Aquatic Contamination

Aquatic Contamination

Tolerance and Bioremediation

Edited by

Rouf Ahmad Bhat
Gowhar Hamid Dar
Fernanda Maria Policarpo Tonelli
Saima Hamid

WILEY

This edition first published 2024

Registered Office(s)
John Wiley & Sons, Inc., 111 River Street, Hoboken, NJ 07030, USA
John Wiley & Sons Ltd, The Atrium, Southern Gate, Chichester, West Sussex, PO19 8SQ, UK

For details of our global editorial offices, customer services, and more information about Wiley products visit us at www.wiley.com.

Wiley also publishes its books in a variety of electronic formats and by print-on-demand. Some content that appears in standard print versions of this book may not be available in other formats.

Library of Congress Cataloging-in-Publication Data applied for:

Names: Bhat, Rouf Ahmad, 1981– editor. | Dar, Gowhar Hamid, editor. | Tonelli, Fernanda Maria Policarpo, editor. | Hamid, Saima, editor.
Title: Aquatic contamination : tolerance and bioremediation / edited by Rouf A. Bhat, Gowhar H. Dar, Fernanda M.P. Tonelli, Saima Hamid.
Description: Hoboken, NJ : Wiley, 2024. | Includes index.
Identifiers: LCCN 2023012172 (print) | LCCN 2023012173 (ebook) | ISBN 9781119989288 (hardback) | ISBN 9781119989295 (adobe pdf) | ISBN 9781119989301 (epub)
Subjects: LCSH: Water–Pollution. | Bioremediation.
Classification: LCC TD420 .A68 2024 (print) | LCC TD420 (ebook) | DDC 628.1/68–dc23/eng/20230330
LC record available at https://lccn.loc.gov/2023012172
LC ebook record available at https://lccn.loc.gov/2023012173

Cover Design: Wiley
Cover Image: Courtesy of Rouf Bhat

Set in 9.5/12.5pt STIXTwoText by Straive, Pondicherry, India

This book is dedicated to the mothers of all the editors of this book.

Contents

About the Book

More than two-thirds of earth is covered with aquatic habitats that play a key role in stabilizing the global environment and providing a wide variety of services to increase human needs. Nevertheless, anthropogenic activities are rapidly destroying the quality of both fresh and marine waters globally due to excessive use of chemicals and fertilizers, and pollution from suburban and industrial areas eventually makes their way into the aquatic world.

To preserve the integrity of aquatic environs is the greatest challenge in current era because more than 80% of urban wastewater, millions of tons of heavy metals, toxic chemicals, sludge, fertilizers, and other waste are being directly discharged into inland waters. There are huge issues involved with technological facilitation and need large investment funds.

This book presents the broader spectrum of biological applicability with better understanding of cellular mechanisms for remediation of aquatic contaminants. Besides, it focuses on practices involved in molecular and genetic approach, necessary to achieve target of bioremediation and phytoremediation to solve the global water contamination problems. Such approaches pave the way for the utilization of biological assets to design new, efficient, and environmentally sound remediation strategies by inculcating genomic techniques at cellular and molecular levels with model assessment.

The content of the book provides a comprehensive background for those who are interested to look upon all perspectives of contamination of aquatic environ. It covers various research aspects that are being carried globally to understand the simulation models in assessment of xenobiotic, role of genomics, transgenic plants, and microbial enzymes for degradation and removal of toxic substances in aquatic environs. In general, it will be very much helpful to the environmentalists, microbiologists, biotechnologists, and scientists and enlighten them on various modern technologies for remediation of contaminants in aquatic ecosystems.

Key features

- The book covers every interaction between plant, metal, and microbes including influence of biotic and abiotic factors.
- It discusses comprehensive details of molecular mechanism from assimilation to detoxification levels.
- It includes the enzymatic approaches of potential plants that act as hyperaccumulators for contaminants in aquatic environs.
- It contains sustainable tools such as transgenic plants to manipulate important functional microbial genes to achieve higher certainty of bioremediation.
- It provides advancement tools and models such as microarrays and simulation models for the complete assessment of xenobiotic compounds from cellular to degradation hierarchies.

About the Editors

Dr. Rouf Ahmad Bhat (PhD) has pursued his doctorate at Sher-e-Kashmir University of Agricultural Sciences and Technology Kashmir (Division of Environmental Sciences) and presently is working in the Department of School Education, Government of Jammu and Kashmir. Dr. Bhat has been teaching graduate and postgraduate students of environmental sciences for the past 3 years. He is author of more than 55 research articles (h-index 25; i-index 35; total citation 2000) and 40 book chapters and has published more than 32 books with international publishers (*Springer*, *Elsevier*, *CRC Press Taylor and Francis*, *Apple Academic Press*, *John Wiley*, and *IGI Global*). He has specialization in limnology, toxicology, phytochemistry, and phytoremediation. Dr. Bhat has presented and participated in numerous state, national, and international conferences, seminars, workshops, and symposium. Besides, he has worked as an associate environmental expert in World Bank-funded Flood Recovery Project and also the environmental support staff in the Asian Development Bank (ADB)-funded development projects. He has received many awards, appreciations, and recognition for his services to the science of water testing, air, and noise analysis. He has served as an editorial board member and a reviewer of reputed international journals. Dr. Bhat is still writing and experimenting with diverse capacities of plants for use in aquatic pollution remediation.

Dr. Gowhar Hamid Dar (PhD) is currently working as an assistant professor in environmental science, Sri Pratap College, Cluster University Srinagar, Department of Higher Education (J&K). He has a PhD in environmental science with specialization in environmental microbiology (fish microbiology, fish pathology, industrial microbiology, taxonomy, and limnology). He has been teaching postgraduate and graduate students for the past many years at Postgraduate Department of Environmental Science, Sri Pratap College, Cluster University Srinagar. He has more than 70 research articles (h-index 15; i-index 20; total citation >700) in international and national journals of repute and more than 20 books with international publishers (*Springer*, *Elsevier*, *CRC Press Taylor and Francis*, *Apple Academic Press*, *John Wiley*, and *IGI Global*) to his credit. Moreover, he is supervising a number of students for the completion of degrees (PhD/Masters). He has been working on the isolation, identification, and characterization of microbes for a decade to understand their utility for humans. His research is based to understand the pathogenic behavior of microbes and to particularly understand the impact of pollution on development of diseases in fish fauna for past several years. He has received many awards and appreciations for his services toward science and development. Besides, he also acts as a member of various research and academic committees. Further, Dr. Dar is principal investigator and coprincipal investigator for different R&D projects sanctioned by Government of India and Government of Jammu and Kashmir.

Dr. Fernanda Maria Policarpo Tonelli (PhD) is an Assistant Professor at Federal University of São João del Rei, Divinópolis, MG, Brazil, where she teaches graduate students. She specializes in biotechnology/molecular biology and has been studying nanomaterials' green synthesis protocols and applications such as delivery and remediation, as same as allelopathic potential of secondary metabolites from plants. The research projects resulted in 10 patent applications, 3 of them already analyzed and granted. She has published five books, authored 14 scientific articles and more than 30 book chapters with international publishers. She has presented and participated in many national and international conferences and has also had the opportunity to contribute to the organization of various scientific events. Dr. Tonelli has also dedicated herself to the promotion of science and technology through co-funding a non-governmental organization. Dr. Tonelli is active in scientific advocation groups for women. Her efforts as a researcher have been recognized with various awards, including For Women in Science Brazil (awarded by L'Oréal Brasil, UNESCO in Brazil, and Brazilian Academy of Sciences) and Under 30 Brazil (Forbes) as well as various certificates of merit.

Dr. Saima Hamid holds doctorate as well as master's degree in environmental science from the University of Kashmir, India. She has qualified UGC-NET exams multiple times and other state level exams. Her research area includes plant molecular biology, stress adaptations, bioactive compound isolation and characterization, climate change, and microbial biotechnology. She has published more than 30 research and review articles in highly reputed journals such as *Elsevier*, *Frontier's in Plant Sciences*, *Springer*, *MDPI*, *Wiley*, and other peer reviewed journals. She is the cofounder of CERD Foundation and member of various international organizations that focus on plant adaptations and climate change. She is an international forum speaker from six years and columnist of various international magazines and local newsletters in different countries such as Island Chief of Maldives, Sri Lanka, Bangladesh, and many others.

Preface

Environmental pollution in the current scenario is a serious threat to all forms of living creatures. The water scarcity is a reality, although it covers the major part of Earth's surface, helping to stabilize ecosystems. Industrialization does not consider sustainability principles and has established a concerning scenario worldwide. Anthropogenic activities have promoted contamination in aquatic ecosystems that leads to deterioration of freshwater qualities and hence makes it inadequate to be used by living beings. Inorganic, organic substances, and biological contamination released from different sources pose severe threat to quality aquatic ecosystems and make it unfit for dependent organisms.

The growing population, however, needs to receive clean water supply to live and maintain good health. So, it is necessary to develop efficient strategies to diminish and preferably stop pollution generated from domestic, industrial, and agricultural sources.

The book is dedicated to discussing the problems associated with contaminated aquatic environs and addressing strategies to preserve and restore the integrity of aquatic environs. Biological materials such as enzymes, microorganisms, and plants can be used in bioremediation protocols to offer low-cost and eco-friendly technologies to deal with contaminants of diverse chemical nature. The nanotechnology field can also offer materials that can be useful tools to promote polluted environments' restoration. The aim of the book is to offer to undergraduate students and researchers a multidisciplinary and comprehensive approach to the theme, from a biotechnological point-of-view.

We are extremely grateful to all the authors who have contributed chapters in this project and to the project handling team of John Wiley & Sons, Inc., for their generous cooperation and for publishing this book.

Rouf Ahmad Bhat
Gowhar Hamid Dar
Fernanda Maria Policarpo Tonelli
Saima Hamid

1

Emerging Techniques for Treatment of Wastewater

Naseema A. Wani[1], Nazir A. Malik[2], Younas R. Tantary[3], Ishrat Jan[4], Tawseef Ahmad[5], and Mohammad S. Wani[6]

[1] Department of Botany, Punjabi University, Patiala, Punjab, India
[2] Department of Botany, Dolphin PG College of Science and Agriculture, Chandigarh, Punjab, India
[3] Department of Botany, Government Degree College, Tangdhar, Jammu and Kashmir, India
[4] Department of Zoology, University of Kashmir, Srinagar, Jammu and Kashmir, India
[5] Department of Biotechnology, Punjabi University, Patiala, Punjab, India
[6] Department of Agriculture, Government Mohindra College, Patiala, Punjab, India

1.1 Introduction

Water is exceptionally a prerequisite for sustaining earthly existence. While 70% of the earth is made up of water, it is surprisingly accessible to only less than 1%. The world population will increase by up to nine million by 2050 with the current population growth rate that may trigger serious freshwater shortages in the immediate future. It is predicted that access to clean water for human needs will be challenging around the globe by 2030, as natural freshwater resources may be under great strain (Wichelns et al. 2015). Regrettably, 97% of the world's surface water is saltwater; two-thirds of the remaining 3% have frozen because 1% of the world's water source is not equally dispersed, and this scarcity is a major trouble in developing nations (Smith 2009). The most important prerequisite of each and every human being is the clean and potable water, but at the same time, usage of this is the utmost responsibility and concern of today's world. Some areas are supplied with water for this purpose despite constant rainfall. There is a shortage in this regard in some areas. In all regions across the globe, this creates a major obstacle in managing water distribution. Water is a predominant factor in the environment which influences human health and other living organisms. The physical and chemical facets of water quality have become a problem with the rise in population, as wastewater from various sources poses a serious threat. Globally, billions of people face the big issue of proper sanitation and clean water supply (WHO 2013). The rapid development of human society throughout the globe in the form of population growth, urbanization, and industrial development has led to an increase in the need for clean and safe drinking water on the one hand and the formation of wastewater on the other. In both the developed world and the developing world, human actions have contributed to water resource pollution by discharging harmful chemical substances from factories, industrial waste, agricultural land, etc., raising some additional load on accessible water supplies and generating large amounts of wastewater (Li et al. 2013). This wastewater approaching from various anthropogenic places is released in a very identical manner into distinct water bodies, which include streams, seas, oceans, and estuaries where it pollutes and significantly damages the aquatic vegetation and fauna. Over the years, various regulations, protocols, and procedures associated with the treatment and release of wastewater into water bodies have been developed and applied around the world and reused to address the continuing challenge of drinking water. But, alternatively, apart from all the rules and regulations, there is another sad truth that a substantial quantity of wastewater has been abandoned untreated or dealt with using strategies that are ineffective and discharged into the environment which subsequently ends in the degradation of the environment. These inadequate sanitation services result in many waterborne diseases (Bixio et al. 2006). According to the WHO, about 80% of the diseases in the developing world are

Aquatic Contamination: Tolerance and Bioremediation, First Edition. Edited by Rouf Ahmad Bhat, Gowhar Hamid Dar, Fernanda Maria Policarpo Tonelli, and Saima Hamid.

water-related, due to low water quality and lack of sanitation, and this is worse in rural areas (WHO 2013). Lack of sanitation contaminates watercourses around the world and is one of the most important sources of water contamination. Every year there are three million deaths from diarrheal diseases due to *Escherichia coli*, Salmonella and Cholera bacterial infections, as well as parasites and viruses. The number of children dying of diarrhea in the 1990s was much greater than the number of conflict-ridden victims after World War II (Smith 2009). Furthermore, it is reported that approximately four million humans around the globe have little or no accessibility to safe and sanitized water sources and every year huge numbers of people die from waterborne diseases (Montgomery and Elimelech 2007; Malato et al. 2009).

In order to solve these issues, wastewater treatment is needed to reprocess wastewater in a useful way from distinctive sources. Wastewater treatment approaches that are currently in use have been validated to be effective in the past, but there is a desperate need to reconsider and develop current policies and procedures and create new wastewater management systems if you want to lessen the environmental risk. Another prime advantage of wastewater treatment is that it can be utilized for consuming purposes that will significantly reduce the burden on natural freshwater supplies from increasing water requirements. The World Health Organization reported that 1.1 billion people did not have access to safe drinking water in 2015. So, purifying wastewater and keeping it safe to drink can help reduce the need for drinking water.

1.2 Composition of Untreated Wastewater and Its Effect on Water Bodies

Defining wastewater really is difficult. Water consideration in its natural state, i.e. water from streams, lakes, and rivers, contains a wide range of minerals, nutrients, and suspended and dispersed substances. The water that is employed for one reason may not be suitable for another. Therefore, to explain the purity of water, there is a list of criteria set by various countries for different water groups. In order to make sure of the availability of clean water, changes are made regularly on the basis of the databases accessible across the planet from certain water and health regulatory authorities. Any degradation in water quality due to the involvement of human activities is water pollution. Anthropogenic and ecological waste both together contribute with regard to the switch in the natural constitution of water. Wastes discharged into water bodies of one country do not remain restricted to specific limits, and this leads to an increase in conflict between bordering nations. Therefore, it is an international concern and that is why it needs global support. Every segment of society has been experiencing an increasing need for sustainable use of resources. Wastewater is any water that has been influenced by human consumption. Wastewater is used water from any mechanical, residential, or arming exercises, stormwater, surface overflow, or any sewer inflow or sewer penetration (Almuktar et al. 2018). Because of the water global scarcity issues, the situation is fundamental to consider nonconventional sources of water for satisfying the increase in the need for freshwater. Wastewater is considered an acceptable solution suitable for addressing the scarcity of water resources caused by multiple factors such as population progression (Bichai et al. 2012; Almuktar and Scholz 2016). However, the unbelievable diversity in wastewater sources intrusive of organic and inorganic elements generates the recycling of that water condition to ordered checking to evaluate incoming hazards affecting the overall environment (FAO 2003). Only 20% of the total wastewater produced is treated before it is delivered into the environment (UNESCO 2012). Satisfactory recycling of wastewater is important to look after the health of the community, the environment, and water resources.

Wastewater normally originates from various anthropogenic places that lead to various forms of contaminants and pollutants present in it counting organic, inorganic, and biological origin (Das et al. 2014). Inorganic contaminants and heavy metal pollutants, viz. copper, cadmium, nickel, and zinc, other chemical toxins and pollutants, such as colorants, detergents, hydrocarbons, and biodegradable substances, are contained in drinking water. Also contained in wastewater are biological pollutants, such as bacteria, microorganisms (pathogenic and nonpathogenic), fungi, and viruses (UN Water 2015). All of the three wide variety of contaminants and pollutants discussed if left untreated or not properly handled affects and impacts the environment in a dangerous way that leads to the degradation of the environment and its related problems. The level of impact depends on the contaminant and pollutant type, form, and concentration.

As discussed earlier, wastewater is a complex mixture of many contaminants and pollutants discharged from domestic, industrial, and agricultural places and, when left untreated, these contaminants and pollutants will pollute the water bodies and produce a significant hazard to the water's natural environment.

Biological pollutants of wastewater by pathogenic or nonpathogenic sources from various sources directly or indirectly contaminate groundwater, which can contribute to infection outbreaks. Untreated wastewater from various places is of major concern due to the long-term effect of water resources on the aquatic environment.

1.2.1 Effect on River Water

Globally, river water is the main source of freshwater for humans, and any pollution of any sort would directly impact the human population. Since river water is an important source of freshwater, it also acts as an important waste disposal center to prevent the treatment of wastewater generated from various sources. These all have a direct effect on river ecology, affecting environmental cycles (Sharma et al. 2019). Wastewater changes the natural environment of rivers. An excessive quantity of toxins and pollutants leads to nutrient accumulation in the rivers resulting in eutrophication (Kanu and Achi 2011). A large amount of growth in river water results in the reduction of oxygen which interferes with the growth of natural aquatic organisms and can lead to the destruction of aquatic life. Industrial and mining effluents containing heavy metals such as arsenic, cadmium, and lead in rivers may result in the addition of these metals that can further lead to biomagnifications of these organic compounds in organisms. Industrial wastewater also produces acid waste resulting in the acidification of the river water by increasing its pH and temperature and thereby destroying the aquatic environment (Akpor and Muchie 2011; UN Water 2015). The presence of hazardous substances will adverse the aquatic health. This is further worsened by the excessive population of waterborne pathogens. Using microbiologically polluted water for household and other uses is harmful to human health and the entire environment. The prolonged exposure of rivers to untreated wastewater can forever alter their composition and properties, which influence normal aquatic life growth and survival. Suspended solids in rivers reduce sunlight penetration and consequently decrease photosynthesis, preventing the growth of aquatic plants and bacteria that sustain and stabilize rivers (Akpor and Muchie 2011).

1.2.2 Effect on Marine Environment

Discharge of untreated wastewater into marine ecosystems, for instance, oceans and seas, has a negative impact on the survival and growth rate of the sea species. The degree of the decline is predicated on the solubility, concentration, and retention of heavy metals, pesticides, insecticides, pharmaceuticals, microorganisms, fossil fuels, hazardous chemicals, and their residues in the contaminants and pollutants found in wastewater. They all may have a worrying effect on the life of the sea. This may build up or bioconcentrate in the species. Wastewater being inaugurated into seawater is detrimental to aquatic life because of increased toxicity. This enhanced toxicity may cause acute and lasting destruction to marine life and will eventually lead to the disruption of the biological cycle (Ren et al. 2017). To investigate the major effect of these toxins and pollutants on aquatic life, much research has been done on one of the diverse, in addition to the most delicate, fragments of marine ecosystems, the coral reefs. The consequences of numerous researches have revealed that the association between wastewater and marine environment destruction is quite consequential and is creating a serious threat to marine life. Industrial wastewater, sewage wastewater, and fertilizer waste are damaging the coastal regions and coral reefs to a greater degree and are also changing the entire water quality of oceans and seas and this in turn is hampering the survival and growth rates of marine life. Another primary point noted was that the polluted sites spread their contamination to unpolluted areas through water currents in oceans and seas, therefore troubling remote areas. Another issue that has a huge impact is the decalcification of calcium carbonate structures contained in the aquatic system. Biogenic calcification is an important process in the marine ecosystem process in which various calcium carbonate structures are being produced that enable several marine organisms to evolve. Calcium carbonate structures in marine organisms has many advantages such as predator defense and mechanical support (Andersson and Gledhill 2013). With the aid of calcium carbonate structures surrounding them, these coral reefs can develop in shallow ocean waters, thereby aiding in providing habitat to other organisms. Acidification of oceans causes erosion and decalcification of these structures leading to the degradation of coral structures. There is a correlation between the calcification and decalcification process that conserves the coral reef biodiversity. Ocean acidification indicates the pH of marine water is transferring toward a low alkaline state which is creating significant damage to the aquatic life (Abbasi and Abbasi 2011). One probable reason for acidification may be the disposal of commercial wastewater comprising toxic metals and others chemicals. These acids can pollute the water and stay there for a significant period of time, changing the average pH of the oceanic waters. The coastal waters encountering contaminated wastewater have elevated levels of nutrient content that leads to algal bloom and decreases the oxygen content in water leading to an unbalanced growth of aquatic organisms and also changing the ratio of coral reef communities. The ratio of coral cover and fish population diminishes with the increasing pollution. It was also observed that due to the low quality and high nutrient content in coastal waters, the amount of dead coral is higher and, at the same time, the proportion of algae increases.

Another dominant aspect which is being taken into account for deteriorating aquatic life is the nutrient enrichment of marine water. Untreated sewage and fertilizer waste are contributing to the nutrient enhancement of marine water in a massive way and damaging aquatic life. Nutritional enhancement results in a strong increase in the concentration of inorganic compounds such as phosphate, nitrate, and ammonium. Prolonged persistence causes chemical imbalances in marine waters and increases algal growth, decreasing photosynthetic activity and oxygen levels of aquatic plants, thus consequently influencing their growth. The nutritional enhancement also results in coral reef bleaching, which leads to the loss of coral cover around the globe (Wiedenmann et al. 2013).

1.3 Strategies to Treat Wastewater

With the fast growth in human population and industrialization, the challenges of acquiring water resources and disposing of wastewater will become increasingly more and more difficult. People around the globe express trouble in this sector and assembled various meetings and plans for environmental protection. Various intellectuals expressed profound concern over the continued deterioration of natural resources and proposed remedies by introducing different goals.

Most importantly, one such idea is ecological understanding that directs to exploring the accessible environmental information and implementing those concepts for designing new projects to accomplish the target of sustainable growth without harming the environment and balance the needs of future generations. The essence of ecological understanding can also be implemented in plotting wastewater treatment methods by making use of soil, plants, and microorganisms. Many such strategies are presently being used like aquatic ponds, wetlands, trickling filters, activated sludge method, and rotating biological contactor (RBC) techniques. Treatment refers to the reduction of value contaminants or pollutants in wastewater until they are released into the atmosphere with low negative impacts. The method of treatment differs considerably and depends on factors such as technology complexity, affordability, cost, efficiency, and the type of wastewater involved such as industrial, agricultural, or urban. There are several standards and rules developed by specific countries in order to preserve the quality of water and industries related to acceptable amounts of toxic components in their waste.

The drinking water standards are divided into two categories: primary and secondary. The primary requirements are legal and designed to govern potentially dangerous substances such as organic and inorganic chemicals and microbes in order to guarantee a safe public water supply. On the other hand, the secondary requirements, which are not enforced by law, mainly involve color and odor. So it is not a matter of great concern.

Technology is rapidly evolving and is indirectly disrupting natural wealth, especially water. The byproducts and waste produced by these activities add xenobiotic components that are challenging to deteriorate by natural processes. Therefore, there is a need for improved treatment strategies to deal with this ever-changing waste matter (Sharma et al. 2019). The type of wastewater treatment techniques relays on the kind of environment where the wastewater will later be released and what kind of effect it would have on that specific environment (Laugesen et al. 2010). The wastewater treatments are classified into different stages depending on the procedure involved and the methodology used to meet the required standards. The easiest kind of categorization is mainly based on natural and advanced methods. But globally the most acceptable and widely studied methods of treatment schemes were divided into three main classes: primary, secondary, and tertiary schemes (Figure 1.1). These strategies often comprise various forms of physical, chemical, and biological treatment methods.

The treatment of wastewater is done to make it acceptable for release or reuse into the system again. The types and degrees of the treatment process used depend on the efficiency of the wastewater disposal and the intended use of treated water. For example, the water for the use of domestic consumption should not contain any type of microbial contamination such as pathogens but may contain tracer amounts of calcium and magnesium salts. However, boiler water may contain microorganisms and even pathogen-causing diseases but calcium and potassium salts (hardness) are not allowed as they can cause the scaling of the reactor and prevent heat transfer. Wastewater released into the water bodies requires less intense treatment as compared to the water which has to be reused. The purpose of a modern wastewater treatment system is to generate clear, free-from-suspended solids, safe water that should be free from disease-causing microorganisms.

1.3.1 Primary Treatment

The primary treatment highlights physicochemical and mechanical approaches which do not have a significant impact on the chemical oxygen demand (COD) and the biological oxygen demand (BOD) of the effluent. It is highly intended to reduce the bulkiness of the effluent with the aid of removing pollutants which could interfere with the further operating

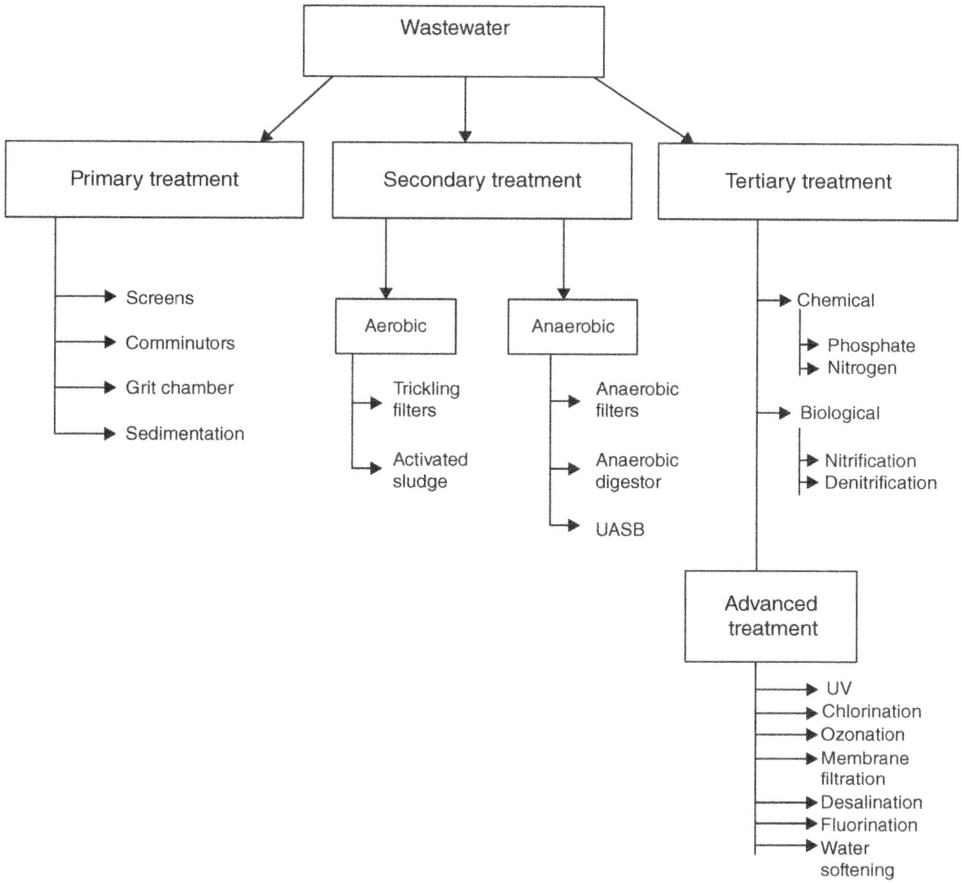

Figure 1.1 Scheme for handling wastewater.

facility of the sewage treatment plant. The purpose of primary treatment in particular is the removal of organic and inorganic solvents through sedimentation, as well as the removal of scum materials by skimming. During primary treatment about half of the biochemical oxygen demands, three-fourths of the overall suspended solids, and the greater part of the oil and grease are separated. Several organic phosphorus, nitrogen, and heavy metals incorporated into solids are often separated throughout primary sedimentation. However, the colloidal and dissolved constituents are unaffected. Grit such as sand and heavy particles with high settling velocity are separated to avoid the rush of alternating wastewater treatment plants. Chemicals can be added to clarifiers to increase the rate of settling in order to increase the efficiency of the process. The primary treatment is meant for removing only those contaminants that have the tendency to float or be resolved under the influence of gravity. Therefore, there is a need to separate those contaminants which are in a suspended form and are mostly composed of oxygen-demanding substances. Figure 1.2 shows the simplified diagram of the primary treatment.

Figure 1.2 Simplified diagram of the primary treatment.

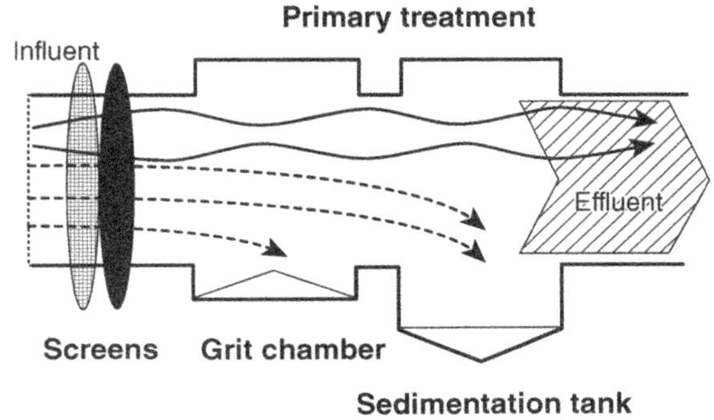

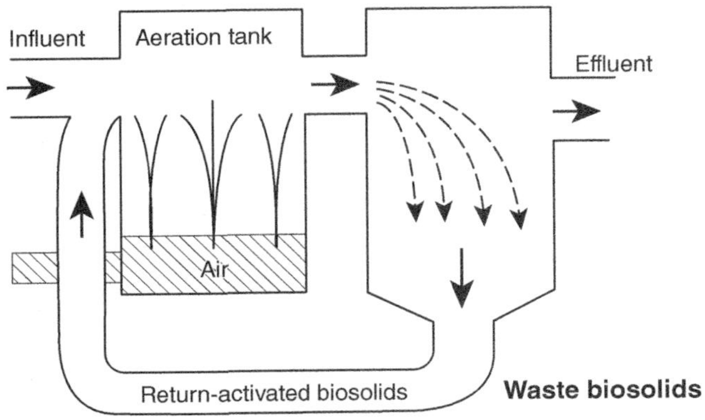

Figure 1.3 Simplified illustration of the secondary treatment.

1.3.2 Secondary Treatment

The primary treatment is accompanied by the secondary treatment, which is commonly known as biological treatment. Generally, secondary treatments include elimination of dissolved biodegradable and organic compounds that use biological treatments. The most popularly used biological treatment processes are tricky filters that are an unchangeable growth system and an activated sludge program that is a suspended growth system. Both systems are virtually used for the successful removal of total suspended solids (TSS) and dissolved BOD. A typical simplified plan of secondary treatment is revealed in Figure 1.3.

The biological treatment depends on the functioning of the microbial community. So, it is interesting the results can be obtained by providing the necessary growth conditions required for the existing community. Several parameters are considered to monitor the effect of processes such as residence time, hydraulic retention time, food-to-microbe ratio, rate of organic load, and rate of recycling. The most widely used activated sludge process is one of the oldest and main methods being used by the wastewater treatment industry.

1.3.3 Activated Sludge

Presently, the activated sludge system is probably the most successful and versatile method for the wastewater treatment process. Activated sludge designates scientific treatment methods that take advantage of an adjourned development of organisms to remove suspended solids and BOD. It relies on the rule that excess wastewater for air circulation to shape flocs of microorganisms decreases the organic matter and is separated by sedimentation (Narmadha and Kavitha 2012). The technique includes aeration and settles down the tanks and other accessories, reoccurrence and waste pumps, blowers and mixers for air circulation, and the device for flow measurement (Liu et al. 2011; Wu et al. 2012). Continuing the attention of active microorganisms in the reservoir, a portion of the activated sludge is reused (Xu et al. 2015). The activated sludge technique inhabits a small place as incorporated with a trickling filter and has better quality of waste. The only disadvantage is the significant amount of BOD in one end of the reservoir. However, on the other end, the microorganisms are more active physiologically until the mixer is entirely activated in the activated sludge system.

In the inactivated sludge method, various modifications have been made in order to get better results. By doing several modifications in the conventional type of facility like extended aeration, step aeration, mechanical aeration, and contact stabilization the efficiency of the activated sludge system can be increased. It is possible to modify the process by judiciously augmenting the sludge with acclimatized microbes and optimization of other parameters related to the process due to the development in technology. Apart from generating energy, these systems have the facilities to recover important resources from wastewater. A few years ago, research has been done in this area and therefore it is possible to retrieve many valuable products, such as fibers, cellulose biopolymers, and bioplastics (Guest et al. 2009; Lin et al. 2010; Ruiken et al. 2013). A general type of activated sludge system is shown in Figure 1.4 (Obuobie et al. 2006).

1.3.4 Trickling Filter

This is a developmental process wherein the microorganisms in charge of treatment are linked to a dormant pressing material. It is made of a circular reservoir with which it is a loaded product (volcanic, rock, or engineered material). Wastewater

Figure 1.4 An activated device of sludge.

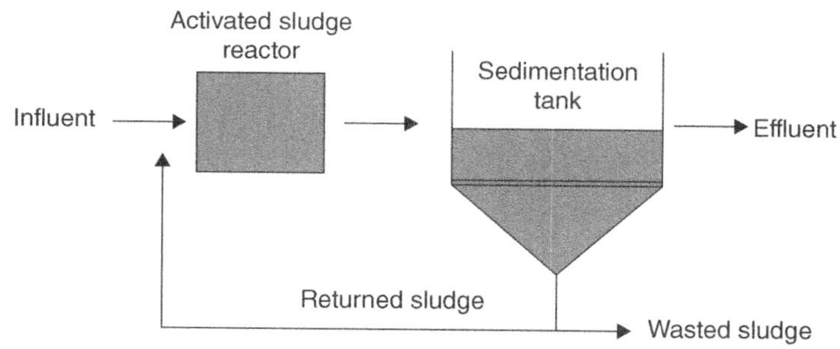

Figure 1.5 Main parts of a trickling filter.

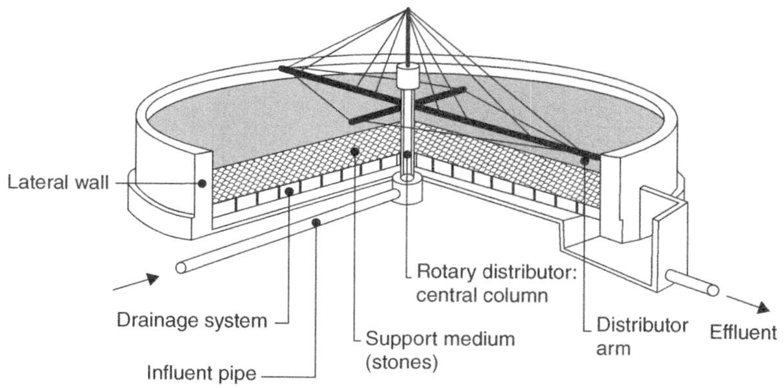

is pumped from above and seeps via filter media that allow organic material to be impregnated in wastewater by populated microorganisms such as aerobic, anaerobic, algae, facultative bacteria, protozoa, fungi referred to the stream as a biological film, or slime layer (about the thickness from 0.1 to 0.2 mm). Trickling filters are effective in which waste eminence such as suspended solids and BOD elimination is significant. The method is less complicated compared to activated sludge method but requirements for repair and operation necessities are higher because of the energy consumption (Fahad et al. 2019). This process is definitely worth it compared with prosperous, densely populated areas which have good treatment of wastewater and better water system. And it is suitable for the treatment of gray water. Moreover, it requires more area contrasted with certain other technologies and has impending for filter flies and odor (Liu et al. 2011). The main components of a trickling filter are shown in Figure 1.5 (Xu et al. 2015).

1.3.5 Rotating Biological Contactors

One of the natural methods for biological wastewater treatment is aerobic RBC. RBC is a different type of wastewater treatment method which associates benefits of fixed biological film, such as high concentration of biomass, short hydraulic retention time, fast operation, minimal energy cost, and insensitivity to hazardous material shock loads. Consequently, the aerobic RBC reactor is commonly used in the treatment of domestic and commercial wastewater (Amoatey and Bani 2011; Vianna et al. 2012). The RBC reactor schematic diagram is shown in Figure 1.6.

1.3.6 Oxidation Ditches

Oxidation Ditches (OD) are usually comprehensive mixing techniques however can be revised. Normal OD treatment methods constitute a single or multichannel arrangement inside a ring or oval. Initial treatment for instance grit elimination and bar screens usually leads to an oxidation ditch. Principal resolving proceeding to an oxidation ditch may also be used and tertiary filters might be needed following elucidation, reliant on the sewage necessities. Disinfection is necessary and reaeration could be required prior to final absolution. In the ditch oxygen transfer, circulation and aeration are provided by horizontally or vertically attached aerators (Zhang et al. 2014). The flow is aerated and blended with recovered sludge from a secondary clarifier in the oxidation ditch. In order to facilitate microbial growth, the mixing phase

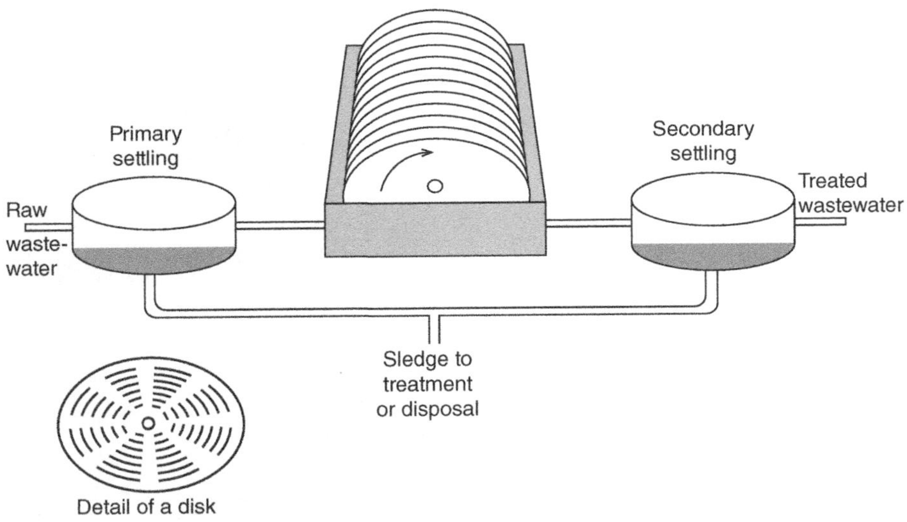

Figure 1.6 Diagram of an RBC reactor.

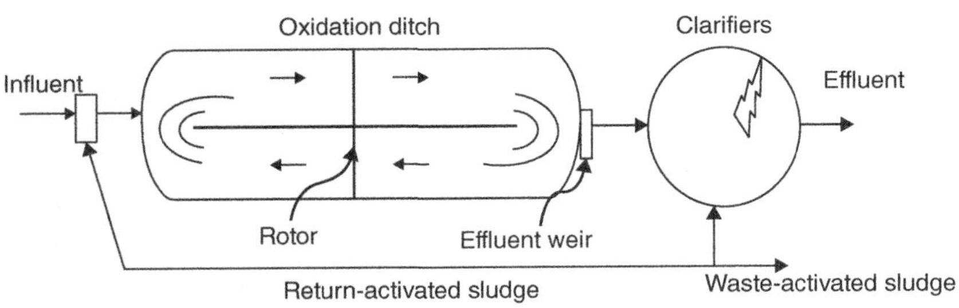

Figure 1.7 Oxidation ditch.

incorporates oxygen into the blended liquor and the drive velocity confirms microorganisms' contact with the powerful liquor (US-EPA 2000). Figure 1.7 shows the diagram of oxidation ditch treatment (Amoatey and Bani 2011).

The secondary effluent consequently collected may be dealt with again by modern methods such as tertiary treatment followed by membrane filtration and disinfection to regain good-quality potable water. Secondary treatments have been additionally divided into aerobic and anaerobic processes. The most extensively used aerobic processes have already been discussed earlier such as activated sludge, trickling filters, RBCs, and oxidation ditches. Among the anaerobic methods are anaerobic digestors, anaerobic filters, and upflow anaerobic sludge blanket reactor. Anaerobic methods are more beneficial than aerobic methods in terms of more energy production and biomass production in the form of methane gas possessing high calorific value (Metcalf and Eddy 2003).

1.4 Tertiary Treatment

Tertiary processing or advanced processing is used where it is required to remove particular components of wastewater that cannot be eliminated by secondary treatment. The tertiary treatment technique commonly known as effluent polishing is used to remove plant nutrients such as nitrogen and phosphorus which are responsible for the eutrophication of water bodies. This technique includes a diversity of processes performed on the effluent discharge from the secondary waste treatment process so that the water can be reused. It is employed when primary and secondary treatment cannot attain all that is needed. Suspended solids dissolved inorganic materials and organic compounds are eliminated by tertiary wastewater treatment. The tertiary treatment technique makes use of physicochemical and biological processes. Nitrogen removal can be done by the biological process which employs nitrification and denitrification with the application of additional aeration and nitrifying and denitrifying bacteria. Another method includes ammonia stripping where pH of water is first raised to alkaline range leading to the formation of ammonium ions followed by rapid flow of water at high speed through the long towers leading up to the removal of nitrogen in the form of ammonia gas. The removal of phosphorus is supported by chemical treatment facilitated by the addition of an alkaline chemical, alum, which results in precipitation (Nathanson 2000). The objective of the tertiary treatment is to eliminate nonbiodegradable toxic organic contaminants like polychlorinated biphenyls, chlorophenols, and other synthetic pollutants.

1.5 Natural Processes for Wastewater Management

Natural treatment strategies include the proper use of natural systems such as plants, soil, and microorganisms where pollutants and contaminants are eliminated without the help of any chemical. Natural treatment takes advantage of the following techniques to combat the wastewater epidemic.

1.5.1 Constructed Wetlands (CWs)

Wetlands constitute marshy areas which are composed of swampy forest areas and salt marshes and they can be artificial, natural, permanent, or temporary. The naturally existing wetlands are abundant in microflora and can be used for wastewater treatment. The constructed Wetlands (CWs) are near-natural techniques and are fully engineered wetlands seen as the future of the wastewater treatment industry for which many newest investigations have been dynamically established in the past decades (Kayombo et al. 2000; Vymazal 2014). CWs are regions with a considerable depth possessing water in them and supporting the growth of different life forms including microorganisms as well as various plant forms such as reeds and sedges. CWs have been shown to make sure some benefits over conventional techniques. CWs are cost-effective, east to operate and maintain, decrease the cost of the waste collection process, absorb low energy, promise appropriate reprocessing, can be worked on several levels, and increase visual importance to the atmosphere (Chen et al. 2008; Foladori et al. 2015). This method seeks to use the natural power of microbes as well as plants to remove pollutants. Plant-microbe interactions play a major role in this remediation of pollutants. The roots of plants provide an environment for the growth of bacteria, provide sufficient nutrients for growth, and enrich water bodies with oxygen, which is prepared by photosynthesis. The removal or degradation of contaminants present in the wastewater is carried out by the bacteria which are residing in the roots. These systems are similar to oxidation ponds to some extent in their functioning.

The most extensively used CW systems are subsurface water flow systems and surface water flow systems. Surface systems have a depth of about two feet possessing an impermeable layer of soil at the bottom (Sanks 1976). A typical type of CW (Amoatey and Bani 2011) is shown in Figure 1.8 (subsurface flow system) and Figure 1.9 (free water surface

Figure 1.8 Constructed wetland (subsurface flow system).

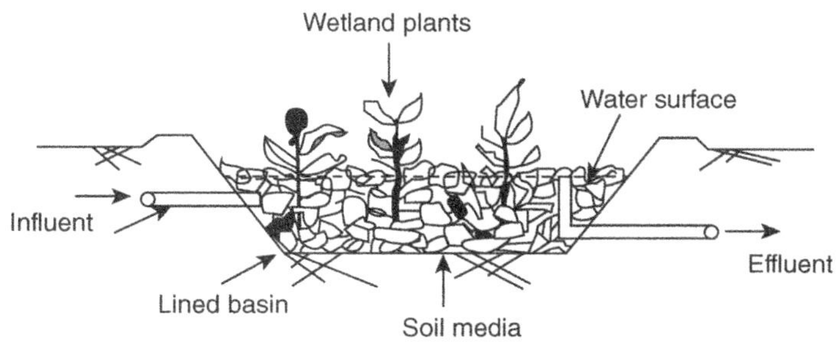

Figure 1.9 Constructed wetland (free water surface system).

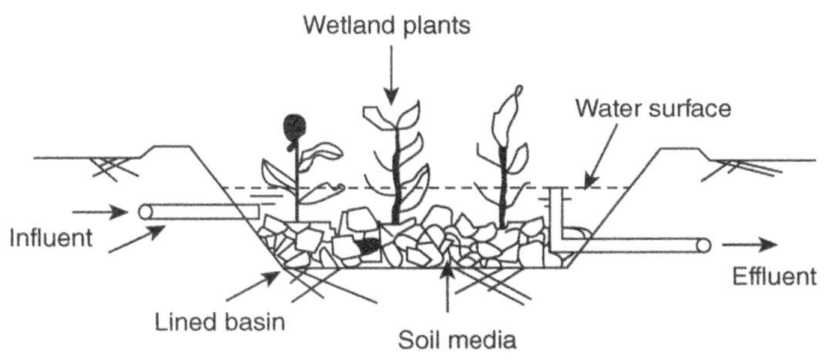

system). Waste water goes through the plant material in the case of subsurface structures and is specifically targeted at removing dissolved pollutants from waste water, while the water to be processed passes through the roots and stems of the growing plants in the surface systems.

CWs for the treatment of wastewater can be segregated as per the life-form of the prevailing macrophyte, into systems of floating-leaved, free-floating, rooted-emerging, and underwater macrophytes (Brix and Schierup 1989). The subsurface system is further divided according to the flow direction into horizontal and vertical (Vymazal and Kröpfelová 2008). The usage of CWs with both vertical and horizontal flow is increasing worldwide for treatment of domestic wastewater (Liu et al. 2005). The CW from a horizontal flow can come up with a reliable secondary degree of treatment pertaining to TSS and BOD but in addition to adequate oxygenation and longer hydraulic retention time, they are rarely successful in removing nitrogen (Liu et al. 2005). Contrarily, vertical flow CWs (VFCWs) are quite constructive not merely for the elimination of TSS and BOD and also for nitrification surprisingly at high loading rate (Prochaska et al. 2007). Thus, vertical flow CWs are getting increasingly accepted (Molle et al. 2006) and are presently under extensive research to optimize the parameters of their basic design (Brix and Arias 2005). In addition, wetland plants are an essential component of the CWs' treatment process. They perform significant part in the CWs due to their ability to remove and degrade nutrients and other contaminants. Many pollutants, for instance, COD, BOD, total Kjeldahl nitrogen (TKN), TSS, total phosphorus (TP), and metals may be eliminated from wastewater with CWs through plant uptake, substrate adsorption, microbial degradation, biological predation, and filtration by the packed material (Saeed and Sun 2011). A large number of CWs exhibit high elimination efficiency of >80% for TSS and BOD (Newman et al. 2000; Vymazal 2002; Wang et al. 2005). They found higher removal of pollutants excluding TKN, TSS, and NO_3. They recommended developing countries use CWs.

The wetlands may be considered the future of the wastewater industry as they are eco-friendly, cost-effective, and have higher potential to deteriorate/remove toxins which remain untreated in conventional wastewater facilities (Sharma et al. 2019). One such pollutant is the pharmaceutical waste produced from domestic sources. The major awareness was paid to the degradation of effluents and the xenobiotic compounds being produced by different industries. This part of the waste is often overlooked as it is being disposed of along with regular domestic waste. The main issue related to them is that they vary a lot depending on their chemical nature i.e. hydrophilic, lipophilic, hydrophobic, reducible, and oxidizable. They are regarded to be pseudo-persistent pollutants as they are added to the environment at a constant rate and, consequently, create instant damage to aquatic organisms and are therefore very hard to remove by any single approach (Fent et al. 2006). They are even more toxic as they sit there in the effluent due to the lack of any treatment plan designed for their removal. Therefore, wetlands are used to eliminate these pollutants by the use of microbial and phytoremediation potential of microbes and plants. A major problem related to the use of CWs is the demand for a large area of land as compared to conventional treatments. This is a systematic method for the treatment of wastewater in rural or remote areas with a small population (Wu et al. 2015).

1.5.2 Aquatic Plant Ponds

This type of treatment makes use of floating type of aquatic plants including duckweed, water hyacinth, Elodea, and water clover (Starkl et al. 2013). Aquatic plant system has been considered one of the processes for wastewater recycling and recovery. The main objectives of using this system focus on nutrient removal and stabilization of the waste (Kanabkaew and Puetpaiboon 2004). In general, the removal of pollutants and contaminants from the environment using plants is known as phytoremediation. The phytoremediation method has already gained important scientific significance because of its versatility and the degree of polluted areas (Chandra and Yadav 2011; Rahman et al. 2016). The aquatic vegetation or wetlands are used extensively to remove heavy metals and contaminants from wastewater over the past 50 years (Marbaniang and Chaturvedi 2015). Usually, three kinds of aquatic macrophytes, namely emergent, submerged, and free-floating, are utilized in CWs, which are preferred to native plants (Valipour and Ahn 2016). Categorically, floating aquatic plants grow speedily in water bodies, thereby helping in removing nitrogen, metals, and TSS due to their high accumulation and uptake ability, and also decrease BOD remarkably (UN Economic Social Commission for Western Asia 2003). In addition, plants provide shade on the surface of the water to slow the growth of algae. Using aquatic plants such as water hyacinth is an inexpensive and, at the same time, eco-friendly approach as compared to other technologies that involve the use of chemicals (Rezania et al. 2015). The potential of water hyacinth was explored to treat wastewater approaching from domestic sources by (Rezania et al. 2015). Several aquatic weeds such as *Eichhornia crassipes* and *Lemna minor* were discovered to have the capability of decontaminating organic pollutants and to take inorganic nutrient salts, for instance, phosphorus and nitrogen compounds. A wide variety of wetland plant species demonstrate their ability to extract metals from different kinds of wastewaters, such as *Eichhornia* sp., viz. *E. azurea*, *E. paniculata*, *E. diversifolia*, and *E. crassipes*; *Salvinia* sp., viz. *S. auriculata*, *S. molesta*, and *S. minima*; *Pistia stratiotes*, and *L. minor*; *Azolla* sp., underwater species like *Myriophyllum* sp. and *Potamogeton* sp. Emergent species such as *Limnocharis flava*, *Typha* sp. *Scirpus* sp. *Cyperus* sp., *Phragmites* sp., and *Spartina* sp. etc. (Soda et al. 2012; Rodríguez and Brisson 2015).

The aquatic macrophytes frequently used for heavy metal remediation in aquatic system were *L. minor*, *P. stratiotes*, and *E. crassipes*. These three floating water aquatic plants are at the forefront of the phytoremediation process (Rezania et al. 2016).

Research over the past few decades has shown that some floating plants like *E. crassipes*, *P. stratiotes*, *L. minor*, *Centella asiatica*, and *Trapa natans* have major effects on eutrophic water purification (Reddy and Debusk 1987; Jianbo et al. 2008). Therefore, the macrophyte reservoir can also be utilized as low-cost wastewater treatment facilities with benefits such as maintenance costs of operating, energy savings, elimination of broad-spectrum contaminates, and the regeneration of the aquatic plants as animal foods or production of green manure and the recycling of wastewater as resources (Badr El-Din and Abdel-Aziz 2018). Various aquatic plants have been tested to remove heavy metals present in wastewater such as selenium, copper, and cadmium. In addition to this aquatic plant, duckweed is also a very popular option for the wastewater treatment due to its phytoremediation ability and its potential to convert the nutrients present in wastewater into useful compounds.

1.6 Emerging or Advanced Techniques for the Treatment of Wastewater

Three main factors have primarily guided the creation and implementation of water treatment technologies: the announcement of new water quality standards, the discovery of new rarer contaminants and the expense. For the first 75 years of this century, granular media filtration, chemical filtering, and chlorination were the lone treatment methods utilized in municipal water treatment. Nevertheless, over the past two decades, there has been a significant shift in the water treatment industry in which water utilities have been used as a starting point to seriously examine alternative treatment technologies to the traditional chlorination/filtration treatment approach. Traditional treatment techniques like filtration and chlorination can remove the contaminants and other physical particles but also causes the formation of a toxic sludge resulting in its difficult disposal (Ferroudj et al. 2013; Anjum et al. 2016). Some of these emerging or advanced technologies are discussed here.

1.6.1 Wastewater Treatment by Means of Nanoparticles

Speedy urbanization has yielded the significant trouble of water shortage on earth and wastewater treatment is the lone recycling process that can overcome this problem. Advancements in technology have brought us to nanotechnology that has immense potential for the treatment of wastewater. Nanoparticles are materials with different properties and nanoscale size of 1–100 nm, which gives an advantage of a large surface area (Zhang et al. 2008). Biologists are seeing a huge amount of potential in nanoparticles in the treatment of environmental pollution. One such area is the treatment of wastewater using metal nanoparticles. A large number of metals are used to synthesize nanoparticles, viz. iron oxide, gold, TiO_2, and silver nanoparticles. These nanoparticles have been proven to be very effective in removing heavy metal contamination and, to some extent, powerful radioactive compounds present in the wastewater (Li et al. 2013; Gehrke et al. 2015). These nanoparticles are beneficial as they have a somewhat negative impact on the environment, low solubility, and high adsorption properties, making them the most promising, efficient, and safe technologies for the treatment of wastewater (Gupta et al. 2015; Anjum et al. 2016). Adsorption and removal of impurities using nanoparticles are eliminated by external adsorption and then their intraparticle diffusion along the walls (Li et al. 2013). Silver nanoparticles have various applications for the treatment of wastewater ranging from therapeutic, drug delivery systems, antimicrobial activity, etc. (Ahamed et al. 2010). Due to their antimicrobial activity, they have been used to remove microorganisms present in wastewater. Zinc oxide nanoparticles with porous structures have been used to remove various types of heavy metal pollutants such as mercury, lead, nickel, arsenic, and cadmium (Singh et al. 2013). Manganese oxide nanoparticles close to other nanoparticles provide a large surface area for the adsorption of metal pollutants in wastewater. Manganese oxide-based nanoparticles are very effective in the adsorption and removal of arsenic in wastewater (Wang et al. 2011; Anjum et al. 2016).

Water contamination by toxic metal ions Pb^{2+}, Cr^{4+}, Hg^{2+}, Co^{2+}, Cr^{3+}, Ni^{2+}, CD^{2+}, Cu^{2+}, Ag^{1+}, As^{5+}, and As^{3+} has become a serious environmental and public health crisis. To accomplish ecological detoxification, different techniques such as precipitation, adsorption, reverse osmosis, electrochemical treatments, ion exchange, evaporation, membrane filtration, flotation, biosorption, and oxidation processes are used. Among these, with the aid of magnetic nanoparticles, adsorption is an effective but efficient way to extract toxic metal ions and bacterial pathogens from water. In the treatment of water sediments, charged particles, heavy metals, nanostructured materials such as carbon nanotubes, nanostructured iron zeolite, silver-impregnated cyclodextrin nanocomposites, magnetic nanoparticles, carbon-iron nanoparticles, photocatalytic titania nanoparticles, nanofiltration membranes and functionalized silica nanoparticles, and nanofiltration membranes can be used for sediment removal (Bapusaheb and Patil 2015). Our earth is rich in various types of iron oxides, such as hematite, maghemite, and magnetite. Iron oxides are excellent substances used in the formation of nanoparticles and have proven to be a new and novel technology for wastewater treatment. Iron oxides have significant affinity for toxic metal pollutants such as Co^{2+}, Cr^{3+}, Pb^{2+}, Cd^{2+}, and As^{3+} (Gupta et al. 2015). Since iron oxide-based nanoparticles have magnetic properties, these can be easily segregated and retrieved from the mixture under the influence of magnetic field (Kunduru et al. 2017). Because of their magnetic field potential, they are also known as magnetic nanoparticles. Magnetic nanoparticles have shown to be a promising in situ technology for water treatment as these can be applied directly to the contaminated site and the contaminants can be easily removed without any type of damage to the environment (Jung et al. 2004). Specifically, the utilization of magnetic nanoparticles as water treatment adsorbents offers an easy way for removing and separating pollutants by using outdoor magnetic forces. Therefore, the utilization of these nanoparticles in different potentials for the stability of iron oxide particles by surface change is desirable.

1.6.2 Activated Carbon

In perfecting the traditional form of treatment, activated carbon (AC) used as a form of absorption medium plays a major role. In addition, AC technology is one of the most effective and safe technologies for the treatment of organic water pollutants. AC is a widely used material in various fields due to its excellent adsorbent properties. These structures are assigned by finely developed internal pore structure, high surface area, and different functional groups, such as carboxyl, carbonyl, and phenols (Bhatnagar et al. 2013). The functional groups can be modified as per the requirement present on the AC by various surface treatments such as base, acid treatment (Shen et al. 2008a,b), microwave (Ania et al. 2005), ozone (Zaror 1997), impregnation (Bhatnagar et al. 2013), and chemical treatment using anionic and cationic surfactants (Gao et al. 2009). In Philadelphia, United States, the first water treatment plant was installed in 1930, using granular AC adsorption tanks (Wenjie et al. 2006). During the 1960s–1970s, developed Western nations began to use AC technologies in drinking water treatment to improve the elimination of organic pollutants. At that time, prechlorination was the commonly used first step in AC treatment. The development of the microorganism was hindered as the inflow of the carbon layer included free chlorine, and there was no apparent biological activity indicated by the carbon layer. It was Yang (2003) who documented the use of AC for the removal of pollutants. Further studies are needed to look at the specific type of contaminants by using selective AC in wastewater by choosing specific industrial effluent. The biggest obstacle in using AC is its high cost, but further research in this field and the application of economically feasible methods can make it a viable alternative for wastewater treatment.

1.6.3 Carbon Nanotubes

Nanotechnology has a variety of purposes in various fields including food industry, medical sciences, electronics, drug delivery systems, and the environment. By the utilization of nanotechnology, many nanostructures have been created such as nanoadsorbents, nanoparticles, nanotubes, and zeolites with remarkable properties and are well used in various domains (Das et al. 2014). Carbon nanotube is one such nanostructure which is having the potential of treating and purifying wastewater on a large scale. Carbon nanotubes are nanoscale structures with a large surface area and have adsorption sites for the adsorption of contaminants present in wastewater. These are cylindrical structures made of graphite which are allotropes of carbon forms. It can be of two kinds: multiwalled carbon nanotubes and single-walled carbon nanotubes. Multiwalled nanotubes have multilayered graphite sheets while single-walled carbon nanotubes are made up of a single layer of graphite sheet (Das et al. 2014; Gehrke et al. 2015). These nanotubes are effective in adsorbing toxic heavy metals as well as metal ions and organic contaminants, even insecticides, dyes, pharmaceutics, etc., and biological and inorganic contaminants. These adsorbed contaminants enter the carbon nanotubes by developing various kinds of interactions like electrostatic interactions, hydrogen bonding, and van der Waals' interactions with the variety of functional groups existing on the adsorbing matrix such as −OH and COOH. Carbon nanotubes have been extensively used for adsorbing heavy metals such as Pb^{2+}, Cd^{2+}, and Cu^{2+} (Li et al. 2013). Carbon nanotubes are very effective in removing heavy metal contamination from wastewater. Previous studies have shown that carbon nanotubes with their high oxidation density by various functional groups are very efficient in the adsorption and removal of heavy metals such as lead, cadmium, copper, and zinc (Li et al. 2013; Lu et al. 2006). Another potential use of carbon nanotubes has been found for their use as photocatalysts and other nanomaterials to form a hybrid to remove contaminants. These can be used in combination with zinc oxide nanomaterials, titanium oxide nanomaterials, etc. This combination of nanomaterials and carbon nanotubes increases the properties of carbon nanotubes manyfolds. It has enhanced surface area for electrical conductivity, adsorption, total durability, and photostability that make them ideal for water purification. However, carbon nanotubes have great potential for the treatment of wastewater, but the disadvantage of their high production costs does not allow their industrial use on a large scale (De Volder et al. 2013; Kunduru et al. 2017).

1.6.4 Microbial Fuel Cells

Wastewater that is considered to be useless can actually be a very useful resource when managed and treated properly. In order to explore the undiscovered potential for wastewater, many environmentalists and scientists are trying to focus on microbial fuel cell (MFC) technology. MFCs have appeared as an encouraging technology in wastewater treatment that includes a combination of both electrochemical and biological processes (Ahn and Logan 2010).

Because of the following potential benefits microbial cell technology (Figure 1.10) can be proven to be a promising technology in the treatment of wastewater including the clean energy generation in the form of biomass and electricity. In addition, technology produces less environmental impact and proves to be a sustainable method for the wastewater treatment. Furthermore, in real-time data production and collection, this technology is very sensitive (Li et al. 2014). Microbial cell technology has been studied for the generation of energy from the inorganic and organic wastes found in wastewater by transforming it into electricity.

The major defect of MFC technology is its efficiency in producing energy and operating costs. Extensive research is still needed for the technology to be used on a large scale with minimal energy consumption and high energy production to achieve sustainable wastewater treatment. Li et al. (2014) reviewed that microbial cell technology is more efficient in the sense that it uses less energy compared to other treatment processes as it can be more efficient at room temperature and

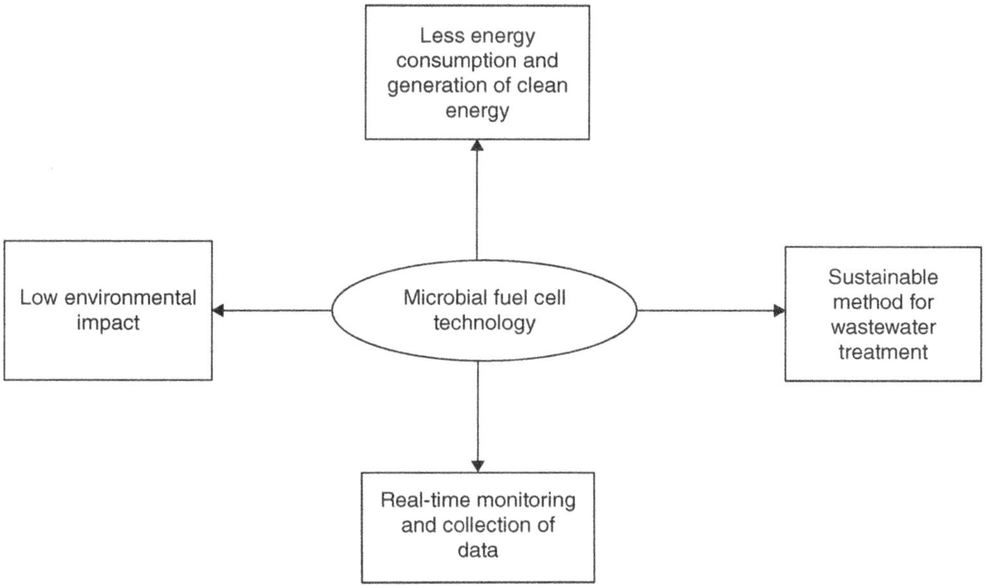

Figure 1.10 Microbial fuel cell technology benefits.

consequently, lower energy demand will be there for temperature maintenance. This technology is continuously tested for different types of pollutants. One such study by Adelaja et al. (2017) evaluated the effectiveness of MFC technology against hydrocarbons such as benzene and phenanthrene. The hydrocarbon degradation process was evaluated for 155 days by MFCs cultured according to in situ and ex situ conditions. In situ research showed the removal of benzene and phenanthrene and electricity generation up to 6.75 mW/m^2. In the case of ex situ method, proper hydrocarbons were removed, suggesting that MFC technology could prove to be a promising technology for the cleaning up of effluents coming from refineries or wastewater contaminated by hydrocarbons. Using this technology, the removal of inorganic compounds such as nitrogen can also be done.

MFCs can offer the solution to some of the problems that conventional treatment of wastewater is facing. They allow the restoration of energy in the wastewater while reducing the input capacity and excess sludge production (Rabaey and Verstraete 2005). MFC technology is used to treat and purify the environment's vital part of water. The technology is able to accurately remove various contaminants like organic, various metals, and sulfur-based pollutants while maintaining water quality levels so that they can be reused for different anthropogenic purposes. In addition to being efficient, MFC technology releases less CO_2, thus producing less sludge and also providing fewer greenhouse gases (Li et al. 2014). This technique generates energy right away from the organic materials in which the microbial biofilm oxidizes organic waste and shifts electrons to the electrode. Various studies have been conducted in recent times to look at the effectiveness of the technology even with certain modifications or in combination with other hybrid techniques across the earth. For example, in a study done by Sevda et al. (2013), MFCs attached to an air cathode were used against compostable sewage water collected from a treatment plant in Belgium to generate electricity. The samples used were original wastewater and centrifuged and half-diluted wastewater, and the machine worked for 160 days. The overall generating power is approximately 762 mV and the maximum COD purification of approximately 60% is recorded by the original wastewater followed by centrifuged and half-diluted wastewater samples. The results of the study revealed that air cathode microbial cell technology is a promising way to generate electricity and treat wastewater. A review by Pandey et al. (2016) focused on the potentiality of some microorganisms to purify wastewater and, at the same time, generate electricity using different substrates adding to various microbial cell technologies as various types of pollutants are added daily to the water by anthropogenic activities. However, this technology is very much at its primary stage and therefore its effectiveness and environmental impact are yet to be established and will be standardized for this method.

Large-scale system protocols are desperately needed but there might be current new problems and weaknesses that need to be resolved systematically (Dong et al. 2015; Ge et al. 2015). The future of MFC technology is filled with promising possibilities and difficulties that require bigger discoveries and developments in engineering and science to resolve a number of obstacles and to establish advanced technologies for processing wastewater treatment centered on bioelectrochemical concepts.

1.6.5 Fenton's Method in Combination with Cavitation

Various oxidation processes have been used to reduce various kinds of contaminants present in wastewater. It can deteriorate or convert chemical pollutants with long chains into short-chain compounds that are easier to handle than conventional methods (Gogate and Pandit 2004a,b; Bagal and Gogate 2014). Fenton's procedure is one such method for oxidation and is being used to treat wastewater with toxic inorganic and organic pollutants. From hydrogen peroxide, it produces hydroxyl radicals. In this process, the ferrous ion acts as a catalyst. This process has the advantage of removing a broad variety of pollutants like insecticides, pesticides, and pollutants from the plastic, refinery, and rubber industries (Pera-Titus et al. 2004). However, in order to carry out an efficient process, various parameters must have to be strictly monitored such as temperature, catalyst amount, and the concentration of hydrogen and pH should be maintained within the range of 2–4 to acquire the highest productivity of the process. The maintenance of all these parameters and the high cost of hydrogen peroxide from time to time prove to be dangerous, and also its effectiveness in many compounds such as acetic acid and carbon tetrachloride is still not well established. Considering these undesirable effects, some modifications in Fenton's process are done to improve the efficiency of waste removal. Fenton's process was used in combination with the cavitation process in one of the modifications. In this combination, the Fenton process is used by ultrasound waves to produce large amounts of hydroxyl radicals that will enhance the conversion rate and degradation of organic compounds present in wastewater (Bagal and Gogate 2014). Cavitation in addition to the effect of ultrasound waves also takes away the contaminants with the aid of pyrolysis as other organic compounds are unaffected by hydroxyl radicals.

1.6.6 Molecularly Imprinted Polymers (MIPs)

The available treatment methods cannot remove the specific type of pollutants which are present in a very lesser concentration, thus getting away from the treatment process and beginning to accumulate in the environment. As discussed in the treatment of wetlands, certain pharmaceutical components are in this category. To deteriorate such components, a specific approach is required. One such option is the use of a selective emerging technique, i.e. *molecularly imprinted polymers (MIPs)*. In order to form a complex the template is linked to the monomers carrying particular functional groups that will be polymerized along with a selective crosslinker to form MIPs. The elimination of templates leads to the production of cavities that have complementary shape and functional groups (Advincula 2011; Shen et al. 2012; Huang et al. 2015). MIPs have an advantage over other techniques in terms of cost-effectiveness, stability at high temperature and pressure, and specificity (Bui and Haupt 2010). They have been reported to be used in a variety of studies such as separation, quantification, and catalysis (Strikovsky et al. 2003; Byun et al. 2010; Sueyoshi et al. 2010). The adsorption capacity of MIPs was reported for the removal of the various pollutants present in trace amounts such as estrogen compounds (Zhongbo and Hu 2008), polyaromatic hydrocarbons (Krupadam et al. 2010), and Diclofenac (Dai et al. 2011). MIPs also find their use as catalysts. The use of MIP-coated TiO_2 as a photocatalyst was to efficiently remove 2-nitrophenol and 4-nitrophenol, which was selective and more efficient than conventional TiO_2 as reported by Shen et al. (2008b). The currently available MIPs still face many problems such as template removal and binding affinity (Huang et al. 2015) that can be overcome by using MIP-based composites using nanomaterials, magnetic particles, and quantum dots (Chen et al. 2014). These composites have the properties of both the MIP and the other matrix-related components. In addition, the adsorption sites are located on the surface of the polymers providing better cohesion and selectivity against the corresponding analyte. The composite of MIPs and Fe_3O_4 has been used successfully in the determination of Bisphenol-A (BPA) and estrogens in a water sample by Guo et al. (2011) and the elimination of estrogens via MIP/TiO_2 nanotubes was studied by Zhang et al. (2013). Such studies are needed to exploit the potential of MIPs in wastewater treatment to identify specific contaminants that can be eliminated by other available alternatives.

1.6.7 Ceramic Foam Fabrication Techniques for Wastewater Treatment

The wastewater process has become more complex recently because of the number of times pollutants are discharged to a water source. Besides the traditional wastewater remedy plant, membrane technology turns into an alternative method of sanitization of wastewater. Ceramic membranes receive amazing interest over polymeric membranes because of excessive chemical, thermal, and mechanical balances which are capable of functioning under intense operating conditions (Ciora and Liu 2003; de Souza et al. 2012). Other than that, ceramic membranes show several advantages such as it is unambiguous in controlling the pore intensity, higher persistence, and achievement. The potency of the membrane is manipulated by techniques of its structural residences for separation for the period of the filtration manner and can be dealt with throughout the fabrication of the membrane (Nor et al. 2018). The surface of the membrane which is in touch with the fluid acquires a strong layer which determines the porosity of the membrane. The active layer is full of holes from special diameters to support the preferred molecules to escape through the membrane.

Ceramic foam is a group of exceptionally porous components needed for a broad extent of mechanical operations, unequivocally as sponges and film for the approach to wastewater analysis. There is another catalogue of ceramic materials that has been taken advantage of as fired films, for example, alumina (Al_2O_3), zirconia (ZrO_2), silicon dioxide (SiO_2), and titanium dioxide (TiO_2) (Fakhfakh et al. 2010; Zawrah et al. 2014). Titanium dioxide is normally applied as a top layer because of its photocatalytic action, compound solidness, and long water transition (Li et al. 2013). Photocatalytic debasement is significant in evacuating natural contamination. The performance of sanitation of water film can be broadened by coupling the detachment procedure with photocatalytic action. Ceramic foams are made up of cell structures made out of a three-dimensional system of swaggers (Luyten et al. 2009). It is lightweight equipment that occur as cell structures, swaggers, and windows (Luyten et al. 2003). Ceramic foams are permeable and weak materials, including enormous void cells, with direct measurements in the scope of 10 μm to 5 mm (Sharafat et al. 2006). In any case, a strong material containing cavity openings or interstices can likewise be viewed as permeable material. These foams accompany porosity that is closed, fully open, or somewhat interconnected. It may be delivered from an expansive scope of clay materials, explicitly both oxide and nonoxide, which incorporates silicates. Metallic oxides, aluminosilicates, and carbides are known to be feasible applications in various territories (Dhara and Bhagarva 2003; Sharafat et al. 2006). The application has ceramic qualities, for instance, low thickness, low warm conductivity, temperature stability, and large protection from substance assault (Blachou et al. 1992; Han et al. 2002). Foams are either closed-cell or open-cell structures. They comprise a get-together of sporadically molded polyhedral cells or motley, associated with one another with a strong side for open cells and countenances for closed cells. Closed-cell ceramic materials are generally utilized for warm protection and fire insurance materials while open-cell ceramic foams are chiefly utilized for liquid metal filtration, hot gas filtration, and diesel motor fume channels (Sharafat et al. 2006; Peng et al. 2000). Additionally, potential utilization of this foam has incorporated a layer, sponges, exhaust system, protection, biomedical gadgets, and as center materials for sandwich development (Woyansky and Scott 1999; Zhu et al. 2002). Cell ceramics are large-scale or small-scale cells with a sort of foam. Full-scale cell artistic foams are fired bodies with an enormous volume of porosity, regularly running from around 70% to over 90% of the all-out volume. However, there is a contrast in the cell's overall morphology as well as in the foam's properties (Colombo and Hellmann 2002). Large-scale cell foams with cell size ranges between 100 and 600 μm and a mass thickness of 0.25–0.58 g/cm^3, contingent upon the handling parameters that were created utilizing a direct foaming methodology (Colombo and Bernardo 2003). In contrast with large-scale cell clay froth, small-scale cell plastics have cell densities more prominent than 109 cells/cm^3 and completely developed cells little than 30–50 μm. Then, small-scale cell foams with a cell size of around 8 μm, and mass thickness of about 0.31–0.48 g/cm^3, contingent upon the material, temperature, weight, and different parameters are produced using traditional plastic-framing advancements (Martini-Vvedensky et al. 1984; Colombo and Bernardo 2003). The instances of ceramic foams have been broadly concentrated, for example, silicon carbide, alumina, and halfway balanced out zirconia (Montanaro et al. 1998). Alumina (Al_2O_3) is a customary adsorbent for substantial metals. It is a significant material that can be set up in a few distinct stages. The stage incorporates γ-Al_2O_3 and α-Al_2O_3. The structure of γ-Al_2O_3 is foreseen to be more adsorptive dynamic than α-Al_2O_3 (Li et al. 2008).

Ceramic foam can be created utilizing different strategies, including replication strategy (Schmidt et al. 2001), starch solidification (Lyckfildt and Ferreira 1998), foaming strategy (Minnear 1992), and gel-throwing of foam (Sepulveda 1997). A different strategy was accounted for and utilized in the readiness of permeable clay bolster, for example, polymeric sponge technique, pore-forming agent, and a foaming agent (Sharmiwati et al. 2014; Li et al. 2014b). Among them is the polymer replication technique which delivers high porosity, high pore network, and more prominent pore size of permeable fired (Woyansky and Scott 1999). The ceramic membrane film can be set up in single or various layers relying upon the applications. In multilayer clay layer (MLCM), permeable help is typically used to improve the exhibition of the film because of its superb mechanical and synthetic properties (Li et al. 2008).

1.6.7.1 Replication Method

Replication method (polymeric sponge) includes the screening of open-cell polymeric foam with clay slurry proceeded by wearing via sintering process out of polymeric foam, which produces fired foam that is generated using unique polymeric foam (Woyansky and Scott 1999). These froths obtained by utilizing this procedure have less mechanical aspect and greater penetrability.

1.6.7.2 Starch Consolidation

Starch consolidation has developed a remarkable position since it is inexpensive, nonpoisonous, earth-amicable, and displays imperfection-free burnout between 300 and 600 °C (Zivcova et al. 2009). Handling of starch union is set up by presenting starch, for example, nourishment quality, fine rice flour, and also as a gelling specialist, and fired powder into refined water. At that point, the blends are blended by mixing, throwing, tapping, coagulation, remoulding, drying, and sintering food grade, fine rice flour, and others as a gelling specialist, and fired powder.

1.6.7.3 Direct Foaming Technique

Direct foaming technique appears with artistic slurry, that is, clay suspension, foaming polymerization, expulsion from the shape (remolding), drying, and sintering. Elevated flexibility and the capacity to create exceptionally permeable green bodies, which are similarly solid, and can without much of a stretch, withstand machining (Sepulveda and Binner 1999).

1.6.7.4 Gel-Throwing

Gel-throwing of foam has set up for effortlessness and capacity to deliver an elevated level of uniformity just as green body aspect, which is subsequently highly machinable. The creation procedure commences with a blending of a colloidal earthenware suspension comprising water solvent monomers and a frothing operator. After the froth arrangement, the suspension is quickly felled by methods for the polymerization of the monomers, giving unbending clay froth. The green body is then dehydrated and sintered, carrying about fired foam with almost round pores and extremely thick swaggers (Ortega et al. 2003).

1.7 Conclusion

Throughout the years, the nature of water has been stumbling primarily due to anthropogenic exercises, rapid industrialization, impromptu urbanization, population development, and the inefficient use of natural water supplies. Besides, the raising understanding of the value of achieving results because of the current environmental policies has driven research communities to develop effective, environmentally friendly, and economically achievable techniques that have the potential of removing pollutants from the water and at the same time, protecting the well-being of an influenced community. There are many techniques for eliminating the contaminants from sewage water and treating the wastewater for reuse. Different studies have been implemented to solve the problem of global water scarcity and, in this case, the treatment of wastewater is one among them.

Currently, there is a wide selection of methods available to remove contaminants from wastewater varying from principles based on either anthropogenic or natural and the combination of these two methods. Promising progress has been made in this area, and much research is being done in this area on the growing problems upraised by different areas of the world in environmental fortification. The present study aims to review current and traditional approaches available so far for wastewater treatment. The biggest obstacle related to most pollutants is their accumulation in the environment. Such contamination stays in the treated water due to the lack of required treatment techniques in conventional methods. One of the reasons for accumulation is that they are sometimes converted to secondary form and therefore difficult to assess their presence and treatment. A basic understanding of the type of pollutants present in different wastes will greatly assist in increasing the efficiency of existing techniques. The efficacy of the treatments can be increased by looking at the existing lacunae with the available methods, and the necessary amendments should be made, as well as by adopting hybrid methods for better results. Another important area of research is related to the movement of possible transformations, contaminants, and accumulations in various environmental conditions. The advancements made in wastewater technologies and studies worldwide should be shared across the globe in order to create an information database that will help in strengthening the research and develop more eco-friendly and innovative methods or ways to break through small obstacles. Regulatory agencies should be strengthened and strong legislation should be enacted on defaulters. Water management is a very important issue and, as such, it can be resolved through the combined efforts of all sectors such as research, public, industrial, and nongovernmental and governmental organizations. There is a need to constantly evaluate the effectiveness of existing methods and add new treatment strategies simultaneously to have the desired effect in this area.

References

Abbasi, T. and Abbasi, S.A. (2011). Ocean acidification: the newest threat to the global environment. *Crit. Rev. Env. Sci. Technol.* 41 (18): 1601–1663.

Adelaja, O., Keshavarz, T., and Kyazze, G. (2017). Treatment of phenanthrene and benzene using microbial fuel cells operated continuously for possible in situ and ex situ applications. *Int. Biodeterior. Biodegrad.* 116: 91–103.

Advincula, R.C. (2011). Engineering molecularly imprinted polymer (MIP) materials: developments and challenges for sensing and separation technologies. *Korean J. Chem. Eng.* 28 (6): 1313–1321.

Ahamed, M., AlSalhi, M.S., and Siddiqui, M.K.J. (2010). Silver nanoparticle applications and human health. *Clin. Chim. Acta* 411 (23): 1841–1848.

Ahn, Y. and Logan, B.E. (2010). Effectiveness of domestic wastewater treatment using microbial fuel cells at ambient and mesophilic temperatures. *Bioresour. Technol.* 101 (2): 469–475.

Akpor, O.B. and Muchie, B. (2011). Environmental and public health implications of wastewater quality. *Afr. J. Biotechnol.* 10 (13): 2379–2387.

Almuktar, S.A. and Scholz, M. (2016). Experimental assessment of recycled diesel spill-contaminated domestic wastewater treated by reed beds for irrigation of Sweet Peppers. *Int. J. Environ. Res. Public Health* 13: 208. https://doi.org/10.3390/ijerph13020208 PMid: 26861370, PMCid: PMC4772228.

Almuktar, S.A., Abed, S.N., and Scholz, M. (2018). Wetlands for wastewater treatment and subsequent recycling of treated effluent: a review. *Environ. Sci. Pollut. Res.* 25 (24): 235–295. https://doi.org/10.1007/s11356-018-2629-3. PMid: 29959736, PMCid: PMC6096557.

Amoatey, P. and Bani, R. (2011). Chapter 20: wastewater management. In: *Waste Water - Evaluation and Management* (ed. F.S.G. Einschlag). Intech https://doi.org/10.5772/2051.

Andersson, A.J. and Gledhill, D. (2013). Ocean acidification and coral reefs: effects on breakdown, dissolution, and net ecosystem calcification. *Annu. Rev. Mar. Sci.* 5: 321–348.

Ania, C.O., Parra, J.B., Menendez, J.A., and Pis, J.J. (2005). Effect of microwave and conventional regeneration on the microporous and mesoporous network and on the adsorptive capacity of activated carbons. *Microporous Mesoporous Mater.* 85 (1): 7–15.

Anjum, M., Miandad, R., Waqas, M. et al. (2016). Remediation of wastewater using various nano-materials. *Arab. J. Chem.* 12 (8): 4897–4919.

Badr El-Din, S.M. and Abdel-Aziz, R.A. (2018). Potential uses of aquatic plants for wastewater treatment. *J. Microbiol. Biotechnol. Rep.* 2 (2): 47–48.

Bagal, M.V. and Gogate, P.R. (2014). Wastewater treatment using hybrid treatment schemes based on cavitation and Fenton chemistry: a review. *Ultrason. Sonochem.* 21 (1): 1–14.

Bapusaheb, B. and Patil, T. (2015). Wastewater treatment using nanoparticles. *J. Adv. Chem. Eng.* 5: 131. https://doi.org/10.4172/2090-4568.1000131.

Bhatnagar, A., Hogland, W., Marques, M., and Sillanpää, M. (2013). An overview of the modification methods of activated carbon for its water treatment applications. *Chem. Eng. J.* 219: 499–511.

Bichai, F., Polo-López, M.I., and Fernández Ibañez, P. (2012). Solar disinfection of wastewater to reduce contamination of lettuce crops by Escherichia coli in reclaimed water irrigation. *Water Res.* 46: 6040–6050. https://doi.org/10.1016/j.watres.2012.08.024. PMid: 22981489.

Bixio, D., Thoeye, C., De Koning, J. et al. (2006). Wastewater reuse in Europe. *Desalination* 187: 89–101.

Blachou, V., Goula, D., and Philippopoulos, C. (1992). Wet milling of alumina and preparation of slurries for monolithic structures impregnation. *Ind. Eng. Chem. Res.* 31: 364–369.

Brix, H. and Arias, C.A. (2005). The use of vertical flow constructed wetlands for on-site treatment of domestic wastewater: new Danish guidelines. *Ecol. Eng.* 25: 491–500.

Brix, H. and Schierup, H.H. (1989). The use of macrophytes in water pollution control. *AMBIO* 18: 100–107.

Bui, B.T.S. and Haupt, K. (2010). Molecularly imprinted polymers: synthetic receptors in bioanalysis. *Anal. Bioanal. Chem.* 398 (6): 2481–2492.

Byun, H.S., Youn, Y.N., Yun, Y.H., and Yoon, S.D. (2010). Selective separation of aspirin using molecularly imprinted polymers. *Sep. Purif. Technol.* 74 (1): 144–153.

Chandra, R. and Yadav, S. (2011). Phytoremediation of Cd, Cr Cu, Mn, Fe, Ni, Pb and Znfrom aqueous solution using *Phragmites cummunis*, *Typha angustifolia* and *Cyperous esculentus*. *Int. J. Phytorem.* 13: 580–591.

Chen, Z., Chen, B., Zhou, J. et al. (2008). A vertical subsurface-flow constructed wetland in Beijing. *Commun. Nonlinear Sci. Numer. Simul.* 13 (9): 1986–1997. https://doi.org/10.1016/j.cnsns.2007.02.009.

Chen, H., Kong, J., Yuan, D., and Fu, G. (2014). Synthesis of surface molecularly imprinted nanoparticles for recognition of lysozyme using a metal coordination monomer. *Biosens. Bioelectron.* 53: 5–11.

Ciora, R.J. and Liu, P.K.T. (2003). Ceramic membranes for environmental related applications. *Fluid/Part. Sep. J.* 15 (1): 51–60.

Colombo, P. and Bernardo, E. (2003). Macro and micro-cellular porous ceramics from preceramic polymers. *J. Comput. Sci. Technol.* 63: 2353–2359.

Colombo, P. and Hellmann, J.R. (2002). Ceramic foams from preceramic polymers. *Mater. Res. Innovations* 6: 260–272.

Dai, C.M., Geissen, S.U., Zhang, Y.L. et al. (2011). Selective removal of diclofenac from contaminated water using molecularly imprinted polymer microspheres. *Environ. Pollut.* 159 (6): 1660–1666.

Das, R., Hamid, S.B.A., Ali, M.E. et al. (2014). Multifunctional carbon nanotubes in water treatment: the present, past and future. *Desalination* 354: 160–179.

De Volder, M.F., Tawfick, S.H., Baughman, R.H., and Hart, A.J. (2013). Carbon nanotubes: present and future commercial applications. *Science* 339 (6119): 535–539.

Dhara, S. and Bhagarva, P. (2003). A simple direct casting route to ceramic foams. *J. Am. Ceram. Soc.* 86: 1645–1650.

Dong, Y., Qu, Y., He, W. et al. (2015). A 90-liter stackable baffled microbial fuel cell for brewery wastewater treatment based on energy self-sufficient mode. *Bioresour. Technol.* 195: 66e72.

Fahad, A., Mohamed, R.M.S., Radhi, B., and Al-Sahari, M. (2019). Wastewater and its treatment techniques: an ample review. *Indian J. Sci. Technol.* 12 (25): 1–13. https://doi.org/10.17485/ijst/2019/v12i25/146059.

Fakhfakh, S., Baklouti, S., and Baklouti Bouaziz, S.J. (2010). Preparation, characterization and application in BSA solution of silica ceramic membranes. *Desalination* 262: 188.

FAO (2003). User's Manual for Irrigation with Treated Wastewater. Food and Agriculture Organization (FAO) of the United Nations. FAO Regional Office of the Near East, Cairo.

Fent, K., Weston, A.A., and Caminada, D. (2006). Ecotoxicology of human pharmaceuticals. *Aquat. Toxicol.* 76 (2): 122–159.

Ferroudj, N., Nzimoto, J., Davidson, A. et al. (2013). Maghemite nanoparticles and maghemite/silica nanocomposite microspheres as magnetic Fenton catalysts for the removal of water pollutants. *Appl. Catal. B* 136: 9–18.

Foladori, P., Bruni, L., and Tamburini, S. (2015). Bacteria viability and decay in water and soil of vertical subsurface flow constructed wetlands. *Ecol. Eng.* 82: 49–56. https://doi.org/10.1016/j.ecoleng.2015.04.058.

Gao, R., Hu, Z., Chang, X. et al. (2009). Chemically modified activated carbon with 1-acylthiosemicarbazide for selective solid-phase extraction and preconcentration of trace Cu (II), Hg (II) and Pb (II) from water samples. *J. Hazard. Mater.* 172 (1): 324–329.

Ge, Z., Wu, L., Zhang, F., and He, Z. (2015). Energy extraction from a large-scale microbial fuel cell system treating municipal wastewater. *J. Power Sources* 297: 260e264.

Gehrke, I., Geiser, A., and Somborn-Schulz, A. (2015). Innovations in nanotechnology for water treatment. *Nanotechnol. Sci. Appl.* 8: 1.

Gogate, P.R. and Pandit, A.B. (2004a). A review of imperative technologies for wastewater treatment I: oxidation technologies at ambient conditions. *Adv. Environ. Res.* 8 (3): 501–551.

Gogate, P.R. and Pandit, A.B. (2004b). A review of imperative technologies for wastewater treatment II: hybrid methods. *Adv. Environ. Res.* 8 (3): 553–597.

Guest, J.S., Skerlos, S.J., Barnard, J.L. et al. (2009). A new planning and design paradigm to achieve sustainable resource recovery from wastewater. *Environ. Sci. Technol.* 43 (16): 6126–6130.

Guo, W., Hu, W., Pan, J. et al. (2011). Selective adsorption and separation of BPA from aqueous solution using novel molecularly imprinted polymers based on kaolinite/Fe_3O_4 composites. *Chem. Eng. J.* 171 (2): 603–611.

Gupta, V.K., Tyagi, I., Sadegh, H. et al. (2015). Nanoparticles as adsorbent; a positive approach for removal of noxious metal ions: a review. *Sci. Technol. Dev.* 34: 195–214.

Han, Y.S., Li, J.B., Chen, Y.J., and Wei, Q.M. (2002). A study on the factors involved in collapse of macroporous α-Al_2O_3 structure. *J. Mater. Process. Technol.* 128: 313–317.

Huang, D.L., Wang, R.Z., Liu, Y.G. et al. (2015). Application of molecularly imprinted polymers in wastewater treatment: a review. *Environ. Sci. Pollut. Res.* 22 (2): 963–977.

Jianbo, L.U., Zhihui, F.U., and Zhaozheng, Y.I.N. (2008). Performance of a water hyacinth (*Eichhornia crassipes*) system in the treatment of wastewater from a duck farm and the effects of using water hyacinth as duck feed. *J. Environ. Sci.* 20: 513–519.

Jung, J.Y., Chung, Y.C., Shin, H.S., and Son, D.H. (2004). Enhanced ammonia nitrogen removal using consistent biological regeneration and ammonium exchange of zeolite in modified SBR process. *Water Res.* 38 (2): 347–354.

Kanabkaew, T. and Puetpaiboon, U. (2004). Aquatic plants for domestic wastewater treatment: lotus (*Nelumbo nucifera*) and Hydrilla (*Hydrilla verticillata*) systems. *Aquatic* 26 (5): 749–756.

Kanu, I. and Achi, O.K. (2011). Industrial effluents and their impact on water quality of receiving rivers in Nigeria. *J. Appl. Technol. Environ. Sanit.* 1 (1): 75–86.

Kayombo, S., Mbwette, T.S.A., Katima, J.H.Y. et al. (2000). Waste Stabilisation Ponds and Constructed Wetlands Design Manual. UNEP-IETC and DANIDA Publication.

Krupadam, R.J., Khan, M.S., and Wate, S.R. (2010). Removal of probable human carcinogenic polycyclic aromatic hydrocarbons from contaminated water using molecularly imprinted polymer. *Water Res.* 44 (3): 681–688.

Kunduru, K.R., Nazarkovsky, M., Farah, S. et al. (2017). Nanotechnology for water purification: applications of nanotechnology methods in wastewater treatment. *Water Purif.* 33–74.

Laugesen, C.H., Fryd, O., Koottatep, T., and Brix, H. (2010). *Sustainable Wastewater Management in Developing Countries: New Paradigms and Case Studies from the Field.* Reston, VA: ASCE Press.

Li, G.L., Bai, X., Yu, S. et al. (2012). Urban phosphorus metabolism through food consumption: The case of China. *Journal of Industrial Ecology.* 16:588–599.

Li, J.D., Shi, Y.L., Cai, Y.Q. et al. (2008). Adsorption of di-ethyl-phthalate from aqueous solutions with surfactant-coated nano/microsized alumina. *J. Chem. Eng.* 140: 214–220.

Li, Q., Qu, X., and Alvarez, P.J.J. (2013). Applications of nanotechnology in water and wastewater treatment. *Water Res.* 47: 3931–3946.

Li, W.W., Yu, H.Q., and He, Z. (2014a). Towards sustainable wastewater treatment by using microbial fuel cells-centered technologies. *Energy Environ. Sci.* 7 (3): 911–924.

Li, F., Kang, Z., Huang, X. et al. (2014b). Preparation of zirconium carbide foam by direct foaming method. *J. Eur. Ceram. Soc.* 34: 3513–3520.

Lin, Y., de Kreuk, M., Van Loosdrecht, M.C.M., and Adin, A. (2010). Characterization of alginate-like exopolysaccharides isolated from aerobic granular sludge in pilot-plant. *Water Res.* 44 (11): 3355–3364.

Liu, W., Dahab, M.F., and Surampalli, R.Y. (2005). Nitrogen transformations modeling in subsurface-flow constructed wetlands. *Water Environ. Res.* 77: 246–258.

Liu, J.J., Wang, X.C., and Fan, B. (2011). Characteristics of PAHs adsorption on inorganic particles and activated sludge in domestic wastewater treatment. *Bioresour. Technol.* 102 (9): 5305–5311. https://doi.org/10.1016/j.biortech.2010.12.063. PMid:21220202.

Lu, C., Chiu, H., and Liu, C. (2006). Removal of zinc (II) from aqueous solution by purified carbon nanotubes: kinetics and equilibrium studies. *Ind. Eng. Chem. Res.* 45 (8): 2850–2855.

Luyten, J., Mullens, S., Cooymans, J. et al. (2003). New processing techniques of ceramic foams. *J. Adv. Eng. Mater.* 5: 715–718.

Luyten, J., Mullens, S., Cooymans, J. et al. (2009). Different methods to synthesize ceramic foams. *J. Eur. Ceram. Soc.* 29: 829–832.

Lyckfildt, O. and Ferreira, J.M.F. (1998). Processing of porous ceramics by starch consolidation. *J. Eur. Ceram. Soc.* 18: 131–140.

Malato, S., Fernandez-Ibanez, P., Maldonado, M.I. et al. (2009). Decontamination and disinfection of water by solar photocatalysis: recent overview and trends. *Catal. Today* 147: 1–59. https://doi.org/10.1016/j.cattod.2009.06.018.

Marbaniang, D. and Chaturvedi, S.S. (2015). Chapter 4: aquatic macrophytes as a tool for phytoremediation of heavy metals. In: *Biology, Biotechnology and Sustainable Development*, 62–85. India: Research India Publications.

Martini-Vvedensky, J.E., Suh, N.P., and Waldman, F.A. (1984). Microcellular closed cell foams and their method of manufacture. US Patent 4, 473,665.

Minnear, W.P. (1992). Processing of foamed ceramics. In: *Ceramic Transactions, Forming Science and Technology for Ceramics* (ed. M.J. Cima), 149–156. Westerville, OH: American Ceramic Society.

Molle, P., Liénard, A., Grasmick, A., and Iwema, A. (2006). Effect of reeds and feeding operations on hydraulic behavior of vertical flow constructed wetlands under hydraulic overloads. *Water Res.* 40: 606–612.

Montanaro, L., Jorand, Y., Fantozzi, G., and Negro, A. (1998). Ceramic foams by powder processing. *J. Eur. Ceram. Soc.* 18: 1339–1350.

Montgomery, M.A. and Elimelech, M. (2007). Water and sanitation in developing countries: including health in the equation. *Environ. Sci. Technol.* 41: 17–24. https://doi.org/10.1021/es072435t.

Narmadha, D. and Kavitha, M.S. (2012). Treatment of domestic waste water using natural flocculants. *Int. J. Life Sci. Biotechnol. Pharm. Res.* 1 (3): 206–213.

Nathanson, J.A. (2000). *Basic Environmental Technology: Water Supply, Waste Management, and Pollution Control*. Prentice-Hall.

Newman, J.M., Clausen, J.C., and Neafsey, J.A. (2000). Seasonal performance of a wet-land constructed to process milk house wastewater in Connecticut. *Ecol. Eng.* 14: 181–198.

Nor, M.A.A.M., Zaine, S.N.N.M., Hasmizam, R.M.D., and Asmadi, A. (2018). Development of ceramic membrane for wastewater filtration. *Int. J. Adv. Sci. Eng. Technol.* 6 (1): 2321–8991, ISSN(e): 2321-9009.

Obuobie, E., Keraita, B.N., Danso, G. et al. (2006). Sanitation and urban wastewater management-book chapter in irrigated urban vegetable farming in Ghana: characteristics, benefits and risks. https://ageconsearch.umn.edu/bitstream/137958/2/Irrigated_Urban_Vegetable_Production_in_Ghana.pdf.

Ortega, F.S., Valenzuela, F.A.O., Scuracchio, C.H., and Pandolfelli, V.C. (2003). Alternative gelling agents for the gelcasting of ceramic foams. *J. Eur. Ceram. Soc.* 23: 75–80.

Pandey, P., Shinde, V.N., Deopurkar, R.L. et al. (2016). Recent advances in the use of different substrates in microbial fuel cells toward wastewater treatment and simultaneous energy recovery. *Appl. Energy* 168: 706–723.

Peng, H.X., Fan, Z., Evans, J.R.G., and Busfield, J.J.C. (2000). Microstructure of ceramic foams. *J. Eur. Ceram. Soc.* 20: 807–813.

Pera-Titus, M., García-Molina, V., Baños, M.A. et al. (2004). Degradation of chlorophenols by means of advanced oxidation processes: a general review. *Appl. Catal. B Environ.* 47 (4): 219–256.

Prochaska, C.A., Zouboulis, A.I., and Eskridge, K.M. (2007). Performance of pilot-scale vertical-flow constructed wetlands, as affected by season, substrate, hydraulic load and frequency of application of simulated urban sewage. *Ecol. Eng.* 31: 57–66.

Rabaey, K. and Verstraete, W. (2005). Microbial fuel cells: novel biotechnology for energy generation. *Trends Biotechnol.* 23 (6): 291e298.

Rahman, M.A., Reichman, S.M., De Filippis, L. et al. (2016). Phytoremediation of toxic metals in soils and wetlands: concepts and applications. In: *Environmental Remediation Technologies for Metal-Contaminated Soils* (ed. H. Hasegawa, I. Rahman, and M. Rahman), 161–195. Japan: Springer.

Reddy, K.R. and DeBusk, T.A. (1987). State of the art utilization of aquatic plants in water pollution control. *Water Sci. Technol.* 19 (10): 61–79.

Ren, Z., Chon, T.S., Xia, C., and Li, F. (2017). The monitoring and assessment of aquatic toxicology. *Biomed. Res. Int.* 2017: 9179728.

Rezania, S., Ponraj, M., Talaiekhozani, A. et al. (2015). Perspectives of phytoremediation using water hyacinth for removal of heavy metals, organic and inorganic pollutants in wastewater. *J. Environ. Manage.* 163: 125–133.

Rezania, S., Taib, S.M., Din, M.F.M. et al. (2016). Comprehensive review on phytotechnology: heavy metals removal by diverse aquatic plants species from wastewater. *J. Hazard. Mater.* 318: 587–599.

Rodríguez, M. and Brisson, J. (2015). Pollutant removal efficiency of native versus exotic common reed (*Phragmites australis*) in North American treatment wetlands. *Ecol. Eng.* 74 (2015): 364–370.

Ruiken, C.J., Breuer, G., Klaversma, E. et al. (2013). Sieving wastewater–cellulose recovery, economic and energy evaluation. *Water Res.* 47 (1): 43–48.

Saeed, T. and Sun, G. (2011). A comparative study on the removal of nutrients and organic matter in wetland reactors employing organic media. *Chem. Eng. J.* 171: 439–447.

Sanks, R.L. (1976). *Land Treatment and Disposal of Municipal and Industrial Wastewater.* Ann Arbor Science.

Schmidt, H., Koch, D., Grathwohl, G., and Colombo, P. (2001). Micro and macroporous ceramics from preceramic precursors. *J. Am. Ceram. Soc.* 84: 2252–2255.

Sepulveda, P. (1997). Gelcasting of foams for porous ceramics. *J. Am. Ceram. Soc.* 76: 61–65.

Sepulveda, P. and Binner, J.G.P. (1999). Processing of cellular ceramics by foaming and in situ polymerisation of organic monomers. *J. Eur. Ceram. Soc.* 19: 2059–2066.

Sevda, S., Dominguez-Benetton, X., Vanbroekhoven, K. et al. (2013). High strength wastewater treatment accompanied by power generation using air cathode microbial fuel cell. *Appl. Energy* 105: 194–206.

Sharafat, S., Ghoniem, N., Sawan, M., and Ying Williams, A.B. (2006). Breeder foam: an innovative solid breeder material for fusion application. *J. Am. Ceram. Soc.* 81: 455–460.

Sharma, N., Singh, A., and Batra, N. (2019). Modern and emerging methods of wastewater treatment. In: *Ecological Wisdom Inspired Restoration Engineering, EcoWISE* (ed. V. Achal and A. Mukherjee), 223–247. Singapore: Springer Nature https://doi.org/10.1007/978-981-13-0149-0_13.

Sharmiwati, M.S., Mizan, R.M., and Noorhelinahan, A.B. (2014). Preparation and characterization of ceramic sponge for water filter. *Int. J. Sci. Technol. Res.* 3: 6.

Shen, W., Li, Z., and Liu, Y. (2008a). Surface chemical functional groups modification of porous carbon. *Recent Pat. Chem. Eng.* 1 (1): 27–40.

Shen, X., Zhu, L., Liu, G. et al. (2008b). Enhanced photocatalytic degradation and selective removal of nitrophenols by using surface molecular imprinted titania. *Environ. Sci. Technol.* 42 (5): 1687–1692.

Shen, X., Zhu, L., Wang, N. et al. (2012). Molecular imprinting for removing highly toxic organic pollutants. *Chem. Commun.* 48 (6): 788–798.

Singh, S., Barick, K.C., and Bahadur, D. (2013). Fe_3O_4 embedded ZnO nanocomposites for the removal of toxic metal ions, organic dyes and bacterial pathogens. *J. Mater. Chem. A* 1 (10): 3325–3333.

Smith, A. (2009). Nanotechnology: an answer to the world's water crisis. *Chem. Int.* 31: 12–14.

Soda, S., Hamada, T., Yamaoka, Y. et al. (2012). Constructed wetlands for advanced treatment of wastewater with acomplex matrix from a metal-processing plant: bioconcentration andtranslocation factors of various metals in Acorus gramineus and Cyperusalternifolius. *Ecol. Eng.* 39 (2012): 63–70.

de Souza, J.S., Vieira, T.M., Barbosa, E.S. et al. (2012). Numerical study of oil/water separation by ceramic membranes in the presence of turbulent flow. *Adv. Chem. Eng. Sci.* 02 (02): 257–265.

Starkl, M., Amerasinghe, P., Essl, L. et al. (2013). Potential of natural treatment technologies for wastewater management in India. *J. Water Sanit. Hyg. Dev.* 3 (4): 500–511.

Strikovsky, A., Hradil, J., and Wulff, G. (2003). Catalytically active, molecularly imprinted polymers in bead form. *React. Funct. Polym.* 54 (1): 49–61.

Sueyoshi, Y., Fukushima, C., and Yoshikawa, M. (2010). Molecularly imprinted nanofiber membranes from cellulose acetate aimed for chiral separation. *J. Membr. Sci.* 357 (1): 90–97.

Tchobanoglous, G., Burton, F.L., Stensel, H.D., and Metcalf, E. (2003) Wastewater Engineering: Treatment and Reuse, 4th ed. McGraw-Hill Series in Civil and Environmental Engineering.

UN Economic Social Commission for Western Asia (2003). Governance for Sustainable Development in the Arab Region: Institutions and Instruments for Moving beyond and Environmental Management Culture. United Nations, New York.

UN Water (2015). Wastewater Management-A UN-Water Analytical Brief. New York.

UNESCO (2012). Managing Water under Uncertainty and Risk. The United Nations World Water Development Report 4. United Nations Educational, Scientific and Cultural Organization, Paris.

US-EPA (2000). "Oxidation Ditches" Wastewater Technology Fact Sheet. Washington, DC.

Valipour, A. and Ahn, Y.H. (2016). Constructed wetlands as sustainable ecotechnologiesin decentralization practices: a review. *Environ. Sci. Pollut. Res.* 23: 180–197.

Vianna, M.R., de Melo, G.C., and Neto, M.R. (2012). Wastewater treatment in trickling filters using LUFFA cyllindrica as biofilm supporting medium. *J. Urban Environ. Eng.* 6 (2): 57–66. https://doi.org/10.4090/juee.2012.v6n2.057066.

Vymazal, J. (2002). The use of sub-surface constructed wetlands for wastewater treatment in the Czech Republic: 10 years experience. *Ecol. Eng.* 18: 632–646.

Vymazal, J. (2014). Constructed wetlands for treatment of industrial wastewaters: a review. *Ecol. Eng.* 73: 724–751. https://doi.org/10.1016/j.ecoleng.2014.09.034.

Vymazal, J. and Kröpfelová, L. (2008). *Wastewater Treatment in Constructed Wetlands with Horizontal Sub-Surface Flow.* Dordrecht, The Netherlands: Springer.

Wang, X., Bai, X., and Wang, B. (2005). Municipal wastewater treatment with pond-constructed wetland system: a case study. *Water Sci. Technol.* 51: 325–329.

Wang, H.Q., Yang, G.F., Li, Q.Y. et al. (2011). Porous nano-MnO_2: large scale synthesis via a facile quick-redox procedure and application in a supercapacitor. *New J. Chem.* 35 (2): 469–475.

Wenjie, H., Weiguang, L., Xiaojian, Z. et al. (2006). *Novel Technology for Drinking Water Safety [M].* China Architecture & Building Press.

Wichelns, D., Drechsel, P., and Qadir, M. (2015). Wastewater: economic asset in an urbanizing world. In: *Wastewater*, 3–14. Netherlands: Springer.

Wiedenmann, J., D'Angelo, C., Smith, E.G. et al. (2013). Nutrient enrichment can increase the susceptibility of reef corals to bleaching. *Nat. Clim. Change* 3 (2): 160–164.

World Health Organization (WHO) (2013). Research for Universal Coverage: World Health Report.

Woyansky, S. and Scott, J.C.E. (1999). Processing of ceramic foams. *Am. Ceram. Soc. Bull.* 71: 1674–1682.

Wu, Y., Li, T., and Yang, L. (2012). Mechanisms of removing pollutants from aqueous solutions by microorganisms and their aggregates: a review. *Bioresour. Technol.* 107: 10–18. https://doi.org/10.1016/j.biortech.2011.12.088. PMid: 22257855.

Wu, H., Zhang, J., Ngo, H.H. et al. (2015). A review on the sustainability of constructed wetlands for wastewater treatment: design and operation. *Bioresour. Technol.* 175: 594–601.

Xu, G., Yang, X., and Spinosa, L. (2015). Development of sludge-based adsorbents: preparation, characterization, utilization and its feasibility assessment. *J. Environ. Manage.* 151: 221–232. https://doi.org/10.1016/j.jenvman.2014.08.001. PMid: 25577702.

Yang, R.T. (2003). *Adsorbents: Fundamentals and Applications.* Wiley.

Zaror, C.A. (1997). Enhanced oxidation of toxic effluents using simultaneous ozonation and activated carbon treatment. *J. Chem. Technol. Biotechnol.* 70 (1): 21–28.

Zawrah, M.F., Khattab, R.M., Lamey, G. et al. (2014). Effect of CTAB as a foaming agent on the properties of alumina ceramic membranes. *Ceram. Int.* 40: 5299.

Zhang, L., Gu, F.X., Chan, J.M. et al. (2008). Nanoparticles in medicine: therapeutic applications and developments. *Clin. Pharmacol. Ther.* 83 (5): 761–769.

Zhang, W., Li, Y., Wang, Q. et al. (2013). Performance evaluation and application of surface-molecular-imprinted polymer-modified TiO_2 nanotubes for the removal of estrogenic chemicals from secondary effluents. *Environ. Sci. Pollut. Res.* 20 (3): 1431–1440.

Zhang, D.Q., Jinadasa, K.B.S.N., Gersberg, R.M. et al. (2014). Application of constructed wetlands for wastewater treatment in developing countries-a review of recent developments (2000–2013). *J. Environ. Manage.* 141: 116–131. https://doi.org/10.1016/j.jenvman.2014.03.015. PMid:24784754.

Zhongbo, Z. and Hu, J. (2008). Selective removal of estrogenic compounds by molecular imprinted polymer (MIP). *Water Res.* 42 (15): 4101–4108.

Zhu, X., Jiang, D., and Tan, S. (2002). The control of slurry rheology in the processing of reticulated ceramic foams. *Mater. Res. Bull.* 37: 541–553.

Zivcova, Z., Gregorova, E., Pabst, W. et al. (2009). Thermal conductivity of porous alumina ceramics prepared using starch as a pore-forming agent. *J. Eur. Ceram. Soc.* 29: 347–353.

2

Aquatic Ecosystems and Health Threats: Case Study on the Nickel Pollution in Gölbaşı Lake in Hatay – Turkiye

Volkan Altay[1], Büşra Kara[2], Ibrahim E. Yalcin[3,4], and Munir Ozturk[5]

[1] *Department of Biology, Faculty of Science and Arts, Hatay Mustafa Kemal University, Hatay, Turkiye*
[2] *Department of Plant Protection, Institute of Pure and Applied Sciences, Hatay Mustafa Kemal University, Hatay, Turkiye*
[3] *Department of Biology, Institute of Pure and Applied Sciences, Hatay Mustafa Kemal University, Hatay, Turkiye*
[4] *Department of Civil Engineering, Faculty of Engineering and Natural Sciences, Bahcesehir University, Istanbul, Turkiye*
[5] *Centre for Environmental Studies and Botany Department, Ege University, Izmir, Turkiye*

2.1 Introduction

The current ecological conditions of the environment provide important contributions to the structure, function, and service delivery of all ecosystems. Several books, book chapters, and articles have been published on the functions of different ecosystems including coastal areas and wetlands (Altay and Ozturk 2012; Ozturk et al. 2021). The aquatic ecosystems are attracting much attention as sustainable models for future ecosystem studies, especially the effect of multiple global environmental changes on these as they are highly fragile.

Aquatic ecosystems are grouped mainly as "marine" and "freshwater" ecosystems. The former cover more than 70% of our globe being represented by the oceans, estuaries, coral reefs, and coastal ecosystems. The latter cover less than 1% of the earth's surface, including lotic, lentic, and wetland ecosystems (Barange et al. 2010; Bashir et al. 2020).

Generally, we come across the toxic effects resulting from many pollutants in aquatic ecosystems. Among the contaminants leading toward this toxicity are heavy metals, detergents, pesticides, fertilizers, microfibers, plastics, other chemicals, and radioactive elements. The major role in this connection is played by industrial solvents, paints, volatile organic compounds, and drugs as well as those originating from anthropogenic impacts and industrial wastes (Hampel et al. 2015; Bhat et al. 2017; Bashir et al. 2020).

Aquatic ecosystems have played an important role in the development of civilizations. These ecosystems possess a natural tendency to overcome pollution with their buffering power, therefore it is important to maintain their quality. This is one of the important environmental issues of the day (Demirak et al. 2006; Fernandesa et al. 2007; Hampel et al. 2015; Mateo-Sagasta et al. 2017; Bashir et al. 2020).

2.2 Threats to the Health of Aquatic Ecosystems

Unlike other environments, aquatic habitats, especially freshwater ecosystems, may be exposed to more pollution due to the use of water in industrial processes and the discharge of wastewater from industry and urban areas into this environment. All these have a direct impact on their biodiversity (Araujo et al. 2021; Ceschin et al. 2021; Ding et al. 2021; Servadio et al. 2021; Rakib et al. 2022). These ecosystems support an enormous diversity of life and are critical for human survival (Othman et al. 2021). With the increase in the world population, the demand for water use is increasing day by day. This situation, especially as a result of anthropogenic activities, pollution, stress, and changing climatic conditions are disrupting the natural water cycle, therefore putting all aquatic ecosystems, especially freshwater ones under tremendous stress (Erasmus et al. 2022; Reddy et al. 2022). The effect of harmful pollutants on aquatic life is indisputably constituting one of the most important environmental problems. Therefore, a detailed study of aquatic ecosystems will play an important role

Aquatic Contamination: Tolerance and Bioremediation, First Edition. Edited by Rouf Ahmad Bhat, Gowhar Hamid Dar, Fernanda Maria Policarpo Tonelli, and Saima Hamid.

in revealing the degree of environmental pressure and demonstrating the effectiveness of reinstatement actions (Akhtar et al. 2021; Kolarova and Napiórkowski 2021).

Some of the threats leading to pollution in aquatic ecosystems are outlined here:

2.2.1 Threats from Agricultural Practices: Agrochemicals and Other Chemicals

The global population increase has increased the demand for food. This has led to the expansion of agricultural areas for increasing crop production together with an increase in the application of the amount of agricultural chemicals (hormones, pesticides, fertilizers, herbicides, etc.) to increase crop productivity (Schwarzenbach et al. 2010; Mateo-Sagasta et al. 2017; Bashir et al. 2020; Mushtaq et al. 2020). Unsustainable use of agricultural chemicals has led toward their accumulation in the rivers, lakes, and aquifers in nearby areas, with factors such as runoff, direct drift, and seepage from agricultural as well as urban and industrial areas (Rathore and Nollet 2016; Bashir et al. 2020; Haq et al. 2021).

The application of excessive amounts of organic fertilizers and/or synthetic chemical fertilizers in agricultural areas causes aquatic pollution. Moreover, intensive open storage and/or storage in agricultural areas also is washed away following heavy rainfall and cause pollution in existing aquatic ecosystems through surface runoff. The excess accumulation of nitrogenous and phosphate fertilizers leaching into groundwaters end up with their pollution as well as pollution of the surface water bodies following surface runoff. These cause an overgrowth in these ecosystems, especially aquatic plants, resulting in eutrophication due to nutrient enrichment over time. This is undesirable for the health of sustainable aquatic ecosystems (Mateo-Sagasta et al. 2017; Bashir et al. 2020).

The other potential threat to aquatic ecosystems is pesticides including insecticides, herbicides, and fungicides. In many different countries, these chemicals are widely used in agricultural fields. These easily move toward the aquatic ecosystems, thereby playing an active role in polluting our water resources. The chemicals involved in these applications constitute a very dangerous group because they contain carcinogens and other toxic substances which can kill aquatic life. These can at the same time reach humans through the food chain and can cause fatal cases due to their toxic levels (Schreinemachers and Tipraqsa 2012; Mateo-Sagasta et al. 2017; Bhat et al. 2018; Bashir et al. 2020).

In arid and semi-arid regions, irrigation in agricultural areas leads to salt accumulation in the soil (Ozturk et al. 2014). These accumulated salts are easily transported to aquatic habitats by drainage waters and cause salinization (Mateo-Sagasta et al. 2017; Bashir et al. 2020; Parnian et al. 2022). High levels of salt in aquatic environments can cause changes in the geochemical cycles of key elements such as carbon, iron, nitrogen, phosphorus, and sulfur. Salinization in freshwater biota also causes significant negative changes at the species and community level (Herbert et al. 2015; Mateo-Sagasta et al. 2017).

Chemicals such as antibiotics, vaccines, growth promoters, and hormones, which are frequently used in agricultural lands to increase yield and quality, do appear as functional pollutants in aquatic ecosystems. Moreover, the reuse of wastewater for irrigation purposes in agricultural areas is another threatening factor, effective in the spread of existing pollution in aquatic environments (Mateo-Sagasta et al. 2017; Bashir et al. 2020; Haq et al. 2021).

2.2.2 Popular Anthropogenic Threats: Sewages from Industrial, Municipal, and Household Wastes

These constitute a potential threat after entering the ecosystem from the bath, washing machine, kitchen, and fecal waste. These constitute the largest volume of waste discharged into aquatic ecosystems (Das and Acharya 2003; Tukura et al. 2009; Bashir et al. 2020). The most important negative effect of sewage is based on the composition and concentration of pollutants and the volume and frequency of wastewater entering the system (Akpor and Muchie 2011; Bhat et al. 2017; Bashir et al. 2020). Some researchers have reported that 58% of the wastewater originating from urban areas and 81% originating from industrial areas are discharged directly into aquatic ecosystems without treatment and/or insufficient treatment, which causes pollution of up to 73% of existing aquatic ecosystems (Vargas-Gonzalez et al. 2014). Keeping this in view, as long as necessary precautions are not taken, this percentage value is expected to increase even more with the rapid increase in urbanization in parallel with the increase in population density. The pollution from sewers causes a decrease in water quality and the usability of aquatic ecosystems. There is the precipitation of pollution-rich substances in sediments, the occurrence of eutrophication, changes in the distribution of species in the biota, and the loss of some species (Oberdorff et al. 2002; Bhat et al. 2017; Bashir et al. 2020).

2.2.3 Eutrophication: A Threat to Water Quality Deterioration, Economic Extinction of Fisheries, and a Risk to Public Health

The agricultural, industrial, and sewage disposals lead to nutrient enrichment resulting from anthropogenic activities in both freshwater and coastal marine ecosystems around the world. The point source discharges from human activities and limitations of nitrogen and phosphorus nutrients accelerate the rate and extent of eutrophication through nonpoint loading in such ecosystems (Schindler 1974; Carpenter et al. 1998; Bhat et al. 2017; Bashir et al. 2020). The main criterion in eutrophication is characterized by excessive plant and algal growth because of their increased presence in the environment; for this purpose, one or more limiting growth factors are needed for photosynthesis, e.g. sunlight, carbon dioxide, and nutrients (Schindler 2006). This can lead to more dire consequences over time. In particular, as a result of eutrophication in aquatic ecosystems, the light penetration in the water is limited by the excessive growth of algae, followed by a reduction in the vegetative growth, which is interrupted over time, ending up with the death of plants in coastal areas (Bashir et al. 2020). In aquatic ecosystems exposed to eutrophication, potential drinking water sources serving humans also face dramatic consequences; these also affect fisheries and recreational water bodies (Carpenter et al. 1998; Bhat et al. 2017).

2.2.4 Plastics as an Ignored Global Threat

Among the anthropogenic pressures on the ecosystems under question, an accumulation of plastic waste is also becoming the most notable lately. The use of plastics in human society provides significant benefits, but these easily accumulate in aquatic ecosystems, due to their unsustainable use and inappropriate waste management (Andrady and Neal 2009; Barnes et al. 2009; Bashir et al. 2020). These are dumped in huge volumes on beaches, lakes, navigation channels, and other water bodies (Lechner et al. 2014). A report by GESAMP (2010) has fully emphasized that 70–80% of marine litter, most of which is plastic, comes from internal sources and spreads from rivers to the sea and/or oceans. The potential sources of plastics include wastewater treatment plants; runoffs from urban, agricultural, touristic, and industrial areas; shipping activities; sewage sludge; beach litter; fisheries; and ports as per the studies conducted by different researchers (Zubris and Richards 2005; Norén 2007; GESAMP 2010; Claessens et al. 2011; Leslie et al. 2012). In the marine environment, plastics of various sizes, classes, and origins (particularly single-use plastic bags and microbeads) are ubiquitous and affect an aesthetic problem as well as the large number of species that become entangled in or ingest plastics (Perkins 2015).

Depending on the ecological conditions of the environment, plastics may break down to sizes smaller than 5 mm over time, resulting in microplastics. Microplastics, which are smaller in size, can be found in aquatic ecosystems and food web complexes. From plankton to fish, birds, and even mammals, there may be a tendency to take up microplastics. In view of this, although microplastics are permanent in the aquatic ecosystem, they also cause bioaccumulation in the aquatic biota with the toxic compounds found in them (Oehlmann et al. 2009; Teuten et al. 2009; Ashton et al. 2010; Zarfl and Matthies 2010; Bakir et al. 2012; Koelmans et al. 2013; Rochman 2013; Wright et al. 2013).

2.2.5 A Toxic Threat to All Life Forms: Oil Spills

Liquid petroleum hydrocarbons too cause environmental pollution after being discharged into the environment as a result of anthropogenic activities (oil tankers, facilities, nontank ships, pipelines, and other sources) in marine ecosystems. These are called "oil spills" (Bashir et al. 2020), which too are included among the hazardous waste originating from the damages in ships, shipping accidents, as well as other industrial and mining activities (Bartha and Bossert 1984; Bashir et al. 2020). Knapp and Van de Velden (2011) have also reported that marine oil spills from ships can occur as a result of shipwrecks and/or operations and/or deliberate dumping of oily wastes into the oceans. For this reason, oil spills are accepted as the most common organic pollutant of aquatic ecosystems, with highly toxic effects on all life forms because these damage both aquatic as well as terrestrial organisms (Margesin and Schinnur 1997; Alló and Loureiro 2013).

2.2.6 Threats Posed by Nonecofriendly Practices in Aquatic Ecosystems: Land Reclamation, Reservoir, and Dam Constructions

These include nonenvironmentally friendly practices, particularly land reclamation for food production and infrastructure construction, irrigation-water extraction for industrial and domestic uses, and separation from main rivers by dam construction (river floodplains) (Junk et al. 2013).

Globally, in regions with very high population density, land reclamation (especially to obtain arable land) and drainage events contribute negatively to both land degradation and pollution in aquatic ecosystems (especially wetland and coastal ecosystems). Such environmentally unfriendly practices cause landslides with an increase in the precipitation flow rate in the regions; this needs to know and understand the ecological structure of the relevant regions (Junk et al. 2013). Moreover, the practices like river straightening, channel opening, agricultural land expansions, and civil constructions, especially around the wetlands, cause serious irreversible damages (Sousa Jr. et al. 2011; Junk et al. 2013).

The construction of reservoirs for storing some of the annual runoff for irrigation, hydropower generation, flood control, and water supply is also a major threat to wetlands; the reservoir and embankment constructions change the discharge pattern and sediment load of rivers, by cutting the longitudinal connection, and are therefore considered a major threat as they negatively change the ecological conditions (Junk et al. 2013).

2.2.7 Other Deadly Threats in Aquatic Ecosystems: Acidic Rains and Radioactive Wastes

Thermal power plants emit large amounts of SO_2 and NO_2 into the atmosphere, which react with rainwater, reducing its pH below 5. As a result of this, it causes the formation of a well-known phenomenon – the acid rains, which can have a negative impact on the existing biota within the scope of the hydrological cycle, especially in aquatic ecosystems (Hamid et al. 2020).

Radioactive materials, radioactive deposits, and radioactive minerals from nuclear power plants potentially constitute nuclear waste. These wastes pose a serious threat to water, soil, and all living organisms, even in very small concentrations. In fact, since the half-life of some dangerous radioactive elements in these wastes lasts for thousands of years, they appear as a potential danger which can cause many deadly diseases, especially cancer, in living beings for generations (Hamid et al. 2020).

2.2.8 A Threat with a Very High Input Rate into Aquatic Ecosystems: Heavy Metals

Heavy metals are naturally occurring elements with a density greater than $5\,g/cm^3$, widely distributed in the environment due to natural processes and anthropogenic activities (Ghori et al. 2019; Mohan et al. 2021). The penetration of heavy metals as natural constituents of the earth's crust is ubiquitous (Maanan et al. 2015; Zaynab et al. 2022). However, industrial, military, domestic, medical, technological, and agricultural processes and practices around the globe can maximize their accumulation effects on the environment (Ozturk et al. 2017; Ghori et al. 2019; Caparrós et al. 2022). Although there are a few heavy metals that are necessary for life, high intakes can lead to toxic and carcinogenic effects causing hypertension; atherosclerosis; cardiovascular diseases; endocrine dysfunction; kidney failure; infertility; and many chronic and acute diseases including bones, deformities, hyperglycemia, diabetes, and inflammation have a wide range of devastating health consequences, including neurological diseases, autoimmune diseases, prostate dysfunction, osteomalacia, and osteoporosis (Aendo et al. 2019; Sofoulaki et al. 2019; Chen et al. 2021; Nag et al. 2022; Ozturk et al. 2022).

The aquatic ecosystems contaminated with heavy metals can get accumulated up to a highly toxic concentration because they can easily accumulate in water reservoirs, are easily transferred between existing aquatic organisms in the food chain, cannot be easily removed from the water, and may thus result in a high degree of pollution in aquatic ecosystems (Xu and Yang 1996; Loska and Wiechuła 2003; Harguinteguy et al. 2014; Bashir et al. 2020). Therefore, these are serious toxic substances with highly pathogenic risks to the health of ecosystems. Cadmium, chromium, cobalt, copper, mercury, lead, and nickel are the leading ones which cause environmental pollution (Cui et al. 2021; Rostami et al. 2021; Li et al. 2022; Parnian et al. 2022).

2.2.8.1 A Case Study on Nickel Pollution

Nickel was first isolated in the eighteenth century and is among the most abundant elements in the earth's crust, where it occurs as a free metal or with iron in igneous rocks. In the periodic table, its atomic number is 28 and atomic weight is 58.71 (Thyssen 2011; Chowdhury et al. 2021; Ferreira et al. 2021; Begum et al. 2022). As a member of the transition series in the periodic table, it is readily soluble in dilute oxidizing acids, although it tends to resist corrosion by air, water, and alkali (Tran et al. 2022). In general, naturally occurring nickel concentrations in soil and surface waters are less than 22 ppm and 200 ppb, respectively (Chen et al. 2009).

Nickel is also released into the environment from anthropogenic activities such as metal mining, smelting, fossil fuel burning, vehicle emissions, domestic and industrial waste disposal, fertilizer application, and organic fertilizers (El-Naggar et al. 2021; Said 2022). It is mainly used as a raw material in the metallurgical and electroplating industries, as a catalyst in the chemical and food industries, and as a component of electric batteries (Lee and Lee 2021; Yadav et al. 2021). It can be found in high concentrations in aquatic environments with common applications in many industrial processes such as electroplating, coin manufacturing, battery manufacturing, forging, porcelain enamel, copper sulfate production, and power stations (Mu et al. 2018; Bigos et al. 2021; Pang et al. 2022). The chronic toxicity of nickel in humans and the environment is well known (Das et al. 2019). While nickel can be useful as an activator of some enzyme systems in trace amounts, high concentrations can cause lung and bone cancer (Zhu et al. 2019; Yoneda et al. 2021; de Oliveira et al. 2022). At very high exposure levels, nickel salts are known to be carcinogenic (Son 2020).

The presence of its ions above the permissible limits in aquatic ecosystems and wastewater poses a significant threat to the environment and public health (Anchidin-Norocel et al. 2021). Considering toxicity, incremental accumulation in the food chain and persistence in the ecosystem, wastewater containing nickel, like other heavy metals, needs to be treated appropriately before being discharged into receiving waters (Punia et al. 2021). Chemical precipitation, membrane filtration, ion exchange, and co-precipitation/adsorption are applied among various treatment methods to remove these ions from water and wastewater (Costa et al. 2022; Singh et al. 2022).

An attempt is made here to evaluate the threats causing global water pollution, especially nickel because of its toxic effects. The existing Gölbaşı Lake in the Hatay province of Turkiye has been chosen as the pilot study area. The water samples were taken from different locations of the lake, along with the sediments, and the total nickel content in the rhizome, petiole, and leaves of *N. alba* and *N. lutea* plant species was investigated. The results enlighten the extent of heavy metal pollution using these plants as bioindicators.

2.3 Data Analysis

2.3.1 Study Area

Gölbaşı Lake (Kırıkhan-Hatay) is located in the Eastern Mediterranean basin, near the Amik plain in the easternmost part of Adana State in the Turkish Mediterranean Region (Figure 2.1). Its geographical features have led to the formation of different wetland ecosystems. This freshwater ecosystem has been taken under protection as a natural protected area in 1992 (Doğaner 2001; Özşahin 2011). The lake basin, which opens to the Amik plain from the west, is located in a position surrounded by swamps from the northwest, and bedrock (limestone and basalt) elevations from the east, south, and partly

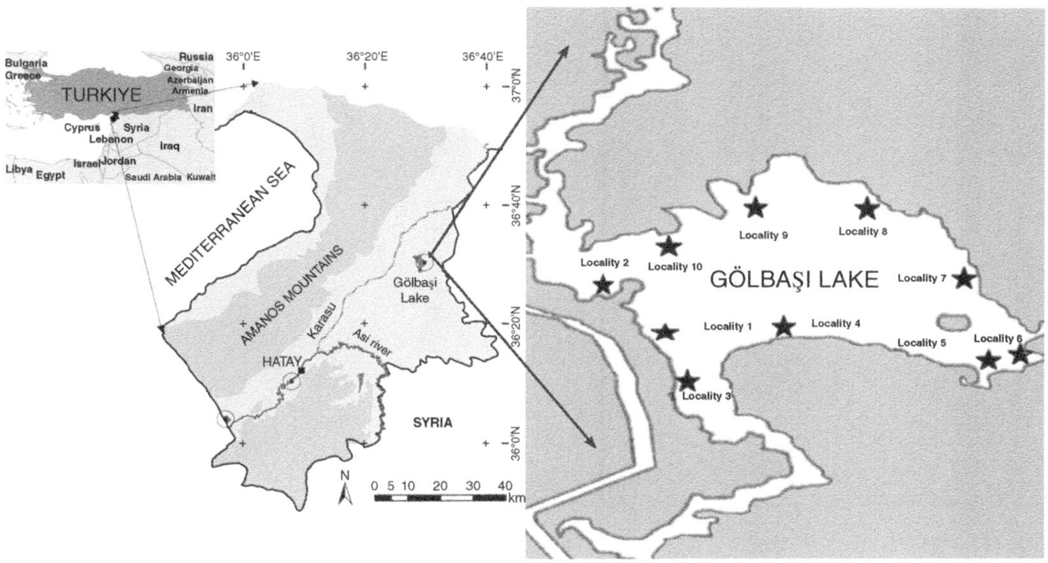

Figure 2.1 Location map showing the sample collection localities from the research area. *Source:* modified from Türkmen and Ciminli (2011); Atahan et al. (2015).

Figure 2.2 The photographic views of the studied localities. (a) locality 4; (b) locality 3; (c) locality 2; (d) locality 3; (e) locality 1; (f) locality 6.

from the north. According to Özşahin (2011), the lake is fed by the spring waters coming out of the skirts of these elevations. The settlements around the lake benefit from it primarily in the form of agricultural irrigation, drinking and utility water, and fishing. This situation is responsible for the developments in the lake area (Figure 2.2). Therefore, this lake is very important for agriculture and aquaculture for the local people.

2.3.2 Water, Soil, and Plant Materials

N. lutea (L.) and *N. alba* (L.) species (Figure 2.3) belonging to the Nymphaeaceae family were selected as plant materials. In all, 10 localities from the research area were selected, taking into account the population density around the lake (Figures 2.1 and 2.4). The rhizome, petiole, and leaf samples of these species were collected together with the sediment and water samples. The samples were brought to the laboratory for comparative analysis (Figure 2.5), kept in an oven at 80 °C for 48 hours to dry. The plant parts of the samples were dried in an oven to remove moisture, powdered with a high-speed grinder (Lavion) to ensure homogeneity. These sediment samples were transferred into glass Petri dishes, dried in an oven at 80 °C for 48 hours and passed through a steel sieve with a pore diameter of 2 mm and made ready for weighing (Figure 2.6).

2.3.3 Determination of Nickel Concentrations

0.2500 g of the plant and sediment samples were taken using an analytical balance (Precisa XB 220A SCS) and transferred to Teflon Wessels of Berghof – MSW2 brand microwave device. For plant parts, 8 ml 65% Merck brand HNO_3, for sediment samples 6 ml 65% Merck brand HNO_3, 3 ml 37% Merck brand HCl, and 2 ml 48% Merck brand HF were used. The dissolution process was carried out to prepare the samples. These were transferred to 50 ml sterile falcon tubes using ultrapure water and filtered with a blue band Whatman filter. The water samples taken from the lake were transferred to sterile 50 ml falcon tubes by filtering through blue band filter paper for elemental analysis. The filtered samples were diluted with 1% HNO_3 into a new falcon tube and made ready for elemental analysis. The elemental analysis was carried out with an inductively coupled plasma-optical emission spectrometry (ICP-OES, PerkinElmer – Optima 7000DV) instrument using nickel standard solutions (1000 mg/L – Merck).

2.4 Results from the Study

N. alba was observed in the 1, 2, 3, 9, and 10 coded localities and *N. lutea* in 4, 5, 6, 7, and 8 coded localities, during two different vegetation periods (autumn and spring). The samples were collected from each locality. These included water, sediment, and plant parts (rhizome, petiole, and leaf) taken from their distribution areas and nickel content was examined (Table 2.1).

Nickel values in **water samples** taken from the localities where *N. alba* is distributed are the lowest in the 10th locality, whereas the 9th locality contains the highest value (in both autumn and spring seasons) (Table 2.1). Although it has the same order in both seasons, the current nickel values in the water samples from smallest to largest are recorded as follows: on the basis of the locality $10 < 3 < 2 < 1 < 9$ (Table 2.1). In the water samples, where the *N. lutea* was distributed, the 6th locality contains the highest value during the autumn as well as spring seasons. The current nickel values in other locations varied on a seasonal basis (Table 2.1). It has been observed that the current concentration values of this heavy metal in water are higher in the autumn season than spring season in all localities (Table 2.1).

In the **sediment samples** taken from *N. alba* localities, in both autumn and spring seasons, nickel values were the lowest in the 10th locality, while the 9th locality contained the highest value (Table 2.1). Although it has the same order in both seasons, the current nickel values in the sediment samples as smallest to largest on the basis of the locality were as follows; $10 < 3 < 2 < 1 < 9$ (Table 2.1). In *N. lutea*, on the other hand, although the ranking of values varies on a seasonal basis, nickel values in both autumn and spring seasons are the lowest in the 8th locality, while the 6th locality shows the highest value (Table 2.1). The current concentration values of nickel in the sediment are much higher in the autumn season compared to the spring in all localities (Table 2.1).

In the **rhizome** samples of ***N. alba*** collected from different localities in the autumn and spring seasons, the nickel values were the lowest in the 10th locality, while the 9th locality depicted the highest value. If nickel values present in the rhizomes are evaluated on the basis of localities, their order from smallest to largest is as follows: $10 < 3 < 2 < 1 < 9$ (Table 2.1). In the **petioles** of this species, nickel values in both autumn and spring seasons are the lowest in the 9th locality, but the

Figure 2.3 *Nuphar lutea* (right) and *Nymphaea alba* (left).

10th locality showed the highest value. An evaluation of the values present in the **leaves** on the basis of localities reveals their order from smallest to largest as $10 < 3 < 2 < 1 < 9$. Nickel values of *N. lutea* detected in all plant parts collected from different localities in both the autumn and spring seasons varied. As in the water and sediment samples, the nickel values detected in the plant parts of both species were higher in the autumn season at all locations (Table 2.1).

Chen et al. (2009) have provided acceptable references for maximum nickel values in surface waters and these are less than 200 μg/L. As per these authors, the acceptable reference maximum nickel values in the sediments are less than 22 mg/kg (Chen et al. 2009). According to Sobati-Nasab et al. (2021), acceptable reference nickel values in the plant species have been reported to be around 0.01–5 mg/kg dry weight. A comparison with the limit values accepted as a reference; the data determined in our studies shows that the sediments in all locations are well below the acceptable limit values. However, acceptable limit values exceeded in surface waters and plant parts (especially in rhizomes and leaves) (Table 2.1). Therefore, the nickel values in the water samples taken from Gölbaşı Lake exceed the limit values, posing a threat in terms of aquatic ecosystem pollution.

The aquatic plants used here may serve as bioindicators for nickel through active and passive absorption (Bose et al. 2008; Vodyanitskii and Shoba 2015; Bonanno et al. 2018). Some plants distributed in aquatic ecosystems can accumulate high levels of heavy metals from water and sediment, the reason being their well-developed roots and rhizomes. They appear to show high tolerance to toxicity, with high-yielding biomass, and stable structures (Milošković et al. 2013; Rezania et al. 2016). The heavy metal concentrations are lower in the water samples examined in the study area compared to the sediment and related parts of the plants investigated. The high concentrations of heavy metals in the plant samples, especially in the rhizome and leaves, are in full consistence with other studies (Guilizzoni 1991; Duman et al. 2006; Grudnik 2010).

Nickel is mainly used as a raw material in the metallurgical and electroplating industries, as a catalyst in the chemical and food industries, and as a component of electric batteries (Lee and Lee 2021; Yadav et al. 2021). As stressed earlier, it is released into the environment from different anthropogenic activities (El-Naggar et al. 2021; Said 2022). Therefore, the application of intensive organic fertilizer application, sprinkling of chemical control agents, disposal of domestic wastes, burning of fossil fuels, and vehicle emissions around the agricultural lands surrounding the aquatic ecosystems encountered frequently as anthropogenic activities should be decreased (Figure 2.7).

The presence of nickel ions in water and wastewater above the permissible limits poses a significant threat to the environment and public health (Anchidin-Norocel et al. 2021). The chronic toxicity of nickel in humans as well as in the environment is well known (Das et al. 2019). Nickel may be useful as an activator of some enzyme systems in trace amounts, but its high concentrations can cause lung and bone cancer (Zhu et al. 2019; Yoneda et al. 2021; de Oliveira et al. 2022). At very high exposure levels, its salts are known to be carcinogenic (Son 2020). Moreover, with an increase in nickel pollution, its high concentrations are frequently encountered rather than a deficiency in plants (Lopez et al. 2021). Toxic effects of high nickel concentrations in plants frequently reported include inhibition of mitotic activities, reduced plant growth, and adverse effects on fruit yield and quality (Haseeb et al. 2022; Roccotiello et al. 2022). Excessive nickel concentration also

Figure 2.4 General views of the seasonally investigated localities in the Gölbaşı Lake area. (a) locality 1 (autumn); (b) locality 1 (spring); (c) locality 2 (autumn); (d) locality 2 (spring); (e) locality 3 (autumn); (f) locality 3 (spring); (g) locality 4 (autumn); (h) locality 4 (spring); (i) locality 5 (autumn); (j) locality 5 (spring); (k) locality 6 (autumn); (l) locality 6 (spring); (m) locality 7 (autumn); (n) locality 7 (spring); (o) locality 8 (autumn); (p) locality 8 (spring); (q) locality 9 (autumn); (r) locality 9 (spring); (s) locality 10 (autumn); (t) locality 10 (spring).

Figure 2.5 Samples collected from the localities. (a) recording location parameters; (b–d) collection of plant samples; (e) water sampling procedures; (f) transfer of water and sediment samples to storage containers; (g) transfer of collected samples from the locations to the laboratory; (h) preparation of sediment samples for elemental analysis; (i) separation of rhizome, petiole, and leaf parts of the samples in the laboratory.

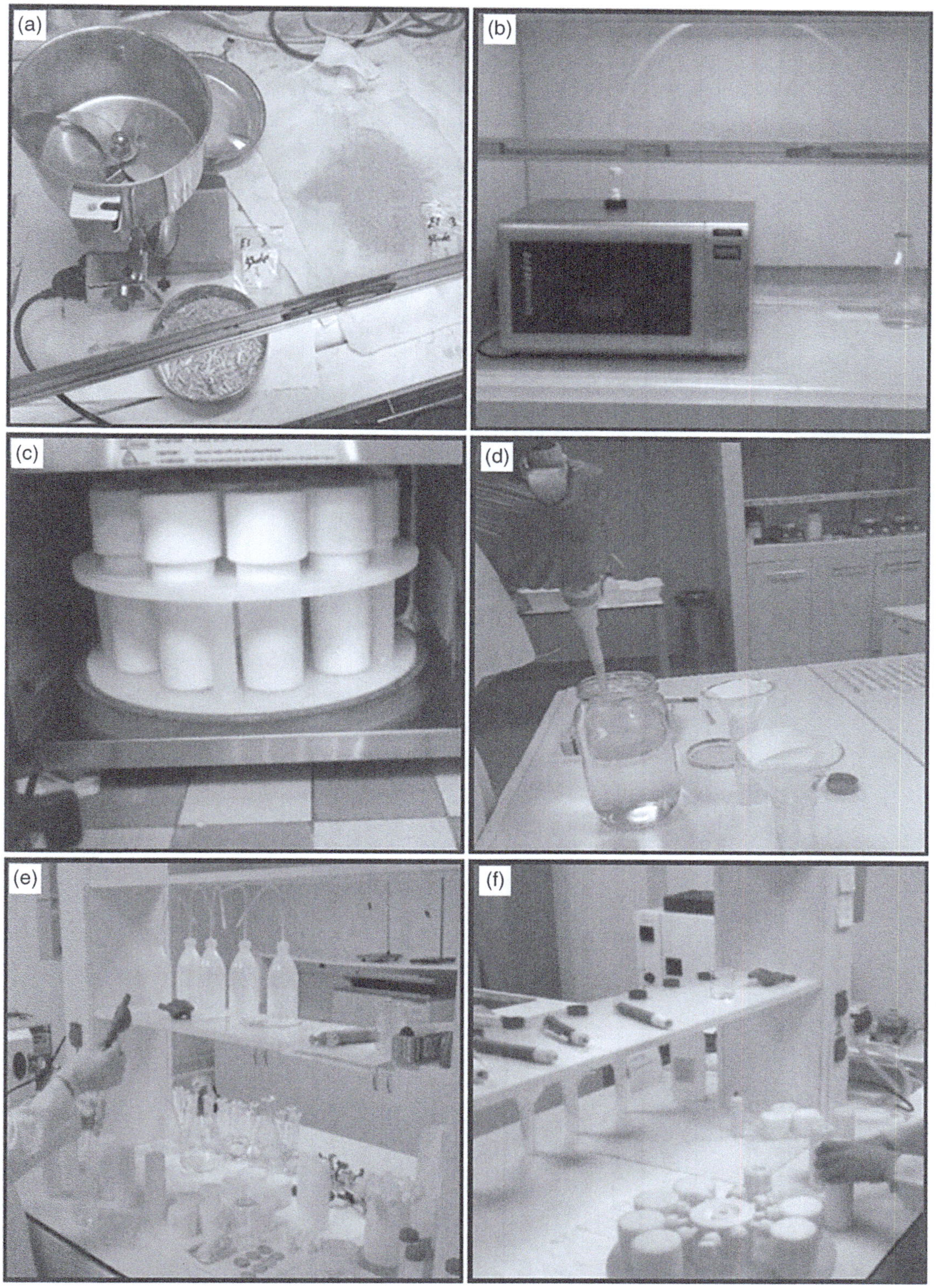

Figure 2.6 Sample preparations for analysis. (a) grinding of oven-dried plant parts dried using a high-speed grinder; (b) Berghof MWS2 brand device used for digestion of samples; (c) microwave Teflon Wessels used for digestion; (d) preparing water samples for elemental analysis following filtration; (e) preparation of plant/sediment samples for digestion; (f) placement of Teflon Wessels in their sockets for digestion.

Table 2.1 Nickel concentrations in plant parts (mg/kg), sediments (mg/kg), and water (μg/L) in the Gölbaşı Lake ecosystem. (Bold numbers refer to min. and max. values).

		Leaf		Petiole		Rhizome		Sediment		Water	
Species	Locality no	Autumn	Spring	Autumn	Spring	Autumn	Spring	Autumn	Spring	Autumn	Spring
Nymphaea alba	1	8.135 ± 0.012	7.200 ± 0.008	2.993 ± 0.002	2.553 ± 0.003	6.676 ± 0.008	5.919 ± 0.007	3.521 ± 0.005	0.607 ± 0.011	457.363 ± 5.054	400.606 ± 5.732
	2	7.849 ± 0.009	6.855 ± 0.007	2.806 ± 0.003	2.457 ± 0.002	6.664 ± 0.009	5.919 ± 0.008	3.374 ± 0.006	0.605 ± 0.012	442.172 ± 6.217	393.336 ± 4.621
	3	7.540 ± 0.010	6.571 ± 0.009	2.838 ± 0.003	2.454 ± 0.003	6.299 ± 0.007	5.441 ± 0.006	3.212 ± 0.004	0.577 ± 0.009	424.799 ± 7.392	387.697 ± 4.983
	9	**8.345 ± 0.012**	7.296 ± 0.007	3.077 ± 0.005	2.633 ± 0.004	7.031 ± 0.008	6.094 ± 0.009	3.640 ± 0.005	0.635 ± 0.007	482.243 ± 5.004	412.656 ± 7.437
	10	7.223 ± 0.008	6.380 ± 0.009	2.735 ± 0.004	**2.364 ± 0.003**	6.067 ± 0.009	5.377 ± 0.011	3.202 ± 0.007	**0.555 ± 0.010**	421.583 ± 6.832	**370.883 ± 4.886**
Nuphar lutea	4	8.521 ± 0.011	7.350 ± 0.007	3.072 ± 0.005	**2.693 ± 0.005**	7.110 ± 0.011	6.253 ± 0.013	3.681 ± 0.005	0.659 ± 0.011	478.891 ± 5.821	423.085 ± 6.037
	5	8.774 ± 0.012	7.685 ± 0.009	3.281 ± 0.004	2.866 ± 0.003	7.291 ± 0.010	6.515 ± 0.009	3.697 ± 0.004	0.656 ± 0.013	498.088 ± 4.994	426.294 ± 8.115
	6	**8.939 ± 0.014**	7.904 ± 0.008	3.291 ± 0.006	2.844 ± 0.004	7.443 ± 0.009	6.410 ± 0.010	**3.926 ± 0.006**	0.693 ± 0.010	**510.013 ± 7.374**	441.668 ± 5.243
	7	8.419 ± 0.017	7.474 ± 0.007	3.175 ± 0.005	2.839 ± 0.004	7.105 ± 0.010	6.236 ± 0.011	3.670 ± 0.005	0.668 ± 0.009	496.891 ± 6.883	432.873 ± 6.925
	8	8.223 ± 0.011	7.500 ± 0.006	3.160 ± 0.004	2.780 ± 0.005	7.109 ± 0.012	6.142 ± 0.013	3.633 ± 0.007	0.630 ± 0.007	480.462 ± 7.047	421.524 ± 7.275

Figure 2.7 Anthropogenic factors causing pollution in the Gölbaşı Lake. (a–d) daily tourism activities in the vicinity of the lake during spring and summer seasons, and disposal of domestic waste; (e) transit access roads located very close to the lake; (f–g) agricultural lands surrounding the lake; (h and i) a marble quarry near the lake.

affects the structural and anatomical dynamics of plants (Saraiva et al. 2021). Higher concentrations have been shown to reduce vegetative growth such as plant height and overall biomass production in various agricultural crops (Rehman et al. 2022).

2.5 Conclusions

Deterioration and/or damage to aquatic ecosystems is/are mostly due to human activities, especially agricultural activities, as well as other effects originating from the density of transit transportation roads. The anthropogenic activities causing water pollution include mainly the discharge of solid, agricultural, and domestic wastes in the Gölbaşı Lake area. The chemical fertilizers and pesticides used for agricultural pests on the agricultural lands around the lake should be checked carefully by the relevant agricultural directorates. The use of chemicals containing heavy metals should be limited. Precautions should also be taken urgently for the regulation of domestic wastes which cause pollution in the lake and its vicinity. As part of the eco-sustainability of the lake, it is imperative that heavy metals are treated effectively before these wastewaters are discharged into the lake ecosystem.

Other anthropogenic effects which cause heavy metal pollution in this aquatic ecosystem should be reduced, and awareness-raising activities carried out for local people within the scope of environmental education practices. These activities should be started urgently with the support of relevant official institutions. The size of the pollution should be controlled by repeating analysis in the research area at regular intervals (at least twice a year).

The aquatic natural vegetation that develops in the Gölbaşı Lake and its immediate surroundings must need to be protected. The sustainability of this aquatic ecosystem draws attention in terms of its ability to act as a filter in terms of heavy metal pollution and its attractive landscape features. In light of this evaluation, both these species evaluated here appear as important bioindicators for aquatic ecosystems. They have the potential to act as filters in terms of heavy metal pollution.

References

Aendo, P., Thongyuan, S., Songserm, T., and Tulayakul, P. (2019). Carcinogenic and non-carcinogenic risk assessment of heavy metals contamination in duck eggs and meat as a warning scenario in Thailand. *Sci. Total Environ.* 689: 215–222.

Akhtar, N., Syakir Ishak, M.I., Bhawani, S.A., and Umar, K. (2021). Various natural and anthropogenic factors responsible for water quality degradation: a review. *Water* 13 (19): 2660.

Akpor, O.B. and Muchie, M. (2011). Environmental and public health implications of wastewater quality. *Afr. J. Biotechnol.* 10 (13): 2379–2387.

Alló, M. and Loureiro, M.L. (2013). Estimating a meta-damage regression model for large accidental oil spills. *Ecol. Econ.* 86: 167–175.

Altay, V. and Ozturk, M. (2012). Land degradation and halophytic plant diversity of Milleyha wetland ecosystem (Samandağ-Hatay), Turkey. *Pak. J. Bot.* 44: 37–50.

Anchidin-Norocel, L., Savage, W.K., Gutt, G., and Amariei, S. (2021). Development, optimization, characterization, and application of electrochemical biosensors for detecting nickel ıons in food. *Bios* 11 (12): 519.

Andrady, A.L. and Neal, M.A. (2009). Applications and societal benefits of plastics. *Philos. Trans. R. Soc. Lond. B Biol. Sci.* 364: 1977–1984.

Araujo, D.F., Knoery, J., Briant, N. et al. (2021). Metal stable isotopes in transplanted oysters as a new tool for monitoring anthropogenic metal bioaccumulation in marine environments: the case for copper. *Environ. Pollut.* 290: 118012.

Ashton, K., Holmes, L., and Turner, A. (2010). Association of metals with plastic production pellets in the marine environment. *Mar. Pollut. Bull.* 60: 2050–2055.

Atahan, A., Gül, O., Atahan, M., and Gül, M. (2015). Unusual wintering records of pipits (Aves: Motacillidae) in Hatay, eastern Mediterranean region of Turkey. *Turk. J. Zool.* 39: 74–79.

Bakir, A., Rowland, S.J., and Thompson, R.C. (2012). Competitive sorption of persistent organic pollutants onto microplastics in the marine environment. *Mar. Pollut. Bull.* 64: 2782–2789.

Barange, M., Field, J.G., Harris, R.P. et al. (2010). *Marine Ecosystems and Global Change*. Oxford: Oxford University Press.

Barnes, D.K., Galgani, F., Thompson, R.C., and Barlaz, M. (2009). Accumulation and fragmentation of plastic debris in global environments. *Philos. Trans. R. Soc. Lond. B Biol. Sci.* 364: 1985–1998.

Bartha, R. and Bossert, I. (1984). The treatment and disposal of petroleum refinery wastes. In: *Petroleum Microbiology* (ed. R.M. Atlas), 1–61. New York: Macmillan.

Bashir, I., Lone, F.A., Bhat, R.A. et al. (2020). Concerns and threats of contamination on aquatic ecosystems. In: *Bioremediation and Biotechnology* (ed. K.R. Hakeem et al.), 1–26. Cham: Springer.

Begum, W., Rai, S., Banerjee, S. et al. (2022). A comprehensive review on the sources, essentiality and toxicological profile of nickel. *RSC Adv.* 12 (15): 9139–9153.

Bhat, R.A., Shafiq-ur-Rehman, M.M.A., Dervash, M.A. et al. (2017). Current status of nutrient load in Dal Lake of Kashmir Himalaya. *J. Pharmacog. Phytochem.* 6 (6): 165–169.

Bhat, R.A., Beigh, B.A., Mir, S.A. et al. (2018). Biopesticide techniques to remediate pesticides in polluted ecosystems. In: *Handbook of Research on the Adverse Effects of Pesticide Pollution in Aquatic Ecosystems* (ed. K.A. Wani and K. Mamta), 387–407. Hershey, PA: IGI Global.

Bigos, A., Wolowicz, M., Janusz-Skuza, M. et al. (2021). Citrate-based baths for electrodeposition of nanocrystalline nickel coatings with enhanced hardness. *J. Alloys Compd.* 850: 156857.

Bonanno, G., Vymazal, J., and Cirelli, G.L. (2018). Translocation, accumulation and bioindication of trace elements in wetland plants. *Sci. Total Environ.* 2018: 252–261.

Bose, S., Vedamati, J., Rai, V., and Ramanathan, A.L. (2008). Metal uptake and transport by *Typha angustata* L. grown on metal contaminated waste amended soil: an implication of phytoremediation. *Geoderma* 145: 136–142.

Caparrós, P.G., Ozturk, M., Gul, A. et al. (2022). Halophytes have potential as heavy metal phytoremediators: a comprehensive review. *Environ. Exp. Bot.* 193: 104666.

Carpenter, S.R., Caraco, N.F., Correl, D.L. et al. (1998). Nonpoint pollution of surface waters with phosphorus and nitrogen. *Ecol. Appl.* 8: 559–568.

Ceschin, S., Bellini, A., and Scalici, M. (2021). Aquatic plants and ecotoxicological assessment in freshwater ecosystems: a review. *Environ. Sci. Pollut. Res.* 28 (5): 4975–4988.

Chen, C., Huang, D., and Liu, J. (2009). Functions and toxicity of nickel in plants: recent advances and future prospects. *Clean-Soil Air Water* 37 (4–5): 304–313.

Chen, Y.G., Huang, J.H., Luo, R. et al. (2021). Impacts of heavy metals and medicinal crops on ecological systems, environmental pollution, cultivation, and production processes in China. *Ecotoxicol. Environ. Saf.* 219: 112336.

Chowdhury, A., Naz, A., and Sharma, D. (2021). Geochemistry, extent of pollution, and ecological impact of heavy metal pollutants in soil. *Geochem. Concepts Appl.* 2021: 73–91.

Claessens, M., De Meester, S., Van Landuyt, L. et al. (2011). Occurrence and distribution of microplastics in marine sediments along the Belgian coast. *Mar. Pollut. Bull.* 62: 2199–2204.

Costa, J.M., da Costa, J.G.D.R., and de Almeida Neto, A.F. (2022). Techniques of nickel (II) removal from electroplating industry wastewater: overview and trends. *J. Water Process Eng.* 46: 102593.

Cui, L., Wang, X., Li, J. et al. (2021). Ecological and health risk assessments and water quality criteria of heavy metals in the Haihe River. *Environ. Pollut.* 290: 117971.

Das, J. and Acharya, B.C. (2003). Hydrology and assessment of lotic water quality in Cuttack City, India. *Water Air Soil Pollut.* 150: 163–175.

Das, K.K., Reddy, R.C., Bagoji, I.B. et al. (2019). Primary concept of nickel toxicity–an overview. *J. Basic Clin. Physiol. Pharmacol.* 30 (2): 141–152.

Demirak, A., Yilmaz, F., Levent Tuna, A., and Ozdemir, N. (2006). Heavy metals in water, sediment and tissues of *Leuciscus cephalus* from a stream in southwestern Turkey. *Chemosphere* 63 (9): 1451–1458.

Ding, L., Wang, X., Ouyang, Z. et al. (2021). The occurrence of microplastic in Mu us sand land soils in Northwest China: different soil types, vegetation cover and restoration years. *J. Hazard. Mater.* 403: 123982.

Doğaner, S. (2001). *Türkiye Turizm Coğrafyası*. İstanbul: Çantay Kitabevi Yayınları.

Duman, F., Obali, O., and Demirezen, D. (2006). Seasonal changes of metal accumulation and distribution in shining pondweed (*Potamogeton lucens*). *Chemosphere* 65: 2145–2151.

El-Naggar, A., Ahmed, N., Mosa, A. et al. (2021). Nickel in soil and water: sources, biogeochemistry, and remediation using biochar. *J. Hazard. Mater.* 419: 126421.

Erasmus, J.H., Zimmermann, S., Smit, N.J. et al. (2022). Human health risks associated with consumption of fish contaminated with trace elements from intensive mining activities in a peri-urban region. *Sci. Total Environ.* 825: 154011.

Fernandesa, C., Fontainhas, F.A., Peixotoc, F., and Salgadod, M.A. (2007). Bioaccumulation of heavy metals in *Liza saliens* from the Esmoriz-Paramos coastal lagoon, Portugal. *Ecotoxicol. Environ. Saf.* 66 (3): 426–431.

Ferreira, C.F., de Oliveira, M.M.F., Mansur, E.T., and Rosa, W.D. (2021). The jaguar hydrothermal nickel sulfide deposit: evidence for a nickel-rich member of IOCG-type deposits in the Carajás Mineral Province, Brazil. *J. S. Am. Earth Sci.* 111: 103501.

GESAMP (2010). IMO/FAO/UNESCO-IOC/UNIDO/WMO/IAEA/UN/UNEP. Joint Group of Experts on the Scientific Aspects of Marine Environmental Protection. *Proceedings of the GESAMP International Workshop on Micro-plastic Particles as a Vector in Transporting Persistent, Bio-accumulating and Toxic Substances in the Oceans* (28–30 June 2010). Paris: UNESCO-IOC.

Ghori, N.-H., Ghori, T., Hayat, M.Q. et al. (2019). Heavy metal stress and responses in plants. *Int. J. Environ. Sci. Technol.* 16: 1807–1828.

Grudnik, Z.M. (2010). Seasonal changes in the concentration of some trace elements in macrophyte shoots. *Acta Biol. Slo.* 53: 55–61.

Guilizzoni, P. (1991). The role of heavy metals and toxic materials in the physiological ecology of submerged macrophytes. *Aquat. Bot.* 41: 87–109.

Hamid, S., Yaseen, A., Kamili, A.N., and Yatoo, A.M. (2020). Pollution in aquatic environs: sources and consequences. In: *Bioremediation and Biotechnology*, vol. 4 (ed. R.A. Bhat and K.R. Hakeem), 21–38. Cham: Springer.

Hampel, M., Blasco, J., and Segner, H. (2015). Molecular and cellular effects of contamination in aquatic ecosystems. *Environ. Sci. Pollut. Res.* 22: 17261–17266.

Haq, S., Bharose, R., Ozturk, M. et al. (2021). Impact of treated sewage water on nutrient status of alfisols and vegetable crops. *Not. Bot. Horti Agrobot. Cluj Napoca* 49 (2): 12255–12255.

Harguinteguy, C.A., Cirelli, A.F., and Pignata, M.I. (2014). Heavy metal accumulation in leaves of aquatic plant *Stuckenia filiformis* and its relationship with sediment and water in the Suquía river (Argentina). *Microchem. J.* 114: 111–118.

Haseeb, M., Iqbal, S., Hafeez, M.B. et al. (2022). Phytoremediation of nickel by quinoa: morphological and physiological response. *PLoS One* 17 (1): e0262309.

Herbert, E.R., Boon, P., Burgin, A.J. et al. (2015). A global perspective on wetland salinization: ecological consequences of a growing threat to freshwater wetlands. *Ecosphere* 6 (10): 1–43.

Junk, W.J., An, S., Finlayson, C.M. et al. (2013). Current state of knowledge regarding the world's wetlands and their future under global climate change: a synthesis. *Aquat. Sci.* 75 (1): 151–167.

Knapp, S. and Van de Velden, M. (2011). Global ship risk profiles: safety and the marine environment. *Trans. Res. D Transp. Environ.* 16 (8): 595–603.

Koelmans, A.A., Besseling, E., Wegner, A., and Foekema, E.M. (2013). Plastic as a carrier of POPs to aquatic organisms: a model analysis. *Environ. Sci. Technol.* 47: 7812–7820.

Kolarova, N. and Napiórkowski, P. (2021). Trace elements in aquatic environment. Origin, distribution, assessment and toxicity effect for the aquatic biota. *Ecohydrol. Hydrobiol.* 21 (4): 655–668.

Lechner, A., Keckeis, H., Lumesberger-Loisl, F. et al. (2014). The Danube so colourful: a potpourri of plastic litter outnumbers fish larvae in Europe's second largest river. *Environ. Pollut.* 188: 177–181.

Lee, S.H. and Lee, J.H. (2021). Pulse electroplating of Ni–W alloy for the diffusion barrier in under bump metallurgy. *J. Nanosci. Nanotechnol.* 21 (5): 3016–3019.

Leslie, H.A., Moester, M., de Kreuk, M., and Vethaak, A.D. (2012). Pilot study on emissions of microplastics from wastewater treatment plants. *H2O* 14 (15): 45–47.

Li, Y., Li, P., and Liu, L. (2022). Source identification and potential ecological risk assessment of heavy metals in the topsoil of the Weining plain (Northwest China). *Exposure Health* 14 (2): 281–294.

Lopez, S., Nkrumah, P.N., Echevarria, G. et al. (2021). Bacterial community diversity and functional roles in the rhizosphere of *Rinorea cf. bengalensis* and *Phyllanthus rufuschaneyi* under a nickel concentration gradient. *Plant Soil* 459 (1): 343–355.

Loska, K. and Wiechuła, D. (2003). Application of principal component analysis for the estimation of source of heavy metal contamination in surface sediments from the Rybnik reservoir. *Chemosphere* 51 (8): 723–733.

Maanan, M., Saddik, M., Maanan, M. et al. (2015). Environmental and ecological risk assessment of heavy metals in sediments of Nador lagoon, Morocco. *Ecol. Indic.* 48: 616–626.

Margesin, R. and Schinnur, F. (1997). Efficiency of endogenous and inoculated cold-adapted soil microorganisms for biodegradation of diesel oil in alpine soils. *Appl. Environ. Microbiol.* 63: 2660–2664.

Mateo-Sagasta, J., Zadeh, S.M., Turral, H., and Burke, J. (2017). Water pollution from agriculture: a global review. Food and Agriculture Organization of the United Nations, Rome and the International Water Management Institute on behalf of the Water Land and Ecosystems Research Program, Colombo.

Milošković, A., Branković, S., Simić, V. et al. (2013). The accumulation and distribution of metals in water, sediment, aquatic Macrophytes and fishes of the Gruža reservoir Serbia. *Bull. Environ. Contam. Toxicol.* 90 (5): 563–569.

Mohan, I., Goria, K., Dhar, S. et al. (2021). Phytoremediation of heavy metals from the biosphere perspective and solutions. *Pollut. Water Manage.* 16: 95–127.

Mu, W., Cui, F., Huang, Z. et al. (2018). Synchronous extraction of nickel and copper from a mixed oxide-sulfide nickel ore in a low-temperature roasting system. *J. Clean. Prod.* 177: 371–377.

Mushtaq, W., Mehdizade, M., Siddiqui, M.B. et al. (2020). Phytotoxicity of above – ground weed residue against some crops and weeds. *Pak. J. Bot.* 52 (3): 851–860.

Nag, R., O'Rourke, S.M., and Cummins, E. (2022). Risk factors and assessment strategies for the evaluation of human or environmental risk from metal (loid)s–a focus on Ireland. *Sci. Total Environ.* 802: 149839.

Norén, F. (2007). *Small Plastic Particles in Coastal Swedish Waters*. Sweden: KIMO.

Oberdorff, T., Pont, D., Hugueny, B., and Porcher, J. (2002). Development and validation of a fish based index for the assessment of 'river health' in France. *Freshwater Biol.* 47 (9): 1720–1734.

Oehlmann, J., Schulte-Oehlmann, U., Kloas, W. et al. (2009). A critical analysis of the biological impacts of plasticizers on wildlife. *Philos. Trans. R. Soc. Lond. B Biol. Sci.* 364: 2047–2062.

de Oliveira, J.B., Lavres, J., and van der Ent, A. (2022). In situ analysis of nickel uptake from foliar application in pecan using instrumental μXRF analysis. *J. Soil Sci. Plant Nutr.* 22 (1): 1–9.

Othman, Y.A., Al-Assaf, A., Tadros, M.J., and Albalawneh, A. (2021). Heavy metals and microbes accumulation in soil and food crops irrigated with wastewater and the potential human health risk: a metadata analysis. *Water* 13 (23): 3405.

Özşahin, E. (2011). Gölbaşı (Balık) Gölü'nde (Hatay) meydana gelen değişimin coğrafi analizi. *Turk. Stud.* 6 (1): 1604–1621.

Ozturk, M., Altay, V., Gucel, S., and Guvensen, A. (2014). Halophytes in the East Mediterranean – their medicinal and other economical values. In: *Sabkha Ecosystems: Volume IV: Cash Crop Halophyte and Biodiversity Conservation Tasks for Vegetation Science*, vol. 47 (ed. M.A. Khan, B. Böer, M. Öztürk, et al.), 247–272. Netherlands: Springer.

Ozturk, M., Altay, V., and Karahan, F. (2017). Studies on trace elements distributed in *Glycyrrhiza* taxa in Hatay-Turkey. *Int. J. Plant Environ.* 3 (2): 1–7.

Ozturk, M., Altay, V., Yarcı, C. et al. (2021). Endangered swamp forests in Turkey – an ecological inventory, prospects and challenges. In: *Biodiversity, Conservation and Sustainability in Asia: Volume I: Prospects and Challenges in West Asia and Caucasus* (ed. M. Ozturk et al.), 661–679. Cham: Springer Nature Switzerland AG.

Ozturk, M., Metin, M., Altay, V. et al. (2022). Arsenic and human health: genotoxicity, epigenomic effects, and cancer signaling. *Biol. Trace Elem. Res.* 200: 988–1001.

Pang, Y., Li, L., Wang, Y. et al. (2022). Zinc-induced phase reconstruction of cobalt–nickel double hydroxide cathodes for high-stability and high-rate nickel–zinc batteries. *Chem. Eng. J.* 436: 135202.

Parnian, A., Chorom, M., Jaafarzadeh, N. et al. (2022). Bioremediation of cadmium and nickel from a saline aquatic environment using *Ceratophyllum demersum*. *Chiang Mai J. Sci.* 49 (2): 339–347.

Perkins, S. (2015). Nearly every seabird may be eating plastic by 2050. *Science* 31: https://doi.org/10.1126/science.add1694.

Punia, P., Bharti, M.K., Chalia, S. et al. (2021). Recent advances in synthesis, characterization, and applications of nanoparticles for contaminated water treatment-a review. *Ceram. Int.* 47 (2): 1526–1550.

Rakib, M.R.J., Jolly, Y.N., Begum, B.A. et al. (2022). Assessment of trace element toxicity in surface water of a fish breeding river in Bangladesh: a novel approach for ecological and health risk evaluation. *Toxin Rev.* 41 (2): 420–436.

Rathore, H.S. and Nollet, L.M. (2016). *Handbook of Pesticides: Methods of Pesticide Residues Analysis*. Boca Raton, FL: CRC Press.

Reddy, C.K.V., Golla, V., Badapalli, P.K., and Reddy, N.B.Y. (2022). Evaluation of groundwater contamination for fluoride and nitrate in Nellore Urban Province, southern India: a special emphasis on human health risk assessment (HHRA). *Appl Water Sci* 12 (3): 1–9.

Rehman, S., Mansoora, N., Al-Dhumri, S.A. et al. (2022). Associative effects of activated carbon biochar and arbuscular mycorrhizal fungi on wheat for reducing nickel food chain bioavailability. *Environ. Technol. Innov.* 26: 102539.

Rezania, S., Taib, S.M., Din, M.F.M. et al. (2016). Comprehensive review on phytotechnology: heavy metals removal by diverse aquatic plants species from wastewater. *J. Hazard. Mater.* 318: 587–599.

Roccotiello, E., Nicosia, E., Pierdonà, L. et al. (2022). Tomato (*Solanum lycopersicum* L.) accumulation and allergenicity in response to nickel stress. *Sci. Rep.* 12 (1): 1–15.

Rochman, C.M. (2013). Plastics and priority pollutants: a multiple stressor in aquatic habitats. *Environ. Sci. Technol.* 47: 2439–2440.

Rostami, S., Kamani, H., Shahsavani, S., and Hoseini, M. (2021). Environmental monitoring and ecological risk assessment of heavy metals in farmland soils. *Hum. Ecol. Risk Assess.: Int. J.* 27 (2): 392–404.

Said, I. (2022). Nickel pollution pathways in small ecosystem, Egypt. *Arab. J. Geosci.* 15 (10): 1–8.

Saraiva, M.P., Maia, C.F., da Silva, B.R.S. et al. (2021). 24-Epibrassinolide induces protection against nickel excess in soybean plants: anatomical evidences. *Braz. J. Bot.* 44 (1): 197–205.

Schindler, D.W. (1974). Eutrophication and recovery in experimental lakes: implications for lake management. *Science* 174: 897–899.

Schindler, D.W. (2006). Recent advances in the understanding and management of eutrophication. *Limnol. Oceanogr.* 51: 356–363.

Schreinemachers, P. and Tipraqsa, P. (2012). Agricultural pesticides and land use intensification in high, middle and low income countries. *Food Policy* 37: 616–626.

Schwarzenbach, R.P., Egli, T., Hofstetter, T.B. et al. (2010). Global water pollution and human health. *Annu. Rev. Environ. Resour.* 35: 109–136.

Servadio, J.L., Deere, J.R., Jankowski, M.D. et al. (2021). Anthropogenic factors associated with contaminants of emerging concern detected in inland Minnesota lakes (phase II). *Sci. Total Environ.* 772: 146188.

Singh, S., Benny, C.K., and Chakraborty, S. (2022). An overview on the application of constructed wetlands for the treatment of metallic wastewater. In: *Biodegradation and Detoxification of Micropollutants in Industrial Wastewater* (ed. I. Haq, A.S. Kalamdhad, and M.P. Shah), 103–130. Elsevier.

Sobati-Nasab, Z., Alirezalu, A., and Noruzi, P. (2021). Effect of foliar application of nickel on physiological and phytochemical characteristics of pot marigold (*Calendula officinalis*). *J. Agric. Food Res.* 3: 100108.

Sofoulaki, K., Kalantzi, I., Machias, A. et al. (2019). Metals in sardine and anchovy from Greek coastal areas: public health risk and nutritional benefits assessment. *Food Chem. Toxicol.* 123: 113–124.

Son, Y.O. (2020). Molecular mechanisms of nickel-induced carcinogenesis. *Endocr. Metab. Immune. Disord. Drug Targets* 20 (7): 1015–1023.

Sousa, P.T. Jr., Piedade, M.T.F., and Candotti, E. (2011). Letter to nature. *Nature* 478: 458.

Teuten, E.L., Saquing, J.M., Knappe, D.R. et al. (2009). Transport and release of chemicals from plastics to the environment and to wildlife. *Philos. Trans. R. Soc. Lond. B Biol. Sci.* 364: 2027–2045.

Thyssen, J.P. (2011). Nickel and cobalt allergy before and after nickel regulation–evaluation of a public health intervention. *Contact Dermatitis* 65: 1–68.

Tran, T.T., Moon, H.S., and Lee, M.S. (2022). Separation of cobalt, nickel, and copper from synthetic metallic alloy by selective dissolution with acid solutions containing oxidizing agent. *Miner. Process. Extr. Metall. Rev.* 43 (3): 313–325.

Tukura, B.W., Kagbu, J.A., and Gimba, C.E. (2009). Effects of pH and seasonal variations on dissolved and suspended heavy metals in dam surface water. *Chem. Class J.* 6: 27–30.

Türkmen, M. and Ciminli, C. (2011). Seasonal variations of the metal concentrations in the waters of lake Gölbaşı in northern east mediterranean area of Turkey. *Black Sea J. Sci.* 1 (3): 86–93.

Vargas-Gonzalez, H.H., Arreola-Lizarraga, J.A., Mendoza-Salgado, R.A. et al. (2014). Effect of sewage discharge on trophic state and water quality in a coastal ecosystem of the gulf of California. *Sci. World J.* 2014: 618054.

Vodyanitskii, Y.N. and Shoba, S.A. (2015). Biogeochemistry of carbon, iron, and heavymetals in wetlands (analytical review). *Moscow Univ. Soil Sci. Bull.* 70 (3): 89–97.

Wright, S.L., Thompson, R.C., and Galloway, T.S. (2013). The physical impacts of microplastics on marine organisms: a review. *Environ. Pollut.* 178: 483–492.

Xu, J.L. and Yang, J.R. (1996). *Heavy Metals in Terrestrial Ecosystem*. China Environmental Science Press.

Yadav, M., Goel, G., Hatton, F.L. et al. (2021). A review on biomass-derived materials and their applications as corrosion inhibitors, catalysts, food and drug delivery agents. *Curr. Res. Green Sustainable Chem.* 4: 100153.

Yoneda, T., Hiasa, M., Okui, T., and Hata, K. (2021). Sensory nerves: a driver of the vicious cycle in bone metastasis? *J. Bone Oncol.* 30: 100387.

Zarfl, C. and Matthies, M. (2010). Are marine plastic particles transport vectors for organic pollutants to the Arctic? *Mar. Pollut. Bull.* 60: 1810–1814.

Zaynab, M., Al-Yahyai, R., Ameen, A. et al. (2022). Health and environmental effects of heavy metals. *J. King Saud Univ. Sci.* 34 (1): 101653.

Zhu, Y., Chen, Q.Y., Li, A.H., and Costa, M. (2019). The role of non-coding RNAs involved in nickel-induced lung carcinogenic mechanisms. *Inorganics* 7 (7): 81.

Zubris, K.A. and Richards, B.K. (2005). Synthetic fibers as an indicator of land application of sludge. *Environ. Pollut.* 138: 201–211.

3

Endophytic Fungi and Bacteria: Enhancement of Heavy Metal Phytoextraction

Amauri Ponce-Hernández[1], Javier A. Gómez-Rubio[1], Juan G. Ceballos-Maldonado[1], Domingo Martínez-Soto[2], Margarita Márquez-Vega[3], Alejandro Hernández-Morales[3,4], Juan José Maldonado Miranda[3], and Candy Carranza-Álvarez[3,4]

[1] *Faculty of Chemical Sciences, Autonomous University of San Luis Potosí, San Luis Potosí, Mexico*
[2] *Department of Microbiology, Center for Scientific Research and Higher Education of Ensenada (CICESE), Ensenada, Mexico*
[3] *Faculty of Professional Studies Huasteca Zone, Autonomous University of San Luis Potosí, San Luis Potosí, Mexico*
[4] *Chemical Sciences Faculty, Autonomous University of San Luis Potosí, San Luis Potosí, Mexico*

3.1 Introduction

The rate increase of anthropogenic activities has resulted in higher production of toxic pollutants, such as heavy metals (HMs), which can be accumulated in the environment. The main sources of metals in the environment are industrial and agricultural activities, mainly located in the urban sectors. Since HMs are not biodegradable, they can persist for years in the environment, being available to be absorbed and accumulated by plants and other living beings, entering food chains, and negatively affecting human and animal health. An alternative to help this current problem regarding metals is the application of phytoremediation, a low-cost and environmentally friendly technology that uses tolerant plants to remove contaminants from water, soil, sediment, and air. To carry out the phytoremediation of a contaminated site, the plants use various mechanisms such as phytoextraction, phytostabilization, phytodegradation, phytovolatilization, and rhizofiltration, which can be carried out individually or several of them at the same time in the same plant. Various physicochemical and biological factors are involved in these mechanisms. Physicochemical factors include the physical and chemical compositions of the medium (soil or water) and the species or state of the contaminant. Biological factors include the physiological characteristics of plants and microorganisms present in the rhizosphere. It is currently known that phytoremediation is improved when plants establish interactions with rhizosphere microorganisms, allowing plants better resistance to HMs toxicity, increasing their phytoextraction capacities, and improving their growth and development. These interactions include microorganisms that promote plant growth, such as endophytic bacteria and fungi, which establish a mutualistic coexistence with their host. Endophytic microorganisms play a key role in the growth and development of plants since they can alter the physicochemical characteristics of the soil, the content of minerals and nutrients, the deposition of soil waste, and the absorption of water. Moreover, endophytic plant bacteria and fungi can remove and/or deactivate contaminants such as HMs through biological and chemical reactions. They also contribute to the mobilization of contaminants present in the rhizosphere and the subsequent absorption by plants, which, in turn, release exudates and enzymes that promote biochemical and microbial activities to aid phytoextraction.

3.2 Main Anthropogenic Sources Releasing HMs into the Environment

The term "HMs" includes metals and metalloids that have been linked to pollution and ecotoxicity (Armah et al. 2014). According to Ali and Khan (2018), the term "HMs" refers to natural metals that have an atomic number greater than 20 and a density greater than $5\,g/cm^3$. Their concentrations in the environment, their mobilization rates, and their transport have accelerated considerably since the 1940s (Ernest 1984; Adriano 2001). Nowadays, there are different sources that

Aquatic Contamination: Tolerance and Bioremediation, First Edition. Edited by Rouf Ahmad Bhat, Gowhar Hamid Dar, Fernanda Maria Policarpo Tonelli, and Saima Hamid.

Table 3.1 Average natural contents in selected rocks with HMs (values in mg/kg).

	Ingneous rocks			Sedimentary rocks				
Heavy metal	**Granitic**	**Basaltic**	**Ultramafic**	**Limestone**	**Sandstone**	**Shale**	**Black shale**	**Streams (ppb)**
As	2	2.2	0.7	1	0.5	8	<500	50
Cd	0.13	0.21	0.05	0.3	<0.04	0.3	<240	0.01
Co	4	47	110	0.1	0.3	20	67	0.1
Cr	10	185	2300	11	35	100	<700	1
Cu	20	94	40	4	2	50	<300	7
Fe (%)	1.42	8.6	—	0.38	—	5.1	—	40
Hg (ppb)	0.03	0.09	—	0.04	—	0.4	—	0.07
Mn	400	1500	1200	15	100	850	—	—
Ni	10	145	2000	20	2	60	<300	0.3
Pb	17	7	0.05	9	10	20	<100	1
Sb	0.22	0.6	0.1	0.2	0.05	1.5	<10	0.07
Zn	50	118	60	20	20	85	<2314	20

Source: Adapted from Bradl (2005).

release HMs into the environment, which can be natural or anthropogenic (Su et al. 2022). Natural sources of HMs release into the environment include magmatic rocks, sedimentary and metamorphic rocks, weathering and soil formation, the rock cycle, and the origin of HMs in surface and groundwater, as well as in the atmosphere (Bradl 2005). According to Alloway (2013), the most important rocks considered natural sources of HMs release are granitic, basaltic, ultramafic, sandstone, shale, black shale, and limestone. Table 3.1 shows the concentration of HMs in these types of rocks and the concentration with which they are found in water bodies (Bradl 2005). On the other hand, anthropogenic air emissions have been shown to be the main sources of pollution of atmospheric, terrestrial, and aquatic environments (Cheng et al. 2014). HMs can be released into the environment through gases, particles, aqueous, or solid substances. According to Zhang et al. (2022), the main sources of urban air pollution by HMs are traffic, industrial sources, sources of carbon combustion, and metal smelting. Additionally, a study conducted by Yang et al. (2022) indicated that the main sources that contribute to the increase in high concentrations of HMs in urban surface dust are: industrial sources, traffic sources, coal sources, and natural sources. On the other hand, Bradl (2005) indicated that the main anthropogenic sources that release HMs into the soil, water, and atmosphere are agricultural and industrial activities (Figure 3.1).

3.3 Phytoremediation of HMs

Phytoremediation is a biological method using plants as biological components to treat contaminated environments (Schück and Greger 2020). It is a modern technology for removing pollutants like HMs (Ashraf et al. 2019) from contaminated soil, water and/or sediments (Table 3.2). Since metals are extremely toxic to plants and animals, they are representing one of the most critical environmental concerns (Ngo et al. 2020; Sharma and Rath 2020). Phytoremediation is a widely accepted approach for the decontamination of HMs, in a cost-effective and noninvasive manner, less disturbing, highly efficient, relatively less costly, etc. (Al-Thani and Yasseen 2020).

Plants used for phytoremediation have fast growth and high biomass production, tolerance to HMs, easy harvest, and accumulation of high levels of HMs (Patra et al. 2019; Yan et al. 2020). In this context, phytoremediation involves the absorption and modification of the contaminant through the roots to a less toxic element and the confinement of metals in different tissues such as stems and leaves (Kaur et al. 2018; Prieto et al. 2018). Therefore, it is possible to reduce the presence of HMs in the environment. For example, San Juan et al. (2018) reported the use of *Limnobium laevigatum* and *Ludwigia peploides* for remediation of Pb and Zn in contaminated water, both plants showed a high metal removal rate. Similarly, they observed a high accumulation of zinc in leaves and roots.

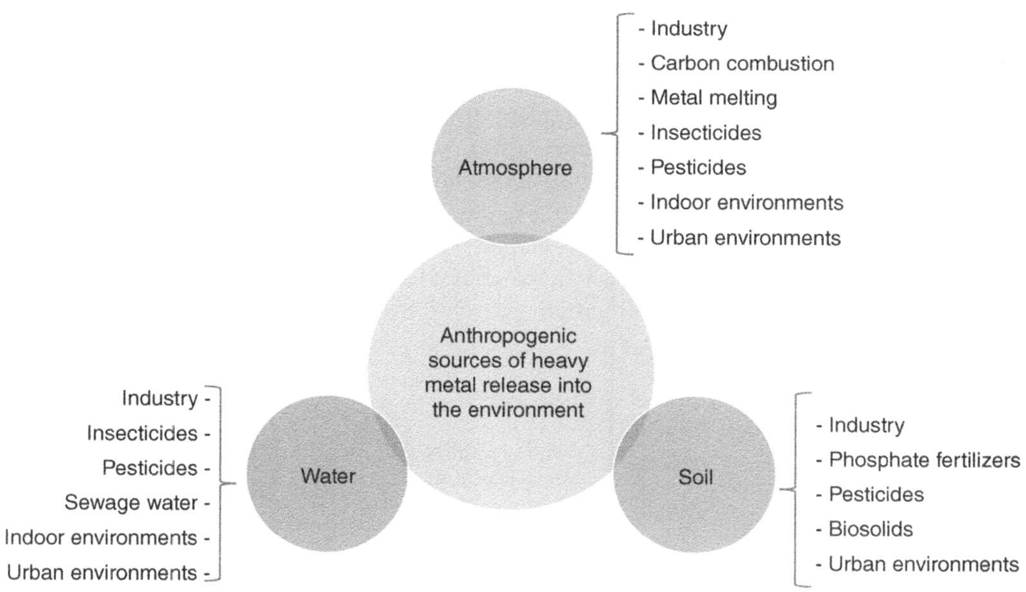

Figure 3.1 Main anthropogenic sources of heavy metal release into the environment. *Source:* Bradl (2005), Briffa et al. (2020), Zhang et al. (2022), Yang et al. (2022).

Table 3.2 Plant species used for phytoremediation.

Phytoremediation mechanisms	Plant	Heavy metal	Reference
Phytovolatilization	*Arundo donax* L.	As	Guarino et al. (2020)
Phytodegradation	*Cynodon dactylon*	Mn, Cu, Fe, Pb	Islam et al. (2015)
Rhizofiltration	*Eichhornia crassipes*	Fe, Cr, Cu, Cd, Zn, Ni, As	Rai (2019)
Phytoextraction	*Xanthium strumarium*	Cd, Pb, Ni, Zn	Khalid et al. (2019)
	Pelargonium hortorum	Pb	Manzoor et al. (2020)
	Polygonum capitatum	Cd, Cu, Cr, Ni	Wu et al. (2021)
	Typha latifolia	Cd	Rolón-Cárdenas et al. (2022)
	Typha domingensis	Zn, Pb, Ni, Cr, Cd	Soudani et al. (2022)
	Scirpus validus *Typha orientalis* *Phragmites australis*	Cd, Cr, Cu, Ni, Pb, Zn, As, Hg	Yan et al. (2022)
	Populus nigra *Salix matsudana* *Sapium sebiferum* *Swida wilsoniana* *Corchorus capsularis* *Hibiscus cannabinus* *Ricinus communis*	Pb, Zn	Tang et al. (2019)
Phytostabilization	*Boehmeria nivea* *Chrysanthemum indicum* *Miscanthus floridulus* *Conyza canadensis* *Rubus setchuenensis* *Senecio scandens* *Arthraxon hispidus*	Cd, Cu, Cr, Ni	Wu et al. (2021)
	Scariola orientalis	Zn, Fe	Mahdavian et al. (2022)

Pilipović et al. (2019) found the potential of *Populus deltoides* Bartr. ex Marsh. clone "Bora" and *Salix viminalis* L. clone "SV068" to phytoextract Zn and Cd in aerial parts, while Cr, Ni, Pb, and Cu were phytostabilized in the roots. On the other hand, the photostabilizer grass *Festuca arundinacea* was found to have the potential for the removal of Cd, Ni, and Pb. These metals were accumulated by the plant roots (Steliga and Kluk 2020). In addition, Yang et al. (2020) examined the uptake capacity of Cd and Zn in *Pennisetum purpureum* plants. Their results showed that *Pennisetum purpureum* accumulated the highest concentration of Cd (197.5 g/ha) and Zn (5023.9 g/ha) in the shoots.

Interestingly, several processes are involved in the metal remediation by plants: translocation, accumulation, distribution, exclusion, and osmoregulation (Dong et al. 2017). Also, the phytoremediation process can be classified as phytoextraction, phytostabilization, phytovolatilization, and rhizodegradation (Dal Corso et al. 2019; Goyal et al. 2020). Other authors include other phytoremediation mechanisms such as rhizofiltration, phytofiltration, and phytostimulation (Kanwar et al. 2020; Nedjimi 2021) (Figure 3.2).

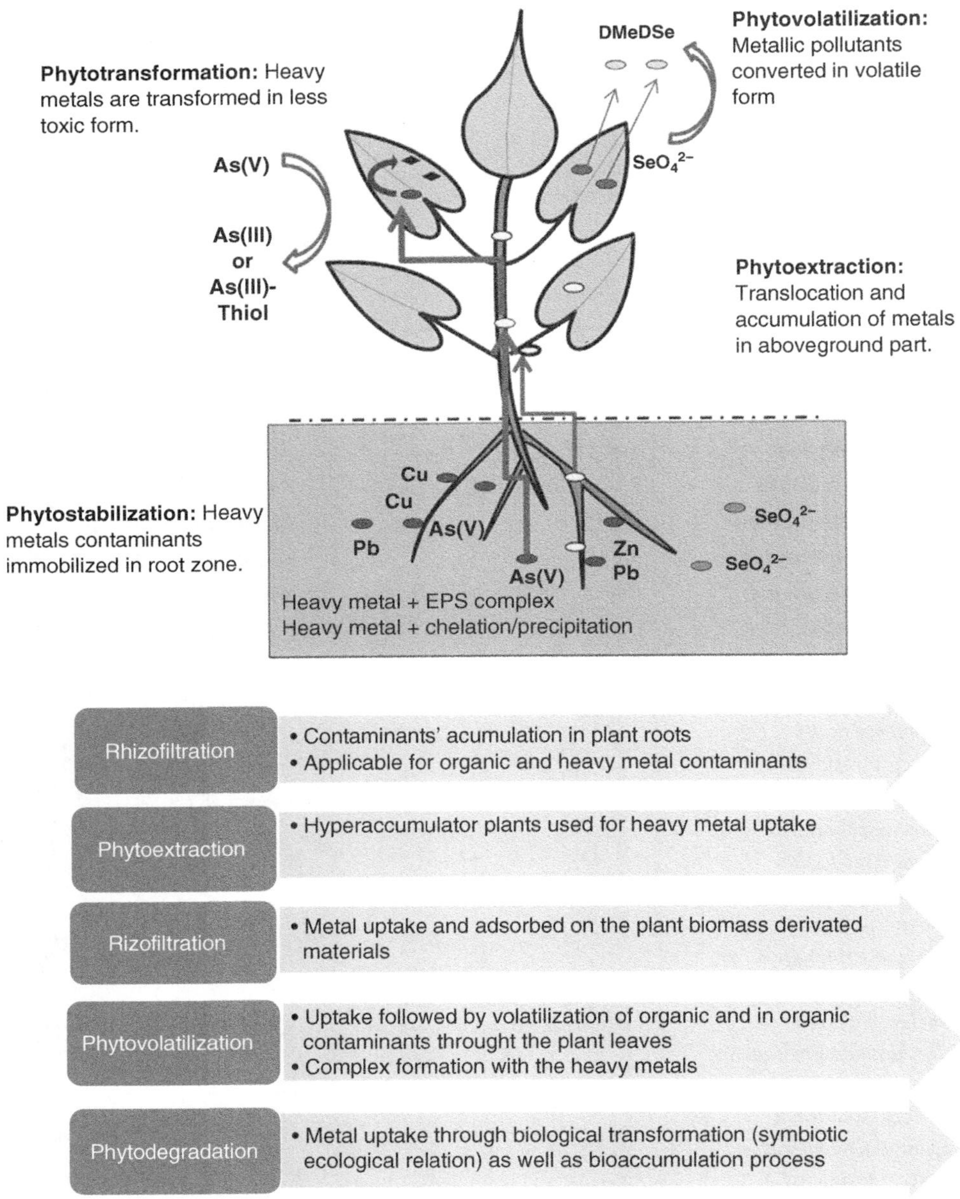

Figure 3.2 Overview of mechanisms of HMs' uptake. *Source:* Shah and Daverey (2020), Kanwar et al. (2020).

Table 3.3 Advantages and disadvantages of phytoremediation (Diarra et al. 2021).

Advantages	Disadvantages
Applicable in situ and ex situ	It is a slow process to effectively remediate contaminated sites
With an affinity to the environment and socially accepted	restricted by root depth and bioavailability of the contaminant
Cheaper than conventional remediation methods	Restricted to sites with low contamination
Solar driven – less energy requirements	Harvested plant biomass requires proper disposal as hazardous waste
Easy to implement and maintain	Incorrect waste management can cause ecological damage
Without environmental damage and improves the functionality of the soil	Possible introduction of invasive species
Suitable for large, contaminated areas	Depends on environmental factors, soil, water, and plants

3.4 Advantages and Disadvantages

Phytoremediation has been considered as an alternative strategy with an ecological and sustainable approach to remove HMs or other toxic compounds from the environment (Bian et al. 2019). This method has been very attractive in recent years. The main advantages are profitability, feasibility, and ecological approach. In addition, phytoremediation has many advantages compared to other technologies to remediate contaminated sites (Ekta and Modi 2018). When phytoremediation is combined with sustainable site management, phytomanagement generates a broader range of benefits that include economic, environmental, and social well-being (Burges et al. 2018). Furthermore, the harvested plant biomass after phytoextraction can be used for bioenergy production (Tian and Zhang 2016; Tripathi et al. 2016). Moreover, this technology can be applied to in situ remediation as well as ex situ experiments for extracting multiple toxic metals such as Pb, Cd, Hg, and Cr.

However, various challenges limit the application and effectiveness of phytoremediation technologies (Diarra et al. 2021). Phytoremediation is a process that requires a lot of time because it depends on biological cycles, and it takes from months to years to remediate contaminated sites (Khalid et al. 2017). In addition, environmental, soil and plant factors, such as climate, growth time, root depth, soil chemical composition, and level of contamination, strongly reduce the capacity of plants that can be used, as well as the effectiveness of phytoremediation (Sheoran et al. 2016). The biomass that is collected after the remediation process has been carried out is classified as hazardous waste. Therefore, they must be disposed of or recycled in accordance with applicable regulations (Shen et al. 2019). Finally, the introduction of non-native species for phytoremediation purposes can generate invasiveness and threaten local biodiversity. Table 3.3 summarizes the advantages and disadvantages of the technologies used in phytoremediation.

3.5 Factors that Increase HMs Phytoremediation

Phytoremediation efficiency by plants can be affected by several factors that influence the HMs' removal. Plant types, environmental conditions, root composition, and elemental species, as well as the soil's physiochemical and biological properties, all influence HM buildup and distribution in plants (Bhat et al. 2022).

3.5.1 Physicochemical

Humans' health can be impacted by HMs-contaminated soil and effective remediation is necessary to solve this environmental problem (Awa and Hadibarata 2020). The phytoavailability of HMs is directly affected by the environmental pH (Kader et al. 2016). The low pH (<6.0) could increase the uptake of metal cations by plants. The metal cations are available and soluble in the environment in an acid medium. In contrast, at high pH, the metal cations precipitate or form complexes affecting their metal availability and phytoextraction (Figure 3.3) (reviewed by Beiyuan et al. 2017; Agrawal and Singh 2018; Rosenfeld et al. 2018).

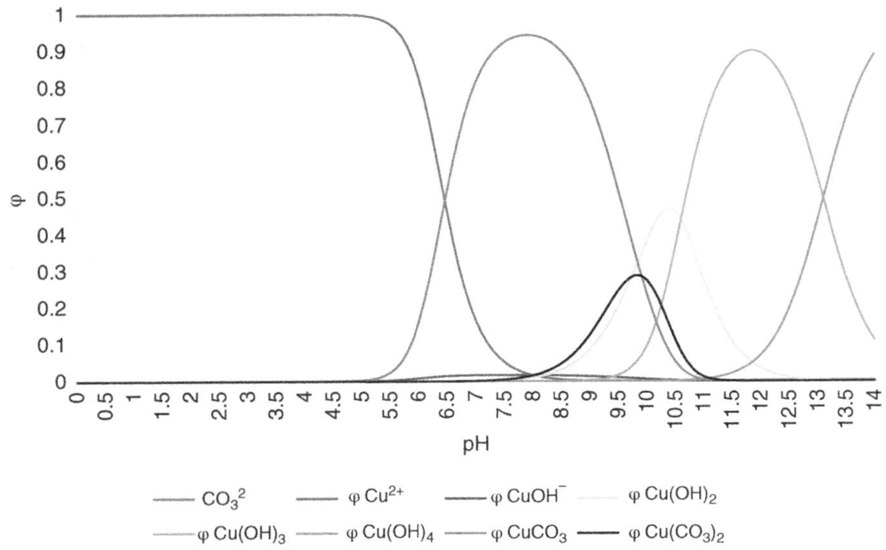

Figure 3.3 Speciation diagram of Cu(II)/OH/CO_3.

Also, the temperature affects the metal bioremediation. Under high temperatures, HMs can be easily removed by plants (Ali et al. 2016; Jiang et al. 2019). For example, Xu et al. (2020) showed that the high temperature increases the efficiency of Cd^{2+}, Cu^{2+}, Pb^{2+}, and Ni^{2+} removal by using *Hovenia acerba*, *Fatsia japónica*, and *Pterocarya stenoptera*. On the other hand, chelating agents are required in alkaline soils with a pH greater than 5.5–6 since, in alkaline soils, the bioavailability of HMs is decreased (Gómez-Garrido et al. 2018). The reduction of the soil pH could be hazardous to physical and biochemical features of the soil (Postigo et al. 2018), affecting the natural microbiota of the environment. However, there are organic acids produced by some microorganisms and plants, which increase the metal bioavailability by reducing the soil pH and increasing the phytoextraction of HMs. For example, gluconic acid is produced by some bacteria (Yang et al. 2018).

3.5.2 Biological

With biotechnological advances, phytoremediation has become an important strategy for ecological restoration (Vishnoi and Dixit 2019). There are biological factors that rise the phytoremediation technics. One of the most basic biological strategies for enhancing phytoremediation is using hyperaccumulator plant species (Muthusaravanan et al. 2020; Singh et al. 2022). For example, Lin et al. (2016) showed that *Arabidopsis thaliana* transgenic plants overexpressing a Zn transporter showed an enhanced Zn and Cd.

In other cases, the use of biological organic amendments increases the extraction of HMs. An example was the use of sugar beet residues to promote the phytoextraction of HM, such as B, Cd, Cr, Fe, Ni, and Zn in *Trifolium repens* (Azcón et al. 2009). Therefore, one of the key features of effective phytoremediation is the use of multiple microorganisms (Thakare et al. 2021). For instance, microbe consortia and hyperaccumulator plants are used to eliminate HM pollutants (Jan and Parray 2016).

3.6 Phytoremediation Mechanisms

Plants can transform or degrade organic pollutants and remove or stabilize inorganic pollutants through specific or various mechanisms. The success of phytoremediation mechanisms depends mainly on the bioavailability of the contaminant, the characteristics of the plant species and the environmental conditions of the contaminated site (Sheoran et al. 2016). Through these mechanisms, plants can accumulate, immobilize, and/or transform pollutants (Banjoko and Eslamian 2016). Within these phytoremediation mechanisms are mainly phytoextraction, phytostabilization, phytodegradation, phytovolatilization, and rhizofiltration. Some authors include other mechanisms that influence phytoremediation such as phytostimulation, phytodeanization, and hydraulic control (Banjoko and Eslamian 2016; Muthusaravanan et al. 2018; Kafle et al. 2022). The five main mechanisms of phytoremediation are explained as follows.

3.6.1 Phytoextraction

Phytoextraction is the process of transferring organic and inorganic contaminants present in some environmental matrix and sediments to plant tissues. In this mechanism, the contaminants are absorbed through the roots followed by translocation; accumulation in the stem, leaves, shoots, or roots of the plants; and, finally, the deposition in the vacuole, cell wall, and cell membrane (Rashid et al. 2014). Phytoextraction has been widely applied to decontaminate sites impacted with metals, metalloids, radionuclides, and nonmetals (Pivetz 2001). For the effective application of phytoextraction, plants should have the following characteristics: (i) tolerance to the contaminants present at toxic levels in situ, (ii) having a high rate of biomass production, and (iii) having a high rate of accumulation of contaminants in their tissues (Vangronsveld et al. 2009).

The phytoextraction potential of hyperaccumulator plants can be determined through the bioconcentration factor (BCF) and the translocation factor (TF). The first indicates the capacity of a plant to accumulate metals. It is the relationship between the metal concentration in the root and the concentration in the environmental matrix. On the other hand, the FT refers to the ability of a plant to translocate metals from the root to the shoot and is defined as the concentration of metal in the stem divided by the concentration of metal in the root. A plant with BCF and FT values >1 has the potential to be used in phytoextraction (Yoon et al. 2006). It has been reported that approximately 500 plant species belonging to at least 45 plant families hyperaccumulate several different metals (Sheoran et al. 2016). The mechanisms involved in the accumulation of HMs include the absorption and formation of a metal-phytochelatin complex inside the cell, followed by the transfer and confinement at the vacuoles (Yadav 2010). Biomass containing high levels of metals is removed from the study site and processed to decrease its weight or volume through composting, drying, compaction, or thermal decomposition. The resulting mass can be used for the re-extraction of trace elements (Suman et al. 2018).

3.6.2 Phytostabilization

In this mechanism, some plants are used to immobilize the contaminants on the study site. Phytostabilization takes place in the rhizosphere and is frequently used to remediate soil, sediment, or sludge (Pivetz 2001). In phytostabilization, contaminants are immobilized by their accumulation in the roots, by adsorption on the root surface or by precipitation in the rhizosphere (Mendez and Maier 2008). Hammond et al. (2018) reported that arsenic (As) was immobilized in the root of *Prosopis juliflora* through binding the As with ferric sulfate, forming an As-tris-thiolate complex in the vacuoles. In addition, contaminants can be captured by binding to the hydroxyl groups of hemicellulose or lignin in the cell wall or by lipids in the plasma membrane (Hussain et al. 2009). In the same way, they can be trapped on the root surface by transport proteins. On the other hand, plant roots can release phytochemicals such as sugars, proteins, or amino acids that conjugate HMs and some organic pollutants from the rhizosphere and can transform them into less toxic compounds (Kafle et al. 2022). In the rhizosphere, metals change their soluble oxidation state to an insoluble state through root-mediated precipitation. An example of this is the precipitation of lead as lead phosphate mediated by the root (Pivetz 2001).

3.6.3 Phytodegradation

In this mechanism, plants degrade pollutants through metabolic processes inside or outside plant tissues through enzymes produced by plants (Muthusaravanan et al. 2018). According to Alsafran et al. (2022), phytodegradation is mainly applied for organic pollutants since it is less effective with inorganic pollutants such as HMs. Therefore, contaminants that can be phytodegraded include compounds such as ammunition, chlorinated solvents, herbicides, insecticides, organic nutrients, antibiotics, and nonsteroidal anti-inflammatory drugs (Gujarathi et al. 2005). These contaminants are decomposed inside or outside plant tissues by the action of enzymes produced by plants. Among these enzymes are hydrolases, nitroreductases, laccases, dehalogenases, peroxidases, and nitrilases (Grzegórska et al. 2020). For phytodegradation to occur within plant tissues, the plant must first take up the contaminants followed by degradation. The degraded products are used by plants as metabolites or nutrients that are incorporated into plant tissues to aid growth or possibly incorporated into lignin or cellulose (Chandra and Kumar 2018). On the other hand, contaminants can be broken down outside the plant before being absorbed. Because plants can release enzymes that metabolize pollutants into the rhizosphere. Once the contaminants are broken down into simpler forms, they are absorbed by the plant. The factors involved in phytodegradation are the efficiency of absorption and concentration of pollutants present in the soil and water (Zayed and Terry 2003).

3.6.4 Phytovolatilization

Phytovolatilization is a process where plants absorb a pollutant and subsequently release it into the environment in a volatile form, which can finally undergo photochemical decomposition (Limmer and Burken 2016). In this bioremediation method, the pollutants are transferred from the initial matrix to the atmosphere. A volatile compound can be the volatile form of a contaminant that was initially nonvolatile or the degradation product of a contaminant (Pivetz 2001). For the phytovolatilization mechanism to be effective, products released into the atmosphere must be less toxic than their initial form. Phytovolatilization has been applied for organic and inorganic contaminants. However, it is mainly applied to volatile organic compounds (VOCs) (Guarino et al. 2020). Among the inorganic pollutants can be found arsenic (As), mercury (Hg), and selenium (Se). Of them, As (V) is absorbed by the roots through specific membrane transporters; once in the plant cell, As (V) is reduced to As (III) form by enzymatic action. As (III) can be methylated by the plant to generate trimethylarsine gas (TMAs (III)), which can volatilize (Yong-Guan and Rosen 2009). Similarly, mercury ion (Hg (II)) can be transformed inside the plant by the enzyme mercuric reductase and generate metallic mercury (Hg (0)), which is less toxic and volatile (Meagher et al. 2000). Finally, plants metabolize various organic and inorganic species of Se such as selenate and selenite to gaseous forms such as methyl-selenate. This route is considered the most adequate for the removal of Se from the soil (Berken et al. 2002).

Moreover, phytovolatilization exists in two forms: direct and indirect phytovolatilization. Direct phytovolatilization uses plants to absorb contaminants from the soil, transform them into volatile compounds, and subsequently release them to the atmosphere by transpiration (Limmer and Burken 2016). In direct phytovolatilization, plants produce physical changes on the surface, which leads to the reduction of contaminants in the soil and water (Chandra and Kumar 2018). On the other hand, indirect phytovolatilization refers to the flow of volatile organic pollutants from the subsoil because of plant roots, such as an increase in soil permeability, a decrease in the water table, chemical transport through hydraulic redistribution, or fluctuations in the water table (Limmer and Burken 2016).

3.6.5 Rhizofiltration

In rhizofiltration, contaminants can be adsorbed around the root (rhizosphere) or absorbed inside the root tissue, concentrated, and precipitated (Abdullahi 2015). Plants must have a broad fibrous root system covered in root hairs with a large surface area to provide effective remediation (Grzegórska et al. 2020). Rhizofiltration reduces the mobility of contaminants and prevents their migration to groundwater. It is an effective and low-cost technique when applied to low concentrations of contaminants and a large volume of water (Chandra and Kumar 2018). A plant suitable for rhizofiltration must have fast-growing roots and be able to remove contaminants from the solution for a long time (Chandra and Kumar 2018). Rhizofiltration works through several physical and biochemical processes, such as adsorption, precipitation, and bioaccumulation. The efficiency of this mechanism resides in the efficiency of the roots to release root exudates (secondary metabolites) in the form of fatty acids, organic acids, amino acids, sugars, enzymes, sterols, growth factors, flavones, nucleotides, and other compounds (Sharma and Juwarkar 2015). Root exudates alter the geochemical conditions of the soil, such as pH, which leads HMs to accumulate in the plant and subsequently precipitate on the root surface, thus minimizing the movement of metals (Yan et al. 2020).

3.7 Microbiota in Plants Used in Phytoremediation

3.7.1 Fungi that Enhance Phytoremediation

The use of accumulator plants is a strategy used in the phytoremediation of soil and water contaminated with HMs. However, most of the plants are slow-growing and produce little biomass (Aguinaga-Ayen 2019). That is why, it is necessary to implement different alternatives to improve the phytoextraction of nonessential metals. One of the promising alternative strategies is the use of plants and microorganisms to hyperaccumulate, detoxify, or remediate soil contaminants. The rhizospheric and endophytic microbiota of plants play an important role in their protection, stability, productivity, and plant development. Also, they contribute to the immobilization of HMs present in the soil outside the rhizosphere of the plant and therefore improve phytostabilization (Figure 3.4) (Sánchez-Suarez 2022). Due to the presence of a chelation system and metal sequestration capabilities, fungi allow the plant to tolerate a greater number of metals by accumulating them in leaves and roots (Yadav et al. 2019) and likewise reduce stress and its toxic effect, providing an improvement in phytoremediation (Reyes-Pérez and Villanueva-Carranza 2021).

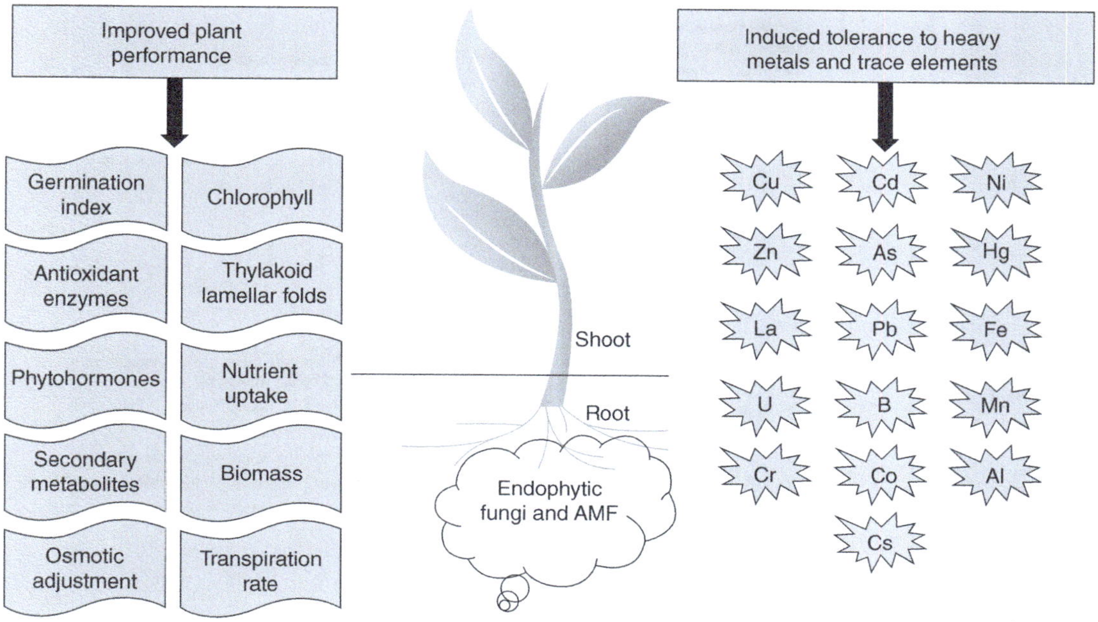

Figure 3.4 Positive effects of endophytic and mycorrhizal fungi on plant development under heavy metal and trace element stress conditions. *Source:* Khalid et al. (2021)/with permission of Elsevier.

Plant-fungus symbiosis is built through communication processes using signal molecules. These mutualistic relationships include arbuscular mycorrhizal fungi (AMF), ectomycorrhizal fungi (EMF), endophytic fungi (EF), and dark septate fungi (DSF), which colonize the roots and internal tissue of the plants and proliferate in the rhizosphere region, helping the translocation of nutrients to the host plant (Smith and Read 2008; Latef et al. 2015). In addition, they use different metabolic routes for the biotransformation of organic substances and different types of pollutant remediation (Stępniewska and Kuźniar 2013).

The fungi species associated with plants have different bioremediation capacities according to their genetic components, environmental factors, and rate growth (Yadav et al. 2019). Mycorrhizal fungi present a greater adsorption capacity for the elimination of contaminants, turning them into a potential tool for phytoremediation (Singh and Gauba 2014).

3.7.2 Mechanisms Involved in Fungal-Mediated Phytoremediation

Fungi growing on soils contaminated with HMs have adopted different molecular and cellular systems to tolerate and survive in such environments (Chiapello et al. 2015; Yadav et al. 2019). Several studies show that a great diversity of bacteria and fungi that are associated with plants produce organic acids that complex HMs, which alter their bioavailability in the soil through acidification, chelation, complexation, precipitation, and oxidation–reduction reactions (Seneviratne et al. 2017). Fungi increase the ability of roots to absorb HMs (Mishra et al. 2017). The absorption and immobilization of HMs occur in the fungal structures that generate the activation of transporters and specific or nonspecific pores that intervene in the plasmatic membrane of plants and fungi, activating the chelation that takes place in the cytosol and the sequestration in plant vacuoles and inside fungal cells. Therefore, the interaction of hyperaccumulator plants with HMs-tolerant fungi could be an effective strategy for phytoremediation (Chang et al. 2016). Figure 3.5 shows a diagram of the various mechanisms involved in the detoxification and remediation of soils contaminated with HMs, which are favored by various compounds generated by fungi (Chang et al. 2016; Franchi et al. 2017).

One of the great advantages of the symbiotic fungus-plant interaction is the production of root exudates, which play an important role in the growth and survival of plants, as well as providing sources of nutrients and energy to microorganisms, as well as changing the bioavailability of metals (Mishra et al. 2017). Table 3.4 shows studies of endophytic or mycorrhizal fungi in a symbiotic relationship with plants and their effect on the phytoremediation processes of soils contaminated by HMs.

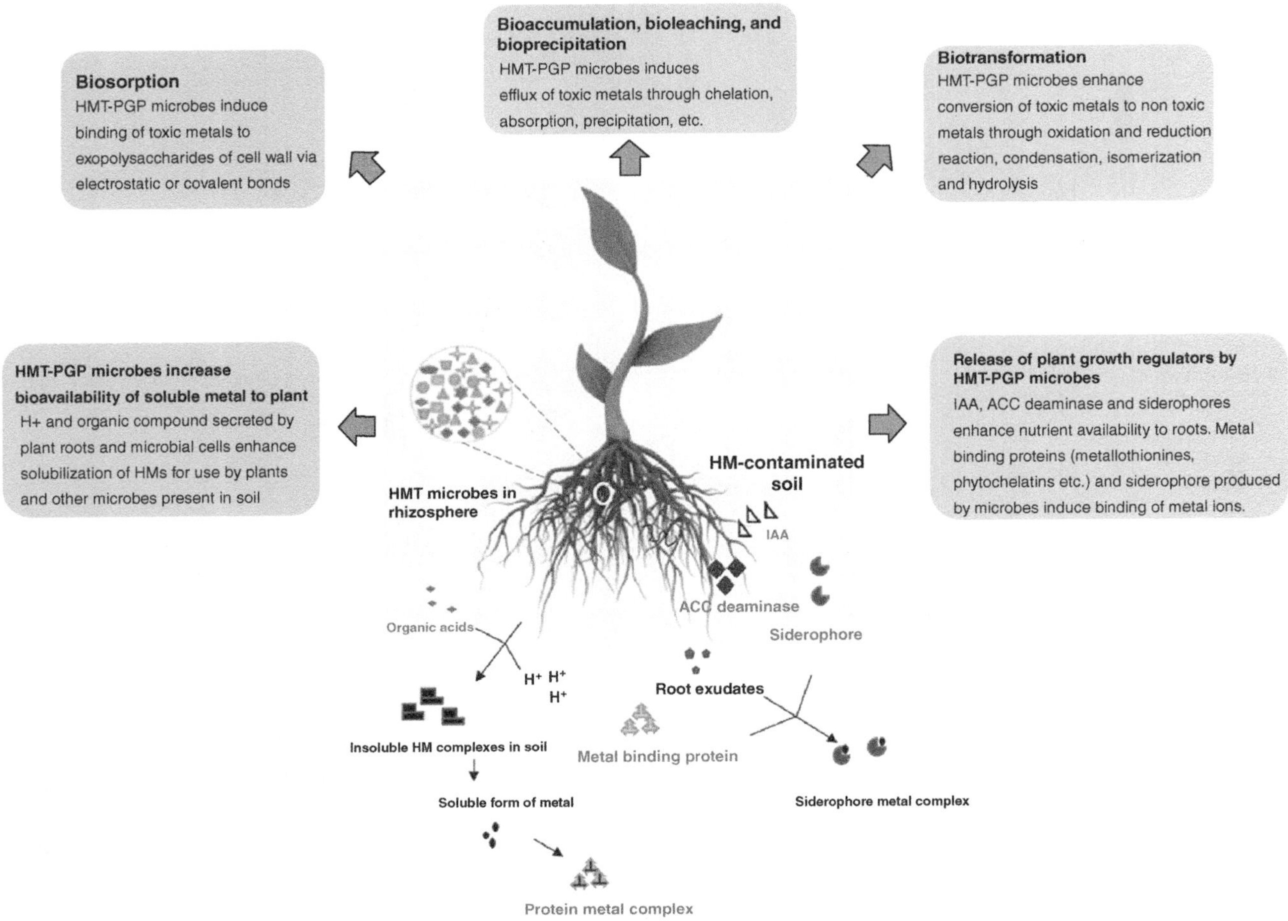

Figure 3.5 Mechanisms of action of rhizosphere microorganisms in the remediation of soils contaminated with HM. HM-tolerant microorganisms present in the rhizosphere can promote plant growth, including the secretion of phytohormones, siderophores, etc. *Source:* Adapted from Mishra et al. (2017).

Table 3.4 Effects of endophytic and mycorrhizal fungi on HMs phytoremediation.

Plant species	Fungus	Metal	Effect	Reference
Lolium arundinaceum	*Neotyphodium coenophialum*	Cd	Increase in biomass, accumulation, and transport of Cd	Ren et al. (2011)
Zea mays	*Paraphaeosphaeria* sp.	Cd	Major resistance to Cd	Emamverdian et al. (2015)
Oryza sativa	*Fusarium oxysporum*	Cr, Cd	Resistance to high levels of Cr and Cd	Ding et al. (2016)
Zea mays L.	*Gaeumannomyces cylindrosporum*	Pb	Improves photosynthesis and resistance to Pb	Ban et al. (2017)
Latuca sativa L.	*Aspergillus fumigatus*	Cr	Reduces Cr toxicity	Bibi et al. (2018)
Trifolium repens L.	*Glomus versiforme*	As	Increase in biomass, antioxidants and phytoextraction of As	Wang et al. (2018)
Heleniatus annuus L.	*Funneliformis caledonium*	Pb, Cd, Cr	Plant growth and decrease of HMs in shoots	Zhang et al. (2019)
Amaranthus hypochondriacus L.	*Rhizophagus intraradices*	Cd	Increased cadmium in roots	Li et al. (2019)
Vetiveria zizanioides	*Glomus mosseae*	Cd, Pb, Cu	Biomass increase and stress reduction	Kafil et al. (2019)
Zea mays L.	*Glomus aggregatum*	Cd, Pb, Cr	Greater root and shoot length and high chlorophyll content	Singh et al. (2019)

3.8 Bacteria that Enhance Phytoremediation

Phytoremediation can be improved by a microorganism's action present in the rhizosphere. Of these microorganisms, bacteria represent the most common type of microorganism in the rhizosphere, with an approximate 95% presence (Glick 2012). Among the wide variety of bacteria present in the rhizosphere are plant growth-promoting bacteria (PGPB). Some reports indicate that PGPB have been isolated from sites contaminated with HMs such as *Agrobacterium* sp., *Alcaligenes* sp., *Arthrobacter* sp., *Azoarcus* sp., *Azospirillum* sp., *Azotobacter* sp., *Bacillus* sp., *Burkholderia* sp., *Corynebacterium* sp., *Enterobacter* sp., *Flavobacterium* sp., *Klebsiella* sp., *Mycobacterium* sp., *Pseudomonas* sp., *Serratia* sp., *Rhizobium* sp., and *Rhodococcus* sp. (Enebe and Babalola 2018; Sharma 2021). HMs-tolerant PGPBs increase plant tolerance to biotic and abiotic stress. They improve plant tolerance to HMs, improve plant detoxification, increase the secretion of enzymes secreted by roots, modify soil pH, and can alter the mobility and bioavailability of metals (Verma and Kuila 2019). Therefore, HMs tolerant PGPBs are an appropriate strategy to improve HMs phytoremediation, reduce the impact of metals on plants, and increase the uptake and accumulation of metals in the plant (Wang et al. 2022). According to Alves et al. (2022), PGPB can be classified according to their functional properties as (i) biofertilizers, improving the availability of nutrients, (ii) phytostimulators, producing hormones that help plant growth and development, (iii) rhizoremediators, degrading organic pollutants, and solubilizing metals, and (iv) biopesticides, producing antibiotics and antifungals that restrict the development of pathogens.

3.8.1 Mechanisms Involved Bacterial-Mediated Phytoremediation

PGPB can promote through direct or indirect mechanisms the growth of plants against stress caused by HMs. Under a direct mechanism, PGPBs promote the production of phytohormones (auxins, cytokinins, gibberellins, ethylene, etc.) or facilitate the absorption of nutrients (nitrogen, phosphorus, potassium, etc.) by the plant (Wang et al. 2022). In the same way, they can produce antibiotics or antifungals that inhibit the growth of pathogens and reduce damage to the plant (Novo et al. 2018). On the other hand, some PGPBs can adsorb HMs through ion exchange or chelation. Furthermore, bacteria are partially responsible for reducing the solubility of metals to immobilize them and/or improving the solubility of metals to remove them from the site (Gadd 2000). For immobilization, PGPBs reduce uptake or translocation through metal precipitation, alkalinization of the medium, and metal complexation (Ma et al. 2015). To improve solubility, bacteria acidify, chelate, and protonate metals. Alternatively, bacteria produce low-molecular-weight organic compounds, biosurfactants, exopolysaccharides, and siderophores that can complex metals and improve their availability in soils (Ullah et al. 2015).

3.9 Conclusion

HMs from anthropogenic sources are a serious problem affecting living organisms. Fortunately, phytoremediation of sites contaminated by HMs is an economic and effective strategy to deal with this problem in nature. Although phytoremediation is a well-studied technique, it is important to study the plants-rhizosphere microorganisms' interactions, such as those established with bacteria and fungi to improve the knowledge of the bioremediation processes and bacteria and fungi roles to improve the plant phytoremediation capacity and their tolerance to toxic environments.

References

Abdullahi, M.S. (2015). Chapter 18: soil contamination, remediation and plants: prospects and challenges. In: *Soil Remediation and Plants Prospects and Challenges* (ed. K.R. Hakeem, M. Sabir, M. Öztürk, and A.R. Mermut), 525–546. Elsevier https://doi.org/10.1016/B978-0-12-799937-1.00018-8.

Adriano, D.C. (2001). *Trace Elements in Terrestrial Environments: Biogeochemistry, Bioavailability, and Risks of Metals*, 2e, vol. 1. India: Springer https://doi.org/10.1007/978-0-387-21510-5.

Agrawal, M. and Singh, A. (2018). Reduction in metal toxicity by applying different soil amendments in agricultural field and its consequent effects on characteristics of radish plants (*Raphanus sativus* L.). *J. Agric. Sci. Technol.* 15: 1553–1564. DOR: 20.1001.1.16807073.2013.15.7.9.8.

Aguinaga-Ayen, M.E. (2019). Inoculación de hongos micorrízicos arbusculares en Amaranthus sp. y su efecto en la biorremediación de suelo contaminado con plomo. Tesis de licenciatura. Universidad Nacional Pedro Ruiz Gallo, Perú. https://hdl.handle.net/20.500.12893/4640.

Ali, H. and Khan, E. (2018). What are heavy metals? Long-standing controversy over the scientific use of the term 'heavy metals' – proposal of a comprehensive definition. *Toxicol. Environ. Chem.* 100: 6–19. https://doi.org/10.1080/02772248.2017.1413652.

Ali, R. M., Hamad, H. A., Hussein, M. M., Malash G. F. (2016). Potential of using green adsorbent of heavy metal removal from aqueous solutions: adsorption kinetics, isotherm, thermodynamic, mechanism and economic analysis, *Ecol. Eng.*, 91, 317–332. https://doi.org/10.1016/j.ecoleng.2016.03.015.

Alloway, B.J. (2013). Chapter 2: sources of heavy metals and metalloids in soils. In: *Heavy Metals in Soil* (ed. B.J. Alloway), 11–50. Netherlands: Springer.

Alsafran, M., Usman, K., Ahmed, B. et al. (2022). Understanding the phytoremediation mechanisms of potentially toxic elements: a proteomic overview of recent advances. *Front. Plant Sci.* 13: 881242. https://doi.org/10.3389/fpls.2022.881242.

Al-Thani, R.F. and Yasseen, B.T. (2020). Phytoremediation of polluted soils and waters by native Qatari plants: future perspectives. *Environ. Pollut.* 259: 113694. https://doi.org/10.1016/j.envpol.2019.113694.

Alves, A.R.A., Yin, Q., Oliveira, R.S. et al. (2022). Plant growth-promoting bacteria in phytoremediation of metal-polluted soils: current knowledge and future directions. *Sci. Total Environ.* 838 (4): 156435. https://doi.org/10.1016/j.scitotenv.2022.156435.

Armah, F.A., Quansah, R., and Luginaah, I. (2014). A systematic review of heavy metals of anthropogenic origin in environmental media and biota in the context of gold mining in Ghana. *Int. Sch. Res. Notices* 2014: 37. https://doi.org/10.1155/2014/252148.

Ashraf, S., Ali, Q., Zahir, Z.A. et al. (2019). Phytoremediation: environmentally sustainable way for reclamation of heavy metal polluted soils. *Ecotoxicol. Environ. Saf.* 174: 714–727. https://doi.org/10.1016/j.ecoenv.2019.02.068.

Awa, S.H. and Hadibarata, T. (2020). Removal of heavy metals in contaminated soil by phytoremediation mechanism: a review. *Water Air Soil Pollut.* 231 (2): 1–15. https://doi.org/10.1007/s11270-020-4426-0.

Azcón, R., Medina, A., Roldán, A. et al. (2009). Significance of treated agrowaste residue and autochthonous inoculates (Arbuscular mycorrhizal fungi and *Bacillus cereus*) on bacterial community structure and phytoextraction to remediate soils contaminated with heavy metals. *Chemosphere* 75: 327–334. https://doi.org/10.1016/j.chemosphere.2008.12.029.

Ban, Y.H., Xu, Z.Y., Yang, Y.R. et al. (2017). Effect of dark septate endophytic fungus *Gaeumannomyces cylindrosporus* on plant growth, photosynthesis and Pb tolerance of maize (*Zea mays* L.). *Pedosphere* 27 (2): 283–292. https://doi.org/10.1016/S1002-0160(17)60316-3.

Banjoko, B. and Eslamian, S. (2016). Chapter 53: phytoremediation. In: *Urban Water Reuse Handbook*, 1e (ed. S. Eslamian), 663–705. Taylor and Francis https://doi.org/10.1201/b19646-65.

Beiyuan, J., Awad, Y.M., Beckers, F. et al. (2017). Mobility and phytoavailability of As and Pb in a contaminated soil using pine sawdust biochar under systematic change of redox conditions. *Chemosphere* 178: 110–118. https://doi.org/10.1016/j.chemosphere.2017.03.022.

Berken, A., Mulholland, M.M., LeDuc, D.L., and Terry, N. (2002). Genetic engineering of plants to enhance selenium phytoremediation. *Crit. Rev. Plant Sci.* 21: 567–582. https://doi.org/10.1080/0735-260291044368.

Bhat, S.A., Bashir, O., Haq, S.A.U. et al. (2022). Phytoremediation of heavy metals in soil and water: an eco-friendly, sustainable, and multidisciplinary approach. *Chemosphere* 303: 134788. https://doi.org/10.1016/j.chemosphere.2022.134788.

Bian, F., Zhong, Z., Zhang, X. et al. (2019). Bamboo – an untapped plant resource for the phytoremediation of heavy metal contaminated soils. *Chemosphere* 246: 25750. https://doi.org/10.1016/j.chemosphere.2019.125750.

Bibi, S., Hussain, A., Hamayun, M. et al. (2018). Bioremediation of hexavalent chromium by endophytic fungi; safe and improved production of *Lactuca sativa* L. *Chemosphere* 211: 653–663. https://doi.org/10.1016/j.chemosphere.2018.07.197.

Bradl, H.B. (2005). Chapter 1: sources and origins of heavy metals. In: *Heavy Metals in the Environment: Origin, Interaction and Remediation. Interface Science and Technology*, vol. 6 (ed. H.B. Bradl), 1–27. Elsevier https://doi.org/10.1016/S1573-4285(05)80020-1.

Briffa, J., Sinagra, E., and Blundell, R. (2020). Heavy metal pollution in the environment and their toxicological effects on humans. *Heliyon* 6: e04691. https://doi.org/10.1016/j.heliyon.2020.e04691.

Burges, A., Alkorta, I., Epelde, L., and Garbisu, C. (2018). From phytoremediation of soil contaminants to phytomanagement of ecosystem services in metal contaminated sites. *Int. J. Phytoremediat.* 20 (4): 384–397. https://doi.org/10.1080/15226514.2017.1365340.

Chandra, R. and Kumar, V. (2018). Phytoremediation: a green sustainable technology for industrial waste management. In: *Phytoremediation of Environmental Pollutants*, vol. 1 (ed. R. Chandra, N.K. Dubei, and V. Kumar), 1–42. Taylor & Francis https://doi.org/10.4324/9781315161549.

Chang, Y.T., Lee, J.F., Liu, K.H. et al. (2016). Immobilization of fungal laccase onto a nonionic surfactant-modified clay material: application to PAH degradation. *Environ. Sci. Pollut. Res.* 23 (5): 4024–4035. https://doi.org/10.1007/s11356-015-4248-6.

Cheng, K., Tian, H.Z., Zhao, D. et al. (2014). Atmospheric emission inventory of cadmium from anthropogenic sources. *Int. J. Environ. Sci. Technol.* 11: 605–616. https://doi.org/10.1007/s13762-013-0206-3.

Chiapello, M., Martino, E., and Perotto, S. (2015). Common and metal-specific proteomic responses to cadmium and zinc in the metal tolerant ericoid mycorrhizal fungus *Oidiodendron maius* Zn. *Metallomics* 7 (5): 805–815. https://doi.org/10.1039/c5mt00024f.

Dal Corso, G., Fasani, E., Manara, A. et al. (2019). Heavy metal pollutions: state of the art and innovation in phytoremediation. *Int. J. Mol. Sci.* 20: 3412. https://doi.org/10.3390/ijms20143412.

Diarra, I., Kotra, K., and Prasad, S. (2021). Application of phytoremediation for heavy metal contaminated sites in the South Pacific: strategies, current challenges and prospects. *Appl. Spectrosc. Rev.* https://doi.org/10.1080/05704928.2021.1904410.

Ding, Z., Wu, J., Jiao, C., and Cao, C. (2016). Isolation of heavy metal-resistant fungi from contaminated soil and co-culturing with rice seedlings. *Afr. J. Microbiol. Res.* 10 (28): 1080–1085. https://doi.org/10.5897/AJMR2016.8126.

Dong, Q., Xu, P.X., and Wang, Z.L. (2017). Differential cadmium distribution and translocation in roots and shoots related to hyper-tolerance between tall fescue and Kentucky bluegrass. *Front. Plant Sci.* 8: 113. https://doi.org/10.3389/fpls.2017.00113.

Ekta, P.E. and Modi, N.R. (2018). A review of phytoremediation. *J. Pharmacogn. Phytochem.* 7: 1485–1489. https://doi.org/10.1016/j.chemosphere.2020.126310.

Emamverdian, A., Ding, Y., Mokhberdoran, F., and Xie, Y. (2015). Heavy metal stress and some mechanisms of plant defense response. *Sci. World J.* 2015: 756120. https://doi.org/10.1155/2015/756120.

Enebe, M.C. and Babalola, O.O. (2018). The influence of plant growth-promoting rhizobacteria in plant tolerance to abiotic stress: a survival strategy. *Appl. Microbiol. Biotechnol.* 102: 7821–7835. https://doi.org/10.1007/s00253-018-9214-z.

Ernest, M. (1984). Introduction on environmental chemistry and global cycles of chromium, nickel, cobalt beryllium, arsenic, cadmium and selenium, and their derivatives. *Toxicol. Environ. Chem.* 8: 9–38. https://doi.org/10.1080/02772248409357038.

Franchi, E., Rolli, E., Marasco, R. et al. (2017). Phytoremediation of a multi contaminated soil: mercury and arsenic phytoextraction assisted by mobilizing agent and plant growth promoting bacteria. *J. Soils Sediments* 17 (5): 1224–1236.

Gadd, G.M. (2000). Bioremedial potential of microbial mechanisms of metal mobilization and immobilization. *Curr. Opin. Biotechnol.* 11 (3): 271–279. https://doi.org/10.1016/S0958-1669(00)00095-1.

Glick, B.R. (2012). Plant growth-promoting bacteria: mechanisms and applications. *Scientifica* 2012: 1–15. https://doi.org/10.6064/2012/963401.

Gómez-Garrido, M., Mora Navarro, J., Murcia Navarro, F.J., and Faz Cano, A. (2018). The chelating effect of citric acid, oxalic acid, amino acids, and Pseudomonas fluorescens bacteria on phytoremediation of Cu, Zn, and Cr from soil using Suaeda vera. *Int. J. Phytorem.* 20: 1033–1042. https://doi.org/10.1080/15226514.2018.1452189.

Goyal, D., Yadav, A., Prasad, M. et al. (2020). Effect of heavy metals on plant growth: an overview. In: *Contaminants in Agriculture* (ed. M. Naeem, A. Ansari, and S. Gill). Cham: Springer https://doi.org/10.1007/978-3-030-41552-5_4.

Grzegórska, A., Rybarczyk, P., Rogala, A., and Zabrocki, D. (2020). Phytoremediation from environment cleaning to energy generation—current status and future perspectives. *Energies* 13 (11): 2905. https://doi.org/10.3390/en13112905.

Guarino, F., Miranda, A., Castiglione, S., and Cicatelli, A. (2020). Arsenic phytovolatilization and epigenetic modifications in *Arundo donax* L. assisted by a PGPR consortium. *Chemosphere* 251: 126310. https://doi.org/10.1016/j.chemosphere.2020.126310.

Gujarathi, N.P., Haney, B.J., and Linden, J.C. (2005). Phytoremediation potential of *Myriophyllum aquaticum* and *Pistia stratiotes* to modify antibiotic growth promoters, tetracycline, and oxytetracycline, in aqueous wastewater systems. *Int. J. Phytoremediat.* 7: 99–112. https://doi.org/10.1080/16226510590950405.

Hammond, C.M., Root, R.A., Maier, R.M., and Chorover, J. (2018). Mechanisms of arsenic sequestration by *Prosopis juliflora* during the phytostabilization of metalliferous mine tailings. *Environ. Sci. Technol.* 52 (3): 1156–1164. https://doi.org/10.1021/acs.est.7b04363.

Hussain, S., Siddique, T., Arshad, M., and Saleem, M. (2009). Bioremediation and phytoremediation of pesticides: recent advances. *Crit. Rev. Environ. Sci. Technol.* 39: 843–907. https://doi.org/10.1080/10643380801910090.

Islam, M., Hossain, M., Khatun, M., and Hossen, M. (2015). Environmental impact assessment on frequency of pesticide uses during vegetable production. *Progress. Agric.* 26: 97–102. https://doi.org/10.3329/pa.v26i2.25962.

Jan, S. and Parray, J.A. (2016). *Approaches to Heavy Metal Tolerance in Plants*, 1–18. Singapore: Springer https://doi.org/10.1007/978-981-10-1693-6_1.

Jiang, M., Liu, S., Li, Y. et al. (2019). EDTA-facilitated toxic tolerance, absorption and translocation and phytoremediation of lead by dwarf bamboos. *Ecotoxicol. Environ. Saf.* 170: 502–512. https://doi.org/10.1016/j.ecoenv.2018.12.020.

Kader, M., Lamb, D.T., Megharaj, M., and Naidu, R. (2016). Sorption parameters as a predictor of arsenic phytotoxicity in Australian soils. *Geoderma* 265: 103–110. https://doi.org/10.1016/j.geoderma.2015.11.019.

Kafil, M., Boroomand Nasab, S., Moazed, H., and Bhatnagar, A. (2019). Phytoremediation potential of vetiver grass irrigated with wastewater for treatment of metal contaminated soil. *Int. J. Phytoremediat.* 21 (2): 92–100. https://doi.org/10.1080/15226514.2018.1474443.

Kafle, A., Timilsina, A., Gautam, A. et al. (2022). Phytoremediation: mechanisms, plant selection and enhancement by natural and synthetic agents. *Environ. Adv.* 8: https://doi.org/10.1016/j.envadv.2022.100203.

Kanwar, V.S., Sharma, A., Srivastav, A.L., and Rani, L. (2020). Phytoremediation of toxic metals present in soil and water environment: a critical review. *Environ. Sci. Pollut. Res.* 27 (36): 44835–44860. https://doi.org/10.1007/s11356-020-10713-3.

Kaur, R., Bhatti, S.S., Singh, S. et al. (2018). Phytoremediation of heavy metals using cotton plant: a field analysis. *Bull. Environ. Contam. Toxicol.* 101: 637–643. https://doi.org/10.1007/s00128-018-2472-8.

Khalid, S., Shahid, M., Niazi, N.K. et al. (2017). A comparison of technologies for remediation of heavy metal contaminated soils. *J. Geochem. Explor.* 182: 247–268. https://doi.org/10.1016/j.gexplo.2016.11.021.

Khalid, N., Noman, A., Aqeel, M. et al. (2019). Phytoremediation potential of *Xanthium strumarium* for heavy metals contaminated soils at roadsides. *Int. J. Environ. Sci. Technol.* 16: 2091–2100. https://doi.org/10.1007/s13762-018-1825-5.

Khalid, M., Saeed, U.R., Hassani, D. et al. (2021). Advances in fungal-assisted phytoremediation of heavy metals: a review. *Pedosphere* 31 (3): 475–495. https://doi.org/10.1016/S1002-0160(20)60091-1.

Latef, A.A.H.A., Hashem, A., Rasool, S. et al. (2015). Assessment of arbuscular mycorrhizal fungi status and heavy metal accumulation characteristics of tree species in a lead–zinc mine area: potential applications for phytoremediation. *Environ. Sci. Pollut. Res.* 22 (17): 13179–13193. https://doi.org/10.1007/s11356-015-4521-8.

Li, X., Chen, A.Y., Chen, X.X. et al. (2019). Effects of β-cyclodextrin on phytoremediation of soil co-contaminated with Cd and BDE-209 by arbuscular mycorrhizal amaranth. *Chemosphere* 220: 910–920. https://doi.org/10.1016/j.chemosphere.2018.12.211.

Limmer, M. and Burken, J. (2016). Phytovolatilization of organic contaminants. *Environ. Sci. Technol.* 50 (13): 6632–6643. https://doi.org/10.1021/acs.est.5b04113.

Lin, Y.F., Hassan, Z., Talukdar, S. et al. (2016). Expression of the ZNT1 zinc transporter from the metal hyperaccumulator *Noccaea caerulescens* confers enhanced zinc and cadmium tolerance and accumulation to Arabidopsis thaliana. *PLoS One* 11 (3): e0149750. https://doi.org/10.1371/journal.pone.0149750.

Ma, Y., Rajkumar, M., Rocha, I. et al. (2015). Serpentine bacteria influence metal translocation and bioconcentration of *Brassica juncea* and *Ricinus communis* grown in multi-metal polluted soils. *Front. Plant Sci.* 5: 757. https://doi.org/10.3389/fpls.2014.00757.

Mahdavian, K., Asadigerkan, S., Sangtarash, M.H., and Nasibi, F. (2022). Phytoextraction and phytostabilization of copper, zinc, and iron by growing plants in Chahar Gonbad Copper Mining Area, Iran. *Proc. Natl. Acad. Sci. India B – Biol. Sci.* 92 (2): 319–327. https://doi.org/10.1007/s40011-021-01345-9.

Manzoor, M., Gul, I., Manzoor, A. et al. (2020). Lead availability and phytoextraction in the rhizosphere of *Pelargonium* species. *Environ. Sci. Pollut. Res.* 27 (32): 39753–39762. https://doi.org/10.1007/s11356-020-08226-0.

Meagher, R.B., Rugh, C.L., Kandasamy, M.K. et al. (2000). Engineered phytoremediation of mercury pollution in soil and water using bacterial genes. In: *Phytoremediation of Contaminated Soil and Water* (ed. N. Terry and G. Bañuelos), 201–219. Boca Raton, FL: Lewis Publishers https://doi.org/10.1201/9780367803148-11.

Mendez, M.O. and Maier, R.M. (2008). Phytostabilization of mine tailings in arid and semiarid environments an emerging remediation technology. *Environ. Health Perspect.* 116: 278–283. https://doi.org/10.1289/ehp.10608.

Mishra, J., Singh, R., and Arora, N.K. (2017). Alleviation of heavy metal stress in plants and remediation of soil by rhizosphere microorganisms. *Front. Microbiol.* 8: 1706. https://doi.org/10.3389/fmicb.2017.01706.

Muthusaravanan, S., Sivarajasekar, N., Vivek, J.S. et al. (2018). Phytoremediation of heavy metals: mechanisms, methods, and enhancements. *Environ. Chem. Lett.* 16: 1339–1359. https://doi.org/10.1007/s10311-018-0762-3.

Muthusaravanan, S., Sivarajasekar, N., Vivek, J.S. et al. (2020). Research updates on heavy metal phytoremediation: enhancements, efficient post-harvesting strategies and economic opportunities. In: *Green Materials for Wastewater Treatment. Environmental Chemistry for a Sustainable World*, vol. 38 (ed. M. Naushad and E. Lichtfouse), 191–222. Cham: Springer https://doi.org/10.1007/978-3-030-17724-9_9.

Nedjimi, B. (2021). Phytoremediation: a sustainable environmental technology for heavy metals decontamination. *SN Appl. Sci.* 3 (3): 1–19. https://doi.org/10.1007/s42452-021-04301-4.

Ngo, H.H., Bui, X.T., Nghiem, L.D., and Guo, W. (2020). Green technologies for sustainable water. *Bioresour. Technol.* 317: 123978. https://doi.org/10.1016/j.biortech.2020.123978.

Novo, L.A.B., Castro, P.M.L., Alvarenga, P., and da Silva, E.F. (2018). Plant growth-promoting rhizobacteria-assisted phytoremediation of mine soils. In: *Bio-Geotechnologies for Mine Site Rehabilitation*, vol. 1 (ed. M.N.V. Prasad, P.J. de Campos Favas, and S.K. Maiti), 281–295. Elsevier https://doi.org/10.1016/B978-0-12-812986-9.00016-6.

Patra, D.K., Pradhan, C., and Patra, H.K. (2019). Chromium bioaccumulation, oxidative stress metabolism and oil content in lemon grass *Cymbopogon flexuosus* (Nees ex Steud.) W. Watson grown in chromium rich over burden soil of Sukinda chromite mine, India. *Chemosphere* 218: 1082–1088. https://doi.org/10.1016/j.chemosphere.2018.11.211.

Pilipović, A., Zalesny, R.S. Jr., Rončević, S. et al. (2019). Growth, physiology, and phytoextraction potential of poplar and willow established in soils amended with heavy-metal contaminated, dredged river sediments. *J. Environ. Manage.* 239: 352–365. https://doi.org/10.1016/j.jenvman.2019.03.072.

Pivetz, B.E. (2001). Phytoremediation of contaminated soil and ground water at hazardous waste sites. United States Environmental Protection Agency EPA 540: 136. https://www.epa.gov/sites/production/files/2015-06/documents/epa_540_s01_500.pdf.

Postigo, C., Martinez, D., Grondona, S., and Miglioranza, K. (2018). Groundwater pollution: sources, mechanisms, and prevention. *Water Resour. Prot* 11: 87–96. https://doi.org/10.1016/B978-0-12-409548-9.09880-8.

Prieto, M., Acevedo, S., Prieto, G., and González, N. (2018). Phytoremediation of soils contaminated with heavy metals. *Biodivers. Int. J.* 2: 362–376. https://doi.org/10.15406/bij.2018.02.00088.

Rai, P.K. (2019). Heavy metals/metalloids remediation from wastewater using free floating macrophytes of a natural wetland. *Environ. Technol. Innovation.* 15: 100393. https://doi.org/10.1016/j.eti.2019.100393.

Rashid, A., Mahmood, T., Mehmood, F. et al. (2014). Phytoaccumulation, competitive adsorption and evaluation of chelators-metal interaction in lettuce plant. *Environ. Eng. Manage. J.* 13 (10): 2583–2592. https://doi.org/10.30638/eemj.2014.288.

Ren, A., Li, C., and Gao, Y. (2011). Endophytic fungus improves growth and metal uptake of *Lolium arundinaceum* Darbyshire ex Schreb. *Int. J. Phytorem.* 13 (3): 233–243. https://doi.org/10.1080/15226511003671387.

Reyes-Pérez, L.K. and Villanueva-Carranza, C.N. (2021). Diseño de una propuesta ambiental para la fitoextracción de metales pesados utilizando el género amaranthus sp. para la restauración de suelos contaminados por la Minera Quiruvilca SA–La Libertad 2020. https://hdl.handle.net/11537/28881.

Rolón-Cárdenas, G.A., Martínez-Martínez, J.G., Arvizu-Gómez, J.L. et al. (2022). Enhanced Cd-accumulation in Typha latifolia by interaction with Pseudomonas rhodesiae GRC140 under axenic hydroponic conditions. *Plants* 11 (11): 1447. https://doi.org/10.3390/plants11111447.

Rosenfeld, C.E., Chaney, R.L., and Martínez, C.E. (2018). Soil geochemical factors regulate Cd accumulation by metal hyperaccumulating *Noccaea caerulescens* (J. Presl & C. Presl) FK Mey in field-contaminated soils. *Sci. Total Environ.* 616: 279–287. https://doi.org/10.1016/j.scitotenv.2017.11.016.

San Juan, M.R.F., Albornoz, C.B., Larsen, K., and Najle, R. (2018). Bioaccumulation of heavy metals in *Limnobium laevigatum* and Ludwigia peploides: their phytoremediation potential in water contaminated with heavy metals. *Environ. Earth Sci.* 77: 404. https://doi.org/10.1007/s12665-018-7566-4.

Sánchez-Suarez, R.I. (2022). Importancia de los HMA (Hongos Micorrícicos Arbusticos) en sistemas agroforestales de la provincia de Los Ríos. Bachelor's thesis, UTB, Babahoyo. http://dspace.utb.edu.ec/handle/49000/13127.

Schück, M. and Greger, M. (2020). Plant traits related to the heavy metal removal capacities of wetland plants. *Int. J. Phytoremediat.* 22 (4): 427–435. https://doi.org/10.1080/15226514.2019.1669529.

Seneviratne, M., Seneviratne, G., Madawala, H.M.S.P., and Vithanage, M. (2017). Role of rhizospheric microbes in heavy metal uptake by plants. In: *Agro-Environmental Sustainability*, vol. 2 (ed. J. Singh and G. Seneviratne), 147–163. Springer https://doi.org/10.1007/978-3-319-49727-3_8.

Shah, V. and Daverey, A. (2020). Phytoremediation: a multidisciplinary approach to clean up heavy metal contaminated soil. *Environ. Technol. Innovation* 18: 100774. https://doi.org/10.1016/j.eti.2020.100774.

Sharma, J.K. and Juwarkar, A.A. (2015). Phytoremediation: general account and its application. In: *Plant Biology and Biotechnology. Vol. II: Plant Genomics and Biotechnology* (ed. B. Bahadur, M.V. Rajam, L. Sahijram, and K.V. Krishnamurthy). New Delhi: Springer https://doi.org/10.1007/978-81-322-2283-5_34.

Sharma, P. (2021). Efficiency of bacteria and bacterial assisted phytoremediation of heavy metals: an update. *Bioresour. Technol.* 328: 124835. https://doi.org/10.1016/j.biortech.2021.124835.

Sharma, P. and Rath, S.K. (2020). Potential applications of fungi in the remediation of toxic effluents from pulp and paper industries. In: *Fungi Bio-Prospects in Sustainable Agriculture, Environment and Nanotechnology*, 193–211. Academic Press https://doi.org/10.1016/B978-0-12-821925-6.00010-1.

Shen, X., Chi, Y., and Xiong, K. (2019). The effect of heavy metal contamination on humans and animals in the vicinity of a zinc smelting facility. *PLoS One* 14: e0207423. https://doi.org/10.1371/journal.pone.0207423.

Sheoran, V., Sheoran, A.S., and Poonia, P. (2016). Factors affecting phytoextraction: a review. *Pedosphere* 26 (2): 148–166. https://doi.org/10.1016/S1002-0160(15)60032-7.

Singh, A. and Gauba, P. (2014). Micoremediación: un tratamiento para la contaminación del suelo por metales pesados. *J. Civ. Eng. Environ. Technol.* 1: 59–61. http://www.krishisanskriti.org/jceet.html.

Singh, G., Pankaj, U., Chand, S., and Verma, R.K. (2019). Fitoextracción asistida por hongos micorrízicos arbusculares de metales tóxicos por Zea mays L. a partir de lodos de curtiduría. *Contaminación de suelos y sedimentos: una revista internacional* 28 (8): 729–746. https://doi.org/10.1080/15320383.2019.1657381.

Singh, B.M., Singh, D., and Dhal, N.K. (2022). Enhanced phytoremediation strategy for sustainable management of heavy metals and radionuclides. *CSCEE* 5: 100176. https://doi.org/10.1016/j.cscee.2021.100176.

Smith, S.E. and Read, D. (2008). The symbionts forming arbuscular mycorrhizas. In: *Mycorrhizal Symbiosis*, 3e, vol. 2 (ed. S.E. Smith and D.J. Read), 13–41. London: Academic Press https://doi.org/10.1016/B978-012370526-6.50003-9.

Soudani, A., Gholami, A., Mohammadi Roozbahani, M. et al. (2022). Heavy metal phytoremediation of aqueous solution by Typha domingensis. *Aquat. Ecol.* 56 (2): 513–523. https://doi.org/10.1007/s10452-022-09945-x.

Steliga, T. and Kluk, D. (2020). Application of Festuca arundinacea in phytoremediation of soils contaminated with Pb, Ni, Cd, and petroleum hydrocarbons. *Ecotoxicol. Environ. Saf.* 194: 110409. https://doi.org/10.1016/j.ecoenv.2020.110409.

Stępniewska, Z. and Kuźniar, A. (2013). Endophytic microorganisms—promising applications in bioremediation of greenhouse gases. *Appl. Microbiol. Biotechnol.* 97 (22): 9589–9596. https://doi.org/10.1007/s00253-013-5235-9.

Su, C., Meng, J., Zhou, Y. et al. (2022). Heavy metals in soils from intense industrial areas in South China: spatial distribution, source apportionment, and risk assessment. *Front. Environ. Sci.* 10: 820536. https://doi.org/10.3389/fenvs.2022.820536.

Suman, J., Uhlik, O., Viktorova, J., and Macek, T. (2018). Phytoextraction of heavy metals: a promising tool for clean-up of polluted environment? *Front. Plant Sci.* 9: 1476. https://doi.org/10.3389/fpls.2018.01476.

Tang, C., Chen, Y., Zhang, Q. et al. (2019). Effects of peat on plant growth and lead and zinc phytostabilization from lead-zinc mine tailing in southern China: screening plant species resisting and accumulating metals. *Ecotoxicol. Environ. Saf.* 176: 42–49. https://doi.org/10.1016/j.ecoenv.2019.03.078.

Thakare, M., Sarma, H., Datar, S. et al. (2021). Understanding the holistic approach to plant-microbe remediation technologies for removing heavy metals and radionuclides from soil. *Curr. Res. Biotechnol.* 3: 84–98. https://doi.org/10.1016/j.crbiot.2021.02.004.

Tian, Y. and Zhang, H. (2016). Producing biogas from agricultural residues generated during phytoremediation process: possibility, threshold, and challenges. *Int. J. Green Energy* 13 (15): 1556–1563. https://doi.org/10.1080/15435075.2016.1206017.

Tripathi, V., Edrisi, S.A., and Abhilash, P.C. (2016). Towards the coupling of phytoremediation with bioenergy production. *Renew. Sust. Energ. Rev.* 57: 1386–1389. https://doi.org/10.1016/j.rser.2015.12.116.

Ullah, A., Heng, S., Munis, M.F.H. et al. (2015). Phytoremediation of heavy metals assisted by plant growth promoting (PGP) bacteria: a review. *Environ. Exp. Bot.* 117: 28–40. https://doi.org/10.1016/j.envexpbot.2015.05.001.

Vangronsveld, J., Herzig, R., Weyens, N. et al. (2009). Phytoremediation of contaminated soils and groundwater: lessons from the field. *Environ. Sci. Pollut. Res.* 16: 765–794. https://doi.org/10.1007/s11356-009-0213-6.

Verma, S. and Kuila, A. (2019). Bioremediation of heavy metals by microbial process. *Environ. Technol. Innovation* 14: 100369. https://doi.org/10.1016/j.eti.2019.100369.

Vishnoi, N. and Dixit, S. (2019). Bioremediation: new prospects for environmental cleaning by fungal enzymes. In: *Recent Advancement in White Biotechnology Through Fungi. Fungal Biology* (ed. A.N. Yadav, S. Singh, S. Mishra, and A. Gupta). Springer https://doi.org/10.1007/978-3-030-25506-0_2.

Wang, F., Adams, C.A., Shi, Z., and Sun, Y. (2018). Combined effects of ZnO NPs and Cd on sweet sorghum as influenced by an arbuscular mycorrhizal fungus. *Chemosphere* 209: 421–429. https://doi.org/10.1016/j.chemosphere.2018.06.099.

Wang, Y., Narayanan, M., Shi, X. et al. (2022). Plant growth-promoting bacteria in metal-contaminated soil: current perspectives on remediation mechanisms. *Front. Microbiol.* 13: 966226. https://doi.org/10.3389/fmicb.2022.966226.

Wu, B., Peng, H., Sheng, M. et al. (2021). Evaluation of phytoremediation potential of native dominant plants and spatial distribution of heavy metals in abandoned mining area in Southwest China. *Ecotoxicol. Environ. Saf.* 220: 112368. https://doi.org/10.1016/j.ecoenv.2021.112368.

Xu, X., Yang, Y., Wang, G. et al. (2020). Removal of heavy metals from industrial sludge with new plant–based washing agents. *Chemosphere* 246: 125816. https://doi.org/10.1016/j.chemosphere.2020.125816.

Yadav, S. (2010). Heavy metals toxicity in plants: an overview on the role of glutathione and phytochelatins in heavy metal stress tolerance of plants. *S. Afr. J. Bot.* 76: 167–179. https://doi.org/10.1016/j.sajb.2009.10.007.

Yadav, A.N., Singh, S., Mishra, S., and Gupta, A. (2019). *Recent Advancement in White Biotechnology Through Fungi*. Springer https://doi.org/10.1007/978-3-030-10480-1.

Yan, A., Wang, Y., Tan, S.N. et al. (2020). Phytoremediation: a promising approach for revegetation of heavy metal-polluted land. *Front. Plant Sci.* 11: https://doi.org/10.3389/fpls.2020.00359.

Yan, X., An, J., Yin, Y. et al. (2022). Heavy metals uptake and translocation of typical wetland plants and their ecological effects on the coastal soil of a contaminated bay in northeast China. *Sci. Total Environ.* 803: 149871. https://doi.org/10.1016/j.scitotenv.2021.149871.

Yang, P., Zhou, X.F., Wang, L.L. et al. (2018). Effect of phosphate-solubilizing bacteria on the mobility of insoluble cadmium and metabolic analysis. *Int. J. Environ. Res. Public Health* 15 (7): 1330. https://doi.org/10.3390/ijerph15071330.

Yang, W., Gu, J., Zhou, H. et al. (2020). Effect of three Napier grass varieties on phytoextraction of Cd- and Zn-contaminated cultivated soil under mowing and their safe utilization. *Environ. Sci. Pollut. Res.* 27: 16134–16144. https://doi.org/10.1007/s11356-020-07887-1.

Yang, X., Eziz, M., Hayrat, A. et al. (2022). Heavy metal pollution and risk assessment of surface dust in the arid NW China. *Int. J. Environ. Res. Public Health* 19: 13296. https://doi.org/10.3390/ijerph192013296.

Yong-Guan, Z. and Rosen, B.P. (2009). Perspectives for genetic engineering for the phytoremediation of arsenic-contaminated environments: from imagination to reality? *Curr. Opin. Biotechnol.* 20 (2): 220–224. https://doi.org/10.1016/j.copbio.2009.02.011.

Yoon, J., Cao, X., Zhou, Q., and Ma, L.Q. (2006). Accumulation of Pb, Cu, and Zn in native plants growing on a contaminated Florida site. *Sci. Total Environ.* 368 (2-3): 456–464. https://doi.org/10.1016/j.scitotenv.2006.01.016.

Zayed, A.M. and Terry, N. (2003). Chromium in the environment: factors affecting biological remediation. *Plant Soil* 249: 139–156. https://doi.org/10.1023/A:1022504826342.

Zhang, F., Liu, M., Li, Y. et al. (2019). Effects of arbuscular mycorrhizal fungi, biochar and cadmium on the yield and element uptake of Medicago sativa. *Sci. Total Environ.* 655: 1150–1158. https://doi.org/10.1016/j.scitotenv.2018.11.317.

Zhang, H., Zhao, Z., Cai, A. et al. (2022). Source apportionment of heavy metals in wet deposition in a typical industry city based on multiple models. *Atmosphere* 13: 1716. https://doi.org/10.3390/atmos13101716.

4

Mechanism of Heavy Metal-Induced Stress and Tolerance

Jose A. Montes-Rocha[1], Angel J. Alonso-Castro[2], and Candy Carranza-Álvarez[3]

[1] *Postgraduate in Chemical Sciences, Autonomous University of San Luis Potosí, San Luis Potosí, México*
[2] *Division of Natural and Exact Sciences, University of Guanajuato, Guanajuato, México*
[3] *Faculty of Professional Studies Huasteca Zone, Autonomous University of San Luis Potosí, Ciudad Valles, San Luis Potosí, México*

4.1 Introduction

Nowadays, pollution by heavy metals has taken on significant importance worldwide, and several countries have taken measures to regulate these elements. Increased activities like mining, industry, agriculture, chemical weathering of rocks, and environmental factors have increased the levels of metals in soil, water, and air. Heavy metals cause toxic effects at various levels of the food chain. There is a great concern in the world about eliminating elevated levels of these elements from the soil, sediment, water, and atmosphere. Essential metals such as manganese (Mn), copper (Cu), iron (Fe), and magnesium (Mg) are necessary for plant growth. Zinc (Zn), molybdenum (Mb), nickel (Ni), and cobalt (Co) function as enzyme cofactors. Nonessential or toxic metals, including arsenic (As), lead (Pb), cadmium (Cd), and mercury (Hg) do not fulfill any biological function. These metals cause harmful effects by displacing other essential metals, causing the generation of reactive oxygen species (ROS), and reducing photosynthesis. Plants are exposed to distinct types of stress, including those caused by metals. Plants growing under heavy metal-induced stress have adapted and developed metal tolerance mechanisms and have acquired the capacity to accumulate metals in roots, stems, leaves, and fruits. Among the main mechanisms are root exudates, sequestration in the cell wall, transport across the membrane, sequestration in the vacuole, and chelation in the xylem and phloem. Another mechanism developed by plants is the chelation of various ligands and compounds responsible for chelating metals and reducing their toxicity in the plant. Some of the most important ligands are glutathione, phytochelatins (PCs), metallothionein (MT), citrate, malate, phenols, carboxylic acids, saponins, and tocopherols. Plant tolerance to heavy metals will be activated depending on the species and type of plants, as well as the chemical, physical, and biological conditions of the study area. In this chapter, we will discuss heavy metal-induced stress (ROS generation) and the main mechanisms of metal tolerance: functional groups for metal binding, transporters, organic acids, amino acids, and amino acid derivatives.

4.2 Heavy Metal-Induced Stress

Plants need low concentrations of micronutrients (Zn, Cu, and Ni) for their essential growth and development. Some of these elements participate as cofactors of several enzymes. However, there are other elements (Cd and Pb) without a beneficial role in plant development and represent a risk since they can be toxic at minimum concentrations. Plants suffering from heavy metal-induced stress show decreased growth, chlorosis, browning of roots, inactivation, and denaturation of enzymes and proteins, altered photosynthesis, respiration and homeostasis, DNA damage, and the stimulation of ROS production (Ghori et al. 2019; Feng et al. 2021).

ROS are generated in plants in chloroplasts, mitochondria, and peroxisomes. These chemical species are highly reactive, unstable, and shortlived. The main ROS are hydrogen peroxide (H_2O_2), singlet oxygen (O_2), superoxide anion (O_2^-), hydroxyl ($HO^{\cdot}$), alkoxyl ($RO^{\cdot}$), and peroxyl ($RO_2^{\cdot}$). Heavy metal stress in plants increases ROS production, leading to affections in

Aquatic Contamination: Tolerance and Bioremediation, First Edition. Edited by Rouf Ahmad Bhat, Gowhar Hamid Dar, Fernanda Maria Policarpo Tonelli, and Saima Hamid.

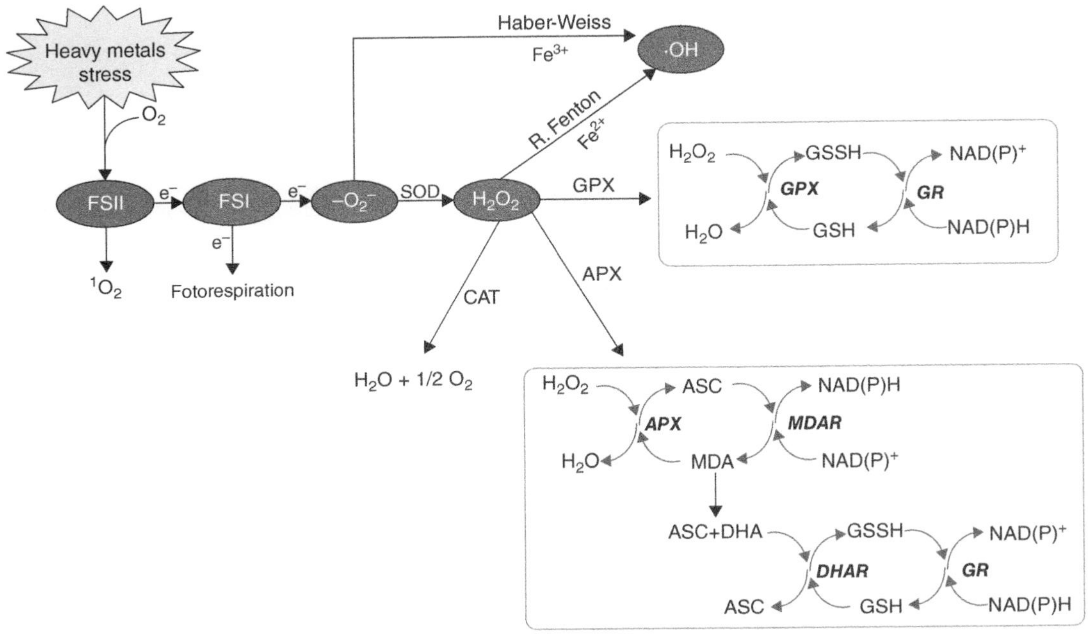

Figure 4.1 ROS generation and stress tolerance mechanisms induced by heavy metals. Phase II photosynthesis (FSII), Phase I photosynthesis (FSI), superoxide anion (O_2^-), hydrogen peroxide (H_2O_2), superoxide dismutase (SOD), Catalase (CAT), Glutathione Peroxidase (GPX), Ascorbate Peroxidase (APX), glutathione disulfide (GSSH), glutathione (GSH), glutathione reductase (GR), nicotinamide-adenine dinucleotide phosphate (NAD(P)H), ascorbate (ASC), monodehydro-ascorbate (MDA), MDA reductase (MDAR), dehydro-ascorbate (DHA), DHA reductase (DHAR). *Source:* Own elaboration based on information from Shahid et al. (2017), Berni et al. (2018), Feng et al. (2021).

metabolic processes, cell death, membrane lipid peroxidation, RNA and DNA damage, enzyme inhibition, and protein oxidation (Shahid et al. 2017; Berni et al. 2018; Yadav et al. 2021). The toxic properties of O_2^- and H_2O_2 are increased with redox-active metals (Fe^{3+} and Cu^{2+}), generating the OH radical via the Haber–Weiss reaction (Khan et al. 2021). The OH radical reacts rapidly with all types of macromolecules, leading to cell death. Plants can eliminate ROS through antioxidant enzymatic activities including catalase (CAT), superoxide dismutase (SOD), ascorbate peroxidase (APX), glutathione peroxidase (GPX), and using glutathione, a nonenzymatic antioxidant protector (Figure 4.1) (Berni et al. 2018; Feng et al. 2021).

4.3 Metal Tolerance Mechanisms

Plants have developed diverse mechanisms of tolerance to heavy metals, varying according to the plant species. These mechanisms give plants the ability to grow and thrive under metal stress. Plants subjected to metal stress have developed internal tolerance mechanisms such as (i) root exudates, (ii) sequestration in the primary and secondary cell wall, (iii) selective transport in the plasma membrane, (iv) vacuole sequestration, (v) chelation by various ligands in the xylem and phloem, and finally (vi) chelation of metals in the cytosol by various ligands.

4.4 Root Exudates

Secondary metabolites secreted by plants to the rhizosphere can alter pH as well as form complexes with metals. The metabolites secreted are mainly complex mixtures of inorganic ions, gaseous molecules, and carbon-based compounds such as amino acids, organic acids, phenolic compounds, sugars, mucilages, and proteins (Feng et al. 2021; Pasricha et al. 2021). The metabolites with the highest detoxification activity are organic acids (low molecular weight compounds), which act on Cd, aluminum (Al), gallium (Ga), Cu, Mn, and Pb. Organic acids are secreted from the apex into the rhizosphere and some of them are ion chelators. The most effective organic acid in detoxification is citric acid, followed by malate and oxalate. Malate secretion is mediated by outflow transporters such as Al^-- and Ga^--activated malate transporter 1 (ALMT), whereas citrate secretion is carried out through multidrug and toxin outflow (MATE).

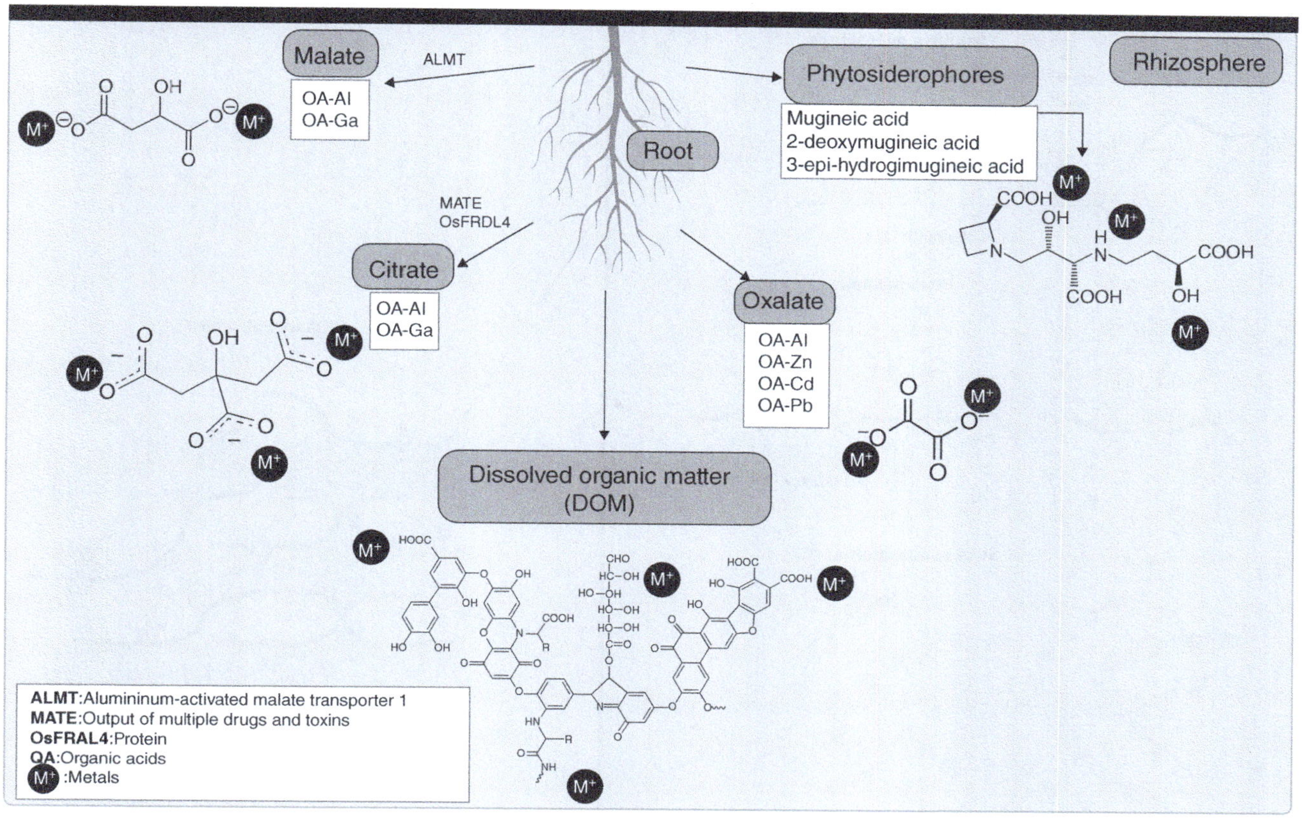

Figure 4.2 Involvement of root exudates in metal tolerance. *Source:* Own elaboration based on information from Gupta and Singh (2017), Chen et al. (2017).

The plasma membrane-localized protein (OsFRDL4) is another citrate-related transporter that participates in Al- and Ga-induced stress. Oxalate is another important chelator of Al, Zn, Cd, and Pb. Chelators can retain metals in the rhizosphere or the apoplast. In the exudate secreted by plants, dissolved organic matter (DOM), acting as a chelating agent, forms complexes with Zn, Cd, and Pb (Chen et al. 2017). Root exudates contain DOM, which can sequester metals such as Cd. The composition of root exudates can change depending on the type of metal-induced stress. DOM is categorized into CHO, CHON, CHOS, and CHONS compounds, with CHOs being the main constituents of DOM (Li et al. 2021; Zhang et al. 2021).

Phytosiderophores (PS) are compounds derived from mugineic acid, including mugineic acid (the first PS discovered), 2-deoxymugineic acid, 3-epihydrogimugineic acid, avenic acid, and 3-epihydroxy-2-deoxymugineic acid. Most of these compounds possess carboxyl, amino, and hydroxyl groups, which can chelate metals (Gupta and Singh 2017). Mugineic acid PS secreted into the rhizosphere are solubilized, chelates iron (Fe^{3+}) forming a complex, and transported to the root by yellow streak (YS) transporters (Carrillo and Borthakur 2021) (Figure 4.2).

4.5 Cellular Wall

Cellulose microfibrils, hemicelluloses, and pectins are the main constituents of the cell wall (Figure 4.3a-d). The most abundant hemicellulose is xyloglucan, a polysaccharide with short side chains of xylose, galactose, and sometimes fucose. Pectins are characterized by galacturonic acid and neutral sugars (rhamnose, galactose, and arabinose). One of the most complex pectins is the complex polysaccharide rhamnogalacturonan, which has a wide variety of side chains in its skeleton. The secondary plant cell wall contains lignin, a phenolic polymer with an irregular and complex pattern of bonds between its aromatic alcohol units, as well as three basic units of alcohol: *p*-coumaryl, coniferyl, and sinapyl (Berni et al. 2018).

The first contact of metals and metalloids in the soil with the plant is through the cell wall. These elements can bind to functional groups through covalent and noncovalent bonds. Metal stress promotes the generation of phenolic compounds

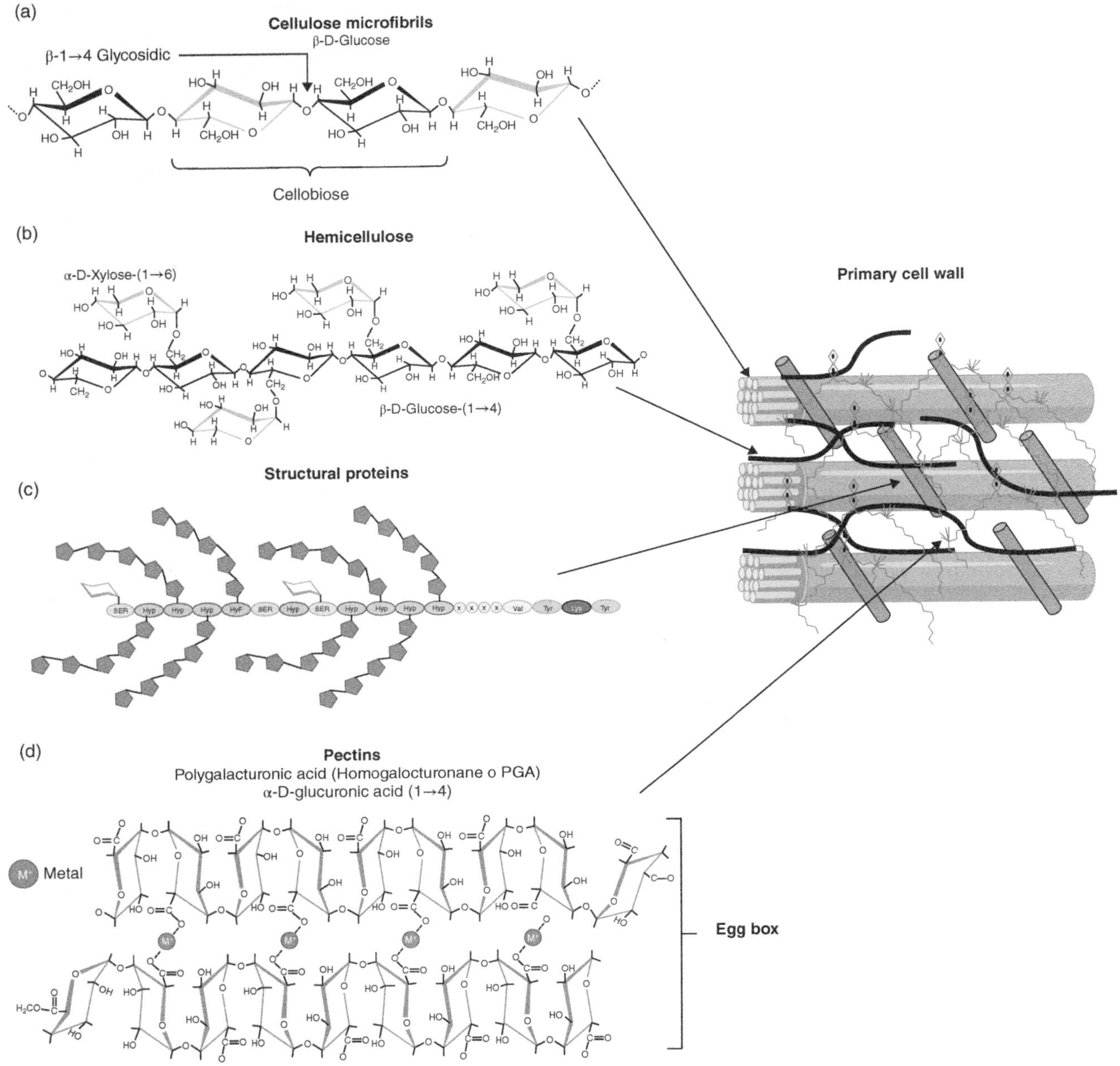

Figure 4.3 Primary wall components involved in metal tolerance. (a) Cellulose microfibrils. (b) Hemicellulose. (c) Structural proteins. (d) Pectins. *Source:* Own elaboration based on information from Anjitha et al. (2021), Meychik et al. (2021).

and lignin in the secondary cell wall. Phenolic compounds and flavonoids can trap metals in the cell wall. Increased pectin methylation and xylogalacturonan content are excellent indicators of metal stress (Anjitha et al. 2021). The thickening of the cell wall is a strategy that helps chelate heavy metals avoiding their absorption. The cell wall has negatively charged sites that can retain heavy metals (Zeroual et al. 2020).

Metal ions can be chelated in acidic peptide polymers such as homogalacturonan and rhamnogalacturonan (Carrillo and Borthakur 2021; Gu and Lan 2021). The cell wall may contain organic acids, amino acids, PCs, MTs, proteins, pectins, and polyphenols that help sequester metals and decrease their toxicity in plants (Pasricha et al. 2021). Cellulose and hemicellulose lack ion exchange functional groups, except for glucuronoarabinoxylans containing glucuronic acid which ionizes at physiological pH. It is important to mention that the main characteristic of pectins is their cation exchange capacity, conferred by polygalacturonic acid (homogalocturonan or PGA) (egg crate) (Figure 4.3d). This polysaccharide is a polymer of acid-α-D-glucuronic acid residues with (1→4) bonds (Pidatala et al. 2018; Jia et al. 2019; Meychik et al. 2021; Luyckx et al. 2021).

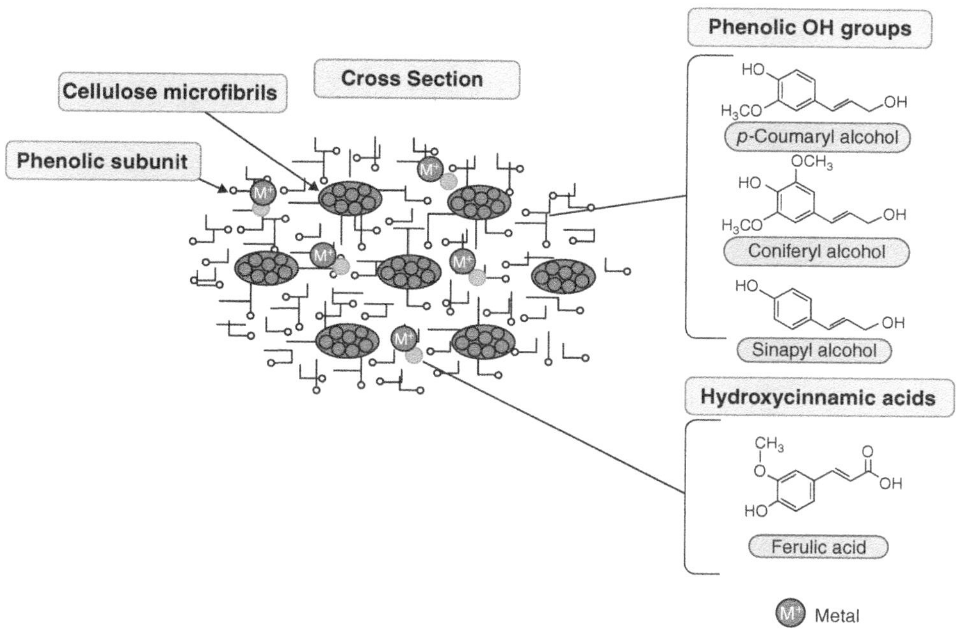

Figure 4.4 Secondary wall components involved in metal tolerance. *Source:* Own elaboration based on information from Anjitha et al. (2021), Meychik et al. (2021).

The secondary wall is composed of cellulose, a few pectins, and proteins with suberin and lignin as the main component. Lignin has two cation exchange groups, hydroxyl groups of phenolic compounds and carboxyl groups of hydroxycinnamic acids (HCA), mainly ferulic acid, which can be found in a free state or bound to the lignin (Figure 4.4) (Berni et al. 2018). The cell wall contains cation exchange groups such as carboxylic and phenolic OH groups, which can trap metals at physiological pH. The ability of the cell wall to retain heavy metals is determined by the charge, cation size, concentration, and pH of the solution. This capacity decreases according to the valence number of the metal (M), $M^{3+} > M^{2+} > M$. The main binding of metals takes place via the carboxyl groups of the PGAs. A divalent metal cation can be trapped by the carboxyl groups present in the wall, as shown in the following reaction: $2R-COOH + M^{2+} \rightarrow (RCOO)_2M + 2H^+$ (Figure 4.4) (Luyckx et al. 2021; Meychik et al. 2021; Wang et al. 2021).

4.6 Plasma Membrane

The plasma membrane is a complex and highly sophisticated functioning system, being the entrance for any element (nutrient or metal). The plasma membrane is severely affected by metals, and the effects vary according to the metal, causing damage through several mechanisms: oxidation and cross-linking of proteins, inhibition of membrane proteins (H^+-ATPase), variation in the composition and fluidity of membrane lipids, and alterations in permeability (Mishra et al. 2022). There are several classes of metal transporter proteins involved in metal tolerance, including CPx-ATPases, Nramps, CDF family, and ZIP family (Niu et al. 2021). Many pumps transport H^+ or Ca^+ and some pumps like the ATP-binding cassette (ABC) family can transport large organic molecules. This type of transport extrudes protons from the cytosol by H^+-ATPases operating in the plasma and vacuolar membrane (Li et al. 2022). Pb transport across the membrane can occur via the Ca^{2+} ATPase system, the Ca^{2+}/H^+ antiporter system, or through Ca^{2+} channels. Cd^{2+} can be transported by IRT1, ZNT1, and AtNramp3. At the plasma membrane, the uptake and transport of Zn^{2+} are mediated by ZIP transporters, which transport Zn^{2+} from the extracellular space to the cytoplasm. ZIP transporters also can transport Cd^{2+}, Mn^{2+}, and Fe^{2+}. Yellow stripe-like protein 1 (YSL) mediates Fe uptake, forming complexes with PS or nicotianamine. The most studied protein of this family is YS1, involved in metal homeostasis through the transport of Zn^{2+}, Cu^{2+}, or Ni^{2+}. Some transporters involved in Ni^{2+} and Mn^{2+} homeostasis are AtYSL4 and AtYSL6. Another transporter is LCT1 involved in the transport of low-affinity cations and Nramp (natural resistance-associated macrophage proteins) is involved in the transport of Cd^{2+} Fe^{2+}. COPT is a transporter involved in Cu^{2+} uptake by plants (Tibbett et al. 2021; Yaashikaa et al. 2022). Another transporter is ZNT1, involved in transporting Zn^{2+} and Cd^{2+} at very high levels in roots and the aerial part of plants. TcNRAMP3 is a transporter related to increased sensitivity and Cd^{2+} content in roots (Figure 4.5) (Pasricha et al. 2021).

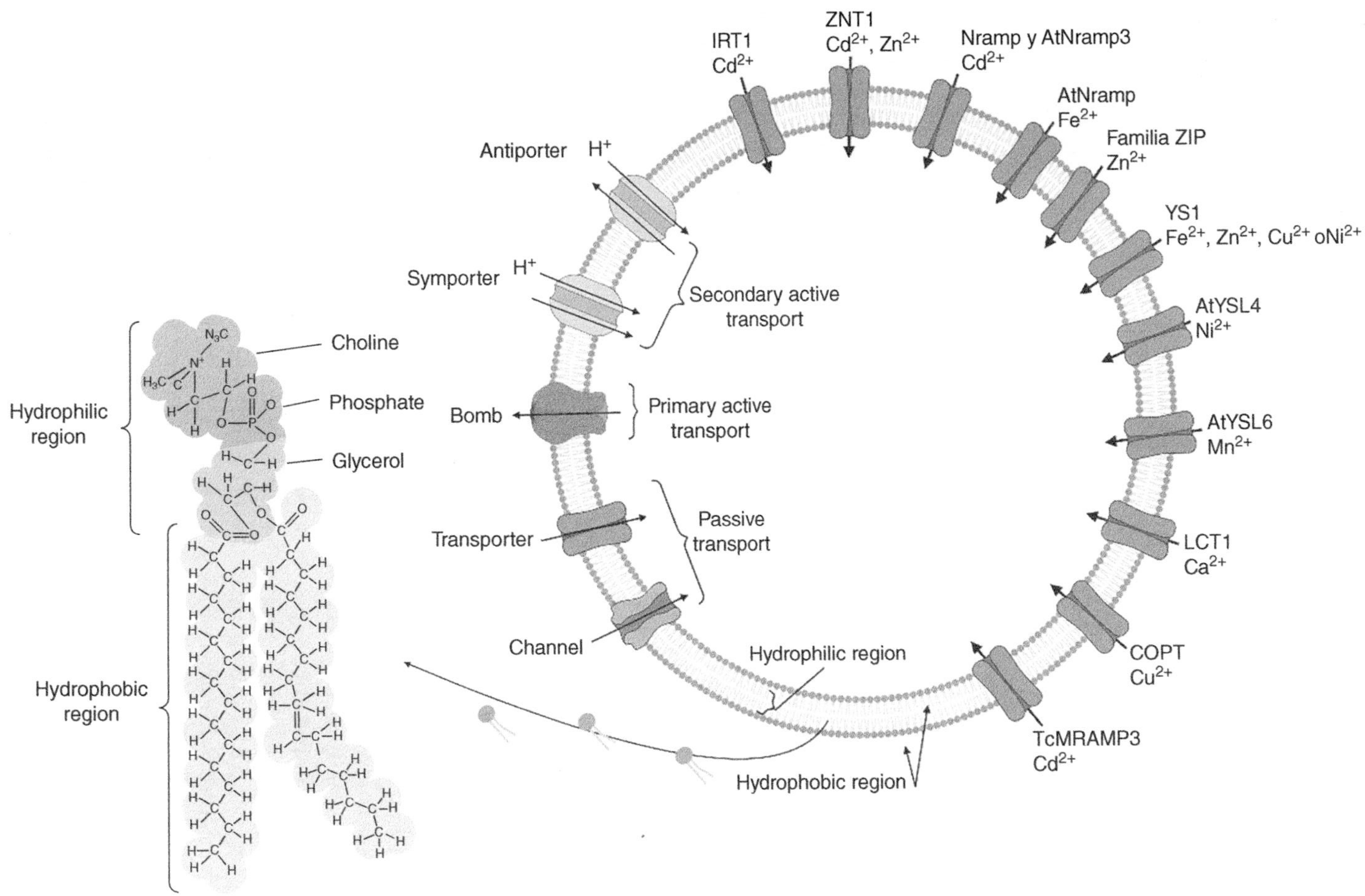

Figure 4.5 Plasma membrane metal transport mechanisms. *Source:* Own elaboration based on information from Pasricha et al. (2021), Tibbett et al. (2021), Yaashikaa et al. (2022).

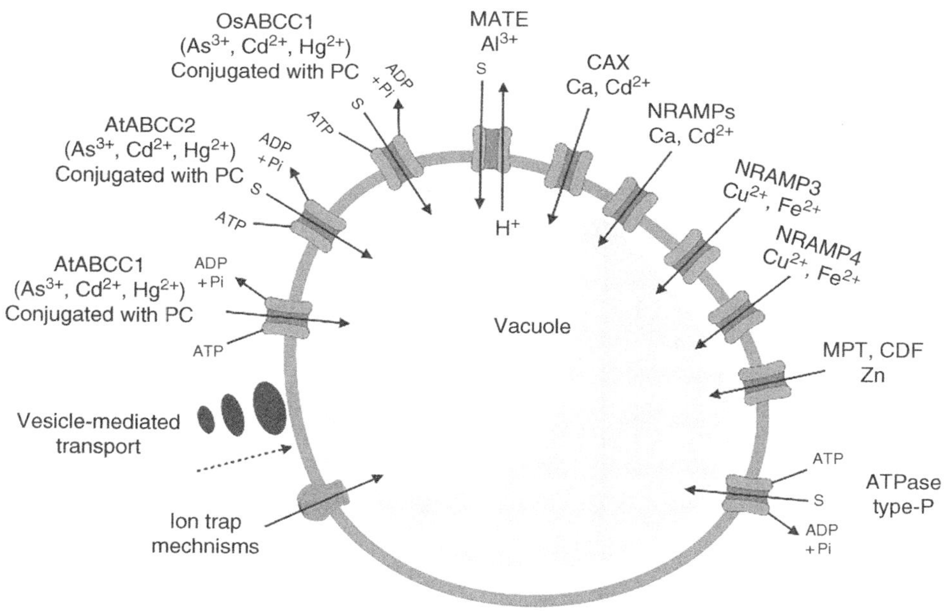

Figure 4.6 Mechanisms of metal transport to the vacuole. *Source:* Own elaboration based on information from Shitan and Yazaki (2019), Sharma et al. (2021a).

4.7 Vacuole

There are three transport mechanisms in the vacuole: ion trap, vesicle-mediated, and protein-dependent transport. ABC transporter proteins are classified into ABCA, ABCB, ABCC, ABCD, ABCE, ABCF, ABCG, and ABCI depending on their structure. Cation exchanger CAX2, the Cd/H^+ antiporter, and the Ca^{2+}/H^+ antiporters (ECA1) assist in Mn uptake and transport into the vacuole, and the P-type ATPase promotes Mn transport into the secretory pathway, and one of the ZIP proteins transports Mg out of the vacuole (Dhir 2021). ABC transporters are ATP-dependent to transport substrates. Such transporters could transport xenobiotics, hormones, primary and secondary metabolites, and ions (Rekha et al. 2021; Koul et al. 2022). The AtABCC1, AtABCC2, and OsABCC1 can transport arsenite cadmium (Cd II), mercury (Hg II), and (As III) in PC-conjugated form to the vacuole (Huang et al. 2021). MATE (multidrug and toxic compound extrusion) transporters contribute to the transport of Fe and Al^{3+}. Protein transporters and metal ATPases (HMA), belonging to the P-type ATPases, are heavy metal transporters (Huang et al. 2021). The Cation/H^+ Exchanger (CAX) transports calcium and contributes to Cd tolerance in the root. The NRAMP transporter is associated with Cd transport in the cytoplasm and vacuolar membrane (Shitan and Yazaki 2019; Pasricha et al. 2021; Sharma et al. 2021a; Yadav et al. 2021; Yaashikaa et al. 2022). The main transporters involved in Zn detoxification are the metal tolerance proteins (MTPs) and the cation diffusion facilitator (CDF). These move them from the cytoplasm to the vacuole. Crosstalk has been demonstrated between the mobilization of vacuolar Cu and Fe groups by the transporters COPT5, Nramp3, and Nramp4 (Figure 4.6) (Pasricha et al. 2021; Tibbett et al. 2021).

4.8 Xylem

The xylem is characterized by the transport of metal ions that move vertically (root to leaf) (Huang et al. 2021). The ions can be transported in bulk or by chromatographic exchange, however, the translocation depends on the valence number of the metal–ligand complex. In both processes, metal transport in the xylem is carried out via the mass flow pathway in the transpiration stream, indicating that the number of metals transported is proportional to the transpiration rate (Rekha et al. 2021; Tibbett et al. 2021). Some examples of transporters in the xylem are histidine that increases due to metal stress derived from Ni, which is in a free state and can bind other metal ions during transport in the xylem.

The xylem has a pH of 5–6.5, which favours the union of metal ions with histidine and citrate, decreasing the toxicity of these elements (Carrillo and Borthakur 2021). Other elements such as Cd and Zn are transported to the xylem by heavy metal ATPases. HMA4 is a key transporter in the loading of Cd and Zn in the xylem (Huang et al. 2021). And in some cases, citric acid and malic acid also participate together with histidine, which sequester Cu and Ni in the xylem (Rekha et al. 2021; Tibbett et al. 2021).

4.9 Phloem

In the phloem, metal translocation occurs laterally using holes in the vessel walls. Metals can enter the phloem sap by selective transport. They enter by mobilization of the xylem vessels of the stem, through the nodes, or through the network of leaf veins. The movement of metals in the phloem is affected by pH, ionic strength, and their organic components in the sap. In the phloem, metals are found in the form of complexes instead of ions. These complexes are formed with nicotianamine (Cu, Fe, and Zn), glutathione, and PCs (Hg, Cd, and Zn) (Figure 4.10). Nicotianamine is highly efficient in the kidnapping of Fe while maintaining a nonreactive status with a consequent decrease in the formation of hydroxyl radicals through the Fenton and Haber-Weiss reactions (Figure 4.7) (Carrillo and Borthakur 2021; Tibbett et al. 2021).

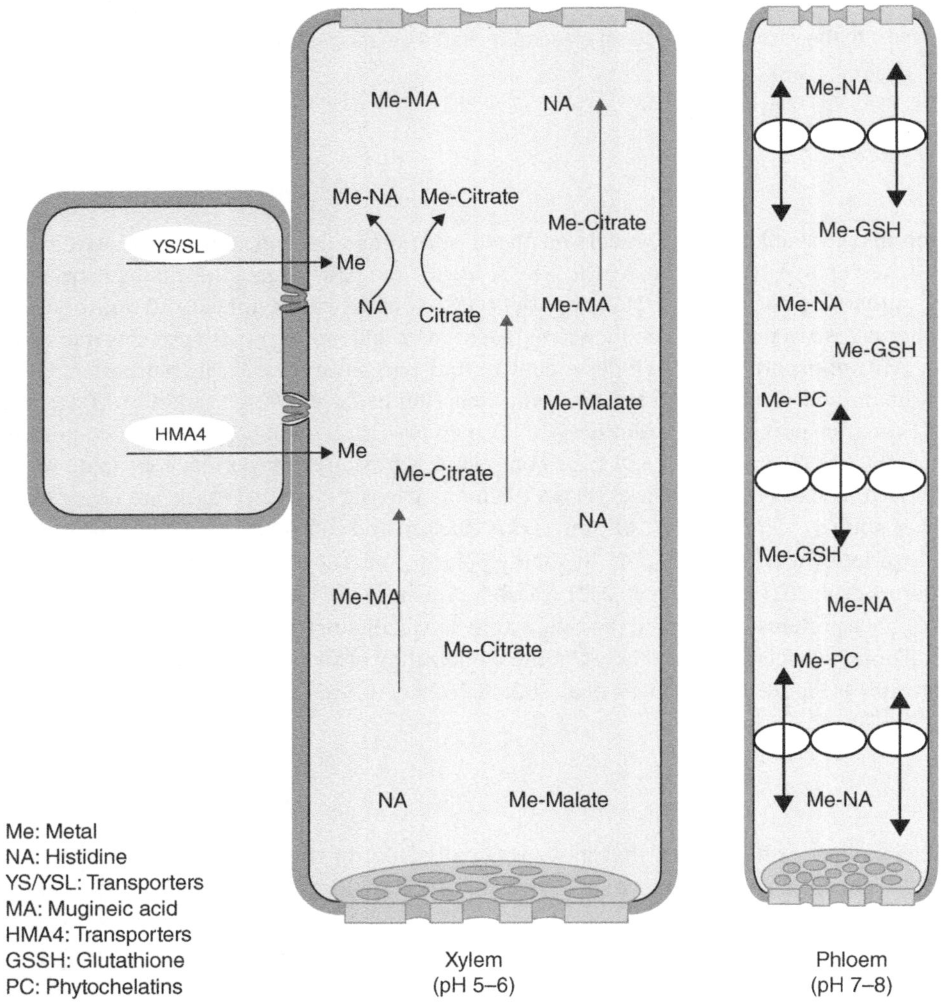

Figure 4.7 Mechanism of metal transport in the xylem and phloem. *Source:* Own elaboration based on information from Carrillo and Borthakur (2021), Tibbett et al. (2021).

4.10 Sequestering of Metals in the Cytosol by Various Ligands

4.10.1 Glutathione (GSH)

Glutathione, a tripeptide (γ-glutamyl-cysteine-glycine) abundant in thiol groups, is involved in the defense against ROS, heavy metal sequestration, and detoxification of xenobiotics (Vezza et al. 2019; Huang et al. 2021). GSH has been identified in the vacuole, endoplasmic reticulum, chloroplast, mitochondria, and cytosol. GSH is the source of nonprotein thiols in most plant cells. The thiol group present in GSH has high reactivity and a nucleophilic nature that forms bonds with metals (Figure 4.8) (Yadav et al. 2021).

GSH presents relative stability, high-water solubility, and reactivity. Therefore, GSH can protect plants from heavy metals-induced stress, endogenous chemicals-induced stress, and oxidative stress. GSH is an antioxidant involved in the reduction of ROS during metal stress. GSH is considered one of the most relevant defenses against ROS. Plants possess a complex antioxidant enzyme system in different compartments. Between 1% and 2% of the O_2 consumed by a plant leads to the generation of O_2^- radical. The main enzyme involved in the reduction–oxidation of O_2^- to H_2O_2 is SOD. Once H_2O_2 is formed, it can be catabolized by CAT or with several types of peroxidases in the ascorbate-glutathione cycle (Kaya et al. 2021). Ascorbate-glutathione cycle is the main route of H_2O_2 removal, and this occurs through the oxidation of GSH and ASC by H_2O_2. The transformation of H_2O_2 to H_2O is carried out by APX accomplished by the oxidation of ASC to monodehydroascorbate (MDA), which can be synthesized by monodehydroascorbate reductase (MDAR) with NADPH as reductant. Glutathione reductase (GR) can regenerate GSH from glutathione disulfide (GSSG), using NAD(P)H as the reducing agent. Another route that detoxifies H_2O_2 to H_2O is through GR using GSH as a reductant. In this way, the oxidized forms of ASC and GSH are transformed to their initial reduced states concluding the cycle and avoiding the accumulation of H_2O_2 (Dobritzsch et al. 2019; Kaya et al. 2020a,b; Ivanov 2021).

GSH detoxifies plants by conjugation with xenobiotics and metals through glutathione S-transferase, with the subsequent transport to the vacuole for protecting cells from harmful effects. However, the high demand for GSH resulting from metal stress leads to a decrease of GSH in the cytosol, impacting the redox potential. Therefore, maintaining the GSH/GSSG ratio is of vital importance for plant survival (Hernández Estévez and Rodríguez Hernández 2020; Khan et al. 2021).

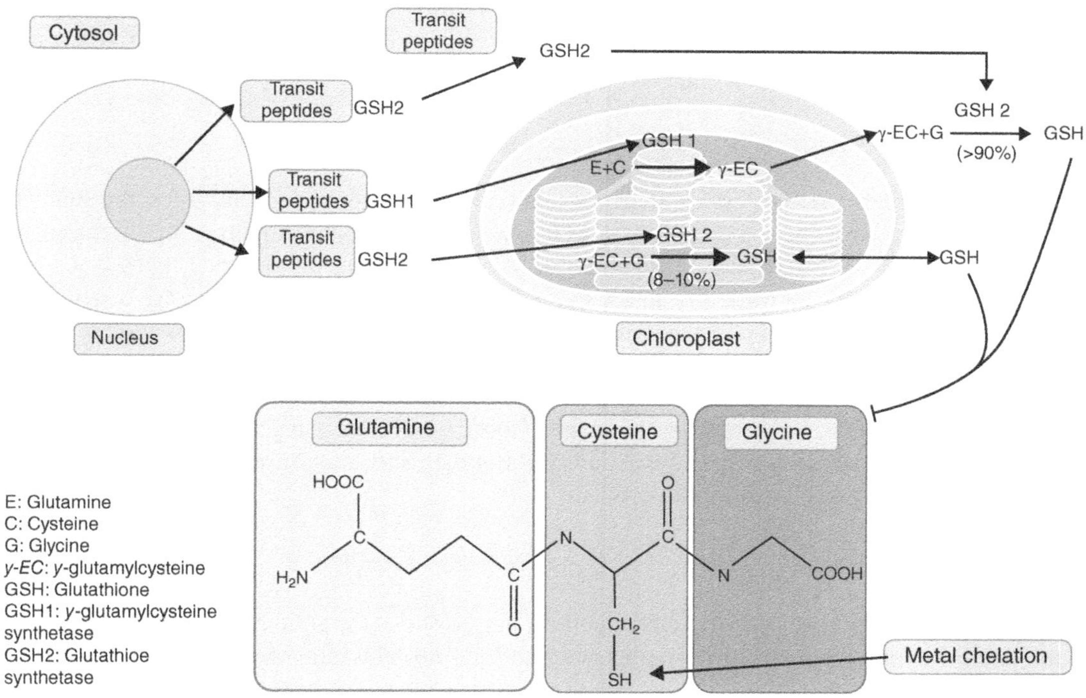

Figure 4.8 Synthesis and composition of Glutathione. *Source:* Own elaboration based on information from Dobritzsch et al. (2019), Kaya et al. (2020a,b), Ivanov (2021), Yadav et al. (2021).

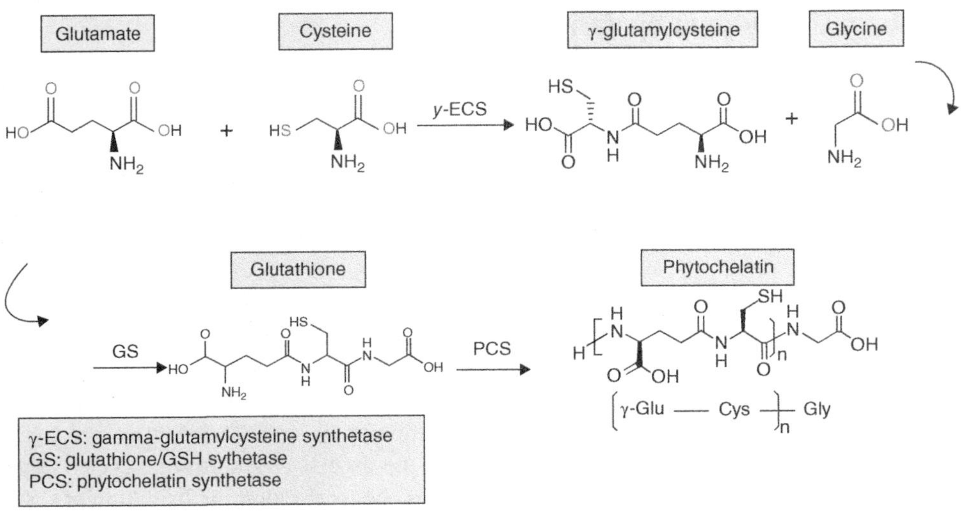

Figure 4.9 Phytochelatin synthesis. *Source:* Own elaboration based on information from Carrillo and Borthakur (2021), Khan et al. (2021), Yadav et al. (2021).

4.10.2 Phytochelatins

PCs are rich in cysteines which contain sulfhydryl groups (Sharma et al. 2021a) and are synthesized under stress caused by metals such as Ag, Zn, Hg, Pb, Ni, and Cu. Their synthesis is carried out through GSH by the activity of 3-glutamyl cysteine dipeptidyl transpeptidase (PCS) (Figure 4.9). PCs participate in heavy metal detoxification by forming mercapeptide bonds and acting as natural chelators by transporting toxic metals from the cytosol to the vacuole (Yadav et al. 2021). PCs is activated by two GSH molecules that bind to the metal generating a thiolate (Cd-GS2). During activation, a γ-Glu-Cys moiety is transferred to a GSH molecule or a PC molecule. PC-metal complexes can be transported to the vacuole by cassette transporters, the most stable form of storage. The activation of PCs is carried out in the following order: As, Sn, Au, Cd, Zn, Pb, Bi, Sb, Hg, Ag, and Cu. PCs play an important role in the tolerance of plants to metals and metalloids (Carrillo and Borthakur 2021; Khan et al. 2021).

4.10.3 Metallothioneins

MTs are polypeptides (4–8 kDa) with high cysteine (Cys) content. Cysteine contains a thiol group that confers the ability to form complexes with ions in the cytoplasm (Sharma et al. 2021a). MTs are classified into three main groups according to the disposition of their Cys residues: Cys-X-Cys, Cys-Cys, and Cys-XX-Cys, where X is any amino acid (Figure 4.10). The organization and distribution of Cys residues provide the ability to bind and chelate metals. MTs decrease heavy metal toxicity through the sequestration of essential and nonessential metals in storage organelles and the regulation of metalloenzymes. Biosynthesis of MT can be induced by metals (Cu, Co, Hg, Au, Cd, Ag, Ni, and Zn), cytotoxic agents, and hormones. MTs can protect cells from oxidants and electrophilic compounds that react with cysteinyl thiolate groups (Anjitha et al. 2021; Khan et al. 2021). There are four types of plant MTs: type I (root), type II (leaf), type III (fruit ripening), and type IV identified and known as early cysteine proteins. Type I and type II are directly associated with Cu accumulation and type III and type IV chelate mainly to Zn (Carrillo and Borthakur 2021).

4.10.4 Other Chelating Compounds

Other compounds that can chelate metals are organic acids, phenols, polyphenols, saponins, and anthocyanins. The organic acids such as malonate, tartrate, oxalate, citrate, malic, aconitate, amino acids, and some of their main derivatives help in the chelation of heavy metals. Organic acids chelate metals in the cytosol decreasing their toxicity. For

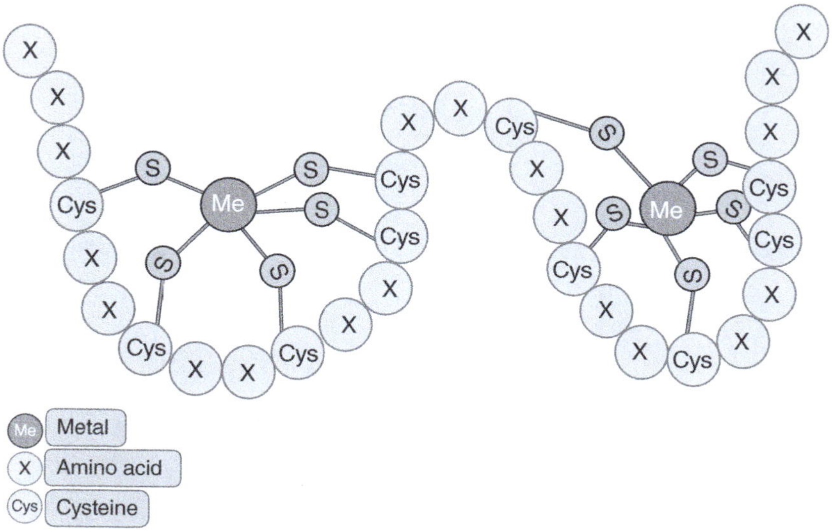

Figure 4.10 Structure of metallothioneins. *Source:* Own elaboration based on information from Anjitha et al. (2021), Carrillo and Borthakur (2021), Khan et al. (2021), Sharma et al. (2021b).

instance, Ni can form complexes with organic acids in the vacuole by decreasing its toxic effect. Citrate has a high capacity to chelate metals (Zn, Al, Ni, Cu, Al, Cd, and Fe Cd), while amino acids and their derivatives chelate metals to a lesser extent. The histidine biosynthetic pathway is the most effective in the chelation and detoxification of Ni in hyperaccumulator plants (Yadav et al. 2021). These amino acids include alanine, glycine, histidine, polyols, proline, sucrose, and trehalose, involved in the reduction of metal stress (Yadav et al. 2021). Nicotianamine forms complexes in the xylem with Ni (Anjitha et al. 2021). Other compounds that can help chelate metals are anthocyanins (cyanidins, petunidins, and delphinidins) because of their adjacent hydroxyl groups. Cyanidin forms the cyanidin-Zn complex by accepting two electrons from Zn^{2+} (Janeeshma and Puthur 2020). Other compounds involved in chelation or complex formation are phenols, alkaloids, and saponins. Tocopherols play a significant role in the reduction of heavy metal-induced stress, as well as in the elimination of ROS, and are also compounds involved in the termination of lipid peroxidation (Yadav et al. 2021). Phenols possess important antioxidant and chelating properties for protection against metallic stress. Polyphenols can chelate Cr, Hg, and Pb, whereas tannin fractions can chelate Cu, Fe, and Zn (Figure 4.11) (Anjitha et al. 2021).

4.11 Considerations

The tolerance mechanisms expressed in plants will be mediated by the type and species of plants, and the chemical, biological, and physical conditions of the study area. Table 4.1 shows scientific studies documenting plants with the capacity to tolerate metals, as well as the main mechanisms of accumulation and, in some cases, the expression of ligands such as organic acids, enzymatic, and nonenzymatic ligands. These plants can be a viable alternative for the phytoremediation of sites impacted by heavy metals.

4.12 Conclusion

Potentially toxic elements (PTE) are present in various matrices in minimal concentrations. The increase of PTE in the environment has been of interest worldwide since the concentrations of these elements have increased considerably. Activities contributing to the increase of ETPs in soil, sediment, water, and air are agriculture, mining, and industry,

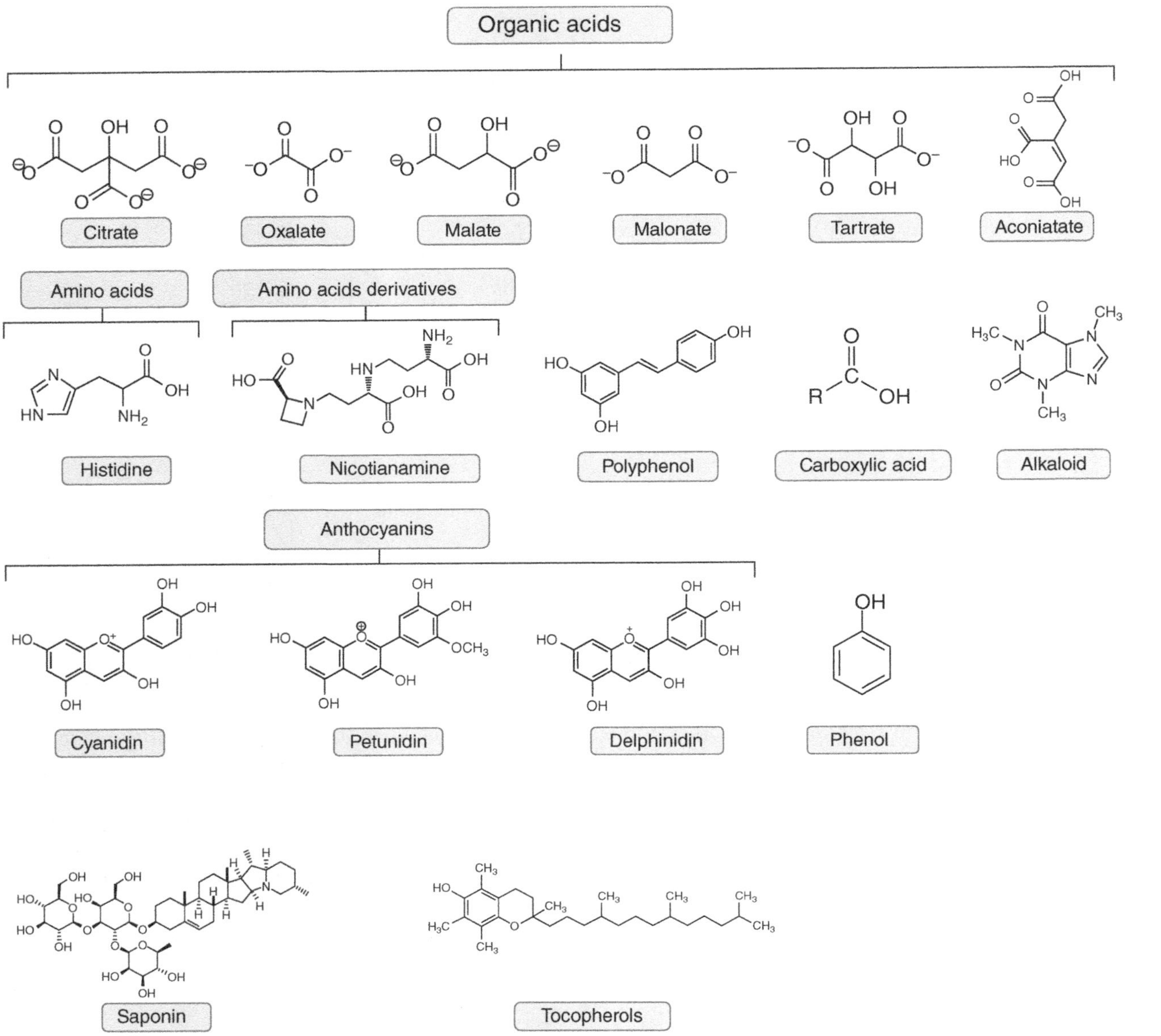

Figure 4.11 Main organic acids involved in metal chelation. *Source:* Own elaboration based on information from Anjitha et al. (2021), Yadav et al. (2021).

directly affecting various levels of the food chain. Plants exposed to these elements have been able to adapt and generate tolerance responses. The response to PTE will depend on the type and species of plant, as well as the chemical, biological, and physical conditions present at the impacted site. The main tolerance mechanisms are (i) root exudates, (ii) sequestration in the primary and secondary cell walls, (iii) selective transport in the plasma membrane, (iv) sequestration in the vacuole, (v) chelation by various ligands in the xylem and phloem, and (vi) chelation of metals in the cytosol. Such mechanisms are not expressed simultaneously as some plants have well developed one mechanism more than the other. Some plants can generate root exudates and others generate the chelation of PTE in the cell wall. It is advisable to continue studying plants associated with sites impacted with PTE to increase the number of species that can be used in phytoremediation, as well as to know the mechanisms involved in EPT tolerance.

Table 4.1 Plant species and main organs involved in metal uptake.

Scientific name	Metal	Plant organs involved in metal uptake	Reference
P. autralis	Copper (Cu) Zinc (Zn) Arsenic (As) Lead (Pb)	Root	Yan et al. (2022)
Typha orientalis	Lead (Pb) Zinc (Zn) Arsenic (As) Copper (Cu)	Root	Yan et al. (2022)
Erigeron annuus	Cadmium (Cd)	Root cell wall Root vacuoles Sheet Root exudates (DOM)	Zhang et al. (2021)
Phragmites sp.	Lead (Pb) Manganese (Mn) Cadmium (Cd)	Root SOD	Ding and Sun (2021)
Phragmites sp.	Lead (Pb) Copper (Cu) Chromium (Cr)	SOD, CAT, GSH	Rai (2021)
Phragmites sp.	Copper (Cu) Zinc (Zn) Lead (Pb)	Root, rhizomes	Nawrot et al. (2021)
Phragmites sp.	Cadmium (Cd) Lead (Pb) Manganese (Mn) Zinc (Zn)	Root, rhizomes	Sharma et al. (2020b)
Lemongras	Lead (Pb) Nickel (Ni)	Root, leaves	Pandey et al. (2020)
Calotropis procera *Pdactylifera*	Chromium (Cr) Cadmium (Cd) Manganese (Mn) Lead (Pb) Nickel (Ni)	Root, leaves	Ullah and Muhammad (2020)
Ricinus communis	Zinc (Zn) Lead (Pb) Iron (Fe)	Root, shoot	Zhang and Zhang (2020)
R. communis	Mercury (Hg) Arsenic (As) Cadmium (Cd)	Root, shoot, and leaves	Sharma et al. (2020a)
Chrysopogon zizanioides	Lead (Pb)	Shoot	Pidatala et al. (2018)
Zea mays	Lead (Pb)	Root	Pidatala et al. (2018)

References

Anjitha, K.S., Sameena, P.P., and Puthur, J.T. (2021). Functional aspects of plant secondary metabolites in metal stress tolerance and their importance in pharmacology. *Plant Stress* 2: 100038. https://doi.org/10.1016/j.stress.2021.100038.

Berni, R., Luyckx, M., Xu, X. et al. (2018). Reactive oxygen species and heavy metal stress in plants: impact on the cell wall and secondary metabolism. *Environ. Exp. Bot.* https://doi.org/10.1016/j.envexpbot.2018.10.0.

Carrillo, J.T. and Borthakur, D. (2021). Methods for metal chelation in plant homeostasis: review. *Plant Physiol. Biochem.* 163: 95–107. https://doi.org/10.1016/j.plaphy.2021.03.045.

Chen, Y.-T., Wang, Y., and Yeh, K.-C. (2017). Role of root exudates in metal acquisition and tolerance. *Curr. Opin. Plant Biol.* 39: 66–72. https://doi.org/10.1016/j.pbi.2017.06.004.

Dhir, B. (2021). Role of transporters of copper, manganese, zinc, and nickel in plants exposed to heavy metal stress. In: *Metal and Nutrient Transporters in Abiotic Stress* (ed. A. Roychoudhury, D.K. Tripathi, and R. Deshmukh), 145–168. Academic Press https://doi.org/10.1016/B978-0-12-817955-0.00008-9.

Ding, Z. and Sun, Q. (2021). Effects of flooding depth on metal(loid) absorption and physiological characteristics of *Phragmites australis* in acid mine drainage phytoremediation. *Environ. Technol. Innovation* 22: 101512. https://doi.org/10.1016/j.eti.2021.101512.

Dobritzsch, D., Grancharov, K., Hermsen, C. et al. (2019). Inhibitory effect of metals on animal and plant glutathione transferases. *J. Trace Elem. Med. Biol.* 57: 48–56. https://doi.org/10.1016/j.jtemb.2019.09.007.

Feng, Z., Ji, S., Ping, J., and Cui, D. (2021). Recent advances in metabolomics for studying heavy metal stress in plants. *TrAC, Trends Anal. Chem.* 143: 116402. https://doi.org/10.1016/j.trac.2021.116402.

Ghori, N.-H., Ghori, T., Hayat, M.Q. et al. (2019). Heavy metal stress and responses in plants. *Int. J. Environ. Sci. Technol.* 16: 1807–1828. https://doi.org/10.1007/s13762-019-02215-8.

Gu, S. and Lan, C.Q. (2021). Biosorption of heavy metal ions by green alga *Neochloris oleoabundans*: effects of metal ion properties and cell wall structure. *J. Hazard. Mater.* 418: 126336. https://doi.org/10.1016/j.jhazmat.2021.126336.

Gupta, C.K. and Singh, B. (2017). Uninhibited biosynthesis and release of phytosiderophores in the presence of heavy metal (HM) favors HM remediation. *Environ. Sci. Pollut. Res.* 24 (10): 9407–9416. https://doi.org/10.1007/s11356-017-8636-y.

Hernández Estévez, I. and Rodríguez Hernández, M. (2020). Plant glutathione S-transferases: an overview. *Plant Genet.* 23: 100233. https://doi.org/10.1016/j.plgene.2020.100233.

Huang, Y., Chen, J., Zhang, D. et al. (2021). Enhanced vacuole compartmentalization of cadmium in root cells contributes to glutathione-induced reduction of cadmium translocation from roots to shoots in pakchoi (*Brassica chinensis* L.). *Ecotoxicol. Environ. Saf.* 208: 111616. https://doi.org/10.1016/j.ecoenv.2020.111616.

Ivanov, A.A. (2021). Chapter 8: role of glutathione in enhancing metal hyperaccumulation in plants. In: *Handbook of Bioremediation, Physiological, Molecular and Biotechnological Interventions* (ed. M. Hasanuzzaman and M.N.V. Prasad), 115–152. Academic Press https://doi.org/10.1016/B978-0-12-819382-2.00008-9.

Janeeshma, E. and Puthur, J.T. (2020). Direct and indirect influence of arbuscular mycorrhizae on enhancing metal tolerance of plants. *Arch. Microbiol.* 202: 1–16. https://doi.org/10.1007/s00203-019-01730-z.

Jia, H., Wang, X., Wei, T. et al. (2019). Accumulation and fixation of Cd by tomato cell wall pectin under Cd stress. *Environ. Exp. Bot.* 167: 103829. https://doi.org/10.1016/j.envexpbot.2019.1038.

Kaya, C., Ashraf, M., Alyemeni, M.N. et al. (2020a). Salicylic acid-induced nitric oxide enhances arsenic toxicity tolerance in maize plants by upregulating the ascorbate-glutathione cycle and glyoxalase system. *J. Hazard. Mater.* 399: 123020. https://doi.org/10.1016/j.jhazmat.2020.123020.

Kaya, C., Ashraf, M., Alyemeni, M.N., and Ahmad, P. (2020b). The role of endogenous nitric oxide in salicylic acid-induced up-regulation of ascorbate-glutathione cycle involved in salinity tolerance of pepper (*Capsicum annuum* L.) plants. *Plant Physiol. Biochem.* 147: 10–20. https://doi.org/10.1016/j.plaphy.2019.11.040.

Kaya, C., Polat, T., Ashraf, M. et al. (2021). Endogenous nitric oxide and its potential sources regulate glutathione-induced cadmium stress tolerance in maize plants. *Plant Physiol. Biochem.* 167: 723–737. https://doi.org/10.1016/j.plaphy.2021.08.030.

Khan, M.I.R., Chopra, P., Chhillar, H. et al. (2021). Regulatory hubs and strategies for improving heavy metal tolerance in plants: chemical messengers, omics and genetic engineering. *Plant Physiol. Biochem.* 164: 260–278. https://doi.org/10.1016/j.plaphy.2021.05.006.

Koul, B., Singh, S., Yousuf-Parray, S. et al. (2022). Biochemical and molecular aspects of heavy metal stress tolerance in plants. In: *Phytoremediation Technology for the Removal of Heavy Metals and Other Contaminants from Soil and Water* (ed. V. Kumar, M.P. Shah, and S.K. Shahi), 205–219. Elsevier https://doi.org/10.1016/B978-0-323-85763-5.00003-9.

Li, Y., Gong, X., Xiong, J. et al. (2021). Different dissolved organic matters regulate the bioavailability of heavy metals and rhizosphere microbial activity in a plant-wetland soil system. *J. Environ. Chem. Eng.* 9 (6): 106823. https://doi.org/10.1016/j.jece.2021.106823.

Li, J., Guo, Y., and Yang, Y. (2022). The molecular mechanism of plasma membrane H^+-ATPases in plant responses to abiotic stress. *J. Genet. Genomics* 49 (8): 715–725. https://doi.org/10.1016/j.jgg.2022.05.007.

Luyckx, M., Hausman, J.-F., Isenborghs, A. et al. (2021). Impact of cadmium and zinc on proteins and cell wall-related gene expression in young stems of hemp (*Cannabis sativa* L.) and influence of exogenous silicon. *Environ. Exp. Bot.* 183: 104363. https://doi.org/10.1016/j.envexpbot.2020.1043.

Meychik, N., Nikolaeva, Y., and Kushunina, M. (2021). The significance of ion-exchange properties of plant root cell walls for nutrient and water uptake by plants. *Plant Physiol. Biochem.* 166: 140–147. https://doi.org/10.1016/j.plaphy.2021.05.048.

Mishra, P., Singh, A., and Roy, S. (2022). Chapter 19: plasma membrane H^+-ATPase in plants. In: *Cation Transporters in Plants* (ed. S.K. Upadhyay), 357–373. Academic Press https://doi.org/10.1016/B978-0-323-85790-1.00012-9.

Nawrot, N., Wojciechowska, E., Pazdro, K. et al. (2021). Uptake, accumulation, and translocation of Zn, Cu, Pb, Cd, Ni, and Cr by *P. australis* seedlings in an urban dredged sediment mesocosm: impact of seedling origin and initial trace metal content. *Sci. Total Environ.* 768: 144983. https://doi.org/10.1016/j.scitote.

Niu, M., Bao, C., Zhan, J. et al. (2021). Plasma membrane-localized protein BcHIPP16 promotes the uptake of copper and cadmium in *planta*. *Ecotoxicol. Environ. Saf.* 227: 112920. https://doi.org/10.1016/j.ecoenv.2021.112920.

Pandey, J., Sarkar, S., Verma, R.K., and Singh, S. (2020). Sub-cellular localization and quantitative estimation of heavy metals in lemongrass plants grown in multi-metal contaminated tannery sludge. *S. Afr. J. Bot.* 131: 74–83. https://doi.org/10.1016/j.sajb.2020.01.034.

Pasricha, S., Mathur, V., Garg, A. et al. (2021). Molecular mechanisms underlying heavy metal uptake, translocation and tolerance in hyperaccumulators-an analysis. *Environ. Challenges* 4: 100197. https://doi.org/10.1016/j.envc.2021.100197.

Pidatala, V.R., Li, K., Sarkar, D. et al. (2018). Comparative metabolic profiling of vetiver (*Chrysopogon zizanioides*) and maize (*Zea mays*) under lead stress. *Chemosphere* 193: 903–911. https://doi.org/10.1016/j.chemosphere.2017.11.087.

Rai, P.K. (2021). Heavy metals and arsenic phytoremediation potential of invasive alien wetland plants *Phragmites karka* and *Arundo donax*: water-energy-food (W-E-F) nexus linked sustainability implications. *Bioresour. Technol. Rep.* 15: 100741. https://doi.org/10.1016/j.biteb.2021.100741.

Rekha, K., Usha, B., and Keeran, N.S. (2021). Role of ABC transporters and other vacuolar transporters during heavy metal stress in plants. In: *Metal and Nutrient Transporters in Abiotic Stress* (ed. A. Roychoudhury, D.K. Tripathi, and R. Deshmukh), 55–76. Academic Press https://doi.org/10.1016/B978-0-12-817955-0.00003-X.

Shahid, M., Dumat, C., Khalid, S. et al. (2017). Foliar heavy metal uptake, toxicity and detoxification in plants: a comparison of foliar and root metal uptake. *J. Hazard. Mater.* 325: 36–58. https://doi.org/10.1016/j.jhazmat.2016.11.063.

Sharma, P., Tripathi, S., and Chandra, R. (2020a). Phytoremediation potential of heavy metal accumulator plants for waste management in the pulp and paper industry. *Heliyon* 6 (7): e04559. https://doi.org/10.1016/j.heliyon.2020.e04559.

Sharma, P., Tripathi, S., Chaturvedi, P. et al. (2020b). Newly isolated *Bacillus* sp. PS-6 assisted phytoremediation of heavy metals using *Phragmites communis*: potential application in wastewater treatment. *Bioresour. Technol.* 320: 124353. https://doi.org/10.1016/j.biortech.2020.12435.

Sharma, P., Pandey, A.K., Udayan, A., and Kumar, S. (2021a). Role of microbial community and metal-binding proteins in phytoremediation of heavy metals from industrial wastewater. *Bioresour. Technol.* 326: 124750. https://doi.org/10.1016/j.biortech.2021.12475.

Sharma, P., Ngo, H.H., Khanal, S. et al. (2021b). Efficiency of transporter genes and proteins in hyperaccumulator plants for metals tolerance in wastewater treatment: sustainable technique for metal detoxification. *Environ. Technol. Innovation* 23: 101725. https://doi.org/10.1016/j.eti.2021.101725.

Shitan, N. and Yazaki, K. (2019). Dynamism of vacuoles toward survival strategy in plants. *Biochim. Biophys. Acta Biomembr.* 1862: 183127. https://doi.org/10.1016/j.bbamem.2019.183127.

Tibbett, M., Green, I., Rate, A. et al. (2021). The transfer of trace metals in the soil-plant-arthropod system. *Sci. Total Environ.* 779: 146260. https://doi.org/10.1016/j.scitotenv.2021.1462.

Ullah, R. and Muhammad, S. (2020). Heavy metals contamination in soils and plants along with the mafic-ultramafic complex (ophiolites), Baluchistan, Pakistan: evaluation for risk and phytoremediation potential. *Environ. Technol. Innovation* 19: 100931. https://doi.org/10.1016/j.eti.2020.100931.

Vezza, M.E., Luna, D.F., Agostini, E., and Talano, M.A. (2019). Glutathione, a key compound for As accumulation and tolerance in soybean plants treated with AsV and AsIII. *Environ. Exp. Bot.* 162: 272–282. https://doi.org/10.1016/j.envexpbot.2019.03.002.

Wang, L., Liu, J., Filipiak, M. et al. (2021). Fast and efficient cadmium biosorption by *Chlorella vulgaris* K-01 strain: the role of cell walls in metal sequestration. *Algal Res.* 60: 102497. https://doi.org/10.1016/j.algal.2021.102497.

Yaashikaa, P.R., Senthil-Kumar, P., Jeevanantham, S., and Saravanan, R. (2022). A review on bioremediation approach for heavy metal detoxification and accumulation in plants. *Environ. Pollut.* 301: 119035. https://doi.org/10.1016/j.envpol.2022.119035.

Yadav, B., Dubey, R., Gnanasekaran, P., and Narayan, O.P. (2021). OMICS approaches towards understanding plant's responses to counterattack heavy metal stress: an insight into molecular mechanisms of plant defense. *Plant Genet.* 28: 100333. https://doi.org/10.1016/j.plgene.2021.100333.

Yan, X., An, J., Yin, Y. et al. (2022). Heavy metals uptake and translocation of typical wetland plants and their ecological effects on the coastal soil of a contaminated bay in Northeast China. *Sci. Total Environ.* 803: 149871. https://doi.org/10.1016/j.scitotenv.2021.1498.

Zeroual, S., El Bakkal, S.E., Mansori, M. et al. (2020). Cell wall thickening in two *Ulva* species in response to heavy metal marine pollution. *Reg. Stud. Mar. Sci.* 35: 101125. https://doi.org/10.1016/j.rsma.2020.101125.

Zhang, H. and Zhang, Y. (2020). Effects of iron oxide nanoparticles on Fe and heavy metal accumulation in castor (*Ricinus communis* L.) plants and the soil aggregate. *Ecotoxicol. Environ. Saf.* 200: 110728. https://doi.org/10.1016/j.ecoenv.2020.110728.

Zhang, H., Heal, K., Zhu, X. et al. (2021). Tolerance and detoxification mechanisms to cadmium stress by hyperaccumulator *Erigeron annuus* include molecule synthesis in root exudate. *Ecotoxicol. Environ. Saf.* 219: 112359. https://doi.org/10.1016/j.ecoenv.2021.112359.

5

Biotechnology for Sustainable Remediation of Contaminated Wastewater

Younis A. Hajam

Department of Life Sciences and Allied Health Sciences, Sant Baba Bhag Singh University, Jalandhar, Punjab, India

5.1 Introduction

In the current era, environmental pollution and global warming have emerged as major global challenges (Rogers 2006; Soler and Sánchez 2014; Pino et al. 2017). The scarcity of water and its deteriorating quality are significant issues that must be addressed, particularly as they impact millions of people in developing countries (Qu et al. 2013; Thines et al. 2017). Unplanned industrial development, population growth, and overexploitation of natural resources are key contributors to water quality degradation (Cundy et al. 2008; Chong et al. 2010; Zeng et al. 2011; Xu et al. 2012), leading to the discharge of heavy metals, pesticide residues, and other harmful compounds into waterways (Xu et al. 2012). This pollution directly or indirectly affects human health and the environment (Yue et al. 2015), making it imperative to manage these pollutants. Water is a valuable natural resource that is essential for life, as all plants and animals require it. Although about 71% of the Earth's surface is covered with water, only 2.5% is freshwater, and it is becoming increasingly polluted with human activity. Water scarcity has negative impacts on the economy, livelihoods, and environment worldwide, affecting an estimated 1.1 billion people who drink polluted water (Narmadha and Kavitha 2012). Polluted water causes 21% of communicable diseases in India, and diarrhea alone was responsible for over 535,000 deaths in 2004. Wastewater treatment systems harbor many microorganisms, such as algae, fungi, bacteria, viruses, and helminths, which can cause disease in humans. Nitrogen and phosphorus are the primary water pollutants, with heavy metals, detergents, and insecticides also contributing. Nitrogen and phosphorus play a significant role in eutrophication and are considered limiting nutrients (Shivajirao 2012). Table 5.1 documents the different common technologies used for treating wastewater and remediating polluted water (Fu and Wang 2011).

These are the most common technologies being used in various industries (Thines et al. 2017). Various aliphatic and aromatic compounds like hydrocarbons, aromatic compounds, and phthalate esters are some of the chemical contaminants that are liberated in the surroundings on a huge scale. In the environment, they circulate through different paths. Some are distributed by direct application and some are distributed as a component of fertilizers, pesticides, and herbicides. Harayama (1997) and Janssen et al. (2001) said that the concentration of toxins relies on (i) the potential of a toxin under both aerobic and anaerobic environments; (ii) dilution in the environment; (iii) the amount present as well as the speed at which the toxins are discharged in the environment; (iv) movement of toxins in a specific environment; and (v) degradation speed. Glazer et al. (1996) describe the process of "bioremediation." They said that it is used to describe the process of using microbes to decompose hazardous ingredients present in the wastes. Over the past decade, there are various authors who have reviewed the subject of biodegradation of organopollutants (Sangodkar et al. 1989; Chaudhry and Chapalamadugu 1991; Van Der Meer et al. 1992; Deo et al. 1994; Johri et al. 1996; Kumar et al. 1996; Janssen et al. 2001). All the untreated pollutants present in the normal milieu are biotransformed and are widely used to understand physiology and microbial ecology for their efficiency in bioremediation (Bouwer and Zehnder 1993; Chen et al. 1999; van Hylckama Vlieg and Janssen 2001; Mishra et al. 2001; Watanabe 2001) (Figure 5.1).

Aquatic Contamination: Tolerance and Bioremediation, First Edition. Edited by Rouf Ahmad Bhat, Gowhar Hamid Dar, Fernanda Maria Policarpo Tonelli, and Saima Hamid.

Table 5.1 Various processes used for the treatment of wastewater.

S. no.	Technology used	Reference
1.	Activated carbon-based adsorption	Srivastava et al. (2011), Zamboulis et al. (2011)
2.	Membrane filtration	Ersahin et al. (2012)
3.	Ion exchange	Gode and Pehlivan (2006), Xing et al. (2007), Motsi et al. (2009)
4.	Coagulation and flocculation	Zhang et al. (2003)
5.	Reverse osmosis	Bodalo-Santoyo et al. (2003)
6.	Flotation and extraction	Rykowska et al. (2008)
7.	Electrochemical treatment	Lu and Astruc (2018)
8.	Advanced oxidations and biosorption	Esser-Kahn et al. (2008), Kavamura and Esposito (2010), Dave et al. (2010)

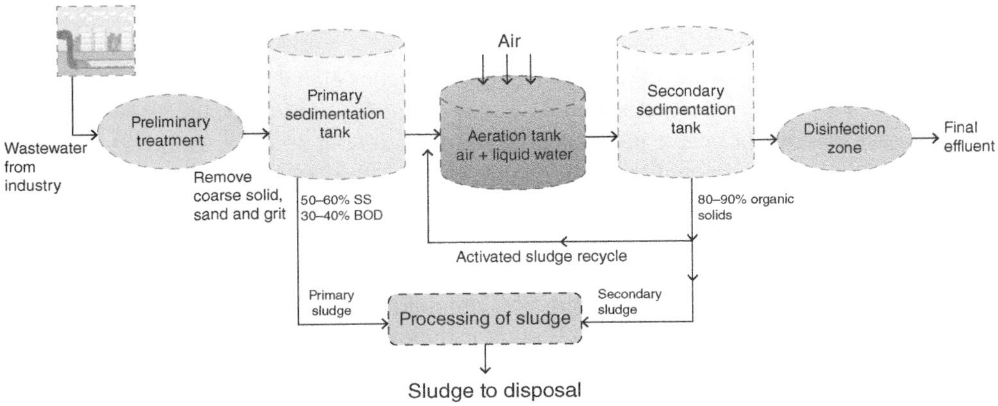

Figure 5.1 The mechanism of removal of wastes from contaminated water.

5.2 Organic Contaminants

Insecticides, household garbage, industrial waste, etc., all contribute to organic contamination. Major health issues like respiratory illnesses, chronic illnesses, and hormonal imbalances are brought on by organic material contamination (Harvey et al. 1984; Ram et al. 1990). When chlorine combines with untreated materials in treated drinking water, trihalomethanes (THMs) are created. Pesticide contamination comes from agricultural practices and other sources related to public hygiene (Younes and Galal-Gorchev 2000; Damalas and Eleftherohorinos 2011). Because of improper dose and application methods, pesticides used in agriculture and public health have a variety of environmental impacts (WHO 2010). People are unaware of the right methods for using or handling these insecticides. Pesticides are created using a variety of chemical procedures. All of these pesticides, which are truly chemicals, might interact with living things' metabolic processes. Humans are subject to a great deal of harm from pesticides; the liver and nervous system are particularly vulnerable. According to Bolognesi (2003), herbicides can potentially cause liver tumors. Volatile organic compounds (VOCs) include degreasers, paints, petrol, and fuel additives (Wehrmann 1996). These also include organic and solvent compounds including vinyl chloride, toluene, benzene, styrene, and trichloroethylene (TCE), among others. Birth defects, cancer, CNS diseases, hepatic and renal damage, reproductive system problems, and birth defects are all brought on by VOCs (Brown et al. 1984). The release of contaminants into the environment leads to non-aesthetic pollution and eutrophication. These non-aesthetic pollutants are converted into dangerous byproducts by a variety of processes that take place in water bodies, such as oxidation, hydrolysis, or other chemical reactions (Pagga and Brown 1986). These substances can also have other negative effects on a person's health or the health of their surroundings. According to Stuart et al. (2012), Lapworth et al. (2012), and other researchers, these substances are referred to as "Emerging Organic Contaminants," a new word.

5.2.1 Inorganic Contaminants

Chemical parameters are used to measure the presence of these contaminants. If there is significant amount of calcium or magnesium components present in the drinking water, then that water is naturally hard and this hardness of water depends upon its geographical nature. Carbonate or noncarbonate hardness is the classification of the hardness of the water. There are many other organic substances like copper, chromium, mercury, fluoride, arsenic, lead, and cyanide that contaminate water resources, which may enter drinking water through industrial discharge and even natural sources (Nriagu 1988). Geological or anthropogenic sources may act as sources of fluoride. Higher fluoride levels in groundwater are due to excess fluorite, fluorspar, cryolite, fluorapatite, ralstonite, etc. (fluoride-bearing minerals) on the Earth's crust. Groundwater is getting overexploited and emerging as a main difficulty of excess fluoride content in water. Further, here are some sources of fluorides that are used to treat hyperthyroidism, which are some pharmaceutical products, viz. medicines, toothpaste, fertilizers, vitamin supplements, insecticides, disinfectants, preservatives, and others. Dental and skeletal fluorosis is actually caused by fluoride and also it is harmful to the development of children and elder people especially. Alzheimer's and many other diseases related to the brain like dementia (Susheela 1999; Fawell et al. 2006; WHO 2008) are also caused by fluoride. After entering the brain, it gives a signal to aluminum to invade through the blood–brain barrier, thereby increasing the chance of brain ailment. Ram Gopal and Ghosh (1985) and Eswar and Devaraj (2011) reported that groundwater in more than 20 countries is concentrated with fluoride (4.0 mg/L by EPA) and in India, 19 states are suffering from acute fluorosis issues. Arsenic is also a major contaminant that can also affect people and the environment. Due to agricultural and industrial practices, arsenic (MCL 0.01 mg/L) gets entered into water supplies unit used for drinking (Smith et al. 2000). In India and Bangladesh, it is the biggest poisoning contaminant (Chatterjee et al. 1995; Khan et al. 2003). Intake of arsenic by using polluted water for drinking, cooking, and many other purposes leads to a disease known as arsenicosis (Chen et al. 1988).

5.2.2 Biological Contaminants

Nature has various microbes, such as algae, fungi, bacteria, and viruses, etc., that act as biological contaminants and play an important role in biological contamination. If these are present in our drinking water, this may lead to various typical health issues (Daschner et al. 1996; Ashbolt 2004). Algae are generally unicellular microscopic organisms found abundantly and take their food from nutrients found in water like phosphorus. Run-off from domestic activities or industries provides nutrients to algae which helps to grow them inside water and excess growth of algae causes many problems. It may spoil taste and odor of water. It also gets in the way of sieves and stimulates growth of slime on the carriers. If humans consume this water, then some algae like blue-green algae liberate toxins inside human body and they damage the liver. These cause hepatotoxicity because these act as hepatotoxins and cause neurotoxicity by acting as neurotoxins (that cause toxicity in the nervous system). These also cause skin toxicity (Hitzfeld et al. 2000; Rao et al. 2002). Bacteria are a large group of unicellular and microscopic organisms. Pathogenic bacteria can cause many problems. They spread infection through water (Inamori and Fujimoto 2009). Various diseases are caused by bacteria like typhoid, gastroenteritis, dysentery, and cholera. Many taste and odor problems may be caused by some nonpathogenic bacteria (Rusin et al. 1997; Nwachcuku and Gerba 2004).

5.2.3 Radiological Contaminants

Radioactive elements cause radiological contaminants. Some industrial waste or rocks' run-off could be the source of radioactive material. Radioactive materials are deposits from natural erosion and may emit radiation. In groundwater treatment, radioisotopes are used, which creates a major issue in comparison to the treatment of surface water because if these radioactive elements (U226, Ra226, Ra228, and Rn228) enter the body along with water, it augments the threat of cancer (Hakl et al. 1995).

5.3 Biotechnology in Environmental Engineering

Various anthropogenic activities result in the spoilage of the natural environment. Nowadays, the most important problem all over the world is environmental pollution. In wastewater treatment, biotechnology plays its role by decomposing waste through biological methods and disposing of solid wastes by composting technique. The biotechnological approaches for the treatment of wastewater treatment include activated sludge, membrane bioreactors (MBRs), anaerobic fixed-film reactor, and anaerobic treatment. Moreover, solid waste-composting techniques, nanofiltration (NF), electrochemical methods, and biosorption are the applications of biotechnology in environmental engineering. These techniques require treatment and degradation of organic substances in advance. Some of the biotechnological approaches for the treatment of wastewater are discussed next.

5.3.1 Activated Sludge

High concentrations of microbial decomposers are used in activated sludge process. By stirring, these are kept in suspension in the form of fine particles. Organic matter is removed from wastewater by this activated sludge process. For removing heavy metals from wastewater, there is the emergence of a recent cost-effective and efficient alternative technique, i.e. biosorption (Zhang et al. 2011). Biomasses in nonliving form, which have heavy metal uptake capacities and suitability, have been studied for the development of biosorbent bases. Bacteria and marine algae were included by He and Chen (2014) and Velan et al. (2005). In biosorption, "bio" means biological components present, like biomaterial or biopolymer, both in sorbent and biosorption, which is a sorption process. In early 1970s, the phenomenon of biosorption was observed from a nuclear power station from where the radioactive elements got released in the wastewater. It was demonstrated in early studies that heavy metals may be eliminated from aqueous medium by biosorption, which is a capable and low-cost technology. Low operating cost, to detoxify heavy metals even at lower concentrations because of having high efficiency, the final disposable amount is spent on biosorbent, and there is no requirement for nutrients (Sheng et al. 2007). These are the advantages of biosorption technology in comparison to other wastewater treatment methods like chemical reduction, ion exchange, precipitation, membrane separation, and biosorption. Biosorbents play a major role in the removal of heavy metal pollutants and these are also rich inorganic legends or functional groups. COOH, phosphate, OH, sulfate as well as amine groups are important functional groups. Agriculture wastes and activated carbon, which are low-cost adsorbents, developed in agriculture industries, have been used in large amounts to extract/eliminate harmful metals from an aqueous medium in the sorption process (Babel and Kurniawan 2003; Zhang et al. 2009). Gupta and Rastogi (2009) said that algal biomass is used as an effective biomaterial. Carboxylic acid, sulfonic acid, and hydroxyl (–OH) groups' presence is confirmed in marine algae polysaccharides, which are liable for metal uptake (Davis et al. 2003). For the sorption process, macroscopic structures that are present in marine algae are used to produce biosorbent particles (Vieira and Volesky 2000). The heavy metals present in wastewater are found to be absorbed by the algae (Figure 5.2).

5.3.2 Anaerobic Treatment

Anaerobic treatment is used for paper waste, distillery effluents, waste from tanneries, and textile waste. Food-processing anaerobic technology was used earlier. This approach of treatment is a very old one, generally for the treatment of wastewater from various factories and reactors including fluidized bed reactor (FBR), anaerobic contact reactor (ACR), upflow anaerobic sludge blanket reactor (UASB), and anaerobic fixed film reactor (AFFR) have been designed. These reactors have operational constraints that are actually of different types but are used very commonly at large scale (Rao et al. 2004). In very small sludges, large quantities of organic wastes can be loaded, and less nutrient and energy supply is required in this case. Moreover, the generation of biogas is an advantage of the anaerobic treatment of industrial waste. Therefore, this process can be used to produce energy (Nandy et al. 2002). Treatment efficiency can be affected by inoculum source and feed pretreatment. Greater hydraulic retention time (HRT) is because of the presence of acidic pH and minimal growth rate. Patel and Madamwar (2000) and Enright et al. (2005) reported that 60–70% of COD elimination potential was attained when an anaerobic biological treatment was given for pharmaceutical wastewater.

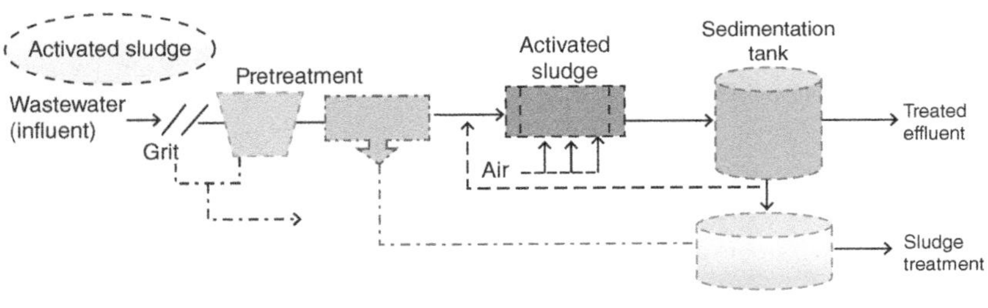

Figure 5.2 The process of activated sludge for the treatment of wastewater.

5.3.3 Aerobic Treatment

With the help of microorganisms organic content in industrial wastewater is decomposed into carbon dioxide and methane through aerobic treatment. An aerobic sequencing batch reactor (ASBR) is used for industrial wastewater treatment. Further an activated sludge process is commonly used, and then a principal settling tank and intermediate retention are also been used. Wastewater derived from pharmaceutical factories is treated through an aerobic biological treatment process and an aerobic treatment in a biological reactor (LaPara et al. 2001; Khan and Mostafa 2011). An aerobic biological treatment plant of integrated laboratory-scale batch type was constructed and an operation was done for the treatment of pharmaceuticals along with the evaluation of its performance. Air enters the tank through an open cylindrical tank in order to aerate the reactor. Approximately, within 15 days, maximum hydraulic retention (HR) can be obtained. When water is collected in a plant every day, the potential of the plant can be analyzed by evaluating the following parameters such as COD, conductivity, TDS, TSS, and pH. Fifteen days after employing HRT, about 75% removal of COD was attained. The air injection is used to increase the oxygenation flow that increases the rate of respiration of microorganisms that enhances the rate of COD elimination from wastewater. Biological treatment approaches are very significant to eliminate organic wastes from water, having higher efficiency such as removal of COD and TSS. The aerobic method of a treatment plant is eco-friendly because of the least use of chemicals, low cost, and ease of operation to remove waste from industrial wastewater.

In addition, these advanced technologies are also adopted to eliminate organic waste from water obtained from herbal pharmaceutical industries such as sequencing batch reactor (SBR) and activated sludge. This combined technology significantly eliminates wastes and other chemicals from water, as documented in Table 5.2.

5.3.4 Anaerobic Fixed Film Reactor (AFFR)

AFFR constitutes a biofilm as a supporting base to attach the biomass in the AFFR. It gives us many advantages like its construction is simple, it eliminates the mechanical mixing, stability is better, and it has the capability to tolerate toxic shock load. Rajeshwari et al. (2000) studied that after a period of starvation, AFFR has rapid recovery potential. In the treatment of industrial wastewater-packing material such as glass bead, red drain clay, polyurethane foam, waste tire rubber, splintered glass (Pérez et al. 1997), polyacryl nitrile acrylamide, corrugated plastic (Garcia et al. 1995), coconut coir, charcoal, and nylon fiber can be used. Packing substance acts as supportive media. It designates the performance of the reactor

Table 5.2 Combined technology.

S. no.	Name of chemical/waste	Percentage	Reference
Name of technique: sequencing batch reactor (SBR)			
1.	Biological oxygen demand (BOD)	85% approximately	Ileri et al. (2003)
2.	Chemical oxygen demand (COD)	87% approximately	Ileri et al. (2003)
3.	Suspended solids	98% approximately	Ileri et al. (2003)
4.	Ammonia	96% approximately	Ileri et al. (2003)
5.	Phosphate	5% approximately	Ileri et al. (2003)
Name of technique: activated sludge			
1.	Reduction in chemical oxygen demand (COD)	68.98–91.02%	Vanerkar et al. (2013)
2.	Reduction in biological oxygen demand (BOD)	76.90–97.26%	Vanerkar et al. (2013)
3.	Reduction in suspended solids	74.61–95.54%	Elmolla et al. (2012)
Name of technique: hydraulic retention (HR)			
1.	Biological oxygen demand (BOD)	94%	Lapara et al. (2001), Khan and Mostafa (2011)
2.	Chemical oxygen demand (COD)	83%	Lapara et al. (2001), Khan and Mostafa (2011)
Name of technique: coupled solar photo-Fenton process			
1.	Chemical oxygen demand (COD)	98%	

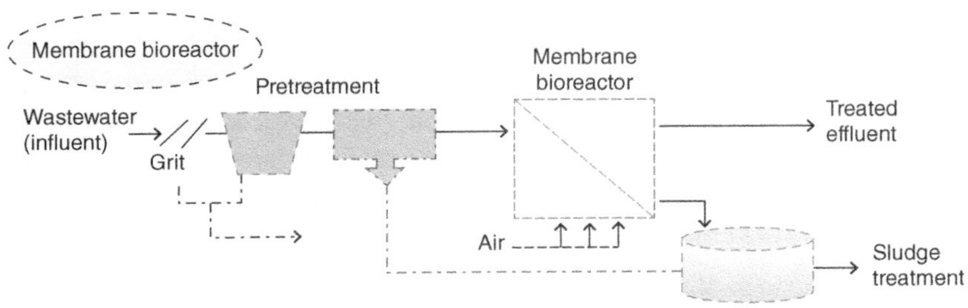

Figure 5.3 Membrane bioreactor for the treatment of wastewater.

(Acharya et al. 2008). High organic load removal is achieved by the suspended form of anaerobic biomass and organic load. AFFR is used in bulk drug industries for the management of wastewater with highly suspended solids. It was concluded that in bulk drug industries, wastewater is having high potential to remove COD (60–70%), BOD (80–90%), TDS, and TSS. Therefore, AFFR technology could be efficient to manage the wastewater obtained from drug industries.

5.3.5 Membrane Bioreactor (MBR)

MBR is mostly used for the treatment, management, and purification of wastewater obtained from industries, households, and municipal sources (Yang et al. 2006). MBR technique involves physical separation of biomass and biological degradation of wastewater compounds and porous membrane filtration is done to purify the wastewater. Peculiar characteristics of MBR include complete elimination of suspended solids, packed plant size, high degradation rate, ease and flexibility, simple operation process, small production of sludge, control over the disinfection and odor, higher ability to retain microorganisms, and intractable as well as poisonous organic and inorganic effluents that are eliminated by MBR. Chang et al. (2008) observed COD and BOD removal of 95% and 99%, respectively, when pharmaceutical wastewater was treated by the MBR. Pharmaceutical wastewater can also be processed by Zee Weed MBR technology (Noble 2006) (Figure 5.3).

5.4 Biological Treatment

Wastewater can also be managed and purified by using biological agents such as microbes and plants. So, it is commonly referred to as biological treatment. The waste is converted into some other forms by the microbes. Primarily, it must be confirmed that the products are converted or degraded into nontoxic end products and they should not be problematic. Numerous factors are responsible for the biodegradation of products, which includes toxicity of compounds, the concentration of compounds, microbial strain efficiency stereochemistry of compounds, degradation conditions, time of retention, and concentration and presence of other compounds. In the removal of xenobiotics, microorganisms play a very important role. These microorganisms also maintain the minimum concentration of a toxic compound in the surroundings (Misal et al. 2011). Formed residues (byproducts) during microbial transformation, these residues become more stable than the original compound, i.e. they have been converted into less toxic compound. Maximum reduction in COD (86.6%) in pharmaceutical industrial wastewater (PIWW) is achieved when a biological treatment approach is applied (Raj and Anjaneyulu 2005). Further, Rosen et al. (1998) reported that microbial consortia such as fungal and bacterial treatment given to pharmaceutical wastewater significantly reduce/remove toxicity. Deegan et al. (2011) reported that both aerobic as well as anaerobic biological treatments can be used to treat PIWW (Table 5.3).

Todd and Josephson (1996) studied the various biological methods used for the treatment of sewage. They reported that 12 principal factors include high exchange rates, mesocosm structure, subecosystems, mineral diversity, nutrient reservoirs, steep gradients, periodic and random pulses, cellular design microbial communities, photosynthetic bases, animal diversity, and biological exchanges. Microbes use wastes as nutrients present in the water to carry out their physiological activities as well as to remove the wastes efficiently. Microbiological processes and hydroponics are involved in the combined treatment of household wastewater. Pant and Adholeya (2007) stated that in biological approaches, mostly fungi and bacteria are used for distillery effluents according to the color of compounds in the molasses.

Table 5.3 Techniques used for the treatment of pharmaceutical industrial wastewater (PIWW).

S. no.	Name of chemical/waste	Percentage	Reference
Name of technique: bioremediation			
1.	COD		Das et al. (2012)
2.	TSS		Das et al. (2012)
3.	TDS		Das et al. (2012)
4.	Sulfate		
Name of technique: membrane aerated biofilm reactor (MABR)			
1.	Chemical oxygen demand (COD)	90% reduction	Wei et al. (2012)
2.	Ammonia nitrogen	98% reduction	Wei et al. (2012)
Name of technique: solar photo-Fenton immobilized biomass reactor			
1.	Reduces dissolved organic carbon (DOC) in pharmaceutical industrial wastewater (PIWW)	95%	Sirtori et al. (2009)
2.	Chemical oxygen demand (COD)	86–91%	Mayabhate et al. (1988)
3.	Phenol	50%	Mayabhate et al. (1988)
Name of technique: sequencing batch reactor (SBR)			
1.	Chemical oxygen demand (COD)	85%	Ng et al. (1989)
2.	Biological oxygen demand (BOD)	93%	Ng et al. (1989)
3.	Relative toxicity in dissolved oxygen (2 mg/l) at various pH (6.62, 6.69, 6.79, and 6.9) for 7, 14, and 21 days	Reduction in BOD, COD, oil, grease, TDS, TSS, and ammonia level. Values were found the same as per the National Environmental Quality Standards (NEQS)	Altaf and Ali (2010)

5.4.1 Bacterial Treatment

In bioremediation, bacterial colonies are used to treat industrial wastewater (Ghosh et al. 2004). *Pseudomonas*, *Enterobacter*, *Streptomonas*, *Aeromonas*, *Acinetobacter*, and *Klebsiella* are some bacterial strains that show up to 44% COD reduction. Fifteen rhizosphere bacteria were isolated by Chaturvedi et al. (2006), which show 85–86% BOD and COD reduction and 76% color reduction within 30 days. For the removal of toxic pollutants, *Escherichia coli bacterial colonies* and methanogenic consortium are used to remove toxic pollutants. Pentachlorophenols are highly dangerous, toxic, mutations-causing, and cancer-causing compounds. Bacterial colonies provide metabolic fuel for complete mineralization of organic compounds needed for the deterioration of organic contaminants (Tewari and Malviya 2002). Phenolic and other complex organic compounds are degraded by using the *Arthrobacter*, *Comamonas*, *Rhodococcus*, and *Ralstonia* bacteria. Kavitha and Beebi (2003) reported that *Clostridium* species helps in fermentation having the potential to decompose resorcinol. Thermophilic *Bacillus thermoglucosidasius* A7 decomposes the phenol at 65 °C through the meta-cleavage pathway (Duffner et al. 2000). *Pseudomonas putida* MTCC 1194 is used to study the kinetics of phenolic and catechol biodegradation (Kumar et al. 2005). Bacteria *Pseudomonas fluorescence* possesses the potential to completely degrade the phenol and catechol, which were 100 and 500 mg/l, respectively (Agarry and Solomon 2008). Das et al. (2012) studied the control of pharmaceutical effluent parameters through bioremediation. Nine distinct points were located in the industry where samples were collected and pollutants were treated by using bacterial consortia. The use of bacterial colonies resulted in the reduction in the following parameters – TDS value decreases from 484–1452 to 68–540 mg/l, total suspended solids 24–84 to 12–56 mg/l, sulfates from 44–1527 to 6–65.8 mg/l, and COD from 1257.9–1542.9 to 113.2–377.6 mg/l (Das et al. 2012). Phototropic bacteria, i.e. *Rhodobacter sphaeroides*, was used as a treatment method for pharmaceutical wastewater (Madukasi et al. 2010). It results in 80% COD attained. In addition to this use of mixed bacterial cultures for PIWW, it has been shown that these mixed bacterial colonies significantly reduce the COD (62% at 30 °C and 38% at 60 °C) (LaPara et al. 2001). *Cyanobacteria* and *Nostoc rivularis* produce a higher metal-binding protein in absorbance of the heavy metals.

N. rivularis has been revealed very efficiently. Wastewater-foaming bacterial species were isolated and sequenced from wastewater by El-Enany and Issa (2000) and Zhong et al. (2015).

5.4.2 Adsorption

In adsorption process, dissolved pollutants are attached to the porous surface of the solid bodies (Jiuhui 2008). Adsorption is a physical process that is also called the surface phenomenon and results in surface energy. The integrant atoms bonding necessity with the material are filled by other atoms. Another adsorbent atom is present on the surface but does not surround the atoms, leading to physically attractive forces. It can originate from van der Waals forces. So, it is called physisorption and can originate from covalent forces. Thus, it is called chemisorption. Chemical and physical processes are joined by adsorbents to eliminate the substances that give color, taste, and odor to water. Adsorbents have very small porous particles and those most preferred are highly microporous and well-controlled (Yang 1997). Jiuhui (2008) and Ali and Gupta (2007) explained that porous solids carry many pores which have small diameters as a fraction of a nanometer, e.g. activated carbon, silica gels, alumina, zeolites, etc., and which are more useful. Activated carbon is the most common adsorbent substance which is very much similar to charcoal. Physical method is used to determine the high porous character. Activated carbon is griped at a very elevated temperature (1300 °C) in order to provide steam in the nonexistence or existence of oxygen and inorganic salts. The carbon is made of petroleum coke, bituminous coal, lignite, wood products, and coconut/peanut shells. Carbon items are oxidized in carbon dioxide and steam at high temperatures. In the carbon structure, pores are generated and microfractures and gases are evacuated. Chemical method is used in some cases for treating carbonaceous matter, phosphoric acid, zinc chloride, and the mixture carbonized (a chemical-activating agent) at high temperatures along with the elimination of activating items through washing with water. The rate of the adsorption process varies and relies on the following factors such as the nature of carbon particles and pore size, surface area, density, hardness, and the nature of pollutants (concentration, hydrophobicity, polarity, and solubility of the pollutants and pollutant attraction with the carbon surface).

5.4.3 Nanofiltration

Heavy metals are eliminated through the process of NF. It provides a solution for various vital problems related with conventional elimination process (Al-Rashdi et al. 2013). The main constitution of NF includes a thin film (composed of a charged group) comparing various artificial polymers (Kotrappanavar et al. 2011). These charged groups help in the elimination of heavy metals. For removal of heavy metal ions, NF membranes have some successful applications. The pore size in NF is typically 1 nm. However, it ranges from ultrafiltration (UF) to RO and is considered a pressure-driven process, and negative or positive charges are present around the NF membrane (Escobar and Van der Bruggen 2011). In the NF membrane, there are mechanisms like Donnan exclusion (charge repulsion) as well as size exclusion. These mechanisms are unique as compared to others (Escobar and Van der Bruggen 2011; Soroko et al. 2011; Gherasim et al. 2013). In multivalent copper ions, the NF membrane is used instead of the UF membrane due to having advantages in comparison to other membranes (Duan et al. 2010). The NF membrane is highly permeable to water, is a less energy-consuming process, and also eliminates heavy metals very efficiently (Zhu et al. 2014).

5.5 Electrochemical Method

In metal finishing and metallurgy, electrochemical procedures are used to isolate metal ions (zinc, copper, silver, and lead) (Fu and Wang 2011; Nasef and Güven 2012; Li et al. 2017). According to Caprarescu et al. (2012), the electrodialysis technique is actually an electrochemical process in which copper ions are moved off the membrane by the action of an electric field. Heavy metal ions are primarily removed using electrocoagulation and electrodialysis. Additionally, the technology of electrodialysis is utilized to separate copper ions from industrial wastewater. When protons are at the anode and hydroxyl ions are at the cathode, an electric field is created. Heavy metals have a stronger affinity and migrate electrically toward the cathode. In contrast, protons from hydroxyl ions undergo greater ionic change (Pedersen et al. 2015). When handled using a flow battery that has a single integrated cell, metal-rich wastewater flows continuously (Al-Saydeh et al. 2017). According to Chang et al. (2010), the characteristics of the membrane employed in electrodialysis operations determine how copper ions are separated. Additionally, Capraresću et al. (2012) indicated that the membrane employed in the electrodialysis process should have chemical and thermal stability in situations where separation requires a high temperature in a solution with a range of pH values. It comprises a core segment with metal-contaminated wastewater and two adjacent sections with continually flowing electrolytes. Time and current density play key roles in determining how effectively heavy metal ions are removed from a process (Pedersen et al. 2015). The bench-scale electrodialysis procedure known as electrodialysis remediation (EDR), which happens by reducing their leaching ability and also increasing the capability of heavy metal removal from the contaminated wastewater, was first introduced by Jensen et al. (2010). The contaminated wastewater is continuously circulated within the feed section. The concentrate and feed parts are divided by the ion-exchange membrane. The ion-exchange membrane regulates the migration of ions from both portions (Akbal and Camcı 2010; Lakshmanan et al. 2010; Rincon and La Motta 2014). The electrocoagulation technique is a suitable technology for removing heavy metals from industrial effluent, such as Cu. To remove copper ions from wastewater generated by industries, electrocoagulation (electrochemical processes) uses electrical current (Jack et al. 2014).

5.5.1 Biologically Active Carbon Filtration

Biologically active carbon filtration is a type of bioremediation technique. In this process, granular activated carbon (GAC) is utilized for the filtration of water. Formation of "biofilm" is a porous tangled mass of slime matrix that contains microbial cells attached to the surface of GAC (substratum) or is dipped in an extracellular microbial organic polymer matrix (Weber Jr et al. 1978; Scholz and Martin 1997; Ghosh et al. 1999; Lawrence and Tong 2005). Bacterial and fungal cells secrete various extracellular materials in the biofilm and create a cohesive, stable matrix within which cells are compacted in a dense matrix (polysaccharides, proteins, nucleic acids, and lipids) (Goodwin and Forster 1985; Lazarova and Manem 1995; Branda et al. 2005). The physiological changes are associated with the task of biofilm and advancement of some genes (Dagostino et al. 1991) or enhancing the concentration of nutrients, O_2, and enzymes within the surroundings of the bacterial cell (Ghosh et al. 1999) or restricting the allocation of toxic or hampering components (Blenkinsopp and Costerton 1991). Hillis (2000) reported water treatment processes for the prevention of chemical disinfection. In source filtration processes, it is less possible that bacteria grow again. The service life can be extended and it eliminates the need for coagulants.

5.5.2 Distillation

Distillation process is one of the widespread techniques for the separation of various components (Veil 2008). Heat is applied in order to separate mixed components from water. The basic principle described that the vapor pressure is raised by the input of heat energy. During the boiling of the liquid mixture, vapor pressure increases and becomes equivalent to its surrounding pressure due to the differences in volatility in the mixture in which distillation occurs. Bacteria gets ruined when water is separated from inorganic elements, viz. lead, calcium, magnesium, etc. Boiling points less than 100 °C cannot separate organics effectively and in the water, these organics can actually become concentrated. For industrial purposes, the distilled water purification technology was used. For removing organic chemicals, distilled water purification technology was not effective. Carbon filter system is used to make water safe for drinking. The carbon filters can rapidly become breeding grounds for bacterial growth. They need to be regularly changed though distilled water is safe because due to distillation they lose their nutrient mineral content. So, it is not healthy to drink because, for drinking purposes, nutrients are essential. And it is also a very slow water purification technology.

5.5.3 Hydrogenation of Nitrates

The most effective catalytic approach for removing nitrate from wastewater is hydrogenation. The reaction must be carried out at low temperatures, which calls for a particularly active catalyst. The nitrate is transformed during this reaction into NO_2, NO, N_2O, N_2O, and N_2. Over hydrogenation results in the side reaction of NH_4 formation, which is undesirable (Soares et al. 2010; Mikami et al. 2006; Berndt et al. 2001). Pd/Cu, Pd/In, and Pd/Sn have been found to be efficient catalysts for nitrate hydrogenation (Deganello et al. 2000; Gao et al. 2003; Mikami et al. 2003); other metals, such as Cu, In, Sn, and Co, are also assisting these processes. The initial reduction process that turns NO_3 into NO_2 is supported by these metals (Soares et al. 2008; Pintar et al. 2004; Arino et al. 2004; Qi et al. 2006). N_2 and ammonium (NH_4), both dangerous aquatic contaminants, are the final byproducts of this catalytic reduction process.

Permeable Reactive Barrier (PRB)

Permeable reactor barrier (PRB) technique is used to remediate polluted groundwater permeably. In the PRB technique, biological reactions such as biodegradation, precipitation, and adsorption are used to eliminate pollutants (Thiruvenkatachari et al. 2008; Obiri-Nyarko et al. 2014). Further, the PRB technique (in situ technique) is used for the remediation of various types of pollutants such as heavy metals and chlorinated compounds present in polluted groundwater. A zero-valent iron creates a semipermeable or permanent reactive barrier sunken within the surroundings of pullulated groundwater (García et al. 2014; Zhou et al. 2014). During the flow of contaminated water through a barrier, pollutants are caught and a chain reaction occurs to clean water (Thiruvenkatachari et al. 2008; Obiri-Nyarko et al. 2014). These barriers are permeable only for water instead of pollutants because barriers are very reactive to trap pollutants, energy consumption is very low, low-cost, and have easy availability and accessibility (De Pourcq et al. 2015). Studies by Obiri-Nyarko et al. (2014) and Liu et al. (2015) reported that effectiveness of the media-type technique depends on some factors which are influenced by the types of pollutants, influences on surroundings and well-being, automatic stability, and price. Nowadays researchers focus on electrokinetic as well as coupling technique for the treatment of various kinds of pollutants (Mena et al. 2015; Ramírez et al. 2015).

5.5.4 Phosphorus Removal Mechanisms

The phosphorus removal mechanism is a type of filtration in which phosphorus remains physically attached to the surface of the filter bed. In different processes such as adsorption, precipitation, or biological processes, the dissolved portion of phosphorus can be achieved (Kim 2014). Inorganic soluble phosphorus is adsorbed and is achieved through a microbial biofilm of the filtering technique. It is reported that phosphorus adsorption is limited due to the porous medium retention capacity in the filter. However, filter depth and hydraulic load limit the elimination of orthophosphate. Moreover, it was also studied that when phosphorus is treated at 50, 100, and 150 cm depths, respectively, $0.27\,m^3/m^2/day^1$, it means phosphorus elimination rates were observed at 43.40%, 55.35%, and 64.61%, respectively. Normal elimination rate of PO_4-P has been observed as 64.61% for $H = 0.27\,m^3/m^2$ and 63% for $H = 0.40\,m^3/m$. These findings showed that the removal of orthophosphate mostly takes place in the upper sand layers. Significant phosphorus reduction needed a minimum water residence time. Phosphorus is retained in the porous medium due to an increase in the water residence time and filter depth. Due to this, the pollutant transport increases through the filter, and in the sand bed, the residence time of wastewater decreases. It was showed that the number of feeding–drainage cycles are closely associated with the reduction of orthophosphate, f (fractionation factor). PO_4-P elimination rates are 61.4%, 71.6%, and 77.2%, respectively, for $f = 1$, 2, and 3, respectively. Brissaud et al. (1999) reported that with an increased inflow rate of water, the f value decreases, and the water retention time might be shorter.

Residual clearance time is a significant contributing factor for infiltration-percolation systems for optimizing the elimination of phosphorus. In infiltration-percolation, phosphorus can be removed in two ways such as assimilating microorganisms and attaching them to the filter material. Moreover, temperature has been considered an important factor that affects the growth of microbes because bacterial growth increases with an increase in temperature and slows down with a decrease in temperature. Electrochemical conditions of the mass filter and the mineralogical composition of the sand filter affect the infiltration-percolation systems (Brahmi et al. 2013). Both sorption and precipitation phenomena are responsible for phosphorus retention in sand filters (Bori Akadar et al. 2014). The removal of phosphorus precipitation is considered to be a major process because sand filters have a limited adsorption potential (Arias and Brix 2005) (Figure 5.4).

5.6 Heavy Metal Treatment

Chromium kinetic mechanism has been studied by using flame atomic absorption spectrometry (Javadian et al. 2013). By using algae, a new industrial heavy metal water treatment method was given. Cs and Sr are eliminated from solutions by chemical modification as well as by testing. Azolla is used as a biosorption material (Ghazvini and Mashkani 2009). In

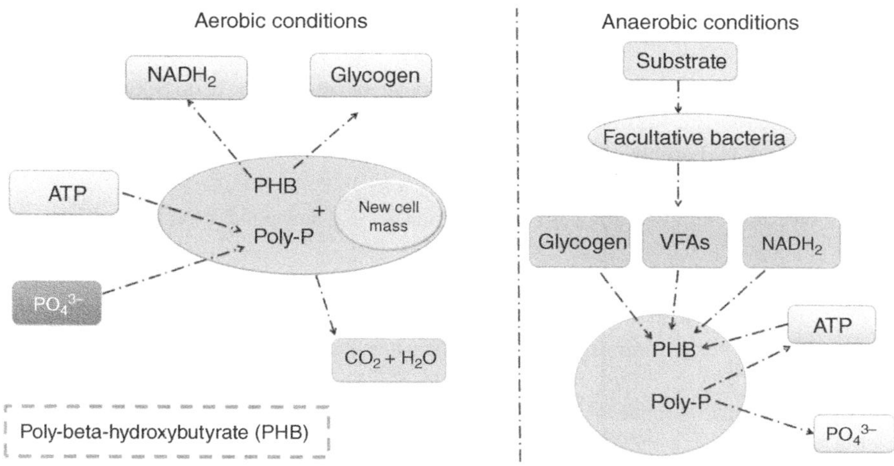

Figure 5.4 The mechanism of removal of phosphorus from the wastewater.

normal growth of algae, toxicity of these metals was studied. X-ray emission and Fourier-transform infrared spectroscopy (FTIR) that induced microparticles were used to study the absorption isotherm. At pH 8.8, the best result was obtained. Chemical modification at the industry level is quite difficult. In batch and continuous reactors, heavy metals are eliminated from solutions through dried algae culture (Dhaouadu et al. 2011). Dried algal powder has potential for the removal of poisonous chromium from water and different solutions.

The macro brown algae *Undaria pinnatifida* is converted into chars and copper metals in solution using both physical and chemical techniques (Park et al. 2013). Slow macroalgal growth rates are irrelevant to industrial scale and char production. Daniel et al. (2005) examined the biomass that contained both bacteria and algae in order to remove heavy metal (cadmium and copper) wastewater obtained from manufacturers.

5.6.1 Vegetated Filter Strips

Planted or indigenous vegetation functions like filter strips, located between the surface of a water body and pollutant-source area that receives runoff. For the treatment of sheet flow from neighboring surfaces such as vegetated filter strips including grassed filter strips, filter strips, and grassed filters are made. Vegetated surfaces are also called vegetated filter strips (Dillaha et al. 1989; Delgado et al. 1995). Biological and chemical processes utilized to trap plant nutrients and pesticides might be simply decayed/altered. Bermuda grass buffer is used for the elimination of chloropyrifos (62–99%), dicamba (90–100%), 2,4D (89–98%), and mecoprop (89–95%). Vegetated drainage ditches are used for the removal of atrazine (98%) and pyrethroid (100%) (Moore et al. 2001).

5.7 Conclusion

Wastewater treatment has become an important issue in recent years because of the increasing demand for limited resources along with water pollution as one of the major risk factors. Discharge of various hazardous materials from industries and manufacturing companies deteriorates the water quality, making it unsafe for drinking purposes and other day-to-day uses. Water is the primary as well as universal solvent for the sustainability of life and agricultural purposes. Sustainable development has three milestones such as environment sustainability, social sustainability, and economic sustainability. Therefore, different techniques are available through which we can dispose of/treat/filter the toxic chemicals from industrial wastes to minimize their adverse effects on health. Further, elimination of toxins from various kinds of waste can be a good approach to a sustainable environment.

References

Acharya, B.K., Mohana, S., and Madamwar, D. (2008). Anaerobic treatment of distillery spent wash–a study on upflow anaerobic fixed film bioreactor. *Bioresour. Technol.* 99 (11): 4621–4626.

Agarry, S.E. and Solomon, B.O. (2008). Kinetics of batch microbial degradation of phenols by indigenous pseudomonas fluorescence. *Int. J. Environ. Sci. Technol.* 5 (2): 223–232.

Akbal, F. and Camcı, S. (2010). Comparison of electrocoagulation and chemical coagulation for heavy metal removal. *Chem. Eng. Technol.* 33 (10): 1655–1664.

Ali, I. and Gupta, V.K. (2007). Advances in water treatment by adsorption technology. *Nat. Protoc.* 1 (6): 2661–2668.

Al-Rashdi, B.A.M., Johnson, D.J., and Hilal, N. (2013). Removal of heavy metal ions by nanofiltration. *Desalination* 315: 2–17.

Al-Saydeh, S.A., El-Naas, M.H., and Zaidi, S.J. (2017). Copper removal from industrial wastewater: a comprehensive review. *J. Ind. Eng. Chem.* 56: 35–44.

Altaf, M.S. and Ali, T.A. (2010). Waste water treatment using sequential batch reactor and development of microbiological method for the analysis of relative toxicity. *Pak. J. Nutr.* 9 (6): 574–576.

Arias, C.A. and Brix, H. (2005). Phosphorus removal in constructed wetlands: can suitable alternative media be identified? *Water Sci. Technol.* 51 (9): 267–273.

Arino, M., Pinna, F., and Strukul, G. (2004). Nitrate and nitrite hydrogenation with Pd and Pt/SnO2 catalysts: the effect of the support porosity and the role of carbon dioxide in the control of selectivity. *Appl. Catal. B Environ.* 53 (3): 161–168.

Ashbolt, N.J. (2004). Microbial contamination of drinking water and disease outcomes in developing regions. *Toxicology* 198 (1–3): 229–238.

Shivajirao, P.A. (2012). Treatment of distillery wastewater using membrane technologies. *Int. J. Adv. Eng. Res. Stud.* 1 (3): 275–283.

Babel, S. and Kurniawan, T.A. (2003). Low-cost adsorbents for heavy metals uptake from contaminated water: a review. *J. Hazard. Mater.* 97 (1–3): 219–243.

Berndt, H., Mönnich, I., Lücke, B., and Menzel, M. (2001). Tin promoted palladium catalysts for nitrate removal from drinking water. *Appl. Catal. B Environ.* 30 (1–2): 111–122.

Blenkinsopp, S.A. and Costerton, J.W. (1991). Understanding bacterial biofilms. *Trends Biotechnol.* 9: 138–143.

Bodalo-Santoyo, A., Gómez-Carrasco, J.L., Gomez-Gomez, E. et al. (2003). Application of reverse osmosis to reduce pollutants present in industrial wastewater. *Desalination* 155 (2): 101–108.

Bolognesi, C. (2003). Genotoxicity of pesticides: a review of human biomonitoring studies. *Mutat. Res.* 543 (3): 251–272.

Bori Akadar, A., Bourioug, M., Mohamed, N. et al. (2014). Evaluation of effectiveness of domestic wastewater treatment by infiltration through sand and pozzolana in PVC columns. *Int. J. Environ. Res.* 8 (3): 515–522.

Bouwer, E.J. and Zehnder, A.J. (1993). Bioremediation of organic compounds—putting microbial metabolism to work. *Trends Biotechnol.* 11 (8): 360–367.

Brahmi, M., Fourti, O., and Abdennaceur, H. (2013). Disinfection of wastewater by infiltration-percolation coupled to UV irradiation in an arid Tunisian area. *Hydrol. Curr. Res.* 4: 155.

Branda, S.S., Vik, Å., Friedman, L., and Kolter, R. (2005). Biofilms: the matrix revisited. *Trends Microbiol.* 13 (1): 20–26.

Brissaud, F., Salgot, M., Bancolé, A. et al. (1999). Residence time distribution and disinfection of secondary effluents by infiltration percolation. *Water Sci. Technol.* 40 (4–5): 215–222.

Brown, H.S., Bishop, D.R., and Rowan, C.A. (1984). The role of skin absorption as a route of exposure for volatile organic compounds (VOCs) in drinking water. *Am. J. Public Health* 74 (5): 479–484.

Caprarescu, S., Purcar, V., and Vaireanu, D.I. (2012). Separation of copper ions from synthetically prepared electroplating wastewater at different operating conditions using electrodialysis. *Sep. Sci. Technol.* 47 (16): 2273–2280.

Chang, C.Y., Chang, J.S., Vigneswaran, S., and Kandasamy, J. (2008). Pharmaceutical wastewater treatment by membrane bioreactor process–a case study in southern Taiwan. *Desalination* 234 (1–3): 393–401.

Chang, J.H., Ellis, A.V., Tung, C.H., and Huang, W.C. (2010). Copper cation transport and scaling of ionic exchange membranes using electrodialysis under electroconvection conditions. *J. Membr. Sci.* 361 (1–2): 56–62.

Chatterjee, A., Das, D., Mandal, B.K. et al. (1995). Arsenic in ground water in six districts of West Bengal, India: the biggest arsenic calamity in the world. Part I. arsenic species in drinking water and urine of the affected people. *Analyst* 120 (3): 917–924.

Chaturvedi, S., Chandra, R., and Rai, V. (2006). Isolation and characterization of *Phragmites australis* (L.) rhizosphere bacteria from contaminated site for bioremediation of colored distillery effluent. *Ecol. Eng.* 27 (3): 202–207.

Chaudhry, G.R. and Chapalamadugu, S. (1991). Biodegradation of halogenated organic compounds. *Microbiol. Mol. Biol. Rev.* 55 (1): 59–79.

Chen, C.J., Kuo, T.L., and Wu, M.M. (1988). Arsenic and cancers. *Lancet* 1: 414–415.

Chen, W., Brühlmann, F., Richins, R.D., and Mulchandani, A. (1999). Engineering of improved microbes and enzymes for bioremediation. *Curr. Opin. Biotechnol.* 10 (2): 137–141.

Chong, M.N., Jin, B., Chow, C.W., and Saint, C. (2010). Recent developments in photocatalytic water treatment technology: a review. *Water Res.* 44 (10): 2997–3027.

Cundy, A.B., Hopkinson, L., and Whitby, R.L. (2008). Use of iron-based technologies in contaminated land and groundwater remediation: a review. *Sci. Total Environ.* 400 (1–3): 42–51.

Dagostino, L., Goodman, A.E., and Marshall, K.C. (1991). Physiological responses induced in bacteria adhering to surfaces. *Biofouling* 4 (1–3): 113–119.

Damalas, C.A. and Eleftherohorinos, I.G. (2011). Pesticide exposure, safety issues, and risk assessment indicators. *Int. J. Environ. Res. Public Health* 8 (5): 1402–1419.

Daniel, C., Alfano, D., Venditto, V. et al. (2005). Aerogels with a microporous crystalline host phase. *Adv. Mater.* 17 (12): 1515–1518.

Das, M.P., Bashwant, M., Kumar, K., and Das, J. (2012). Control of pharmaceutical effluent parameters through bioremediation. *J. Chem. Pharm. Res.* 4 (2): 1061–1065.

Daschner, F.D., Rüden, H., Simon, R., and Clotten, J. (1996). Microbiological contamination of drinking water in commercial household water filter system. *Eur. J. Clin. Microbiol. Infect. Dis.* 15 (3): 233–237.

Dave, N., Chan, M.Y., Huang, P.J.J. et al. (2010). Regenerable DNA-functionalized hydrogels for ultrasensitive, instrument-free mercury (II) detection and removal in water. *J. Am. Chem. Soc.* 132 (36): 12668–12673.

Davis, T.A., Volesky, B., and Mucci, A. (2003). A review of the biochemistry of heavy metal biosorption by brown algae. *Water Res.* 37 (18): 4311–4330.

De Pourcq, K., Ayora, C., García-Gutiérrez, M. et al. (2015). A clay permeable reactive barrier to remove Cs-137 from groundwater: column experiments. *J. Environ. Radioact.* 149: 36–42.

Deegan, A.M., Shaik, B., Nolan, K. et al. (2011). Treatment options for wastewater effluents from pharmaceutical companies. *Int. J. Environ. Sci. Technol.* 8 (3): 649–666.

Deganello, F., Liotta, L.F., Macaluso, A. et al. (2000). Catalytic reduction of nitrates and nitrites in water solution on pumice-supported Pd–Cu catalysts. *Appl. Catal. B Environ.* 24 (3–4): 265–273.

Delgado, A.N., Periago, E.L., and Viqueira, F.D.F. (1995). Vegetated filter strips for wastewater purification: a review. *Bioresour. Technol.* 51 (1): 13–22.

Deo, P.G., Karanth, N.G., Gopalakrishna, N., and Karanth, K. (1994). Biodegradation of hexachlorocyclohexane isomers in soil and food environment. *Crit. Rev. Microbiol.* 20 (1): 57–78.

Dhaouadu, H., Zakhama, S., and M'henni, F. (2011). Nonlinear modelisation of heavy metal removal from aqueous solution using *Ulva lactuca* algae. *Bioresour. Technol.* 102 (2): 786–796.

Dillaha, T.A., Reneau, R.B., Mostaghimi, S., and Lee, D. (1989). Vegetative filter strips for agricultural nonpoint source pollution control. *Trans. ASAE* 32 (2): 513–0519.

Duan, J., Lu, Q., Chen, R. et al. (2010). Synthesis of a novel flocculant on the basis of crosslinked Konjac glucomannan-graft-polyacrylamide-co-sodium xanthate and its application in removal of Cu^{2+} ion. *Carbohydr. Polym.* 80 (2): 436–441.

Duffner, F.M., Kirchner, U., Bauer, M.P., and Müller, R. (2000). Phenol/cresol degradation by the thermophilic bacillus thermoglucosidasius A7: cloning and sequence analysis of five genes involved in the pathway. *Gene* 256 (1–2): 215–221.

El-Enany, A.E. and Issa, A.A. (2000). Cyanobacteria as a biosorbent of heavy metals in sewage water. *Environ. Toxicol. Pharmacol.* 8 (2): 95–101.

Elmolla, E.S., Ramdass, N., and Chaudhun, M. (2012). Optimization of sequencing batch reactor operating conditions for treatment of high-strength pharmaceutical wastewater. *J. Environ. Sci. Technol.* 5 (6): 452–459.

Enright, A.M., McHugh, S., Collins, G., and O'Flaherty, V. (2005). Low-temperature anaerobic biological treatment of solvent-containing pharmaceutical wastewater. *Water Res.* 39 (19): 4587–4596.

Ersahin, M.E., Ozgun, H., Dereli, R.K. et al. (2012). A review on dynamic membrane filtration: materials, applications and future perspectives. *Bioresour. Technol.* 122: 196–206.

Escobar, I. and Van der Bruggen, B. (ed.) (2011). *Modern Applications in Membrane Science and Technology*. American Chemical Society.

Esser-Kahn, A.P., Iavarone, A.T., and Francis, M.B. (2008). Metallothionein-cross-linked hydrogels for the selective removal of heavy metals from water. *J. Am. Chem. Soc.* 130 (47): 15820–15822.

Eswar, P. and Devaraj, C.G. (2011). Water defluoridation: field studies in India. *Indian J. Dent. Adv.* 3 (2): 526–533.

Fawell, J., Bailey, K., Chilton, J. et al. (2006). *Fluoride in Drinking-Water*. IWA publishing.

Fu, F. and Wang, Q. (2011). Removal of heavy metal ions from wastewaters: a review. *J. Environ. Manage.* 92 (3): 407–418.

Gao, W., Guan, N., Chen, J. et al. (2003). Titania supported Pd-Cu bimetallic catalyst for the reduction of nitrate in drinking water. *Appl. Catal. B Environ.* 46 (2): 341–351.

Garcia, A., Rivas, H.M., Figueroa, J.L., and Monroe, A.L. (1995). Case history: pharmaceutical wastewater treatment plant upgrade SmithKline Beecham pharmaceuticals company. *Desalination* 102 (1–3): 255–263.

García, Y., Ruiz, C., Mena, E. et al. (2014). Removal of nitrates from spiked clay soils by coupling electrokinetic and permeable reactive barrier technologies. *J. Chem. Technol. Biotechnol.* 9 (90): 1719–1726.

Ghazvini, P.T.M. and Mashkani, S.G. (2009). Biotechnological potential of Azolla filiculoides for biosorption of Cs and Sr: application of micro-PIXE for measurement of biosorption. *Bioresour. Technol.* 100 (6): 1915–1921.

Gherasim, C.V., Cuhorka, J., and Mikulášek, P. (2013). Analysis of lead (II) retention from single salt and binary aqueous solutions by a polyamide nanofiltration membrane: experimental results and modelling. *J. Membr. Sci.* 436: 132–144.

Ghosh, U., Weber, A.S., Jensen, J.N., and Smith, J.R. (1999). Granular activated carbon and biological activated carbon treatment of dissolved and sorbed polychlorinated biphenyls. *Water Environ. Res.* 71 (2): 232–240.

Ghosh, M., Verma, S.C., Mengoni, A., and Tripathi, A.K. (2004). Enrichment and identification of bacteria capable of reducing chemical oxygen demand of anaerobically treated molasses spent wash. *J. Appl. Microbiol.* 96 (6): 1278–1286.

Glazer, A.N., Nikaido, H., and Brown, A.J. (1996). Microbial biotechnology: fundamentals of applied microbiology. *Trends Biotechnol.* 14 (6): 215–215.

Gode, F. and Pehlivan, E. (2006). Removal of chromium (III) from aqueous solutions using Lewatit S 100: the effect of pH, time, metal concentration and temperature. *J. Hazard. Mater.* 136 (2): 330–337.

Goodwin, J.A.S. and Forster, C.F. (1985). A further examination into the composition of activated sludge surfaces in relation to their settlement characteristics. *Water Res.* 19 (4): 527–533.

Gopal, R. and Ghosh, P.K. (1985). Fluoride in drinking water-its effects and removal. *Def. Sci. J.* 35 (1): 71–88.

Gupta, V.K. and Rastogi, A. (2009). Biosorption of hexavalent chromium by raw and acid-treated green alga *Oedogonium hatei* from aqueous solutions. *J. Hazard. Mater.* 163 (1): 396–402.

Hakl, J., Hunyadi, I., Varga, K., and Csige, I. (1995). Determination of radon and radium content of water samples by SSNTD technique. *Radiat. Meas.* 25 (1–4): 657–658.

Harayama, S. (1997). Polycyclic aromatic hydrocarbon bioremediation design. *Curr. Opin. Biotechnol.* 8 (3): 268–273.

Harvey, R.W., Smith, R.L., and George, L. (1984). Effect of organic contamination upon microbial distributions and heterotrophic uptake in a Cape Cod, Mass., aquifer. *Appl. Environ. Microbiol.* 48 (6): 1197–1202.

He, J. and Chen, J.P. (2014). A comprehensive review on biosorption of heavy metals by algal biomass: materials, performances, chemistry, and modeling simulation tools. *Bioresour. Technol.* 160: 67–78.

Hillis, P. (2000). Membrane technology in water and wastewater treatment. *International Conference on Membrane Technology in Water and Wastewater Treatment (2000, Lancaster University)*. Royal Society of Chemistry.

Hitzfeld, B.C., Höger, S.J., and Dietrich, D.R. (2000). Cyanobacterial toxins: removal during drinking water treatment, and human risk assessment. *Environ. Health Perspect.* 108 (suppl 1): 113–122.

van Hylckama Vlieg, J.E.T. and Janssen, D.B. (2001). Formation and detoxification of reactive intermediates in the metabolism of chlorinated ethenes. *J. Biotechnol.* 2 (85): 81–102.

Ileri, R., Sengil, I.A., Kulac, S., and Damar, Y. (2003). Treatment of mixed pharmaceutical industry and domestic wastewater by sequencing batch reactor. *J. Environ. Sci. Health A* 38 (10): 2101–2111.

Inamori, Y. and Fujimoto, N. (2009). Water quality and standards—vol. II, microbial/biological contamination of water. In: *Encyclopaedia of Life Support Systems (EOLSS)*.

Jack, F., Bostock, J., Tito, D. et al. (2014). Electrocoagulation for the removal of copper from distillery waste streams. *J. Inst. Brew.* 120 (1): 60–64.

Janssen, D.B., Oppentocht, J.E., and Poelarends, G.J. (2001). Microbial dehalogenation. *Curr. Opin. Biotechnol.* 12 (3): 254–258.

Javadian, H., Ahmadi, M., Ghiasvand, M. et al. (2013). Removal of Cr (VI) by modified brown algae Sargassum bevanom from aqueous solution and industrial wastewater. *J. Taiwan Inst. Chem. Eng.* 44 (6): 977–989.

Jensen, P.E., Ferreira, C.M., Hansen, H.K. et al. (2010). Electroremediation of air pollution control residues in a continuous reactor. *J. Appl. Electrochem.* 40 (6): 1173–1181.

Jiuhui, Q.U. (2008). Research progress of novel adsorption processes in water purification: a review. *J. Environ. Sci.* 20 (1): 1–13.

Johri, A.K., Dua, M., Tuteja, D. et al. (1996). Genetic manipulations of microorganisms for the degradation of hexachlorocyclohexane. *FEMS Microbiol. Rev.* 19 (2): 69–84.

Kavamura, V.N. and Esposito, E. (2010). Biotechnological strategies applied to the decontamination of soils polluted with heavy metals. *Biotechnol. Adv.* 28 (1): 61–69.

Kavitha, G.V. and Beebi, S.K. (2003). Biodegradation of phenol in a packed bed reactor using peat media. *Asian J. Microbiol. Biotechnol. Environ. Sci.* 5 (2): 157–159.

Khan, M.Z.H. and Mostafa, M.G. (2011). Aerobic treatment of pharmaceutical wastewater in a biological reactor. *Int. J. Environ. Sci.* 1 (7): 1797–1805.

Khan, M.M.H., Sakauchi, F., Sonoda, T. et al. (2003). Magnitude of arsenic toxicity in tube-well drinking water in Bangladesh and its adverse effects on human health including cancer: evidence from a review of the literature. *Asian Pac. J. Cancer Prev.* 4 (1): 7–14.

Kim, B. (2014). Becoming of phosphorus in filters planted with reeds: study of its retention/release and influencing factors. Doctoral dissertation, INSA Lyon.

Kotrappanavar, N.S., Hussain, A.A., Abashar, M.E.E. et al. (2011). Prediction of physical properties of nanofiltration membranes for neutral and charged solutes. *Desalination* 280 (1–3): 174–182.

Kumar, S., Mukerji, K.G., and Lai, R. (1996). Molecular aspects of pesticide degradation by microorganisms. *Crit. Rev. Microbiol.* 22 (1): 1–26.

Kumar, A., Kumar, S., and Kumar, S. (2005). Biodegradation kinetics of phenol and catechol using pseudomonas putida MTCC 1194. *Biochem. Eng. J.* 22 (2): 151–159.

Lakshmanan, D., Clifford, D.A., and Samanta, G. (2010). Comparative study of arsenic removal by iron using electrocoagulation and chemical coagulation. *Water Res.* 44 (19): 5641–5652.

LaPara, T.M., Nakatsu, C.H., Pantea, L.M., and Alleman, J.E. (2001). Aerobic biological treatment of a pharmaceutical wastewater: effect of temperature on COD removal and bacterial community development. *Water Res.* 35 (18): 4417–4425.

Lapworth, D.J., Baran, N., Stuart, M.E., and Ward, R.S. (2012). Emerging organic contaminants in groundwater: a review of sources, fate and occurrence. *Environ. Pollut.* 163: 287–303.

Lawrence, K. and Tong, D. (2005). Feasibility of using biologically activated carbon for treatment of gaseous H2S. *J. Inst. Eng. Singapore* 45: 15–23.

Lazarova, V. and Manem, J. (1995). Biofilm characterization and activity analysis in water and wastewater treatment. *Water Res.* 29 (10): 2227–2245.

Li, J., Wang, X., Wang, H. et al. (2017). Functionalization of biomass carbonaceous aerogels and their application as electrode materials for electro-enhanced recovery of metal ions. *Environ. Sci. Nano* 4 (5): 1114–1123.

Liu, Y., Mou, H., Chen, L. et al. (2015). Cr (VI)-contaminated groundwater remediation with simulated permeable reactive barrier (PRB) filled with natural pyrite as reactive material: environmental factors and effectiveness. *J. Hazard. Mater.* 298: 83–90.

Lu, F. and Astruc, D. (2018). Nanomaterials for removal of toxic elements from water. *Coord. Chem. Rev.* 356: 147–164.

Madukasi, E.I., Dai, X., He, C., and Zhou, J. (2010). Potentials of phototrophic bacteria in treating pharmaceutical wastewater. *Int. J. Environ. Sci. Technol.* 7 (1): 165–174.

Mayabhate, S.P., Gupta, S.K., and Joshi, S.G. (1988). Biological treatment of pharmaceutical wastewater. *Water Air Soil Pollut.* 38 (1–2): 189–197.

Mena, E., Ruiz, C., Villaseñor, J. et al. (2015). Biological permeable reactive barriers coupled with electrokinetic soil flushing for the treatment of diesel-polluted clay soil. *J. Hazard. Mater.* 283: 131–139.

Mikami, I., Sakamoto, Y., Yoshinaga, Y., and Okuhara, T. (2003). Kinetic and adsorption studies on the hydrogenation of nitrate and nitrite in water using Pd-Cu on active carbon support. *Appl. Catal. B Environ.* 44 (1): 79–86.

Mikami, I., Kitayama, R., and Okuhara, T. (2006). Hydrogenations of nitrate and nitrite in water over Pt–promoted Ni catalysts. *Appl. Catal. A Gen.* 297 (1): 24–30.

Misal, S.A., Lingojwar, D.P., Shinde, R.M., and Gawai, K.R. (2011). Purification and characterization of azoreductase from alkaliphilic strain *Bacillus badius*. *Process Biochem.* 46 (6): 1264–1269.

Mishra, V., Lal, R., and Srinivasan. (2001). Enzymes and operons mediating xenobiotic degradation in bacteria. *Crit. Rev. Microbiol.* 27 (2): 133–166.

Moore, M.T., Bennett, E.R., Cooper, C.M. et al. (2001). Transport and fate of atrazine and lambda-cyhalothrin in an agricultural drainage ditch in the Mississippi Delta, USA. *Agric. Ecosyst. Environ.* 87 (3): 309–314.

Motsi, T., Rowson, N.A., and Simmons, M.J.H. (2009). Adsorption of heavy metals from acid mine drainage by natural zeolite. *Int. J. Miner. Process.* 92 (1–2): 42–48.

Nandy, T., Shastry, S., and Kaul, S.N. (2002). Wastewater management in a cane molasses distillery involving bioresource recovery. *J. Environ. Manage.* 65 (1): 25–38.

Narmadha, D. and Kavitha, M.S. (2012). Treatment of domestic wastewater using natural flocculants. *Int. J. Life Sci. Biotechnol. Pharma Res.* 1 (3): 206–213.

Nasef, M.M. and Güven, O. (2012). Radiation-grafted copolymers for separation and purification purposes: status, challenges and future directions. *Prog. Polym. Sci.* 37 (12): 1597–1656.

Ng, W.J., Yap, M.G., and Sivadas, M. (1989). Biological treatment of a pharmaceutical wastewater. *Biol. Wastes* 29 (4): 299–311.

Noble, J. (2006). GE ZeeWeed MBR technology for pharmaceutical wastewater treatment. *Membr. Technol.* 2006 (9): 7–9.

Nriagu, J.O. (1988). A silent epidemic of environmental metal poisoning? *Environ. Pollut.* 50 (1–2): 139–161.

Nwachcuku, N. and Gerba, C.P. (2004). Emerging waterborne pathogens: can we kill them all? *Curr. Opin. Biotechnol.* 15 (3): 175–180.

Obiri-Nyarko, F., Grajales-Mesa, S.J., and Malina, G. (2014). An overview of permeable reactive barriers for in situ sustainable groundwater remediation. *Chemosphere* 111: 243–259.

Pagga, U. and Brown, D. (1986). The degradation of dyestuffs: part II behaviour of dyestuffs in aerobic biodegradation tests. *Chemosphere* 15 (4): 479–491.

Pant, D. and Adholeya, A. (2007). Biological approaches for treatment of distillery wastewater: a review. *Bioresour. Technol.* 98 (12): 2321–2334.

Park, S.H., Cho, H.J., Baek, K. et al. (2013). Removal characteristics of copper by marine macro-algae-derived chars. *Chem. Eng. J.* 217: 205–211.

Patel, H. and Madamwar, D. (2000). Biomethanation of low pH petrochemical wastewater using up-flow fixed-film anaerobic bioreactors. *World J. Microbiol. Biotechnol.* 16 (1): 69–75.

Pedersen, K.B., Ottosen, L.M., Jensen, P.E., and Lejon, T. (2015). Comparison of 2-compartment, 3-compartment and stack designs for electrodialytic removal of heavy metals from harbour sediments. *Electrochim. Acta* 181: 48–57.

Pérez, M., Romero, L.I., Nebot, E., and Sales, D. (1997). Colonisation of a porous sintered-glass support in anaerobic thermophilic bioreactors. *Bioresour. Technol.* 59 (2–3): 177–183.

Pino, F., Mayorga-Martinez, C.C., Merkoçi, A., et al. (2017). Nanomaterials-based platforms for environmental monitoring. *Past, present and future challenges of biosensors and bioanalytical tools in analytical chemistry: a tribute to professor Marco Mascini, 77.*

Pintar, A., Batista, J., and Muševič, I. (2004). Palladium-copper and palladium-tin catalysts in the liquid phase nitrate hydrogenation in a batch-recycle reactor. *Appl. Catal. B Environ.* 52 (1): 49–60.

Qi, G., Yang, R.T., and Rinaldi, F.C. (2006). Selective catalytic reduction of nitric oxide with hydrogen over Pd-based catalysts. *J. Catal.* 237 (2): 381–392.

Qu, X., Alvarez, P.J., and Li, Q. (2013). Applications of nanotechnology in water and wastewater treatment. *Water Res.* 47 (12): 3931–3946.

Raj, D.S.S. and Anjaneyulu, Y. (2005). Evaluation of biokinetic parameters for pharmaceutical wastewaters using aerobic oxidation integrated with chemical treatment. *Process Biochem.* 40 (1): 165–175.

Rajeshwari, K.V., Balakrishnan, M., Kansal, A. et al. (2000). State-of-the-art of anaerobic digestion technology for industrial wastewater treatment. *Renew. Sust. Energ. Rev.* 4 (2): 135–156.

Ram, N.M., Christman, R.F., and Cantor, K.P. (1990). *Significance and Treatment of Volatile Organic Compounds in Water Supplies*. CRC Press.

Ramírez, E.M., Jiménez, C.S., Camacho, J.V. et al. (2015). Feasibility of coupling permeable bio-barriers and electrokinetics for the treatment of diesel hydrocarbons polluted soils. *Electrochim. Acta* 181: 192–199.

Rao, P.V., Gupta, N., Bhaskar, A.S., and Jayaraj, R. (2002). Toxins and bioactive compounds from cyanobacteria and their implications on human health. *J. Environ. Biol.* 23 (3): 215–224.

Rao, A.G., Naidu, G.V., Prasad, K.K. et al. (2004). Anaerobic treatment of wastewater with high suspended solids from a bulk drug industry using fixed film reactor (AFFR). *Bioresour. Technol.* 93 (3): 241–247.

Rincon, G.J. and La Motta, E.J. (2014). Simultaneous removal of oil and grease, and heavy metals from artificial bilge water using electro-coagulation/flotation. *J. Environ. Manag.* 144: 42–50.

Rogers, K.R. (2006). Recent advances in biosensor techniques for environmental monitoring. *Anal. Chim. Acta* 568 (1–2): 222–231.

Rosen, M., Welander, T., Löfqvist, A., and Holmgren, J. (1998). Development of a new process for treatment of a pharmaceutical wastewater. *Water Sci. Technol.* 37 (9): 251–258.

Rusin, P.A., Rose, J.B., Haas, C.N., and Gerba, C.P. (1997). Risk assessment of opportunistic bacterial pathogens in drinking water. In: *Reviews of Environmental Contamination and Toxicology*, 57–83. New York, NY: Springer.

Rykowska, I., Wasiak, W., and Byra, J. (2008). Extraction of copper ions using silica gel with chemically modified surface. *Chem. Pap.* 62 (3): 255–259.

Sangodkar, U.M.X., Aldrich, T.L., Haugland, R.A. et al. (1989). Molecular basis of biodegradation of chloroaromatic compounds. *Acta Biotechnol.* 9 (4): 301–316.

Scholz, M. and Martin, R.J. (1997). Ecological equilibrium on biological activated carbon. *Water Res.* 31 (12): 2959–2968.

Sheng, P.X., Ting, Y.P., and Chen, J.P. (2007). Biosorption of heavy metal ions (Pb, Cu, and Cd) from aqueous solutions by the marine alga Sargassum sp. in single-and multiple-metal systems. *Ind. Eng. Chem. Res.* 46 (8): 2438–2444.

Sirtori, C., Zapata, A., Oller, I. et al. (2009). Decontamination industrial pharmaceutical wastewater by combining solar photo-Fenton and biological treatment. *Water Res.* 43 (3): 661–668.

Smith, A.H., Lingas, E.O., and Rahman, M. (2000). Contamination of drinking-water by arsenic in Bangladesh: a public health emergency. *Bull. World Health Organ.* 78: 1093–1103.

Soares, O.S.G., Órfão, J.J., and Pereira, M.F.R. (2008). Activated carbon supported metal catalysts for nitrate and nitrite reduction in water. *Catal. Lett.* 126 (3–4): 253–260.

Soares, O.S.G.P., Órfão, J.J.M., Ruiz-Martínez, J. et al. (2010). Pd–Cu/AC and Pt–Cu/AC catalysts for nitrate reduction with hydrogen: influence of calcination and reduction temperatures. *Chem. Eng. J.* 165 (1): 78–88.

Soler, L. and Sánchez, S. (2014). Catalytic nanomotors for environmental monitoring and water remediation. *Nanoscale* 6 (13): 7175–7182.

Soroko, I., Bhole, Y., and Livingston, A.G. (2011). Environmentally friendly route for the preparation of solvent resistant polyimide nanofiltration membranes. *Green Chem.* 13 (1): 162–168.

Srivastava, V., Weng, C.H., Singh, V.K., and Sharma, Y.C. (2011). Adsorption of nickel ions from aqueous solutions by nano alumina: kinetic, mass transfer, and equilibrium studies. *J. Chem. Eng. Data* 56 (4): 1414–1422.

Stuart, M., Lapworth, D., Crane, E., and Hart, A. (2012). Review of risk from potential emerging contaminants in UK groundwater. *Sci. Total Environ.* 416: 1–21.

Susheela, A.K. (1999). Fluorosis management programme in India. *Curr. Sci.* 77 (10): 1250–1256.

Tewari, L. and Malviya, P. (2002). Biodegradation of catechol by fluorescent pseudomonas for sustainable environment. *J. Sci. Ind. Res.* 61 (1): 70–74.

Thines, R.K., Mubarak, N.M., Nizamuddin, S. et al. (2017). Application potential of carbon nanomaterials in water and wastewater treatment: a review. *J. Taiwan Inst. Chem. Eng.* 72: 116–133.

Thiruvenkatachari, R., Vigneswaran, S., and Naidu, R. (2008). Permeable reactive barrier for groundwater remediation. *J. Ind. Eng. Chem.* 14 (2): 145–156.

Todd, J. and Josephson, B. (1996). The design of living technologies for waste treatment. *Ecol. Eng.* 6 (1–3): 109–136.

Van Der Meer, J.R., De Vos, W.M., Harayama, S., and Zehnder, A.J. (1992). Molecular mechanisms of genetic adaptation to xenobiotic compounds. *Microbiol. Mol. Biol. Rev.* 56 (4): 677–694.

Vanerkar, A.P., Satyanarayan, S., and Dharmadhikari, D.M. (2013). Full scale treatment of herbal pharmaceutical industry wastewater. *Int. J. Chem. Phys. Sci.* 2: 52–62.

Veil, J.A. (2008). Thermal distillation technology for management of produced water and frac flowback water. *United States Department of Energy, Argonne National Laboratory Water Technology Brief, 1.*

Velan, M., Vijayaraghavan, K., Jegan, J., and Palanivelu, K. (2005). Biosorption of copper, cobalt and nickel by marine green alga *Ulva reticulata* in a packed column. *Chemosphere* 60 (3): 419–426.

Vieira, R.H. and Volesky, B. (2000). Biosorption: a solution to pollution? *Int. Microbiol.* 3 (1): 17–24.

Watanabe, K. (2001). Microorganisms relevant to bioremediation. *Curr. Opin. Biotechnol.* 12 (3): 237–241.

Weber, W. Jr., Pirbazari, M., and Melson, G. (1978). Biological growth on activated carbon: an investigation by scanning electron microscopy. *Environ. Sci. Technol.* 12 (7): 817–819.

Wehrmann, H.A. (1996). Ground-water contamination by volatile organic compounds: site characterization, spatial and temporal variability. *ISWS Contract Report CR 591.*

Wei, X., Li, B., Zhao, S. et al. (2012). Mixed pharmaceutical wastewater treatment by integrated membrane-aerated biofilm reactor (MABR) system–a pilot-scale study. *Bioresour. Technol.* 122: 189–195.

WHO (2008). *Guidelines for Drinking-Water Quality. Recommendations*, 3rde, vol. 1. Geneva: World Health Organization.

WHO (2010). International code of conduct on the distribution and use of pesticides: guidelines for the registration of pesticides (No. WHO/HTM/NTD/WHOPES/2010.7). World Health Organization.

Xing, Y., Chen, X., and Wang, D. (2007). Electrically regenerated ion exchange for removal and recovery of Cr (VI) from wastewater. *Environ. Sci. Technol.* 41 (4): 1439–1443.

Xu, X.J., Chen, C., Wang, A.J. et al. (2012). Enhanced elementary sulfur recovery in integrated sulfate-reducing, sulfur-producing rector under micro-aerobic condition. *Bioresour. Technol.* 116: 517–521.

Yang, R.T. (1997). *Gas Separation by Adsorption Process*. London: Imperial College Press.

Yang, W., Cicek, N., and Ilg, J. (2006). State-of-the-art of membrane bioreactors: worldwide research and commercial applications in North America. *J. Membr. Sci.* 270 (1–2): 201–211.

Younes, M. and Galal-Gorchev, H. (2000). Pesticides in drinking water—a case study. *Food Chem. Toxicol.* 38 (1): S87–S90.

Yue, D., Qian, X., and Zhao, Y. (2015). Photocatalytic remediation of ionic pollutant. *Sci. Bull.* 60 (21): 1791–1806.

Zamboulis, D., Peleka, E.N., Lazaridis, N.K., and Matis, K.A. (2011). Metal ion separation and recovery from environmental sources using various flotation and sorption techniques. *J. Chem. Technol. Biotechnol.* 86 (3): 335–344.

Zeng, W., Yang, Y., Li, L. et al. (2011). Effect of nitrite from nitritation on biological phosphorus removal in a sequencing batch reactor treating domestic wastewater. *Bioresour. Technol.* 102 (12): 6657–6664.

Zhang, P., Hahn, H.H., and Hoffmann, E. (2003). Different behavior of iron (III) and aluminum (III) salts to coagulate silica particle suspension. *Acta Hydrochim. Hydrobiol.* 31 (2): 145–151.

Zhang, Y., Deng, L., Qin, J. et al. (2009). Biosorption of Cr (VI) from aqueous solutions by nonliving green algae Cladophora albida. *Miner. Eng.* 22 (4): 372–377.

Zhang, Y., Wu, H.X., Wang, T.J. et al. (2011). Granulation of Fe–Al–Ce hydroxide nano-adsorbent by immobilization in porous polyvinyl alcohol for fluoride removal in drinking water. *Powder Technol.* 209 (1–3): 92–97.

Zhong, Y., Zhao, J., Wang, D. et al. (2015). An efficient process for wastewater treatment to mitigate free nitrous acid generation and its inhibition on biological phosphorus removal. *Sci. Rep.* 5: 8602.

Zhou, D., Li, Y., Zhang, Y. et al. (2014). Column test-based optimization of the permeable reactive barrier (PRB) technique for remediating groundwater contaminated by landfill leachates. *J. Contam. Hydrol.* 168: 1–16.

Zhu, W.P., Sun, S.P., Gao, J. et al. (2014). Dual-layer polybenzimidazole/polyethersulfone (PBI/PES) nanofiltration (NF) hollow fiber membranes for heavy metals removal from wastewater. *J. Membr. Sci.* 456: 117–127.

6

Novel Trends of Biotechnology in Wastewater Treatment

Anjani K. Upadhyay[1], *Kazi N. Hasan*[1], *Apratim Chakraborty*[1], *and Manisha Priyadarshini*[2]

[1] *School of Biotechnology, Kalinga Institute of Industrial Technology, Bhubaneswar, Odisha, India*
[2] *MITS School of Biotechnology, Bhubaneswar, Odisha, India*

6.1 Introduction

During the 1960s, the scientific community got acquainted with the term "water pollution". This resulted in a rapid increase of global awareness and since then wastewater treatment techniques have played a key role in the purification of contaminated water, safe disposal, or recycling of wastes, thereby providing solutions to the widespread crisis related to polluted water (Rittmann 2010). It has thus emerged as a key player in achieving sustainable development. But it is yet to address many more challenges.

Wastewater can be defined as water that is contaminated by impurities, often toxic materials due to human use, thus also known as used water. A range of activities (agricultural, industrial, and household among others) generate wastewater. It is broadly classified into three types – industrial, domestic, and storm sewage. Domestic sewage carries used water from houses and apartments; it is also called sanitary sewage. Industrial sewage is used for water from manufacturing or chemical processes (Archis Ambulkar 2021).

Wastewater is highly detrimental to human health. Exposure to it might cause acute health problems. Pathogenic microorganisms are known to be causative agents of most water-related infections like typhoid, cholera, and diarrhea among others. The diseases caused by bacteria, viruses, and protozoa are the most common health hazards associated with untreated waters. It also contains heavy metals and harmful chemicals that adversely affect both the environment and human health by causing diseases (Akpor et al. 2014). Industrial wastes may be toxic, ignitable, corrosive, or reactive and will lead to dangerous environmental outcomes if not properly administered.

Until recently, culture-dependent methods, such as plate count or most probable number (MPN) counting were widely used for the enumeration and detection of bacteria being relevant to biological wastewater treatment performance. The earliest methods of wastewater treatment were waste stabilization ponds, better known as oxidation ponds or lagoons. Till the end of the 1970s, wastewater treatment operations were mainly focused on three aspects – (i) the process to remove suspended materials from wastewater, (ii) the treatment of organics which are biodegradable by preferably negating biological oxygen demand (BOD), (iii) to get rid of pathogenic microorganisms (Rajasulochana and Preethy 2016).

From the early 1970s to about the 1990s, wastewater treatment focused on aesthetic and environmental concerns. The earlier tasks of reduction and removal of BOD, suspended solids, and pathogenic microorganisms were continued, but at larger levels. Removal of nutrients such as nitrogen and phosphorus also began to be addressed, particularly in some of the streams and lakes. The process of membrane separation also commenced in this period, with simple symmetric membranes acting merely as a filtration medium. Asymmetric membranes came up, allowing for increased flux, as did membranes with diverse materials, better characteristics, and improved mechanical properties. The toxic and intractable nature of physicochemical processes like distillery wastewater (DWW) treatment was the primary reason for its preference for wastewater treatment. But it is highly organic and sludge content caused major setbacks. The DWW treatment by

Aquatic Contamination: Tolerance and Bioremediation, First Edition. Edited by Rouf Ahmad Bhat, Gowhar Hamid Dar, Fernanda Maria Policarpo Tonelli, and Saima Hamid.

biological processes was limited due to temporal variations in the loading rate and the inhibitive nature of phenolics and melanoidins present in the wastewater. As a result, anaerobic or fungal modes of treatment became widespread (Phuong Thu and Michèle 2016; Johnson et al. 2019).

Scientists, though, looked at more cost-effective techniques as energy consumption is one of the largest expenses for the operation of a wastewater treatment plant (WWTP), and providing adequate oxygen without excessive aeration is energy-consuming in spite of changing influent conditions. With increasing population, urbanization, and industrialization, regulatory pressure to remove nutrients such as nitrogen and phosphorus is more than ever, and it requires more complicated processes, thus leading to more costs. Along with these, sludge residue generated during physical, chemical, and biological treatment, and its disposal is a major challenge for wastewater treatment. But activated sludge treatment is costly not only in its construction but also in acquiring large areas of land for settling tanks and aeration basins, not to mention the huge footprint it demands (Syron and Heffernan 2017).

6.2 The Nascent Organic Methods

6.2.1 Key Roles Played by Enzymes in Wastewater Treatment

Each strain of a microorganism produces an enormous number of hydrolyzing, oxidizing, and reducing biocatalysts which have a huge importance in adequately treating industrial and municipal wastewater (Pandey et al. 2017). Enzymatic treatment is getting more familiar than other conventional methods as it can be operated over a vast range of salinity, temperature, pH, etc. Also, the effects of loading shock and acclimatization with biomass are minimal, which simplifies and eases the process of controlling (Varga et al. 2019). Enzymes like lipases and lysozymes help in advanced sludge dewatering (Boran et al. 2019; Liu et al. 2019a), amylases and cellulases aid in the reduction of waste chemical discharge in water (Maharajaa et al. 2019; Verma et al. 2020). Reactive remediation technology using nanoparticles stands exemplary against the pernicious influence of contaminants in the water (Thakur and Kumar 2020). An enzyme impregnated into a thick cage of nanoparticles known as single enzyme nanoparticles (SEN) helps in the degradation of a variety of compounds such as phenols, polyaromatic dyes, and pesticides which are malignant in nature and ultimately leads to the remediation of the contaminated waterbody (Watlington 2005). Enzymes are biodegradable and thus can easily be regenerated and can be used in the next catalytic cycle concussing enzymatic wastewater treatment to be a biologically sustainable method (Meng et al. 2019).

There are several enzymes (Table 6.1) which are crucial factors in the repertoire of wastewater treatment procedures.

6.2.2 Use of Oligonucleotide Probes to Treat Water

The utilization of oligonucleotide hybridization probes has gotten boundless in microbial biology and ecological microbiology, thus possessing vital criteria to be part of wastewater treatment (Liu et al. 2019b). A DNA microarray of a minute scale containing a lot of oligonucleotide tests in which 16S rRNAs are focused from a few gatherings of nitrifying microorganisms was produced for the observation of plant samples for wastewater treatment (Yang et al. 2020b). To evacuate nitrogen, nitrification is primarily applied in municipal and industrial WWTPs. The key cycle of nitrogen evacuation is nitrification performed generally in the city and modern WWTPs. It is termed "nitrification" because it alludes to the vigorous, successive oxidation of alkali to nitrite and of nitrite to nitrate. Particularly, the chemolithoautotrophic prokaryotes catalyze the mentioned two stages: ammonia-oxidizing microorganisms (AOM), alkali-oxidizing archaea (AOA), and nitrite-oxidizing microbes (NOM) (Table 6.2). Oligonucleotide probes w.r.t. fluorescence in situ hybridization (FISH) are frequently used to detect and quantify nitrifiers and ammox organisms (Xie et al. 2020).

6.2.3 The Implication of Efficient Microbial Fuel Cell

Microbial fuel cells (MFCs) utilize organisms to be legitimately empowered from biochemical energy and offer spotlessly sustainable, and conceivably self-sufficient vitality and a novel methodology for wastewater treatment (Al-Sayed et al. 2020). According to the acknowledgment, the power density reductions occur with expanding scale, the profusing endeavors to challenge MFC scaling is still an ongoing process (Aelterman et al. 2006; Gálvez et al. 2009; Kim et al. 2011). The fundamental difficulties for a versatile plan are expanding power, expanding power recuperation, and diminishing material expenses. As compared to batch culture frameworks, the continuous flow is more reasonable for the practical utilization of

Table 6.1 List of enzymes and their role in wastewater treatment.

Enzymes	Applications	References
Alkyl sulfatase	Surfactant degradation	Bargh et al. (2020)
Amylase	Starch hydrolysis and production of glucose	Zou et al. (2019)
Cellulolytic enzymes (cellulase cellobiase, etc.)	Hydrolysis of cellulosic sludge from pulp and paper to produce sugars and alcohol and hydrolysis of cellulose in municipal solid waste to sugars and other energy sources	Champreda et al. (2019)
Chitinase	Bioconversion of shellfish waste to N-acetyl glucosamine	Amick et al. (2019)
Chloroperoxidase	Oxidation of phenolic compounds	Swain et al. (2020)
Cyanidase	Cyanide decomposition	Dalal et al. (2012)
Cyanide hydratase	Cyanide hydrolysis	Howden and Preston (2009)
L-Galactono-lactone oxidase	Conversion of galactose from whey hydrolysis to L-ascorbic acid	Adinehbeigi et al. (2020)
Laccase	Removal of phenols, decolourization of Kraft bleaching effluents, binding of phenols and aromatic amines with humus	Degani and Goldblat (2020)
Lactases	Dairy waste processing and production of value-added products	Singh et al. (2019)
Lignin peroxidase	Removal of phenols and aromatic compounds, decolourization of Kraft bleaching effluents	Biko et al. (2020)
Lipase	Improved sludge dewatering	Boran et al. (2019)
Lysozyme	Improved sludge dewatering	Liu et al. (2019a)
Mn-peroxidase	Oxidation of monoaromatic phenols and aromatic dyes	Chowdhary et al. (2019)
Parathion hydrolase	Hydrolyzation of organophosphate pesticides	Tian et al. (2013)
Pectin lyase	Pectin degradation	Yadav et al. (2009)
Peroxidase	Removal of phenols and aromatic amines, decolourization of Kraft bleaching effluents, sludge dewatering	Zhang et al. (2019)
Phosphatase	To mitigate effects of heavy metals	Kwiatek and Carman (2020)
Proteases	Solubilization of fish and meat remains improving sludge dewatering	Kwiatek and Carman (2020)
Tyrosinase	Removal of phenols	Yang et al. (2020c)

wastewater treatment. But it gets discontinuous at times. Different MFCs interact in a ceaseless stream requiring the units to be fluidically joined to an inflow and an outflow stream (Munoz-Cupa et al. 2020).

To advance the improvement optimization of MFCs in wastewater treatment, exploration toward the future should be accentuated on: (i) metabolic component; to additionally comprehend the metabolic system, dynamic microorganisms will be added electrochemically to choose high electrochemical action of microorganisms; build a conductive thick biofilm and improve the activity conditions; (ii) plan and engineering reactor, the structure and design of MFCs will legitimately choose the utilization of MFCs in wastewater treatment; (iii) MFC slews, the voltage inversion and short ionic outcome are as yet enormous boundaries in the manner of functional application as a result of bioanalyzed anode responses in the MFC; (iv) usage of power assortment, the advancement of power assortment and usage framework will quicken the business utilization of MFCs; (v) synergistically impact over other innovation of wastewater treatment, a joint venture may quicken the use of MFCs in wastewater treatment (He et al. 2017).

6.2.4 Algal Bacterial System – A Lost Cost Solution to Solve Wastewater Problems

The microalgal technology or phytoremediation incumbent on providing a rich source of many valuable products (biofuels and nutraceuticals) as microalgae are ubiquitous photoautotrophic microorganisms, which are able to deal with umpteen situations (Diniz et al. 2017). Developing the use of microalgae with wastewater for supplement expulsion and profitability

Table 6.2 Oligonucleotide probes used to detect chemolithoautotrophic prokaryotes in nitrifying activated sludge.

Chemolithoautotrophic prokaryotes	Probe name	Target	Sequence (5′-3′)	[a]FA (%)	Competitor oligonucleotide	Reference
AOB	Nso190	*Betaproteobacterial ammonia-oxidizing bacteria*	CGA TCC CCT GCT TTT CTC C	55	none	
	NmV	*Nitrosococcus mobilis*	TCC TCA GAG ACT ACG CGG	35	none	
	Cluster 6a192	*Nitrosomonas oligotropha lineage (Cluster 6a)*	CTT TCG ATC CCC TAC TTT CC	35	CTT TCG ATC CCC TGC TTC C	Adamczyk et al. (2003)
NOB	Ntspa1151	*Nitrospira sublineage II*	TTC TCC TGG GCA GTC TCT CC	35	none	Maixner et al. (2006)
	Nspmar62	*Nitrospira marina-related Nitrospira*	GCC CCG GAT TCT CGT TCG	40	none	Foesel et al. (2008)
AO	Amx368	*all anammox bacteria*	CCT TTC GGG CAT TGC GAA	15	none	
	Apr820	*A. propionicus*	AAA CCC CTC TAC CGA GTG CCC	40	none	Kartal et al. (2007)
	BS820	*Scalindua wagneri*	TAA TTC CCT CTA CTT AGT GCC C	40	none	

[a] FA ¼ Formamide.

of biomass, especially utilizing lake framework, and photobioreactors were demonstrated viable for the preliminary development of microalgae on a larger scale (Yang et al. 2020a). The vitality of microalgal bacterial colligation is one of the significant methodologies equipped for eliminating toxins from wastewater as a reciprocation approach, such as the utilization of carbon dioxide produced throughout bacterial breath as a source of carbon by microalgae, and the usage of oxygen delivered during photosynthesis performed by microalgae for energy yield. Though, extracellular substances and enzymes released by bacteria happen to be toxic to the algal community (Abinandan et al. 2015). Nevertheless, microalgal innovation for wastewater treatment is cost-effective, conservative, and successful. Certain kinds of restrictions located in the current situation are the mishaps to take the technology to the business level (Abinandan et al. 2018).

6.2.5 Nano Chitosan – Reinventing Bioavailability

In randomly distributed D-glucosamine and N-acetyl-D-glucosamine, there is a linear polysaccharide called chitosan, which is derived from chitin and also known as the second most abundant polysaccharide in the entire world. Huge biocompatibility and low toxicity are one of the most important criteria of chitosan (Goy et al. 2016), which has a broader spectrum of activity against bacteria, viruses, and fungi. The chelating properties of nano chitosan having several dicationic metal ions (Ni^{+}, Zn^{2+}, Co^{2+}, Fe^{2+}, Mg^{2+}, and Cu^{2+}) made it a promising player for wastewater treatment (Sahu et al. 2019). Chitosan nanoparticles are expected to be the most promising materials of decision for water treatment for their little size, enormous surface region, quantum size impacts, and nonappearance of diffusion impediments. Nano chitosan aced due to their quick adsorption rates and adsorption limits as well (Sharma et al. 2009; Kalia and Averous 2011; Abou Kana et al. 2013; Liu et al. 2013; Mahgoub and Samaras 2014; Perera and Rajapakse 2014; Vakili et al. 2014). Nano chitosans are used on the surface of water storage tanks and are also embedded in the membranes to treat the contaminants present in the water. Wastewater cleansing utilizing nanosized chitosan-based adsorbents is thought of as a substitute innovation for the energy-intensive and costly activated charcoal-based adsorption (Abdelmouleh et al. 2004; Vismara et al. 2009; Olivera et al. 2016).

6.2.6 Impact of Antibiotic Levels on Treating Wastewater

Water mediums always have an abundance of pathogenic bacteria which are mainly released with wastewater (Baquero et al. 2008). A considerable lot of such microorganisms bear antibiotic resistance genes, which ultimately consolidate into hereditary versatile stages (plasmids, transposons, and integrons) fit for diffusing among water and soil bacterial clumps (Ghernaout et al. 2011a,b; Osińska et al. 2020). Antimicrobial synthetics like sulfonamides, macrolides, trimethoprim, cephalosporins, and fluoroquinolones can be seen at extraordinarily unique levels in enacted slop treatment. During the year, the antibiotic charge is connected to the adjustment in yearly utilization data, being more significant in the winter (Wang et al. 2020).

The antimicrobial level of wastewater is a component of its partition coefficient; nonetheless, the trial of slop execution to agricultural land with fluoroquinolone field demonstrated is delayed by the strength of such synthetic compounds, regardless of whether with confined portability into the earth (Sukul and Spiteller 2007). Treating wastewater may eliminate practically 80% of fluoroquinolones or antibiotic medications before they come in streams, and are inclined to photodegradation (Gulkowska et al. 2008). The development in instrumental investigative science, utilizing electrophoretic and chromatographic procedures, similar tandem mass spectrometry of liquid chromatography, permits us to uncover a few different sorts of antimicrobials at levels of nanograms/l, after extraction of the solid phase (Voigt Alexander et al. 2019). Antibiotic activity in effluent and wastewater can smoothly be monitored using commercially available test kits. Hence, antibiotic level determination helps in the execution of wastewater treatment efficiently (Ghernaout and Elboughdiri 2020).

6.2.7 Advanced Oxidation Processes

The effluents having pollutants, for example, manufactured synthetic chemicals, colors, obstinate natural wastes, substantial metals, and so forth are released to the closest water bodies in either the presence or absence of preliminary medicines. This causes genuine harm to the dissolved oxygen (DO) level and natural parity of the biological system of the closeby water bodies which are receiving contaminants. Advanced oxidation processes (AOPs) quickly turned into the picked innovation for its numerous applications, for example, natural toxin decimation as toxicity decreases, and improvement of biodegradable BOD/COD expulsion just like odor and color evacuation (Garrido-Cardenas et al. 2020). In AOPs, solid-oxidizing agents, for example, hydroxyl radicals are delivered at room temperature, which are viable for the expulsion of synthetic substances from industrial wastewater (Iqbal et al. 2015). Fenton reaction and ozone oxidation process coupled with biotreatment methods are most promising for AOPs due to their high efficiency and cost-effectiveness (Pang and Abdullah 2013).

6.2.7.1 Fenton's Reaction along with the Biotreatment Process

Organophosphorus insecticides, pesticides, and chemicals are considered to be wastewater alleviation containing poisonous and also nonbiodegradable products that stay in the earth even after the conventional processing of exposed wastewater (Unuofin 2020). In the field of AOPs, the Fenton reaction (Fe^{2+}/H_2O_2) and Fenton-like reactions (Fe^{3+}/H_2O_2) played a very important role in the treatment of nonbiodegradable wastewater (Badawy et al. 2006). Experiments showed that there are two sound limitations in Fenton systems (Fenton reaction + Fenton-like reaction), i.e. (i) a huge iron sludge production, and (ii) in their application, a slow ferric-ion reduction by H_2O_2 (Fan et al. 2019). Illumination of the framework by UV rays was recommended to defeat the restrictions of the Fenton framework. The photoreduction of different ferric species adds to the creation of ferrous particles and radical species (Lueder et al. 2019). Now, the so-named bio-electro-Fenton incorporation of MFCs to create H_2O_2 for Fenton oxidation is ascending as an urgent procedure to lessen the cost connected to H_2O_2 buying and transportation (Wang et al. 2016). An ideal position of the bio-electro-Fenton has been discovered where they might have been controlled in the following years, especially for decreasing unmanageable natural contaminants. Differentiated from the traditional electro-Fenton methods, gigantically the bio-electro-Fenton arrangement lessened the uses on the treatment of wastewater regarding electric vitality utilization and running costs. The arrangements of bio-electro-Fenton are starting to be an adaptable strategy giving a novel answer for developing natural issues connected with wastewater treatment (Ghernaout et al. 2020).

6.2.7.2 Ozonation Coupled with the Biotreatment Process

For the treatment of trace recalcitrant natural mixed content of wastewater, for example, organohalogens, natural pesticides, surfactants, and dyes are being taken care of with a combination of variegated methods related to ozonolysis, viz. O_3/VUV (ozone/vacuum ultraviolet), O_3/H_2O_2/UV (ozone/hydrogen peroxide/ultraviolet), and UV/H_2O_2 (ultraviolet/hydrogen peroxide) (Kaplan et al. 2020). Ozonation has been authenticated to be exceptionally successful in water sanitization. Ozonation is utilized to meet drinking water treatment in the United States Environmental Protection Agency (USEPA) guidelines for the inactivation of infections, Cryptosporidium, and Giardia (USEPA 2003). Recently, ozonation has picked up consideration for its capacity of oxidizing endocrine-disrupting chemicals (EDCs) and drugs in both drinking water and wastewater. The microbial sanitization and contaminant oxidation are conjoined to make ozonation an alluring elective for the cutting-edge wastewater treatment (Wert et al. 2007).

6.2.8 Moving Bed Biofilm Reactor

Progresses in wastewater treatment technology have encouraged expanding the contamination evacuation productivity and meeting rigid effluent guidelines. In any case, there are as yet numerous difficulties confronted in wastewater treatment measures, particularly in the removal of nutrient and trace organic substances, which require improving existing wastewater treatment measures for accomplishing higher removal efficiency (Nguyen et al. 2016). In light of this, the moving bed biofilm reactor (MBBR) could be utilized to cultivate the recuperation of wastewater resources (Leyva-Díaz et al. 2019). In an MBBR cycle, biomass for the most part becomes connected to plastic components, called transporters, as a biofilm. These components continue moving inside the bioreactor because of the unsettling brought about by air circulation in an oxygen-consuming reactor, or by stirrers in anoxic/anaerobic mechanical reactors. MBBR innovation shows significant focal points comparable to regular cycles for wastewater treatment. Considering this, this cycle has necessarily less space as a higher surface zone for mass exchange, higher natural stacking rates at lower water-powered maintenance times (HRTs), and higher profluent quality with respect to natural issues and supplements (Leyva-Díaz et al. 2013; Ødegaard 2017). MBBR has probably the highest rate of nitrogen evacuation while MBBR differs between high and medium-high for genuine wastewater and synthetic wastewater as well (Chaali et al. 2018). As of now, this technology faces a few difficulties, for example, high vitality costs due to air circulation for transporters blending reasons, conceivable development of stale zones inside the reactor that impede the total blending, and capital expenses related to reactor development and transporters (Di Biase et al. 2019).

6.2.9 Film Technologies

As per the concern and according to the environmental research a definite similitude amount of wastewater is directly generated to the environment without being treated properly. This deformation of wastewater treatment leads to the formation of a curb for the supply of usable water for human convenience. A healthy source of water is germane for energy, nutrients, and other required uses (Nambikkattu et al. 2020). Film technologies are nothing but various methods which are based on the treatment and purification of wastewater to make them useful for human welfare. Till now, various studies have been done on this particular concern and multiple techniques have been manifesting to more rich levels (Karami et al. 2020). There are three classified methods – physical, chemical, and biological, which are significantly proven to be the most effective in the field of wastewater management. However, the physical methods which include adsorption, membrane filtration, media filtration (sand, gravel, and walnut shell), evaporation, distillation, and air flotation are broadly categorized under film techniques for the treatment of wastewater.

The thin film composite (TFC) membrane technique is one of the techniques under film technology, which is mainly executed by the process of membrane separation. TFC membranes with superior antifouling properties are more likely to be done for wastewater contaminated with highly concentrated oil (Drioli and Giorno 2016). The technique involves the firm attachments between its thin polyamide layer and its thick sublayer, measures the transition of glass, melting, crystallization, and decomposition and temperature of the selective layer (Karami et al. 2020). Another most effective and well-applied technique which has application in the horticultural field is the nutrient film technique (NFT). It is a reported data according to the vegetation filter research that the contaminated water with zero cultural substrates and primary contaminant treatment are supposed to undergo this NFT technique for the filtration purpose. Nutrient-based horticultural plants are emphasized as a positive and effective great output of this technique. The agglomerated organic pollutants are removed and root zones are bedspread with the wastewater after the liberation of pollutants reached over 24 hours. The roots not only possess supremacy for the filtration of the wastewater but also provide a microhabitat for nitrifying and heterotrophic bacteria, which lead to a noticeable and sharp decrease in wastewater pollution (Lv et al. 2020).

Copper zinc tin sulfide (CZTS) is another technique employed for the degradation of pollutants from wastewater, which acquires a thin photoactive film containing multiple titanium dioxide (TiO_2) aggregates on the surface of this heterostructure. In this process, two types of semiconductors are used, i.e. type-n for the TiO_2 layer and type-p for copper zinc layer, and the UV radiation under which dye and pollutant removal is done falls under solar-simulated radiation (Vis + UV). This TiO_2 layer helps protect the narrow bandgap semiconductor of the thin film from chemical or physical interactions with water and allows visible light irradiation for the removal of methylene blue dye when the TiO_2 layer comes in direct contact with water. The present investigation states that methylene blue removal is efficient for this CZTS heterostructure and the addition of 3% of UV makes this process more inclined toward the removal of deposited pollutants on the thin film (Covei et al. 2019).

For the removal of suspended solids and biologically degradable organic components, some of the membrane technologies such as ultrafiltration (UF), nanofiltration (NF) and reverse osmosis (RO) are completely advantageous. NF and RO are used as a secondary treatment in a considerable situation if the UF process is found incapacious (Kasemset et al. 2013; Yalcinkaya et al. 2020). In addition to this, forward osmosis (FO) also has its significant place for its less energy requirement, lower membrane-fouling frequency, and its easy removal with a higher water recovery rate in comparison to pressure-driven membrane processes (Martinetti et al. 2009; Mi and Elimelech 2010; Song et al. 2015). The use of membrane filtration technology could be effective and ideal for the purification of water and its small-scale treatment systems, but these techniques are still under investigation for a better assessment (Zhou et al. 2006).

6.2.10 Channelizing Technologies Electrochemically

The electrochemical technology has been emphasizing its potential in the treatment of wastewater for the past two decades, considering the increasing demand for drinking water supply and stringent environmental regulations on wastewater management (Chaplin 2019). The proposal to use electricity for water treatment was first made in the United Kingdom in 1889 (Strokach 1975). The challenges of population growth and climate change have resulted in a scarcity of essential resources such as food, energy, and drinking water (Baker 2018). Various electrochemical techniques, including electro-coagulation, electro-flocculation, electro-flotation, electro-deposition, electro-oxidation, and electro-disinfection, have been extensively explored. Among these, the electrooxidation technique relies on materials that minimize electrode kinetics overpotential through electrocatalysis processes (Patel et al. 2013). Electrocoagulation is widely used for treating industrial, rural, and sewage water, as well as for drinking water in small cities. While the technique requires optimization and process tuning, it efficiently removes even the smallest colloidal particles and a wide range of contaminants. Electro-flotation involves the use of tiny bubbles generated by electrolysis to float pollutants on the water surface. The effectiveness of pollutant removal depends on bubble size, as smaller bubbles provide a larger surface area for particle attachment. Electro-flocculation, on the other hand, agglomerates dissolved stable particles to form larger sizes. It involves two phases: coagulation, where repulsive forces between particles are reduced, and flocculation, where metal ions are added to induce bubble formation at the cathode (Mickova 2015a, b). Electrodialysis, a technique focused on converting saline water into potable water, relies on ion-exchange membranes and an electric field to separate salt ions. Although it is beneficial for drinking water storage, there are limitations in handling macromolecules and fouling membranes (Kumar and Pan 2020). The extensive use of electrochemical technologies faces challenges such as low power output, high cost, short electrode material life span, and increased maintenance, making it economically less suitable for wastewater treatment (Mahvi et al. 2011). Nevertheless, electrochemical technology has been studied for over a century and has proven effective in metal recovery and treating heavy metal effluents (Chen 2004).

6.2.11 Fabrication of an Aquaporin-Based Forward Osmosis Membrane

FO is considered natural osmosis and is a membrane-driven process in contrast with pressure-based membrane processes resembling RO and NF. In this process, the osmotic pressure gradient is used to drive water across a semipermeable membrane, from the feed solution to the draw solution (from higher to lower chemical concentration). It is thus coming up as a reliable purification and desalination technique which is used for water supply, energy generation, and food processing. Owing to the requirement of low hydraulic pressure, FO has a number of advantages such as shortage energy input, inferior fouling tendency, and facile fouling discharge, as well as dignified water recovery and less accumulation of contaminants (Ding et al. 2015).

Aquaporins (AQPs) are an enlarged family used as water-transport proteins in cells, which evinces desirable characteristics like high permeabilities (three billion water molecules per second per subunit) and superior specificity (allowing the passage of water but blocking entry of all other molecules) (Ding et al. 2015; Luo et al. 2017). Their standout features and sublimate-water transport competence are tempting for probable incorporation with the FO membrane for better application. AQP having biomimetic membranes streamlining a new route for the prolongation of membrane-based separation technology agglutinated a bioimitation membrane by akinetic AQP Z-unified, block copolymer vesicles on a substrate membrane through covalent bonding, followed by impaction through a command interpreter through the layer-by-layer polydopamine-histidine-coating process. Exact methods, in which proteoliposomes in a selective layer were fully encapsulated, by cross-linking a polyelectrolyte with a membrane. By a complex rehydration and protuberant method, liposomes, with a diameter of about 100 nm, were prepared AQP Z-containing proteoliposomes, and the AQP Z solution was added into the liposome solution (Maurel et al. 2015; Giwa et al. 2017; Warsinger et al. 2018).

Then AQZ-fabricated membranes were applied on a phase-inversion method, and a PSf porous support layer was agglutinated, resulting in a homogeneous intoxicate solution, by submersion into a coagulant bath loaded with fresh DI water at room temperature to inflame phase separation. The PDA-coated membrane was preactivated and proteoliposomes were added to the system. SLBs were selected as AQP Z-incorporated DOPE SLB membranes as they have higher solute and water flux. To measure the size and zeta potential of liposomes and proteoliposomes was used to measure by the characterization of proteoliposomes Ze. Here, liposomes show a much lower shrinkage rate (k) than proteoliposomes, while AQP Z-containing proteoliposomes exhibit much higher permeability than liposomes that AQP Z possesses excellent permeability and may increase water flux in novel membranes. To understand the effect of AQP Z on biomimetic FO membranes. The rough surface is determined by atomic force microscopy (AFM) increased in addition to proteoliposomes. The easing membrane hydrophilicity, porosity, average surface pore diameter, mechanical strength, and thermal stability are interlinked with the surface roughness. AQPs assent the carriage of up to a billion water molecules per second when a substantially high osmotic potential exists across the membrane and in addition to proteoliposomes, the water permeability deepens. This also provides extended membrane stability.

In conclusion, the addition of AQPs contributed to the enhanced water flux and salt rejection. The synthetic protocol of the prepared ABM-1 was found beneficial in reducing the formation of defects and the membrane exhibited a better salt abjure and chemical stability than those of ABM-2 (Liang et al. 2019).

6.2.12 Biosensors: Biomorphosis in a Chip

Biosensors have been applied for the identification of important organic chemicals, heavy metals, and microorganisms in polluted water to mitigate acute health problems like damage to the endocrine system, and induction of cytotoxicity and genotoxicity. The development of biosensing instruments is in great demand for their high sensitivity, selectivity, reliability, and low cost and possesses the ability for real-time response. Biosensors are also made into small systems that permit the evolution of mobile sensors which can lead to effluent onsite. Various kinds of biosensors are categorized into (i) electrochemical, (ii) optical, (iii) piezoelectric, and (iv) thermal biosensors (Korostynska et al. 2013; Fathizadeh and Behnia 2020). A usual mechanism used in optical biosensors for keeping track of organic matter is based on immune analytical methods, which require fluorescently labeled antibodies with a high frequency to the target component. This method, known as fluorimetry is often used for signal transducing in optical biosensors of organic pollutants (Tschmelak et al. 2005). Electrochemical bioreceptors comprising enzymes, antibodies, and whole cells are also used to determine a wide range of organic pollutants. The most promising technique in this field is the amperometric biosensors which employ an amperometric transduction technique, based on horseradish peroxidase (HRP) bioreceptor to be characterized as tenacious organic pollutants in wastewater samples (Salvador et al. 2020). A large series of Clark-type oxygen sensors work on the basis of a simple principle, which comprises an immobilized microorganism on a biomembrane, coupled to an amperometric probe to measure the concentrations of DO (Sonawane et al. 2020). MFC-based biosensors have been immensely used not only for checking DO, BOD, and COD but also in regard to identification and abstracting different organic pollutants in water samples (Zhou et al. 2017). Piezoelectric biosensors are mainly built for the asserting of heavy metals. With the covalent attachment of metallothionein (MT) onto a quartz crystal, a piezoelectric biosensor was designed to monitor Zn^{2+} and Cd^{2+} in aqueous media (Saber and Pişkin 2003). Several microbial biosensors were empanelled to emit bioluminescent signals in the presence of a low concentration of heavy metals (Olaniran et al. 2011; Jouanneau et al. 2012; Martin-Betancor et al. 2015). The bacterial biosensors use recombinant *Escherichia coli* strains that were improved based on their fluorescent protein expression. When these bacteria are in connection with the heavy metals, fluorescent proteins show induced expression to report the hazardous number of HMs in wastewater samples (Raja and Selvam 2011; Ravikumar et al. 2012).

Despite the mastery of biosensing methods, some challenging steps must be addressed to see commercial implementation and development of bacterial biosensors for environmental wastewater paragons. Some of the forsakes include the need for sample pretreatment and reach to multiplexing microorganism detection in a single sample (Bazin et al. 2017); (Theron et al. 2010). There are efficient surface plasmon resonance (SPR)-based devices fortunately established for identifying pathogens in an aqueous medium, but the complexity, cost, and large size of appliances limit their application.

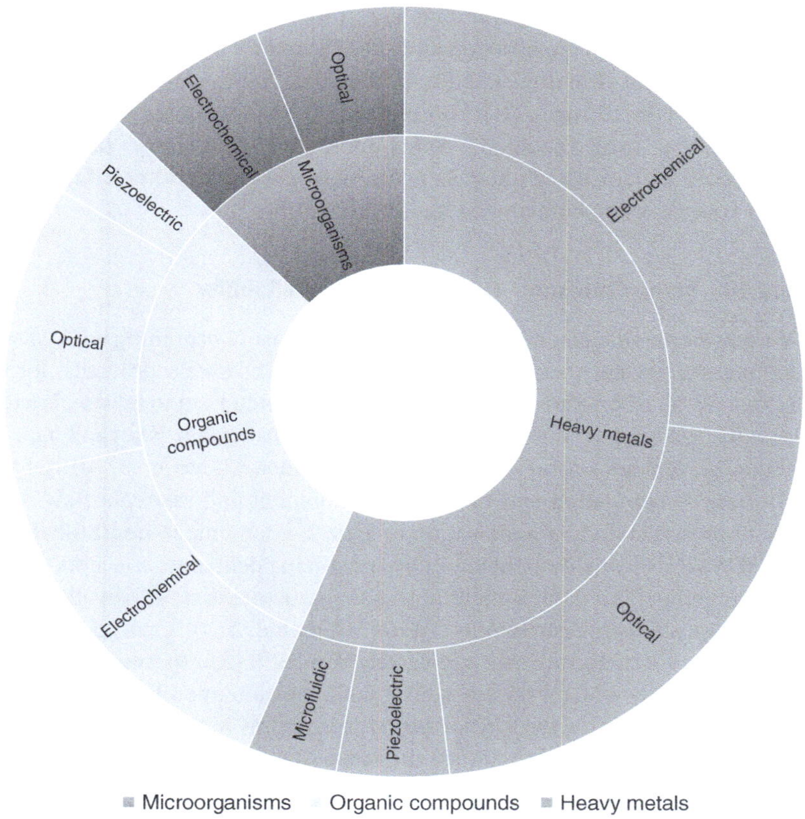

Figure 6.1 Contribution of biosensors in wastewater treatment.

The pie chart (Figure 6.1) exhibits the dynamic application of biosensors in the impelling research and development arena to find out the effects of organic matter, heavy metals, and microorganisms in wastewater treatment.

6.2.13 Anaerobic Sludge Technology

Aerobic granular sludge (AGS) is a recently developed type of biofilm reactor used for wastewater treatment. It consists of dense microbial clusters called aerobic granules, which contain millions of microorganisms per gram of biomass. Unlike activated sludge flocs, a wide range of bacterial species in AGS do not coagulate under reduced hydrodynamic forces and have a higher settling velocity (De Kreuk et al. 2005). The aerobic granulation process involves several stages: (i) microorganism interaction leading to the production of clumps through hydrodynamic, diffusion, gravitational, or thermodynamic forces; (ii) formation of multicellular aggregates through primary attraction, including physical (van der Waals) forces, attraction between opposite charges, and thermodynamic forces; (iii) maturation of the microbial clusters through the conversion of extracellular polymers, cell assembly, and metabolic changes, resulting in organized structures; and (iv) formation and stability of the three-dimensional structure of the microbial aggregate through applied hydrodynamic shear forces (He et al. 2019). This sequential process transforms activated sludge into compact clusters, which eventually take the form of granular sludge and mature into granules (Bassin 2018). Column reactors are inoculated with sludge and operated with bubble aeration and short settling times for cultivation. In these bubble column reactors, the growth of AGS is promoted to form densely packed aggregates, which are later selected through settling durations typically ranging from 2 to 10 minutes. Manipulating sequencing batch reactors with short settling times allows the selection of microbial aggregates that settle quickly, while sludge with poor settling characteristics is removed. This bioreactor manipulation promotes the growth and selection of microbial aggregates, leading to the rapid formation of granular sludge. AGS is separated from the influent water in the same reactor tank through sedimentation. The major characteristics of AGS that contribute to enhanced treatment with a smaller footprint include (i) large particle size (millimeter-scale) of sludge particles; (ii) preservation of diverse redox microenvironments within the granules, including oxic, anoxic, and anaerobic regions; (iii) high biomass retention; (iv) precise control of metabolic reactions through dissolved oxygen (DO) management; and

(v) metabolic cooperation among autotrophic and heterotrophic microorganisms (Nancharaiah and Reddy 2018). The extracellular polymeric substances (EPS) matrix is a crucial component of AGS, playing a key role in microorganism aggregation and structural stability (Chen et al. 2007; Adav et al. 2008). AGS reactors provide opportunities for removing organic carbon, nitrogen (N), phosphorus (P), microcontaminants such as pharmaceuticals and personal care products, persistent or toxic xenobiotic compounds, and metal ions. Various xenobiotic compounds including phenol, para-nitrophenol, chlorinated phenols, pentachlorophenol, pyridine, phthalic acids, and esters can be biologically degraded in AGS systems (Nancharaiah and Reddy 2018).

6.2.14 Nano Chitosan – Reinventing Bioavailability

Wastewater management is a mind-boggling measure beginning with wastewater generation, trailed by the sewage organization, wastewater treatment, and reintegration of excess products, for example, supplements and sludge, into nature (Cossio et al. 2020). Thus, the water cycle, from catchment to release, is a division that includes significant speculation and activity and support costs. Specifically, sewage treatment is a challenge for governments since they are thinking about financial, ecological, and social angles (Hernández-Chover et al. 2018). Expanding requests on water supplies – effectively stressed by population and financial development and instream flow necessities – imply numerous zones are thinking about preservation and reallocation to address economic issues (Stillwell and Webber 2013). These days a sufficient plan for WWTPs is being thought about by actualizing deliberate concepts where theoretical and numerical devices can be utilized together. This commitment proposes a structure that utilizes all-out cost, expended vitality, and recovered wastewater as manageability measurements (Padrón-Páez et al. 2020). On a good note, low- and lower-middle-income countries extensively try to transform waste aggregated from WWTP into nutrients and biofertilizers (Salguero-Puerta et al. 2019). The linear economy which basically consists of expulsion–modification–dispensation–trading–use is an unsustainable process which is streamlining the way for a spheric economy in WWTPs (Morató et al. 2017). Accomplishing an economical wastewater management cycle requires the recognition of site-explicit elements that sway wastewater composition, accessible and productive treatment advancements, the affability of the accepting condition, the hierarchical structure, and the accessibility of ability, also detailed comprehension of how they are interlinked (Munier 2005).

6.3 Forthcoming Technologies/Incubating Ideas: Theory of Existential Growth

As a responsible race on earth, consumption of necessary resources like water should be maintained discreetly. So, as it is a greater concern for a decade and as water is considered to be an inherently renewable source, the upcoming goals should have to include proper functionality of water supplies at the same time while maintaining energy stability, minimal chemical intake, and causative nutrient management. As we are aware, increasing industrialization and urbanization is a generic phenomenon in human societies, which not only treats us to use natural resources but is also protracted in the round of generated waste materials from various sources too (Ghangrekar et al. 2020). MFC technology is considered an upcoming process which is going to be promising in the coming decades. Basically, the mechanism states that the use of this process is to convert the chemical energy present in the bonds of organic matter to electrical energy with the help of electroactive microorganisms (Du et al. 2007). Apart from this, graphene-based materials basically act as adsorbents for variable pollutants from water. Although it is strongly effective, somehow 2D nanomaterials for water purification are in their infancy as because of the initial stage of development of many of these materials, technical issues and production issues have some impact on their execution in industrial processes and somewhere renounce over the application. Overall, ongoing development in promising industrial technologies and techniques has begun to encourage emerging developments and growth in the consumer use of 2D materials (Khalid et al. 2020). Also, the utilization of agricultural remnant or biowaste as an inferior cost of raw adsorbents can suppress environmental degradation, help in treating complex inorganic impurities from wastewater, and will surely increase the system of wastewater treatment by waste (Singh et al. 2020).

The treatment of polluted water is the most challenging and brutal issue of the environment and water ecosystem to make it more useful for people who are straightly and even indirectly dependent on water resources (Reddy and Lee 2013). Many techniques and methods for wastewater treatment have been emphasized worldwide but as per the population and need for large-scale inexpensive treatments, many are yet to come (Singh et al. 2020).

6.4 Conclusion: Progression of Trending Technologies in Water Science

Emerging techniques which can possibly propel sheltered ways and practical innovations to reuse influx ought to be a worldwide need. Rising advancements, for example, online sensors with real-time feedback, will surely assume a significant function in water reuse soon. As more novel techniques emerge, the obstacles related to wastewater treatment are being evaded. Many scientific organizations have identified innovative technologies like anaerobic sludge, which have been effective in the extraction of toxic metal ions and xenobiotic compounds from wastewater while the use of AQP-fabricated FO membrane has shown improved water flux and salt rejection. AOP plays a role in natural toxin depletion, biodegradability evolution, BOD/COD expulsion just as smell and color evacuation. Membrane technologies have also shown positive results in regulating wastewater toxicity. Film and electrochemical technologies comprise a number of innovations which serve multiple purposes. The improvement of biosensor-manufacturing techniques' incorporation with nanoparticles results in higher accuracy and precise detection in many labs to harness sustainable methodologies. In the end, water supply and waste management have to be done proficiently by the authority as clean and clear water has a precious value toward global immunity.

Acknowledgments

We thank Ms. Lakshmi Upadhyay and Ms. Mousumi Meghamala Nayak for their valuable suggestions and for helping us in structuring and formatting the manuscript.

References

Abdelmouleh, M., Boufi, S., Belgacem, M.N. et al. (2004). Modification of cellulosic fibres with functionalised silanes: development of surface properties. *Int. J. Adhes. Adhes.* 24 (1): 43–54.

Abinandan, S., Bhattacharya, R., and Shanthakumar, S. (2015). Efficacy of Chlorella pyrenoidosa and Scenedesmus abundans for nutrient removal in rice mill effluent (paddy-soaked water). *Internat. J. Phytorem.* 17: 377–381.

Abinandan, S., Subashchandrabose, S.R., Venkateswarlu, K., and Megharaj, M. (2018). Nutrient removal and biomass production: advances in microalgal biotechnology for wastewater treatment. *Crit. Rev. Biotechnol.* 38 (8): 1244–1260.

Abou Kana, M.T., Radi, M., and Elsabee, M.Z. (2013). Wastewater treatment with chitosan nano-particles. *Int. J. Nanotechnol. Appl.* 3: 39–50.

Adamczyk, J., Hesselsoe, M., Iversen, N. et al. (2003). The isotope array, a new tool that employs substrate-mediated labeling of rRNA for determination of microbial community structure and function. *Appl. Environ. Microbiol.* 69: 6875–6887.

Adav, S.S., Lee, D.J., and Tay, J.H. (2008). Extracellular polymeric substances and structural stability of aerobic granule. *Water Res.* 42 (6–7): 1644–1650.

Adinehbeigi, K., Shaddel, M., Khalili, S., and Zakeri, A. (2020). Suramin could block the activity of Arabinono-1, 4-lactone oxidase enzyme from Leishmania donovani: structure-based screening and molecular dynamics analyses. *Trans. R. Soc. Trop. Med. Hyg.* 114 (3): 162–172.

Aelterman, P., Rabaey, K., Clauwaert, P., and Verstraete, W. (2006). Microbial fuel cells for wastewater treatment. *Water Sci. Technol.* 54 (8): 9–15.

Akpor, O.B., Otohinoyi, D.A., Olaolu, D.T., and Aderiye, B.I. (2014). Pollutants in wastewater effluents: impacts and remediation processes. *Int. J. Environ. Res. Earth Sci.* 3 (3): 050–059.

AlSayed, A., Soliman, M., and Eldyasti, A. (2020). Microbial fuel cells for municipal wastewater treatment: from technology fundamentals to full-scale development. *Renew. Sust. Energ. Rev.* 134: 110367.

Archis Ambulkar (2021). *Encyclopædia*. Britannica, Inc.

Amick, A.K., Liu, Q., Gautam, S. et al. (2019). Measurement of Chitinase activity in biological samples. *J. Vis. Exp.* 150: e60159.

Badawy, M.I., Ghaly, M.Y., and Gad-Allah, T.A. (2006). Advanced oxidation processes for the removal of organophosphorus pesticides from wastewater. *Desalination* 194 (1–3): 166–175.

Baker, A. (2018). Cape Town is 90 days away from running out of water. *Time*. http://time.com/5103259/cape-town-water-crisis/ (accessed 11 April 2019).

Baquero, F., Martínez, J.-L., and Cantón, R. (2008). Antibiotics and antibiotic resistance in water environments. *Curr. Opin. Biotechnol.* 19: 260–265.

Bargh, J.D., Walsh, S.J., Isidro-Llobet, A. et al. (2020). Sulfatase-cleavable linkers for antibody-drug conjugates. *Chem. Sci.* 11 (9): 2375–2380.

Bassin, J.P. (2018). Aerobic granular sludge technology. In: *Advanced Biological Processes for Wastewater Treatment* (ed. M. Dezotti, G. Lippel, and J.P. Bassin), 75–142. Cham: Springer.

Bazin, I., Seo, H.B., Suehs, C.M. et al. (2017). Profiling the biological effects of wastewater samples via bioluminescent bacterial biosensors combined with estrogenic assays. *Environ. Sci. Pollut. Res.* 24 (1): 33–41.

Biko, O.D.V., Viljoen-Bloom, M., and Van Zyl, W.H. (2020). Microbial lignin peroxidases: applications, production challenges and future perspectives. *Enzym. Microb. Technol.* 141: 109669.

Boran, R., Ugur, A., Sarac, N., and Ceylan, O. (2019). Characterisation of *Streptomyces violascens* OC125-8 lipase for oily wastewater treatment. *3 Biotech.* 9 (1): 5.

Syron, E. & Heffernan, B. (2017). OxyMem, the flexiable MABR. *Proceedings of the Water Environment Federation* (3): 650–656. http://doi.org/10.2175/193864717821494150.

Chaali, M., Naghdi, M., Brar, S.K., and Avalos-Ramirez, A. (2018). A review on the advances in nitrifying biofilm reactors and their removal rates in wastewater treatment. *J. Chem. Technol. Biotechnol.* 93: 3113–3124.

Champreda, V., Mhuantong, W., Lekakarn, H. et al. (2019). Designing cellulolytic enzyme systems for biorefinery: from nature to application. *J. Biosci. Bioeng.* 128 (6): 637–654.

Chaplin, B.P. (2019). The prospect of electrochemical technologies advancing worldwide water treatment. *Acc. Chem. Res.* 52 (3): 596–604.

Chen, G. (2004). Electrochemical technologies in wastewater treatment. *Sep. Purif. Technol.* 38 (1): 11–41.

Chen, M.Y., Lee, D.J., and Tay, J.H. (2007). Distribution of extracellular polymeric substances in aerobic granules. *Appl. Microbiol. Biotechnol.* 73: 1463–1469.

Chowdhary, P., Shukla, G., Raj, G. et al. (2019). Microbial manganese peroxidase: a ligninolytic enzyme and its ample opportunities in research. *SN Appl. Sci.* 1 (1): 45.

Cossio, C., Norrman, J., McConville, J. et al. (2020). Indicators for sustainability assessment of small-scale wastewater treatment plants in low and lower-middle income countries. *Environ. Sustainability Indic.* 100028.

Covei, M., Perniu, D., Bogatu, C., and Duta, A. (2019). CZTS-TiO_2 thin film heterostructures for advanced photocatalytic wastewater treatment. *Catal. Today* 321: 172–177.

Dalal, S., Pandey, R.R., and Dubey, R.C. (2012). Bacterial degradation of phenol and cyanide from industrial wastewater. In: *Bioremediation of Pollutants* (ed. D.K. Maheshwari and R.C. Dubey), 1–23. New Delhi: IK International Pub. Com.

De Kreuk, M.K., McSwain, B.S., Bathe, S. et al. (2005). Discussion outcomes. In: *Aerobic Granular Sludge*, Water and Environmental Management Series (ed. S. Bathe, M. De Kreuk, B. McSwain, and N. Schwarzenbeck), 155–169. London: IWA Publishing.

Degani, O. and Goldblat, Y. (2020). Potential role of laccases in the relationship of the maize late wilt causal agent, *Magnaporthiopsis Maydis*, and its host. *J. Fungi* 6 (2): 63.

Di Biase, A., Kowalski, M.S., Devlin, T.R., and Oleszkiewicz, J.A. (2019). Moving bed biofilm reactor technology municipal wastewater treatment: a review. *J. Environ. Manage.* 247: 849–866.

Ding, W., Cai, J., Yu, Z. et al. (2015). Fabrication of an aquaporin-based forward osmosis membrane through covalent bonding of a lipid bilayer to a microporous support. *J. Mater. Chem. A* 3 (40): 20118–20126.

Diniz, G.S., Silva, A.F., Araújo, O.Q., and Chaloub, R.M. (2017). The potential of microalgal biomass production for biotechnological purposes using wastewater resources. *J. Appl. Phycol.* 29 (2): 821–832.

Drioli, E. and Giorn, L. (2016). Encyclopedia of membranes. Springer.

Du, Z., Li, H., and Gu, T. (2007). A state of the art review on microbial fuel cells: a promising technology for wastewater treatment and bioenergy. *Biotechnol. Adv.* 25 (5): 464482.

Fan, P., Li, L., Sun, Y. et al. (2019). Selenate removal by Fe0 coupled with ferrous iron, hydrogen peroxide, sulfidation, and weak magnetic field: a comparative study. *Water Res.* 159: 375–384.

Fathizadeh, S. and Behnia, S. (2020). Control of a DNA based piezoelectric biosensor. *J. Phys. Soc.* 89 (2): 024004.

Foesel, B.U., Gieseke, A., Schwermer, C. et al. (2008). Nitrosomonas Nm143-like ammonia oxidizers and Nitrospira marina-like nitrite oxidizers dominate the nitrifier community in a marine aquaculture biofilm. *FEMS Microbiol. Ecol.* 63: 192–204.

Gálvez, A., Greenman, J., and Ieropoulos, I. (2009). Landfill leachate treatment with microbial fuel cells; scale-up through plurality. *Bioresour. Technol.* 100 (21): 5085–5091.

Garrido-Cardenas, J.A., Esteban-García, B., Agüera, A. et al. (2020). Wastewater treatment by advanced oxidation process and their worldwide research trends. *Int. J. Environ. Res. Public Health* 17 (1): 170.

Ghangrekar, M.M., Das, S., and Tiwari, B.R. (2020). Integration of bioelectrochemical systems with other existing wastewater treatment processes. In: *Integrated Microbial Fuel Cells for Wastewater Treatment* (ed. R. Abbassi, A.K. Yadav, F. Khan, and V. Garaniya), 229–248. Butterworth-Heinemann.

Ghernaout, D. and Elboughdiri, N. (2020). Antibiotics resistance in water mediums: background, facts, and trends. *Appl. Therm. Eng.* 4 (1): 1.

Ghernaout, D., Ghernaout, B., and Naceur, M.W. (2011a). Embodying the chemical water treatment in the green chemistry – a review. *Desalination* 271: 1–10.

Ghernaout, D., Naceur, M.W., and Aouabed, A. (2011b). On the dependence of chlorine by-products generated species formation of the electrode material and applied charge during electrochemical water treatment. *Desalination* 270 (1–3): 9–22.

Ghernaout, D., Elboughdiri, N., and Ghareba, S. (2020). Fenton Technology for Wastewater Treatment: dares and trends. *Open Access Libr.* 7: 1–26.

Giwa, A., Hasan, S.W., Yousuf, A. et al. (2017). Biomimetic membranes: a critical review of recent progress. *Desalination* 420: 403–424.

Goy, R.C., Morais, S.T.B., and Assis, O.B.G. (2016). Evaluation of the antimicrobial activity of chitosan and its quaternized derivative on *E. coli* and *S. aureus* growth. *Rev. Bras* 26 (1): 122–127.

Gulkowska, A., Leung, H.W., So, M.K. et al. (2008). Removal of antibiotics from wastewater by sewage treatment facilities in Hong Kong and Shenzhen. *China Water Res.* 42: 395–403.

He, L., Du, P., Chen, Y. et al. (2017). Advances in microbial fuel cells for wastewater treatment. *Renew. Sust. Energ. Rev.* 71: 388–403.

He, Q., Chen, L., Zhang, S. et al. (2019). Hydrodynamic shear force shaped the microbial community and function in the aerobic granular sequencing batch reactors for low carbon to nitrogen (C/N) municipal wastewater treatment. *Bioresour. Technol.* 271: 48–58.

Hernández-Chover, V., Bellver-Domingo, Á., and Hernández-Sancho, F. (2018). Efficiency of wastewater treatment facilities: the influence of scale economies. *J. Environ. Manage.* 228: 77–84.

Howden, A.J.M. and Preston, G.M. (2009). Nitrilase enzymes and their role in plant–microbe interactions. *Microb. Biotechnol.* 2 (4): 441–451.

Iqbal, M., Ahmad, M.Z., Bhatti, I.A. et al. (2015). Cytotoxicity reduction of wastewater treated by advanced oxidation process. *Int. J. Chem.* 1 (1): 53–59.

Johnson, I., Ali, M.A.S., and Kumar, M. (2019). Cyanobacteria/microalgae for distillery wastewater treatment-past, present and the future. In: *Microbial Wastewater Treatment* (ed. M.P. Shah and S. Rodriguez-Couto), 195–236. Elsevier.

Jouanneau, S., Durand, M.J., and Thouand, G. (2012). Online detection of metals in environmental samples: comparing two concepts of bioluminescent bacterial biosensors. *Environ. Sci. Technol.* 46 (21): 11979–11987.

Kalia, S. and Averous, L. (2011). *Biopolymers: Biomedical and Environmental Applications*, vol. 70. John Wiley & Sons.

Kaplan, A., Mamane, H., Lester, Y., and Avisar, D. (2020). Trace organic compound removal from wastewater reverse-osmosis concentrate by advanced oxidation processes with $UV/O_3/H_2O_2$. *Materials* 13 (12): 2785.

Karami, P., Khorshidi, B., McGregor, M. et al. (2020). Thermally stable thin-film composite polymeric membranes for water treatment: a review. *J. Clean. Prod.* 250: 119447.

Kartal, B., Rattray, J., van Niftrik, L.A. et al. (2007). Candidatus "Anammoxoglobus propionicus" a new propionate oxidizing species of anaerobic ammonium oxidizing bacteria. *Syst. Appl. Microbiol.* 30: 39–49.

Kasemset, S., Lee, A., Miller, D.J. et al. (2013). Effect of polydopamine deposition conditions on fouling resistance, physical properties, and permeation properties of reverse osmosis membranes in oil/water separation. *J. Membr. Sci.* 425-426: 208–216.

Khalid, Z.B., Nasrullah, M., Nayeem, A. et al. (2020). Application of 2D graphene-based nanomaterials for pollutant removal from advanced water and wastewater treatment processes. In: *Adapting 2D Nanomaterials for Advanced Applications*, 191. 217: American Chemical Society.

Kim J R, Rodríguez J, Hawkes F R, Dinsdale R M, Guwy A J, Premier G C (2011) Increasing power recovery and organic removal efficiency using extended longitudinal tubular microbial fuel cell (MFC) reactors. Lakhveer Singh & Durga Madhab Mahapatra *Energy Environ. Sci.* 4(2):459–465

Korostynska, O., Mason, A., and Al-Shamma'a, A.I. (2013). Monitoring pollutants in wastewater: traditional lab based versus modern real-time approaches. In: *Smart Sensors for Real-Time Water Quality Monitoring*, Smart Sensors, Measurement and Instrumentation, vol. 4 (ed. S. Mukhopadhyay and A. Mason), 1–24. Berlin, Heidelberg: Springer.

Kumar, A. and Pan, S.Y. (2020). *Opportunities and Challenges for Renewable Energy Integrated Water-Energy Nexus Technologies.* Water-Energy Nexus.

Kwiatek, J.M. and Carman, G.M. (2020). Yeast phosphatidic acid phosphatase Pah1 hops and scoots along the membrane phospholipid bilayer. *J. Lipid Res.* 61 (8): 1232–1243.

Leyva-Díaz, J.C., Calderón, K., Rodríguez, F.A. et al. (2013). Comparative kinetic study between moving bed biofilm reactor–membrane bioreactor and membrane bioreactor systems and their influence on organic matter and nutrients removal. *Biochem. Eng. J.* 77: 28–40.

Leyva-Díaz, J.C., Monteoliva, A., Martín-Pascual, J. et al. (2019). Moving bed biofilm reactor as an alternative wastewater treatment process for nutrient removal and recovery in the circular economy model. *Bioresour. Technol.* 299: 122631.

Liang, Z., Yun, Y., Wang, M. et al. (2019). Performance evaluation of interfacial polymerisation-fabricated aquaporin-based biomimetic membranes in forward osmosis. *RSC Adv.* 9 (19): 10715–10726.

Liu, B., Wang, D., Yu, G., and Meng, X. (2013). Adsorption of heavy metal ions, dyes and proteins by chitosan composites and derivatives – a review. *J. Ocean Univ. China* 12 (3): 500–508.

Liu, G., Li, X., Ma, L. et al. (2019a). Enhancement of excess sludge hydrolysis and decomposition by combined lysozyme and rhamnolipid pretreatment. *Bioresour. Technol.* 289: 121703.

Liu, Y., Ngo, H.H., Guo, W. et al. (2019b). The roles of free ammonia (FA) in biological wastewater treatment processes: a review. *Environ. Int.* 123: 10–19.

Lueder, U., Jørgensen, B.B., Kappler, A., and Schmidt, C. (2019). Fe (III) photoreduction producing Feaq2+ in oxic freshwater sediment. *Environ. Sci. Technol.* 54 (2): 862–869.

Luo, W., Xie, M., Song, X. et al. (2017). Biomimetic aquaporin membranes for osmotic membrane bioreactors: membrane performance and contaminant removal. *Bioresour. Technol.* 24: 62–68.

Lv, Y., Pan, J., Huo, T. et al. (2020). Enhance the treatment of low strength wastewater at low temperature with the coexistence system of AnAOB and heterotrophic bacteria: performance and bacterial community. *Sci. Total Environ.* 714: 136799.

Maharajaa, P., Priyadharshinib, S.V., Akshayab, R. et al. (2019). Removal of glycosaminoglycans present in tannery saline soak wastewater using integrated biological reactor and amylase immobilised reactor. *Desalin. Water Treat.* 156: 189–203.

Mahgoub, S. and Samaras, P. (2014). Nanoparticles from biowastes and microbes: focus on role in water purification and food preservation. *2nd International Conference on Sustainable Solid Waste Management.*

Mahvi, A.H., Ebrahimi, S.J., Mesdaghinia, A. et al. (2011). Performance evaluation of a continuous bipolar electrocoagulation/electrooxidation- electrofluotation (ECEO-EF) reactor designed for simultaneous removal of ammonia and phosphate from wastewater effluent. *J. Hazard. Mater.* 192: 1267–1274.

Maixner, F., Noguera, D.R., Anneser, B. et al. (2006). Nitrite concentration influences the population structure of Nitrospira-like bacteria. *Environ. Microbiol.* 8: 1487–1495.

Martin-Betancor, K., Rodea-Palomares, I., Munoz-Martin, M.A. et al. (2015). Construction of a self-luminescent cyanobacterial bioreporter that detects a broad range of bioavailable heavy metals in aquatic environments. *Front. Microbiol.* 6: 186.

Martinetti, C.R., Childress, A.E., and Cath, T.Y. (2009). High recovery of concentrated RO brines using forward osmosis and membrane distillation. *J. Membr. Sci.* 331: 31–39.

Maurel, C., Boursiac, Y., Luu, D.T. et al. (2015). Aquaporins in plants. *Physiol. Rev.* 95: 1321–1358.

Meng, F., Yang, A., Zhang, G. et al. (2019). Effects of dissolved oxygen on key enzyme activities during photosynthetic bacteria wastewater treatment. *Process Biochem.* 76: 165–170.

Mi, B. and Elimelech, M. (2010). Organic fouling of forward osmosis membranes: fouling reversibility and cleaning without chemical reagents. *J. Membr. Sci.* 348: 337–345.

Mickova, I.L. (2015a). Advanced electrochemical technologies in wastewater treatment part I: electrocoagulation. *Am. Sci. Res. J. Eng. Technol. Sci.* 14 (2): 233–257.

Mickova, I.L. (2015b). Advanced electrochemical technologies in wastewater treatment. Part II: electro-flocculation and electro-flotation. *Am. Sci. Res. J. Eng. Technol. Sci.* 14 (2): 273–294.

Morató, J., Tollin, N., Jiménez, L. et al. (2017). *Situación y Evolución de la Economía Circular en España.* Madrid, Spain: Fundación COTEC para la Innovación.

Munier, N. (2005). *Introduction to Sustainability.* Dordrecht, The Netherlands: Springer.

Munoz-Cupa, C., Hu, Y., Xu, C.C., and Bassi, A. (2020). An overview of microbial fuel cell usage in wastewater treatment, resource recovery and energy production. *Sci. Total Environ.* 754: 142429.

Nambikkattu, J., Kaleekkal, N.J., and Jacob, J.P. (2020). Metal ferrite incorporated polysulfone thin-film nanocomposite membranes for wastewater treatment. *Environ. Sci. Pollut. Res.* 28: 1–13.

Nancharaiah, Y.V. and Reddy, G.K.K. (2018). Aerobic granular sludge technology: mechanisms of granulation and biotechnological applications. *Bioresour. Technol.* 247: 1128–1143.

Nguyen, N.C., Chen, S.-S., Nguyen, H.T. et al. (2016). Innovative sponge-based moving bed–osmotic membrane bioreactor hybrid system using a new class of draw solution for municipal wastewater treatment. *Water Res.* 91: 305–313.

Ødegaard, H. (2017). New applications for MBBR and IFAS systems. *Front. Wastewater Treat. Model. Lect. Notes Civ. Eng.* 4: 499–507.

Olaniran, A.O., Hiralal, L., and Pillay, B. (2011). Whole-cell bacterial biosensors for rapid and effective monitoring of heavy metals and inorganic pollutants in wastewater. *J. Environ. Monit.* 13 (10): 2914–2920.

Olivera, S., Muralidhara, H.B., Venkatesh, K. et al. (2016). Potential applications of cellulose and chitosan nanoparticles/composites in wastewater treatment: a review. *Carbohydr. Polym.* 153: 600–618.

Osińska, A., Korzeniewska, E., Harnisz, M. et al. (2020). Small-scale wastewater treatment plants as a source of the dissemination of antibiotic resistance genes in the aquatic environment. *J. Hazard. Mater.* 381: 121221.

Padrón-Páez, J.I., Almaraz, S.D.L., and Román-Martínez, A. (2020). Sustainable wastewater treatment plants design through multiobjective optimization. *Comput. Chem. Eng.* 140: 106850.

Pandey, K., Singh, B., Pandey, A.K. et al. (2017). Application of microbial enzymes in industrial waste water treatment. *Int. J. Curr. Microbiol. App. Sci.* 6 (8): 1243–1254.

Pang, Y.L. and Abdullah, A.Z. (2013). Fe^{3+} doped TiO_2 nanotubes for combined adsorption sonocatalytic degradation of real textile wastewater. *Appl. Catal. B* 129: 473–481.

Patel, P.S., Bandre, N., Saraf, A., and Ruparelia, J.P. (2013). Electro-catalytic materials (electrode materials) in electrochemical wastewater treatment. *Procedia Eng.* 51: 430–435.

Perera, U.M.S.P. and Rajapakse, N. (2014). Chitosan nanoparticles: preparation, characterization, and applications. In: *Seafood Processing by-Products*, 371–387. New York: Springer.

Phuong Thu, L. and Michèle, B. (2016). Carbon and nitrogen removal from glucose-glycine melanoidins solution as a model of distillery wastewater by catalytic wet air oxidation. *Hazard. Mater.* 310: 108–116.

Raja, C.E. and Selvam, G.S. (2011). Construction of green fluorescent protein based bacterial biosensor for heavy metal remediation. *Int. J. Environ. Sci. Technol.* 8: 793–798.

Rajasulochana, P. and Preethy, V. (2016). Comparison on efficiency of various techniques in treatment of waste and sewage water–a comprehensive review. *Resour. Eff. Technol.* 2 (4): 175–184.

Ravikumar, S., Ganesh, I., Yoo, I.K., and Hong, S.H. (2012). Construction of a bacterial biosensor for zinc and copper and its application to the development of multifunctional heavy metal adsorption bacteria. *Process Biochem.* 47 (5): 758–765.

Reddy, D.H.K. and Lee, S. (2013). Application of magnetic chitosan composites for the removal of toxic metal and dyes from aqueous solutions. *Adv. Colloid Interf. Sci.* 201202: 6893.

Rittmann, B.E. (2010). Environmental biotechnology in water and wastewater treatment. *Environ. Eng. Sci.* 136 (4): 348–353.

Saber, R. and Pişkin, E. (2003). Investigation of complexation of immobilized metallothionein with Zn (II) and cd (II) ions using piezoelectric crystals. *Biosens. Bioelectron.* 18 (8): 1039–1046.

Sahu, J.N., Karri, R.R., Zabed, H.M. et al. (2019). Current perspectives and future prospects of nano-biotechnology in wastewater treatment. *Sep. Purif. Rev.* 50: 1–20.

Salguero-Puerta, L., Leyva-Díaz, J.C., Cortés-García, F.J., and Molina-Moreno, V. (2019). Sustainability indicators concerning waste management for implementation of the circular economy model on the University of Lome (Togo) campus. *Int. J. Environ. Res. Public Health* 16 (12): 2234.

Salvador, J., Kopper, K., Miti, A. et al. (2020). Multiplexed Immunosensor based on the Amperometric transduction for monitoring of marine pollutants in sea water. *Sensors* 20 (19): 5532.

Sharma, Y.C., Srivastava, V., Singh, V.K. et al. (2009). Nano-adsorbents for the removal of metallic pollutants from water and wastewater. *Environ. Technol.* 30 (6): 583–609.

Singh, M., Friesen, C., and Singh, V. (2019). Lactase deficiency and colonic mucosal eosinophilia (P19-014-19). *Curr. Develop. Nutr.* 3 (Supplement_1): nzz049-P19.

Singh, S., Wasewar, K.L., and Kansal, S.K. (2020). Low-cost adsorbents for removal of inorganic impurities from wastewater. In: *Inorganic Pollutants in Water* (ed. P. Devi, P. Singh, and S.K. Kansal), 173–203. Elsevier.

Sonawane, J.M., Ezugwu, C.I., and Ghosh, P.C. (2020). Microbial fuel cell-based biological oxygen demand sensors for monitoring wastewater: state-of-the-art and practical applications. *ACS Sens.* 5 (8): 2297–2316.

Song, X., Wang, L., Tang, C.Y. et al. (2015). Fabrication of carbon nanotubes incorporated double-skinned thin film nanocomposite membranes for enhanced separation performance and antifouling capability in forward osmosis process. *Desalination* 369: 1–9.

Stillwell, A.S. and Webber, M.E. (2013). Novel methodology for evaluating economic feasibility of low-water cooling technology retrofits at power plants. *Water Policy* 15 (2): 292–308.

Strokach, P.P. (1975). The prospects of using anodic dissolution of metal for water purification. *Electrochem. Ind. Process. Biol.* 4: 55.

Sukul, P. and Spiteller, M. (2007). Fluoroquinolone antibiotics in the environment. *Rev. Environ. Contam. Toxicol.* 191: 131–162.

Swain, S.S., Paidesetty, S.K., Padhy, R.N., and Hussain, T. (2020). Isoniazid–phytochemical conjugation: a new approach for potent and less toxic anti-TB drug development. *Chem. Biol. Drug Des.* 96 (2): 714–730.

Thakur, A.K. and Kumar, M. (2020). Reappraisal of permeable reactive barrier as a sustainable groundwater remediation technology. In: *Contaminants in Drinking and Wastewater Sources* (ed. M. Kumar, D. Snow, R. Honda, and S. Mukherjee), 179–207. Singapore: Springer.

Theron, J., Eugene Cloete, T., and de Kwaadsteniet, M. (2010). Current molecular and emerging nanobiotechnology approaches for the detection of microbial pathogens. *Crit. Rev. Microbiol.* 36 (4): 318–339.

Tian, J., Wang, P., Huang, L. et al. (2013). Improving the thermostability of methyl parathion hydrolase from Ochrobactrum sp. M231 using a computationally aided method. *Appl. Microbiol. Biotechnol.* 97 (7): 2997–3006.

Tschmelak, J., Proll, G., and Gauglitz, G. (2005). Optical biosensor for pharmaceuticals, antibiotics, hormones, endocrine disrupting chemicals and pesticides in water: assay optimization process for estrone as example. *Talanta* 65 (2): 313–323.

Unuofin, J.O. (2020). Garbage in garbage out: the contribution of our industrial advancement to wastewater degeneration. *Environ. Sci. Pollut. Res. Int.* 27 (18): 22319–22335.

USEPA (2003). Long term 2 enhanced surface water treatment rule toolbox guidance manual (DRAFT). EPA 815-D-03-009

Vakili, M., Rafatullah, M., Salamatinia, B. et al. (2014). Application of chitosan and its derivatives as adsorbents for dye removal from water and wastewater: a review. *Carbohydr. Polym.* 113: 115–130.

Varga, B., Somogyi, V., Meiczinger, M. et al. (2019). Enzymatic treatment and subsequent toxicity of organic micropollutants using oxidoreductases-a review. *J. Clean. Prod.* 221: 306–322.

Verma, N., Kumar, V., and Bansal, M.C. (2020). Valorization of waste biomass in fermentative production of Cellulases: a review. *Waste Biomass Valorization* 12: 613–640.

Vismara, E., Melone, L., Gastaldi, G. et al. (2009). Surface functionalization of cotton cellulose with glycidyl methacrylate and its application for the adsorption of aromatic pollutants from wastewaters. *J. Hazard. Mater.* 170 (2–3): 798–808.

Voigt Alexander, M., Skutlarek, D., Timm, C. et al. (2019). Liquid chromatography-tandem mass spectrometry as a fast and simple method for the determination of several antibiotics in different aqueous matrices. *Environ. Chem.* 17 (1): 54–74.

Wang, N., Zheng, T., Zhang, G., and Wang, P. (2016). A review on Fenton-like processes for organic wastewater treatment. *J. Environ. Chem. Eng.* 4: 762–787.

Wang, J., Chu, L., Wojnárovits, L., and Takács, E. (2020). Occurrence and fate of antibiotics, antibiotic resistant genes (ARGs) and antibiotic resistant bacteria (ARB) in municipal wastewater treatment plants: an overview. *Sci. Total Environ.* 744: 140997.

Warsinger, D.M., Chakraborty, S., Tow, E.W. et al. (2018). A review of polymeric membranes and processes for potable water reuse. *Prog. Polym. Sci.* 81: 209–237.

Watlington, K. (2005). *Emerging Nanotechnologies for Site Remediation and Wastewater Treatment*. Washington, DC: Environmental Protection Agency.

Wert, E., Rosarioortiz, F., Drury, D., and Snyder, S. (2007). Formation of oxidation byproducts from ozonation of wastewater. *Water Res.* 41 (7): 1481–1490.

Xie, J., Zhao, Y., Yu, L. et al. (2020). Molecular karyotyping of Siberian wild rye (Elymus sibiricus L.) with oligonucleotide fluorescence in situ hybridization (FISH) probes. *PLoS One* 15 (1): e0227208.

Yadav, P.K., Singh, V.K., Yadav, S. et al. (2009). In silico analysis of pectin lyase and pectinase sequences. *Biochem. Mosc.* 74 (9): 1049–1055.

Yalcinkaya, F., Boyraz, E., Maryska, J., and Kucerova, K. (2020). A review on membrane technology and chemical surface modification for the oily wastewater treatment. *Materials* 13 (2): 493.

Yang, J., Li, Z., Lu, L. et al. (2020a). Model-based evaluation of algal-bacterial systems for sewage treatment. *J. Water Process Eng.* 38: 101568.

Yang, Y., Li, M., Hu, Z. et al. (2020b). Deep insights into the green nitrogen removal by anammox in four full-scale WWTPs treating landfill leachate based on 16S rRNA gene and transcripts by 16S rRNA high-throughput sequencing. *J. Clean. Prod.* 276 (3): 124176.

Yang, S. et al. (2020c). Substrate-Photocaged enzymatic Fluorogenic probe enabling sequential activation for light-controllable monitoring of intracellular Tyrosinase activity. *Anal. Chem.* 92 (10): 7194–7199.

Zhang, P., Sun, D., Cho, A. et al. (2019). Modified carbon nitride nanozyme as bifunctional glucose oxidase-peroxidase for metal-free bioinspired cascade photocatalysis. *Nat. Commun.* 10 (1): 1–14.

Zhou, P., Brown, G.M., and Gu, B. (2006). Membrane and other treatment technologies – pros and cons. In: *Perchlorate*, 389–404. Boston, MA: Springer.

Zhou, T., Han, H., Liu, P. et al. (2017). Microbial fuels cell-based biosensor for toxicity detection: a review. *Sensors* 17 (10): 2230.

Zou, W., Schulz, B.L., Tan, X. et al. (2019). The role of thermostable proteinaceous α-amylase inhibitors in slowing starch digestion in pasta. *Food Hydrocoll.* 90: 241–247.

7

Role of Free-Floating Macrophytes in the Abatement of Disturbed Environments

Wajiha Anum[1], *Umair Riaz*[2], *Ghulam Murtaza*[3], *Syed Ali Zulqadar*[4], *Laila Shahzad*[5]

[1] *Department of Agronomy, Regional Agricultural Research Institute, Bahawalpur, Pakistan*
[2] *Department of Soil & Environmental Sciences, MNS University of Agriculture, Multan*
[3] *Institute of Soil & Environmental Sciences, University of Agriculture, Faisalabad*
[4] *Soil and Water Testing Laboratory for Research, Bahawalpur-63100, Pakistan*
[5] *Sustainable Development Study Center GC University, Lahore, Pakistan*

7.1 Introduction

Macrophytes are aquatic plants that grow on water or near water bodies. Macrophytes act as one of the significant biological components in freshwater ecosystems (Villa et al. 2017). They are divided into four categories as emerged, submerged, free-floating, and floating left, based on their adaptations (Rejmankova 2011). Emerged macrophytes are mostly seen in shallow, stagnant water bodies, and along with the upper coastal areas at a depth of 1–1.5 m (Rejmankova 2011). As adapted to anaerobic conditions, their rhizomes and root structures thrive below the aquatic sediments (Hasan and Rina 2009) while they keep their aerial photosynthetic structure above the water (Ali et al. 2019). They are further divided into two groups as erect emergent (e.g. *Phragmites* and *Typha*) and creeping-emergent like *Hydrocotyle* spp., *Ludwigia* spp., *Myriophyllum aquaticum*, *Alternanthera phyloxeroides*, and *Nasturtium aquaticum* (Rejmankova 2011).

7.2 Nutrient Equilibrium

Macrophytes are well known for nutrient cycling by transferring them from sediment to water either by active or passive means (Carignan and Kalff 1980; Esteves and Camargo 1986; Camargo et al. 2003). They use two ways for nutrient cycling: first, they retain the solids and nutrients in their roots and leaves (Meerhoff et al. 2003; Pott and Pott 2003), and second, they reduce the nutrient flow by the motion of waves or wind (Madsen et al. 2001).

They also provide algae and bacteria (free-living or attached to macrophytes) with nutrients like nitrogen and phosphorus along with organic carbon (Burkholder and Wetzel 1989; Anesio et al. 2003; Stets and Cotner 2008). Many macrophytes also produce refractory fibrous material, which decomposes slowly (Bianchini Jr 2003), ultimately returning carbon amounts to the sediments. All of these factors result in lesser erosion and shore stabilization. The physicochemical properties of the water column are highly influenced by macrophytes (Thomaz and Cunha 2010).

7.3 Importance of Free-Floating Macrophytes in Ecosystem Structure and Function

Certain free-floating macrophytes [*Solms*, *Pistia stratiotes* L., *Eichhornia crassipes* (Mart.) and *Salvinia molesta* (Mitchell)], considered the most important aquatic weeds (Henry-Silva et al. 2008), are native to tropical America and have pantropical distribution (Cook 1989). They are abundantly present in Brazil both in polluted and nonpolluted aquatic ecosystems (Thomaz and Bini 1998; Henry-Silva and Camargo 2005). Moreover, they are categorized into small-, medium-, and large-sized taxa like *Azolla* and *lemna*, *salvinia* spp, *E. crassipes*, and *P. stratiotes*, respectively (Meerhoff 2006). Free-floating macrophytes include plants with highly reduced morphology (Rejmankova 2011).

Aquatic Contamination: Tolerance and Bioremediation, First Edition. Edited by Rouf Ahmad Bhat, Gowhar Hamid Dar, Fernanda Maria Policarpo Tonelli, and Saima Hamid.

Free-floating species are naturally occurring in their native habitat (e.g. neotropical river-floodplain system). In natural areas, they do not cause problems because of their natural adaptations to the area. As in a few other cases, factors like water-level fluctuations and damages are caused by native fungal species, and pests take control over their populations. However, in other non-native areas, the lack of such factors contributes to their success and abundant populations (Thomaz et al. 2009).

7.4 How Toxins are Added to the Environment

With the advancement in industrialization, a significant concern is the presence of highly toxic material in waste. These toxins are thrown into the water bodies intentionally, and a sufficient part is also added to the waterways, air, and terrestrial environment in one or the other. Inorganic fertilizers are applied to field crops, fruits, and vegetables to ensure adequate yields. However, certain factors like the oversupply of phosphate fertilizers also disturb the levels of Pb, Zn, Ni, Cr, and Cd (Nagajyoti et al. 2010; Uddin 2017).

Mismanagement of agricultural lands by overapplication of fertilizers (heavy metals) and pesticides is becoming a hazardous threat to the environment (Sardar et al. 2013; Czarnecki and Düring 2015). High application rates of inorganic fertilizers also enhance the concentrations of lead, mercury, iron, and cadmium (Ali et al. 2020). Transportation is another means of increasing heavy metal concentration in the environment. Maintenance work generates contaminants in groundwater. Moreover, brake abrasion, tread wear, and corrosion are significant sources of heavy metal contamination, thus affecting the environment (Sanchez-Martin et al. 2000). Among them, water bodies are profoundly affected by the release of wastewater from a different point of origin. The water bodies have high amounts of heavy metals in them, thus altering the natural ecosystems.

7.5 Role of Aquatic Plants in Water Bodies

Aquatic plants are known to efficiently remove heavy metals (Dhir et al. 2009). Various studies suggest that the living or nonliving biomass of macrophytes plays an important role in the removal and monitoring of heavy metals (Lee et al. 1998; Schneider and Rubio 1999; Hu et al. 2003; Kamal et al. 2004). All aquatic macrophytes (free-floating, submerged, and emergent) can impound heavy metals (Maine et al. 2001).

A complicated relationship is present between the free-floating plants with aquatic communities with little evidence being present. However, their presence also influences fish behavior differently (Meerhoff et al. 2003; Abdel-Tawwab 2006). They also decrease the phytoplankton biomass (Bicudo et al. 2007) and zooplankton abundance (Meerhoff et al. 2003; Abdel-Tawwab 2006) while increasing heterotrophic and mixotrophic microbial plankton components (Sinistro et al. 2006).

7.5.1 Mode of Action-Free-Floating Plants

Free-floating plants take metals from their roots (Sen et al. 1987; Maine et al. 2004) and translocate them to other plant parts. A passive process (adsorption) takes place when plant parts are in direct contact with the medium resulting in metal accumulation in aerial parts of plants (Maine et al. 2004). The most advantageous parameter of aquatic plants is their higher above-ground biomass having the ability to accumulate higher amounts of metals (Vymazal 2016). Roots act as initial sites of uptake and the concentrations are higher in roots as compared to the aerial parts (Rezania et al. 2016). It is because they protect the photosynthetic apparatus in the above aerial parts of plants (Drążkiewicz and Baszyński 2005). There is evidence that free-floating plants like *E. crassipes*, *S. molesta*, and *P. stratiotes* have a higher ability of phosphorus absorption as compared to nitrogen (de Tezanos Pinto and O'Farrell 2014). In an experiment, free-floating plants in tanks removed total P (80%) and phosphate (74%), while less total N (44%), ammonia (55%), and nitrate (55%) (Henry-Silva et al. 2008). Various studies demonstrated the role of free-floating aquatic plants in their ability to accumulate heavy metals (Table 7.1).

Table 7.1 Free-floating macrophytes for accumulating various heavy metals.

Plant species	Metals	References
Eicbbornia crassipes	Cr, Cu, Zn, Cd, Pt	Delgado et al. (1993), Farago and Parsons (1994), Low et al. (1994)
Lemna trisulca	Cd	Prasad et al. (2001)
Pistia stratiotes	Zn	Odjegba and Fasidi (2004)
Lemna gibba	As, Uranium	Mkandawire et al. (2004), Mkandawire and Dudel (2005)
Lemna polyrhiza	Pb, Zn, Ni	Sharma and Gaur (1995)
Lemna minor	Pb, Ni, Cu, Cr	Axtell et al. (2003), Kara (2004), Ater et al. (2006)
Spirodela polyrhiza	Cr, As	Appenroth et al. (2000), Rahman et al. (2007)
Wolffia globosa	Cd, Cr	Boonyapookana et al. (2002)
Azolla filiculoides	Pb	Benaroya et al. (2004)
Azolla pinnata	Zn, Cr	Jain et al. (1990), Noraho and Gaur (1996), Arora et al. (2006)
Salvinia minima	Cr, Ni, Cd, As	Srivastav et al. (1994), Olguín et al. (2002), Hoffmann et al. (2004)
Spirodela polyrhiza	As	Rahman et al. (2007)
Salvinia herzogii	Cr	Maine et al. (2004)
Pistia stratiotes	Cr	Maine et al. (2004)
Eichhornia, Lemna, Spirodela, Azolla, Hydrilla	As	Rahman and Hasegawa (2011)

7.6 Phytoremediation

Bioremediation is being used widely in the world. This practice makes use of naturally occurring organisms to control/break down hazardous substances into less toxic or nontoxic substances. The widely accepted techniques for the purification of hazardous substances are bioaccumulation, biosorption, and phytoremediation (Vijayaraghavan and Balasubramanian 2015). In phytoremediation, green plants take/absorb/accumulate the pollutants mostly through their roots and translocate them into the other plant's parts (Sharma et al. 2015).

Phytoremediation has gained popularity among researchers because of its characteristics like fast growth, phytoplankton reduction, high nutrient, and metal accumulation, and a reduction in microbial growth. All of these factors are in favor as it requires little input and provides multiple benefits. In this regard, *Spirodela* and *Lemna* species are more evaluated for reclaiming municipal, industrial, and metal-contaminated wastewater. These species also fulfill biological oxygen demand, total nitrogen and phosphorus, dissolved salts, total suspended salts, and electrical conductivity (Ali et al. 2016).

Various types of pollutants like metals, persistent organic pollutants, and pesticides may be removed through sludges, contaminated soils, water, and sediments (Bhatia and Goyal 2014; Bauddh et al. 2015). Phytoremediation makes use of various species of plants for cleaning soil and water (Sharma et al. 2015).

7.6.1 Duckweeds

Similarly, duckweeds are tiny plants that come under the categorization of free-floating macrophytes commonly found in small moving water bodies. Duckweeds comprise five genera, and almost all of them are used in phytoremediation studies for municipal and industrial water. Because of their fast growth, ability to reduce microbial and phytoplankton populations and high ability for nutrient and heavy metal accumulation. Duckweeds are whole plants measuring only about 1 mm present in the genus *Wolfia* (Sculthorpe 1967). They do not have embedded roots in the sediments and absorb nutrients entirely from the water with their leaves and reproductive organs being aerial or floating on the water surface (Rejmankova 2011). Similarly, duckweeds also have a phenomenon to counter the toxic effects of heavy metals when metal concentration increases above their tolerance levels. In such cases, they produce antioxidant enzymes, organic acid synthesis, and production of proline. The tolerance level is dependent upon several factors like water ph, electrical conductivity, temperature, and metal concentration (Ali et al. 2016). However, the issue that prevails is related to the safe disposal of metal-contaminated duckweeds because if the accurate method is not adopted, it may further pollute the environment. A few methods suggested are anaerobic digestion, incineration, hydrolyzation, and carbonization are the need of the hour. Duckweeds, when incorporated at point pollution sources and followed through safe disposal, can be a promising solution for maintaining the natural ecosystem (Ali et al. 2016).

7.6.2 Water Hyacinth

Water hyacinth (*E. crassipes*) is another free-floating macrophyte, native to South America that belongs to the family *Pontederiaceae* (Martins et al. 2002). It efficiently tolerates polluted environments. It also extracts metal nutrients like Pb, Cd, and Hg. Water hyacinth is known for its property for absorbing elements like Hg; on exposure, plant growth is not affected to a greater extent. At higher concentrations, the plants undergo chlorosis without affecting their growth (de Souza and Silva 2019). Research has shown that for elements such as Hg, it has excellent absorption properties, with only the characteristic of chlorosis in the plant when exposed, without drastically affecting its growth over a long period with a certain quantity of mercury, counting on this attribute even in the remediation of water with high levels of K and NO^{3-} (Caldelas et al. 2009). The relationships between biotic components and their interactions with the physicochemical environment may establish positive feedbacks that autostabilize the ecosystem under free-floating plants (FFP) dominance (O'Farrell et al. 2009). Many studies reveal that free-floating macrophytes influence nutrient dynamics (Camargo and Esteves 1995). Scheffer et al. (2003) state that water bodies dominated by free-floating plants are capable of promoting a self-stabilizing state.

7.6.3 Free-Floating Macrophytes and Its Their Relationship with Other Aquatic Ecosystems

The most important influence of free-floating macrophytes is on phytoplankton biomass (de Tezanos Pinto and O'Farrell 2014) as well as diversity because of their dominance in competition for light (Roijackers et al. 2004; de Tezanos Pinto et al. 2007; O'Farrell et al. 2009). The lowest light-adapted phytoplankton group is cyanobacteria, and the second most efficient are diatoms (Schwaderer et al. 2011). Mixotrophs, and similarly, cryptophytes, mixotrophs, and euglenoids, also maintain cells at low lights (Jones 2000). Complete cover of duckweed characterizes phytoplankton communities with low volume, small size, elongated organism, and high surface-to-volume ratios (O'Farrell et al. 2007); such traits help in efficient survival under low light conditions. Under the dominance of free-floating plants, a lower biomass of periphyton, picoplankton, and submerged plants is observed (Portielje and Roijackers 1995; Janes et al. 1996; Izaguirre et al. 2010, 2012) due to lesser incoming light. Fish showed a differential habitat use according to species, size class, and feeding habits. Dominant omnivore-planktivores, particularly the smallest size classes and preferred submerged plants. In contrast, omnivore-piscivores were significantly associated with free-floating plants (Meerhoff et al. 2003).

7.7 FFPs as Bioabsorbants

Nowadays macrophytes are also used as absorbents. In this technique, the dried and dead macrophytes remove the heavy metals from the contaminated water as a result of industrial activities. The practices of removing metals derived from textile, storage batteries, electroplating, mining, lead smelting, glass, and ceramic industries are most preferably done by using dead macrophytes because of their lower cost in wastewater treatment (Miretzky et al. 2004, 2006). For example, research revealed that dead biomass of *Spirodela polyrhiza* effectively absorbs cadmium, nickel, and lead at 44.90, 13.96, and 6.15 mg/g respectively, based on Langmuir coefficients under diluted metal conditions (Rai and Tripathi 2007).

7.7.1 FFPs as Decomposers

Free-floating macrophytes also play a crucial role in decomposition and carbon sequestration in wetland conditions where nutrient input is low. According to Kim and Rejmánková (2005), a faster decomposition rate is observed for submersed, freely floating, and floating leaved macrophytes as compared to tall emergent. And it is because of their soft tissues (Rejmankova 2011).

7.7.2 Climate Change and its Influence on FFP

Climate change is impacting, altering, and in the long run, shifting the entire ecosystem. In this regard, carbon dioxide (CO_2) levels are also elevated. Aquatic macrophytes survive on CO_2 and HCO_3. Free-floating macrophytes utilize the inorganic form of CO_2 mainly from the air while submerged species use carbon from sediments, water, air, and HCO_3 (Hossain et al. 2017). The rise in CO_2 will increase photosynthesis that will lead to intensification in the growth of aquatic plants that mainly use CO_2 long form. However, Bhupinder Dhir (2015) claimed that the survival of other species might be influenced negatively. Industrial activities give rise to the introduction of heavy metals into the environment, posing a threat to terrestrial as well as aquatic ecosystems. Directly it reduces the quality of water, causes harmful effects to the water fauna, and indirectly affects human health. Heavy metals persist in the food chain and cause different ailments in humans. Below the 2.0 mg Pb^{2+}/l concentration, there was no impact on the growth of aquatic macrophytes. According to the experimental data, both adsorption and bioaccumulation mechanisms are present. In the first 24 hours, a significant lead concentration reduction from the lead-doped hydroponic medium was verified, indicating that the primary process was the biosorption one. After the first day of the experiment, lead bioaccumulation was also observed and checked by the lead concentration in the leaves. From the concentration reduction of phosphorus and a simultaneous increase of lead, on the other hand, a competition between the phosphorus macronutrient and lead for plant growth was observed, when high concentrations of lead in roots are present. Assuming that the lead removal rate is proportional to both bioaccumulation capacity on the macrophytes and the lead concentration in a liquid phase, the nonstructural kinetics models have shown good results when fitting the lead uptake experimental data in all the investigated cases. The aquatic macrophyte *Salvinia biloba* Raddi has shown an excellent metal removal capacity during the long experiment. This plant can be considered a viable alternative biosorbent for use in the metal removal treatment process for industrial effluents in wetlands. Lead is probably among the most dangerous metals, and its exposure can cause serious ailments (Tello Zevallos et al. 2018) related to the central nervous system, reproductive system, hematopoietic disorders, renal disorders, and cause neurotoxic effects even at deficient levels of 20 μg/dL (Kalia and Flora 2005).

7.7.3 Problematic Free-Floating Macrophytes

Free-floating macrophytes also become noxious for the community. A few species of free-floating macrophytes like *E. crassipes*, *Azolla pinnata*, and *S. molesta* and *pistastratioites* are widespread in tropics and subtropics. Some also take place as invasive ones on a global scale, for example, *Trapanatans*, *Lythrum salicaria*, *Myriophyllum spicatum* and *Potamogeton crispus*. However, some are only problematic in some specific regions like *hydrilla verticillata* (North America) and *Elodea* (Europe) (Rejmankova 2011). When a macrophyte plant species turns its status as invasive, it has great potential to alter the food web leading to the extinction of native species and influencing biogeochemistry (Rejmankova 2011). Apart from the positive impacts, higher quantities of *E. crassipes* also alter the ecosystem set up by producing leachates (Sharma et al. 1996) that reduce the phytoplankton population. As a result, the aquatic animals are deprived of food, thus leading to ecosystem disruptions. However, submerged macrophytes have a greater impact on the physic-chemistry of water as compared to free-floating plants. A study suggested that biomass and free-floating macrophytes cover are the two factors which govern the bottom-up mechanisms in Lake Rodó. In this case, the bottom-up could have dampened because of high nutrient concentrations. The authors claim that this mechanism may be more reliable in other pristine conditions (Meerhoff et al. 2003).

References

Abdel-Tawwab, M. (2006). Effect of free-floating macrophyte, *Azolla pinnata* on water physico-chemistry, primary productivity, and the production of Nile Tilapia, *Oreochromisniloticus* (L.), and Common Carp, *Cyprinus carpio* L., in fertilized earthen ponds. *J. Appl. Aquac.* 18 (1): 21–41.

Ali, Z., Waheed, H., Kazi, A.G. et al. (2016). Duckweed: an efficient hyperaccumulator of heavy metals in water bodies. In: *Plant Metal Interaction*, 411–429. Elsevier.

Ali, F., Jilani, G., Fahim, R. et al. (2019). Functional and structural roles of wiry and sturdy rooted emerged macrophytes root functional traits in the abatement of nutrients and metals. *J. Environ. Manage.* 249: 109330.

Ali, S., Abbas, Z., Rizwan, M. et al. (2020). Application of floating aquatic plants in phytoremediation of heavy metals polluted water: a review. *Sustainability* 12 (5): 1927.

Anesio, A.M., Abreu, P.C., and Biddanda, B.A. (2003). The role of free and attached microorganisms in the decomposition of estuarine macrophyte detritus. *Estuar. Coast. Shelf Sci.* 56 (2): 197–201.

Appenroth, K.J., Bischoff, M., Gabryś, H. et al. (2000). Kinetics of chromium (V) formation and reduction in fronds of the duckweed *Spirodela polyrhiza*—a low frequency EPR study. *J. Inorg. Biochem.* 78 (3): 235–242.

Arora, A., Saxena, S., and Sharma, D.K. (2006). Tolerance and phytoaccumulation of chromium by three *Azolla* species. *World J. Microbiol. Biotechnol.* 22 (2): 97–100.

Ater, M., Ali, N., and Kasmi, H. (2006). Tolerance and accumulation of copper and chromium in two species of duckweed: *Lemna minor* L. and *Lemna gibba* L. *Rev. Sci. Eau* 19 (1): 57–67.

Axtell, N.R., Sternberg, S.P., and Claussen, K. (2003). Lead and nickel removal using *Microspora* and *Lemna minor*. *Bioresour. Technol.* 89 (1): 41–48.

Bauddh, K., Singh, K., Singh, B., and Singh, R.P. (2015). *Ricinuscommunis*: a robust plant for bio-energy and phytoremediation of toxic metals from contaminated soil. *Ecol. Eng.* 84: 640–652.

Benaroya, R.O., Tzin, V., Tel-Or, E., and Zamski, E. (2004). Lead accumulation in the aquatic fern *Azolla filiculoides*. *Plant Physiol. Biochem.* 42 (7–8): 639–645.

Bhatia, M. and Goyal, D. (2014). Analyzing remediation potential of wastewater through wetland plants: a review. *Environ. Prog. Sustain. Energy* 33 (1): 9–27.

Bianchini, I. Jr. (2003). Models of growth and decomposition of aquatic macrophytes. In: *Ecology and Management of Aquatic Macrophytes* (ed. S.M. Thomaz and L.M. Bini), 85–126. Maringá: Eduem.

Bicudo, D.D.C., Fonseca, B.M., Bini, L.M. et al. (2007). Undesirable side-effects of water hyacinth control in a shallow tropical reservoir. *Freshw. Biol.* 52 (6): 1120–1133.

Boonyapookana, B., Upatham, E.S., Kruatrachue, M. et al. (2002). Phytoaccumulation and phytotoxicity of cadmium and chromium in duckweed *Wolffia globosa*. *Int. J. Phytorem.* 4 (2): 87–100.

Burkholder, J.M. and Wetzel, R.G. (1989). Epiphytic microalgae on natural substrata in a hardwater lake: seasonal dynamics of community structure, biomass and ATP content. *Archivfür Hydrobiologie* 83 (1): 1–56.

Caldelas, C., Iglesia-Turiño, S., Araus, J.L. et al. (2009). Physiological responses of *Eichhornia crassipes* [Mart.] *Solms* to the combined exposure to excess nutrients and hg. *Braz. J. Plant Physiol.* 21 (1): 1–12.

Camargo, A.F.M. and Esteves, F.D.A. (1995). Biomass and productivity of aquatic macrophytes in Brazilian lacustrine ecosystems. *Limnol. Br.* 137–149.

Camargo, A.F.M., Pezzato, M.M., and Henry-Silva, G.G. (2003). Limiting factors to the primary production of aquatic macrophytes. In: *Ecology and Management of Aquatic Macrophytes* (ed. S.M. Thomaz and L.M. Bini), 59–83. Maringá: Eduem.

Carignan, R. and Kalff, J. (1980). Phosphorus sources for aquatic weeds: water or sediments? *Science* 207 (4434): 987–989.

Cook, C.D.K. (1989). Origin, autecology, and spread of some of the world's most troublesome aquatic weeds. In: *Aquatic Weeds: The Ecology and Management of Nuisance Aquatic Vegetation* (ed. A.H. Pieterse and K.J. Murphy), 31–38. Oxford Science Publications.

Czarnecki, S. and Düring, R.A. (2015). Influence of long-term mineral fertilization on metal contents and properties of soil samples taken from different locations in Hesse, Germany. *Soil* 1 (1): 23.

Delgado, M., Bigeriego, M., and Guardiola, E. (1993). Uptake of Zn, Cr and Cd by water hyacinths. *Water Res.* 27 (2): 269–272.

Dhir, B. (2015). Status of aquatic macrophytes in changing climate: a perspective. *J. Environ. Sci. Technol.* 8 (4): 139.

Dhir, B., Sharmila, P., and Saradhi, P.P. (2009). Potential of aquatic macrophytes for removing contaminants from the environment. *Crit. Rev. Environ. Sci. Technol.* 39 (9): 754–781.

Drążkiewicz, M. and Baszyński, T. (2005). Growth parameters and photosynthetic pigments in leaf segments of Zea mays exposed to cadmium, as related to protection mechanisms. *J. Plant Physiol.* 162 (9): 1013–1021.

Esteves, F.A. and Camargo, A.F.M. (1986). On the role of aquatic Macrophytes in nutrient storage and cycling. *Acta Limnol. Bras.* 1: 273–298.

Farago, M.E. and Parsons, P.J. (1994). The effects of various platinum metal species on the water plant *Eichhornia crassipes* (MART.) *Solms*. *Chem. Speciat. Bioavailab.* 6 (1): 1–12.

Hasan, M.R. and Rina, C. (2009). *Use of algae and aquatic macrophytes as feed in small-scale aquaculture: a review (No. 531).* Food and Agriculture Organization of the United Nations (FAO).

Henry-Silva, G.G. and Camargo, A.F.M. (2005). Ecological interactions between floating aquatic macrophytes *Eichhornia crassipes* and *Pistia stratiotes*. *Hoehnea* 32 (3): 445–452.

Henry-Silva, G.G., Camargo, A.F., and Pezzato, M.M. (2008). Growth of free-floating aquatic macrophytes in different concentrations of nutrients. *Hydrobiologia* 610 (1): 153.

Hoffmann, T., Kutter, C., and Santamaria, J. (2004). Capacity of *Salvinia minima* baker to tolerate and accumulate As and Pb. *Eng. Life Sci.* 4 (1): 61–65.

Hossain, K., Yadav, S., Quaik, S. et al. (2017). Vulnerabilities of macrophytes distribution due to climate change. *Theor. Appl. Climatol.* 129 (3–4): 1123–1132.

Hu, M.J., Wei, Y.L., Yang, Y.W., and Lee, J.F. (2003). Immobilization of chromium (VI) with debris of aquatic plants. *Bull. Environ. Contam. Toxicol.* 71 (4): 0840–0847.

Izaguirre, I., Pizarro, H., de Tezanos Pinto, P. et al. (2010). Macrophyte influence on the structure and productivity of photosynthetic picoplankton in wetlands. *J. Plankton Res.* 32 (2): 221–238.

Izaguirre, I., Sinistro, R., Schiaffino, M.R. et al. (2012). Grazing rates of protists in wetlands under contrasting light conditions due to floating plants. *Aquat. Microb. Ecol.* 65 (3): 221–232.

Jain, S.K., Vasudevan, P., and Jha, N.K. (1990). *Azolla pinnata* R. Br. and *Lemna minor* L. for removal of lead and zinc from polluted water. *Water Res.* 24 (2): 177–183.

Janes, R.A., Eaton, J.W., and Hardwick, K. (1996). The effects of floating mats of *Azolla filiculoides Lam.* and *Lemnaminuta Kunth* on the growth of submerged macrophytes. In: *Management and Ecology of Freshwater Plants*, 23–26. Dordrecht: Springer.

Jones, R.I. (2000). Mixotrophy in planktonic protists: an overview. *Freshw. Biol.* 45 (2): 219–226.

Kalia, K. and Flora, S.J. (2005). Strategies for safe and effective therapeutic measures for chronic arsenic and lead poisoning. *J. Occup. Health* 47 (1): 1–21.

Kamal, M., Ghaly, A.E., Mahmoud, N., and Cote, R. (2004). Phytoaccumulation of heavy metals by aquatic plants. *Environ. Int.* 29 (8): 1029–1039.

Kara, Y. (2004). Bioaccumulation of copper from contaminated wastewater by using *Lemna minor*. *Bull. Environ. Contam. Toxicol.* 72 (3): 467–471.

Kim, J.G. and Rejmánková, E. (2005). Decomposition of macrophytes and dynamics of enzyme activities in subalpine marshes in Lake Tahoe basin, USA. *Plant Soil* 266 (1–2): 303–313.

Lee, C.L., Wang, T.C., Hsu, C.H., and Chiou, A.A. (1998). Heavy metal sorption by aquatic plants in Taiwan. *Bull. Environ. Contam. Toxicol.* 61 (4): 497–504.

Low, K.S., Lee, C.K., and Tai, C.H. (1994). Biosorption of copper by water hyacinth roots. *J. Environ. Sci. Health Part A* 29 (1): 171–188.

Madsen, J.D., Chambers, P.A., James, W.F. et al. (2001). The interaction between water movement, sediment dynamics and submersed macrophytes. *Hydrobiologia* 444 (1–3): 71–84.

Maine, M.A., Duarte, M.V., and Suñé, N.L. (2001). Cadmium uptake by floating macrophytes. *Water Res.* 35 (11): 2629–2634.

Maine, M.A., Suñé, N.L., and Lagger, S.C. (2004). Chromium bioaccumulation: comparison of the capacity of two floating aquatic macrophytes. *Water Res.* 38 (6): 1494–1501.

Martins, D., Velini, E.D., Negrisoli, E., and Tofoli, G.R. (2002). Controle químico de *Pistia stratiotes*, *Eichhornia crassipes* e *Salvinia molesta* em caixas d'água. *Planta Daninha* 20 (SPE): 83–88.

Meerhoff, M. (2006). The structuring role of macrophytes on trophic dynamics in shallow lakes under a climate-warming scenario. Doctoral dissertation, Faculty of Science, University of Aarhus.

Meerhoff, M., Mazzeo, N., Moss, B., and Rodríguez-Gallego, L. (2003). The structuring role of free-floating versus submerged plants in a subtropical shallow lake. *Aquat. Ecol.* 37 (4): 377–391.

Miretzky, P., Saralegui, A., and Cirelli, A.F. (2004). Aquatic macrophytes potential for the simultaneous removal of heavy metals (Buenos Aires, Argentina). *Chemosphere* 57 (8): 997–1005.

Miretzky, P., Saralegui, A., and Cirelli, A.F. (2006). Simultaneous heavy metal removal mechanism by dead macrophytes. *Chemosphere* 62 (2): 247–254.

Mkandawire, M. and Dudel, E.G. (2005). Accumulation of arsenic in *Lemna gibba* L. (duckweed) in tailing waters of two abandoned uranium mining sites in Saxony, Germany. *Sci. Total Environ.* 336 (1–3): 81–89.

Mkandawire, M., Taubert, B., and Dudel, E.G. (2004). Capacity of *Lemna gibba* L. (duckweed) for uranium and arsenic phytoremediation in mine tailing waters. *Int. J. Phytorem.* 6 (4): 347–362.

Nagajyoti, P.C., Lee, K.D., and Sreekanth, T. (2010). Heavy metals, occurrence and toxicity for plants: a review. *Environ. Chem. Lett.* 8: 199–216.

Noraho, N. and Gaur, J.P. (1996). Cadmium adsorption and intracellular uptake by two macrophytes, *Azolla pinnata* and *Spirodela polyrhiza*. *Arch. Hydrobiol.* 136 (1): 135–144.

Odjegba, V.J. and Fasidi, I.O. (2004). Accumulation of trace elements by *Pistia stratiotes*: implications for phytoremediation. *Ecotoxicology* 13 (7): 637–646.

O'Farrell, I., de Tezanos Pinto, P., and Izaguirre, I. (2007). A pattern of morphological variability in phytoplankton in response to different light conditions. *Hydrobiologia* 578: 65–77.

O'Farrell, I.N.É.S., De Tezanos Pinto, P., Rodriguez, P.L. et al. (2009). Experimental evidence of the dynamic effect of free-floating plants on phytoplankton ecology. *Freshw. Biol.* 54 (2): 363–375.

Olguín, E.J., Hernández, E., and Ramos, I. (2002). The effect of both different light conditions and the pH value on the capacity of *Salvinia minima* baker for removing cadmium, lead and chromium. *Acta Biotechnol.* 22 (1–2): 121–131.

Portielje, R. and Roijackers, R.M.M. (1995). Primary succession of aquatic macrophytes in experimental ditches in relation to nutrient input. *Aquat. Bot.* 50 (2): 127–140.

Pott, V.J. and Pott, A. (2003). Dynamics of aquatic vegetation in the Pantanal. In: *Ecology and Management of Aquatic Macrophytes* (ed. S.M. Thomaz and L.M. Bini), 145–162. Maringá: Eduem.

Prasad, M.N.V., Malec, P., Waloszek, A. et al. (2001). Physiological responses of *Lemna trisulca* L. (duckweed) to cadmium and copper bioaccumulation. *Plant Sci.* 161 (5): 881–889.

Rahman, M.A. and Hasegawa, H. (2011). Aquatic arsenic: phytoremediation using floating macrophytes. *Chemosphere* 83 (5): 633–646.

Rahman, M.A., Hasegawa, H., Ueda, K. et al. (2007). Arsenic accumulation in duckweed (*Spirodela polyrhiza* L.): a good option for phytoremediation. *Chemosphere* 69 (3): 493–499.

Rai, P.K. and Tripathi, B.D. (2007). Heavy metals adsorption characteristics of free floating aquatic macrophyte *Spirodela poyrhhiza*. *J. Environ. Res. Dev.* 148 (1–4): 75–84.

Rejmankova, E. (2011). The role of macrophytes in wetland ecosystems. *J. Ecol. Environ.* 34 (4): 333–345.

Rezania, S., Taib, S.M., Din, M.F.M. et al. (2016). Comprehensive review on phytotechnology: heavy metals removal by diverse aquatic plants species from wastewater. *J. Hazard. Mater.* 318: 587–599.

Roijackers, R., Szabó, S., and Scheffer, M. (2004). Experimental analysis of the competition between algae and duckweed. *Arch. Hydrobiol.* 160 (3): 401–412.

Sanchez-Martin, M.J., Sanchez-Camazano, M., and Lorenzo, L.F. (2000). Cadmium and lead contents in suburban and urban soils from two medium-sized cities of Spain: influence of traffic intensity. *Bull. Environ. Contam. Toxicol.* 64 (2).

Sardar, K., Ali, S., Hameed, S. et al. (2013). Heavy metals contamination and what are the impacts on living organisms. *GJEMPS* 2 (4): 172–179.

Scheffer, M., Szabo, S., Gragnani, A. et al. (2003). Floating plant dominance as a stable state. *Proc. Natl. Acad. Sci.* 100 (7): 4040–4045.

Schneider, I.A.H. and Rubio, J. (1999). Sorption of heavy metal ions by the nonliving biomass of freshwater macrophytes. *Environ. Sci. Technol.* 33 (13): 2213–2217.

Schwaderer, A.S., Yoshiyama, K., de Tezanos Pinto, P. et al. (2011). Eco-evolutionary differences in light utilization traits and distributions of freshwater phytoplankton. *Limnol. Oceanogr.* 56 (2): 589–598.

Sculthorpe, C.D. (1967). *Biology of Aquatic Vascular Plants*. London: Edward Arnold Ltd.

Sen, A.K., Mondal, N.G., and Mandal, S. (1987). Studies of uptake and toxic effects of Cr (VI) on *Pistia stratiotes*. *Water Sci. Technol.* 19 (1–2): 119–127.

Sharma, S.S. and Gaur, J.P. (1995). Potential of *Lemna polyrrhiza* for removal of heavy metals. *Ecol. Eng.* 4 (1): 37–43.

Sharma, A., Gupta, M.K., and Singhal, P.K. (1996). Toxic effects of leachate of water hyacinth decay on the growth of *Scenedesmus obliquus* (Chlorophyta). *Water Res.* 30 (10): 2281–2286.

Sharma, S., Singh, B., and Manchanda, V.K. (2015). Phytoremediation: role of terrestrial plants and aquatic macrophytes in the remediation of radionuclides and heavy metal contaminated soil and water. *Environ. Sci. Pollut. Res.* 22 (2): 946–962.

Sinistro, R., Izaguirre, I., and Asikian, V. (2006). Experimental study on the microbial plankton community in a south American wetland (lower Paraná River basin) and the effect of the light deficiency due to the floating macrophytes. *J. Plankton Res.* 28 (8): 753–768.

de Souza, C.B. and Silva, G.R. (2019). Phytoremediation of effluents contaminated with heavy metals by floating aquatic macrophytes species. In: *Biotechnology and Bioengineering* (ed. E. Jacob-Lopes and L. Queiroz), 137–153. IntechOpen.

Srivastav, R.K., Gupta, S.K., Nigam, K.D.P., and Vasudevan, P. (1994). Treatment of chromium and nickel in wastewater by using aquatic plants. *Water Res.* 28 (7): 1631–1638.

Stets, E.G. and Cotner, J.B. (2008). Littoral zones as sources of biodegradable dissolved organic carbon in lakes. *Can. J. Fish. Aquat. Sci.* 65 (11): 2454–2460.

Tello Zevallos, W., Matías Salvatierra, L., Loureiro, D.B. et al. (2018). Evaluation of the autochthonous free-floating macrophyte *Salviniabiloba Raddi* for use in the phytoremediation of water contaminated with lead. *Desalin. Water Treat.* 103: 282–289.

de Tezanos Pinto, P. and O'Farrell, I. (2014). Regime shifts between free-floating plants and phytoplankton: a review. *Hydrobiologia* 740 (1): 13–24.

de Tezanos Pinto, P., Allende, L., and O'Farrell, I. (2007). Influence of free-floating plants on the structure of a natural phytoplankton assemblage: an experimental approach. *J. Plankton Res.* 29 (1): 47–56.

Thomaz, S.M. and Bini, L.M. (1998). Ecologia e manejo de macrófitas aquáticas em reservatórios. *Acta Limnol. Bras.* 10: 103–116.

Thomaz, S.M. and Cunha, E.R.D. (2010). The role of macrophytes in habitat structuring in aquatic ecosystems: methods of measurement, causes and consequences on animal assemblages' composition and biodiversity. *Acta Limnol. Bras.* 22 (2): 218–236.

Thomaz, S.M., Esteves, F.A., Murphy, K.J. et al. (2009). Aquatic macrophytes in the tropics: ecology of populations and communities, impacts of invasions and human use. *Trop. Biol. Conserv. Manage.* 4: 27–60.

Uddin, M.K. (2017). A review on the adsorption of heavy metals by clay minerals, with special focus on the past decade. *Chem. Eng. J.* 308: 438–462.

Vijayaraghavan, K. and Balasubramanian, R. (2015). Is biosorption suitable for decontamination of metal-bearing wastewaters? A critical review on the state-of-the-art of biosorption processes and future directions. *J. Environ. Manage.* 160: 283–296.

Villa, P., Pinardi, M., Tóth, V.R. et al. (2017). Remote sensing of macrophyte morphological traits: implications for the management of shallow lakes. *J. Limnol.* 76 (s1): 109–126.

Vymazal, J. (2016). Concentration is not enough to evaluate accumulation of heavy metals and nutrients in plants. *Sci. Total Environ.* 544: 495–498.

8

Enzymatic Approach for Phytoremediation

Anjali Pathak[1], *Mahendra K. Gupta*[1], *Mir S. Rabani*[1], *Shivani Tripathi*[1], *Sadhna Pandey*[2], *Charu Gupta*[1], *and Meenakshi Shrivastav*[1]

[1] *Microbiology Research Lab, School of Studies in Botany, Jiwaji University, Gwalior, Madhya Pradesh, India*
[2] *Department of Botany, Govt. K.R.G. PG Autonomous College, Gwalior, Madhya Pradesh, India*

8.1 Introduction

The overall state of the surroundings and the standard of life on the earth are inextricably linked. Unfortunately, as science, technology, and industry developed a massive amount of trash – from radioactive waste to raw sewage – is released or discarded into the ecosystem, posing a major threat to human survival on the earth. Hence, these pollutants impact humans, other living creatures, soil, and even the biogeochemical cycle of the ecosystem. The main causes of pollutant toxicity are their stability, limited solubility, and resistance to a variety of physical, chemical, and biological breakdown routes (Eibes et al. 2015). Several studies show various physical and chemical approaches that have evolved for the cleaning up of waste disposal systems like high-temperature ignition and chemical decomposition (e.g. base-catalyzed dechlorination and UV oxidation). But these procedures are very complicated, expensive, and unpopular. Therefore, due to the limitations of these approaches, a feasible biological method called "phytoremediation" is used as an alternative (Karigar and Rao 2011). The term "phyto" means "of a plant or connected to plants," while remediation is defined as the process of restoring damage to the environment. Phytoremediation can be described as the utilization of plants to remove environmental pollutant or detoxify them to contribute harmless substances. There are various sorts of pollution, but environmental contamination is one of the most harmful. Moreover, due to various anthropogenic activities, environmental pollution has become a major issue that can harm human health, crop production, and the ecosystem (Conesa et al. 2012). According to Sparks et al., pollutants such as heavy metals, radioisotopes, nitrates, inorganic acids, organic chemicals, etc., can come from a variety of sources, including waste materials, explosives, chemical pesticides, and fertilizers, acidic breakdown, as well as radioactive fallout (Spark et al. 1995). To eliminate contaminations, in situ and ex situ remediation approaches can be applied. Both methods are heavily reliant on physical and chemical processes. Exhumation is followed via heat treatment, chemical removal, and solidification before disposal in a landfill are examples of ex situ procedures. Volatilization by air venting, surfactant leaching, vitrification (pollutants are stiffened with an electric current), and isolation and confinement using a physical barrier all come in in situ procedures. Remediation processes are typically very expensive; estimates for cleaning up large-scale commercial sites start at $20,000, plus an additional fee of $40–70 per cubic yard of dirt, depending on the circumstance. Environmental cleanup and pollution control account for about 2% of the US gross domestic product (Spark et al. 1995). Synthetic fertilizers containing harmful metals and commercially cultivated fruits are two examples of environmental contamination. Heavy metals like lead and cadmium are abundant in tobacco. Throughout the last century, industrialization operations have contaminated vast surface regions throughout the industrialized world, as well as most of the developing world. In Europe, for example, over a million acres, or nearly 20% of the entire land surface, are afflicted by toxic soil pollution and deterioration (Vidali 2001). Several studies have revealed the use of creatures such as animals and plants in bioremediation to detoxify toxins. Assimilated arsenic contaminants have been absorbed by genetically modified plants (*Arabidopsis thaliana*), hence leading to arsenic pollution. Two bacterial genes are present in these

Aquatic Contamination: Tolerance and Bioremediation, First Edition. Edited by Rouf Ahmad Bhat, Gowhar Hamid Dar, Fernanda Maria Policarpo Tonelli, and Saima Hamid.

plants, one of which transforms arsenate into arsenite and the other of which binds to arsenite and stores it in vacuoles. Phytoextraction, phytostabilization, phytotransformation, and phytovolatilization are some of the other processes that play a part in plant remediation of pollutants (Leung 2004). Remediation plays a pivotal role in which microbes participate and enzymatically transform contaminants into nonhazardous compounds. Nevertheless, the effectiveness of the detoxification of the pollutants depends on the existence of favorable conditions for the microorganism's growth, development, and activity. By turning contaminants into nonhazardous compounds with the aid of enzymes, microbes play a significant role in cleanup. Because enzymes accelerate all chemical reactions on pollutants, they are the most efficient bioremediation agents. Enzyme specificity is usually broad enough to allow them to work on an ample range of molecules with similar patterns. Enzymes can also be modified through bioengineering in order to increase their longevity and efficacy in particular situations or with particular substrates. The benefits of this approach for lowering various contaminants are therefore highlighted in this chapter, along with a review of the most important enzymes involved in phytoremediation and an explanation of how they work.

8.2 Mechanism and Types of Phytoremediation

Depending on the type of plant species and the condition of the environment, plants can remove organic and inorganic pollutants from soil and water through a variety of techniques (Anton and Mathe-Gaspar 2005; Antoniadis et al. 2017). Plants' interference with environmental factors (i.e. water, air, and soil) as well as the number of microbes plays a critical role in the depletion of several toxic pollutants (Dary et al. 2010). Several studies reveal that the pollutant, the species of plants, and the soil all affect how well cleanup works. The feasibility of remediation is significantly influenced by biomass production of plants and metabolism, as well as soil properties like pH, electric conductivity, the amount of organic matter, microbial activity, and other soil additives (Anton and Mathe-Gaspar 2005; Guidi Nissim et al. 2018). There are a variety of phytoremediation processes that can be utilized to remove heavy metals from polluted soils and water. These techniques include phytostabilization (involves plants to minimize heavy metal bioavailability in soil), phytoextraction (involves plant extracts and removes heavy metals from soil and water), phytovolatilization (usually includes plants to absorb heavy metal from soil and releases it into the atmosphere as a volatile component), phytofiltration, or immobilization (it involves hydroponic systems [i.e. cultivated plants utilized to remove metal ions from aqueous waste and groundwater] Ernst 2005, Marques et al. 2009). Phytodegradation as well as rhizodegradation are two additional phytoremediation methods for degrading organic contaminants from soil. The most often utilized phytoremediation approaches for treating soil and water contaminated with heavy metals include phytostabilization, bioaugmentation, phytovolatilization, and phytofiltration (Figure 8.1).

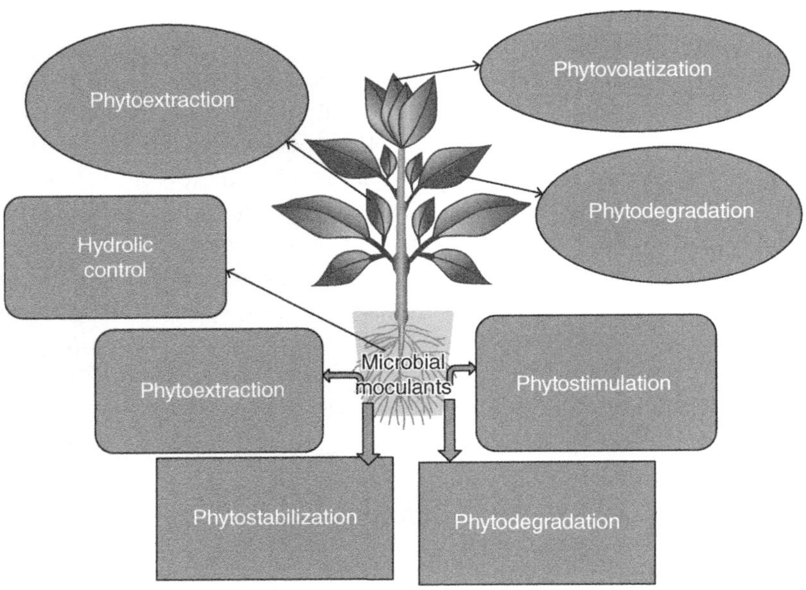

Figure 8.1 Mechanisms of phytoremediation with microbial assistance. *Source:* c.f Alori and Fawole (2017) / Springer Nature.

8.2.1 The Kinetics of Plant Enzymes

Enzyme kinetics is the process through which enzymes catalyze chemical processes. In this, there are 298 smart bioremediation technologies for assessment of the reaction and the reaction rate. In addition to metabolism and drug effect measurement, this is crucial for measuring the activity of enzymes. Enzymes can mimic proteins and change how a specific substrate functions. A variety of substrates for biochemical reactions can attach to the enzyme's active site, causing it to change into a particular product (Biasutti et al. 2008). Michaelis-Menten kinetics can be used to explain the kinetics of plant enzymes. For instance, first-order degradation occurs when the substrate concentration is relatively low; nevertheless, it is only possible at increasing substrate concentrations. Bacteria can be used as a growth substrate, and substrate growth is greatest when it is readily available, claims Monod kinetics.

Furthermore, it has been found that if there is little substrate present, bacteria have limited ability to degrade, which can prevent the biodegradation of polluted locations. Consequently, combining bacteria and plants can be used to solve the problem of environmental pollution (Trapp et al. 2007).

8.2.2 Soil Enzymes Facilitating Phytoremediation

Enzymes are known as biological catalysts that play a role in metabolism and biochemical reactions. Enzymes are employed to treat contaminants in the environment and act as organ catalysis in a variety of activities, as outlined later.

i) **Oxidoreductase**: It takes part in the oxidation process that reduces substrates by transferring oxygen from molecular oxygen with the help of cofactors such as NADP/FAD/NADPH. It reduces the toxicity of contaminants that can be remediated by microbes and higher plants, using oxidoreductase enzymes as a catalyst. By delivering ions from the substrate to the substance, it aids in the breaking of chemical bonds, and the pollution is ultimately oxidized to undisruptive organic molecules. In a soil ecosystem, this enzyme participates in the humification of phenolic compounds as well as the breakdown of lignin. Furthermore, its binding methods can also be used to detoxify xenobiotic substances. Several plant families, like Fabaceae and Solanaceae, produce oxidoreductases, that aids in soil detoxification (Vidali 2001; Husain 2006). Several studies revealed that plants belonging to the Fabaceae, Gramineae, and Solanaceae families have been identified to produce oxidoreductases, which assist in the oxidative breakdown of soil components. The three substances that have been the main targets of organic pollutant phytoextraction are petroleum hydrocarbons, explosives, and chlorinated solvents (Newman et al. 1998; Durán and Esposito 2000) (Figure 8.2).

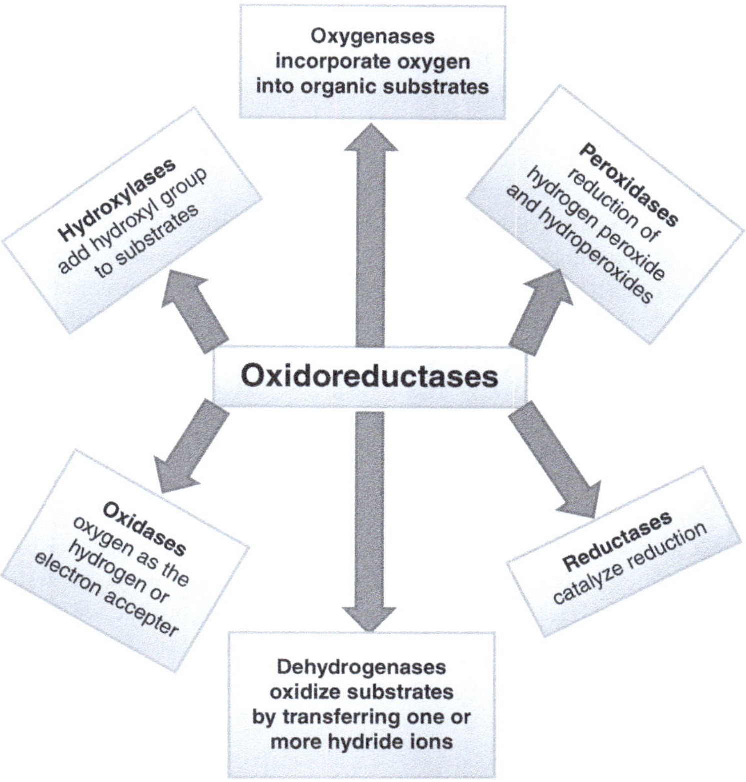

Figure 8.2 Categorization of oxidoreductase. *Source:* Kareem (2020) / IntechOpen / CC BY 3.0).

Multiple soil microorganisms (oxygenases, reductases, dehalogenases, dehydrogenases, and hydroxylases) are imperative in the remediation of environmental contaminants that have spread across the ecosystem as a result of human activity. Oxidase enzymes are particularly essential among these enzymes since they are engaged in the earliest stages of breakdown and catalyze and degrade aromatic molecules. The reaction in this process is catalyzed by the incorporation of oxygen ions into the substrates. Two main types of oxygenases that have been reported are monooxygenases, which add one oxygen molecule, and dioxygenases, which add two oxygen molecules (Arora et al. 2010; Karigar and Rao 2011).

ii) **Monooxygenases**: Monooxygenase enzyme is a single oxygen molecule that is incorporated into the substrate. It is divided into two subclasses based on the cofactors: flavin-dependent monooxygenases and P450 monooxygenases (like CYP102 from *Bacillus megaterium* BM3). The first subclass's prosthetic group is flavin, which is stimulated by the coenzymes NADP or NADPH, whereas the other subclass's prosthetic group is heme. Numerous studies claim that monooxygenases are cofactor-dependent biocatalysts employed in bioremediation. Certain enzymes are cofactor-independent and rely solely on molecular oxygen to function. ActVAorf6 monooxygenase, TcmH, and other enzyme monooxygenases regulate a variety of processes, including aromatic and aliphatic biodegradation, volatilization, nitrogen fixation, ammonification, dehalogenation, transformation, and hydroxylation. The role of methane monooxygenase in the decomposition of hydrocarbons is best known (Lock et al. 2017; Sirajuddin and Rosenzweig 2017; Syed et al. 2013; Urlacher et al. 2004).

iii) **Dioxygenases**: The numerous parts of these enzyme systems connect molecule oxygen to the substrate. These enzymes decompose the fragrant chemicals that are harmful to the environment. Based on how the enzyme works, it can be classified into two subclasses: hydroxylation dioxygenases and cleavage dioxygenases. The cleavage enzyme catalyzes the cleavage of an aromatic ring with two or more hydroxyl groups, whereas the hydroxylation enzyme adds two oxygen atoms to the substrate. Dioxygenases that catalyze the cleavage of intradiol (ortho cleavage with FeIII) and estradiol (meta cleavage with FeII and MnII) are further split into two types. (Xenia and Refugio 2016; Fulekar 2017; Muthukamalam et al. 2017).

iv) **Laccases**: These are blue multicopper oxidases that function as biosensors and catalyze substrate oxidation. Different microorganisms have the ability to create intra- and extracellular laccases, which are thought to be responsible for the oxidation of lignins, polyamines, polyphenols, aryl diamines, ortho- and paradiphenols, aminophenols, and other compounds. Laccases help in wastewater treatment by decolorizing and detoxifying industrial effluents. They are useful in the textile, pulp, and paper industries because they affect both phenolic and nonphenolic organic compounds, which are major ecological pollutants (Nigam 2013; Viswanath et al. 2014). Laccases target the methoxy groups of phenolic and methoxy-phenolic acids as well as oxidizing and decarboxylating them (demethylation). These enzymes are responsible for lignin depolymerization that produces different ranges of phenols. Furthermore, the chemicals serve as microbial nutrition and are repolymerized to humic substances through the enzyme laccase (Kim et al. 2002). Also, they are involved directly in xenobiotic breakdown and bioremediation. Further, the thermophilic laccase enzyme was isolated from desert plant species like Cereus spp. and Opuntia spp. and utilized by the pulping industry (Nigam 2013; Viswanath et al. 2014).

v) **Peroxidases**: All across the world, peroxidases (EC 1.11.1.7) have been discovered. These enzymes can be found in plants, animals, and microorganisms. These were all formed from a multitude of sources, including cyanobacteria, yeast, bacteria, and fungi. The degradation of environmental pollutants, animal feedstock, chemical foodstuffs, agricultural, and paper industries, dyes degradation, paper/pulp manufacturing lignin decomposition, photodegradation, sewage disposal, and biosensors are all accomplished by the use of these microbial enzymes. They help plants build their cell walls, process auxin, lengthen their cells, and establish defense systems. There are two categories of proteins: heme and nonheme. Bansal and Kanwar reported two groups of Hemeperoxidases (i.e. prokaryotes and eukaryotes) (Bansal and Kanwar 2013; Falade et al. 2016). Based on the comparison, peroxidases are categorized into three classes.
Category I: Category I enzymes include bacterial gene-duplicated catalase peroxidases, ascorbate peroxidase (APX) from plants, and yeast cytochrome c peroxidase
Category II: Fungal peroxidases, specifically manganese peroxidase (Mnp) and lignin peroxidase (LiP), are included in Category II and are responsible for the breakdown of lignin in wood.
Category III: Plant peroxidases, which are responsible for the formation of cell walls and lignifications, can be found in Category III and include enzymes from horseradish, barley, or soybean (Falade et al. 2016).

8.2.3 Classification of Peroxidases

In accordance with their source and mode of action, peroxidases can be categorized into various groups. The peroxidases LiP, manganese-dependent peroxidase (MnP), and versatile peroxidase (VP) have received the most attention due to their potent ability to eliminate hazardous substances in nature.

i) **Microbial lignin peroxidases**: In addition to destroying the cell wall of lignin-containing plants, these heme proteins were produced by white-rot fungi during secondary metabolism.
ii) **Microbial manganese peroxidases**: This is an extracellular heme enzyme from the lignin-eroding basidiomycetes fungi, which undergoes a stepwise reaction to change Mn^{2+} into the oxidant Mn^{3+} Mn^{2+} increases the synthesis of MnP and serves as an MnP substrate. The subsequent Mn^{3+} chelate oxalate is small enough to reach locations where even the enzyme is unable to access them, like lignin or other such substances that are embedded deep within the earth's soil but are occasionally inaccessible to enzymes (Ten and Teunissen 2001).
iii) **Microbial versatile peroxidases**: Methoxybenzenes, Mn^{2+}, horseradish peroxidase, LiP, and other phenolic volatile precursors can all be directly oxidized by VP enzymes. VP has an extremely broad substrate selectivity compared to certain other peroxidases and a propensity to oxidize substrates in the exclusion of manganese. VP has also been shown to be able to oxidize model dimers of both phenolic and nonphenolic lignin (Ruiz-Duenas et al. 2007).

8.2.4 Arsenate Reductase

Arsenate reductases are enzymes that are present in numerous species and catalyze the fundamental reduction reaction. The operons that provide arsenic resistance in an organism are made up of three genes in general. They have the following names in *Escherichia coli*: (i) arsC, which codes for both the reductase enzyme; (ii) arsB, which codes for the arsenite transporter; and (iii) arsR, which codes for a gene-repressor (Lopez-Maury et al. 2003). Despite the fact that the genes in this Operon are labeled differently in different species, their functions are very identical (Fernandes and Holmgren 2004). Arsenate reductase has a pair of cysteine residues in its active site that are required for catalysis. One of these cysteine residues is part of the P-loop (Cys-X5-Arg); generally, this cysteine residue forms a thioester bond with As(V), and the arginine residue helps stabilize the intermediate (Doucleff and Terry 2002).

8.2.5 Hydrolytic Enzymes Involved in Phytoremediation

In the bioremediation of soil pollutants, several hydrolytic enzymes play a crucial role. Hydrolytic enzymes break apart the hazardous compound's main chemical bonds, which lessens their toxicity. This method works well for the degradation of oil spills as well as pesticides like carbamate and organophosphate. Organochlorine insecticides like DDT and heptachlor easily break down in anaerobic settings but are constant in well-aerated soil (Williams 1977; Lal and Saxena 1982; Vasileva-Tonkova and Galabova 2003).

1) **Microbial lipase**: The enzyme lipase split down lipids formed by a broad range of bacteria, animals, and plants. Lipase activity was responsible for the significant reduction in total hydrocarbon from polluted soils. Triacylglycerols are hydrolyzed by lipases, which are widely distributed enzymes, into glycerol and free fatty acids. With reactions including esterification, aminolysis, hydrolysis, and others, it interacts with organic soil contaminants and aids in cleaning up contaminants from contaminated soil. It is an evidence of the degradation of soil hydrocarbon pollutants that soil hydrocarbon contaminants are being broken down. Although lipase has numerous uses in chemical, aesthetic, culinary, paper, pulp, and other sectors, it is expensive to produce (Nigam 2013; Okino-Delgado et al. 2017).
2) **Microbial cellulase**: Enzyme typically, cellulases are a mixture of several enzymes. The hydrolysis process involves at least three major cellulase groups. The first is endoglucanase (EG, endo- 1,4-D-glucanohydrolase), which attacks low crystallinity areas in cellulose and generates free chain ends. Exoglucanase or cellobiohydrolase (CBH, 1,4-bD-glucan cellobiohydrolase) degrades the cellulose molecule further by removing cellobiose units from the free chain ends. And the last one is glucosidase, which hydrolyzes cellobiose to glucose units. Some auxiliary enzymes are also present in addition to the primary enzymes. Cellulases degrade cellulose during the enzymatic hydrolysis process, converting it into reducing sugars that yeasts or bacteria can ferment to produce ethanol (Sun and Cheng 2002).

3) **Microbial protease**: These hydrolytic enzymes break down peptide bonds found in the primary structure of polypeptides and other proteins. They are eminent industrial enzymes that are derived from plants, animals, and microbes. They play a major role in a variety of industries, including pharmaceutical, textile, feed, and food. Protease has potential applications in the management of trash from some processing companies and household tasks. Additionally, it is also useful for industries that prepare poultry (Bhunia and Basak 2014).
4) **Microbial amylase**: An extracellular enzyme called alpha-amylases releases oligosaccharides by rupturing the 1,4-glucosidic bond in starch molecules. Maltose's second 1,4-glycosidic bond is likewise broken via amylase, which is produced by both bacteria and plants. Amylases are essential enzymes because they are specifically used in the manufacturing industries for the production and conversion of starch, according to 296 Smart Bioremediation Technologies. These enzymes are classified as glycoside hydrolases and specifically act on disaccharides (sucrose) and polysaccharides (starch). Such enzymes have found wide applications in the paper, food, sugar, pharmaceutical, and starch liquefaction sectors (Gopinath et al. 2017).
5) **Microbial pullulanase**: Pullulanases, also known as Dextrin 6-glucanohydrolases, are produced by numeral microbes, including *Geobacillus Stearothermophilus*, *Klebsiella* spp., and *Bacillus* spp. For its role as a bioprocessor of starch, it is extremely significant. The enzyme pullulanase is used in a variety of commercial applications as well as genetic engineering (Lee et al. 2017; Karigar and Rao 2011.

8.3 Conclusion

The process of phytoremediation cleans the environment without endangering it. The selection of the most viable and ideal agricultural crops, as well as the most creative, is crucial to the accomplishment of this process. Found to be potential substances that are important to this process and many of them work effectively. With the aid of a better understanding of enzymatic potential processes meant for eliminating pollutants from the surrounding soil of roots, as well as the relationship among plants, microorganisms, and chemical contaminants, the use of phytoremediation will be broadened. Cleanup can be aided by various mechanisms found in plants and bacteria. Additionally, a lot of plants function as hyperaccumulators to remove particular hazardous metals from polluted surroundings that are dependent on the industry. Because of its inexpensive processing costs, well-established methods, and ability to break down hazardous heavy metals, this method is favorable. In addition to its benefits, phytoremediation also has significant drawbacks, including phototoxicity and climatic and geological restrictions.

References

Alori, E.T. and Fawole, O.B. (2017). Microbial inoculants-assisted phytoremediation for sustainable soil management. In: *Phytoremediation* (ed. A. Ansari, S. Gill, R. Gill, et al.), 1–17. Cham: Springer https://doi.org/10.1007/978-3-319-52381-1_1.

Anton, A. and Mathe-Gaspar, G. (2005). Factors affecting heavy metal uptake in plant selection for phytoremediation. *Z. Naturforsch. C J. Biosci.* 60: 244–246.

Antoniadis, V., Levizou, E., Shaheen, S.M. et al. (2017). Trace elements in the soil-plant interface: phytoavailability, translocation, and phytoremediation – a review. *Earth-Sci. Rev. 171*: 621–645.

Arora, P.K., Srivastava, A., and Singh, V.P. (2010). Application of monooxygenases in dehalogenation, desulphurization, denitrification and hydroxylation of aromatic compounds. *J. Biorem. Biodegrad.* 1: 18.

Bansal, N. and Kanwar, S.S. (2013). Peroxidase(s) in environment protection. *Sci. World J.* 2013: 714639.

Bhunia, B. and Basak, B. (2014). A review on application of microbial protease in bioremediation. In: *Industrial and Environmental Biotechnology* (ed. K. Pramanik and J.K. Patra), 217–228. New Delhi, India: Studium Press (India) Pvt. Ltd.

Biasutti, M.A., Albuin, E.B., Silber, J.J. et al. (2008). Kinetics of the reactions catalyzed by the enzymes in solutions of surfactants. *Adv. Colloid Interf. Sci.* 136 (12): 1–24.

Conesa, H.M., Evangelou, M.W., Robinson, B.H., and Schulin, R. (2012). A critical view of current state of phytotechnologies to remediate soils: still a promising tool? *Sci. World J. 2012*: 173829.

Dary, M., Chamber-Pérez, M.A., Palomares, A.J., and Pajuelo, E. (2010). "In situ" phytostabilisation of heavy metal polluted soils using *Lupinus luteus* inoculated with metal resistant plant-growth promoting rhizobacteria. *J. Hazard. Mater.* 177 (1–3): 323–330.

Doucleff, M. and Terry, N. (2002). Pumping out the arsenic. *Nat. Biotechnol.* 20: 1094.

Durán, N. and Esposito, E. (2000). Potential applications of oxidative enzymes and phenoloxidase-like compounds in wastewater and soil treatment: a review. *Appl. Catal. B* 28 (2): 83–99.

Eibes, G., Arca-Ramos, A., Feijoo, G. et al. (2015). Enzymatic technologies for remediation of hydrophobic organic pollutants in soil. *Appl. Microbiol. Biotechnol.* 99 (21): 8815–8829.

Ernst, W.H. (2005). Phytoextraction of mine wastes–options and impossibilities. *Chem. Erde* 65: 29–42. https://doi.org/10.1016/j.chemer.2005.06.001.

Falade, A.O., Nwodo, U.U., Iweriebor, B.C. et al. (2016). Lignin peroxidase functionalities and prospective applications. *Microbiol. Open* 6 (1): e00394.

Fernandes, A.P. and Holmgren, A. (2004). Glutaredoxins: glutathione dependent redox enzymes with functions far beyond a simple thioredoxin backup system. *Antioxid. Redox Signal.* 6: 63.

Fulekar, M.H. (2017). Microbial degradation of petrochemical waste-polycyclic aromatic hydrocarbons. *Bioresour. Bioprocess.* 4: 28.

Gopinath, S.C.B., Anbu, P., Arshad, M.K.M. et al. (2017). Biotechnological processes in microbial amylase production. *Biomed. Res. Int.* 1272193: https://doi.org/10.1155/2017/1272193.

Guidi Nissim, W., Palm, E., Mancuso, S. and Azzarello, E. (2018). Trace element phytoextraction from contaminated soil: a case study under Mediterranean climate. *Environ. Sci. Pollut. Res.*, 25, pp. 9114–9131. https://doi.org/10.1007/s11356-018-1197-x

Husain, Q. (2006). Potential applications of the oxidoreductive enzymes in the decolorization and detoxification of textile and other synthetic dyes from polluted water: a review. *Crit. Rev. Biotechnol.* 26 (4): 201221.

Kareem, H.M. (2020). Oxidoreductases: significance for humans and microorganism. In: *Oxidoreductase* (ed. M.A. Mansour), 123–130. IntechOpen https://doi.org/10.5772/intechopen.93961.

Karigar, C.S. and Rao, S.S. (2011). Role of microbial enzymes in the bioremediation of pollutants: a review. *Enzyme Res.* 111: 805187. https://doi.org/10.4061/2011/805187.

Kim, J.S., Park, J.W., Lee, S.E., and Kim, J.E. (2002). Formation of bound residues of 8-hydroxybentazon by oxidoreductive catalysts in soil. *J. Agric. Food Chem.* 50 (12): 3507–3511.

Lal, R. and Saxena, D.M. (1982). Accumulation, metabolism and effects of organochlorine insecticides on microorganisms. *Microbiol. Rev.* 46 (1): 95–127.

Lee, C.W., Jang, S.H., and Chung, H.S. (2017). Improving the stability of cold-adapted enzymes by immobilization. *Catalysts* 7: 112. https://doi.org/10.3390/catal7040112.

Leung, M. (2004). Bioremediation: technique for cleaning up a mess. *Biotech. J.* 2: 18–22.

Lock, M., Nichol, T., Murrell, J.C., and Smith, T.J. (2017). Mutagenesis and expression of methane monooxygenase to alter regioselectivity with aromatic substrates. *FEMS Microbiol. Lett.* 13: 364.

Lopez-Maury, L., Florencio, F.J., and Reyes, J.C. (2003). Arsenic sensing and resistance system in the cyanobacterium *Synechocystis* sp. strain PCC 6803. *J. Bacteriol.* 185: 5363–5371.

Marques, A.P., Rangel, A.O., and Castro, P.M. (2009). Remediation of heavy metal contaminated soils: phytoremediation as a potentially promising clean-up technology. *Crit. Rev. Environ. Sci. Technol.* 39: 622–654. https://doi.org/10.1080/10643380701798272.

Muthukamalam, S.K., Sivagangavathi, S., Dhrishya, D., and Rani, S.S. (2017). Characterization of dioxygenases and biosurfactants produced by crude oil degrading soil bacteria. *Environ. Microbiol.* 48 (4): 637647.

Newman, L.A., Doty, S.L., Gery, K.L. et al. (1998). Phytoremediation of organic contaminants: a review of phytoremediation research at the University of Washington. *J. Soil Contam.* 7 (4): 531–542.

Nigam, P.S. (2013). Microbial enzymes with special characteristics for biotechnological applications. *Biomolecules* 3: 597–611. https://doi.org/10.3390/biom3030597.

Okino-Delgado, C.H., Prado, D.Z., Facanali, R. et al. (2017). Bioremediation of cooking oil waste using lipases from wastes. *PLoS One* 12 (10): e0186246. https://doi.org/10.1371/journal.pone.018624.

Ruiz-Duenas, F.J., Morales, M., Pérez-Boada, M. et al. (2007). Manganese oxidation site in *Pleurotus eryngii* versatile peroxidase: a site-directed mutagenesis, kinetic, and crystallographic study. *Biochemistry* 46 (1): 66–77.

Sirajuddin, S. and Rosenzweig, A.C. (2017). Enzymatic oxidation of methane. *Biochemistry* 54 (14): 2283–2294.

Spark, K.M., Wells, J.D., and Johnson, B.B. (1995). Characterizing trace metal adsorption on kaolinite. *Eur. J. Soil Sci. 46* (4): 633–640.

Sun, Y. and Cheng, J. (2002). Hydrolysis of lignocellulosic materials for ethanol production: a review. *Bioresour. Technol.* 83 (1): 1–11.

Syed, K., Porollo, A., Miller, D., and Yadav, J.S. (2013). Rational engineering of the fungal P 450 monooxygenase CYP5136A3 to improve its oxidizing activity toward polycyclic aromatic hydrocarbons. *Protein Eng. Des. Sel.* 26 (9): 553–557.

Ten, H. and Teunissen, P.J.M. (2001). Oxidative mechanisms involved in lignin degradation by white-rot fungi. *Chem. Rev.* 101 (11): 3397–3413.

Trapp, S., Ücisik, A.S., Romano, P.D., and Larsen, M. (2007). The role of plants and bacteria in phytoremediation-kinetic aspects. In: *Bioremediation of Soils Contaminated with Aromatic Compounds*, 41–49. Dordrecht: Springer.

Urlacher, V.B., Lutz-Wahl, S., and Schmid, R.D. (2004). Microbial P450 enzymes in biotechnology. *Appl. Microbiol. Biotechnol.* 64: 317–325.

Vasileva-Tonkova, E. and Galabova, D. (2003). Hydrolytic enzymes and surfactants of bacterial isolates from lubricant-contaminated wastewater. *Z. Naturforsch.* 58 (1–2): 87–92.

Vidali, M. (2001). Bioremediation: an overview. *Pure Appl. Chem.* 73 (7): 1163–1172.

Viswanath, B., Rajesh, B., Janardhan, A. et al. (2014). Fungal laccases and their applications in bioremediation. *Enzyme Res.* 2014: 163242. https://doi.org/10.1155/2014/163242.

Williams, P.P. (1977). Metabolism of synthetic organic pesticides by anaerobic microorganisms. *Residue Rev.* 66: 63–135.

Xenia, M.E. and Refugio, R.V. (2016). Microorganisms metabolism during bioremediation of oil contaminated soils. *J. Biorem. Biodegrad.* 7: 2. https://doi.org/10.4172/2155-6199.1000340.

9

Phyto-Metalloproteins and Restoration of Freshwater Ecosystems

Ekta B. Jadhav[1], Shefali[2], Varad Nagar[3], Vinay Aseri[3], Poonam Kumari[3], Vanisha Godara[3], Sneha Lohar[3], Kumud K. Awasthi[4], Garima Awasthi[4], and Mahipal S. Sankhla[5]

[1] *Department of Chemistry & Toxicology, Government Institute of Forensic Science, Aurangabad, Maharashtra, India*
[2] *Department of Zoology, DPG Degree College, Gurugram, Haryana, India*
[3] *Department of Forensic Science, Vivekananda Global University, Jaipur, Rajasthan, India*
[4] *Department of Life Sciences, Vivekananda Global University, Jaipur, Rajasthan, India*
[5] *Department of Forensic Science, University Center for Research and Development (UCRD), Chandigarh University, Mohali, Punjab, India*

9.1 Introduction

Freshwater ecosystems comprise around 2.8% of the total water on the earth and most of its part is occupied by frozen polar ice caps, icebergs, and glaciers. Thus, less than 1% of freshwater resources are available for direct human requirements, which are mainly found in lakes, reservoirs, rivers, streams, and underground aquifers (Qadri and Bhat 2020). Although the freshwater ecosystems on the earth are concise, these resources diminish fast with the continuous non-stop process of degradation in the form of contaminants due to the factors like population explosion, urbanization, and industrialization (Tripathi and Pandey 2009). Emerging pollutants such as organic waste, pathogens, fertilizers, and pesticides including toxic heavy metals from some natural sources and most notably from anthropogenic sources are the most universal contaminants of serious environmental concern as they arise water quality challenges globally with serious threats to human health and ecosystems (UNESCO, UN-Water 2020). Heavy metal constitutes a vital chunk of industrial hazardous waste and arise obvious concerns as they are nondegradable, persistent and toxic in the environment as serious ecological ramifications on aquatic ecosystems including freshwater ecosystems, which may impart hazardous effects on human health as well as aquaculture (Jumbe and Nandini 2009; Sankhla et al. 2016; Qadri and Bhat 2020; Sonone et al. 2021). Heavy metals are insoluble, high-density metal(loid)s having atomic numbers above 20, which may be either essential or nonessential and regularly found in nature (Bolan and Duraisamy 2003). At high concentrations, both essential and nonessential heavy metals are toxic to plants and thus to other organisms, and their accumulation threatens the quality and health of freshwater ecosystems and related organisms (Parihar et al. 2019; Wani et al. 2020; Yadav et al. 2020; Awasthi et al. 2022). Mostly the organisms depending on those ecosystems are subjected to various toxic effects of heavy metals (Vanisree et al. 2022). Potential industrial and agricultural entities are the notable sources of release of metals in the environment, which are primarily releasing Fe, Zn, Pb, Cd, Ni, Cu, and Cr into various ecosystems on the earth including freshwater resources (Yahya et al. 2018; Mushtaq et al. 2020). Freshwater is a crucial resource for nature, and humans, and a biodiversity hotspot. Thus, they are economically and ecologically essential (Strayer and Dudgeon 2010). The river is one of the major freshwater ecosystems and river water is significantly used for drinking and other domestic purposes (Singh et al. 2022). But river water and groundwater are also highly vulnerable to heavy metal contamination with changing metal concentration according to temperature, and seasonal and climatic variations and such metal-contaminated water affects human health by harming the cardiovascular system, gastrointestinal tract, nervous system, endocrine glands, kidneys, liver, lungs, and bones (Kumar et al. 2014; Sankhla and Kumar 2019; Sankhla et al. 2019a,b, 2020). As freshwater ecosystems are much more vulnerable to disturbances due to contamination than other ecosystems, the restoration of such ecosystems is an alarming issue and receiving much attention worldwide (Palmer et al. 2007; Hwang et al. 2014). However, there are many conventional physicochemical methods for the restoration of contaminated water bodies but these treatments can be inept, and expensive, and some of them are not

Aquatic Contamination: Tolerance and Bioremediation, First Edition. Edited by Rouf Ahmad Bhat, Gowhar Hamid Dar, Fernanda Maria Policarpo Tonelli, and Saima Hamid.

eco-friendly, i.e. disturbing the surrounding environment. Thus, these disadvantages are the driving force behind the exploration of unconventional remediation methods (Bhattacharya and Chakraborty 2019). There is a need to move from conventional approaches to new sustainable methodologies, which need only biological resources that are cost-effective as well as eco-friendly for the restoration of metal-contaminated water ecosystems. The researches have been conducted in this way using plants as phytoremediation for the uptake, accumulation, and detoxification of heavy metal from metal-polluted effluents (Ghosh and Singh 2005; Fosso-Kankeu and Mulaba-Bafubiandi 2014). Based on the type and site of contamination, phytoremediation methods are ex situ and in situ, which are achieved with plants which bind, extract, and remediate pollutants from the environment in eco-friendly and economically feasible manner and show promising results for the removal of metals, even present in very low concentrations (Ojuederie and Babalola 2017; Skinder et al. 2020). Phytoremediation involves plants, which have the potential for metal uptake, accumulation, and mainly tolerance that are depending on the synthesis of metalloproteins, mainly metallothioneins (MTs) or phytochelatins (PCs) in response to toxic levels of heavy metals. These metalloproteins in these biosorbent plants or hyperaccumulators (phytometalloproteins) play a very essential role in the adsorption and uptake of metals, their accumulation, detoxification, and, hence, restoration of freshwater ecosystems (Fosso-Kankeu and Mulaba-Bafubiandi 2014).

9.2 Phytoremediation

Phytoremediation is a green clean-up process, which involves the use of the potential of plants for treating environmental contaminants, preserving a clean environment and alternatively maintaining the ecological balance (Fosso-Kankeu and Mulaba-Bafubiandi 2014; Singh et al. 2017).

9.2.1 Phytoremediation Mechanisms

Phytoremediation can be applied in different directions for various reasons for the uptake or accumulation of toxic metals inside the plant via rot cells and based on these parameters, it is classified as rhizofiltration, phytoextraction, phytostabilization, phytodegradation, and phytovolatilization. When plants are employed for bringing down the bioavailability, mobility, and reactivity of contaminants such as toxic metals in the environmental ecosystems, that process of phytoremediation is referred to as phytostabilization. The process is classified as phytodegradation as plants are utilized for the transformation of toxic organic contaminants into nontoxic forms by the degradation of organic pollutants with the help of enzymes. Rhizofiltration involves the absorption, uptake, precipitation, and accumulation of toxic heavy metals from contaminated effluents. Phytovolatilization class involves the process of uptake of contaminants particularly from soil and makes them volatile for their subsequent release into the atmosphere. Phytoextraction or phytoaccumulation, in which hyperaccumulator plants are used to uptake, transport, and accumulate heavy metals and radionuclides into the aboveground harvestable parts (Ghosh and Singh 2005; Rai 2009; Ali et al. 2013). Here, the phytoextraction method is mainly taken into consideration, as this phytoremediation process is used by the plants for the removal of heavy metal pollutants from contaminated water using metalloproteins (Fosso-Kankeu and Mulaba-Bafubiandi 2014).

9.2.2 Advantages and Disadvantages of Phytoremediation for Restoration of Freshwater Ecosystems

9.2.2.1 Advantages

Other than major advantages such as environment-friendliness and cost-effectiveness, phytoremediation techniques have vast assets. Phytoremediation is suitable for the decontamination of ecosystems with various organic and inorganic pollutants and even for sites with shallow and low levels of contaminants. It is applicable to the remediation of effluent in both in situ/ex situ environments and in situ applications are able to decrease the spread of contaminants via water. To apply such methods there is no need for highly specialized experts or any high-cost equipment (Ghosh and Singh 2005; Farraji et al. 2016).

9.2.2.2 Disadvantages

As phytoremediation is a completely natural process, it is a slow and seasonally effective treatment for clean-up and phytoextraction efficiency may decline under increasing metal concentrations. In most metal hyperaccumulator plants, their remediation efficiency is generally restricted by their slow growth and development rate as well as low biomass and thus several planting and harvesting are required for the decontamination of particular effluent. Hyperaccumulators may have a selective unique accumulation of specifically one metallic element and can be restricted to the number of extraction target metals and only limited depth can be reached by the roots. As there is a high probability and thus risk of accumulation of pollutants in fruit and other edible parts of absorbent plants, the threat of food chain contamination is higher in case of mismanagement and lack of proper care, which is a cause of concern (Ghosh and Singh 2005; Bhargava et al. 2012; Farraji et al. 2016).

9.3 Role of Metalloproteins in Phytoremediation

9.3.1 Heavy Metal Removal from Metal-Contaminated Sites

Heavy metals like zinc and copper are essential micronutrients for various plants for their physiological processes, which are particularly based on copper and zinc-dependent enzymes. Whereas, other nonessential heavy metal ions, such as mercury, cadmium (Cd), and lead can be toxic to plants due to their high reactivity. Like all living organisms, plants have evolved with the mechanisms to control, tolerate, and regulate the uptake, translocation, accumulation, and detoxification of these essential and nonessential heavy metals (Cobbett and Goldsbrough 2002). Plants synthesize metalloproteins in response to the stress of toxic heavy metals. In such a way, plants are able to tolerate metal stress from the complexation and/or sequestration of those heavy metal ions inside the cell organelles of plants, particularly vacuoles by spontaneously expressing such phytometalloproteins or as a response to stress (Fosso-Kankeu and Mulaba-Bafubiandi 2014). Plants usually use mechanisms of chelation or complexation and sequestration of metals with specific chelators or ligands for detoxification and to deal with toxic metal stress. The uptaken toxic metals via root cells are detoxified by their chelation or complexation with organic acids, amino acids, MTs, and PCs, which are widely distributed in plants. Both best-characterized metalloproteins, MTs and PCs form stable metal–ligand complexes with extracted metals in the cytosol and/or that complexes are then sequestered into the vacuole, which restricts the translocation and reactivity of toxic metals to aboveground organs and protects the plant from heavy metal damage (Goldsbrough 2000; Cobbett and Goldsbrough 2002; Hall 2002; Bhargava et al. 2012).

9.3.2 Metalloproteins as Sensitive Biomarkers for Heavy Metal Toxicity

A biomarker is an indicator, which can sense just observable or measurable changes including behavioral changes, in the range of molecular to the physiological level during a biochemical or biochemical response. Biomarkers have the potential to indicate just exposure as well as toxic effects of particular environmental stressors via their biosensing characteristics and responses (Cossu-Leguille and Vasseur 2013). Phytometalloproteins present in aquatic plants can act as biomarkers to indicate toxicity of heavy metals in a metal-contaminated freshwater ecosystem (Joshi et al. 2016). MTs provide defense against the toxicity of essential and nonessential metals by their detoxification. Various heavy metals such as Cu, Hg Cd, and Pb induce the biosynthesis of MTs as a defense mechanism of the plant against metal stress. Thus, the production of MTs serves as a sensitive biomarker for sensing the ecotoxicological consequence of essential and nonessential toxic metals such as Cd, copper, and mercury (Dabrio et al. 2002). Different modern, up-to-date techniques such as HPLC-inductively coupled with plasma mass spectrometry (HPLC-ICP-MS), HPLC-electrospray ionization-tandem mass spectrometry (HPLC/ESI-MS/MS), and RP-HPLC coupled with fluorescence detection (Nischwitz et al. 2003; Alhama et al. 2006) used for analysis for the analysis. The increasing production and thus concentration of MTs has served to be a biomarker of heavy metal toxicity or exposure (Cataldo et al. 2011). The antioxidant glutathione (GSH) serves a major role as a biomarker, which defends against metal exposure and toxicity associated with oxidative stress (Joshi et al. 2016). PC also served as a probable biomarker for metal exposure or biomonitoring trace metal contamination in both aquatic and terrestrial ecosystems as numerous studies have reported the straight connection between PC synthesis and heavy metal exposure and toxicity (Pawlik-Skowronska 2002; Ferrat et al. 2003; Sun et al. 2005). The PC production in aquatic macrophytes, such as *Lemnaaequinoctialis*, *Pistia stratiotes* L., *Hydrilla verticillata* (l.f.) *Royle*, and *Vallisneria spiralis* also help investigate even low and moderate levels of Cd toxicity in polluted freshwater as a sensitive biomarker (Wang et al. 2009; Wang et al. 2010). Similarly, *Ceratophyllum demersum* L., a submerged aquatic macrophyte, synthesizes PCs in response to lead accumulation and thus acts as a potential biomarker for biomonitoring of lead-contaminated water systems (Mishra et al. 2006).

9.4 Use of Phytometalloproteins for Remediation of Contamination and Restoration of Freshwater Ecosystems

Various conventional methodologies such as chemical precipitation, oxidation, and reduction, reverse osmosis, ultrafiltration, coagulation-flocculation, ion exchange, electrochemical treatment, etc., are used for the removal of heavy metal contaminants. Although these methodologies have specific benefits, they are not cost-effective and environmentally friendly (EPA 1997; Volesky 2001; Rai 2009). Boyd (1970), Stewart (1970), Wooten and Dodd (1976), and Conwell Jr. et al. (1977) are some of the pioneers, who revealed the capacity of aquatic plants to remove nutrients while some of the researchers have proved the major role of aquatic plants in removing organic pollutants from aquatic ecosystems (Seidal 1976; Wolverton and McDonald 1976; Wolverton and Mckown 1976). Since this emerging and innovative approach is a green clean-up technology, plants are solar-driven and make this methodology an eco-friendly as well as cost-effective tool to achieve a sustainable environment. The Greek word "phyto," i.e. plant and the Latin word "remedium," i.e. remediation or restoring balance conjugated to form the term "phytoremediation," which consists of reducing contaminants in polluted air, water, or soil by using plants. These naturally occurring plants sometimes improve their potential; genetically modified plants have the potential to uptake, accumulate, transport, degrade, or sequester various pollutants such as toxic metals, explosives, pesticides, and crude oil (Flathman and Lanza 1998; Prasad and Freitas 2003). The use of large vascular plants having greater biomass, faster development rate, and potential to adapt to a huge number of environmental factors can help overcome all the limitations of the phytoremediation process. *Azolla*, the water fern has many advantages such as the potential to develop in nitrogen deficit sites, free-floating habitat, tolerance against a huge range of contaminants, and uptake and accumulation of various toxic metals from polluted environments, and thus, in future, *Azolla* can be a more promising green phytoremediation tool (Arora et al. 2006; Umali et al. 2006) (Table 9.1).

Table 9.1 Efficient HM-hyperaccumulators for phytometalloproteins-mediated freshwater restoration.

Hyperaccumulator	Type of plant	Types of heavy metal(oid)s removed	References
Pistia stratiotes (water cabbage or water lettuce)	Free-floating perennial macrophyte	Accumulation and sequestration of As, Cr, Mn, Zn	Sytar et al. (2016)
Eichhornia crassipes (water hyacinth)	Free-floating, perennial, freshwater-hydrophyte	Accumulation of Cd, Cu, Zn, Fe, Pb, and Cr Removal of high concentration of Co, Cd, and As	Sytar et al. (2016), Sukumaran (2013)
Egeria densa (Brazilian waterweed)	Submerged aquatic macrophyte	Detoxification of Cd	Malec et al. (2009)
Spirodela polyrhiza (giant duckweed)	Fast-growing, freshwater-floating plant	Accumulation of B, Cu, As	Sytar et al. (2016), Mishra and Tripathi (2008)
Typha latifolia (cattails)	Perennial herbaceous-freshwater plant	Removal of Co, Cu Al Cd, and As	Sukumaran (2013)
Lemna minor (lesser duckweed)	Freshwater-floating plant	Removal of Pb, B, Cd, As, and Cr Potential Cd accumulator	Sytar et al. (2016)
Eleocharis acicularis (dwarf hairgrass or needle spikerush)	Hair-thin aquarium plant	Accumulation of Fe, Cu, Pb, Mn, Cr, Cd, As, and Zn	Ha et al. (2009), Sakakibara et al. (2011)
Ludwigia palustris (creeping primrose)	Aquatic freshwater plant	Removal of Cu, Fe, Hg, and Zn Accumulation Cu and Hg	Kamal et al. (2004)
Mentha aquatica (water mint)	Herbaceous, perennial-aquatic plant	Removal of Cu, Fe, Hg, and Zn Tolerance of As	Kamal et al. (2004)

9.5 Heavy Metal Uptake from Contaminated Water

9.5.1 Metal Uptake

Metal uptake processes are conducted by both living and dead cells with different mechanisms. Dead cells use passive mechanisms to adsorb metals using their dead biomass and this phenomenon is called bio-sorption. However, the active, i.e. energy-consuming mechanisms of metal uptake by living cells with their living biomasses and consequently sequester that metals are referred to as bioaccumulation (Ghori et al. 2016). Biosorption is a term referring to the plants and microorganisms, which take up, remove, and restore toxic heavy metals from contaminated water or effluents by using mechanisms of physicochemical interactions of heavy metals with cellular constituents of particular plants, which involve absorption, ion exchange, adsorption, complexation, and precipitation (Kapoor and Viraraghavan 1998). This mechanism carries the removal of contaminants which may be organic or inorganic; soluble or insoluble components using nonliving biomass or biological material due to the high force of attraction between the two (Volesky and Holan 1995; Ah Ahemad and Malik 2011). Biosorption is a passive process which involves the metabolism-independent accumulation of metals and consists of two phases (Ashkenazy et al. 1997):

1) Solid phase: biomass/sorbent/biosorbent material
2) Liquid phase i.e. solvent, usually water containing a dissolved constituent, i.e. sorbet

However, bioaccumulation is a comparatively slower process, which involves metabolism-dependent intracellular uptake of metal(oid)s via organelles of living organisms. Plants, which are involved in phytoaccumulation must have the potential to accumulate and store heavy metals in their organs such as leaves and roots, particularly in cell organelles such as vacuoles. The metal uptake initiates from the intake of heavy metals into the root cells. Then they are transported via the xylem to aboveground parts, i.e. shoots and aerial parts of the plant and stored into harvested parts such as leaves (Mahmood 2010).

As metals are available for uptake, they enter into roots via two pathways (Ghori et al. 2016):

1) **Apoplastic pathway**: pathway, where soluble metal fractions pass via the intercellular spaces without entering inside the cells.
2) **Symplastic pathway**: energy-consuming pathway, where nonessential metals travel through the cytoplasm.

There are variations in the transport pathways due to the use of various mechanisms for the uptake of different heavy metal ions by various plant species (Lombi et al. 2002; Lu et al. 2009).

9.5.2 Phytoextraction or Phytoaccumulation Mechanism

Phytoextraction, also known as phytoaccumulation, refers to the uptake of pollutants, mainly toxic metal ions from contaminated soil, sediments, and water via root cells and the transport that heavy metals to the tissues of shoots particularly leaves (Kumar et al. 1995). As compared with nonhyperaccumulators, hyperaccumulators have the ability to uptake 50–500 times higher metals without any adverse effect on their own growth (Mahar et al. 2016). These plants specifically belong to rare species of little population size with finite distribution in the ecosystem and usually small and slow-growing (Pollard et al. 2002).

Thus, hyperaccumulation involves the following steps (Ghori et al. 2016):

1) Uptake of metals from water or soil via roots and their translocation across the plasma membrane of root cells
2) Xylem loading and translocation from root to shoot
3) Detoxification and sequestration of metals or metal complexes mostly in vacuoles.

To date, less than 0.2% of all angiosperms, i.e. 45 plant families consisting of more than 450 plant species have been known to have potential as hyperaccumulators, which majorly involved Brassicaceae, Euphorbiaceae, Asteraceae, Fabaceae, Lamiaceae, Caryophyllaceae, Poaceae, and Violaceae families (Bhargava et al. 2012; Bolan et al. 2014; Khalid et al. 2016). *P. stratiotes* L., a free-floating macrophyte water lettuce is successfully used for the treatment of metal-polluted freshwater ecosystems as it has the potential to produce PCs in response to metal exposure, mainly Cd. Also, it has capacity to uptake other metals such as zinc, copper, and manganese (Wang et al. 2010). Similarly, *V. spiralis*, a rooted-submerged aquatic macrophyte also helps with the removal of Cd metal, even at its low concentration in water (Wang et al. 2009).

9.5.3 Factors Influencing the Ability to Uptake Heavy Metals

Many factors affect heavy metal uptake, which can be related to plant biomass, age, microbial colonization, metal size and translocation to various parts, and other environmental conditions (Singh et al. 2017). Plant species, i.e. hyperaccumulator or not, its age, external environment factors, size, and concentration of the metals and its translocation to different parts are the crucial factors that affect the uptake and accumulation of heavy metals in the plants (Singh et al. 2017). The major factors that can influence the metal uptake are:

9.5.3.1 The Plant Species

As all plant species have not adapted to the potential of metal uptake, the uptake can be affected by characteristic plant species. The selection of suitable, i.e. contaminant-specific hyperaccumulator species, which can easily accumulate heavy metals and also can tolerate its stress to detoxify all the toxicity are capable of successful phytoaccumulation for restoration of contaminated water and effluents (Burken and Schnoor 1996; Tu et al. 2004; Rodriguez et al. 2005).

9.5.3.2 The Root Zone

In phytoremediation, the initial and major role of uptake is exploited by root cells or root zone. It has a fundamental capacity to absorb contaminants such as toxic heavy metals from contaminated sites, and store or metabolize them inside the tissue. Thus, morphological characteristic and adaptation of the root zone serve a vital role in the metal uptake mechanism (Merkl et al. 2005).

9.5.3.3 pH

pH of contaminated water seems to have a major influence over the metal uptake process. The metal solubility and nature of biomass binding sites are influenced by the pH of contaminated water. Hence, at low pH values, the biosorption of metals like Ni, Cu, Co, Cd, and Zn is reduced (Greene and Darnall 1990; Deng and Wang 2012). But, as the metal uptake potential can decrease with the increasing pH, it is very important to get the highest metal uptake by deciding the optimum pH (Shamim 2018).

9.5.3.4 Temperature

Temperature variation can affect heavy metals uptake efficiency during phytoaccumulation. However, during biosorption, within 20–35 °C or even at higher temperatures it remains unaffected. But, living cells may damage permanently due to these high temperatures and the uptake of metal also can decrease (Ahalya et al. 2003; Goyal et al. 2003). The adsorption mechanism is usually exothermic and its extent of uptake increases with decreasing temperature (White et al. 1997). Alternatively, temperature also affects the development of plants and consequently the length of the root (Merkl et al. 2005).

9.5.3.5 The Metal Affinity to the Biosorbent

The binding affinity of metal-binding groups depends on surface charges and permeability of the biomass, which can be affected by different physical/chemical pretreatments. To increase the metal uptake potential, the affinity of metal binding components can be manipulated by using pretreatment techniques for the biomass with acid, alkali detergents as well as heat (Zouboulis et al. 1997; Shamim 2018).

9.5.3.6 The Biomass Concentration

The concentration of biomass in contaminated water or effluent affects the specific metal uptake (Modak and Natarajan 1995). The metal uptake is restricted at lower concentrations of biomass due to a decrease in biosorbent concentration leading to limit the access of binding sites to the metal ions, whereas, lower biomass concentration allows the interface of binding sites (Fourest and Roux 1992; Nuhoglu and Malkoc 2005).

9.5.3.7 Initial Concentration of Metal Ions

To manage all mass transfer resistance of metal ions between the solid and aqueous phases, the initial concentration of metals gives an essential driving force (Ahalya et al. 2003). The increasing initial concentration of metals leads to an increase in metal adsorption via biomass. The low initial concentration of metal can help get the optimum removal percentage of metal. Hence, the initial concentration of metal can increase the rate of metal uptake at given concentration of biomass (Shamim 2018).

9.6 Phytometalloproteins in Remediation of Contaminated Freshwater Ecosystems

The redox potential, total and available concentrations, cation exchange capacity, pH, the presence of Al, Fe, and Si oxide, texture, organic matter, aeration, and climatic conditions of metal-polluted effluent or sediment influence over metal uptake (Kabata-Pendias 2011). Plants adapt different mechanisms for the uptake and translocation of contaminants, which are connected with metal distribution, storage, sequestration, and detoxification of metals in harvestable parts, particularly leaves and roots. As adaptive responses, plants shift the expression of the transporter genes, which are mainly accountable for the uptake, transportation, complexation, and/or sequestration of metals for different concentrations of metal ions (Memon and Schröder 2006). These transporter genes design various membrane transporters, organic acids, and/or amino acids, which are responsible for the translocation of metal ions from roots to shoots of plants (Jabeen et al. 2009). The tolerance mechanism in plants for metal toxicity or deficiency is expressed either by accumulating metal into plant tissues or excluding the metal for further transport (Baker 1987). The restricted uptake and accumulation of toxic metal ions decide the exclusion mechanism via:

1) increasing the binding potential of metal ions to the cell wall
2) allowing the transport of metal ions via the plasma membrane by limiting their absorption via adaptation of the ion channels to prevent their intake into the cytoplasm; and/or active efflux pump ejection of the metal ions from the cell and forming metal–ligand complexes outside the plasma membrane (Tong et al. 2004).

Metal accumulation and tolerance to high metal(loid) concentrations inside the plant cell are regulated via:

1) Metal ion inactivation by forming metal-chelate complexes, where PC, GSH, amino acids (AA), and organic acids (OA).
2) Metal compartmentalization in cell organelles, mainly vacuoles and/or sequestration of metals by the proteins involved in metal transport to activation site, i.e. chaperons (Tong et al. 2004; Singh et al. 2016).

Inside the cell cytoplasm, As forms a complex with PCs, i.e. (As-PCs) or with GSH, i.e. (As-GSH), which is then translocated to the vacuole by membrane transporter, particularly ABC transporter (Tripathi et al. 2007). Various aquatic free-floating, emergent, and submerged macrophytes, such as *Lemna aequinoctialis*, *P. stratiotes* L. (water lettuce), *Hydrilla vertcillate* (l.f.) *Royle*, and *V. spiralis* are identified as potential hyperaccumulators that play a major role in toxic metal removal from metal-contaminated water systems (Wang et al. 2009).

9.6.1 Metal Tolerance: Response of Metalloproteins to Heavy Metal Stress in Plants

Plants are classified into three categories based on their resistance and tolerance strategy to toxic heavy metals (Singh et al. 2017):

1) **HM-sensitive plants**: Does not show resistance to uptaken toxic metals via root cells and consequently their biochemical machinery collapse due to increased concentration of heavy metal in water.
2) **HM-resistant plants/excluders**: Prevents the accumulation of toxic metals inside the cells by using mechanisms, which restrict the transport of metals from soil to root cells.
3) **Hyperaccumulators**: Developed with various cellular and molecular adaptations, which actively take up heavy metals from water and accumulate them within their own parts.

These hyperaccumulator plants with a unique mechanism can tolerate heavy metal stress and thus be able to restore freshwater systems (Chaudhary et al. 2016). The hyperaccumulators involve three biochemical mechanisms to resist heavy metal stress (Kumar et al. 2016):

1) Cell wall binding
2) Active transport of ions into vacuoles
3) Chelation or complexation of metals, i.e. via metalloproteins

The metal uptake through phytoextraction is facilitated by the metal adsorption via root cells and hyperaccumulator plants have the capacity to tolerate these adsorbed heavy metals. This capacity of tolerance determines the toxic metal accumulation potential of plants. Extracellular and/or intracellular chelation and complexation by high-affinity ligands is the main process, which reduces the concentration of free reactive metals and consequently restricts their reactivity and solubility within the cytosol and thus associated with hyperaccumulation and detoxification of heavy metal water pollutants within plant cells. The metal chelating ligands, i.e. PCs and MTs in the plants play an important role in plant metal tolerance and thus respond to heavy metal stress within plants (Goldsbrough 2000; Mello-Farias and Chaves 2008; Fosso-Kankeu and Mulaba-Bafubiandi 2014).

9.6.1.1 Metal Tolerance Regulation via Phytochelatins

PCs play a significant role in mechanisms of regulation and tolerance of toxic metals as they bind to multiple heavy metals and/or metalloids and form PC-metal complexes with its free sulfur present in cysteine. The detoxification of toxic heavy metals involves cytosolic strategies and exogenous strategies through PCs from the uptake via roots and metal binding until its transport to shoots. Exogeneous strategies include specific metal occurrence sites, exudation, pH of the rhizosphere, metal stabilization by binding to the cell walls and cytosolic strategies include complex (PCs-metal) formation, sequestration, and its transformation into less toxic and comparatively stable forms, compartmentalization, and storage in cell organelles (Dago et al. 2014; Polak et al. 2014; Shine et al. 2015). This stabilization of metal ions due to exogenous strategies consequently reduces the toxicity of heavy metals and also affects their transportation to different cellular compartments. PC–metal complexes formed in plants reduce the binding potential of toxic metal to the cell wall leading to tolerance for even high levels of metal ions and also helping their detoxification by compartmentalization. These PC-metal complexes are stable and less toxic than free metal ions (Sunitha et al. 2013; Sharma et al. 2016).

9.6.1.2 Metal Tolerance Strategies via Metallothioneins

MTs are the chelators or phytometalloproteins, which have a thiol group present in cysteine and thus strong affinity toward heavy metals are involved in mechanisms of tolerance, detoxification, and metabolism of toxic heavy metals. MTs defend plants against both essential and nonessential heavy metal stress by reducing their reactivity by forming stable MT-metal complexes, i.e. metalloproteins. MTs serve a major role in maintaining the homeostasis uptaken of toxic metal ions, their sequestration by forming MT-metal complexes and defending the plant parts and biochemical mechanisms against intracellular oxidative stress (Sunitha et al. 2013; Sharma et al. 2016).

9.6.1.3 Metal Tolerance via Organic Acids and Amino Acids

One of the mechanisms of metal tolerance is also handled by various metal-binding organic acids and amino acids due to their reactivity and binding capacity with different metal ions as potential ligands of heavy metals. These ligands are not only involved in heavy metal tolerance but also in their transport through the xylem, and sequestration into vacuoles (Eutrópio et al. 2016).

9.6.2 Hyperaccumulation and Detoxification: Complexation and Sequestration of Metals by Metalloproteins Within Plants

Hyperaccumulation pattern is assumed to initiate from metal uptake via roots to their complexation and compartmentalization into leaves, particularly vacuoles, on the basis of the hypothesis of tolerance. The detoxification of heavy metals based on PCs is a protective mechanism involved in both lower plants and higher vascular plants mainly for sequestration and removal of accumulated Cd, which can be successfully employed for phytoremediation of water polluted with Cd metal ions. The factors such as plant species, water pH, temperature, variety of toxic metals, and their bioavailability influence the rate of hyperaccumulation (Chaudhary et al. 2016; Kumar et al. 2016). Hyperaccumulator plants have a genetic disposition for a unique mechanism which allows them a rapid and active translocation of heavy metals to the aboveground organs via the xylem and accumulate high concentrations of metals such as chromium, cobalt, Cd, selenium, lead, mercury, and silver. These plants having the capacity for hyperaccumulation can be used for the phytoremediation of metal-polluted water ecosystems for the removal of heavy metal ions (Fosso-Kankeu and Mulaba-Bafubiandi 2014). Coontail, i.e. *C. demersum* L., a submerged free-floating aquatic plant, induces metal tolerance and detoxification for lead by the synthesis of PCs (Mishra et al. 2006).

9.7 Genetically Engineered or Modified Metalloproteins for Improved Remediation of Contaminated Water

Phytoremediation is a green technology to clean up metal and/or metalloid-contaminated water and sediments as it is both eco-friendly and cost-effective. To improve the process of phytoremediation, enhance the ability of hyperaccumulators, and promote phytoaccumulation from metal-contaminated water, transgenic approaches can be employed for the accumulation of metals, mainly Cd, Pb, Cu, As, and Se. These approaches involved the improved production of enzymes of sulfur metabolism, metal transporters, and production of metal-binding proteins or metal-detoxifying chelators, i.e. PCs and MTs (Das et al. 2016). To enhance the removal potential of hyperaccumulators and improve their tolerance to heavy metal(oid)s with an understanding of the contribution of metal-binding proteins such as MTs and PCs for the phytoremediation process, it is important to genetically modify that metal-binding protein of hyperaccumulators. Misra and Gedamu were the first researchers, who experimentally reported the expression of MTs introduced into transgenic plants in 1989. They introduced the human gene MT-II into oilseed rape and tobacco plants, which significantly improved the metal tolerance of these genetically modified plants, particularly to Cd metal ion (Misra and Gedamu 1989). Afterward, D'Souza et al. used bacterial glutathione reductase for the transformation of the Indian mustard in its cytoplasm and chloroplast, which enhanced the tolerance of that mustard toward Se (D'Souza et al. 2000). Moreover, another study was conducted using the introduction of Glutathione-S-transferase gene from tobacco plant into *Arabidopsis thaliana*, which significantly enhanced the tolerance of transgenic *Arabis thaliana* to metals, particularly, Al and Cu (Ezaki et al. 2000). There are three genetically engineering approaches to improve metal uptake:

1) To improve the number of metal-uptake sites
2) To alter the specificity of the metal-uptake system
3) To increase the intracellular metal-binding sites.

The bacterial gene merA was transferred to *A. thaliana* and tobacco plants for their metal tolerance to Hg a. The vacuolar transporter gene CAX-2 from *A. thaliana* was used to transform tobacco plants for enhancing the accumulation potential of Mn, Ca, and Cd (Arazi et al. 1999).

9.8 Conclusion

Heavy metal contamination of freshwater ecosystems is a global issue of environmental concern and dangerous ecological risk. Phytoremediation technologies employing mechanisms of metalloproteins can be helpful for biomonitoring and restoration of freshwater ecosystems as well as effluents. Efficient hyperaccumulators such as *P. stratiotes* (water lettuce), *L. minor* (lesser duckweed), and *d crassipes* (water hyacinth) are employed for the removal of toxic heavy metals due to their metalloproteins-mediated mechanisms of uptake, accumulation, transportation, sequestration, compartmentalization, and detoxification of metals within a plant's cell organelles. Phytometalloproteins include PCs and MTs as major metal-binding ligands, which are mostly present in roots and leaves of hyperaccumulators and bind with various toxic heavy metals in water and effluents. The concentration of these metalloproteins in plants decides the toxicity of water, effluents, sediments, or soil, where they are growing. Thus, phytometalloproteins serve a major role as potential HM-biomarkers for estimating metal toxicity. Understanding the abilities and limits of freshwater ecosystems to respond to human-generated pressures is becoming a central issue for cultures and a challenge for science. This article explores trends in alterations to freshwater ecosystems, discusses the ecological consequences of biophysical alterations expected to occur in the next 20–30 years, and identifies some of the major scientific challenges and opportunities to effectively address the changes. Understanding the abilities and limits of freshwater ecosystems to respond to human-generated pressures is becoming a central issue for cultures and a challenge for science. This article explores trends in alterations to freshwater ecosystems, discusses the ecological consequences of biophysical alterations expected to occur in the next 20–30 years, and identifies some of the major scientific challenges and opportunities to effectively address the changes.

References

Ah Ahemad, M. and Malik, A. (2011). Bioaccumulation of heavy metals by zinc resistant bacteria isolated from agricultural soils irrigated with wastewater. *Bacteriol. J.* 2: 12–21.

Ahalya, N., Ramachandra, T.V., and Kanamadi, R.D. (2003). Biosorption of heavy metals. *Res. J. Chem. Environ.* 7: 71–79.

Alhama, J., Romero-Ruiz, A., and Lo'pez-Barea, J. (2006). Metallothionein quantification in clams by reversed-phase high-performance liquid chromatography coupled to fluorescence detection after monobromobimane derivatization. *J. Chromatogr. A* 1107: 52–58.

Ali, H., Khan, E., and Sajad, M.A. (2013). Phytoremediation of heavy metals—concepts and applications. *Chemosphere* 91 (7): 869–881.

Arazi, T., Sunkar, B., and Fromm, H.A. (1999). Tobacco plasma membrane calmodulin binding transporter confers Ni^{+} tolerance and Pb^{2+} hypersensitivity in transgenic plants. *Plant J.* 20: 171–182.

Arora, A., Saxena, S., and Sharma, D.K. (2006). Tolerance and phytoaccumulation of chromium by three *Azolla* species. *World J. Microbiol. Biotechnol.* 22: 97–100.

Ashkenazy, R., Gottlieb, L., and Yannai, S. (1997). Characterization of acetone-washed yeast biomass functional groups involved in lead biosorption. *Biotechnol. Bioeng.* 55: 1–10.

Awasthi, G., Nagar, V., Mandzhieva, S. et al. (2022). Sustainable amelioration of heavy metals in soil ecosystem: existing developments to emerging trends. *Fortschr. Mineral.* 12 (1): 85.

Baker, A.J.M. (1987). Metal tolerance. *New Phytol.* 106: 93–111.

Bhargava, A., Carmona, F.F., Bhargava, M., and Srivastava, S. (2012). Approaches for enhanced phytoextraction of heavy metals. *J. Environ. Manage.* 105: 103–120.

Bhattacharya, T. and Chakraborty, S. (2019). Eco-restoration potential of vegetation for contaminated water bodies. In: *Phytomanagement of Polluted Sites* (ed. V.C. Pandey and K. Bauddh), 425–446. Elsevier.

Bolan, N.S. and Duraisamy, V. (2003). Role of inorganic and organic soil amendments on immobilisation and phytoavailability of heavy metals: a review involving specific case studies. *Soil Res.* 41 (3): 533–555.

Bolan, N., Kunhikrishnan, A., Thangarajana, R. et al. (2014). Remediation of heavy metal (loid)s contaminated soils to mobilizeor to immobilize. *J. Hazard. Mater.* 266: 141–166.

Boyd, C.E. (1970). Vascular aquatic plants for mineral nutrient removal from pollutant water. *Econ. Bot.* 24: 95–103.

Burken, J.G. and Schnoor, J.L. (1996). Phytoremediation: plant uptake of atrazine and role of root exudates. *J. Environ. Eng.* 122 (11): 958–963.

Cataldo, J.R., Hidalgo, M.E.L., Neaman, A.H., and Gaete, O. (2011). Use of molecular biomarkers in *Eisenia foetida* to assess copper toxicity in agricultural soils affected by mining activities. *J. Soil Sci. Plant Nutr.* 11: 57–70.

Chaudhary, K., Jan, S., and Khan, S. (2016). Heavy metal ATPase (HMA2, HMA3, and HMA4) genes in hyperaccumulation mechanism of heavy metals. In: *Plant Metal Interaction* (ed. P. Ahmed), 545–556. Elsevier.

Cobbett, C. and Goldsbrough, P. (2002). Phytochelatins and metallothioneins: roles in heavy metal detoxification and homeostasis. *Annu. Rev. Plant Biol.* 53 (1): 159–182.

Conwell, D.A. Jr., Zoltek, J., Patrinely, C.D. et al. (1977). Nutrient removal by water hyacinths. *J. Water Pollut. Control Fed.* 49: 57–65.

Cossu-Leguille, C. and Vasseur, P. (2013). Aquatic biomarkers. In: *Encyclopedia of Aquatic Ecotoxicology* (ed. J.F. Ferard and C. Blaise), 49–66. Springer.

Dabrio, M., Rodŕiguez, A.R., Bordin, G. et al. (2002). Recent developments in quantification methods for metallothionein. *J. Inorg. Biochem.* 88: 123–134.

Dago, A., Gonzalez, I., Arino, C. et al. (2014). Evaluation of mercury stress in plants from the Almadén mining district by analysis of phytochelatins and their Hg complexes. *Environ. Sci. Technol.* es405619y.

Das, S., Dash, H.R., and Chakraborty, J. (2016). Genetic basis and importance of metal resistant genes in bacteria for bioremediation of contaminated environments with toxic metal pollutants. *Appl. Microbiol. Biotechnol.* 100: 2967–2984. https://doi.org/10.1007/s00253-016-7364-4.

Deng, X. and Wang, P. (2012). Isolation of marine bacteria highly resistant to mercury and their bioaccumulation process. *Bioresour. Technol.* 121: 342–347.

D'Souza, M.P., Pilon-Smits, E.A.H., and Terry, N. (2000). The physiology and biochemistry of selenium volatilization by plants. In: *Phytoremediation of Toxic Metals: Using Plants to Clean up the Environment* (ed. I. Raskin and B.D. Ensley), 171–188. John Wiley & Sons.

EPA (Environmental Protection Agency) (1997). Electrokinetic laboratory and field processes applicable to radioactive and hazardous mixed waste in soil and ground water. EPA 402/R-97/006. Washington, DC.

Eutrópio, F.J., Ramos, A.C., da Silva Folli-Pereira, M. et al. (2016). Heavy metal stress and molecular approaches in plants. In: *Plant Metal Interaction* (ed. P. Ahmed), 531–543. Elsevier.

Ezaki, B., Gardner, R.C., Ezaki, Y., and Matsumoto, H. (2000). Expression of aluminium induced genes transgenic arabidopsis plants can ameliorate aluminium stress and/or oxidative stress. *Plant Physiol.* 122: 657–665.

Farraji, H., Zaman, N.Q., Tajuddin, R., and Faraji, H. (2016). Advantages and disadvantages of phytoremediation: a concise review. *Int. J. Environ. Sci. Technol.* 2: 69–75.

Ferrat, L., Pergent-Martini, C., and Romeo, M. (2003). Assessment of the use of biomarkers in aquatic plants for the evaluation of environmental quality: application to seagrasses. *Aquat. Toxicol.* 65: 187–204.

Flathman, P.E. and Lanza, G.R. (1998). Phytoremediation: current views on an emerging green technology. *J. Soil Contam.* 7: 415–432.
Fosso-Kankeu, E. and Mulaba-Bafubiandi, A. (2014). Implication of plants and microbial metalloproteins in the bioremediation of polluted waters: a review. *Phys. Chem. Earth. A/B/C* 67: 242–252.
Fourest, E. and Roux, J.C. (1992). Heavy metal biosorption by fungal mycelial by-product, mechanisms and influence of pH. *Appl. Microbiol. Biotechnol.* 37: 399–403.
Ghori, Z., Iftikhar, H., Bhatti, M.F. et al. (2016). Phytoextraction: the use of plants to remove heavy metals from soil. In: *Plant Metal Interaction* (ed. P. Ahmed), 385–409. Elsevier.
Ghosh, M. and Singh, S. (2005). A review on phytoremediation of heavy metals and utilization of it's by products. *Asian J. Energy Environ.* 6 (4): 18.
Goldsbrough, P. (2000). Metal tolerance in plants: the role of phytochelatins and metallothioneins. In: *Phytoremediation of Contaminated Soil and Water* (ed. N. Terry and G. Banuelos), 221–233. CRC Press LLC.
Goyal, N., Jain, S.C., and Banerjee, U.C. (2003). Comparative studies on the microbial adsorption of heavy metals. *Adv. Environ. Res.* 7: 311–319.
Greene, B. and Darnall, D.W. (1990). Microbial oxygenic photoautotrophs (cyanobacteria and algae) for metal-ion binding. *Microbial. Miner. Recovery* 277–302.
Ha, N.T.H., Sakakibara, M., and Sano, S. (2009). The phytoremediation of Sb, As, Cu, and Zn from contaminated water by the aquatic macrophyte *Eleocharis acicularis*. *Clean–Soil, Air, Water* 37 (9): 720–725.
Hall, J.L. (2002). Cellular mechanisms for heavy metal detoxification and tolerance. *J. Exp. Bot.* 53: 1–11.
Hwang, S., Lee, S., and Yoo, B. (2014). Ecological conservation and the restoration of freshwater environments in Korea. *Paddy Water Environ.* 15: 1–5.
Jabeen, R., Ahmad, A., and Iqbal, M. (2009). Phytoremediation of heavy metals: physiological and molecular mechanisms. *Bot. Rev.* 75: 339–364. https://doi.org/10.1007/s12229-009-9036-x.
Joshi, R., Pareek, A., and Singla-Pareek, S.L. (2016). Plant metallothioneins: classification, distribution, function, and regulation. In: *Plant Metal Interaction* (ed. P. Ahmed), 239–261. Elsevier.
Jumbe, A.S. and Nandini, N. (2009). Heavy metals analysis and sediment quality values in urban lakes. *Am. J. Environ. Sci.* 5 (6): 678.
Kabata-Pendias, A. (2011). *Trace Elements in Soils and Plants*, 4the. Boca Raton, FL; London; New York, NY; Washington, DC: CRC Press LLC.
Kamal, M., Ghaly, A.E., Mahmoud, N., and Cote, R. (2004). Phytoaccumulation of heavy metals by aquatic plants. *Environ. Int.* 29 (8): 1029–1039.
Kapoor, A. and Viraraghavan, T. (1998). Biosorption of heavy metals on *Aspergillus Niger*: effect of pretreatment. *Bioresour. Technol.* 63: 109.
Khalid, S., Shahid, M., Niazi, N.K. et al. (2016). A comparison of technologies for remediation of heavy metal contaminated soils. *J. Geochem. Explor.* 182: 247–268.
Kumar, P.B.A.N., Dushenkov, V., Motto, H., and Raskin, I. (1995). Phytoextraction: the use of plants to remove heavy metals from soils. *Environ. Sci. Technol.* 29: 1232–1238.
Kumar, R., Tripathi, R.M., and Gupta, A.K. (2014). Seasonal variation of heavy metal concentration in water of river Yamuna, Allahabad, Uttar Pradesh, India. *Int. J. Curr. Microbiol. App. Sci.* 3 (7): 945–949.
Kumar, R., Mishra, R.K., Mishra, V. et al. (2016). Detoxification and tolerance of heavy metals in plants. In: *Plant Metal Interaction* (ed. P. Ahmed), 335–359. Elsevier.
Lombi, E., Tearall, K.L., Howarth, J.R. et al. (2002). Influence of iron status on cadmium and zinc uptake by different ecotypes of the hyperaccumulator *Thlaspi caerulescens*. *Plant Physiol.* 128 (4): 1359–1367.
Lu, L., Tian, S.K., Yang, X.E. et al. (2009). Cadmium uptake and xylem loading are active processes in the hyperaccumulator *Sedum alfredii*. *J. Plant Physiol.* 166 (6): 579–587.
Mahar, A., Wang, P., Ali, A. et al. (2016). Challenges and opportunities in the phytoremediation of heavy metals contaminated soils: a review. *Ecotoxicol. Environ. Saf.* 126: 111–121.
Mahmood, T. (2010). Phytoextraction of heavy metals-the process and scope for remediation of contaminated soils. *Soil Environ.* 29 (2): 91–109.
Malec, P., Maleva, M.G., Prasad, M.N.V., and Strzałka, K. (2009). Identification and characterization of Cd-induced peptides in *Egeria densa* (water weed): putative role in Cd detoxification. *Aquat. Toxicol.* 95 (3): 213–221.
Mello-Farias, P.C. and Chaves, A.L.S. (2008). Biochemical and molecular aspects of toxic metals phytoremediation using transgenic plants. In: *Transgenic Approach in Plant Biochemistry and Physiology* (ed. M.E. Tiznado-Hernandez, R. Troncoso-Rojas, and M.A. Rivera-Dominguez), 253–266. Kerala, India: Research Signpost.

Memon, A.R. and Schröder, P. (2006). Implications of metal accumulation mechanisms to phytoremediation. *Environ. Sci. Pollut. Res.* 16: 162–175. https://doi.org/10.1007/s11356-008-0079-z.

Merkl, N., Schultze-Kraft, R., and Infante, C. (2005). Phytoremediation in the tropics–influence of heavy crude oil on root morphological characteristics of graminoids. *Environ. Pollut.* 138 (1): 86–91.

Mishra, V.K. and Tripathi, B.D. (2008). Concurrent removal and accumulation of heavy metals by the three aquatic macrophytes. *Bioresour. Technol.* 99 (15): 7091–7097.

Mishra, S., Srivastava, S., Tripathi, R.D. et al. (2006). Lead detoxification by coontail (*Ceratophyllumdemersum* L.) involves induction of phytochelatins and antioxidant system in response to its accumulation. *Chemosphere* 65 (6): 1027–1039.

Misra, S. and Gedamu, L. (1989). Heavy metal tolerant transgenic *Brassica napus* L. and *Nicotiana tabacum* L. plants. *Theor. Appl. Genet.* 78: 161–168.

Modak, J.M. and Natarajan, K.A. (1995). Biosorption of metals using nonliving biomass-a review. *Miner. Metall. Process.* 12: 189–196.

Mushtaq, N., Singh, D.V., Bhat, R.A. et al. (2020). Freshwater contamination: sources and hazards to aquatic biota. In: *Fresh Water Pollution Dynamics and Remediation* (ed. H. Qadri, R.A. Bhat, M.A. Mehmood, and G.H. Dar), 27–50. Singapore: Springer.

Nischwitz, V., Michalke, B., and Kettrup, A. (2003). Identification and quantification of metallothionein isoforms and superoxide dismutase in spiked liver extracts using HPLC-ESI-MS offline coupling and HPLC-ICP-MS online coupling. *Anal. Bioanal. Chem.* 375: 145–156.

Nuhoglu, Y. and Malkoc, E. (2005). Investigations of nickel (II) removal from aqueous solutions using tea factory waste. *J. Hazard. Mater.* 127: 120.

Ojuederie, O.B. and Babalola, O.O. (2017). Microbial and plant-assisted bioremediation of heavy metal polluted environments: a review. *Int. J. Environ. Res. Public Health* 14 (12): 1504.

Palmer, M., Allan, J.D., Meyer, J., and Bernhardt, E.S. (2007). River restoration in the twenty-first century: data and experiential knowledge to inform future efforts. *Restor. Ecol.* 15 (3): 472–481.

Parihar, K., Sankhla, M.S., and Kumar, R. (2019). Water quality status of Yamuna river and its toxic effects on humans. *Environ. Anal. Eco. Stud.* 6 (1): 597–601. https://doi.org/10.31031/EAsES.2019.06.000628.

Pawlik-Skowronska, B. (2002). Correlations between toxic Pb effects and production of Pb-induced thiol peptides in the microalga *Stichococcus bacillaris*. *Environ. Pollut.* 119: 119–127.

Polak, N., Read, D.S., Jurkschat, K. et al. (2014). Metalloproteins and phytochelatin synthase may confer protection against zinc oxide nanoparticle induced toxicity in *Caenorhabditis* elegans. *Comp. Biochem. Physiol. C* 160: 75–85.

Pollard, A.J., Powell, K.D., Harper, F.A. et al. (2002). The genetic basis of metal hyperaccumulation in plants. *Crit. Rev. Plant Sci.* 21: 539–566.

Prasad, M.N.V. and Freitas, H. (2003). Metal hyperaccumulation in plants—biodiversity prospecting for phytoremediation technology. *Electron. J. Biotechnol.* 6: 285–321.

Qadri, H. and Bhat, R.A. (2020). The concerns for global sustainability of freshwater ecosystems. In: *Fresh Water Pollution Dynamics and Remediation* (ed. H. Qadri, R.A. Bhat, M.A. Mehmood, and G.H. Dar), 1–13. Singapore: Springer.

Rai, P.K. (2009). Heavy metal phytoremediation from aquatic ecosystems with special reference to macrophytes. *Crit. Rev. Environ. Sci. Technol.* 39 (9): 697–753.

Rodriguez, L., Lopez-Bellido, F.J., Carnicer, A. et al. (2005). Mercury recovery from soils by phytoremediation. In: *Environmental Chemistry* (ed. E. Lichtfouse), 197–204. Berlin, Heidelberg: Springer.

Sakakibara, M., Ohmori, Y., Ha, N.T.H. et al. (2011). Phytoremediation of heavy metal-contaminated water and sediment by *Eleocharis acicularis*. *Clean–Soil, Air, Water* 39 (8): 735–741.

Sankhla, M.S. and Kumar, R. (2019). Contaminant of heavy metals in groundwater & its toxic effects on human health & environment. *Int. J. Environ. Sci. Nat. Res.* 18 (5): 555996. https://doi.org/10.19080/IJESNR.2019.18.555996.

Sankhla, M.S., Kumari, M., Nandan, M. et al. (2016). Heavy metals contamination in water and their hazardous effect on human health-a review. *Int. J. Curr. Microbiol. App. Sci.* 5 (10): 759–766.

Sankhla, M.S., Kumar, R., and Biswas, A. (2019a). Dynamic nature of heavy metal toxicity in water and sediments of Ayad river with climatic change. *Int. J. Hydrogen* 3 (5): 339–343. https://doi.org/10.15406/ijh.2019.03.00197.

Sankhla, M.S., Kumar, R., and Prasad, L. (2019b). Zinc impurity in drinking water and its toxic effect on human health. *Indian Internet J. Forensic Med. Toxicol.* 17 (4): 84–87. https://doi.org/10.5958/0974-4487.2019.00015.4.

Sankhla, M.S., Kumar, R., and Prasad, L. (2020). Variation of chromium concentration in Yamuna river (Delhi) water due to change in temperature and humidity. *J. Seybold Rep.* 15 (9): 293–299.

Seidal, K. (1976). Macrophytes and water purification. In: *Biological Control for Water Pollution* (ed. J. Tourbier and R.W. Pierson), 109–121. Philadelphia, PA: Pennsylvania University Press.

Shamim, S. (2018). Biosorption of heavy metals. *Bios* 2: 21–49.

Sharma, R., Bhardwaj, R., Handa, N. et al. (2016). Responses of phytochelatins and metallothioneins in alleviation of heavy metal stress in plants: an overview. In: *Plant Metal Interaction* (ed. P. Ahmed), 263–283. Elsevier.

Shine, A.M., Shakya, V.P., and Idnum, A. (2015). Phytochelatin synthase is required for tolerating metal toxicity in a basidiomycete yeast and is a conserved factor involved in metal homeostasis in fungi. *Fungal Biol. Biotechnol.* 2 (Pii): 3.

Singh, S., Parihar, P., Singh, R. et al. (2016). Heavy metal tolerance in plants: role of transcriptomics, proteomics, metabolomics, and ionomics. *Front. Plant Sci.* 6: 1143. https://doi.org/10.3389/fpls.2015.01143.

Singh, H., Verma, A., Kumar, M. et al. (2017). Phytoremediation: a green technology to clean up the sites with low and moderate level of heavy metals. *Austin Biochem.* 2 (2): 1012.

Singh, A., Sharma, A., Verma, R.K. et al. (2022). Heavy metal contamination of water and their toxic effect on living organisms. In: *The Toxicity of Environmental Pollutants* (ed. D. Dorta and D.P. De Oliveira), 1–19. Intechopen.

Skinder, B.M., Uqab, B., and Ganai, B.A. (2020). Bioremediation: a sustainable and emerging tool for restoration of polluted aquatic ecosystem. In: *Fresh Water Pollution Dynamics and Remediation* (ed. H. Qadri, R.A. Bhat, M.A. Mehmood, and G.H. Dar), 143–165. Singapore: Springer.

Sonone, S.S., Jadhav, S., Sankhla, M.S., and Kumar, R. (2021). Water contamination by heavy metals and their toxic effect on aquaculture and human health through food chain. *Lett. Appl. NanoBioSci.* 10 (2): 2148–2166.

Stewart, K.K. (1970). Nutrient removal potential of various aquatic plants. *Hyacinth Control J.* 8: 34–35.

Strayer, D.L. and Dudgeon, D. (2010). Freshwater biodiversity conservation: recent progress and future challenges. *J. N. Am. Benthol. Soc.* 29 (1): 344–358.

Sukumaran, D. (2013). Phytoremediation of heavy metals from industrial effluent using constructed wetland technology. *Appl. Ecol. Environ. Sci.* 1 (5): 92–97.

Sun, Q., Wang, X.R., Ding, S.M., and Yuan, X.F. (2005). Effects of exogenous organic chelates on phytochelatins production and its relationship with Cd toxicity in wheat (*Triticum aestivum* L.) under Cd stress. *Chemosphere* 60: 22–31.

Sunitha, M.S.L., Prashant, S., Kumar, S.A. et al. (2013). Cellular and molecular mechanisms of heavy metal tolerance in plants: a brief overview of transgenic plants over- expressing phytochelatin synthase and metallothionein genes. *Plant Cell Biotech. Mol. Biol.* 14: 33–48.

Sytar, O., Brestic, M., Taran, N., and Zivcak, M. (2016). Plants used for biomonitoring and phytoremediation of trace elements in soil and water. In: *Plant Metal Interaction* (ed. P. Ahmed), 361–384. Elsevier.

Tong, Y.P., Kneer, R., and Zhu, Y.G. (2004). Vacuolar compartmentalization: a second–generation approach to engineering plants for phytoremediation. *Trends Plant Sci.* 9: 8–9. https://doi.org/10.1016/j.tplants.2003.11.009.

Tripathi, A.K. and Pandey, S.N. (2009). *Water Pollution*. New Delhi: APH Publishing.

Tripathi, R.D., Srivastava, S., Mishra, S. et al. (2007). Arsenic hazards: strategies for tolerance and remediation by plants. *Trends Biotechnol.* 25: 158–165. https://doi.org/10.1016/j.tibtech.2007.02.003.

Tu, S., Ma, L.Q., Fayiga, A.O., and Zillioux, E.J. (2004). Phytoremediation of arsenic-contaminated groundwater by the arsenic hyperaccumulating fern *Pteris vittata* L. *Int. J. Phytorem.* 6 (1): 35–47.

Umali, L.J., Duncan, J.R., and Burgess, J.E. (2006). Performance of dead *Azolla* filiculoides biomass in biosorption of Au from wastewater. *Biotechnol. Lett.* 28: 45–49.

UNESCO UN-Water (2020). *United Nations World Water Development Report 2020: Water and Climate Change*. Paris: UNESCO.

Vanisree, C.R., Sankhla, M.S., Singh, P. et al. (2022). Heavy metal contamination of food crops: transportation via food chain, human consumption, toxicity and management strategies. In: *Environmental Impact and Remediation of Heavy Metals* (ed. H.M. Saleh and A.I. Hassan). IntechOpen.

Volesky, B. (2001). Detoxification of metal-bearing effluents: biosorption for the next century. *Hydrometallurgy* 59: 203–216.

Volesky, B. and Holan, Z. (1995). Biosorption of heavy metals. *Biotechnol. Prog.* 11: 235–250.

Wang, C., Sun, Q., and Wang, L. (2009). Cadmium toxicity and phytochelatin production in a rooted-submerged macrophyte *Vallisneria spiralis* exposed to low concentrations of cadmium. *Environ. Toxicol. Int. J.* 24 (3): 271–278.

Wang, C., Wang, L.Y., and Sun, Q. (2010). Response of phytochelatins and their relationship with cadmium toxicity in a floating macrophyte *Pistia stratiotes* L. at environmentally relevant concentrations. *Water Environ. Res.* 82 (2): 147–154.

Wani, K.A., Manzoor, J., Dar, A.A. et al. (2020). Heavy metal intrusion and accumulation in aquatic ecosystems. In: *Fresh Water Pollution Dynamics and Remediation* (ed. H. Qadri, R.A. Bhat, M.A. Mehmood, and G.H. Dar), 83–104. Singapore: Springer.

White, C., Sayer, J.A., and Gadd, G.M. (1997). Microbial solubilization and immobilization of toxic metals: key biogeochemical processes for treatment of contamination. *FEMS Microbiol. Rev.* 20: 503–516.

Wolverton, B.C. and McDonald, R.C. (1976). Don't waste waterweeds. *New Sci.* 71: 318–320.

Wolverton, B.C. and Mckown, M.M. (1976). Water hyacinth for removal of phenols from polluted waters. *Aquat. Bot.* 30: 29–37.

Wooten, J.W. and Dodd, D.J. (1976). Growth of water hyacinth in treated sewage effluent. *Econ. Bot.* 30: 29–37.

Yadav, H., Kumar, R., and Sankhla, M.S. (2020). Residues of pesticides and heavy metals in crops resulting in toxic effects on living organism. *J. Seybold Rep.* 15 (7): 1527–1541.

Yahya, A.N., Mohamed, S.K., and Mohamed, A.G. (2018). Environmental pollution by heavy metals in the aquatic ecosystems of Egypt. *Open Acc. J. Toxicol.* 3 (1): 555603.

Zouboulis, A.L., Matis, K.A., and Hancock, I.C. (1997). Biosorption of metals from dilute aqueous solutions. *Sep. Purif. Meth.* 26: 255–295.

10

Phytoremediation: The Way Forward

Muatasim Jan[1,2], Tawseef A. Mir[1,2], and Rakesh K. Khare[2]

[1] *School of Life and Allied Sciences, BFIT Group of Institutions, Dehradun, Uttarakhand, India*
[2] *Centre of Research for Ethnobotany, Government Model Science College, Jiwaji University, Gwalior, India*

10.1 Introduction

Increased industrialization and the role of human-related activities have led to a dramatic increase in the discharge of anthropogenic chemicals into the environment over the last few decades and have resulted in the contamination of soils and various other environmental problems throughout the world because of their detrimental effects on the ecosystem, soil productivity, and socioeconomic well-being (Conesa et al. 2012; Waseem et al. 2014). On one side, solid municipal waste has caused serious problems in the disposal and has become a serious concern (Ramachandra and Bachamanda 2007; Vidyasagaran et al. 2014a,b). On the other side, due to the problems like decreasing petroleum reserves, an increase in fossil fuel prices, global warming, and the impact of environmental pollution, there is a need for suitable alternative sources of energy. As a result of various natural (saline seeps and geological erosions) and anthropogenic actions (industrialization, construction, and mining), soil, surface, and ground water may become polluted due to various toxic chemicals. These contaminants may be organic or inorganic in nature, and come from the sources like waste materials, pesticides, explosives, fertilizers, radioactive fallout, and pharmaceuticals (Sparks 1995).

All these pollutants are not degraded in natural conditions and remain accumulated in the ecosystem for a long period of time, posing a great threat toward the stabilization of the environment. Inorganic pollutants, on the other hand, have highly toxic effects and have a negative impact on ecosystems, and due to their mutagenic nature, DNA damage can occur in fauna including humans (Baudouin et al. 2002). Indigenous technologies are costly, labor-intensive, destructive, and not eco-friendly. Hence, these are not adequate enough to remediate the contaminated substrate properly (Pandey et al. 2015a). Therefore, environmental problems as well as solid waste pollution need attention toward the remediation and detoxification of hazardous contaminants in a better way. Keeping in view these facts, the phytoremediation technique has attained greater attention during the past few years as an effective, cost-effective, and eco-friendly technique to ensure a clean, green, and hazard-free environment (Vidyasagaran and Kumar 2016, 2017). However, factors like soil nature, chelating agent, the bioavailability of the plant, environmental conditions, plant ability, effluent medium, and effluent characteristics influence the process of phytoremediation (Figure 10.1). The term "phytoremediation" is formed from Greek "phyto" meaning plant and Latin "remedium" meaning to correct or to eradicate the problem. Phytoremediation is purely a natural environmentally friendly biological process used for the remediation of soil, sediments, surface and ground water by extraction, degradation, stabilization, and volatilization of toxic metals, organic wastes, and radionuclides (Kirkham 2006; Zhang et al. 2013). This process is the man-engineered utilization of various plant species and some associated microbes to remove or immobilize environmental pollutants present in soil, water, sediments, etc. This plant-based remediation process is simple, low-cost, and low-labor-intensive, broadly accepted, eco-friendly, reliable, sustainable, and an encouraging technique, where some plants of socioeconomic and ecological importance are utilized to remediate the contaminated locations (Pandey et al. 2015a, 2016). This technology is used to remediate a wider angle of contaminants and is even ten times less expensive than other available indigenous remediation techniques (Nascimento and Xing 2006). With the processes of bioaccumulation, translocation, and degradation, plants have an inherent capability to neutralize both organic

Aquatic Contamination: Tolerance and Bioremediation, First Edition. Edited by Rouf Ahmad Bhat, Gowhar Hamid Dar, Fernanda Maria Policarpo Tonelli, and Saima Hamid.

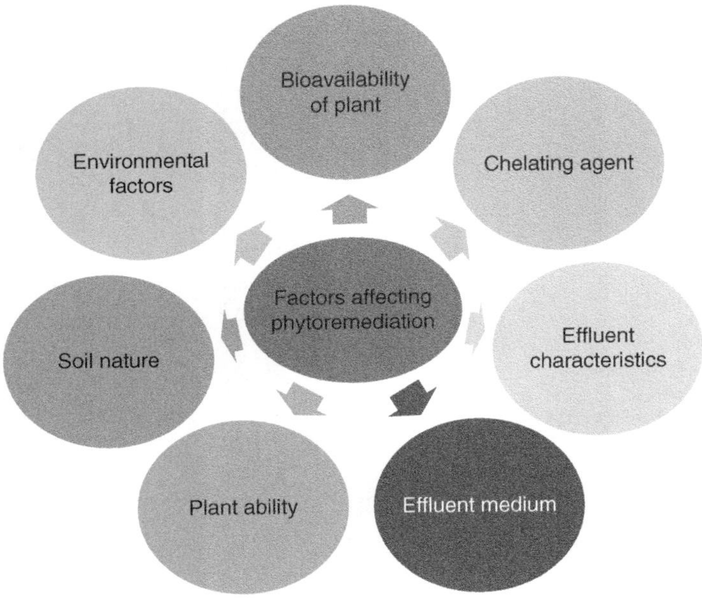

Figure 10.1 Factors influencing phytoremediation.

and inorganic pollutants and act as a biological sink for toxic contaminants. Plants like *Viola calaminaria* and *Thlaspi caerulescens* are regarded as the first plant species reported to have the capability of accumulating heavy metals in their leaves with high concentration (Baumann 1885). In addition, a number of plant species grown on metal-containing soils have been reported to have the capability of accumulating a large number of metals in different parts of the plant body without posing an adverse impact on their growth and development (Reeves and Brooks 1983). It is purely a new technique used for the remediation of contaminated locations through various trees, shrubs, and herbs. An increase in the polluted aquatic and terrestrial ecosystems due to the discharge of various contaminants by a number of industries is a burning issue across the globe. Waste dump sites are well defined, neglected, and uninhabited land areas possessing a major concentration of contaminants (Russ 2000). In the phytoremediation process, several types of vegetation including grasses, trees, and aquatic plant species are utilized to purify the particular site. It can be defined as the natural capability of several plant species which are also known as bioaccumulators, to accumulate, translocate, destroy, or make the pollutants less harmful. It is the collection of valuable plant-centered techniques that utilizes both natural and genetically modified plant species in order to purify the polluted areas (Flathman and Lanza 1998). The phytoremediation is some 300-year-old practice of waste-water remediation, but the novel idea of using plants and associated microorganisms to remediate heavy metal-contaminated sites was reported for the first time in 1983 (Blaylock 2008). The term phytoremediation was coined by Dr. Ilya Raskin (Raskin et al. 1994). Researchers and phytoremediation industries across the globe are busy in remediation of contaminated sites. However, it is the time to use sustainable phytoremediation. Sustainable phytoremediation, broadly termed as phytomanagement, is the process whereby contaminated sites are remediated by utilizing the economically and socioeconomically valuable native plant species (Pandey and Singh 2015). It is the process that will help increase the phytoremediation business as well as environmental remediation. This idea needs to be explored and different stakeholders including researchers, practitioners, policymakers, and other phytoremediation industries are needed to strengthen our knowledge regarding phytoremediation in order to develop its businesses. Therefore, it is important to conduct awareness programs such as workshops, symposia, and seminars on native plants which are economically important so as to use them in the process of remediation of different contaminated locations across the globe.

10.2 Need for Phytoremediation

It was thought that the land and resources of the land are plentiful and will never finish. But due to our carelessness and negligence, most of the resources are not available and a lot more have been contaminated to the extent that they are not in the condition to be used for survival needs. Anthropogenic activities like the use of chemical fertilizers, pesticides, industrial effluent discharges, etc., have led us to face crises in terms of resources (Kumari et al. 2013). On the one hand, the use of chemical fertilizers, herbicides, and other such chemicals has improved yield and productivity, but on the other hand, an enormous concentration of heavy metals has been accumulated in the soil and also got entry into the food chain. Furthermore, the release of hazardous pollutants from anthropogenic sources has caused the contamination of natural resources, which ultimately led to clean water scarcity and loss of soil fertility (Chen et al. 2013). These anthropogenic activities have also led to the loss of biodiversity. Biodiversity plays a crucial part in the pure and clean ecosystem development, which ultimately plays a greater part in providing various economic benefits to man and his environment. But rapid population growth and increased human intervention have threatened this biodiversity. All the major processes like the formation of crude oil and soil, disposal of waste, nitrogen fixation, drug manufacturing, and fruits and pollen dispersal are accomplished by biodiversity (Yerima et al. 2012). It is a big concern as the number of polluted sites are increasing rapidly and are a major challenge to tackle with. To overcome these problems, a much better alternative than the conventional remediation process is needed to completely destroy these contaminants without affecting the stability of the environment. Therefore, the phytoremediation process has evolved a great technique to tackle these contaminants by utilizing a number of plant species (Salt et al. 1998). With phytoremediation, these contaminants are accumulated, degraded, or transformed into less harmful products by the plant body with low risk and enhanced economy (Zouboulis and Katsoyiannis 2005).

10.3 Phytoremediation Approaches

The different approaches used to attain phytoremediation are divided into five different categories, including phytoextraction, phytostabilization, phytodegradation, rhizofiltration, and phytovolatilization (Table 10.1).

10.3.1 Phytoextraction

Phytoextraction is the process of contaminant uptake by the roots of plants and their translocation to plant biomass, including shoots, leaves, and woody tissues. It is basically the uptake of pollutants mainly hazardous metals and radionuclides (Brooks 1998). The process of phytoextraction, also called phytoaccumulation or phytomining, involves the use of plant species or algae in order to eliminate contaminants from soil, sediments, water, or any other contaminated site into harvestable plant biomass. Over the last couple of decades, this technology has attained popularity across the globe. This process is being used mostly for the extraction of heavy metals as compared to organics (Baker et al. 1991). A living plant absorbs toxic pollutants through its roots and accumulates them into root biomass or transports them to aerial parts including stems or leaves, and it continues to absorb contaminants until it is harvested. When the plants are harvested, trace amounts of contaminants still remain in the substrate. So, in order to attain a substantial clean-up, the harvest cycle needs to be repeated through several crops. Therefore, the cleaned substrate can be used for further economic processes (Kumar et al. 1995). Those plant species that have an inherent capability to abide by higher amounts of nonessential pollutants are mostly utilized in this process (Van Nevel et al. 2007; Hasanuzzaman and Fujita 2012). If the vegetation is incinerated, the ash remaining must be disposed of properly. However, the ash volume should not be more than 10% of the volume that would be created if the contaminated substrate is used for the treatment itself. The phytoremediation process works best in removing the metals – zinc, copper, and nickel because the majority of the plant species usually have a high affinity to absorb greater amounts of metallic contaminants (Prasad 2004). The plants used in the process of phytoextraction must have some of the essential characteristics, including healthy roots, high translocation, and bioconcentration factor, fast growth, high contaminant tolerance, high biomass production, high rate of assimilation, and easily harvestable and high biomass-producing capability (EPA 2000). It was documented that the plant species from a number of families (Asteraceae, Brassicaceae, Lamiaceae, Euphorbiaceae, and Scrophulariaceae) are good for this process of contaminant removal (Baker and Brooks 1989). Phytoextraction involves the utilization of plants in a biological process, which is useful in removing contaminants from polluted sites without having adversative effects on the properties of the remediation substrate (Yanai et al. 2006).

10.3.2 Phytostabilization

It is the process of using plant species to limit or stop the translocation and bioavailability of various pollutants in soil, water, and sediments by using physical, chemical, or biological manipulation (Jadia and Fulekar 2009). By using certain selective plants, this process is a low-cost in situ technique with very less invasion properties (Jadia and Fulekar 2009). Phytostabilization is mainly employed in the removal of contaminants from soil, sediments, and sludge and is dependent

Table 10.1 Strategies of phytoremediation with description, benefits, and weakness.

Techniques	Description	Benefits	Weaknesses	References
Phytoextraction	The plant species simply uptake contaminants and accumulate in parts including leaves, roots, and stem. Example: *Helianthus annuus*, *Brassica juncea*, etc.	a) Not so expensive b) Contaminants removed permanently	a) Hyperaccumulators bear low growth with low biomass and root system b) Biomass harvested requires to be disposed of properly	Kumar et al. (1995), Henry (2000), Prasad (2004)
Phytostabilization	The availability and mobility of contaminants are limited to the root system. Example: Grasses, *Populus alba*, and *Salix alba*	a) Hazardous material do not needs disposal b) Reduces erosion attempts and water available in the system	a) Some contaminants accumulate there in the site b) Constant nursing is needed	Smith and Bradshaw (1979)
Phytodegradation	Here plants degrade contaminants by using hydrolytic enzymes and metabolites. Examples: *Populus alba*, *Salix alba*, *Medicago sativa*, and *Sorghum helepense*	a) This process is not dependant on the presence of microbes in the rhizosphere b) Enzymes are employed for degradation	a) Used only for the degradation of organic pollutants b) Does not work well in deep contamination	Burken and Schnoor (1998), Schröder et al. (2007)
Rhizofiltration	Plant roots have the capability to uptake, store, or precipitate contaminants in wastewater by root systems. Examples: *Pistia stratiotes*, *Echhornia crassipes*, *Lemna minor*, and *Hydrocotyle umbellate*	a) Mostly aquatic plants are utilized more than terrestrial plants b) Heavy metals do not translocate to shoots	a) Optimum pH needs to be maintained b) Plants are grown in controlled conditions and then transferred to radiation sites	Dushenkov et al. (1995), Henry (2000), Prasad and de Oliveira Freitas (2003)
Phytovolatilization	Contaminants are converted to their volatile forms and released into the atmosphere through stomata. Examples: *Populus alba*, *Salix alba*, grasses, and legumes	a) Contaminants can be converted into less hazardous forms b) Contaminants released in the atmosphere are in less toxic form	a) Vegetation used can accumulate contaminants b) Low levels of contaminants are reported in tissues of plants	Bañuelos et al. (1999), Burken and Schnoor (1998), Prasad (2004)

on the capability of plant roots to restrict the mobility and bioavailability of contaminants in soil (Mueller et al. 1999; USDA Forest Service 2000). It occurs through absorption, precipitation, or through a reduction in metal valence. The principle goal of plants is to reduce the concentration of water which percolates through the soil matrix, resulting in the hazardous leachate formation to restrict soil erosions and the distribution of hazardous chemical substances to other sites (Beerti and Cunningham 2000). However, contaminant remaining in the soil is a major disadvantage of it and needs regular monitoring. This process focuses on the long-period stabilization and suppression of the contaminants. For example, the presence of plants reduces the erosion caused by wind and the root system restricts the erosion caused by water, immobilizes contaminants by accumulation or adsorption, and provides an area around the roots where the contaminant can precipitate and get stabilized. Apart from phytoextraction, this process works chiefly on seizing contaminants in the substrate near roots but not in the tissues of the plant. In this way, pollutants/contaminants are not available to wildlife, livestock, and human coverage; for example, the utilization of vegetation to reduce and stabilize the mine tailings (Mueller et al. 1999). According to the current status of field practices and research, phytostabilization chiefly focuses on metal (Cr, Cd, As, Hg, Zn, and Cu) stabilization (EPA 1997, 2004). Therefore, certain changes in soil are required for its work on metal solubility and mobility (Aggarwal and Goyal 2007). In addition to the stabilization of metals, phytostabilization also focuses on the contaminated sites of low-level radioactive substances. However, the appraisal of leaching and accumulation of radionuclides is important (Lasat et al. 1998; Watt et al. 2002). According to Ari et al. (2013), dioxane was successfully removed when the trees grown on closed municipal waste sites were properly irrigated with the groundwater having 1,4-dioxane and several other volatile compounds.

10.3.3 Phytodegradation

Phytodegradation uses the enzymatic activity of plant species to break down the toxic contaminants into simple and less toxic materials in the rhizosphere before their translocation to the root system or sometimes after the translocation in the root system. This influential decontamination process is sometimes also termed phytotransformation. Whenever phytodegradation is occurring in the rhizosphere, plant roots are there to release the required enzymes (Meagher et al. 2000; Rugh 2001). In addition, certain microbes which are associated with the root system of plants prove helpful to carry out this practice (Garbisu and Alkorta 2001). In contrast, if the dilapidation of contaminants occurs within the roots, the plant uptakes the contaminants and translocates them in their original form to a place of transformation without any adverse effects on the cells of the plant (Singh et al. 2009; Hasanuzzaman and Fujita 2012). A number of pollutants (herbicides, pesticides, insecticides, and inorganic and chlorinated solvents) can be processed through phytodegradation (Thompson et al. 1998; Pivetz 2001; Gerhardt et al. 2009). Some of the common enzymes of the plant utilized include nitrilase, phosphatase, dehalogenase, oxidoreductase, and nitroreductase (EPA 2000). In this technique, compound contaminants are broken down into simpler ones and utilized by plants for faster growth. The enzymatic metabolic activities in plant roots, stem, and leaf can transform and make contaminants accumulated from soil, water, and air less toxic. Several plant species have the capability to degrade pollutants within the plant system into less harmful compounds, e.g. popular tree (*Salix* spp.) acts for Atrazine and TNT (2,4,6-trinitrtoluene). It has the capability to degrade Atrazine into less harmful compounds like Anneline and dehydroxylated and dealkylated products and the explosive TNT is degraded into 2,4-diamino-6-dinitrotoluene. Similarly, the plants *Avena sativa* and *Beta vulgaris* are suitable species for the degradation of TNT (2,4,6-trinitrotoluene) and GTN (glyceryl trinitrate), respectively. Besides this, the enzyme pentaerythritol tetranitrate reductase (PETR) released by transgenic tobacco is able to degrade GTN and TNT (Gong et al. 1999).

10.3.4 Rhizofiltration

The absorption of pollutants from the rhizosphere to the roots of the plant is termed rhizofiltration. This type of remediation is employed for the decontamination of aquatic bodies by using aquatic and land plant species. In this process, hydroponically cultivated plants are used to accumulate toxic substances or excess nutrients through absorption, concentration, and precipitation. Plant species are grown either in the polluted sites (in situ) or away from the in situ environment in ex situ conditions. Plant roots absorb the contaminants until saturation reaches, and then these plant species are mowed with roots. Since the pollutants are accumulated inside the root system without being translocated to shoots, this paves low chances for atmospheric contamination (Dushenkov et al. 1995). Several plants show the ability to uptake contaminants for various reasons including sequestration, drought resistance, disposal by leaf abscission, and defense against various pathogens and herbivores. Some species can even accumulate an extraordinary amount of contaminants than others, which has led environmental researchers to use these plant species for degrading various contaminants of soil and water (Ye et al. 2011). This process can be used for the decontamination of metals like Cr, Cd, Co, Cu, Ag, Hg, Mn, Ni, Pb, Mo, Zn, and other radioactive elements (Cook et al. 2009; Juwarkar et al. 2010). Some of the plant species including *Brassica juncea*, *Helianthus annuus*, *Nicotiana tabacum*, and *Spinacia oleracea* are some essential plant species used in this particular practice. Plants grown in dry areas have been used widely as their distinguishing feature of the fibrous root system and fast growth (Henry 2000). These plant species are utilized as floating rafts in in situ situations while in ex situ situations, engineered tanks or pipe systems are used to grow these plants. It has been revealed that the introduction of microbes in the rhizosphere has stimulated the contaminant absorption (Entry et al. 1997; Turnau et al. 2006). Due to anthropogenic and industrial activities, this technique is assuming a lot of importance for the decontamination of polluted water. Panda (1996) worked for the utilization of aquatic plants for rhizofiltration of Zn and Ni in a liquid culture medium. Plants like *Pistia stratiotes* and *Echhornia crassipes* are found to be efficient hyperaccumulators of Zn and Ni. To reduce Cr concentration in tannery effluent by more than 50%, *Hydrilla vericillata* and *Nymphea* sp. were found to be effective. A free-floating water fern *Azolla* is present in varied conditions of polluted water bodies (Singh and Mahapatra 2000). This fern has the capacity to accumulate Co, Cr, Hg, Zn, Ni, Cd, Cu, U, and As without any harmful effects on its growth (Young-Huang and Wei-Zhen 1985).

10.3.5 Phytovolatilization

The process of remediation is where contaminants are extracted by plant species using their roots, transform them into less hazardous materials, and then release them into the external environment through their leaves. Roots absorb the soluble pollutants and translocate them to other parts of the plant in altered or unaltered form until they reach the leaves where they are volatilized into the atmosphere through the process of transpiration. This is basically an evapotranspiration method and chiefly volatilizes the Se, Hg, and organic solvent pollution (Bizily et al. 1999; Bañuelos et al. 2000; Karami and Shamsuddin 2010). It has been reported that Tritium, an isotope, is remediated by phytovolatilization using woody plant species (Murphy 2001). Climatic conditions and the genetic capability of plants being used for the process of phytovolatilization plays a crucial role in its efficiency. At the same time, the concentration of contaminants at the contaminated site plays a pivotal part in the absorption, translocation, and chemical fate of the contaminants (Meagher 2000). According to the Environmental Protection Agency (EPA), United States, plants like *Medicago sativa, Brassica napus, and Brassica juncea* are good phytovolatilizing agents (EPA 2000). While this technique is used to degrade a number of hazardous pollutants, it does not resolve the problem. It simply takes the pollutants from the substrate and ultimately releases them into the atmosphere through stomata. Mercury, selenium, and arsenic are the major phytovolatilized metals. Mercuric ion by the process of phytovolatilization is converted to a less toxic elemental form of mercury in transgenic Arabidopsis. The selenium metal, on the other hand, can also be phytovolatilized to dimethyl selenide by the action of microbes and algae. The Indian mustard has also been shown to decrease the concentration of selenium to its low toxic levels (Bañuelos and Meek 1990; Bañuelos et al. 1997). Burken and Schnoor (1999) used the hybrid *Populus alba* trees for the distribution and volatilization of selected contaminants. Hansen et al. (1998) were of the idea that constructed wetlands are extremely effective in order to remove selenium from wastewater contaminated with selenite. They recorded a high rate of selenium volatilization from five different vegetated wasteland sites.

10.4 Hyperaccumulation

Hyperaccumulation is a process in which metal contaminants are absorbed by plants from the contaminated soils and stored to a level higher than in polluted/contaminated soil (Brooks 1998). Approximately, 500 plant species have been reported that can be used as hyperaccumulator species (McGrath and Zhao 2003; Sarma 2011). The main cause behind this distinctive feature of several plant species is believed to be the defense mechanism of these plants. By absorbing and storing toxic substances in their cells, these plant species are able to guard themselves against a number of pathogenic microbes and herbivores. The absorption potential of different hyperaccumulator plants is dependent on the surrounding environment and also on the genes present in them and the regulation and expression of these genes (Rascio and Navari-Izzo 2011). The mechanism of metal accumulation starts with the absorption of metals from soil. Then there occurs the formation of metallocomplex with the assistance of some important proteins. Then these contaminants are translocated from roots to

Table 10.2 List of metal and metalloid hyperaccumulators.

Plant	Family	Contaminants treated	References
Arabidopsis thaliana	Brassicaceae	Mn, Pb, Cu	Lasat (2002)
Astragalus racemosus	Fabaceae	Se	Vallini et al. (2005)
Populus clones	Salicaceae	Soil contaminated with trace elements	Ciadamidaro et al. (2017)
Azolla pinnata	Azollaceae	Cd	Rai (2008)
Bidens pilosa	Asteraceae	Cd	Sun et al. (2009)
Piptatherum miliaceum	Poaceae	Pyritic tailings	Moreno-Barriga et al. (2017)
Brassica juncea	Brassicaceae	Cd, Ni	Saraswat and Rai (2009)
Cardminopsis halleri	Brassicaceae	Zn, Cd	Sun et al. (2007)
Clerodendrum infortunatum	Lamiaceae	Cu	Chipeng et al. (2010)
Dorycnium pentaphyllum	Fabaceae	Mine soil	Kohler et al. (2016)
Euphorbia cheiradenia	Euphorbiaceae	Pb	Chehregani and Malayeri (2007)
Fagopyrum esculentum	Polygonaceae	Pb, Al, Cd	Baker and Brooks (1989)
Pinus halepensis	Pinaceae	Mine tailings	Martinez-Oró et al. (2017)
Populus sp.	Salicaceae	Contaminated sites with Boron	Robinson et al. (2007)
Pteris vittata	Pteridaceae	Cr, Se, As	Ma et al. (2001)
Sedum alfredii	Crassulaceae	Pb, Zn, Cd	Li et al. (2005), Sun et al. (2007)
Solanum nigrum	Solanaceae	Cd	Zhang et al. (2011)
Sonchus asper	Asteraceae	Zn, Pb	Yanqun et al. (2005)
Thlaspi caerulescens	Brassicaceae	Zn, Ni, Cd	Lombi et al. (2001)
Sarcocornia fruticosa	Amaranthaceae	Saline, strongly acidic eutrophic wetlands	González-Alcaraz et al. (2013)

shoot through the apoplast and symplast pathways. After translocation, these contaminants are distributed, sequestered, and then stored in metabolically less active cells (Kotrba et al. 2009). Some of the important hyperaccumulator plant species are given in Table 10.2 with the metals they target upon. Broadly, it is now clear that the plant species with a higher ability for acting as a hyperaccumulator grows at a very slow speed and produces less biomass in comparison to other plants, which is the major disadvantage of these plants. After metallocomplex formation, the transportation of metals is an extremely significant step in the process of hyperaccumulation, which is performed by ZIP family genes; genes from other families like MATE, HMA, MTP, and YSL are important for this process as well (Pagliano et al. 2006; Poschenrieder et al. 2006). As it is now known which of the genes or gene families are responsible for this process of hyperaccumulation, they can now be transferred to the fast-growing plants with which the practice of metal hyperaccumulation can be enhanced (De Souza et al. 1998). The basic practice of hyperaccumulation is depicted in flowchart shown in Figure 10.2. The potential capability of the plant to act as a hyperaccumulator can be resolved by knowing the plants' bioconcentration and translocation factors which include bioconcentration factor (BCF) and translocation (TF). The plants which have TF and BCF values higher than one are appropriate for the process of phytoremediation, particularly phytoextraction (Fitz and Wenzel 2002). According to Padmavathiamma and Li (2007), not less than 450 species of angiosperms belonging to 45 different families (Asteraceae, Brassicaceae, Violaceae, Caryophyllaceae, Cunoniaceae, Cyperaceae, Flacourtiaceae, Fabaceae, Poaceae, Euphorbiaceae, and Lamiaceae) have been reported to have characteristics of BCF and TF. Despite not being like that, several species of the genus Lupinus of the family Fabaceae have been indicated to have a tolerance to metals similar to that of plants used as hyperaccumulators (Zornoza et al. 2002; Lambers et al. 2013). For instance, all the environmental factors must be taken into consideration while determining the phytoremediation techniques. It is essential to work out if and up to what extent the used remediation approach is going to affect the interspecific relations, particularly if it is affecting the rest of the crops. Considering that the plant species that were grown for the phytoremediation process should be subjected to polluted soil or water conditions, other crops are not able to grow in these situations because of the toxic effects they cannot withstand and will likely reach the competition problems (Stankovic and Devetakovic 2016).

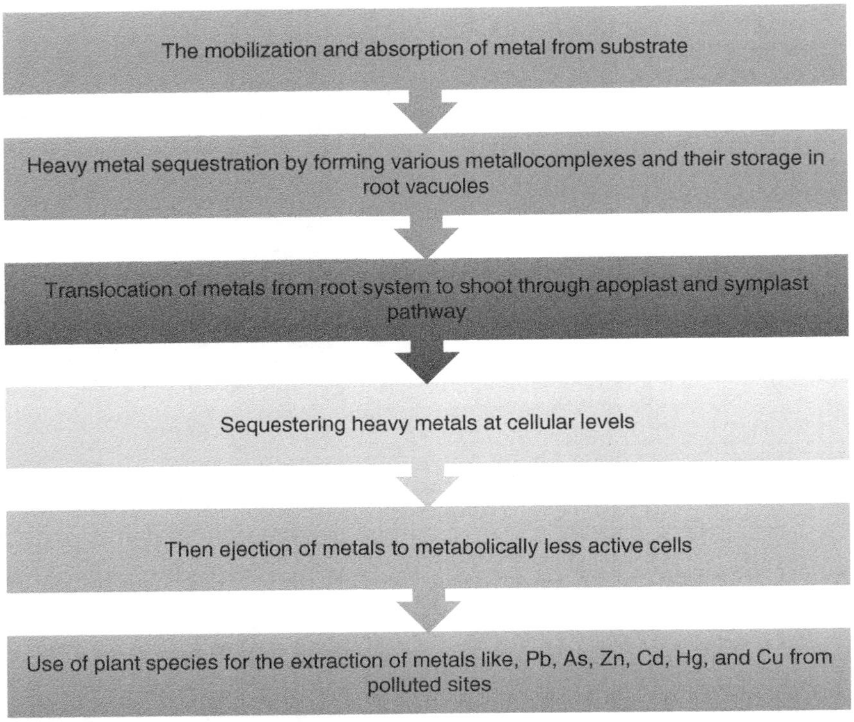

Figure 10.2 Flow chart showing the basic process of metal hyperaccumulation (Kotrba et al. 2009).

10.5 Genetically Engineered Plants and Phytoremediation

Genetically modified plant species are playing a pivotal role in the removal and degradation of a number of contaminants. Plants like *Sesbania grandiflora* of the family Fabaceae and *Arabidopsis thaliana* of the family Brassicaceae have been genetically modified to show the expression of rabbit cytochrome p450 2E1 enzyme so as to increase their activity toward the removal of contaminants including trichloroethylene (TCE) and dichlorodiphenyltrichloroethane (DDT). The effective generation of plants like *Sesbania* and *Arabidopsis* was confirmed by utilizing p450 2E1-specific PCR and western blotting, respectively. It was shown by gas chromatography analysis that the cuttings of transgenic *Sesbania* and *Arabidopsis* when visible to TCE and DDT in a hydroponic system accumulated more TCE and DDT in comparison to plants transformed with empty vector. In addition, these genetically modified plant species (*Sesbania and Arabidopsis*) proved to be more efficacious in the breakdown of TCE and DDT with an enormous surge in the metabolism of TCE. It was revealed that in the two independent lines of *Arabidopsis thaliana*, DDT was metabolized about fourfold higher than that detected in other nontransformed plant species. Similarly, *Sesbania grandiflora* cuttings removed about 51–90% of DDT compared with other nontransformed plants which were able to remove only 3%. The stability of rabbit cytochrome p450 2E1 was proved using third-generation *Arabidopsis* plants that showed a greater potential toward the elimination of two essential contaminants, TCE and DDT as compared with the control (Mouhamad et al. 2012).

10.6 Multiple Benefits of Phytoremediation from Ecological to Socioeconomic

10.6.1 Aesthetic Value

In addition to remediation, phytoremediation comes with a number of intact benefits by increasing the aesthetic of polluted sites. Ornamentally important plant species have more visual benefits than other plant species including energy crops, aromatic crops, timber trees, grasses, and other such plant species, which are utilized for phytoremediation. Most of the essential economic crops including energy crops deliver visual benefits on a short-term basis but disappear after the crop is harvested (Skarback and Becht 2005). In place of biomass-based crops, seed-based crops provide an aesthetic environment because the harvestable part of the plant is the seed. But we also need to plant energy crops in a successive and rotational manner. *Jatropa*, *Ricinus*, *Pogamia*, etc., are seed-based energy crops and plants including *Miscanthus*, *Panicum virgatum*, *Phalaris arundinacea*, *Populus*, *Salix*, etc., are some of the energy crops being used in the phytoremediation process. Aroma-based crop phytoremediation provides the intact benefit of aroma in the environment where it is used. Therefore, these plant species show more benefits in comparison to other nonaromatic plants in phytoremediation (Pandey and Singh 2015).

10.6.2 Biodiversity Conservation

Besides a cost-effective technique for the degradation of toxic materials, the plant utilization for cleaning polluted lands also enhances the floral as well as faunal diversity of the area by attaining the climax condition. The monoculture plantation technique has mostly been used in the process of phytoremediation. According to Ehrlich and Pringle (2008), because of the high propagation rate within a wide range of ecosystems, these plants can prove invasive and may threaten the endemic population of the area. But proper and controlled ecological modeling can help control this invasion (Doren et al. 2009). Currently, the native plants which have the remediating potential are being given preference for the phytoremediation process (Farnsworth and Meyerson 2003; Rowe et al. 2009; Pandey et al. 2015a). As the plants start growing and phytodiversity stabilizes on the site, it provides shelter to large animals like reptiles, amphibians, mammals, and birds as well (Sage et al. 2006). It is now evident that the utilization of various potential plant species for the process of phytoremediation at a site in an engineered manner not only enhances the phytodiversity but also provides support to faunal diversity. Therefore, it is also helpful in biodiversity conservation (Jaradat 2010).

10.6.3 Produced Biomass from Phytoremediation

As a big volume of biomass is produced in the phytoremediation process of heavy metals at contaminated sites, it can provide additional income when used properly. These biomasses can be used in a better way for wood or energy purposes. Some of the derived biomasses like *Tectona grandis*, *Azadirachta indica*, *Salix* spp., *Eucalyptus tereticornis*, and *Gmelina arborea*, from phytoremediation can be utilized for furniture and also for many industrial purposes (Maiti et al. 2005; Lewandowski et al. 2006; Shukla et al. 2011). Several high-yielding herbaceous plants and other hyperaccumulator plant species have been reported to be used for phytomanagement of contaminated soils and their efficacy has been confirmed (Mench et al. 2010). However, using this heavy metal-containing biomass is an essential matter for restricting the re-entry of metals into the environment. In addition, many of the European field trials using *Nicotiana tabacum*, *Salix alba*, *Betula utilis*, and *Populus alba* are meant to produce valuable biomass for the recovery of energy (Kidd et al. 2015).

10.6.4 Economic Benefits of Phytoremediation

If the phytoremediation industries utilize the plants like essential oil-yielding aromatic herbs in the remediation of heavy metal-polluted locations, they will be able to enhance their economy in terms of biodiesel and other essential oils. For remediating polluted areas with the help of the most promising biodiesel crops including *Helianthus annuus*, *Jatropa curcas*, *Pongamia pinnata*, and *Ricinus communis* and the most essential oil-containing crops include *Lavandula augustfolis*, *Mentha piperita*, *Cymbopogon martini*, and *Ocimum basilicum*. Crops have been recommended for sustainable remediation because of low risk (Pandey et al. 2012a; Pandey and Singh 2015).

10.6.5 Recovery of a Metal

Phytoremediation provides an opportunity to enhance the financial possibility of its programs and to decrease disposal risks through the use of metal-containing biomass of plants in energy and metal recovery with the burnt process. It is possible to recover several essential metals like Se, Zn, Fe, and Ni from phytoremediation of phytomining and biofortification for further use (Pandey 2013a). The process by which the metals are recovered from polluted locations by utilizing high biomass-yielding plant species is termed phytomining. On the other hand, biofortification is described as the process of enhancing the nutritional quality of edible crops and vegetables. Processes like agronomic practices, modern biotechnology, and indigenous plant breeding can be used for this. The process of linking phytoremediation with biofortification is a novel idea to confront malnutrition and other environmental remediation (Pandey et al. 2016). Approximately, 300 plants have been recognized for phytoremediation as Zn and Ni hyperaccumulators as these metals are the most accumulated contaminants for different plants (hyperaccumulators). For example, *Alyssum lesbiacum* and *Arabidopsis halleri* act as hyperaccumulators for Ni and Zn, respectively (McNair et al. 2000; Cluis 2004). The consequences and financial threats of phytoremediation are decided by the selection of plant species and target metal. Plants with a high yield of biomass and high value-added metals are supposed to yield more revenue (Jiang et al. 2015).

10.7 Phytoremediation-Theoretical Aspects

Theoretically, in comparison to physical and chemical approaches, the phytoremediation technique is quite simple, low-cost, and widely accepted in large areas. In addition, this technique comes with an enormous number of socioeconomic and ecological benefits. Therefore, it is totally a new technique used for the remediation of contaminated sites by utilizing trees, grasses, and herbs for organic or inorganic pollutants. The increase in terrestrial and aquatic contaminated sites due to the increasing release of contaminants is a burning issue all over the globe (Russ 2000). A few decades ago, the primary aim of this technique (phytoremediation) was only the remediation of pollutants, for which several varieties of plant species were used. A few edible plant species (*Brassica juncea*, *Helianthus annuus*, *Zea mays*, and *Brasica napus*) are very prominent in phytoremediation. Due to the high potential toward heavy metal accumulation as confirmed by many experiments, the Brassicaceae family has attained greater attention and has attracted a number of phytoremdiators (Vamerali et al. 2010). In comparison to *Brassica napus*, *Brassica juncea* shows a high ability toward the accumulation of Cd in aerial parts and it is three times better than that of *Brassica napus*. In addition to the accumulation of Cd, *Brassica juncea* has a higher ability toward the removal of Pb and Se (Salt et al. 1998). Furthermore, it has the ability to accumulate contaminants like Zn from the soil as compared to *T. caerulescens* (which is well known for Zn accumulation) due to its capability to generate ten times more biomass in comparison to *T. caerulscens* (Gisbert et al. 2006). Several nonedible crops have also been studied for remediating metals, metalloids, and other organics, for which *Salix alba*, *Populus alba*, *Lemna minor*, and *Pteris vittata* are widely utilized. To absorb different contaminants (As, Pb, Cu, Co, Zn, Ni, Se, Cs, Cr, and Cd), numerous hyperaccumulators have been reported. These hyperaccumulator species bear some distinguishing features such as the capability to absorb and store metals and metalloids in aerial parts (Prasad and de Oliveira Freitas 2003). Not less than 400 hyperaccumulator species are known to be reported for remediation of various heavy metals and metalloids including Ni, Cd, As, Mn, and Zn. Some of the hyperaccumulator plant species with their accumulation potential are: *T. caerulescens* for Zn (51,600) and Cd (18,000); *Psychotria douarrei* for Ni (47,500); *Thlaspi rotundifolium* for Pb (8200); *Astragalus racemosus* for Se (14,900); *Ipomea alpine* for Cu (12,300); and *Pteris vittata* for As (20,000) (Oh et al. 2013).

With the help of bio input and physicochemical processes, the efficacy of the phytoremediation technique can be enhanced either by increasing bioavailability or by decreasing the hazardous impact of metals and metalloids (Prasad 2011). Besides, remediation by employing microbes has attained popularity as an efficient tool to remediate polluted locations over the last 20 years. As phytoremediation is considered a time-consuming practice, it can be enhanced by knowing the plant-microbe interactions that, in turn, would be helpful in improving plant growth, establishment, and health by helping in the sequestration of hazardous metals and metalloids. It has been reported that plant growth-enhancing bacteria (rhizobacteria) and other endophytes play a pivotal role to accelerate phytoremediation (Ma et al. 2011; Mir et al. 2022; Rabani et al. 2022). These plant-associated microbes including growth-enhancing microorganisms have the capability to produce certain metabolites including siderophores, organic acids, 1-aminocyclopropane-1-carboxylic acid deaminase, indole-3-acetic acid (IAA), etc., that have been reported to be involved in a number of biogeochemical processes which function in a rhizospheric zone of contaminated areas. Some of the notable functions are cell elongation, metal detoxification, nutrient acquisition, and mitigation of biotic and abiotic stresses. These rhizospheric microbes accelerate the mobility of metals. Hence, the pollutant degradation by phytoremediation could be accelerated by plant-associated microbes, where few practices are involved including translocation, solubilization, transformation, precipitation, immobilization, chelation, volatilization, and complexation (Rajkumar et al. 2012). On the other hand, the hyperaccumulation capability of plant species for metals can also be enhanced by the addition of different chelating agents. It may also prove helpful in simplifying the bioavailability of metals through the formation of metallocomplexes (Freitas et al. 2014). Chelating agents can be divided into two categories, viz. natural and synthetic. Oxalic acid, aminoacids, citric acid, gallic acid, gluconic acid, homocitric acid, and 2,3-dihydroxy benzoic acid are categorized into natural or organic chelating agents, while ethylenediamine tetra-acetic acid, diethyltriamine penta acetic acid, ethylenediamine-N,N0-disuccinic acid, and ethylenediamineN,N0-bis(2-hydroxyphenylacetic acid) are included in synthetic category. Chelation is a process in which a definite organic molecule is attached to a mineral ion at more than one site forming a compound structure. With different efficacies, both the categories (natural and synthetic) of chelating agents hasten heavy metal absorption from contaminated areas (Kaushik 2015).

A potential tool for the improvement of phytoremediation by plants is genetic engineering, as it enhances the capability of the plant to withstand, absorb, and metabolize different types of contaminants. The phytoremediation process involving genetically engineered plants has been experimented on on a large scale in laboratories and greenhouses (Sarangi et al. 2009; Van Aken 2009; Maestri and Marmiroli 2011; Kabouw et al. 2012). The essential approaches of genetic engineering that have been employed in phytoremediation include the overexpression of genes which are native to modified species, the transformation of plant species with genes taken from other plant species, and the inoculation of genes from microbes to plants. A number of researchers have used transgenic plants for phytostabilization and phytoaccumulation of different metals including Cu, Cr, Cd, Ni, B, Hg, As, Se, Pb, and Zn (Bañuelos et al. 2005; Bañuelos et al. 2007; Doty 2008; Sarangi et al. 2009; Vamerali et al. 2010). One of the successful events of phytoremediation utilizing genetically modified plant species was of Se by utilizing transgenic *Brassica juncea*, where the plant used the mechanism of gene overexpression facilitating sulfate and selenite reduction and glutathione synthesis (Bañuelos et al. 2007). This process of increased phytoremediation, using genetically engineered plants has been analyzed properly by changing the structure and metabolism of roots, reducing oxidative stress, introduction of genes for chelation, overexpression of genes associated with metal transport, and increased production of biomass (Sarangi et al. 2009; Vamerali et al. 2010). Phytoremediation is considered a suitable approach toward the remediation of inorganic and organic contaminant-polluted sites. However, heavy metal accumulation in different parts of the plant could be hazardous to both flora and fauna. Several plants have been reported as having much more importance in the case of phytoremediation and most of the researchers have used edible crops for this process, but such crops should not be used as they may transfer the metals accumulated by them to humans. So, to mitigate this problem, a new holistic approach to phytomanagement needs to be followed for sustainable phytoremediation. Pandey et al. (2015a) stressed the use of those plant species that are natural colonizers and economically and ecologically valuable for the phytoremediation of contaminated sites toward sustainable phytoremediation. Following this direction, a number of naturally occurring plants on contaminated and waste dump sites have been reported to have great sustainable phytoremediating potential toward the construction of a multifunctional ecosystem. A plant candidate described as suitable for sustainable phytoremediation should have some features, such as fast propagation, rapid development, high biomass production, extensive roots, ability to accumulate contaminants, and tolerance to harsh conditions, and should be inedible (Pandey et al. 2011, 2012a). Also, the plant species must be perennial in nature, economically beneficial, and should also have the capacity to initiate ecological succession. Some of the indirect benefits that sustainable phytoremediation provides include carbon sequestration, substrate quality enrichment, aesthetic value, and conservation of biodiversity (Pandey 2002, 2013a). Various researchers suggest the use of introduced plant species for this type of phytoremediation process after overlooking the input costs and maintenance (Vamerali et al. 2010). Sustainability can be attained in phytoremediation if environmentally and economically significant plant species like aromatic plants and energy crops are screened among the vegetation. For example, *Saccharum spontaneum* and *Saccharum munja* are environmentally and economically essential plants, which have the capability of limiting metal toxicity. Furthermore, *Vetiveria zizaniodes*, *Ocimum basilicum*, and *Cymbopogon flexuosus* are aromatic plants having a high stress-tolerant nature. In addition, several aromatic types of grass have been reported to have phytoremediation properties (Lal et al. 2013). Several energy crops including *Ricinus communis*, *Miscanthus giganteus*, and *Jatropha curcas* have been identified to have the potential of phytoremediation for various contaminant sites (Pandey et al. 2012b, 2013b; Nsanganwimana et al. 2014). These plants are perennial and ecologically appropriate for the phytoremediation of contaminated sites. Fortunately, there is no possible hazard behind the use of these plant species. Therefore, the plant species having such interesting features as being natural colonists, unpalatable and perennial in nature, besides having socioeconomic values could boost phytoremediation-based industries (Pandey et al. 2015a). In conclusion, phytoremediation of contaminated sites is the need of the hour for proper nurturing and sustaining ecosystem services.

10.8 Phytomanagement: A New Paradigm

Over the last few decades, phytoremediation has attained popularity due to several important issues like sustainability, environmental and socioeconomic benefits, less effort, and low risk in order to achieve a sustainable goal (Pandey and Singh 2015; Cundy et al. 2016). Sustainable phytoremediation or simply termed phytomanagement is beneficial in many ways, including the least input and low risk through the selection of specific plant candidates, chiefly native plant species, which are having environmental (CO_2 sequestration property, substrate improvement, seepage control, and remediation of pollutants), aesthetical, recreational value, economical value (biodiesel, biogas, biomass, essential oils, furniture wood, fiber, dyes, and alkaloids), and various other benefits (Pandey and Singh 2012; Kennen and Kirkwood 2015; Pandey et al. 2016). Pandey et al. (2015a) also stressed the socioeconomic significance of various plant species used for the phytoremediation of fly ash (FA) deposits. He stated that the plants which have socioeconomic values must be endorsed for the restoration of such sites in India. These species are used for religious ceremonies and fodder (*Cynodon dactylon*), for making mats, baskets, brooms, ropes, and huts (*Typha latifolia* and *Ipomoea carnea*) and also as fuelwood (*Acacia nilotica*) for the improvement of livelihoods of adjoining communities. Biochar enriched with zinc, which has been produced from the biomass grown on zinc, can serve as a release fertilizer that will provide income to adjoining societies (Evangelou et al. 2014). Similarly, wood and bioenergy are obtained from the polluted sites (Van Slycken et al. 2013a,b). Thus, it is evident that phytomanagement (sustainable phytoremediation) is a novel paradigm involving the use of economically valuable plant species aimed at the management of contaminated sites with a number of additional benefits (Pandey et al. 2015a). For the successful implementation of this phytoremediation idea, the selection of appropriate plant species for a suitable polluted site is a crucial step. The phytomanagement process will be more successful in particularly contaminated sites if the plant species are perennial, indigestible by animals, and indeed natural colonizers (Pandey et al. 2016). Furthermore, the phytomanagement potential can be increased by employing those plant species having characteristics like rapid growth accompanied by huge biomass production, uncomplicated and active propagation, extensive root system, metal and metalloid build-up ability, tolerant to biotic and abiotic stress, and the capability to recover substrate (Pandey et al. 2012a). Several plant species which can be used for the remediation of polluted sites with numerous ecological and socioeconomic advantages include *Saccharum munja*, *Saccharum spontaneum*, *Ricinus communis*, *Juncea curcas*, *Arundo donax*, and a lot more (Papazoglou et al. 2005; Pandey et al. 2012b; Olivares et al. 2013).

Pioneering vegetation with diverse ecological and socioeconomic functions and diverse life forms (grasses for swift growth and tree for soil erosion) need to be used for effective phytomanagement and to ensure sustainability on a long-term basis (Párraga-Aguado et al. 2013, 2014). This process will help the local bodies enjoy a range of direct and indirect benefits. Furthermore, if phytomanagement is being coupled with biofortification, it may prove viable to tackle soil remediation as well as malnutrition of mineral nutrients including Se, Zn, Fe, and I (Bañuelos and Dhillon 2011; Wu et al. 2015). According to Bañuelos and Dhillon (2011). Se-biofortification provides novel commodities (e.g. Se-rich fodder for livestock and vegetables and edible grains for human use) to growers. But it is mandatory to monitor these phytoproducts of phytoremediation or biofortification for possible risks. The biodiversity of plant species has a fundamental role in phytomanagement besides the choice of plants. For the effective phytomanagement of contaminated sites, monocultures are not considered appropriate because of their sensitive nature toward biotic and abiotic stress conditions. Therefore, all these issues should be taken into consideration as to ensure economic, soil, and ecological balance (Evangelou et al. 2012). The use of multiple plant species, particularly value-added, may prove helpful in phytomanagement of contaminated sites, mainly those contaminated with organic pollutants (Ma et al. 2013). Phytomanagement, as supported by many scientific research papers, is emerging as a new paradigm all over the world with its benefit in sustainable site management and economic return (Cundy et al. 2016; Ciadamidaro et al. 2017; Martinez-Oró et al. 2017; Kaewtubtim et al. 2018). The word "phytomanagement" is used in the context of remediation of contaminated sites as a new way to regulate ecosystem services to a greater end (Table 10.3). It is evident that economically valuable native species will be more effective in the

Table 10.3 Plants used in the process of sustainable phytoremediation/phytomanagement with the country location.

Plant species	Approach used	Pollutant treated	Location	Reference
Brassica spp.	Phytomanagement	Heavy metal-contaminated land	Italy	Vamerali et al. (2014)
Helianthus annuus	Phytomanagement	Agricultural land contaminated with Cd and Zn	Switzerland	Fässler et al. (2010)
Salix spp.	Phytomanagement	Agricultural land contaminated with Cd and Zn	Belgium	Van Slycken et al. (2013b)
Lactuca sativa	Phytomanagement	Cu, Zn, As, Cr, Pb	Europe	Quintela-Sabarís et al. (2017)
Sarcocornia fruiticosa	Phytomanagement	Acidic saline eutrophic wetlands contaminated by mine discharge	Spain	González-Alcaraz et al. (2013)
Lasiurus scindicus	Phytomanagement	Pb	India	Sharma and Pandey (2017)
Avicennia marina	Phytomanagement	Radionuclides, heavy metals	Thailand	Kaewtubtim et al. (2018)
Populus alba	Phytomanagement	Contaminants from mine spill	Spain	Domínguez et al. (2008)
Brassica juncea	Phytomanagement	Soil contaminated with Cu	Spain	Rodríguez-Vila et al. (2014)
Piptatherum miliaceum	Phytomanagement	Mine dumps in semiarid regions	Spain	Parraga-Aguado et al. (2015)

process of phytomanagement because of their inherent capacity to tolerate various biotic and abiotic stresses including metal toxicity, water scarcity, poor soil quality, and salinity along with other biotic conditions that halt the remediation and rehabilitation approaches. Thus, economically valued plant species should be used now and again for the remediation of contaminated sites.

10.9 Future Prospects

The outcome of the phytoremediation technique depends on the capability of plant species to tolerate the contaminants which are to be removed. So, it is essential to determine the concentration of contaminants that can be accumulated, degraded, and not accompanied by any injury or critical stress on the metabolic processes of the plant species used. There is a need for more research to determine the heavy metal concentration in any contaminated site so as to use the appropriate plant species for the remediation of soil, water, and sediments (Nan et al. 2013) and also utilize diverse microbial species (Karimzadeh et al. 2012). PCR amplification and DNA fingerprinting may prove helpful in choosing a particular plant variety of microbial strain to be utilized for the remediation of specific pollutants from various contaminated sites. As the pollutants and their byproducts can prove hazardous to human beings and other living organisms together with plants, a comprehensive monitoring of this process should be performed to know if the metabolites discharged are still toxic (Schröder et al. 2008). The most important challenge before phytoremediation is whether the process is being implemented successfully (Harris et al. 2013). The quality of the products obtained during the phytoremediation process should be assessed properly. The phytoremediation of contaminated soils needs to be taken into consideration so as to convert these low-productive lands into high-productive ones. If attention is paid, even these soils have the potential to be used for agricultural purposes. Hence, a proper set of cropping systems has to be developed to allow safe and sustainable production on polluted sites which restricts the transferal of hazardous contaminants to the food chain (Kohler et al. 2016). Phytoremediation can provide good returns if we invest. It can replace conventional methods of remediation and is being accepted properly throughout the world because of its low risk, economic benefits as assessed by various stakeholders, and other benefits to the environment.

10.10 Conclusions

Mostly in the developed countries, phytoremediation has accomplished several advancements during the past couple of decades, but it is yet to achieve popularity among practitioners and scientific societies. To attain this popularity and for successful phytoremediation, plant species have to be selected keeping in consideration several characteristics like high-value-added, high-growth, perennial, unpalatable to livestock, and well-developed root system. In addition, sustainable phytoremediation/phytomanagement is also needed because of its multiple ecological and socioeconomic benefits. Indeed, phytoremediation is an important innovative technique, whereby remediation is obtained at a low cost in an eco-friendly manner. So, this technique of remediation needs much attention in the present world.

References

Aggarwal, H. and Goyal, D. (2007). Phytoremediation of some heavy metals by agronomic crops. *Dev. Environ. Sci.* 5: 79–98.

Ari, M., Ferro, J.K., James, C. et al. (2013). Phytoremediation of 1,4-dioxane-containing recovered groundwater. *Int. J. Phytorem.* 15 (10): 911–923.

Baker, A.J.M. and Brooks, R.R. (1989). Terrestrial higher plants which hyperaccumulated metallic elements-a review of their distribution, ecology and phytochemistry. *Biorecovery* 1: 81–126.

Baker, A.J.M., Reeves, R.D., and Mc Grath, S.P. (1991). In situ decontamination of heavy metal polluted soils using crops of metal-accumulating plants: a feasibility study. In: *In Situ Bioreclamation* (ed. R.L. Hinchee and R.F. Olfenbuttel), 600–605. Boston: Butterworth-Heinemann.

Bañuelos, G.S. and Dhillon, K.S. (2011). Developing a sustainable phytomanagement strategy for excessive selenium in western United States and India. *Int. J. Phytorem.* 13 (1): 208–228.

Bañuelos, G.S. and Meek, D.W. (1990). Accumulation of selenium in plants grown on selenium treated soil. *J. Environ. Qual.* 19: 772–777.

Bañuelos, G.S., Ajaw, H.A., Mackey, B. et al. (1997). Evaluation of different plant species used for phytoremediation of high soil selenium. *J. Environ. Qual.* 26 (3): 639–646.

Bañuelos, G.S., Shannon, M.C., Ajwa, H. et al. (1999). Phytoextraction and accumulation of boron and selenium by poplar (*Populus*) hybrid clones. *Int. J. Phytorem.* 1 (1): 81–96.

Bañuelos, G.S., Zambrzuski, S., and Mackey, B. (2000). Phytoextraction of selenium from soils irrigated with selenium-laden effluent. *Plant Soil* 224 (2): 251–258.

Bañuelos, G., Terry, N., Leduc, D.L. et al. (2005). Field trial of transgenic Indian mustard shows enhanced phytoremediation of selenium contaminated sediment. *Environ. Sci. Technol.* 39: 1771–1777.

Bañuelos, G., Leduc, D.L., Pilon-Smits, E.A.H. et al. (2007). Transgenic Indian mustard overexpressing selenocysteine lyase or selenocysteine methyl transferase exhibit enhanced potential for selenium phytoremediation under field conditions. *Environ. Sci. Technol.* 41: 599–605.

Baudouin, C., Charveron, M., Tarrouse, R. et al. (2002). Environmental pollutants and skin cancer. *Cell Biol. Toxicol.* 18: 341–348.

Baumann, A. (1885). Das Verhalten von Zinksatzengegen Pflanzen und im Boden. *Landwirtsch. Vers. Stn.* 31: 1–53.

Beerti, W.R. and Cunningham, S.D. (2000). *Phytoremediation of Toxic Metals: Using Plants to Clean Up the Environment-Wiley-Interscience*, 71–88. New York, NY: John Wiley and Sons, Inc.

Bizily, S.P., Rugh, C.L., Summers, A.O. et al. (1999). Phytoremediation of methyl mercury pollution: merB expression in *Arabidopsis thaliana* confers resistance to organomercurials. *Proc. Natl. Acad. Sci. U. S. A.* 96: 808–6813.

Blaylock, M. (2008). *Phytoremediation of Contaminated Soil and Water: Field Demonstration of Phytoremediation of Lead Contaminated Soils.* Boca Raton, FL: Lewis Publishers.

Brooks, R.R. (1998). General introduction. In: *Plants that Hyperaccumulate Heavy Metals* (ed. R.R. Brooks), 1–14. Wallingford: CABI.

Burken, J.G. and Schnoor, J.L. (1998). Predictive relationships for uptake of organic contaminants by hybrid poplar trees. *Environ. Sci. Technol.* 32 (21): 3379–3385.

Burken, J.G. and Schnoor, J.L. (1999). Distribution and volatilization of organic compounds following uptake by hybrid poplar trees. *Int. J. Phytorem.* 1 (2): 139–151.

Chehregani, A. and Malayeri, B.E. (2007). Removal of heavy metals by native accumulator plants. *Int. J. Agric. Biol.* 9: 462–465.

Chen, J., Qing-Xuan, X., Yi, S. et al. (2013). Phytoremediation of organic polluted soil. *J. Biorem. Biodegrad.* 4: 1–3.

Chipeng, F.K., Hermans, C., Colinet, G. et al. (2010). Copper tolerance in the cuprophyte *Haumaniastrum katangense* (S. Moore) PA Duvign. & Plancke. *Plant Soil* 328: 235–244.

Ciadamidaro, L., Girardclos, O., Bert, V. et al. (2017). Poplar biomass production at phytomanagement sites is significantly enhanced by mycorrhizal inoculation. *Environ. Exp. Bot.* 139: 48–56.

Cluis, C. (2004). Junk-greedy greens: phytoremediation as a new option for soil decontamination. *BioTeach J.* 2: 61–67.

Conesa, H.M., Evangelou, M.W.H., Robinson, B.H. et al. (2012). A critical view of current state of phytotechnologies to remediate soils: still a promising tool? *Sci. World J.* 2012: 173829. https://doi.org/10.1100/2012/173829.

Cook, L.L., Inouye, R.S., and McGonigle, T.P. (2009). Evaluation of four grasses for use in phytoremediation of contaminated arid land soil. *Plant Soil* 324: 169–184.

Cundy, A.B., Bardos, R.P., Puschenreiter, M. et al. (2016). Brownfields to green fields: realising wider benefits from practical contaminant phytomanagement strategies. *J. Environ. Manage.* 184: 67–77.

De Souza, M.P., Pilon-Smits, E.A.H., Lytle, C.M. et al. (1998). Rate-limiting steps in selenium assimilation and volatilization by Indian mustard. *Plant Physiol.* 117: 1487–1494.

Domínguez, M.T., Marañón, T., Murillo, J.M. et al. (2008). Trace element accumulation in woody plants of the Guadiamar Valley, SW Spain: a large-scale phytomanagement case study. *Environ. Pollut.* 152: 50–59.

Doren, R.E., Richards, J.H., and Volin, J.C. (2009). A conceptual ecological model to facilitate understanding the role of invasive species in large-scale ecosystem restoration. *Ecol. Indic.* 9: 150–160.

Doty, S.L. (2008). Enhancing phytoremediation through the use of transgenics and endophytes. *New Phytol.* 179: 318–333.

Dushenkov, S., Kumar, V., Motto, H. et al. (1995). Rhizo-filtration: the use of plants to remove heavy metals from aqueous streams. *Environ. Sci. Technol.* 29: 1239–1245.

Ehrlich, P.R. and Pringle, R.M. (2008). Where does biodiversity go from here? A grim business-as-usual forecast and a hopeful portfolio of partial solutions. *Proc. Natl. Acad. Sci.* 105: 11579–11586.

Entry, J.A., Watrud, L.S., Manasse, R.S. et al. (1997). Phytoremediation and reclamation of soils contaminated with radionuclides. In: *Phytoremediation of Soil and Water Contaminants. Symposium* (ed. E.L. Kruger, T.A. Anderson, and J.R. Coats) Series No. 664, 299–306. Washington, DC: American Chemical Society.

EPA (1997). Status of in situ phytoremediation technology. Recent Developments for In Situ Treatment of Metal Contaminated Soils, PA-542-R-97-004. US Environmental Protection Agency. http://www.epa.gov.
EPA (2000). Introduction to phytoremediation. National Risk Management Research Laboratory. US Environmental Protection Agency. http://www.clu http://in.org/download/remed/introphyto.
EPA (2004). *Cleaning up the Nation's Waste Sites: Markets and Technology Trends*, 39. Washington, DC: Office of Solid Waste and Emergency Response.
Evangelou, M.W.H., Conesa, H.M., Robinson, B.H. et al. (2012). Biomass production on trace elementcontaminated land: a review. *Environ. Eng. Sci.* 29: 823–839.
Evangelou, M.W., Brem, A., Ugolini, F. et al. (2014). Soil application of biochar produced from biomass grown on trace element contaminated land. *J. Environ. Manage.* 146: 100–106.
Farnsworth, E.J. and Meyerson, L.A. (2003). Comparative ecophysiology of four wetland plant species along a continuum of invasiveness. *Wetlands* 23: 750–762.
Fässler, E., Robinson, B.H., Stauffer, W. et al. (2010). Phytomanagement of metal-contaminated agricultural land using sunflower, maize and tobacco. *Agric. Ecosyst. Environ.* 136: 49–58.
Fitz, W.J. and Wenzel, W.W. (2002). Arsenic transformations in the soil-rhizosphere-plant system: fundamentals and potential application to phytoremediation. *J. Biotechnol.* 99: 259–278.
Flathman, P.E. and Lanza, G.R. (1998). Phyteoremediation: current views on an emerging green technology. *Soil Sediment Contam. Int. J.* 71: 415–432.
Freitas, E.V., Nascimento, C.W., and Silva, W.M. (2014). Citric acid-assisted phytoextraction of lead in the field: the use of soil amendments. *Water Air Soil Pollut.* 225: 1796. https://doi.org/10.1007/s11270-013-1796-6.
Garbisu, C. and Alkorta, I. (2001). Phytoextraction: a costeffective plant-based technology for the removal of metals from the environment. *Bioresour. Technol.* 77 (3): 229–236.
Gerhardt, K.E., Huang, X.D., Glick, B.R. et al. (2009). Phytoremediation and rhizoremediation of organic soil contaminants: potential and challenges. *Plant Sci.* 176: 20–30.
Gisbert, C., Clemente, R., Navarro-Aviñó, J. et al. (2006). Tolerance and accumulation of heavy metals by *Brassicaceae* species grown in contaminated soils from Mediterranean regions of Spain. *Environ. Exp. Bot.* 56: 19–26.
Gong, P., Wilke, B.M., and Fleischmann, S. (1999). Soil-based phytotoxicity of TNT (2,4,6-trinitrotoluene) to terrestrial higher plants. *Arch. Environ. Contam. Toxicol.* 36 (2): 152–157.
González-Alcaraz, M.N., Conesa, H.M., and Álvarez-Rogel, J. (2013). Phytomanagement of strongly acidic, saline eutrophic wetlands polluted by mine wastes: the influence of liming and *Sarcocornia fruticosa* on metals mobility. *Chemosphere* 90: 2512–2519.
Hansen, D., Duda, P.J., Zayed, A. et al. (1998). Selenium removal by constructed wetlands: role of biological volatilization. *Environ. Sci. Technol.* 32 (5): 591–597.
Harris, H.H., Pickering, I.J., and George, G.N. (2013). The chemical form of mercury is fish. *Science* 301 (5637): 1203–1203.
Hasanuzzaman, M. and Fujita, M. (2012). Heavy metals in the environment: current status, toxic effects on plants and possible phytoremediation. In: *Phytotechnologies: Remediation of Environmental Contaminants* (ed. N.A. Anjum, M.A. Pereira, I. Ahmad, et al.), 7–73. Boca Raton, FL: Taylor and Francis/CRC Press.
Henry, J. (2000). *An Overview of Phytoremediation of Lead and Mercury*, 3–9. Washington, DC: NNEMS Report.
Jadia, C.D. and Fulekar, M.H. (2009). Phytoremediation of heavy metals: recent techniques. *Afr. J. Biotechnol.* 8: 921–928.
Jaradat, A.A. (2010). Genetic resources of energy crops: biological systems to combat climate change. *Aust. J. Crop Sci.* 4 (5): 309–323.
Jiang, Y., Lei, M., Ducan, L. et al. (2015). Integrating phytoremediation with biomass valorisation and critical element recovery: a UK contaminated land perspective. *Biomass Bioenergy* 83: 328–339.
Juwarkar, A.A., Singh, S.K., and Mudhoo, A. (2010). A comprehensive overview of elements in bioremediation. *Rev. Environ. Sci. Biotechnol.* 9: 215–288.
Kabouw, P., van Dam, N.M., van der Putten, W.H. et al. (2012). How genetic modification of roots affects rhizosphere processes and plant performance. *J. Exp. Bot.* 63: 3475–3483.
Kaewtubtim, P., Meeinkuirt, W., Seepom, S. et al. (2018). Phytomanagement of radionuclides and heavy metals in mangrove sediments of Pattani Bay, Thailand using *Avicennia marina* and *Pluchea indica*. *Mar. Pollut. Bull.* 127: 320–333.
Karami, A. and Shamsuddin, Z.H. (2010). Phytoremediation of heavy metals with several efficiency enhancer methods. *Afr. J. Botechnol.* 9 (25): 3689–3698.
Karimzadeh, L., Heilmeier, H., and Merkel, B.J. (2012). Effect of microbial siderophore DFO-B on Cd accumulation by *Thlaspic aerulescens* hyperaccumulator in the presence of zeolite. *Chemosphere* 88 (6): 683–687.

Kaushik, P. (2015). Use of natural organic and synthetic chelating agents for efficient phytoremediation. *Int. J. Enhanc. Res. Sci. Technol. Eng.* 4 (6): 99–101.
Kennen, K. and Kirkwood, N. (2015). *Phyto: Principles and Resources for Site Remediation and Landscape Design*, 346. London: Routledge.
Kidd, P., Mench, M., Álvarez-López, V. et al. (2015). Agronomic practices for improving gentle remediation of traceelement-contaminated soils. *Int. J. Phytorem.* 17: 1005–1037.
Kirkham, M.B. (2006). Cadmium in plants on polluted soils: effects of soil factors, hyperaccumulation, and amendments. *Geoderma* 137: 19–32.
Kohler, J., Caravaca, F., Azcon, R. et al. (2016). Suitability of the microbial community composition and function in a semiarid mine soil for assessing phytomanagement practices based on mycorrhizal inoculation and amendment addition. *J. Environ. Manage.* 169: 236–246.
Kotrba, P., Najmanova, J., Macek, T. et al. (2009). Genetically modified plants in phytoremediation of heavy metal and metalloid soil and sediment pollution. *Biotechnol. Adv.* 27 (6): 799–810.
Kumar, N., Dushenkov, V., Motto, H. et al. (1995). Phytoextraction: the use of plants to remove heavy metals from soils. *Environ. Sci. Technol.* 29: 1232–1238.
Kumari, R., Kaur, I., and Bhatnagar, A.K. (2013). Enhancing soil health and productivity of *Lycopersicon esculentum* mill. Using *Sargassum johnstonii Setchell* and *Gardner* as a soil conditioner and fertilizer. *J. Appl. Phycol.* 25 (4): 1225–1235.
Lal, K., Yadav, R.K., Kaur, R. et al. (2013). Productivity, essential oil yield, and heavy metal accumulation in lemon grass (*Cymbopogon flexuosus*) under varied wastewater-groundwater irrigation regimes. *Ind. Crop. Prod.* 45: 270–278.
Lambers, H., Clements, J.C., and Nelson, M.N. (2013). How a phosphorus-acquisition strategy based on carboxylate exudation powers the success and agronomic potential of lupines (*Lupinus*, Fabaceae). *Am. J. Bot.* 100 (2): 263–288.
Lasat, M.M. (2002). Phytoextraction of toxic metals: a review of biological mechanisms. *J. Environ. Qual.* 31: 109–120.
Lasat, M.M., Fuhrmann, M., Ebbs, S.D. et al. (1998). Phytoremediation of a radiocesium-contaminated soil: evaluation of cesium-137 bioaccumulation in the shoots of three plant species. *J. Environ. Qual.* 27: 165–169.
Lewandowski, I.M., Schmidt, U., Londo, M. et al. (2006). The economic value of the phytoremediation function-assessed by the example of cadmium remediation by willow (*Salix* spp.). *Agric. Syst.* 89 (1): 68–89.
Li, T.Q., Yang, X.E., Jin, X.F. et al. (2005). Root responses and metal accumulation in two contrasting ecotypes of *sedum alfredii* Hance under lead and zinc stress. *J. Environ. Sci. Health* 40 (5): 1081–1096.
Lombi, E., Zhao, F.J., Dunham, S.J. et al. (2001). Phytoremediation of heavy metal-contaminated soils: natural hyperaccumulation versus chemically enhanced phytoextraction. *J. Environ. Qual.* 30: 1919–1926.
Ma, L.Q., Komar, K.M., Tu, C. et al. (2001). A fern that hyperaccumulates arsenic: a hardy, versatile, fast-growing plant helps to remove arsenic from contaminated soils. *Nature* 409: 579–579.
Ma, Y., Prasad, M.N.V., Rajkumar, M. et al. (2011). Plant growth promoting rhizobacteria and endophytes accelerate phytoremediation of metalliferous soils. *Biotechnol. Adv.* 29: 248–258.
Ma, T.T., Teng, Y., Luo, Y.M. et al. (2013). Legumegrass intercropping phytoremediation of phthalic acid esters in soil near an electronic waste recycling site: a field study. *Int. J. Phytorem.* 15: 154–167.
Maestri, E. and Marmiroli, N. (2011). Transgenic plants for phytoremediation. *Int. J. Phytorem.* 13: 264–279.
Maiti, S.K., Nandhini, S., and Das, M. (2005). Accumulation of metals by naturally growing herbaceous and tree species in iron ore tailings. *Int. J. Environ. Stud.* 62 (5): 593–603.
Martinez-Oró, D., Parraga-Aguado, I., Querejeta, J.I. et al. (2017). Importance of intra- and interspecific plant interactions for the phytomanagement of semiarid mine tailings using the tree species *Pinus halepensis*. *Chemosphere* 186: 405–413.
McGrath, S.P. and Zhao, F.J. (2003). Phytoextraction of metals and metalloids from contaminated soils. *Curr. Opin. Biotechnol.* 14: 277–282.
McNair, M.R., Tilstone, G.H., and Smith, S.S. (2000). The genetics of metal tolerance and accumulation in higher plants. In: *Phytoremediation of Contaminated Soil and Water* (ed. N. Terry and G. Bañuelos), 235–250. Boca Raton, FL: Lewis Publishers.
Meagher, R.B.P. (2000). Phytoremediation of toxic elemental and organic pollutants. *Curr. Opin. Plant Biol.* 3: 153–162.
Meagher, R.B.P., Rugh, C.L., Kandasamy, M.K. et al. (2000). Engineered phytoremediation of mercury pollution in soil and waters using bacterial genes. In: *Phytoremediation of Contaminated Soil and Water* (ed. N. Terry and G. Bañuelos), 201–219. Boca Raton, FL: Lewis Publishers.
Mench, M., Lepp, N., Bert, V. et al. (2010). Successes and limitations of phytotechnologies at field scale: outcomes, assessment and outlook from COST action 859. *J. Soils Sediments* 10: 1039–1070.

Mir, T.A., Jan, M., and Rabani, M.S. (2022). Microbial intervention for degradation of agricultural wastes. In: *Environmental Biotechnology: Sustainable Remedation of Contamination in Different Environs* (ed. R.A. Bhat, M.A. Dervash, K.R. Hakeem, and K.Z. Masoodi), 25. Apple Academic Press.

Moreno-Barriga, F., Faz, A., Acosta, J.A. et al. (2017). Use of *Piptatherum miliaceum* for the phytomanagement of biochar amended Technosols derived from pyritic tailings to enhance soil aggregation and reduce metal (loid) mobility. *Geoderma* 307: 159–171.

Mouhamad, R., Ghanem, I., Malik, A.O. et al. (2012). Phytoremediation of trichloroethylene and dichlorodiphenyltrichloroethane-polluted water using transgenic *Sesbania grandiflora* and *Arabidopsis Thaliana* plants harboring rabbit cytochrome P450 2E1. *Int. J. Phytorem.* 14 (7): 656–668.

Mueller, B., Rock, S., Gowswami, D. et al. (1999). Phytoremediation decision tree. Prepared by Interstate Technology and Regulatory Cooperation Work Group, 1–36.

Murphy, C.E.J. (2001). *An Estimate of the History of Tritium Inventory in Wood Following Irrigation with Tritiated Water*. Aiken, SC: Westinghouse Savannah River Company https://www.srs.gov/general/srs-home.

Nan, H., Jifang, Z., Dexin, D. et al. (2013). Screening of native hyperaccumulators at the huayuan river contaminated by heavy metals. *Biorem. J.* 17 (1): 21–29.

Nascimento, C.W.A. and Xing, B. (2006). Phytoextraction: a review on enhanced metal availability and plant accumulation. *Sci. Agric.* 63: 299–311.

Nsanganwimana, F., Pourrut, B., Mench, M. et al. (2014). Suitability of *Miscanthus* species for managing inorganic and organic contaminated land and restoring ecosystem services: a review. *J. Environ. Manage.* 143: 123–134.

Oh, K., Li, T., Cheng, H. et al. (2013). Development of profitable phytoremediation of contaminated soils with biofuel crops. *J. Environ. Prot.* 4: 58–64.

Olivares, A.R., Carrillo-González, R., González-Chávez, M.C.A. et al. (2013). Potential of castor bean (*Ricinus communis* L.) for phytoremediation of mine tailings and oil production. *J. Environ. Manage.* 114: 316–323.

Padmavathiamma, P.K. and Li, L.Y. (2007). Phytoremediation technology: hyper-accumulation metals in plants. *Water Air Soil Pollut.* 184: 105–126.

Pagliano, C., Raviolo, M., Vecchia, F.D. et al. (2006). Evidence for PSIIdonor- side damage and photoinhibition induced by cadmium treatment on rice (*Oryza sativa* L.). *J. Phytochem. Photobiol. B* 84: 70–78.

Panda, A.K. (1996). Rhizofiltration of Zn and Ni. *Indian J. Environ. Health* 38: 51–53.

Pandey, D.N. (2002). Sustainability science for mine-spoil restoration. *Curr. Sci.* 83: 792–793.

Pandey, V.C. (2013a). Book review on "phytoremediation and biofortification". *Environ. Eng. Manage. J.* 12 (4): 851–852.

Pandey, V.C. (2013b). Suitability of *Ricinus communis* L. cultivation for phytoremediation of fly ash disposal sites. *Ecol. Eng.* 57: 336–341.

Pandey, V.C. and Singh, B. (2012). Rehabilitation of coal fly ash basins: current need to use ecological engineering. *Ecol. Eng.* 49: 190–192.

Pandey, V.C. and Singh, N. (2015). Aromatic plants versus arsenic hazards in soils. *J. Geochem. Explor.* 157: 77–80.

Pandey, V.C., Singh, J.S., Singh, R.P. et al. (2011). Arsenic hazards in coal fly ash and its fate in Indian scenario. *Resour. Conserv. Recycl.* 55: 819–835.

Pandey, V.C., Singh, K., Singh, R.P. et al. (2012a). Naturally growing *Saccharum munja* on the fly ash lagoons: a potential ecological engineer for the revegetation and stabilization. *Ecol. Eng.* 40: 95–99.

Pandey, V.C., Singh, K., Singh, J.S. et al. (2012b). *Jatropha curcas*: a potential biofuel plant for sustainable environmental development. *Renew. Sust. Energ. Rev.* 16: 2870–2883.

Pandey, V.C., Pandey, D.N., and Singh, N. (2015a). Sustainable phytoremediation based on naturally colonizing and economically valuable plants. *J. Clean. Prod.* 86: 37–39.

Pandey, V.C., Bajpai, O., and Singh, N. (2016). Energy crops in sustainable phytoremediation. *Renew. Sust. Energ. Rev.* 54: 58–73.

Papazoglou, E.G., Karantounias, G.A., Vemmos, S.N. et al. (2005). Photosynthesis and growth responses of giant reed (*Arundo donax* L.) to the heavy metals Cd and Ni. *Environ. Int.* 31 (2): 243–249.

Párraga-Aguado, I., Álvarez-Rogel, J., González Alcaraz, M.N. et al. (2013). Assessment of metal(loid)s availability and their uptake by *Pinus halepensis* in a Mediterranean forest impacted by abandoned tailings. *Ecol. Eng.* 58: 84–90.

Párraga-Aguado, I., Querejeta, J.I., González Alcaraz, M.N. et al. (2014). Usefulness of pioneer vegetation for the phytomanagement of metal(loid)s enriched tailings: grasses vs. shrubs vs. trees. *J. Environ. Manage.* 133: 51–58.

Parraga-Aguado, I., Gonzalez-Alcaraz, M.N., Schulin, R. et al. (2015). The potential use of *Piptatherum miliaceum* for the phytomanagement of mine tailings in semiarid areas: role of soil fertility and plant competition. *J. Environ. Manage.* 158: 74–84.

Pivetz, B.E. (2001). *Ground Water Issue: Phytoremediation of Contaminated Soil and Ground Water at Hazardous Waste Sites*, 1–36. Washington, DC: United States Environmental Protection Agency.

Poschenrieder, C., Tolrà, R., and Barceló, J. (2006). Can metals defend plants against biotic stress? *Trends Plant Sci.* 11: 288–295.

Prasad, M.N.V. (2004). Phytoremediation of metals and radionuclides in the environment: the case of natural hyperaccumulators, metal transporters, soil-amending chelators and transgenic plants. In: *Heavy Metal Stress in Plants from Biomolecules to Ecosystems* (ed. Prasad), 345–391. Berlin, Heidelberg: Springer.

Prasad, M.N.V. (2011). A state-of-the-art report on bioremediation, its applications to contaminated sites in India. Ministry of Environment & Forests, Government of India.

Prasad, M.N.V. and de Oliveira Freitas, H.M. (2003). Metal hyperaccumulation in plants biodiversity prospecting for phytoremediation technology. *Electron. J. Biotechnol.* 6: 285–321.

Quintela-Sabarís, C., Marchand, L., Kidd, P.S. et al. (2017). Assessing phytotoxicity of trace element-contaminated soils phytomanaged with gentle remediation options at ten European field trials. *Sci. Total Environ.* 599 (600): 1388–1398.

Rabani, M.S., Hameed, I., Mir, T.A. et al. (2022). Microbial-assisted phytoremediation. In: *Phytoremediation: Biotechnological Strategies for Promoting Invigorating Environs* (ed. R.A. Bhat, F.M.P. Tonelli, G.H. Dar, and K. Hakeem), 91–114. Academic Press.

Rai, P.K. (2008). Technical note: phytoremediation of Hg and Cd from industrial effluents using an aquatic free floating macrophyte *Azolla* pinnata. *Int. J. Phytorem.* 10: 430–439.

Rajkumar, M., Sandhya, S., Prasad, M.N.V. et al. (2012). Perspectives of plant-associated microbes in heavy metal phytoremediation. *Biotechnol. Adv.* 30: 1562–1574.

Ramachandra, T.V. and Bachamanda, S. (2007). Environmental audit of municipal solid waste management. *Int. J. Environ. Technol. Manage.* 7 (3/4): 369–391.

Rascio, N. and Navari-Izzo, F. (2011). Heavy metal hyperaccumulating plants: how and why do they do it? And what makes them so interesting? *Plant Sci.* 180 (2): 169181.

Raskin, I., Kumar, P.B.A.N., Dushenkov, S. et al. (1994). Bioconcentration of heavy metals by plants. *Curr. Opin. Biotechnol.* 5: 285–290.

Reeves, R.D. and Brooks, R.R. (1983). Hyperaccumulation of lead and zinc by two metallophytes from a mining area of Central Europe. *Environ. Pollut.* 31: 277–287.

Robinson, B.H., Green, S.R., Chancerel, B. et al. (2007). Poplar for the phytomanagement of boron contaminated sites. *Environ. Pollut.* 150 (2): 225–233.

Rodríguez-Vila, A., Covelo, E.F., Forjan, R. et al. (2014). Recovering a copper mine soil using organic amendments and phytomanagement with *Brassica juncea* L. *J. Environ. Manage.* 147: 73–80.

Rowe, R.L., Street, N.R., and Taylor, G. (2009). Identifying potential environmental impacts of large-scale deployment of dedicated bioenergy crops in the UK. *Renew. Sust. Energ. Rev.* 13: 71–290.

Rugh, C.L. (2001). Mercury detoxification with transgenic plants and other biotechnological breakthroughs for phytoremediation. *In Vitro Cell. Dev. Biol.* 37: 321–332.

Russ, T.H. (2000). *Redeveloping Brownfields*. New York: McGraw-Hill.

Sage, R.B., Cunningham, M., and Boatman, N. (2006). Birds in willow short rotation coppice compared to other arable crops in Central England and are view of bird census data from energy crops in the UK. *IBIS* 148: 184–197.

Salt, D.E., Smith, R.D., and Raskin, I. (1998). Phytoremediation. *Annu. Rev. Plant Physiol. Plant Mol. Biol.* 49: 643–668.

Sarangi, B.K., Kalve, S., Pandey, R.A. et al. (2009). Transgenic plants for phytoremediation of arsenic and chromium to enhance tolerance and hyperaccumulation. *Transgenic Plant J.* 3: 57–86.

Saraswat, S. and Rai, J.P.N. (2009). Phytoextraction potential of six plant species grown in multimetal contaminated soil. *Chem. Ecol.* 25 (1): 1–11.

Sarma, H. (2011). Metal hyperaccumulation in plants: a review focusing on phytoremediation technology. *J. Environ. Sci. Technol.* 4 (2): 118–138.

Schröder, P., Navarro-Aviñó, J., Azaizeh, H. et al. (2007). Using phytoremediation technologies to upgrade waste water treatment in Europe. *Environ. Sci. Pollut. Res.* 14 (7): 490–497.

Schröder, P., Daubner, D., Maier, H. et al. (2008). Phytoremediation of organic xenobiotics–glutathione dependent detoxification in Phragmites plants from European treatment sites. *Bioresour. Technol.* 99: 7183–7191.

Sharma, P. and Pandey, S. (2017). Phytomanagement of heavy metal lead by fodder grass *Lasiurus scindicus* in polluted soil and water of Dravyawati River. *Am. J. Environ. Sci.* 13 (2): 167–171.

Shukla, O.P., Juwarkar, A.A., Singh, S.K. et al. (2011). Growth responses and metal accumulation capabilities of woody plants during the phytoremediation of tannery sludge. *Waste Manage.* 31 (1): 115–123.

Singh, P.K. and Mahapatra, J.K. (2000). *The Changing Scenario in Plant Species*, 402–422. Allied Publishers.

Singh, S., Thorat, V., Kaushik, C.P. et al. (2009). Potential of *Chromolaena odorata* for phytoremediation of 137Cs from solution and low level nuclear waste. *J. Hazard. Mater.* 162: 743–745.

Skarback, E. and Becht, P. (2005). Land scape perspective on energy forest. *Biomass Bioenergy* 28: 151–159.

Smith, R.A.H. and Bradshaw, A.D. (1979). The use of metal tolerant plant populations for the reclamation of metalliferous wastes. *J. Appl. Ecol.* 16 (2): 595–612.

Sparks, D.L. (1995). *Environmental Soil Chemistry*. San Diego: Academic Press.

Stankovic, D. and Devetakovic, J. (2016). Application of plants in remediation of contaminated sites. *Reforesta* 1: 300–320.

Sun, Q., Ye, Z.H., Wang, X.R. et al. (2007). Cadmium hyperaccumulation leads to an increase of glutathione rather than phytochelatins in the cadmium hyperaccumulator *Sedum alfredii*. *J. Plant Physiol.* 164: 1489–1498.

Sun, Y.B., Zhou, Q.X., Wang, L. et al. (2009). Cadmium tolerance and accumulation characteristics of *Bidens pilosa* L. as a potential Cd-hyperaccumulator. *J. Hazard. Mater.* 161: 808–814.

Thompson, P.L., Ramer, L.A., and Schnoor, J.L. (1998). Uptake and transformation of TNT by hybrid poplar trees. *Environ. Sci. Technol.* 32: 975–980.

Turnau, K., Orlowska, E., Ryszka, P. et al. (2006). Role of mycorrhizal fungi in phytoremediation and toxicity monitoring of heavy metal rich industrial wastes in southern Poland. In: *Soil and Water Pollution Monitoring, Protection and Remediation* (ed. I. Twardowska). Berlin: Springer, Verlag.

USDA Forest Service (2000). Water and the forest service. FS-660, Washington, DC, 26.

Vallini, G., Di Gregorio, S., and Lampis, S. (2005). Rhizosphereinduced selenium precipitation for possible applications in phytoremediation of se polluted effluents. *Z. Naturforsch. C* 6: 349–356.

Vamerali, T., Bandiera, M., and Mosca, G. (2010). Field crops for phytoremediation of metal-contaminated land: a review. *Environ. Chem. Lett.* 8: 1–17.

Vamerali, T., Bandiera, M., Lucchini, P. et al. (2014). Long-term phytomanagement of metal-contaminated land with field crops: integrated remediation and biofortification. *Eur. J. Agron.* 53: 56–66.

Van Aken, B. (2009). Transgenic plants for enhanced phytoremediation of toxic explosives. *Curr. Opin. Biotechnol.* 20: 231–236.

Van Nevel, L., Mertens, J., Oorts, K. et al. (2007). Phytoextraction of metals from soils: how far from practice? *Environ. Pollut.* 150 (1): 34–40.

Van Slycken, S., Witters, N., Meers, E. et al. (2013a). Safe use of metal contaminated agricultural land by cultivation of energy maize (Zea mays). *Environ. Pollut.* 178: 375–380.

Van Slycken, S., Witters, N., Meiresonne, L. et al. (2013b). Field evaluation of willow under short rotation coppice for phytomanagement of metal-polluted agricultural soils. *Int. J. Phytorem.* 15: 677–689.

Vidyasagaran, K. and Kumar, V. (2016). Use of municipal garbage as potting Media in the Nursery Production of *Ailanthus triphysa* (Dennst.) Alston seedlings. *Indian J. Ecol.* 43 (2): 54–57.

Vidyasagaran, K. and Kumar, V. (2017). Evaluation of municipal garbage as a component of potting media for economically important timber species seedlings for afforestation in tropics. *J. Environ. Biol.* 38: 7.

Vidyasagaran, K., Ajeesh, R., and Kumar, V. (2014a). Use of municipal garbage for the production of quality *Swietenia macrophylla* King (Mahogany) Seedlings. *Nat. Environ. Pollut. Technol.* 13 (4): 705–712.

Vidyasagaran, K., Kumar, V., and Ajeesh, R. (2014b). Utilization of municipal garbage as component potting media for the production of teak (*Tectona grandis* L.f.) seedlings in the nursery. *Ecoscan* 8 (3&4): 215–219.

Waseem, A., Arshad, J., Iqbal, F. et al. (2014). Pollution status of Pakistan: a retrospective review on heavy metal contamination of water, soil and vegetables. *Biomed. J. Int.* 813206: 1–29.

Watt, N.R., Willey, N.J., Hall, S.C. et al. (2002). Phytoextraction of 137Cs: the effect of soil 137Cs concentration on 137Cs uptake by Beta vulgaris. *Acta Biotechnol.* 22: 183–188.

Wu, Z., Bañuelos, G.S., Lin, Z.Q. et al. (2015). Biofortification and phytoremediation of selenium in China. *Front. Plant Sci.* 6: 136. https://doi.org/10.3389/fpls.2015.00136.

Yanai, J., Zhao, F.J., McGrath, S.P. et al. (2006). Effect of soil characteristics on Cd uptake by the hyperaccumulator *Thlaspi caerulescens*. *Environ. Pollut.* 139 (1): 167–175.

Yanqun, Z., Yuan, L., Jianjun, C. et al. (2005). Hyperaccumulation of Pb, Zn and Cd in herbaceous grown on lead_zinc mining area in Yunnan, China. *Environ. Int.* 31: 755–762.

Ye, W.L., Khan, M.A., McGrath, S.P. et al. (2011). Phytoremediation of arsenic contaminated paddy soils with *Pteris vittata* markedly reduces arsenic uptake by rice. *Environ. Pollut.* 159: 3739–3743.

Yerima, M.B., Umar, A.F., Shinkafi, S.A. et al. (2012). Bioremediation of hydrocarbon pollution: a sustainable means of biodiversity conservation. *J. Environ. Prot. Sustainable Dev.* 3: 43–50.

Young-huang, W. and Wei-zhen, X. (1985). The tolerance and concentration capacity of *Azolla* to 11 metal ions. *Azolla Utilization, Proceedings of Workshop on Azolla use*, Fuzhou, Fujian, China, 282.

Zhang, X., Xia, H., Li, Z. et al. (2011). Identification of a new potential Cd-hyperaccumulator *Solanum photeinocarpum* by soil seed bank metal concentration gradient method. *J. Hazard. Mater.* 189: 414–419.

Zhang, H., Tian, Y., Wang, L. et al. (2013). Ecophysiological characteristics and biogas production of cadmium-contaminated crops. *Bioresour. Technol.* 146: 628–636.

Zornoza, P., Vázquez, S., Esteban, E. et al. (2002). Cadmium-stress in nodulated white lupine: strategies to avoid toxicity. *Plant Physiol. Biochem.* 40: 1003–1009.

Zouboulis, A.I. and Katsoyiannis, I.A. (2005). Recent advances in the bioremediation of arsenic-contaminated groundwater. *Environ. Int.* 31 (2): 213–219.

11

Biotechnological Advancements in Phytoremediation

Venkatesh Chunduri, Payal Kapoor, Anita Kumari, Aman Kumar, Saloni Sharma, Natasha Sharma, Satveer Kaur, and Monika Garg

National Agri-Food Biotechnology Institute, Mohali, India

11.1 Introduction

For the sake of economic growth, various anthropogenic activities including agricultural waste disposal, industrial, and military operations have resulted in environmental pollution due to the excessive energy requirement. The continuous release of heavy metals from metallurgical furnaces (Lee et al. 2006; Govarthanan et al. 2013), radioactive nuclides (Mclean and Abbe 2008), and septage/wastewater pollutants (from the industrial, municipal, and domestic origin) (Mapanda et al. 2005; Deng et al. 2007; Robinson et al. 2011) into the ecosystems has resulted in harmful impacts on plants, microflora, aquatic ecosystem and life-supporting functions such as immobilization, mineralization of nutrients, and processes like nitrification. This has primarily affected public health as well as the health of the biosphere in an undesirable manner (Batayneh 2011). Furthermore, the increase in soil pollution has caused soil infertility. Hence, in the present scenario, the conservation of soil and the environment along with human well-being has become one of the major challenges.

In the past few decades, the concept of remediation has emerged, which involves the removal of harmful and deleterious chemicals from the environment and ecosystems. Conventional remediation approaches include physical and chemical methods to remediate contaminated sites. Physical methods include soil excavation, disposal to landfills, soil washing, air spray and pyrometallurgical processes, while chemical treatment involves redox potential mechanisms to detoxify and decrease the mobility of metal contaminants in wastewater. These chemical treatments can be performed in situ by injecting the chemicals into the groundwater but these treatments can further introduce contaminants into the groundwater (Mulligan et al. 2001). Hence, these methods cause damage to the environment and have adverse effects on biodiversity and soil health. Besides, they are highly expensive and difficult to implement. So, there is an urgent requirement for cheap and efficient methods for the cleanup of the contaminated biosphere. To overcome these problems, plants can be used for the removal of pollutants from the environment. Plants employ various remediation strategies to combat many types of pollution and the process is termed "phytoremediation" (Pilon-Smits and Freeman 2006; Doty et al. 2007). Plants can absorb, metabolize, and transport nutrients and contaminants from the soil, water, and growth matrix. In this process, plants are grown in contaminated sites to remove contaminants or to accelerate the immobilization or decomposition (degradation) of pollutants. Unlike conventional remediation processes, phytoremediation is a natural, cost-efficient, and environmentally safe method in which no synthetic inputs are used.

11.2 Types of Phytoremediation

A lot of phytoremediation methods have been used for the treatment of various types of contaminants. A summary of different types of phytoremediation processes was shown in Figure 11.1.

Aquatic Contamination: Tolerance and Bioremediation, First Edition. Edited by Rouf Ahmad Bhat, Gowhar Hamid Dar, Fernanda Maria Policarpo Tonelli, and Saima Hamid.

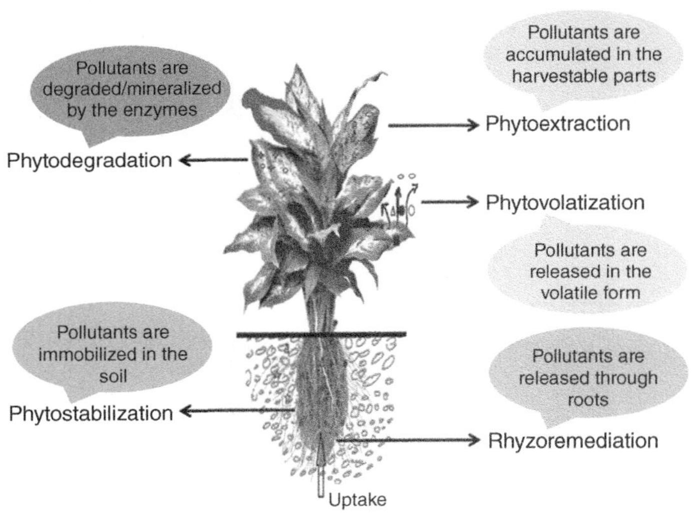

Figure 11.1 Various mechanisms involved in phytoremediation.

11.2.1 Phytoextraction or Phytoaccumulation

It is the phenomenon in which plants are used to remove heavy metal contamination from soil and water through their uptake into the aerial parts of the plant. For this purpose, hyperaccumulator plants are best suited. They are capable of translocating and concentrating high number of heavy metal contaminants into the shoots and roots.

This phenomenon is also called phytomining. These hyperaccumulator plants are planted on the contaminated sites and disposed of or processed by drying, ashing, or composting. The site is replanted until the contaminants reached acceptable levels. The phytoextraction method is used for the treatment of inorganic contaminants.

11.2.2 Phytodegradation or Phytotransformation

In this process, plants degrade the contaminants by the use of enzymes or through metabolic processes. This method is widely used for the treatment of organic contaminants present in soils, sediments, groundwater, and surface water. The degradation process can take place within the plant or inside the rhizosphere with the help of rhizospheric bacteria. In this method, plant or bacterial enzymes transform the organic contaminants completely or partially into inorganic compounds or intermediates (McCutcheon and Schnoor 2003). Some enzymes present in the plants can degrade and transform military supply waste explosives, polychlorinated solvents, and herbicides (Mukhopadhyay and Maiti 2010; Jeevanantham et al. 2019). Enzymes used for the degradation of organic pollutants are peroxidases, peroxygenases, laccases, phosphatases, nitroreductases, and dehalogenases.

11.2.3 Phytovolatilization

Phytovolatization is a process in which plants uptake the contaminants from the soil and translocate them to the leaves where they get transformed into volatile compounds and released into the environment through transpiration. Phytovolatilization can occur with contaminants present in soil, sediments, or water. This method is mainly used for the removal of metal contaminants, primarily Hg. Tobacco plants are known to volatilize inorganic contaminants such as Hg. Plant roots absorb toxic methylmercury from the soil, change their chemical composition, and release less toxic forms of Hg in the environment. The method of phytovolatilization takes place through leaves (ITRC 2009; Mukhopadhyay and Maiti 2010). Besides these, two approaches have been used in the ecosystem to remove Hg from the soil systems. Transgenic plants containing specific metal-detoxifying bacterial genes have been proven to be effective in providing resistance and enhancing the volatilization capacity of Hg. Indian mustard and musk grass have been used for the extraction and accumulation of selenium (Se) (Cherian and Oliveira 2005).

11.2.4 Rhizoremediation/Rhizodegradation, Microbe-Assisted Phytoremediation, and Rhizosphere Bioremediation

Rhizodegradation is described as the degradation of contaminants by the plants in the soil rhizosphere. Microflora of soil plays an important role degradation process, as they utilize these contaminants as a source of nutrients and energy and convert them to harmless compounds (Truu et al. 2015). The degradation of contaminants utilizes both microbial as well as plant enzymes. This approach is mostly used for the treatment of organic contaminants released from various processes like agricultural amendments and petroleum processing.

11.2.5 Phytostabilization

Phytostabilization is the mechanism by which plants reduce the mobility of the contaminants in the soil. Immobilization of contaminants is accomplished by adsorption and accumulation inside the plant root zone (exudates) or precipitation in the rhizosphere. This mechanism is carried out by bacteria or other soil microorganisms present around the plant root system. Phytostabilization is the ability of the plants to retain contaminants that reduce the mobility and bioavailability of contaminants in the soil making them less toxic. Phytostabilization takes place through the process of adsorption, precipitation, complex formation, or metal chelation. Phytostabilization takes advantage of the plant extrudes such as chemoattractants which help in rhizosphere microbial nutrition, thereby changing the soil chemistry and environment (Broeckling et al. 2008; Girkin et al. 2018). These exudates also help in regulating the pH of the soil (Yan et al. 1996) and the chelation of metals (Dakora and Phillips 2002). Phytostabilization is mainly utilized in the treatment of contaminated land areas affected by mining activities of metals like lead (Pb), arsenic (As), cadmium (Cd), chromium (Cr), copper (Cu), and zinc (Zn) (Yadav et al. 2014).

11.2.6 Rhizofiltration

Rhizofiltration is the mechanism by which plants adsorb, precipitate, or concentrate metals from contaminated water into and onto the plant roots. This method has been used for the treatment of surface and groundwater, sewage sludge, and agricultural, industrial, and municipal discharge (Mani and Kumar 2013). Both rhizofiltration and phytoextraction processes are identical, but in rhizofiltration, plants primarily act upon polluted water rather than soil. Biogeochemical processes taking place in exudates help in the precipitation of contaminants in roots and water reservoirs. Root exudates can change metals' chemical form and promote their transport and bioavailability. Common sunflower, Indian mustard, tobacco, rye, spinach, and corn have been used to remove Pb from contaminated water (Dushenkov et al. 1995; Soni and Jain 2014). Rhizofiltration has been used for the treatment of Pb, Cu, Ni, Cd, Zn, and Cr, which are stored in the roots.

11.3 Types of Pollutants

There are two major classes of contaminants: organic and inorganic.

11.3.1 Organic Pollutants

The main cause of organic contaminants is rapid industrialization. Industries like pharmaceutical, dying, petroleum, fertilizers, explosives and chemicals release the majority of the hazardous organic byproducts leading to an increase in the concentration of environmental organic compound contamination. These are mobile and enter the food chain causing serious health issues. Organic contaminants include pharmaceutical wastes/byproducts like antibiotics (cypermethrin (CYP)), petroleum hydrocarbons (benzene, toluene, ethylbenzene, and xylene (BTEX)), explosives such as trinitrotoluene (TNT) and cyclotrimethylenetrinitramine or research department explosive (RDX), chlorinated solvents (trichloroethylene (TCE)), herbicides/pesticides (i.e. atrazine, chlorpyrifos, and 2,4-D) and halogenated hydrocarbons like TCE (Dhankher et al. 2012). In general, detoxification occurs in three steps. Addition of a reactive group to the pollutant (hydroxyl, amino, or sulfhydryl group). In the second step, another compound such as sugar moiety is conjugated into the reactive group. Finally, the conjugated pollutant is sequestered into the vacuole or integrated into cell wall components, making the compound less toxic (Dhankher et al. 2012).

11.3.2 Inorganic Pollutants

Inorganic pollutants include heavy metals (Hg, Cd, Cr, Cu, Zn, Se, and Pb), nonmetallic compounds (As), radionuclides (cesium (Cs), phosphorus (P), uranium (U)) and plant fertilizers (e.g. nitrate and phosphate). Plants require some of these metals for various metalloproteins and enzymes as cofactors that are required for normal growth and development. However, at higher concentrations, metals become toxic leading to oxidative stress by producing free radicals and reactive oxygen species causing plant death (Jones et al. 1991; Cho and Park 2000). The major cause of inorganic contamination is mining and nuclear research and changes in human agricultural practices. These heavy metals are in contrast to many organic pollutants, as they are nondegradable and conserved in the soil (Wade et al. 1993). The majority of heavy metals are immutable and present in nature as charged ions and cannot be degraded or destroyed biologically and depend on plants and their transporters for reduction or oxidation, uptake, and translocation.

11.4 Naturally Available Plant Species for Phytoremediation

Factors like a long time (several years), weather conditions, soil quality, and pollutant phytotoxicity often impede phytoremediation. Therefore, various strategies like sustainable agronomic practices, the use of plant growth-promoting microorganisms (PGMOs) and genetic engineering may be advantageous over the conventional practices of phytoremediation (Gerhardt et al. 2017).

The plant system exhibits highly versatile enzymatic machinery to detoxify pollutants and xenobiotics. The plant xenobiotic metabolism is also termed as "green liver" owing to its resemblance to the mammalian detoxification system (Hills et al. 2007). Phytoremediation constitutes an array of techniques involving metal chelation and storage in plant vacuoles followed by extraction, enzymatic degradation, and volatilization. Various remedial strategies play a crucial role in phytoremediation. For instance, the rhizoremediation strategy involves the release of root exudates into the soil, thereby promoting the symbiotic plant-microbe interaction. Once the pollutants are absorbed by the roots, they get accumulated in root cells. Occasionally, they get translocated to the shoot-through xylem with the aid of plant membrane transporters. This process is widely involved in the treatment of inorganic pollutants by plants. In the shoot, with the help of membrane transporter proteins, the pollutants can be exported from the xylem into leaf cells where the pollutants are destined to get transformed enzymatically or get sequestered often in vacuoles or cell walls or mineralized or volatilized by various enzymes. The pollutants may be assimilated to other parts of the plants such as young leaves, roots, or reproductive organs via phloem where they experience further modifications. So, it is evident from the active plant processes involved that the potential of the plant species to remediate different pollutants is determined by the factors like the abundance of transporters and enzymes, their microbial associates, and also their transpiration rate (Hao et al. 2014). Generally, an ideal plant for phytoremediation must possess certain qualities like rapid growth, high potential for chelating or sequestering heavy metals, ability to produce large biomass, endurance to harsh environments/contaminants, and extensive root system. Plants suited for phytoremediation must also be capable of producing enzymes with antioxidant activities, viz. superoxide dismutase (SOD), ascorbate peroxidase (APX), glutathione peroxidase (GPX), chloramphenicol acetyltransferase (CAT), glutathione reductase (GR), and other antioxidants, viz. cysteine, glutathione (GSH) along with phytochelatins (PCs), peroxidases, veratryl alcohol, oxalate, quinol oxidases, and ligninolytic enzymes (Larson 1988; Stahl and Aust 1995) so that they can efficiently detoxify the harmful contaminants present in the soil or any other environment. Choosing a desirable species of phytoremediation is one such big task. Many scientists in their study have chosen selected plant species for their experiments.

11.5 Phytoremediation of Organic Pollutants

The phytoremediation process by *Bassia scoparia* has reduced the toxicity of petroleum hydrocarbons (Moubasher et al. 2015). In another study, ionic fluid like 1-butyl-3-methylimidazolium bromide was degraded into relatively harmless byproducts by *Lolium perenne* L (Habibul et al. 2019). Poly- and perfluoroalkyl substances (PFAS) are a diverse class of synthetic organofluorine compounds, that includes perfluoroalkyl carboxylic acids (PFCAs), perfluoroalkyl sulfonic acids (PFSAs), and their precursor compounds (Zhang et al. 2019). These are widely used in cosmetics, greases and lubricants, paints, polishes, fire-fighting applications, herbicides, insecticide formulations, and as adhesives (Zhang et al. 2019).

Exposure to PFAS to *Juncus effuses* has significantly removed them from the contaminated soils (Zhang et al. 2019). Other potent organic contaminants are pesticides. Their consumption around the world increased nearly twice in 2016 from 1990 (FAOSTAT 2018). They become a serious concern to the environment because they are persistent and mobile, which means they enter directly into the food chain and other ecological systems (California Department of Pesticide Regulation 2006). CYP is a synthetic pyrethroid insecticide used to control a wide range of insects (Jin et al. 2010). *Plantago major* is one of the potential plant species that can remove CYP from the soil so it may be used as a curative model (Aioub et al. 2019). A new study reported the occurrence of polybrominated diphenyl ethers and dechlorane plus, a class of potential organic pollutants in the plant tissues of mangroves. Their results suggested that 26 species of mangroves can be used in the phytoremediation of these organic pollutants (Qiu et al. 2019).

11.5.1 Phytoremediation of Inorganic Pollutants

Heavy metal phytoremediation methods include phytostabilization, phytoextraction, or rhizofiltration as these are biologically nondegradable, but they can be taken up by plants and make them nontoxic by changing their oxidation state. Phytoremediation of heavy metal-contaminated soils can also be improved by nitrogen fertilization (Arias et al. 2010). It enhances plant growth and development and synthesizes various kinds of proteins to detoxify the heavy metals in the plants (Chang et al. 2013). Other than nitrogen fertilization, acidifying soils also increase the bioavailable heavy-metal concentration in soils (Pedron et al. 2009). Limited studies have shown the phytoremediation of multimetal contaminated soils. One such example is *Azolla* species, which have shown a reduced multimetal concentration of Cd, Cu, Zn, and Ni in soils after treatment in Anzali wetlands (Talebi et al. 2019).

The contamination in water and soil is a growing problem worldwide and in India (West Bengal) and Bangladesh, where rice is flooded with As-contaminated water, leading to a high accumulation of As III in rice (Meharg 2004). The mechanism of As phytoremediation has been well studied. Several studies support the neutralization of As in contaminated waters and soil. *Hydrilla verticillata* detoxifies the As through the synthesis of PCs and increased activity of antioxidant enzymes (Srivastava et al. 2011). Many other plant species have shown phytostabilization of As in several contaminated soils and waters (Table 11.1).

Boron (B) is an essential element for plant growth, but it can be toxic when present in excess (Zhao et al. 2019). *Chrysopogon zizanioides* L., *Gypsophila sphaerocephala*, and *Puccinellia distans* were observed to have high B tolerance (Babaoğlu et al. 2004; Stiles et al. 2010; Xin and Huang 2017). Terrestrial plant species such as *Ambrosia trifida* L., *Phytolacca americana*, and *Commelina communis* and aquatic species like *Lemna gibba* and *Iris wilsonii* were reported to accumulate B in the tissue (Marín and Oron 2007; Zhao et al. 2019).

Cd is listed as number 7 (out of 275) in the priority list of hazardous materials (Substance Priority List | ATSDR 2017). It is highly toxic and can cause diseases like pulmonary emphysema, osteoporosis/osteomalacia, and renal tubular dysfunction in humans (Satarug and Nazar 2010). Previous studies have mentioned some important Cd hyperaccumulators and potential candidates of Cd in soil and aquatic environment, for example, *Lactuca sativa*, *Pentas lanceolata*, *Wolffia globosa*, and *Zea mays* (Bi et al. 2010; Chang et al. 2013; Xie et al. 2012; Hechmi et al. 2013).

Cr is a nonessential micronutrient for plant growth, but wastewater discharges from pigment manufacturing, wood preserving, and electroplating dye and tanning industries have higher concentrations of Cr. Cr is highly toxic, mobile, and has a long residence time. It causes severe health risks to aquatic animals, and human beings, and also affects plant growth and development (Chandra and Kulshreshtha 2004). Previous studies have used *L. gibba* and *Ceratophyllum demersum* for Cr phytoremediation (Abdallah 2012).

Manganese (Mn) is an essential trace element for plants as it helps in maintaining the chloroplast structure, participating in photosynthesis and also helps in activating enzymes. But at higher concentrations, it can lead to plant death (Delhaize et al. 2007), and it is very harmful to humans (Huang et al. 2019). Previous research on plant species belonging to *P. americana*, *Broussonetia papyrifera*, *Solanum nigrum*, and *Xanthium strumarium* showed good absorption ability of Mn in the remediation of contaminated soil (Huang et al. 2019; Pan et al. 2018; Xue et al. 2008).

Hg is one of the major pollutants released into the environment from volcanic eruptions, mining, and thermometer production (Zhao et al. 2018). It causes serious damage to humans at very low concentrations (Wang et al. 2018). Plant species belonging to *Limnocharis flava* and *Acorus calamus* L. have been used for the removal of Hg from the contaminated constructed wetlands (Marrugo-Negrete et al. 2017; Singh et al. 2019).

Se is one of the essential micronutrients for humans and livestock but can be toxic to aquatic animals and waterfowl when they are in higher concentration. Selenium bioaccumulation in lower trophic-level organisms can result in

Table 11.1 List of hyperaccumulator phytoremediators used against various inorganic pollutants.

Species	Contaminants	References
Anadenanthera peregrina	As	Gomes et al. (2019)
Callitriche stagnalis	As	Robinson et al. (2006)
Ceratophyllum demersum	As	Robinson et al. (2006)
Elodea canadensis	As	Robinson et al. (2006)
H. verticillata	As	Srivastava et al. (2007, 2011)
L. gibba	As	Mkandawire and Dudel (2005)
Myriophyllum propinaquum	As	Robinson et al. (2006)
Pistia stratiotes	As	Campos et al. (2019)
P. vittata	As	Lombi et al. (2002)
Spirodela polyrhiza	As	Zhang et al. (2011)
Chrysopogon zizanioides	B	Xin and Huang (2017)
G. sphaerocephala	B	Babaoğlu et al. (2004)
P. distans	B	Stiles et al. (2010)
A. trifida *P. americana* *C. communis*	B	Asada et al. (2006)
L. gibba	B	Marín and Oron (2007)
I. wilsonii	B	Zhao et al. (2019)
Azolla pinneta	Cd	Talebi et al. (2019)
Lactuca sativa	Cd	Bi et al. (2010)
P. lanceolata	Cd	Chang et al. (2013)
W. globosa	Cd	Xie et al. (2012)
Zea mays	Cd	Hechmi et al. (2013)
L. gibba *C. demersum*	Cr	Abdallah (2012)
Azolla pinneta	Cu	Talebi et al. (2019)
Eichhornia crassipes	Pb	De Laet et al. (2019)
Pelargonium	Pb	Arshad et al. (2008)
P. vittata	Pb	Manzoor et al. (2019)
B. papyrifera	Mn	Huang et al. (2019)
P. americana	Mn	Dou et al. (2009)
S. nigrum	Mn	Pan et al. (2018)
Xanthium strumarium	Mn	Xue et al. (2008)
Acorus calamus L	Hg	Singh et al. (2019)
L. flava	Hg	Marrugo-Negrete et al. (2017)
Azolla filiculoides	Nickel (Ni)	Talebi et al. (2019)
Chara sp. *T. angustifolia*	Se	Nattrass et al. (2019)
A. filiculoides	Zn	Talebi et al. (2019)

biomagnification up to 35,000 times greater than aqueous Se concentrations (Lemly 2014). Nattrass et al. (2019) have researched *Chara* species and *Typha angustifolia* for the accumulation of Se from constructed wetlands.

Pb is another heavy metal pollutant placed second among hazardous substances (ATSDR 2017). It is nonbiodegradable having no biological function but can cause pretty much damage to plants (Pourrut et al. 2013).

Phytoremediation by *Pelargonium* and *Pteris vittata* was previously studied for the phytoextraction of Pb from contaminated soils (Arshad et al. 2008; Manzoor et al. 2019). Table 11.1 summarizes the list of various plants which are natural hyperaccumulators.

11.6 Advances in Biotechnological Approaches for Phytoremediation of Different Pollutants

Green plants have the intrinsic mechanisms to decontaminate sites or areas with toxic heavy metals, xenobiotics, and organic and inorganic pollutants. They uptake and sequester both essential and nonessential heavy metals into the cell vacuoles or harvestable parts of the plant. But plant tolerance, uptake, transport, and degradation of contaminants are sometimes hampered by a lack of abundance of membrane transporter proteins and certain enzymes and antioxidants that modify, degrade, or chelate pollutants. Manipulation of the target genes, to design plants which can withstand toxicity caused by pollutants, using recombinant DNA technology is an emerging discipline that augments phytoremediation. R-DNA technology offers a tremendous opportunity to introduce a foreign gene or to modify the expression of the gene relevant to the efficient uptake, transport, and sequestration of the contaminants (Rugh 2004; Cherian and Oliveira 2005; Fasani et al. 2017; Gerhardt et al. 2017). The brief outline of various biotechnological advances in the phytoremediation process was shown schematically in Figure 11.2.

The advent of recombinant DNA technological advances facilitated the amelioration of the plants for enhancing the effectiveness of phytoremediation by transgenic plants, either by promoting the intrinsic expression of the candidate genes or introducing novel genes into the plant. In the context of introducing such genes into the plants, *Agrobacterium tumefaciens*-mediated plant transformation is the most preferred method. Enzymes, metal transporters, and transcription factors, which directly control gene expression, can act as potential targets for manipulation to facilitate phytoremediation. The plant enzymes, including cytochrome P450 monooxygenases, GSH-S-transferases, glucosyl-O-transferases, and acyl-transferases dilute the harmful effects of toxic xenobiotic compounds which are accidentally or deliberately released in the environment. Such detoxifying enzymes have been extensively used for alleviating the xenobiotics and recalcitrant in nature.

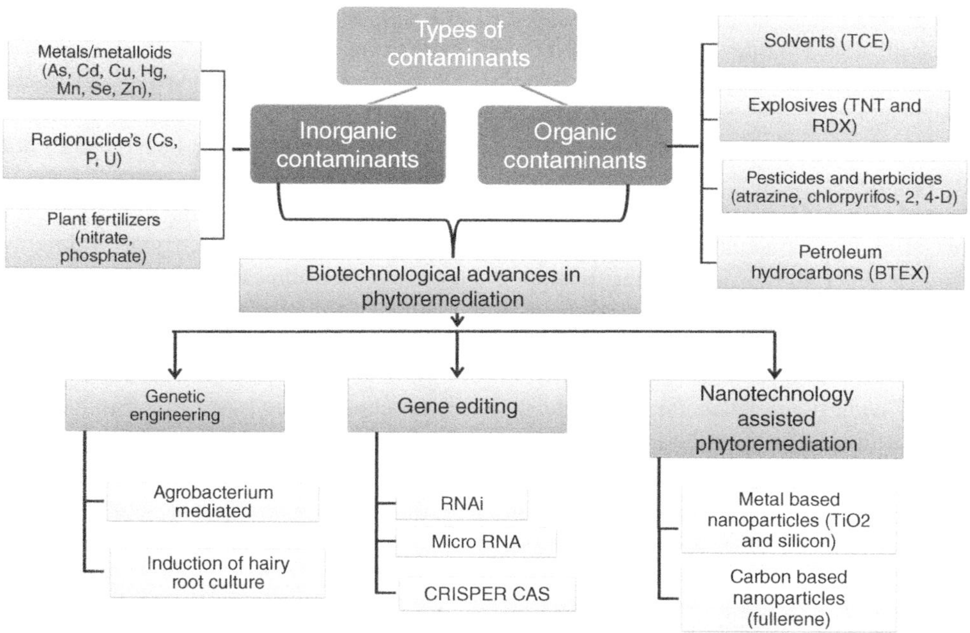

Figure 11.2 Representative diagram showing different types of pollutants and various biotechnological advances in phytoremediation.

11.7 Biotechnology Advances in the Phytoremediation of Inorganic Pollutants

The introduction of specifically designed transgenic plants readily improved the phytoremediation of toxic metals. Membrane transporters play a crucial role in the defense mechanism against inorganic pollutants. So, they can act as the potential target for manipulating the expression patterns to augment the phytoremediation. This approach has been successfully implemented to enhance the accumulation of Ca, Cd, Mn, Pb, and Zn in the harvestable parts of plants. Besides membrane transporters, genes which facilitate the volatilization and chelation of pollutants have also been addressed for improving phytoremediation (Krämer and Chardonnens 2001; Pilon-Smits and Pilon 2002; Rugh 2004; Meagher and Heaton 2005; Eapen et al. 2007). For instance, plant intolerance to the pollutants like heavy metals may be overcome by the overproduction of the transporters mediating the pollutant uptake. Therefore, enhanced metal accumulation in roots and assimilation in the shoots are the successful outcomes of implementing genetic engineering approaches.

Heavy metals and other pollutants usually have the tendency to get concentrated in the food chain. Once entered, they keep on magnifying each level in the food chain. Cd is a detrimental heavy-metal contaminant. It causes phytotoxic effects, hampers the crop yield, and also causes complications to human health (Kaiser 2001; Das et al. 2004). A transmembrane cation diffusion facilitator metal transporter, *OsMTP1* has been characterized and isolated from rice. OsMTP1, a metal transporter gene (*Oryza sativa* L. cv. IR64) was employed to transform the tobacco plant. The transgenic tobacco plants thus obtained exhibited a significant increase in biomass and Cd accumulation as compared to control plants under Cd exposure (Das et al. 2016). Similarly, in *Arabidopsis thaliana*, overexpression of AtATM3 (ATP-binding cassette transporter) conferred enhanced tolerance to Cd and Pd (Bhuiyan et al. 2011). Besides plants, microbes including bacteria, fungi, yeast, etc., also possess genes encoding for metal transporters with potency to promote phytoremediation. For instance, the yeast cadmium factor 1 (*ScYCF1*) from *Saccharomyces cerevisiae* codes for a metal transporter that neutralizes the toxic metals and metalloids by sequestering them into the vacuoles of the yeast. Using this gene, the poplar plant line which is sterile (*Populus alba* x *P. tremula* var. glandulosa) has been transformed to confer resistance against heavy metals like Pb and Cd. These transgenic plants have shown hyperaccumulation of Zn, Cd, and Pb in aerial tissues (Shim et al. 2013).

As is a potent carcinogen as well as toxic to human organs and contaminates soil and water worldwide. Bacteria and yeast have been mined extensively to decipher the mechanism of As detoxification. In *E. coli*, an operon of As resistance is operational, which comprises ArsC gene, which reduces the arsenic to As (III). Further, As (III) export pump, *ArsAB*, then facilitates the expulsion of As (III) out of the cell (Chen et al. 1986; Cervantes et al. 1994). Also, arsenate reductase genes from plant system including grasses like *Holcus lanatus*, rice, *Arabidopsis*, and *P. vittata* have shown the reduction of As concentration in the soil (Bleeker et al. 2006; Dhankher et al. 2006; Ellis et al. 2006; Duan et al. 2007). Besides the reducing enzymes, PCs, metallothioneins (MTs) and the enzymes encoding various GSH enzymes are the principal heavy metal-binding ligands in plant cells which act efficiently against various heavy metals. The PC synthase is an important enzyme which synthesizes PCs. They play an important role in heavy-metal tolerance. They help in chelating heavy metals and reduce their free concentration (Vatamaniuk et al. 1999; Cherian and Oliveira 2005). MTs are the other types of proteins that have a very high affinity toward cations such as Cd, Cu, and Zn (Singh et al. 2003). It has been studied that the overexpression of MT genes improves the phytoremediation efficiency of heavy metals. GSH-encoding enzyme γ-glutamyl cysteine synthetase (GSH) in plants is generally considered the precursor of PCs. Previous studies have reported that altering the expression of GSH in plants increases the tolerance toward heavy metals. Some of the examples of transgenic plants engineered with the PCs, MTs, and GSH have been listed in Table 11.2.

Transgenic phytoremediators for Ni have also been designed to nullify the detrimental effects of the metal. The PgIREG1 gene from the hyperaccumulator *Psychotria gabriellae*, for Ni tolerance, has been characterized as a metal transporter. *PgIREG1* has been proposed to provide tolerance toward Ni in transgenic *Arabidopsis* plants (Merlot et al. 2014).

Arabidopsis and other hyperaccumulator plants have been model plants to study various genes playing a role in scavenging metals and translocating them to the different parts of the plants. But size limitation, slow growth rates, and restricted environment of growth often limit their use as phytoremediators (Meagher and Heaton 2005). Therefore, plants which produce higher biomass with extensive root systems are preferred so that they can mitigate the soil pollution on sites in an effective way. Trees such as willows and *Populus* can grow rapidly and also have an extensive ability to phytoextract Cd, Zn, Cu, Ni, and Pb (Couselo et al. 2010). Therefore, they have been genetically engineered to hypertolerate and hyperaccumulate heavy metals. Table 11.3 summarizes some of the examples in the development of transgenic plants for heavy metal phytoremediation.

Table 11.2 List of the transgenic plants transformed using PCs, MTs, and GSH for improved phytoremediation.

Pollutant	Gene and source	Host organism	Reference
Transgenic plants containing genes encoding phytochelatin synthase for phytoremediation			
As	*CdPCS1* from *Ceratophyllum demersum*	*O. sativa*	Shri et al. (2014)
As, Cd	*AtPCS1* from *Arabidopsis*	*A.thaliana*	Li et al. (2004)
		B. juncea	Gasic and Korban (2007)
Cd	*NnPCS1 from Nelumbo nucifera*	*A. thaliana*	Liu et al. (2012)
Pb	*TaPCS1* from *Triticum aestivum*	*P. tremula × tremuloides*	Couselo et al. (2010)
Pb, Cd		*N. glauca*	Gisbert et al. (2003)
Transgenic plants containing genes encoding metallothioneins for Phytoremediation			
Cd	*CUP1* from yeast	*B.oleracea* var. botrytis	Hesegawa et al. (1997)
		Nicotiana tabacum	Macek et al. (2002)
	MT-I from mouse	*N. tabacum*	Pan et al. (1994)
	MT-II from human	*Brassica napus* and *N. tabacum*	Misra and Gedamu (1989)
Cu	*PsMT$_A$* from *Pisum sativum*	*A. thaliana*	Evans et al. (1992)
Transgenic plants containing genes encoding GSH for phytoremediation			
As, Cd, Cr	*B. juncea*	*γ-glutamylcysteine synthetase (γ-ECS)* or *glutathione synthetase (GS)* from *E. coli*	Reisinger et al. (2008)
Cd	*B. juncea*	*gshI from E.coli*	Zhu et al. (1999)
	P. tremula × P. alba	γ-glutamylcysteine synthetase (γ-ECS) from bacteria	Koprivova et al. (2002)
Cd, Zn	*B. juncea*	Glutamylcysteine synthetase (ECS) or glutathione synthetase (GS)	Bennett et al. (2003)
Hg, TCE	Glutathione S-transferase *(GST)* and human P450 2E1 (*CYP2E1)*	*Medicago sativa*	Zhang et al. (2013)
Se	Cystathionine-γ -synthase (*CGS*)	*B. juncea*	Huysen et al. (2003)
Zn	g-glutamylcysteine synthetase from bacteria	*Populus canescens*	Bittsanszky et al. (2005)

Table 11.3 List of the transgenic plants engineered against heavy metals, xenobiotic compounds, and herbicides.

Pollutants/contaminants	Gene and source organism	Host organism	Reference
I. Transgenic plants for phytoremediation of heavy metals			
Hg	*MerA* and *MerB* from bacterial mercury reductase enzyme	*A. thaliana* L. and *Nicotiana tabacum* L.	Heaton et al. (1998)
	MerA	*Liriodendron tulipifera*	Rugh et al. (1998)
Cd	*CUP1* yeast metallothionein gene originating from *Saccharomyces cerevisiae*	*N. tabacum*	Macek et al. (2002)
	OsMTP1 gene from *O. sativa*	*N. tabacum*	Das et al. (2016)
	ScYCF1 from S. cerevisiae	*P. alba* x *P. tremula*	Shim et al. (2013)
	GOR gene encoding Glutathione reductase from *E. coli*	*B. juncea*	Pilon-Smits et al. (2000)
	Gsh1 encoding for C-glutamylcysteine synthetase from *E. coli*	*P. tremula* x *P. alba*	Noctor et al. (1998)

(*Continued*)

Table 11.3 (Continued)

Pollutants/contaminants	Gene and source organism	Host organism	Reference
Cd and Pb	*AtATM3* from *A. thaliana*	*B. juncea*	Bhuiyan et al. (2011)
	YCF1 from *S. cerevisiae*	*A. thaliana*	Song et al. (2003)
Cu and Ni	Cysteine synthase from spinach	*N. tabacum*	Kawashima et al. (2004)
Se	Cystathionine-y-synthase from *A. thaliana*	*B. juncea*	Huysen et al. (2004)
	Selenocysteine methyltransferase (*SMT*) from *Astragalus bisulcatus*	*A. thaliana* and *B. juncea*	Leduc et al. (2003)
	Mouse (*Mus musculus*) Se-Cys lyase (*SL*)	*Arabidopsis*	Pilon et al. (2003)
	Arabidopsis APS1 gene, encoding a plastidic ATP sulfurylase	*B. juncea*	Pilon-Smits et al. (1999)
Ni	*PgIREG1* from *P. gabriellae*	*A. thaliana*	Merlot et al. (2014)
II. Transgenic plants for phytoremdiation of herbicides			
Alachlor	Glutathione S-transferases (GST) I from maize	*N. tabacum*	Karavangeli et al. (2005)
Atrazine	Human P450 genes CYP1A1, CYP2B6 and CYP2C19	*S. tuberosum*	Inui et al. (2001)
	Atrazine chlorohydrolase from *Pseudomonas* sp.	*N. tabacum* and *M. sativa*	Wang and Samac (2005)
Atrazine and Metolachlor	Human *CYP1A1*, *CYP2B6*, and *CYP2C19*	*O. sativa*	Kawahigashi et al. (2005)
Atrazine, CDNB, metolachlor and phenanthrene	Glutamylcysteine synthetase (*ECS*) and glutathione synthetase (*GS*)	*Brassica juncea*	Flocco et al. (2004)
Chloroacetanilide	Gamma-glutamylcysteine synthetase from bacteria	Poplar	Gullner et al. (2001)
Chlorsulfuron and imazosulfuron	Human CYP2C9 gene	*O. sativa*	Inui et al. (2001)
Chlorotoluron and norflurazon	Human cytochrome P450 monooxygenase gene CYP1A1	*O. sativa*	Kawahigashi (2009)
Oxyfluorfen	Human protoporphyrinogen IX oxidase (*PPO*)	*O. sativa*	Lee et al. (2004)
	Protoporphyrinogen oxidases (protoxes) from *Myxococcus xanthus*	*O. sativa*	Jung et al. (2004)
Sulfonylurea	*CYP2C9* from human	*O. sativa*	Hirose et al. (2005)
III. Transgenic plants for phytoremediation of the other organic compounds			
2,3-dihydroxybipheny and 1-chlorobutane	*DbfB* from *Terrabacter* sp. and *DhaA* from *Rhodococcus* sp.	*A. thaliana* and *N. tabacum*	Uchida et al. (2005)
PETN and glycerol trinitrate (GTN)	Pentaerythritoltetranitrate (PETN) reductase from *Enterobacter cloacae* PB2	*N. tabaccum*	French et al. (1999)
RDX	RDX-degrading cytochrome P450 from *Rhodococcus rhodochrous*	*A. thaliana*	Rylott et al. (2006)
RDX and TNT	*xplA* and *xplB* from *Rhodococcus rhodocous* and nitroreductase gene nfsI from bacteria	*Pascopyrum smithii*	Zhang et al. (2018)
Polychlorinated biphenyls like 4-chlorobiphenyl	Biphenyl chlorophenyl dioxygenase from *Burkholderia xenovorans*	*N. benthamiana*	Mohammadi et al. (2007)
Sulfur dioxide and sulfite	Cysteine (*Cys*) synthase from spinach	*N. tabacum*	Noji et al. (2001)
TCE	Mammalian *P450 2E1* (CYP2E1)	*N. tabacum*	Doty et al. (2000)
	Rabbit cytochrome *P450 2E1*	*Atropa belladonna*	Banerjee et al. (2002)
TCE, chloroform, and benzene	Mammalian cytochrome *P450 CYP2E1*	*P. tremula* x *P. alba*	Doty et al. (2007)
TNT	Nitroreductase gene PnrA from *Pseudomonas putida* JLR11	*P. tremula* x *tremuloides*	Hannink et al. (2001)
	Nitroreductase (*NTR*) from *E* coli	*A. thaliana*	Kurumata et al. (2005)

11.8 Biotechnology Advances in the Phytoremediation of Organic Pollutants

Powerful explosives like TNT, pentaerythritol tetranitrate (PETN), RDX, and nitrate esters (e.g. GTN) are the widespread organic contaminants common in military practices like blasting caps and detonators (Rosenblatt et al. 1991). Among them, TNT and RDX are potent carcinogens and referred to them as "priority pollutants" (Keith and Telliard 1979).

The development of transgenic plants offers a great promising strategy to metabolize the xenobiotics present in the environment. The microbial and mammalian community harbors great metabolic machinery which can be exploited to completely mineralize the recalcitrant or xenobiotics (Aken et al. 2010). Several plants have been modulated to express bacterial as well as mammalian genes to proficiently putrefy the recalcitrant pollutants. However, plants may lack the degradable properties possessed by bacterial species. The bacterial-derived detoxifying enzymes have been exploited extensively for improving phytoremediation.

A bacterial strain, PB2, belonging to *Enterobacter cloacae*, can metabolize the PETN (Obata et al. 1983). The transgenic tobacco plants that encode PETN-reductase have shown efficient metabolism of nitroaromatic explosives such as GTN and TNT as compared to the control plants (French et al. 1999).

Transgenic strategy when applied to trees belonging to the *Populus* genus has shown promising results in the controlled laboratory environment. As discussed earlier, properties like an extensive root system, rapid growth, and elevated transpiration rates might attract biotechnologists to tailor this trees for phytoremediation. Hydroponic culture of transgenic hybrid aspen (*P. tremula* x *tremuloides* var. *Etropole*) harboring a bacterial pnrA gene encoding for nitroreductase enzyme was able to withstand up to 57 mg TNT/L of media. Also, transgenic poplar plants have been a successful remedy against contaminated waters and soil with TNT (Dillewijn et al. 2008).

RDX contamination of water and soil has raised a serious concern owing to its recalcitrant nature. The bacterial strain *Rhodococcus rhodochrous* possesses a gene named xplA which has a cytochrome P450 domain. The *Arabidopsis* plants transformed with the P450 domain of the enzymes in conjugation to flavodoxin redox partner efficiently catalyzed the degradation of RDX in the liquid media (Rylott et al. 2006).

11.9 Implications of Transgenic Plants for Phytoremediation against Herbicides

The s-triazine class of herbicides includes Atrazine which is a predominant pollutant/contaminant of the soil. The plasmid-borne atzA gene, coding for atrazine chlorohydrolase (Martinez et al. 2001) is isolated from the *Pseudomonas* sp. It takes part in the hydrolytic reaction in which atrazine is converted to hydroxyatrazine, which is a nontoxic (Gysin and Knuesli 1960) and comparatively biodegradable form of atrazine (Goswami and Green 1971). This candidate gene was used to transform the tobacco and alpha-alpha plants which were capable of dechlorinating atrazine to the nonherbicidal compound hydroxyatrazine (Wang and Samac 2005). Implications of mammalian genes in enhancing phytoremediation have been well documented by pioneer researchers. A study conducted by Kawahigashi et al. (2005) evaluated transgenic rice plants overexpressing the three genes CYP1A1, CYP2B6, and CYP2C19 coding for cytochrome P450 from humans. They found that the transgenic plants (pIKBACH) reduced the abundance of atrazine and metolachlor in contaminated water and soil. Various enzymes like P450 2E1 (CYP2E1) able to oxidize carbon tetrachloride, benzene, styrene, chloroform, and other chlorinated compounds have been identified in the mammalian system (Guengerich et al. 1991). Tobacco plants transformed with mammalian cytochrome P450 2E1 proficiently oxidized the TCE which is considered the major pollutant of the groundwater (Doty et al. 2000).

Interactions between plants and microbes play a vital role in phytoremediation. A pioneer report led by Mohammadi et al. (2007) documented that plant-microbe interaction could assist in the detoxification of the potentially harmful polychlorinated biphenyls. In this study, *Nicotiana benthamiana* plants were tailored to transiently express the three biphenyl dioxygenase genes (BphAE, BphF, and BphG) from *Burkholderia xenovorans* LB400. The transformed plants were able to catalyze the oxygenation of 4-chlorobiphenyl. Staple food crops like rice being a proficient suitor for phytoremediation metabolizes various herbicides and thus reduce the number of herbicides present in paddy fields and streams (Kawahigashi 2009). Transgenic rice has been developed which contains the protoporphyrinogen oxidase (protox) gene from humans and *Myxococcus Xanthus* and confers the enhanced tolerance of plants against oxyfluorfen, which is a peroxidizing herbicide. Also, it has been found that the activities of catalase, peroxidase, SOD, and GR were upregulated in the transgenic plants upon the introduction of foreign genes (Jung et al. 2004; Lee et al. 2004).

11.10 Nanomaterials-Assisted Phytoremediation

The development of nanotechnology brings new horizons to the phytoremediation of contaminated soil and water (Peng et al. 2012). Nanomaterials are a type of materials having nanometer in dimension (1–100 nm) (Tang et al. 2014). Among them, the most commonly used nanomaterials are carbon-based and metal-based. Many studies have reported their use for the cleaning of groundwater (Mueller and Nowack 2010). Over the past few years, researchers started studying the use of nanomaterials for the improvement of phytoremediation efficiency. *L. perenne*, commonly known as perennial ryegrass, was used in the soil contaminated with Pb by using nanohydroxyapatite, which helps in the direct stabilization of Pb and thus reduces the phytotoxicity and promotes plant growth (Ding et al. 2017; Liang et al. 2017). In another report, nanoscale zero-valent iron (nZVI) was used for the phytoremediation of Pb by ryegrass (Huang et al. 2018). *Isatis cappadocica* Desv., a member of the family *Brassicaceae* was used for As phytoremediation with the help of salicylic acid nanoparticles. The nanoparticles also helped in better absorption of nutrients for plant growth (Souri et al. 2017). Cd was removed from the contaminated soils by using *Glycine max* and TiO_2 nanoparticles, which helps in enhancing the sprouting and increase in photosynthetic rate leading to a five times increase in Cd removal efficiency (Singh and Lee 2016). Tripathi et al. (2015) have used silicon nanoparticles for Cr phytoremediation by *P. sativum*. Studies have also been conducted on the use of nanomaterials in soil contaminated with multimetals. For example, after using nZVI for phytostabilization, the concentration of As, Cd, Pb, and Zn in the roots of *Helianthus annus* was decreased to 50–60% (Vítková et al. 2018).

Research on the use of nanomaterials in the phytoremediation of organic pollutants has also been reported in recent times. Ma and Wang (2010) reported the use of fullerene nanoparticles increased the bioavailability of TCE, thus increasing the uptake by 82% after applying 15 mg/L by *Populus deltoides*. Phytoremediation of TNT from contaminated soil by *Panicum maximum* was studied by Jiamjitrpanich et al. (2012), the addition of TNT-nZVI particles in a ratio of 1/10 in the phytoremediation process removed the prior in just 60 days where the initial TNT concentration is 100 mg/kg. In another phytoremediation study, the removal of endosulfan from the contaminated soil was increased from 81.2% to 100% (*Alpinia calcarata*), from 65.08% to 86.16% (*Ocimum sanctum*), and from 20.76% to 76.28% (*Cymbopogon citrates*) with the assistance of nZVI particles (Pillai and Kottekottil 2016).

11.11 Next-Generation Sequencing and Omics Approach for Improving Phytoremediation

Microarray has become an old practice to study the transcript expression in various tissues as it has the drawback of requiring a reference genome. New emerging technologies like next-generation sequencing (NGS) platforms have rapidly taken over the place of microarray because of their high fidelity and offer de novo sequencing platforms. NGS technologies have led to an explosion in our understanding of plants, microbes, and plant-microbe interactions. The ever-decreasing cost and high resolution offered by NGS platforms make its use possible to address environmental issues like soil pollution as well as water pollution (Bell et al. 2014). The multiomics strategies could identify the potential target that can be manipulated to enhance the production of ligands for metals and metalloids, and convert them to less toxic and volatile forms (Basu et al. 2018). Omics technology can be used efficiently to decipher novel possibilities for improving phytoremediation. Identifying the enzymes, transporters, chelators, etc., through transcriptomics, proteomics, and metabolomics facilitates the biotechnologists to introduce new genes as well as alter the expression of the genes in an organism. The resulting plant phenotype may fasten the phytoremediation (Figure 11.3)

Plant and microbial systems offer a versatile gene pool coding for the diverse enzymatic network of genes responsible for metabolizing, and transforming pollutants and recalcitrants. The dissection of the organism's genome unravels the genomic abilities that can be exploited for phytoremediation. The uptake of metals, chelation, and their distribution and storage inside plants are tightly controlled by a highly sophisticated metal homeostasis network. Therefore, the deduction of knowledge of the natural diversity of plant phenotypes and their correlation with variations at the sequence level can reveal various adaptations of the metal homeostasis networks.

Extensive and exhaustive omics studies have been carried out to understand various processes and signaling pathways responding to environmental stresses and their genetic control (Deshmukh et al. 2014). Omics technologies permit the analysis of microbial community as well as plant community composition and activity which facilitates community modifications or alterations.

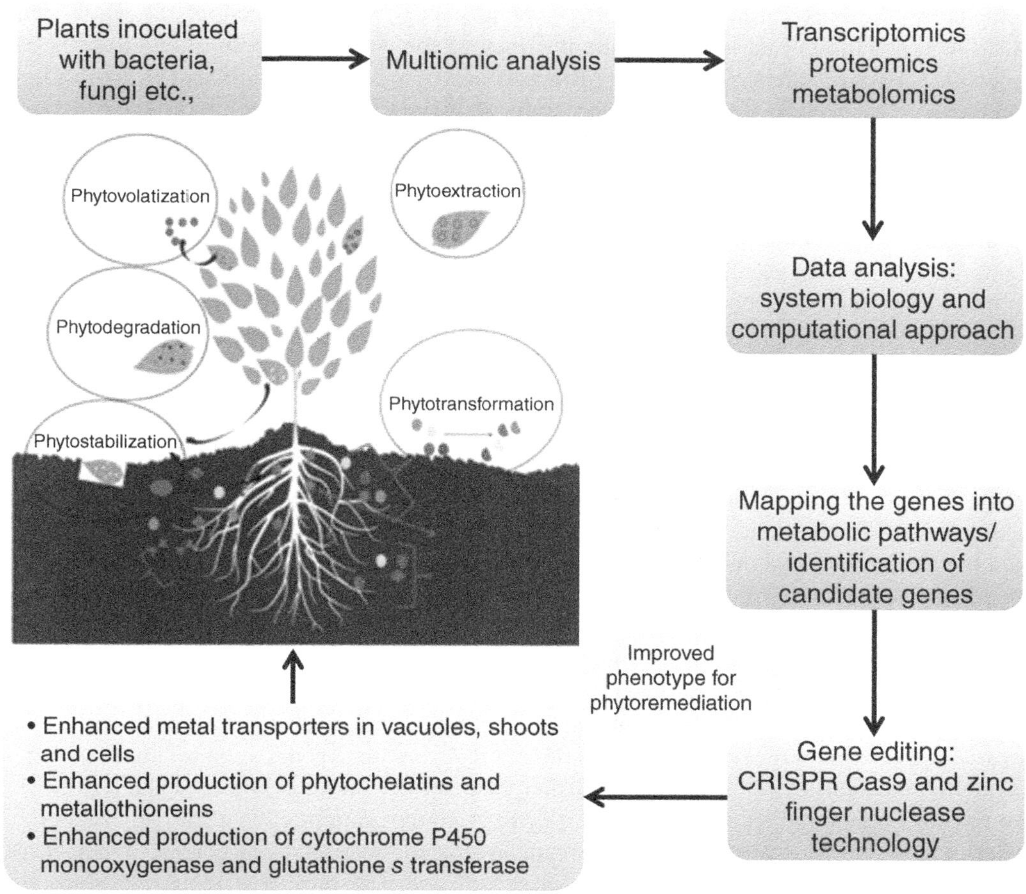

Figure 11.3 Outline for the use of the multiomics approach for a better phytoremediation process.

For instance, the transcriptomics approach has been successfully implemented in probing plant-associated microbial communities and metagenome analysis of environments such as the rhizosphere, and endosphere of the plants through any sequencing strategy like shotgun sequencing of genes. Plant rhizosphere in the soil is thought to be the hotspot for plasmid exchange, and the plasmids are considered to be the cause of the genomic innovations which introduce the adaptive mechanisms against various stress conditions. For instance, it has been suggested that the adaptability of the various microbes to the xenobiotics has been inferred by the mobile genetic elements (MGEs) like plasmids, which play an important role in the establishment of new catabolic pathways in bacteria (Top and Springael 2003; Sentchilo et al. 2013). It can be concluded that the metagenomic analyses can reveal the genetics and taxonomy of a microbial community and allocate functional attributes to them (Fierer et al. 2012). Although the metagenome infers microbial richness, the metatranscriptome accurately represents the functionality of a particular gene which might contribute to being transformed against a pollutant/contaminant. A recent study showed a linear relationship between the atrazine-degrading activity and the abundance of its associated encoding transcript of an undefined microbial community procured from a pilot-scale wastewater treatment plant containing varying fractions of a recombinant strain of *E. coli* capable of hydrolyzing atrazine. This suggests that metatranscriptomics can directly predict microbial community functions (Helbling et al. 2012). A transcriptomic data study of *Geobacter uraniireducens* growing in uranium-contaminated subsurface sediments revealed that many iron–sulfur cluster-binding proteins, c-type cytochromes, as well as dehydrogenases were upregulated (Holmes et al. 2009). The metal uptake systems, transporter genes carrying metals or metalloids, redox enzymes, etc., are the potential targets addressed for phytoremediation. Various reports demonstrated the implications of NGS technology in plants for phytoremediation. For instance, the transcriptome of *P. gabriellae* assisted the researchers to identify and clone the three transporters related to NRAMP and IREG/FPN families by heterologous expression in yeast. Further studies indicated the difference in the expression patterns of PgIREG between *P. gabriellae* and in the related nonaccumulator *P. semperflorens* and concluded that PgIREG1 efficiently transports Ni in vacuoles of *Arabidopsis* cells. The sequencing of the whole plant genome of *A. thaliana* hailed as a revolution and untangled the intricacy of the genome, exposing the enzymes like cytochrome P450 monooxygenase, GSH-S-transferases, glycosyltransferases, and transporters (Schäffner et al. 2002). These enzymes are common to plants as well as the animal kingdom. Many studies have well reviewed the development of transgenics by utilizing such enzymes from the plant as well as mammals to address soil pollution (Doty et al. 2000).

The transcriptome analysis of radish (*Raphanus sativus* L.) taproots identified differentially expressed genes (DEGs) during Pb stress (Wang et al. 2013). This helps us to understand the molecular mechanism underlying the stress induced by various metals and may impart useful information that can promote understanding and designing the processes involving heavy metal tolerance and accumilation in root vegetable crops like a radish. The sole knowledge of genes and proteins is not sufficient for understanding the biological process involved in the metabolism of various compounds. In such case, metabolomics provides a better way to get insights into the metabolic pathways of the plant as well as the microbial population in conjugation to the NGS technology. Plants respond to various environmental perturbations by changing the gene expression (i.e. the transcriptome) and the accumulation of proteins that can degrade contaminants. Mass spectrometry coupled to NGS enables the identification and quantification of an array of the primary and secondary metabolites involved in various biological processes underlying plant-microbe interactions (Deshmukh et al. 2014).

Analyzing the transcriptome of yeast, unwound the potential genes like ScYCF1 and a plasma membrane channel protein (NtCBP4), which, when overexpressed in plants withstand the heavy metals like Cd or Pb (Sunkar et al. 2000; Song et al. 2003). Omics approach has been implicated on various bacteria, such as *Rhodococcus*, *Comamonas*, *Ralstonia*, and *Burkholderia*, to identify the genes responsible for the degradation of polyaromatic hydrocarbons. The genes isolated from *Rhodococcus* bacteria could efficiently metabolize the harmful explosives like TNT and RDX (Uchida et al. 2005; Zhang et al. 2018). These examples demonstrated how the use of transcriptomics technologies assisted to identify the plant growth-promoting microorganisms (PGPMO) genes and alter them to exploit to augment phytoremediation processes (Basu et al. 2018). A report by He et al. (2012) showed that comparative transcriptomics and proteomic studies of developing rhizomes in *Phragmites australis* (common reed) led to the identification of candidate genes and proteins involved in rhizome growth, development and metabolism. The results were further supported by Sauvêtre and Schröder (2015) who demonstrated that the rhizomes and endophytic bacteria of *P. australis* enhanced the accumulation of carbamazepine.

Proteomics help in deciphering the expression of proteins in an organism which is encoded by its genome and also precisely describes the variations (Anderson and Anderson 1998). The proteomic approach is advantageous over transcriptomics as the changes in the genome are not always reflected at protein level (Hossain and Komatsu 2013; Singh et al. 2016). For instance, MHX transporter protein, that transports Zn and Mg was more abundant in *Arabidopsis* even though its corresponding transcript level was minimal (Elbaz et al. 2006). Proteomic analysis has enabled the identification of differentially expressed proteins in response to Cd stress in leaves of *P. americana* (Zhao et al. 2011). Also, it has emerged as a principle approach in identifying the DEGs in Hg^{2+} stress in *Suaeda salsa*. The proteomics approach has successfully identified stress-responsive proteins in the Cd-tolerant bacteria *C. taiwanensis* under Cd stress (Siripornadulsil et al. 2014).

The combinatorial approaches comprising omics knowledge from plant and animals provides a clear understanding of the accessory processes committed to symbiont-mediated decontamination process (Basu et al. 2018). As widely reviewed by Göhre and Paszkowski (2006), the arbuscular mycorrhizal symbiosis has a remarkable contribution toward heavy metal phytoremediation.

11.12 Gene Editing Tools and Phytoremediation

The development of an efficient and precise method that would induce targeted changes has been accomplished by revolutionary gene editing tools like clustered regularly interspaced short palindromic repeat-associated protein-9 system (CRISPR-Cas9). This technology has been successfully implemented in designing many crops with the potential to improve agronomics. Recently, a report has been published, where editing in Maize *Acetolactate Synthase2* (*ALS2*) gene resulted in tolerance to chlorsulfuron (Svitashev et al. 2015). Small noncoding RNAs (snRNAs) can also serve as an alternative tool for improving phytoremediation. They help in the regulation of gene expression in several cellular processes like conferring pathogen resistance and heavy metal stress resistance. siRNAs and miRNAs mediate RNA silencing in plants by the formation of RNA-induced silencing complexes (RISCs) in response to heavy metal stress. The suspension-culture of rice cells and calli, containing a cDNA construct of the gene coding for the subunit of rice GSH-S-transferases, *Os*GST in antisense orientation, led to the detoxification of the herbicide pretilachlor and the metabolism of phenolic compounds. Thus, altering the post-transcriptional pattern of gene expression can be successfully implemented to improve the plant–PGPMO interaction for better phytoremediation of contaminated soil. Besides si-RNA, riboswitches or ribozymes may also offer an approach to modify the PGMOs or the host plant for better phytoremediation (Basu et al. 2018).

11.13 Conclusion

Phytoremediation is a promising technology that has gained wide accountability for various types of environmental pollution caused due to anthropogenic, industrial, and mining interventions. Plants used for phytoremediation efficiently sequester, transfer, stabilize, and degrade the contaminants in the soil. The plants which possess the capacity to accumulate extraordinary levels of toxic metals are termed "hyperaccumulators". Plant species belonging to the *Brassicaceae* family including *Brassica napus* and *Thlaspi caerulescens* (Alpine penny grass) are well-renowned hyperaccumulators and possess extraordinary detoxification machinery to diminish the hazardous pollutants from the environment (Cunningham and Ow 1996). Various reports have convinced *Brassica* spp. including *B. napus* and *B. juncea*, *Alyssum murale*, *Thlaspi rotundifolium*, and *Dicoma niccoliferafor* has been effective in uptake and accumulation of heavy metals like Ni, Cr, As, Cd, Cr, Cu, Pb, and Zn (Dushenkov et al. 1995; Rathore et al. 2013; Ansari et al. 2015; Su et al. 2018). Other plant species *A. thaliana* (Delhaize 1996), *Typha latifolia*, and *P. australis* (Ye et al. 2001) are some of the identified plants suitable for phytoremediation. However, plants like *Arabidopsis* have limited biomass and root system, which impedes phytoremediation efficiency. Therefore, attention has also been paid to tree species, such as willow, poplar, etc. Yet, transgenic plants have gained considerable success mostly under controlled laboratory conditions due to the ethical and regulatory issues associated with GM crops.

Advances in breakthrough biotechnology approaches like NGS have made it feasible to decipher the underlying genetic and biochemical mechanisms involved in scavenging, translocating, and storage by hyperaccumulating species. Mining of genes and their associated protein products, followed by altering the plants has greatly improved the phytoremediation potential. Various omics technologies have facilitated access to the whole genome sequences of an organism and allowed various breakthrough researches like gene editing tools such as CRISPR-Cas9 and zinc-finger nuclease technology (Ahmed et al. 2019). The NGS technology provided accessibility to the whole genome sequence of several plant species and enabled various progressive gene editing tools to edit an organism's genome precisely at predetermined sites. Genome-edited plants will favor public acceptance easily as they can surpass the controlling regulations associated with GM crops (Waltz 2018).

References

Abdallah, M.A.M. (2012). Phytoremediation of heavy metals from aqueous solutions by two aquatic macrophytes, *Ceratophyllum demersum* and *Lemna gibba* L. *Environ. Technol.* 33: 1609–1614. https://doi.org/10.1080/09593330.2011.640354.

Ahmed, I., Sebastain, A., Prasad, M.N.V. et al. (2019). Emerging trends in transgenic technology for phytoremediation of toxic metals and metalloids. In: *Transgenic Plant Technology for Remediation of Toxic Metals and Metalloids* (ed. M.N.V. Prasad), 43–62. Academic Press.

Aioub, A.A.A., Li, Y., Qie, X. et al. (2019). Reduction of soil contamination by cypermethrin residues using phytoremediation with *Plantago major* and some surfactants. *Environ. Sci. Eur.* 31 (1): 26. https://doi.org/10.1186/s12302-019-0210-4.

Aken, B.V., Correa, P.A., and Schnoor, J.L. (2010). Phytoremediation of polychlorinated biphenyls: new trends and promises. *Environ. Sci. Technol.* 44: 2767–2776. https://doi.org/10.1021/es902514d.

Anderson, N.L. and Anderson, N.G. (1998). Proteome and proteomics: new technologies, new concepts, and new words. *Electrophoresis* 19: 1853–1861. https://doi.org/10.1002/elps.1150191103.

Ansari, M.K., Ahmad, A., Umar, S. et al. (2015). Genotypic variation in phytoremediation potential of Indian mustard exposed to heavy metal stress: a hydroponic study. *Int. J. Phytoremed.* 17: 135–144.

Arias, J.A., Peralta-Videa, J.R., Ellzey, J.T. et al. (2010). Effects of *Glomus deserticola* inoculation on *Prosopis*: enhancing chromium and lead uptake and translocation as confirmed by X-ray mapping, ICP-OES and TEM techniques. *Environ. Exp. Bot.* 68: 139–148. https://doi.org/10.1016/j.envexpbot.2009.08.009.

Arshad, M., Silvestre, J., Pinelli, E. et al. (2008). A field study of lead phytoextraction by various scented *Pelargonium* cultivars. *Chemosphere* 71: 2187–2192. https://doi.org/10.1016/j.chemosphere.2008.02.013.

Asada, M., Parkpian, P., and Horiuchi, S. (2006). Remediation technology for boron and fluoride contaminated sediments using green plants. *J. ASTM Int.* 3 (6): 1–7.

Babaoğlu, M., Gezgin, S., Topal, A. et al. (2004). Gypsophila sphaerocephala Fenzl ex Tchihat.: a boron hyperaccumulator plant species that may phytoremediate soils with toxic B levels. *Turk. J. Bot.* 28 (3): 273–278.

Banerjee, S., Shang, T.Q., Wilson, A.M. et al. (2002). Expression of functional mammalian P450 2E1 in hairy root cultures. *Biotechnol. Bioeng.* 77: 462–466. https://doi.org/10.1002/bit.10151.

Basu, S., Rabara, R.C., Negi, S., and Shukla, P. (2018). Engineering PGPMOs through gene editing and systems biology: a solution for phytoremediation? *Trends Biotechnol.* 36: 499–510. https://doi.org/10.1016/j.tibtech.2018.01.011.

Batayneh, A.T. (2011). Toxic (aluminum, beryllium, boron, chromium and zinc) in groundwater: health risk assessment. *Int. J. Environ. Sci. Technol.* 9 (1): 153–162. https://doi.org/10.1007/s13762-011-0009-3.

Bell, T.H., Joly, S., Pitre, F.E. et al. (2014). Increasing phytoremediation efficiency and reliability using novel omics approaches. *Trends Biotechnol.* 32 (5): 271–280. https://doi.org/10.1016/j.tibtech.2014.02.008.

Bennett, L.E., Burkhead, J.L., Hale, K.L. et al. (2003). Analysis of transgenic Indian mustard plants for phytoremediation of metal-contaminated mine tailings. *J. Environ. Qual.* 32 (2): 432–440. https://doi.org/10.2134/jeq2003.0432.

Bhuiyan, M.S.U., Min, S.R., Jeong, W.J. et al. (2011). Overexpression of AtATM3 in *Brassica juncea* confers enhanced heavy metal tolerance and accumulation. *Plant Cell Tissue Organ Cult.* 107 (1): 69–77. https://doi.org/10.1007/s11240-011-9958-y.

Bi, R., Schlaak, M., Siefert, E. et al. (2010). Alternating current electrical field effects on lettuce (*Lactuca sativa*) growing in hydroponic culture with and without cadmium contamination. *J. Appl. Electrochem.* 40: 1217–1223. https://doi.org/10.1007/s10800-010-0094-x.

Bittsanszky, A., Komives, T., Gullner, G. et al. (2005). Ability of transgenic poplars with elevated glutathione content to tolerate zinc (2+) stress. *Environ. Int.* 31: 251–254. https://doi.org/10.1016/j.envint.2004.10.001.

Bleeker, P.M., Hakvoort, H.W., Bliek, M. et al. (2006). Enhanced arsenate reduction by a CDC25-like tyrosine phosphatase explains increased phytochelatin accumulation in arsenate tolerant *Holcus lanatus*. *Plant J.* 45: 917–929. https://doi.org/10.1111/j.1365-313x.2005.02651.x.

Broeckling, C.D., Broz, A.K., Bergelson, J. et al. (2008). Root exudates regulate soil fungal community composition and diversity. *J. Appl. Environ. Microbiol.* 74: 738–744.

California Department of Pesticide Regulation (2006). Pesticide Use Reporting (PUR). Pesticide Use Reporting. www.cdpr.ca.gov/docs/pur/purmain.htm. Accessed 2 July 2019.

Campos, F.V.D., Oliveira, J.A.D., Silva, A.A.D. et al. (2019). Phytoremediation of arsenite-contaminated environments: is *Pistia stratiotes* L. a useful tool? *Ecol. Indic.* 104: 794–801. https://doi.org/10.1016/j.ecolind.2019.04.048.

Cervantes, C., Ji, G., Ramirez, J.L. et al. (1994). Resistance to arsenic compounds in microorganisms. *FEMS Microbiol. Rev.* 15: 355–367. https://doi.org/10.1111/j.1574-6976.1994.tb00145.x.

Chandra, P. and Kulshreshtha, K. (2004). Chromium accumulation and toxicity in aquatic vascular plants. *Bot. Rev.* 70: 313–327. https://doi.org/10.1663/0006-8101(2004)070[0313,caatia]2.0.co;2.

Chang, Y.S., Chang, Y.J., Lin, C.T. et al. (2013). Nitrogen fertilization promotes the phytoremediation of cadmium in *Pentas lanceolata*. *Int. Biodeterior. Biodegrad.* 85: 709–714. https://doi.org/10.1016/j.ibiod.2013.05.021.

Chen, C.M., Misra, T.K., Silver, S. et al. (1986). Nucleotide sequence of the structural genes for an anion pump. The plasmid-encoded arsenical resistance operon. *J. Biol. Chem.* 261: 15030–15038.

Cherian, S. and Oliveira, M.M. (2005). Transgenic plants in phytoremediation: recent advances and new possibilities. *Environ. Sci. Technol.* 39 (24): 9377–9390. https://doi.org/10.1021/es051134l.

Cho, U.H. and Park, J.O. (2000). Mercury-induced oxidative stress in tomato seedlings. *Plant Sci.* 156: 1–9. https://doi.org/10.1016/s0168-9452(00)00227-2.

Couselo, J.L., Navarro-Aviñó, J., and Ballester, A. (2010). Expression of the phytochelatin synthase TaPCS1 in transgenic aspen, insight into the problems and qualities in phytoremediation of Pb. *Int. J. Phytoremed.* 12 (4): 358–370.

Cunningham, S.D. and Ow, D.W. (1996). Promises and prospects of phytoremediation. *Plant Physiol.* 110: 715–719.

Dakora, F.D. and Phillips, D.A. (2002). Root exudates as mediators of mineral acquisition in low nutrient environments. *Plant Soil* 245: 35–47.

Das, H.K., Mitra, A.K., Sengupta, P.K. et al. (2004). Arsenic concentrations in rice, vegetables, and fish in Bangladesh: a preliminary study. *Environ. Int.* 30: 383–387. https://doi.org/10.1016/j.envint.2003.09.005.

Das, N., Bhattacharya, S., and Maiti, M.K. (2016). Enhanced cadmium accumulation and tolerance in transgenic tobacco overexpressing rice metal tolerance protein gene OsMTP1 is promising for phytoremediation. *Plant Physiol. Biochem.* 105: 297–309. https://doi.org/10.1016/j.plaphy.2016.04.049.

De Laet, C., Matringe, T., Petit, E. et al. (2019). *Eichhornia crassipes*: a powerful bio-indicator for water pollution by emerging pollutants. *Sci. Rep.* 9 (1): 7326.

Delhaize, E. (1996). A metal-accumulator mutant of *Arabidopsis thaliana*. *Plant Physiol.* 111 (3): 849–855. https://doi.org/10.1104/pp.111.3.849.

Delhaize, E., Gruber, B.D., Pittman, J.K. et al. (2007). A role for the AtMTP11gene of Arabidopsis in manganese transport and tolerance. *Plant J.* 51: 198–210. https://doi.org/10.1111/j.1365-313x.2007.03138.x.

Deng, L., Su, Y., Su, H. et al. (2007). Sorption and desorption of lead (II) from wastewater by green algae *Cladophora fascicularis*. *J. Hazard. Mater.* 143 (1–2): 220–225. https://doi.org/10.1016/j.jhazmat.2006.09.009.

Deshmukh, R., Sonah, H., Patil, G. et al. (2014). Integrating omic approaches for abiotic stress tolerance in soybean. *Front. Plant Sci.* 5: 244. https://doi.org/10.3389/fpls.2014.00244.

Dhankher, O.P., Rosen, B.P., Mckinney, E.C., and Meagher, R.B. (2006). Hyperaccumulation of arsenic in the shoots of *Arabidopsis* silenced for arsenate reductase (ACR2). *Proc. Natl. Acad. Sci. USA* 103: 5413–5418. https://doi.org/10.1073/pnas.0509770102.

Dhankher, O.P., Pilon-Smits, E.A., Meagher, R.B., and Doty, S. (2012). Biotechnological approaches for phytoremediation. *Plant Biotechnol. Agric.* 309–328: https://doi.org/10.1016/b978-0-12-381466-1.00020-1.

Dillewijn, P.V., Couselo José, L., Corredoira, E. et al. (2008). Bioremediation of 2,4,6-trinitrotoluene by bacterial nitroreductase expressing transgenic aspen. *Environ. Sci. Technol.* 42: 7405–7410. https://doi,org/10.1021/es801231w.

Ding, L., Li, J., Liu, W. et al. (2017). Influence of nano-hydroxyapatite on the metal bioavailability, plant metal accumulation and root exudates of ryegrass for phytoremediation in lead-polluted soil. *Int. J. Environ. Res. Public Health* 14 (5): 532.

Doty, S.L., Shang, T.Q., Wilson, A.M. et al. (2000). Enhanced metabolism of halogenated hydrocarbons in transgenic plants containing mammalian cytochrome P450 2E1. *Proc. Natl. Acad. Sci.* 97 (12): 6287–6291. https://doi.org/10.1073/pnas.97.12.6287.

Doty, S.L., James, C.A., Moore, A.L. et al. (2007). Enhanced phytoremediation of volatile environmental pollutants with transgenic trees. *Proc. Natl. Acad. Sci. USA* 104: 16816–16821. https://doi.org/10.1073/pnas.0703276104.

Dou, C.M., Fu, X.P., Chen, X.C. et al. (2009). Accumulation and detoxification of manganese in hyperaccumulator *Phytolacca americana*. *Plant Biol.* 11 (5): 664–670.

Duan, G.L., Zhou, Y., Tong, Y.P. et al. (2007). A CDC25 homologous from rice functions as an arsenate reductase. *New Phytol.* 174: 31–321. https://doi.org/10.1111/j.1469-8137.2007.02009.x.

Dushenkov, V., Kumar, P.B.A.N., Motto, H. et al. (1995). Rhizofiltration: the use of plants to remove heavy metals from aqueous streams. *Environ. Sci. Technol.* 29: 1239–1245. https://doi.org/10.1021/es00005a015.

Eapen, S., Singh, S., and D'souza, S.F. (2007). Advances in development of transgenic plants for remediation of xenobiotic pollutants. *Biotechnol. Adv.* 25 (5): 442–451. https://doi.org/10.1016/j.biotechadv.2007.05.001.

Elbaz, B., Shoshani-Knaani, N., David-Assael, O. et al. (2006). High expression in leaves of the zinc hyperaccumulator *Arabidopsis halleri* of AhMHX, a homolog of an *Arabidopsis thaliana* vacuolar metal/proton exchanger. *Plant Cell Environ.* 29: 1179–1190. https://doi.org/10.1111/j.1365-3040.2006.01500.x.

Ellis, D.R., Gumaelius, L., Indriolo, E. et al. (2006). A novel arsenate reductase from the arsenic hyperaccumulating fern *Pteris vittata*. *Plant Physiol.* 141: 1544–1554. https://doi.org/10.1104/pp.106.084079.

Evans, K.M., Gatehouse, J.A., Lindsay, W.P. et al. (1992). Expression of pea metallothionein-like gene PsMTA function. *Plant Mol. Biol.* 20: 1019–1028.

FAOSTAT (2018). Food and Agriculture Organization of the United Nations-Statistic Division. http://www.fao.org/faostat/en/#data (accessed 9 July 2020).

Fasani, E., Manara, A., Martini, F. et al. (2017). The potential of genetic engineering of plants for the remediation of soils contaminated with heavy metals. *Plant Cell Environ.* 41: 1201–1232. https://doi.org/10.1111/pce.12963.

Fierer, N., Lauber, C.L., Ramirez, K.S. et al. (2012). Comparative metagenomic, phylogenetic and physiological analyses of soil microbial communities across nitrogen gradients. *ISME J.* 6: 1007–1017. https://doi.org/10.1038/ismej.2011.159.

Flocco, C.G., Lindblom, S.D., Elizabeth, A.H., and Smits, P. (2004). Overexpression of enzymes involved in glutathione synthesis enhances tolerance to organic pollutants in *Brassica juncea*. *Int. J. Phytoremed.* 6: 289–304. https://doi.org/10.1080/16226510490888811.

Food and Agriculture *Organization* Corporate Statistical Database.

French, C.E., Rosser, S.J., Davies, G.J. et al. (1999). Biodegradation of explosives by transgenic plants expressing pentaerythritol tetranitrate reductase. *Nat. Biotechnol.* 17: 491–494. https://doi.org/10.1038/8673.

Gasic, K. and Korban, S.S. (2007). Transgenic Indian mustard (*Brassica juncea*) plants expressing an *Arabidopsis* (AtPCS1) exhibit enhanced As and Cd tolerance. *Plant Mol. Biol.* 64 (4): 361–369.

Gerhardt, K.E., Gerwing, P.D., and Greenberg, B.M. (2017). Opinion: taking phytoremediation from proven technology to accepted practice. *Plant Sci.* 256: 170–185. https://doi.org/10.1016/j.plantsci.2016.11.016.

Girkin, N.T., Turner, B.L., Ostle, N. et al. (2018). Root exudate analogues accelerate CO_2 and CH_4 production in tropical peat. *Soil Biol. Biochem.* 117: 48–55.

Gisbert, C., Ros, R., Haro, A.D. et al. (2003). A plant genetically modified that accumulates Pb is especially promising for phytoremediation. *Biochem. Biophys. Res. Commun.* 303: 440–445.

Göhre, V. and Paszkowski, U. (2006). Contribution of the arbuscular mycorrhizal symbiosis to heavy metal phytoremediation. *Planta* 223: 1115–1122. https://doi.org/10.1007/s00425-006-0225-0.

Gomes, H.I., Mayes, W.M., Whitby, P. et al. (2019). Constructed wetlands for steel slag leachate management: partitioning of arsenic, chromium, and vanadium in waters, sediments, and plants. *J. Environ. Manage.* 24: 330–338.

Goswami, K. and Green, R.E. (1971). Microbial degradation of the herbicide atrazine and its 2-hydroxy analog in submerged soils. *Environ. Sci. Technol.* 5: 426–429.

Govarthanan, M., Lee, K., Cho, M. et al. (2013). Significance of autochthonous *Bacillus* sp. KK1 on biomineralization of lead in mine tailings. *Chemosphere* 90 (8): 2267–2272. https://doi.org/10.1016/j.chemosphere.2012.10.038.

Guengerich, F.P., Kim, D.H., and Iwasaki, M. (1991). Role of human cytochrome P-450 IIE1 in the oxidation of many low molecular weight cancer suspects. *Chem. Res. Toxicol.* 4 (2): 168–179. https://doi.org/10.1021/tx00020a008.

Gullner, G., Kömives, T., and Rennenberg, H. (2001). Enhanced tolerance of transgenic poplar plants overexpressing γ-glutamylcysteine synthetase towards chloroacetanilide herbicides. *J. Exp. Bot.* 52: 971–979. https://doi.org/10.1093/jexbot/52.358.971.

Gysin, H. and Knuesli, E. (1960). Chemistry and herbicidal properties of triazine derivatives. In: *Advances in Pest Control Research*, vol. III (ed. R. Metcalf), 289–358. New York: Wiley.

Habibul, N., Chen, J.-J., Hu, Y.-Y. et al. (2019). Uptake, accumulation and metabolization of 1-butyl-3-methylimidazolium bromide by ryegrass from water: prospects for phytoremediation. *Water Res.* 156: 82–91. https://doi.org/10.1016/j.watres.2019.03.016.

Hannink, N., Rosser, S.J., French, C.E. et al. (2001). Phytodetoxification of TNT by transgenic plants expressing a bacterial nitroreductase. *Nat. Biotechnol.* 19: 1168–1172. https://doi.org/10.1038/nbt1201-1168.

Hao, X., Taghavi, S., Xie, P. et al. (2014). Phytoremediation of heavy and transition metals aided by legume-rhizobia symbiosis. *Int. J. Phytoremed.* 16 (2): 179–202. https://doi.org/10.1080/15226514.2013.773273.

He, R., Kim, M.J., Nelson, W. et al. (2012). Next-generation sequencing-based transcriptomic and proteomic analysis of the common reed, *Phragmites australis* (Poaceae), reveals genes involved in invasiveness and rhizome specifi city. *Am. J. Bot.* 99: 232–247. https://doi.org/10.3732/ajb.1100429.

Heaton, A.C.P., Rugh, C.L., Wang, N.-J., and Meagher, R.B. (1998). Phytoremediation of mercury- and methylmercury-polluted soils using genetically engineered plants. *J. Soil Contam.* 7: 497–509. https://doi.org/10.1080/10588339891334384.

Hechmi, N., Aissa, N.B., Abdennaceur, H., and Jedidi, N. (2013). Phytoremediation potential of maize (*Zea Mays* l.) in co-contaminated soils with pentachlorophenol and cadmium. *Int. J. Phytoremed.* 15: 703–713. https://doi.org/10.1080/15226514.2012.723067.

Helbling, D.E., Ackermann, M., Fenner, K. et al. (2012). The activity level of a microbial community function can be predicted from its metatranscriptome. *ISME J.* 6 (4): 902–904. https://doi.org/10.1038/ismej.2011.158.

Hesegawa, I., Terada, E., Sunairi, M. et al. (1997). Genetic improvement of heavy metal tolerance in plants by transfer of the yeast metallothionein gene (CUP1). *Plant Soil* 196: 277–281.

Hills, M.J., Hall, L., Arnison, P.G. et al. (2007). Genetic use restriction technologies (GURTs): strategies to impede transgene movement. *Trends Plant Sci.* 12 (4): 177–183. https://doi.org/10.1016/j.tplants.2007.02.002.

Hirose, S., Kawahigashi, H., Inoue, T. et al. (2005). Enhanced expression of CYP2C9 and tolerance to sulfonylurea herbicides in transgenic rice plants. *Plant Biotechnol.* 22: 89–96. https://doi.org/10.5511/plantbiotechnology.22.89.

Holmes, D.E., O'neil, R.A., Chavan, M.A. et al. (2009). Transcriptome of *Geobacter uraniireducens* growing in uranium-contaminated subsurface sediments. *ISME J.* 3 (2): 216–230. https://doi.org/10.1038/ismej.2008.89.

Hossain, Z. and Komatsu, S. (2013). Contribution of proteomic studies towards understanding plant heavy metal stress response. *Front. Plant Sci.* 3: 310. https://doi.org/10.3389/fpls.2012.00310.

Huang, D., Qin, X., Peng, Z. et al. (2018). Nanoscale zero-valent iron assisted phytoremediation of Pb in sediment: impacts on metal accumulation and antioxidative system of *Lolium perenne*. *Ecotox. Environ. Safe* 153: 229–237. https://doi.org/10.1016/j.ecoenv.2018.01.060.

Huang, H., Zhao, Y., Xu, Z. et al. (2019). Physiological responses of *Broussonetia papyrifera* to manganese stress, a candidate plant for phytoremediation. *Ecotox. Environ. Safe* 181: 18–25. https://doi.org/10.1016/j.ecoenv.2019.05.063.

Huysen, T.V., Abdel-Ghany, S., Hale, K.L. et al. (2003). Overexpression of cystathionine-γ-synthase enhances selenium volatilization in *Brassica juncea*. *Planta* 218: 71–78. https://doi.org/10.1007/s00425-003-1070-z.

Huysen, T.V., Terry, N., and Pilon-Smits, E.A.H. (2004). Exploring the selenium phytoremediation potential of transgenic Indian mustard overexpressing ATP sulfurylase or cystathionine-γ-synthase. *Int. J. Phytoremed.* 6 (2): 111–118.

Inui, H., Shiota, N., Ido, Y. et al. (2001). Herbicide metabolism and tolerance in the transgenic rice plants expressing human CYP2C9 and CYP2C19. *Pestic. Biochem. Physiol.* 71: 156–169. https://doi.org/10.1006/pest.2001.2574.

ITRC (Interstate Technology & Regulatory Council) (2009). Phytotechnology technical and regulatory guidance and decision trees, revised. PHYTO-3, Washington, DC.

Jeevanantham, S., Saravanan, A., Hemavathy, R. et al. (2019). Removal of toxic pollutants from water environment by phytoremediation: a survey on application and future prospects. *Environ. Technol. Innov.* 13: 264–276. https://doi.org/10.1016/j.eti.2018.12.007.

Jiamjitrpanich, W., Parkpian, P., Polprasert, C. et al. (2012). The tolerance efficiency of *Panicum maximum* and *Helianthus annuus* in TNT-contaminated soil and nZVI-contaminated soil. *J. Environ. Sci. Health A* 47 (11): 1506–1513.

Jin, M., Li, L., Xu, C. et al. (2010). Estrogenic activities of two synthetic pyrethroids and their metabolites. *J. Environ. Sci.* 22: 290–296. https://doi.org/10.1016/s1001-0742(09)60107-8.

Jones, P., Kortenkamp, A., O'Brien, P. et al. (1991). Evidence for the generation of hydroxyl radicals from a chromium (V) intermediate isolated from the reaction of chromate with glutathione. *Arch. Biochem. Biophys.* 286 (2): 652–655. https://doi.org/10.1016/0003-9861(91)90095-z.

Jung, S., Lee, Y., Yang, K. et al. (2004). Dual targeting of *Myxococcus xanthus* protoporphyrinogen oxidase into chloroplasts and mitochondria and high level oxyfluorfen resistance. *Plant Cell Environ.* 27: 1436–1446. https://doi.org/10.1111/j.1365-3040.2004.01247.x.

Kaiser, J. (2001). Environmental health: second look at arsenic finds higher risk. *Science* https://doi.org/10.1126/science.293.5538.2189b.

Karavangeli, M., Labrou, N.E., Clonis, Y.D., and Tsaftaris, A. (2005). Development of transgenic tobacco plants overexpressing maize glutathione S-transferase I for chloroacetanilide herbicides phytoremediation. *Biomol. Eng.* 22: 121–128. https://doi.org/10.1016/j.bioeng.2005.03.001.

Kawahigashi, H. (2009). Transgenic plants for phytoremediation of herbicides. *Curr. Opin. Biotechnol.* 20 (2): 225–230. https://doi.org/10.1016/j.copbio.2009.01.010.

Kawahigashi, H., Hirose, S., Inui, H. et al. (2005). Enhanced herbicide cross-tolerance in transgenic rice plants co-expressing human CYP1A1, CYP2B6, and CYP2C19. *Plant Sci.* 168: 773–781. https://doi.org/10.1016/j.plantsci.2004.10.014.

Kawashima, C.G., Noji, M., Nakamura, M. et al. (2004). Heavy metal tolerance of transgenic tobacco plants over-expressing cysteine synthase. *Biotechnol. Lett.* 26 (2): 153–157. https://doi.org/10.1023/b:bile.0000012895.60773.ff.

Keith, L. and Telliard, W. (1979). ES&T special report: priority pollutants: in perspective view. *Environ. Sci. Technol.* 13 (4): 416–423. https://doi.org/10.1021/es60152a601.

Koprivova, A., Kopriva, S., Jager, D. et al. (2002). Evaluation of transgenic poplars over-expressing enzymes of glutathione synthesis for phytoremediation of cadmium. *Plant Biol.* 4: 664–670. https://doi.org/10.1055/s-2002-37399.

Krämer, U. and Chardonnens, A. (2001). The use of transgenic plants in the bioremediation of soils contaminated with trace elements. *Appl. Microbiol. Biotechnol.* 55 (6): 661–672.

Kurumata, M., Takahashi, M., Sakamoto, A. et al. (2005). Tolerance to, and uptake and degradation of 2,4,6-trinitrotoluene (TNT) are enhanced by the expression of a bacterial Nitroreductase gene in *Arabidopsis thaliana*. *Z. Naturforsch. C* 60: 272–278. https://doi.org/10.1515/znc-2005-3-412.

Larson, R.A. (1988). The antioxidants of higher plants. *Phytochemistry* 27: 969–978. https://doi.org/10.1016/0031-9422(88)80254-1.

Leduc, D.L., Tarun, A.S., Montes-Bayon, M. et al. (2003). Overexpression of Selenocysteine Methyltransferase in Arabidopsis and Indian mustard increases selenium tolerance and accumulation. *Plant Physiol.* 135: 377–383. https://doi.org/10.1104/pp.103.026989.

Lee, Y., Jung, S., and Back, K. (2004). Expression of human protoporphyrinogen oxidase in transgenic rice induces both a photodynamic response and oxyfluorfen resistance. *Pestic. Biochem. Physiol.* 80: 65–74. https://doi.org/10.1016/j.pestbp.2004.06.008.

Lee, C., Li, X., Shi, W. et al. (2006). Metal contamination in urban, suburban, and country park soils of Hong Kong: a study based on GIS and multivariate statistics. *Sci. Total Environ.* 356 (1–3): 45–61. https://doi.org/10.1016/j.scitotenv.2005.03.024.

Lemly, A.D. (2014). Teratogenic effects and monetary cost of selenium poisoning of fish in Lake Sutton, North Carolina. *Ecotox. Environ. Safe* 104: 160–167. https://doi.org/10.1016/j.ecoenv.2014.02.022.

Li, Y., Dhankher, O.P., Carreira, L. et al. (2004). Overexpression of phytochelatin synthase in *Arabidopsis* leads to enhanced arsenic tolerance and cadmium hypersensitivity. *Plant Cell Physiol.* 45 (12): 1787–1797. https://doi.org/10.1093/pcp/pch202.

Liang, S.-X., Jin, Y., Liu, W. et al. (2017). Feasibility of Pb phytoextraction using nano-materials assisted ryegrass: results of a one-year field-scale experiment. *J. Environ. Manage.* 190: 170–175. https://doi.org/10.1016/j.jenvman.2016.12.064.

Liu, Z., Gu, C., Chen, F. et al. (2012). Heterologous expression of a *Nelumbo nucifera* gene enhances cadmium tolerance in *Arabidopsis thaliana*. *Biotechnol. Appl. Biochem.* 166 (3): 722–734.

Lombi, E., Zhao, F.J., Fuhrmann, M. et al. (2002). Arsenic distribution and speciation in the fronds of the hyperaccumulator *Pteris vittata*. *New Phytol.* 156 (2): 195–203.

Ma, X. and Wang, C. (2010). Fullerene nanoparticles affect the fate and uptake of trichloroethylene in phytoremediation systems. *Environ. Eng. Sci.* 27: 989–992. https://doi.org/10.1089/ees.2010.0141.

Macek, T., Macková, M., Pavlíková, D. et al. (2002). Accumulation of cadmium by transgenic tobacco. *Acta Biotechnol.* 22 (1–2): 101–106.

Mani, D. and Kumar, C. (2013). Biotechnological advances in bioremediation of heavy metals contaminated ecosystems: an overview with special reference to phytoremediation. *Int. J. Environ. Sci. Tech.* 11: 843–872. https://doi.org/10.1007/s13762-013-0299-8.

Manzoor, M., Gul, I., Ahmed, I. et al. (2019). Metal tolerant bacteria enhanced phytoextraction of lead by two accumulator ornamental species. *Chemosphere* 227: 561–569. https://doi.org/10.1016/j.chemosphere.2019.04.093.

Mapanda, F., Mangwayana, E., Nyamangara, J. et al. (2005). The effect of long-term irrigation using wastewater on heavy metal contents of soils under vegetables in Harare, Zimbabwe. *Agric. Ecosyst. Environ.* 107 (2–3): 151–165. https://doi.org/10.1016/j.agee.2004.11.005.

Marín, C.M.D.-C. and Oron, G. (2007). Boron removal by the duckweed *Lemna gibba*: a potential method for the remediation of boron-polluted waters. *Water Res.* 41: 4579–4584. https://doi.org/10.1016/j.watres.2007.06.051.

Marrugo-Negrete, J., Enamorado-Montes, G., Durango-Hernández, J. et al. (2017). Removal of mercury from gold mine effluents using *Limnocharis flava* in constructed wetlands. *Chemosphere* 167: 188–192. https://doi.org/10.1016/j.chemosphere.2016.09.130.

Martinez, B., Tomkins, J., Wackett, L.P. et al. (2001). Complete nucleotide sequence and organization of the atrazine catabolic plasmid pADP-1 from *Pseudomonas* sp. strain ADP. *J. Bacteriol.* 183 (19): 5684–5697. https://doi.org/10.1128/jb.183.19.5684-5697.2001.

McCutcheon, S.C. and Schnoor, J.L. (2003). Overview of phytotransformation and control of wastes. In: *Phytoremediation: Transformation and Control of Contaminants* (ed. M.C. SC and J.L. Schnoor), 3–58. New York: Wiley.

McLean, R.I. and Abbe, G.R. (2008). Characteristics of *Crassostrea ariakensis* (Fujita 1913) and *Crassostrea virginica* (Gmelin 1791) in the discharge area of a nuclear power Plant in Central Chesapeake bay. *J. Shellfish Res.* 27 (3): 517–523. https://doi.org/10.2983/0730-8000(2008)27[517,cocafa]2.0.co;2.

Meagher, R.B. and Heaton, A.C.P. (2005). Strategies for the engineered phytoremediation of toxic element pollution: mercury and arsenic. *J. Ind. Microbiol. Biotechnol.* 32: 502–513. https://doi.org/10.1007/s10295-005-0255-9.

Meharg, A.A. (2004). Arsenic in rice – understanding a new disaster for South-East Asia. *Trends Plant Sci.* 9: 415–417. https://doi.org/10.1016/j.tplants.2004.07.002.

Merlot, S., Hannibal, L., Martins, S. et al. (2014). The metal transporter PgIREG1 from the hyperaccumulator *Psychotria gabriellae* is a candidate gene for nickel tolerance and accumulation. *J. Exp. Bot.* 65 (6): 1551–1564. https://doi.org/10.1093/jxb/eru025.

Misra, S. and Gedamu, L. (1989). Heavy metal tolerant transgenic *Brassica napus* L. and *Nicotiana tabacum* L. plants. *Theor. Appl. Genet.* 78: 161–168.

Mkandawire, M. and Dudel, E. (2005). Accumulation of arsenic in Lemna gibba L. (duckweed) in tailing waters of two abandoned uranium mining sites in Saxony, Germany. *Sci. Total Environ.* 336: 81–89. https://doi.org/10.1016/j.scitotenv.2004.06.002.

Mohammadi, M., Chalavi, V., Novakova-Sura, M. et al. (2007). Expression of bacterial biphenyl-chlorobiphenyl dioxygenase genes in tobacco plants. *Biotechnol. Bioeng.* 97: 496–505. https://doi.org/10.1002/bit.21188.

Moubasher, H., Hegazy, A., Mohamed, N. et al. (2015). Phytoremediation of soils polluted with crude petroleum oil using *Bassia scoparia* and its associated rhizosphere microorganisms. *Int. Biodeterior. Biodegrad.* 98: 113–120. https://doi.org/10.1016/j.ibiod.2014.11.019.

Mueller, N.C. and Nowack, B. (2010). Nanoparticles for remediation: solving big problems with little particles. *Elements* 6: 395–400. https://doi.org/10.2113/gselements.6.6.395.

Mukhopadhyay, S. and Maiti, S.K. (2010). Phytoremediation of metal mine waste. *Appl. Ecol. Environ. Res.* 8: 207–222.

Mulligan, C., Yong, R., and Gibbs, B. (2001). Remediation technologies for metal-contaminated soils and groundwater: an evaluation. *Eng. Geol.* 60 (1–4): 193–207. https://doi.org/10.1016/s0013-7952(00)00101-0.

Nattrass, M., Mcgrew, N.R., Morrison, J.I., and Baldwin, B.S. (2019). Phytoremediation of selenium-impacted water by aquatic macrophytes. *J. Arab Soc. Med. Res.* 8: 69–79. https://doi.org/10.21000/jasmr19010069.

Noctor, G., Arisi, A.C.M., Jouanin, L. et al. (1998). Glutathione biosynthesis metabolism and relationship to stress tolerance explored in transformed plants. *J. Exp. Bot.* 49: 623–647.

Noji, M., Saito, M., Nakamura, M. et al. (2001). Cysteine synthase overexpression in tobacco confers tolerance to sulfur-containing environmental pollutants. *Plant Physiol.* 126 (3): 973–980. https://doi.org/10.1104/pp.126.3.973.

Obata, H., Akizuki, D., Tanishita, J. et al. (1983). Isolation and characterisation of pentaerythritol utilising bacterium. *J. Ferment. Technol.* 61 (2): 121–126.

Pan, A., Yang, M., and Tie, F. (1994). Expression of mouse metallothionein-I gene confers cadmium resistance in transgenic tobacco plants. *Plant Mol. Biol.* 24: 341–351.

Pan, G., Liu, W., Zhang, H., and Liu, P. (2018). Morphophysiological responses and tolerance mechanisms of *Xanthium strumarium* to manganese stress. *Ecotox. Environ. Safe* 165: 654–661. https://doi.org/10.1016/j.ecoenv.2018.08.107.

Pedron, F., Petruzzelli, G., Barbafieri, M., and Tassi, E. (2009). Strategies to use phytoextraction in very acidic soil contaminated by heavy metals. *Chemosphere* 75: 808–814. https://doi.org/10.1016/j.chemosphere.2009.01.044.

Peng, J., Sun, Y., Liu, Q. et al. (2012). Upconversion nanoparticles dramatically promote plant growth without toxicity. *Nano Res.* 5: 770–782. https://doi.org/10.1007/s12274-012-0261-y.

Pillai, H.P.S. and Kottekottil, J. (2016). Nano-phytotechnological remediation of endosulfan using zero valent iron nanoparticles. *J. Environ. Prot.* 07: 734–744. https://doi.org/10.4236/jep.2016.75066.

Pilon, M., Owen, J.D., Garifullina, G.F. et al. (2003). Enhanced selenium tolerance and accumulation in transgenic Arabidopsis expressing a mouse Selenocysteine Lyase. *Plant Physiol.* 131: 1250–1257. https://doi.org/10.1104/pp.102.014639.

Pilon-Smits, E.A. and Freeman, J.L. (2006). Environmental cleanup using plants: biotechnological advances and ecological considerations. *Front. Ecol. Environ.* 4 (4): 203–210. https://doi.org/10.1890/1540-9295(2006)004[0203,ecupba]2.0.co;2.

Pilon-Smits, E. and Pilon, M. (2002). Phytoremediation of metals using transgenic plants. *Crit. Rev. Plant Sci.* 21 (5): 439–456.

Pilon-Smits, E.A., Hwang, S., Lytle, C.M. et al. (1999). Overexpression of ATP sulfurylase in Indian mustard leads to increased selenate uptake, reduction, and tolerance. *Plant Physiol.* 119: 123–132. https://doi.org/10.1104/pp.119.1.123.

Pilon-Smits, E.A.H., Zhu, Y.L., Sears, T., and Terry, N. (2000). Overexpression of glutathione reductase in *Brassica juncea*: effects on cadmium accumulation and tolerance. *Physiol. Plant.* 110: 455–460. https://doi.org/10.1034/j.1399-3054.2000.1100405.x.

Pourrut, B., Shahid, M., Douay, F. et al. (2013). Molecular mechanisms involved in Lead uptake, toxicity and detoxification in higher plants. In: *Heavy Metal Stress Plants* (ed. D. Gupta, F. Corpas, and J. Palma), 121–147. Berlin, Heidelberg: Springer https://doi.org/10.1007/978-3-642-38469-1_7.

Qiu, Y.-W., Qiu, H.-L., Zhang, G., and Li, J. (2019). Bioaccumulation and cycling of polybrominated diphenyl ethers (PBDEs) and dechlorane plus (DP) in three natural mangrove ecosystems of South China. *Sci. Total Environ.* 651: 1788–1795. https://doi.org/10.1016/j.scitotenv.2018.10.055.

Rathore, S.S., Kapila, S., Premi, O.P. et al. (2013). Water-use efficiency, productivity, photosynthesis and sustainability of pressurized irrigation systems for Indian mustard (*Brassica juncea* L. Czernj and Cosson) under semi-arid conditions of Rajasthan. *Res. Crops* 14 (1): 140–150.

Reisinger, S., Schiavon, M., Terry, N. et al. (2008). Heavy metal tolerance and accumulation in Indian mustard (*Brassica juncea* L.) expressing bacterial γ-glutamylcysteinesynthetase or glutathione synthetase. *Int. J. Phytoremed.* 10 (5): 440–454.

Robinson, B., Kim, N., Marchetti, M. et al. (2006). Arsenic hyperaccumulation by aquatic macrophytes in the Taupo volcanic zone, New Zealand. *Environ. Exp. Bot.* 58 (1–3): 206–215.

Robinson, C., Brömssen, M.V., Bhattacharya, P. et al. (2011). Dynamics of arsenic adsorption in the targeted arsenic-safe aquifers in Matlab, South-Eastern Bangladesh: insight from experimental studies. *Appl. Geochem.* 26 (4): 624–635. https://doi.org/10.1016/j.apgeochem.2011.01.019.

Rosenblatt, D.H., Burrows, E.P., Mitchell, W.R. et al. (1991). Organic explosives and related compounds. In: *The Handbook of Environmental Chemistry 3 (G)* (ed. O. Hutzinger), 195–234. Berlin: Springer-Verlag KG.

Rugh, C.L. (2004). Genetically engineered phytoremediation: one mans trash is another mans transgene. *Trends Biotechnol.* 22: 496–498. https://doi.org/10.1016/j.tibtech.2004.08.003.

Rugh, C.L., Senecoff, J.F., Meagher, R.B., and Merkle, S.A. (1998). Development of transgenic yellow poplar for mercury phytoremediation. *Nat. Biotechnol.* 16: 925–928. https://doi.org/10.1038/nbt1098-925.

Rylott, E.L., Jackson, R.G., Edwards, J. et al. (2006). An explosive-degrading cytochrome P450 activity and its targeted application for the phytoremediation of RDX. *Nat. Biotechnol.* 24: 216–219. https://doi.org/10.1038/nbt1184.

Satarug, S. and Nazar, S. (2010). Cadmium in food and human health: technologies for environmental restoration and rehabilitation January 15–17, 2010, Phitsanulok, Thailand. *Toxicol. Lett.* 198: 2–6. https://doi.org/10.1016/j.toxlet.2010.02.019.

Sauvêtre, A. and Schröder, P. (2015). Uptake of carbamazepine by rhizomes and endophytic bacteria of *Phragmites australis*. *Front. Plant Sci.* 6: 83. https://doi.org/10.3389/fpls.2015.00083.

Schäffner, A., Messner, B., Langebartels, C. et al. (2002). Genes and enzymes for in-planta phytoremediation of air, water and soil. *Acta Biotechnol.* 22 (1–2): 141–151. https://doi.org/10.1002/1521-3846(200205)22:1/23.0.co;2-7.

Sentchilo, V., Mayer, A.P., Guy, L. et al. (2013). Community-wide plasmid gene mobilization and selection. *ISME J.* 7: 1173–1186. https://doi.org/10.1038/ismej.2013.13.

Shim, D., Kim, S., Choi, Y.I. et al. (2013). Transgenic poplar trees expressing yeast cadmium factor 1 exhibit the characteristics necessary for the phytoremediation of mine tailing soil. *Chemosphere* 90 (4): 1478–1486. https://doi.org/10.1016/j.chemosphere.2012.09.044.

Shri, M., Dave, R., Diwedi, S. et al. (2014). Heterologous expression of *Ceratophyllum demersum*, CdPCS1, in rice leads to lower arsenic accumulation in grain. *Sci. Rep.* 4: 5784.

Singh, J. and Lee, B.-K. (2016). Influence of nano-TiO_2 particles on the bioaccumulation of cd in soybean plants (Glycine max): a possible mechanism for the removal of cd from the contaminated soil. *J. Environ. Manage.* 170: 88–96. https://doi.org/10.1016/j.jenvman.2016.01.015.

Singh, O.V., Labana, S., Pandey, G. et al. (2003). Phytoremediation: an overview of metallic ion decontamination from soil. *Appl. Microbiol. Biotechnol.* 61 (5–6): 405–412.

Singh, S., Parihar, P., Singh, R. et al. (2016). Heavy metal tolerance in plants: role of transcriptomics, proteomics, metabolomics, and ionomics. *Front. Plant Sci.* 6: 1143. https://doi.org/10.3389/fpls.2015.01143.

Singh, R.P., Wu, J., and Fu, D. (2019). Purification of water contaminated with Hg using horizontal subsurface constructed wetlands. *Environ. Sci. Pollut. R* 26: 9697–9706. https://doi.org/10.1007/s11356-019-04260-9.

Siripornadulsil, S., Thanwisai, L., and Siripornadulsil, W. (2014). Changes in the proteome of the cadmium-tolerant bacteria *Cupriavidus taiwanensis* KKU2500-3 in response to cadmium toxicity. *Can. J. Microbiol.* 60 (3): 121–131. https://doi.org/10.1139/cjm-2013-0713.

Song, W.-Y., Sohn, E.J., Martinoia, E. et al. (2003). Engineering tolerance and accumulation of lead and cadmium in transgenic plants. *Nat. Biotechnol.* 21: 914–919. https://doi.org/10.1038/nbt850.

Soni, S. and Jain, S. (2014). A review on phytoremediation of heavy metals from soil by using plants to remove pollutants from the environment. *Int. J. Adv. Res.* 2 (8): 197–203.

Souri, Z., Karimi, N., Sarmadi, M., and Rostami, E. (2017). Salicylic acid nanoparticles (SANPs) improve growth and phytoremediation efficiency of *Isatis cappadocica* Desv., under as stress. *IET Nanobiotechnol.* 11: 650–655. https://doi.org/10.1049/iet-nbt.2016.0202.

Srivastava, S., Mishra, S., Tripathi, R.D. et al. (2007). Phytochelatins and antioxidant systems respond differentially during arsenite and arsenate stress in *Hydrilla verticillata* (L.f.) Royle. *Environ. Sci. Technol.* 41: 2930–2936. https://doi.org/10.1021/es062167j.

Srivastava, S., Shrivastava, M., Suprasanna, P., and Dsouza, S. (2011). Phytofiltration of arsenic from simulated contaminated water using *Hydrilla verticillata* in field conditions. *Ecol. Eng.* 37: 1937–1941. https://doi.org/10.1016/j.ecoleng.2011.06.012.

Stahl, J.D. and Aust, S.D. (1995). Biodegradation of 2,4,6-trinitrotoluene by the white rot fungus *Phanerochaete Chrysosporium*. In: *Biodegradation of Nitroaromatic Compounds*, vol. 49 (ed. J.C. Spain), 117–133. Boston, MA: Springer https://doi.org/10.1007/978-1-4757-9447-2_8.

Stiles, A.R., Bautista, D., Atalay, E. et al. (2010). Mechanisms of boron tolerance and accumulation in plants: a physiological comparison of the extremely boron-tolerant plant species, *Puccinellia distans*, with the moderately boron-tolerant *gypsophila arrostil*. *Environ. Sci. Technol.* 44: 7089–7095. https://doi.org/10.1021/es1016334.

Su, D., Jiang, R., and Li, H. (2018). The potential of oilseed rape and *Thlaspi caerulescens* for phytoremediation of cadmium-contaminated soil. In: *Twenty Years of Research and Development on Soil Pollution and Remediation in China* (ed. Y. Luo and C. Tu), 349–363. Singapore: Springer http://doi.org/10.1007/978-981-10-6029-8_18.

Substance Priority List | ATSDR (2017). Centers for Disease Control and Prevention. https://www.atsdr.cdc.gov/spl. Accessed 2 July 2019.

Sunkar, R., Kaplan, B., Bouché, N. et al. (2000). Expression of a truncated tobacco NtCBP4 channel in transgenic plants and disruption of the homologous *Arabidopsis* CNGC1 gene confer Pb2+ tolerance. *Plant J.* 24 (4): 533–542. https://doi.org/10.1046/j.1365-313x.2000.00901.x.

Svitashev, S., Young, J.K., Schwartz, C. et al. (2015). Targeted mutagenesis, precise gene editing, and site-specific gene insertion in maize using Cas9 and guide RNA. *Plant Physiol.* 169 (2): 931–945. https://doi.org/10.1104/pp.15.00793.

Talebi, M., Tabatabaei, B.E.S., and Akbarzadeh, H. (2019). Hyperaccumulation of Cu, Zn, Ni, and Cd in *Azolla* species inducing expression of methallothionein and phytochelatin synthase genes. *Chemosphere* 230: 488–497. https://doi.org/10.1016/j.chemosphere.2019.05.098.

Tang, W.-W., Zeng, G.-M., Gong, J.-L. et al. (2014). Impact of humic/fulvic acid on the removal of heavy metals from aqueous solutions using nanomaterials: a review. *Sci. Total Environ.* 468-469: 1014–1027. https://doi.org/10.1016/j.scitotenv.2013.09.044.

Top, E.M. and Springael, D. (2003). The role of mobile genetic elements in bacterial adaptation to xenobiotic organic compounds. *Curr. Opin. Biotechnol.* 14 (3): 262–269. https://doi.org/10.1016/s0958-1669(03)00066-1.

Tripathi, D.K., Singh, V.P., Prasad, S.M. et al. (2015). Silicon nanoparticles (SiNp) alleviate chromium (VI) phytotoxicity in *Pisum sativum* (L.) seedlings. *Plant Physiol. Biochem.* 96: 189–198. https://doi.org/10.1016/j.plaphy.2015.07.026.

Truu, J., Truu, M., Espenberg, M. et al. (2015). Phytoremediation and plant-assisted bioremediation in soil and treatment wetlands: a review. *Open Biotechnol. J.* 9 (1): 85–92. https://doi.org/10.2174/1874070701509010085.

Uchida, E., Ouchi, T., Suzuki, Y. et al. (2005). Secretion of bacterial xenobiotic-degrading enzymes from transgenic plants by an apoplastic expressional system: an applicability for phytoremediation. *Environ. Sci. Technol.* 39: 7671–7677. https://doi.org/10.1021/es0506814.

Vatamaniuk, O.K., Mari, S., Lu, Y.P. et al. (1999). At PCS1, a phytochelatin synthase from *Arabidopsis*: isolation and in vitro reconstitution. *Proc. Natl Acad. Sci. USA* 96 (12): 7110–7115.

Vítková, M., Puschenreiter, M., and Komárek, M. (2018). Effect of nano zero-valent iron application on As, Cd, Pb, and Zn availability in the rhizosphere of metal(loid) contaminated soils. *Chemosphere* 200: 217–226. https://doi.org/10.1016/j.chemosphere.2018.02.118.

Wade, M.J., Davis, B.K., Carlisle, J.S. et al. (1993). Environmental transformation of toxic metals. *Occup. Med.* 8: 575–601.

Waltz, E. (2018). With a free pass, CRISPR-edited plants reach market in record time. *Nat. Biotechnol.* 36: 67.

Wang, L., Samac, D.A., Shapir, N. et al. (2005). Biodegradation of atrazine in transgenic plants expressing a modified bacterial atrazine chlorohydrolase (atzA) gene. *Plant Biotechnol. J.* 3 (5): 475–486. https://doi.org/10.1111/j.1467-7652.2005.00138.x.

Wang, Y., Xu, L., Chen, Y. et al. (2013). Transcriptome profiling of radish (*Raphanus sativus* L.) root and identification of genes involved in response to Lead (Pb) stress with next generation sequencing. *PLoS One* https://doi.org/10.1371/journal.pone.0066539.

Wang, M., Zhao, W., Jia, X. et al. (2018). Eco-toxicology effect on *Moina mongolicaDaday* exposed to Cd2, Pb2, and Hg2 by the food chain. *Environ. Sci. Pollut. R* 25: 16024–16036. https://doi.org/10.1007/s11356-018-1800-1.

Xie, W.-Y., Huang, Q., Li, G. et al. (2012). Cadmium accumulation in the rootless macrophyte *Wolffia Globosa* and its potential for phytoremediation. *Int. J. Phytoremed.* 15: 385–397. https://doi.org/10.1080/15226514.2012.702809.

Xin, J. and Huang, B. (2017). Comparison of boron uptake, translocation, and accumulation in reed, cattail, and vetiver: an extremely boron-tolerant plant, vetiver. *Plant Soil* 416: 17–25. https://doi.org/10.1007/s11104-017-3186-0.

Xue, S.G., Ye, S., Zhou, F. et al. (2008). Identity of *Phytolacca americana* L. (Phytolaccaceae), pokeweed: a manganese hyperaccumulator plant. *Acta Ecol. Sin.* 28 (12): 6344–6347.

Yadav BK, Akhtar SM, Panwar J (2014) Rhizospheric plant-microbe interactions: key factors to soil fertility and plant nutrition. Plant Microbes Symbiosis: Applied Facets, Arora, N. (eds) 127–145. Springer, New Delhihttps://doi.org/10.1007/978-81-322-2068-8_6

Yan, F., Schubert, S., and Menge, K. (1996). Soil pH increase due to biological decarboxylation of organic anions. *Soil Biol. Biochem.* 28: 617–624.

Ye, Z.H., Whiting, S.N., Lin, Z.Q. et al. (2001). Removal and distribution of iron, manganese, cobalt and nickel within a Pennsylvania constructed wetland treating coal combustion by-product leachate. *J. Environ. Qual.* 30: 1464–1473. https://doi.org/10.2134/jeq2001.3041464x.

Zhang, X., Hu, Y., Liu, Y. et al. (2011). Arsenic uptake, accumulation and phytofiltration by duckweed (*Spirodela polyrhiza* L.). *J. Environ. Sci.* 23 (4): 601–606.

Zhang, Y., Liu, J., Zhou, Y. et al. (2013). Enhanced phytoremediation of mixed heavy metal (mercury)–organic pollutants (trichloroethylene) with transgenic alfalfa co-expressing glutathione S-transferase and human P450 2E1. *J. Hazard. Mater.* 260: 1100–1107.

Zhang, L., Rylott, E.L., Bruce, N.C., and Strand, S.E. (2018). Genetic modification of western wheatgrass (*Pascopyrum smithii*) for the phytoremediation of RDX and TNT. *Planta* 249: 1007–1015. https://doi.org/10.1007/s00425-018-3057-9.

Zhang, W., Zhang, D., Zagorevski, D.V., and Liang, Y. (2019). Exposure of *Juncus effusus* to seven perfluoroalkyl acids: uptake, accumulation and phytotoxicity. *Chemosphere* 233: 300–308. https://doi.org/10.1016/j.chemosphere.2019.05.258.

Zhao, L., Sun, Y.L., Cui, S.X. et al. (2011). Cd-induced changes in leaf proteome of the hyperaccumulator plant Phytolacca americana. *Chemosphere* 85 (1): 56–66. https://doi.org/10.1016/j.chemosphere.2011.06.029.

Zhao, L., Xu, L., Wu, X. et al. (2018). Characteristics and sources of mercury in precipitation collected at the urban, suburban and rural sites in a city of Southeast China. *Atmos. Res.* 211: 21–29. https://doi.org/10.1016/j.atmosres.2018.04.019.

Zhao, Q., Li, J., Dai, Z. et al. (2019). Boron tolerance and accumulation potential of four salt-tolerant plant species. *Sci. Rep.* 9 (1): 6260. https://doi.org/10.1038/s41598-019-42626-8.

Zhu, Y., Pilon-Smits, E.A.H., and Tarun, A. (1999). Cadmium tolerance and accumulation in indian mustard is enhanced by overexpressing γ-glutamylcysteine synthetase. *Plant Physiol.* 121 (4): 1169–1177. https://doi.org/10.1104/pp.121.4.1169.

12

Phytoremediation of Pesticides and Heavy Metals in Contaminated Environs

Durdana Shah[1], Azra Kamili[1], Nasreena Sajjad[2], Sumira Tyub[1], Gousia Majeed[1], Sabira Hafiz[1], Wasifa Noor[1], Saba Yaqoob[1], and Ishfaq Maqbool[1]

[1] *Plant Tissue Culture Laboratory, Centre of Research for Development, University of Kashmir, Srinagar, Jammu and Kashmir, India*
[2] *Department of Biochemistry, University of Kashmir, Srinagar, Jammu and Kashmir, India*

12.1 Introduction

Soil–plant systems are extremely affected by contaminated soils. Notable water and soil contamination is due to the emission of waste into neighboring environments through the mining process and the release of heavy metals by humans (Cheng 2003). Heavy metals contaminate large land areas around the world as a result of their release from waste incinerators, car exhumes, remains from processes of mining and manufacturing industries, and the usage of sludge or urban composts. The chief reason for heavy metal pollution particularly in agriculture is due to the use of fertilizers and pesticides (Bharagava and Mishra 2018). In food security and green revolution, soil plays an important role and forms an essential faction of our agricultural system. Metallic, organic, and inorganic contaminants have extensively polluted agricultural lands as an outcome of fast industrial development and urbanization (Sivarajasekar et al. 2018), the main sink for heavy metals is soil. Physicochemical and biological assets of agricultural soil are affected due to toxic metal pollution which results in the decrease in land use for agricultural farming, which, in turn, leads to food insufficiency difficulties (Wuana and Okieimen 2011). Immense utilization of pesticides has inflicted health risks on neighboring flora and fauna as well as on humans residing close to sites contaminated by pesticides (Damalas and Eleftherohorinos 2011). The environment as well as biota health is affected due to the build-up of metals in distinct compartments of the environment (Maria et al. 2016). Some vegetables cultivated in coastal areas as well as fish and fishery products are polluted in several ways and through the food chain enter the human body (Rahman et al. 2019). Thus, one of the grave global tasks today is the prevention and decrease of soil and water pollution from heavy metal contamination (Kozminska et al. 2018). In this regard, phytoremediation is a productive route to take off thrash metals from water and soils. Hence, it is regarded as a satisfactory way pertaining to aesthetic, eco-friendly, and economic value (Neha et al. 2017). Phytoremediation employs the capability of plants having high accumulation capability to extract thrash metals from the soil compared to other approaches of phytoremediation like chemical methods and physical methods (Kramer 2005). These plants have the capability of heavy metal accumulation and transferability (Fu et al. 2019) by utilizing the distinctive and particular accumulation capacity of root systems, besides the bioaccumulation, translocation, and pollutant deterioration capacity of the whole plant (Maria et al. 2016). In halophytes, little quantities of heavy metals are transferred to leaves and stems and are mostly concentrated in roots (Windham et al. 2003). In plants like *Arabidopsis thaliana* (Delhaize 1996), *Thlaspicaerulescens* and *Sebera acuminate* (Cunningham and Ow 1996), *Phragmites australis*, and *Typha latifolia* (Ye et al. 2001), it has been reported that heavy metals concentrate in the tissues. Mangrove plants which are also called salt marsh establish in coastal deposits and operate as a medium for biological accumulation and could alter the heavy metals' adhesion frequency for their phytoremediation capability. Plants eliminate metals by way of cation interchange, absorption, filtration, and chemical modifications during root aggregation methods (Wright and Otte 1999). Pesticides are remediated by plants by the elevation of soil rhizosphere

Aquatic Contamination: Tolerance and Bioremediation, First Edition. Edited by Rouf Ahmad Bhat, Gowhar Hamid Dar, Fernanda Maria Policarpo Tonelli, and Saima Hamid.

functioning or through accumulation, adsorption, and volatilization (Xia and Ma 2006). Rhizofiltration, phytostabilization, phytoextraction, phytovolatilization, and phytodegradation are the methods of phytoremediation (Xia and Ma 2006; Ali et al. 2013; Lee 2013). An efficient way of phytoremediation is the recognition and selection of appropriate species of native wild plants that can withstand heavy metals (Brunetti et al. 2009). In wild plants, the metal resistance may primarily rely on their capability of metal elimination (Nouri et al. 2009). The capability of plants to aggregate heavy metals differs by species (Brunetti et al. 2009).

An effective way of phytoremediation is through the selection of genotypes/ecotypes which grow naturally in polluted sites (Muszynska et al. 2015). Presently, in addition to traditional methods of removing contaminated substrate including chemical or mechanical treatment, new possibilities of employing plants (phytoremediation) and microorganisms (bioremediation) have been explored to alter or eliminate harmful components from the surroundings (Gavrilescu et al. 2015). Plant biomass as well as resistance to metals is enhanced due to the connection between useful microorganisms and plants, thus implying that microorganisms like algae, protozoa, bacteria, and fungi, are essential constituents for phytoremediation.

12.2 Mechanism of Phytoremediation by Heavy Metals

In general, plants take metals in the solution form from soil that turns to be a bioavailable pond of nutrients from plants as well as heavy metals. To translocate and store micronutrients, plants have developed highly definite mechanisms. The series of known mechanisms of transport and specified protein set in the plant cell, which mainly takes place in the plasma membrane, which helps in the uptake of ions and translocation of ions, which include (i) proton pumps; (ii) co- and anti-transporters; (iii) and passages (proteins that help in carrying these ions). Every single transportation process is expected to uptake a series of ions. The leaves of a plant evaporate water and function as a pump, which, in turn, helps in the uptake of minerals from the soil into the roots of a plant. This practice is called evapotranspiration, and it is also responsible for the movement of pollutants (heavy metals) to the plant shoot. Microorganisms including fungi and bacteria, present in the rhizosphere, also contribute to mobilizing metal ions as these are closely associated with plants, thus helping in increasing the bioavailable part. Their efficiency to remove inorganic compounds is significantly less than that of organic contaminants (Erdei et al. 2005). The mechanisms which plants use in the uptake of essential nutrients are also used for the application, transfer, and storing of contaminants because their biological properties mimic that of vital nutrients. Therefore, in phytoremediation, nutrient uptake mechanisms serve great importance (US Department of Energy, Plume Focus Area 1994). The phytoremediation mechanism consists of major steps: metal ion uptake by roots of a plant, transfer to the aerial parts of a plant through xylem vessels, metal mobilization in the rhizosphere (phytoimmobilization), bioaccumulation, abilities of the entire plant body for contaminant degradation, sequestration of metal in the tissues of a plant and resistance to heavy metals (Ali et al. 2013). In shoots, metals mostly get accumulated in vacuoles, which seem to be the most significant (US Department of Energy, Plume Focus Area 1994).

In the phytoremediation process, heavy metal tolerance is a precondition because the greater the plant shows tolerance to metal stress, the greater would be the accumulation of metal into the tissues of plants with the least detrimental effects on the fitness and tolerance of a plant. The tolerance potential of metals in plants is reliant on various processes like translocation of ions into the vacuoles, adsorption to the plant cell wall, chelation of proteins with metal ions as well as with peptides (metallothioneins (MTS) and phytochelatins (PCs)), inorganic complexes (sulfides), organic (citrates), and improvement of the active efflux (Memon and Schroder 2009). Heavy metals like arsenic, chromium, lead, cadmium, and numerous radionuclides are absorbed by many plant species from soils. Plants act as "accumulators" as well as "excluders" (Sinha et al. 2004). Accumulators endure even though accumulating toxins into the aerial parts of a plant by biodegrading contaminants into less toxic substances. The excluders limit contaminant absorption into their biomass. The phytoremediation process uses many techniques for heavy metal uptake: phytoextraction, phytovolatilization, phytostabilization, rhizofiltration, and phytotransformation (Figure 12.1).

12.2.1 Techniques of Phytoremediation

Different techniques are there to take advantage of the potential of plants for heavy metal elimination from contaminated soil.

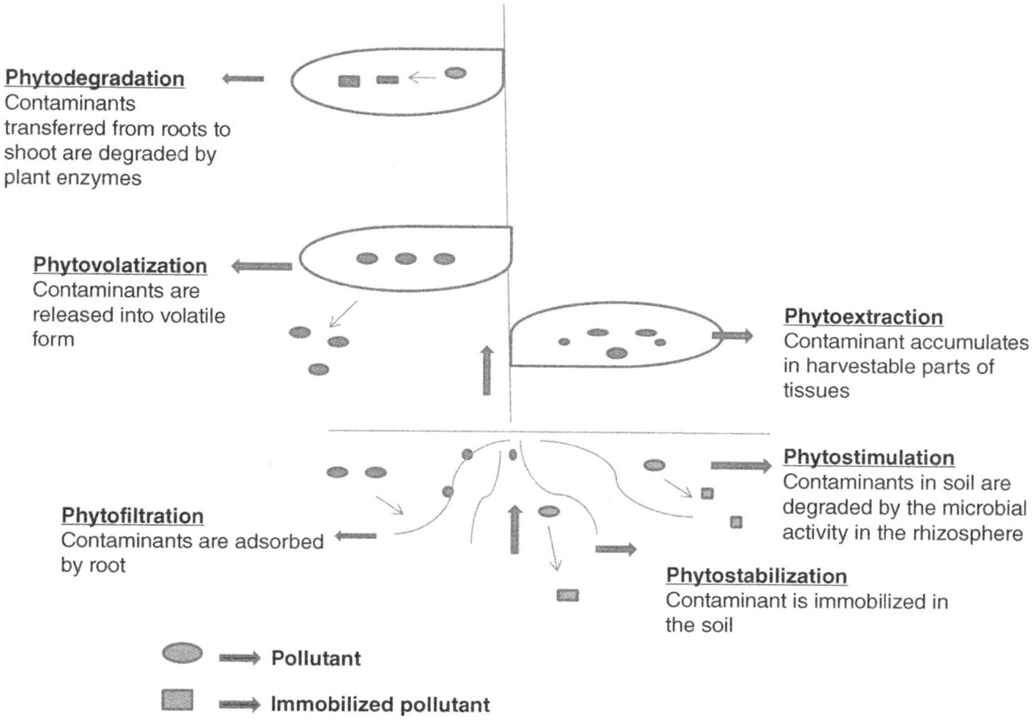

Figure 12.1 Mechanism of phytoremediation.

12.2.2 Phytoextraction

In this process, the uptake and transfer of contaminants from the soil through plants with a high potential to accumulate heavy metal contaminants takes place and later deposit into the harvestable parts (Salt et al. 1998). To decrease the concentration of heavy metals in the soil, repeated harvesting is required. The phytoremediation technique is considered the most prominent and suitable method for the elimination of metal ions and metalloids from polluted soils, but its effectiveness relies on various features like availability of metals, plant species with high phytoextraction potential, metal speciation, and soil properties (Li et al. 2010). Phytoextraction can be a continuous process using plants with high metal-accumulating potential, or an induced method for the usage of chemicals to increase the availability of metal ions in the soil. It is considered to be the utmost effective among many phytoremediation technologies (Kramer 2005). For effective phytoextraction, it is important to select an appropriate plant species. Plants best for phytoextraction should have a potential of metal hyperaccumulation and tolerance and must also have numerous characteristics like (i) above-ground biomass with hasty growth; (ii) widely distributed and far-branched root system; (iii) capability to grow outside its natural habitat; (iv) comparatively easy to grow and foster; (v) sometimes with symbiotic mycorrhizal fungi; (vi) ability to translocate the heavy metals accumulated from roots to shoots; (vii) possible repulsiveness to herbivores to prevent the escape of accumulated metals to the food chain; (viii) plant species with simple life cycles that provide continuous harvest in a single growth period also show immense potential in phytoextraction (Ali et al. 2013).

12.2.3 Phytofiltration

It is the process of inhibition of organic pollutants from blending into the water streams or groundwater using roots for filtration purposes. This technique mainly implicates submerged or floating water plants that by their root systems absorb, adsorb, and remove harmful compounds (including heavy metals) from their environment (Eapen et al. 2003). The technique of phytofiltration can be performed at lower costs by directly growing plants in contaminated water bodies (Suthersan and McDonough 1996). Phytofiltration can be rhizofiltration, blastofiltration, or caulifiltration. In rhizofiltration, plant roots are used and in blastofiltration, seedlings can be utilized, and caulifiltration involves the use of excised plant shoots (Sarma 2011). Some land plants use root biofilters made by their rhizospheric microorganisms for rhizofiltration

(Salt et al. 1995). The efficacy of the rhizofiltration technique is mainly dependent on the plant's physicochemical properties that may favor the bioadsorption process (Olgun and Sanchez-Galvan 2012). Plants suitable for the rhizofiltration technique should have (i) a large and extensive root system that can take up toxic metals from solution for a longer time period; (ii) the capability to tolerate accumulated metals at high concentrations; and (iii) greater biomass production and a restricted root to shoot translocation of the contaminants. Blastofiltration is mostly suited for water remediation taking the benefit of the gradual rise in surface-to-volume ratio that take place after sprouting and those innumerable sprouts are capable of adsorbing and absorbing huge amounts of metals in water bodies (Rawat et al. 2012). The plants with a high root-to-shoot translocation rate of metals actually lessen the outcome benefits of the rhizofiltration technique because the increased translocation rate exposes a greater portion of plants to metal contamination and, consequently, increases contamination risk for other organisms via the food chain.

12.2.4 Phytodegradation/Phytotransformation

This technique employs the practice of using plants and associated microorganisms to degrade/transform contaminants and nutrients from water, sediment, or soil by various plant metabolic processes. This technique is generally used for organic pollutants like herbicides or trichloroethylene and these are transformed into the simpler ones that through various enzymatic reactions can be amalgamated in the plant tissues (Salt et al. 1998; Spaczynski et al. 2012). Plant enzymes like laccases, peroxidases, dehalogenase, nitrilases, and nitroreductase, are utilized in this process to metabolize toxic compounds into less toxic and microorganisms such as bacteria, fungi, and yeasts also support this process (Dobos and Puia 2009). Several plants exhibit the potential of converting toxic metal elements into lesser toxic and more stable forms, for example, Cr which is converted from more mobile and more toxic Cr^{6+} to less toxic and less mobile Cr^{3+} (Shanker et al. 2005). Phytodegradation takes place both in the roots (rhizodegradation) and/or shoots (Bulak et al. 2014). Rhizodegradation is the process of attenuating organic pollutants into lesser toxic forms within the rhizosphere with assistance from soil microbes and root exudates (organic molecules) that support populations of soil microbes. The efficacy of this process can be increased by adding to the contaminated soil the inoculum of specific bacterial strains that have the capability of degrading targeted contaminants. Also, the populations of soil bacteria can be increased by the process of bio stimulation which involves the manipulation of soil nutrients and soil PH levels. Sometimes soil modifications also prove beneficial, such as the addition of chelating agents that facilitate plant metal uptake by breaking bonds and by binding contaminants to soil particles (Miller 1996). By enzyme complexes involved in the plant metabolic cycle, these substances later get metabolized into CO_2 and H_2O (McCutcheon and Schnoor 2003). The plants mostly suited for phytodegradation should possess qualities like (i) advanced root system with enough potential to secrete enzymes for the breakdown of the contaminants; (ii) tolerance to the large number of contaminants present in the soil; (iii) rapid growth; and (iv) a moderately high biomass. Thus, plant-derived enzymes having the ability to degrade harmful contaminants can be effectively utilized for the expansion of future phytoremediation strategies (Salt et al. 1998).

12.2.5 Phytostimulation

This technique involves the stimulation of enzymes existing in the rhizosphere by means of microbes or through fungal exudates which can lead to bioremediation. The metabolism and growth of rhizospheric microflora present in the soil are sustained by the root secretions of plants. This rhizosphere has the ability to diminish diverse contaminants. So, this is also known as rhizospheric biodegradation (Anderson et al. 1994). Uptake of contaminants for decontamination is not required by plants because the chemicals are modified by the enzymes released in the rhizosphere. Purification of areas is also done by enhancing humic components' generation from organic contaminants or by vaporizing contaminants, for example, polynuclear aromatic hydrocarbons (PAHs) by the microorganisms of the rhizosphere (Cunningham and Ow 1996; Dec and Bollag 1994).

12.2.6 Phytovolatilization

This technique involves the uptake of contaminants by plant roots and their conversion and dispense via stomata in a gaseous state into the atmosphere, thus eliminating the soil contaminants (Ghosh and Singh 2005). The evapotranspiration of plants is the driving force of this process. Plants with high evapotranspiration rates are a prerequisite in phytovolatilization. It is very much promising for metals which contain high volatility features, for example, mercury (Hg) and selenium (Se). Mercury (Hg) as a mercuric ion is converted into comparatively not much toxic elemental form and Se could be volatilized by plants when it gets converted into dimethyl selenide [$Se(CH_3)_2$] for release and dilution into the atmosphere (Sharma et al. 2015). This technique is valuable over other phytoremediation techniques as it can remove the contaminant from a site without the need for the harvest/disposal of contaminated plants. Though, its application is still limited because of the fact that it does not involve the complete removal of pollutants from the environment. It simply transfers contaminant from one part of the environment (soil) to another part (atmosphere), and is expected to precipitate with rainfall and thus returns back to the ecosystem (Gill 2014). This forms the phytovolatilization process the most controversial of phytoremediation techniques (Padmavathiamma and Li 2007).

12.2.7 Phytostabilization

This method involves the use of plants to reduce the bioavailability of organic pollutants by inhibiting their movement in soil (Erdei et al. 2005; Erakhrumen and Agbontalor 2007). Thus, phytostabilization is also known as phytoimmobilization as it focuses on holding contaminants in the soil and limits their ability to enter into the food chain and further dispersal into the groundwater. It involves certain mechanisms like adsorption onto roots, absorption, and accumulation by roots, and complex formation or precipitation within the root zone (Erakhrumen and Agbontalor 2007). Generally, it is used for the stabilization of heavy metals in contaminated soil, sediment, sludge, and water (Ghosh and Singh 2005). Since the metals are not really removed from the soil but are only inactivated, thus can be used to manage the pollutants (Vangronsveld et al. 2009). It is a very efficient strategy but requires rapid immobilization to conserve surface and ground waters (Chhotu and Fulekar 2009). Phytostabilization shows an advantage over other phytoremediation techniques as the contaminants are mostly retained in the roots with low translocation to the shoots. So, the demand to treat the aerial parts is lowered (Abreu et al. 2012). Generally, high metal-tolerant plants that can endure drought and high temperature as well (Ernst 2005) are suitable for phytostabilization with the ability to immobilize the contaminants in the rhizosphere and develop an extensive root system that offers good soil colonization and low translocation to aerial parts of plants (Kramer 2005).

12.3 Factors which Affect Uptake Mechanisms

Various factors such as plant species, root zone, vegetative uptake, the addition of chelating agents, and medium properties could affect the uptake mechanism of heavy metals. The uptake performance of plants can be significantly improved after studying these factors in detail.

12.3.1 The Plant Species

Plant species characteristics affect the uptake mechanism (Burken and Schnoor 1996). Plant species are screened and those with higher remediation potential and producing large quantities of biomass are selected (Rodriguez et al. 2005; Prasad and De Oliveira Freitas 2003).

12.3.2 Properties of Medium

Medium properties, e.g. soil pH and organic matter content affect the uptake mechanism. The properties of soil can be modified to improve remediation potential like pH adjustment and the addition of chelators (Prasad and De Oliveira Freitas 2003). For instance, to enhance the uptake of lead by plants, the pH of the soil is adjusted with lime to a level of 6.5–7.0 (Traunfeld and Clement 2001).

12.3.3 The Root Zone

The root zone is of great importance in phytoremediation. It supports the absorption of contaminants and accumulates or metabolizes it within the plant tissue. Plant enzymes released through the roots cause the degradation of contaminants in soil.

12.3.4 Vegetative Uptake

Vegetative uptake is influenced by environmental conditions that affect the tendency of the metal to exist in ionic and plant-available forms, like soil pH, redox potential, and organic matter content (Burken and Schnoor 1996; Fritioff and Greger 2003). The metal uptake by plants also depends on the bioavailability of metals in the water phase, which, in turn, depends on the retention time as well as on the interaction of the metal with elements and other substances in the water.

12.3.5 Addition of Chelating Agent

Biodegradable chelating agents such as nitrilotriacetic acid (NTA) and ethylene diamine tetraacetic acid (EDTA) increase the bioavailability and uptake of heavy metals by plants and also promote uptake ability of the microbial community in and around the plant. This will eventually result in shorter and less expensive remediation periods (Ginneken et al. 2007).

12.4 Strategies for Improved Efficiency of Phytoremediation

Metal metabolism in plants is very complex. Various strategies for making plants suitable to tolerate excessive concentrations of heavy metals and for effective soil clean-up are critically evaluated. The potentially high species with valuable selection particularly for adapted remediation is carried out to produce different variety of plants. Characters of a high potential plant species should include (i) growing with potential in the contaminated area; (ii) potential to restrict the high toxic contaminant to a less toxic compound, and (iii) incorporation of plant with ease in field clean-up. There are a number of strategies which could be used for selected plant tissues to amplify the accumulation of metal, that include transgenesis, complex formation with chelators and ligands, mobilization of metal together by root uptake from the soil, detoxification by deposition in vacuoles, and eventually from far distance translocation to shoots via symplast or apoplast (Nakamura et al. 2014). Genetically modified plants have revealed novel ways in phytoremediation technology by expression of the potential gene and thus help overcome the limitations. Various promising sites of manipulated genes are highlighted as follows.

12.4.1 Overexpression of Genes Encoding Metal Transporters

The genes that encode metal transporters for ions and metal binding ligands (chelators) are widespread in use to amplify the metabolism of metal in plants. The cation diffusion facilitator (CDF) family constituted genes that encode proteins with different types like metal tolerance/transport protein (MTP) that permit cells to eliminate the excess amount of ions from the cytoplasm. High amounts of thiol compounds are produced by plants overexpressing CDF genes and have the ability to chelate the metal ions that are sequestered effectively. This way the plant potential to remediate heavy metals can be increased, thus resulting in hyperaccumulation of toxic ions in plant biomass (Song et al. 2014).

12.5 Metal Chelators Encoded by Overexpression Genes

Proteins associated with metal chelating like MTs and PCs play an essential role in the accumulation and detoxification of metals (Hirata et al. 2005). Thus, by genetic modification of the plants, the ability for metal uptake as well as bioavailability of such metals will be improved (Clemens and Kraemer 2002).

12.5.1 Plants with Phytoremediation Potential

Phytoremediation is believed to be a green alternative solution to the metal pollution problem. Numerous plant species have been identified as potential phytoremediators with enormous potential to remediate metal contaminants (Table 12.1). In addition, genetically modified plants generated for phytoremediation capabilities are also getting more attention (Novakova et al. 2010). These plants have benefits for both abiotic stresses as well as for presence of metals in the environment (Ibanez et al. 2015).

12.6 Origins of Pesticide Entry into Water

Various studies from the past have shown several strong entry sites which differ according to the nature and characteristics of the active substances and the existing agro-climatic conditions (Notenboom et al. 1998). Recently, it has been demonstrated that there are also several other entry routes, in addition to diffuse losses, arising from nonapproved use, poor practice, illegal practices, or abuse (Table 12.2).

Table 12.1 Plants with phytoremediation potential and their role in eliminating heavy metals.

Plant	Contaminant-remediated	Observation	Reference
Lupinus albus	Cd and As	It has a high potential for remediation of As and Cd	Vazquez et al. (2006)
Agrostis capillaris	Pb	It grows in highly polluted Zn/Pb mine wastes and form pyromorphite, thus helping in reducing the escape of lead movement	Cotter-Howells and Capom (1996)
Sedum alfredii	Zn, Cd, and Pb	It has gained importance due to a higher accumulation rate of Pb, Cd, and Zn	Lu et al. (2008)
Thlaspi caerulescens	Zn, and occasionally Cd and Ni	It is capable of hyperaccumulating Zn and occasionally Ni and Cd	Milner and Kochian (2008)
Moringa oleifera	Pb, Cd, and Cr	The seeds of this plant are reported to have the ability to metal uptake from contaminated water for Pb, Cd, Cu, and Cr	
Stevia rebaudiana	As, Cu, Se, and Al	The excised stems from this plant have an ability to accumulate large amounts of Al, Cu, As, and Se	
S. rebaudiana	Cd, Ni, Pb, and Zn	The excised shoots from this plant have ability to accumulate significant amounts of Ni, Pb, Cd, or Zn metals in the leaves	Mesjasz-Przybyłowicz et al. (2004)
Ipomoea aquatic	Cd	In a hydroponic solution, this plant shows a high ability of translocation of Cd to shoots	Wang et al. (2008)
Polypogon monspeliensis	As	This plant is considered an appropriate species for the phytoremediation of As from contaminated soil and water	Ruppert et al. (2013)
Vallisnerian natans	As	It is a submerged aquatic plant and has a high potential for the removal of As	Chen et al. (2014)
Pteris vittata	As	It is capable of hyperaccumulating As which can be effectively harvested	

Table 12.2 Types of pesticides.

Classification of pesticides based on their chemical nature	Chemical names	Persistence	Effect	Uses
Organochlorines	DDT, DDD, Dicofol, Eldrin, Dieldrin	High persistence	Affect membrane function and behavioral development	Pesticides
Organophosphates	Malathion, Dimethoate, Trichlorofan	Nonpersistent	Neuropathy, growth retardation, reproductive toxicity	Insecticides
Carbamates	Carbaryl, Pebulate, Methane	Moderate persistence	Visual disturbances, excessive salivation, weakness	Insecticides
Pyrethroids	Allethrin, Bonthrin, Dimethrin	Moderate persistence	Seizures, paresthesia, CNS depression, increased salivation	Insecticide
Phenyl amides	Barban, Carbetamide, Chloropropane	Non persistent	They affect nucleic acids by inhibiting the activity of RNA polymerase I system	Herbicides
Trazines	Atratone, Chlorazine, Cynazine,	Moderately persistent	Endocrine disruption, skin reactions.	Herbicides
Phthalimide	Captan, Diflotan, Folpet	Nonpersistent	Skin and respiratory tract irritant, teratogen, carcinogen	Fungicides
Benzoic acid	Dicamba, Neptalan, Bromoxynil	Moderate persistence	Muscle spasms, slow heart rate	Herbicide

12.6.1 Diffuse Sources

12.6.1.1 Spray Drift

Spray drift used in plant protection products acts as a strong course to surface water and occurs when a fine spray is made in close space to water surfaces. Some data is available from regular field observations about the volume of the product, which moves into the water. For instance, a soil surface application of pesticides to arable crops results in drift accumulation ranging from 0.3% to 3.5% of the standard field application rate at a distance of 1 m from the treatment area. The buildup of remains below the sediments may result from recurrent applications. Such remains may or may not be of any ecological significance (Helfrich et al. 2009).

12.6.1.2 Volatilization and Precipitation

The supplementary pathway by means of which the pesticides are moved unnoticed from potential sites after their application is volatilization. While most of the losses do not surpass 20% and losses can be as high as 90%. Although, the pesticides that are commonly traced in water bodies are not likely to be of a volatile nature (Taylor 1995).

12.6.2 Surface Runoff/Overland Flow

The soluble remains and residues absorbed by fine particles can move across an application surface when the infiltration capacity of an application surface is surpassed. Such losses are mostly lower than 0.05% from agricultural fields unless there is extreme rainfall within one to two weeks of the application of the applied active substance. Due to particular agro-climatic conditions at the local level, losses can be environmentally significant and those soils which have the capacity to fill or cap, or are present at vertical slopes, can be principally exposed to surface runoff.

12.6.3 Leaching

Leaching is a process through which the transportation of soluble remains of pesticides to ground and surface water occurs. The residues might go direct to underlying groundwater or might be transported laterally to surface waters through the process of leaching. The leaching may result in the loss of applied active substances as high as 5% but commonly these losses are less than 1%. In some cases, there is a loss of applied active substance by leaching, which are as high as 5%, but typically these losses are less than 1%. When the water of soil moves through fast preferential flow, it can be significant losses, through the soil by cracks, fissures, or macrospores since residues have less possibility to dissipate. Residues absorbed by the soil or organic particles can be elated from the soil surface suspended in leachate water in some soil types (Hardy et al. 1998).

12.6.4 Point Sources

The various sources of pollution are farm yard activities like spillages, tank filling, faulty equipment, washing and waste disposal, use of sumps, soaking away, and drainage. Misuse, unintentional spillage, or inadequate handling/storage conditions may lead to severe cases of water contamination. The contamination from back siphoning in the borehole during the tank-filling process may give rise to elevated concentrations (Helweg 1994).

12.6.5 Direct Contamination and Overspray

During the application of pesticides, the overspray of surface waters is not advisable as its effect could be very strong. By using spreaders, the granular formations which are applied might also enter directly into water courses. The direct spillages into surface water are causing widespread contamination.

12.6.6 Consented Discharges

The agricultural and industrial processes can offer a way to lavish residues which can be released into surface waters. Concur varies according to the concentration of residues at the point of discharge and the nature of the receiving water. These point sources can act as ground-laying drivers of water pollution.

12.7 Effects of Pesticides

12.7.1 Effects on Aquatic Ecosystem

The freshwater network comprises assorted gatherings of organisms, for example, fish, amphibians, invertebrates, plants, and microorganisms. Pesticides have immediate and roundabout consequences for these living beings. Direct impacts are brought about by the physiological activity of a pesticide inside an organism laborious to identify while indirect impacts are increasingly complicated to clarify. Competitive advantage for algae in connection to macrophytes might be prompted by an introduction of the two herbicides and insecticides; all things considered, the indirect impacts of these two gatherings of pesticides can act synergistically at the ecosystem level, despite the fact that their direct toxic impacts are applied through diverse modes of activity. Most pesticides connected to agriculture are changed by physical, chemical, and natural procedures into at least one transformation product. Whenever the concentration is sufficient to animate an impact, both parent pesticides and transformation products exert a harmful activity in aquatic life forms.

Benthic communities are of main importance having a role in the movement of ecological contaminants to the next trophic levels (fish, waterfowl, and humans), and have prime position in nutrients, sediments, and contaminant flow (Burton 1992; Horne and Goldman 1994). The build-up of toxic pesticides in the dregs may hit key life forms/forms in this manner causing grave mischief to the ecosystem. Courses on pesticide introduction in aquatic invertebrates contain take-up water from pores, water lying across body walls, and respiratory surfaces, and ingestion of contaminated sediment particles (Power et al. 1992). The dynamic element in the pesticide and residues formed during photochemical and microbial degradation are responsible for the negative impact of pesticides on the aquatic ecosystem (Papendick et al. 1986). The chemical properties like solubility in water, molecule affinity, and instability decide their degradation product in an aquatic environment. Herbicides hinder the take-up of nutrients by wiping out primary producers or plunging their competence. Therefore, the aquatic food chain is hindered at the fundamental degree of primary production (O'Toole and Irvine 2006). Then again, insecticides accelerate the take-up of nutrients, in aquatic ecosystems leading to eutrophication and top-to-down control expulsion of those organisms belonging to higher trophic levels. Raised concentration of pesticide build-ups may hence result in diminished uptake of periphyton/phytoplankton, thereby harming the grazers, which ultimately leads to eutrophication-like indications, for instance, algal blooms. The impacts of aggregate nutrient toxicity and deposits of pesticides could be connected to the trophic levels underneath. At the point when the prey lethality increases and their endurance is repressed, at that point, the existence of the upper trophic level will likewise be adversely influenced giving rise to low biomass, though, may result in species abundance. Multiple times more energy is produced in the case of aerobic respiration contrasted with anaerobic respiration (Peters et al. 2013). Poisoning brought about by raised pesticide concentrations in aquatic ecosystems can promote anaerobic conditions by improving the respiratory oxygen demand. Subsequently, the productivity of the aquatic ecosystem would decrease. Enormous decline in species because of outrageous eutrophication and poisoning caused by excessive use of a pesticide may likewise transfer the piled-up pesticides into the silt, lessening the hazard to pelagic life forms (Breitburg et al. 1999). Macrophytic standing biomass may likewise bolt assemble of pesticide buildups over a vast stretch, particularly in systems deficient in nutrients, following consequent degradation.

12.7.2 Sublethal Effects

All pesticide poisonings do not bring about the early death of an organism. Little "sublethal" dosages of certain pesticides may prompt a shift in behavior, body weight reduction, disabled reproductive potential, inability to dodge predators, and lowered resistance to extreme temperatures. The fishes moving in streams through croplands and orchards are probably going to encounter lower doses of pesticides from continuous field run-off. Repeated pesticide exposure can bring about a decline in the production of fish eggs and bring forth, niche and brood desertion, low protection from infirmities, the diminished weight of the body, hormonal disturbances, and decreased evasion from predators.

Sublethal effects include:

- Loss in weight
- Low resistance to diseases
- Sterility
- Reduction in the production of eggs
- Lowering of attention
- More predator encounters

12.7.3 Habitat Alteration

Due to the negative effect of pesticides on insects and plants, the accessibility to such habitats and other sources of sustenance for aquatic organisms is diminished. The insect population declines due to the application of pesticides as some insect-eating fish may lose their source of food. This forces these organisms to expand their range in search of food exposing them to higher threats of predation. Herbicide splashing can likewise lessen success in reproduction in fish and aquatic organisms. The weedy shallower nursery regions provide abundant food and shelter to young fish. Thus, herbicide application in such weedy nurseries can decrease the shelter and measure of cover that young fish use for feeding and also to avoid predation. Utilizing herbicides for a complete "clean up" of a lake will extensively decrease habitat availability, dissolved oxygen, food abundance, and fish productivity (Helfrich et al. 2009).

12.7.4 Threats to Biodiversity

The negative impact of the unrestrained application of pesticides cannot be ignored. Pesticide accumulation in the food chain is of noteworthy concern as it has a direct influence on raptors and predation. In any case, pesticides can likewise diminish the diversity of various bushes, weeds, and insects on which higher levels depend. Splashing of insecticides, fungicides, and herbicide has been identified as the reason for the decline of some rare species of birds and animals. So, the utilization of pesticides brings about a decrease in biodiversity. The decline in various fish species has been observed with the abuse of pesticides (Scholz et al. 2012). Death of aquatic flora by pesticides brings about seriously lower fish count and, in due course of time, prompts fish to be suffocated with decreased productivity (Helfrich et al. 2009). Levels of pesticides are far more raised in surface water compared to ground water as the surface water is contaminated by spray drift and surface runoff from farmland (Anon 1993). Water bodies are facing threats as the pesticides make their way into ponds, streams, and lakes. Amphibians are mainly influenced by these chemical chemical-rich sullied waters, which ultimately leads to loss of habitat. A widely used pesticide Carbaryl is known to cause lethality in many species of amphibians, while as a herbicide, glyphosate is responsible for causing mortality in juvenile frogs and tadpoles (Relyea and Hoverman 2006). Malathion in minute concentrations has been shown to alter the composition and richness of the periphyton population which has a direct influence on frog tadpole growth. The spraying of herbicide close to nurseries effect the reproductive capability of oceanic live forms (Helfrich et al. 2009).

12.7.5 Effect on Functioning of Freshwater Ecosystem

Freshwater ecosystems provide a significant exhibit of ecosystem services, extending from water for irrigation, clean drinking water, water storage, and a domain for living beings. Invertebrates are an important part of the freshwater food chain and are an important link for the exchange of nutrients and energy. Thus, the application of insecticides can seriously affect services provided by aquatic ecosystems. The insecticide exposure has a direct effect on shredder populations. As this group restrains from feeding, it directly influences the conversion of coarse terrestrial into fine particulates, which is the main feed available for other species for consumption. This process thereby changes the composition of the invertebrate community, nutrient cycling, rate of decomposition, and quality of water on which all the biological services depend.

12.7.6 Effects on Aquatic Food Chain

The effect of pesticides or insecticides has a main role to play in nutrient cycling and decomposition. However, there are also indirect effects of these chemicals on decreasing invertebrate prey. This might affect those species that are highly consumed for food (crawfish and fish). The balance between the trophic levels is disturbed due to the negative impacts posed by insecticides on the base of the food chain. A decrease in reproductive rates, development, and survival can alter all the basic functioning of the ecosystem like cycling of nutrients, decomposition, etc. Thus, the consequences of imbalance in the functioning of aquatic ecosystems are indispensable (Van Dijk et al. 2013).

12.8 Threats to Terrestrial Biodiversity

Pesticides are usually sprayed in liquid form on soil or plants, either infused in soil or as granules. The significant concern of these chemical constituents is their ability to leach into the soil, thereby influencing the microbial flora of the soil. The soil microflora has many functions including nutrient uptake, decomposition, and fertility on which plants rely. Shockingly, the pesticides may have serious results and maybe in the near future, we may have soil devoid of microflora. The soil microorganisms are involved in many activities including fixation of atmospheric nitrogen, but these essential life-sustaining processes could also be influenced by pesticide abuse. Herbicides can also cause damage to the fungal flora of soil, e.g. trifluralin and oryzalin hinder the development of mycorrhizal fungal species. Some pesticides like Oxadiazon may have a negative impact on fungal spores (Chakravarty and Sidhu 1987). Earthworms may face neurotoxic impacts by the pesticide-contaminated soil through soil pore water, leading to physiological damage (Schreck et al. 2008).

12.9 Impacts of Pesticides on Soil Ecosystem Services

Almost all the ecosystem services are organically derived, and the various chemical abuses can cause interruption to non-target living components of soil including the services provided by soil ecosystems. The impacts may include acute to chronic toxicity to various biotic components, thereby inducing behavioral changes, predator-prey interactions, and food web dynamics. All these ultimately affect soil stability and biodiversity (Edwards 2002).

12.9.1 The Effect of Pesticides on Soil Microflora

The expanding number and widespread utilization of these potent synthetic compounds have stirred concern for the possible harmful impacts on soil microorganisms and their activities of significance in soil fertility. Studies on the impact of DDT on the number of soil microorganisms cultivated on agar plates (immunization strategies) show that independent of the nature and origin of the microorganisms, the addition of DDT to growth media altogether influences their populace numbers. Studies by different workers have shown a propensity among microorganisms for resistance from DDT. The most resistant appeared to be fungi, trailed by marginally less resistant actinomycetes and by the least resistant are microbes. It has likewise been demonstrated that a population of bacteria originating from soil already with DDT is equivalent to that originating from polluted soil. The different chemicals utilized as pesticides are capable of repressing at least certain soil microorganisms under specific conditions. Investigations up until this point, in any case, demonstrate that when utilized at prescribed field rates, these toxicants will not essentially influence microbial activities that are significant for soil fertility.

12.9.2 Effect of Pesticides on Soil Fertility and its Control Through Phytoremediation

The biological capacity of soil has been either directly or indirectly affected by the usage of pesticides in order to save crops (Kirk et al. 2004). Pesticide overexposure has led to soil leaching, which influences the microbes living in it (Hayo and Werf 1996). Excessive use of pesticides influences fauna and flora of the soil including microflora and also affects its physiochemical characteristics such as salinity, pH, and alkalinity, thus resulting in reduced soil fertility (Arora et al. 2019). Soil microflora is important in sustaining and increasing important elements and nutrient accumulation such as phosphorus and nitrogen in the soil and is also involved in several different natural procedures of soil. The soil microbes help in the disintegration of organic matter, and nutrient uptake, and hence enhance the fertility of the soil (Mahmood et al. 2016). Change in agricultural soil microbial population framework because of any element is connected to impudence the complete productivity of soil (Cycon and Piotrowska 2007). The incorporation of new pesticides into soil includes various risks. The biological consequences are highly and unexpectedly altered even due to little alterations in the chemical composition. The microbial community uses 2, 4-D for nutrition. Hence, it disintegrates quickly in soil and is a common herbicide example. Although herbicide is formed due to the addition of a single chlorine molecule making 2, 4, 5-T compounds which will withstand disintegration in soil for a long time (Funke and Schulz 1971). An additional characteristic of soil in which microorganisms are fully implicated is nitrogen fixation by leguminous crops. Rhizobium bacteria which are present in the soil cause infection in plant roots as a result of which nodules are made which are responsible for nitrogen fixation (Funke and Schulz 1971).

Soil acts as a final sink of pesticides utilized in public health protection and in agriculture (Magan et al. 2010). Soil fertility relies not only on the texture but also on the biological ability of the soil (Lo 2010). Soil health and efficiency are greatly influenced by the aggregation of pesticides in soil. Soil functions as a depository of several microbial types. Pesticides' consequence on the microbial divergence of soil and capability to disintegrate pesticides in the soil is not clear yet (Tao and Yang 2011).

1) Likewise, pesticides influencing bacterial groups with a crucial role in the nitrogen transformation of soil have been emphasized. Nitrogen-fixing rhizobium in symbiosis with leguminous plants is one such case (Revellin et al. 1992; Ramos and Ribeiro 1993). Nitrogen-fixing free-living bacteria Azospirillum and Azotobacter are other examples (Banerjee and Banerjee 1987; Martinez Toledo et al. 1988). Therefore, pesticide use gravely influences the nitrifying bacteria playing an important role in the nitrogen cycling of soil (Doneche et al. 1993; Banerjee and Banerjee 1987; Martinez Toledo et al. 1992a,b). Doneche et al. (1993) reported that the fungicide called mancozeb executed a continuing obstructive aftermath (three months) in natural soil on aerobic fixers of dinitrogen. In the investigation of mancozeb consequences, the ammonium-oxidizing groups of nitrifiers were hindered in enduring in soils examined (Doneche et al. 1993).
2) Unfortunately, continuous use of pesticides has severe results on soil microorganisms and these organisms may become extinct with time and soil may lose its fertility. The complete count of bacteria, protozoa, fungi, and algae may enhance or decline to rely mainly on the pesticide harmfulness and capacity as an origin of nutrients. For the conversion of atmospheric nitrogen to nitrates, many microbes of soil are engaged. The bacteria-dependent nitrification and denitrification processes have been disorganized by Chlorothalonil and dinitrophenyl fungicides (Lang and Cai 2009). Bacteria, present in the soil, implicated in the ammonia-to-nitrite conversion are hindered by triclopyr herbicide (Pell et al. 1998). 2,4-D hinders the ammonia to nitrates conversion bacteria present in the soil (Frankenberger et al. 1991) while the activity and growth of nitrogen-fixing bacteria which is free-living in the soil are decreased by Glyphosate, which is a nonselective herbicide (Santos and Flores 1995). Fungal species present in the soil are also highly harmed by herbicides as the development of symbiotic mycorrhizal fungi which is symbiotic and that assist in the metabolism of nutrients, are hindered by pesticides trifluralin and oryzalin. In soil productivity, earthworms show an important part and act as bioindicators of soil pollution. Prior researches had examined the influence of pesticides on earthworms and when integrated into pesticide-polluted soil exhibited a 50% reduction in the motion of earthworms and when pesticides are kept over the surface of the soil, 90% reduction takes place. Earthworm survival is highly significant for agro-ecosystem sustainability. However, earthworms could be despoiled by the rigorous usage of pesticides (Nugroho et al. 2014). Schreck et al. (2008) researched that earthworms, fungicides, and/or insecticides result in neurotoxicity and physiological impairment was caused due to continuous exposure. The damage in DNA results because of Glyphosate and chlorpyrifos in earthworms. Consuming capability and sustainability of earthworms are influenced by glyphosates (Casabé et al. 2007). Goulson studied the hindrances of neonicotinoids on the life of animals and the environment. He studied that neonicotinoids have a capability to accumulate in the soil. Thus, earthworms, for example, species of *Eiseniafetida* are eradicated by them (Goulson 2013). Reinecke and Reinecke (2007) had studied earthworm biomass and cholinesterase activity influenced by pesticides. The authors established that as a result of sporadic exposure (azinphos-methyl) and chronic (chlorpyrifos) earthworms were influenced deleteriously by the pesticides. Other studies also revealed the weight of the body and sperm feasibility was reduced in an adult's species of *Eiseniafetidam* as a result of subjection to malathion. The organism also had an objectionable impact on development and reproduction by the chlorpyrifos contact, and the considerable decrease in the generation of cocoons also resulted due to the subjugation of cypermethrin (Yasmin and D'Souza 2010). Off-target organisms are affected by pesticides administered in the soil and the nearby metabolism needed for the fertility of the soil is damaged (Andrea et al. 2001). Environmental protection agencies are very watchful and prior to going for any chemical to work as a pesticide, they put certain checks (Gallegos-Saliner et al. 2008; Balderacchi et al. 2014). However, certain checks such as the consequence of pesticides on micronutrients (Senesi 1992) are still not appropriately addressed (Shirzadi et al. 2008). Metal depauperating is among the determinants that influence the fertility of the soil. It is performed by microorganisms naturally and different natural chelating factors that occur in the environment naturally (Tullberg 2010). Current researchers have distressed that the overabundance of pesticides can be the possible reason for devastating microflora that is probably deteriorating the fertility of the

soil (Wasim et al. 2009). Additional research spotted the appalling cut in the metal ions percentage of the soil of the world (copper 3%, zinc 49%, molybdenum 11%, iron 12%, boron 33%, and manganese 5% (Alloway 2008; Schroeder et al. 2013) resulting in a drop in the value of soil (Racke et al. 1997). In the soil ecosystem, the microbial biomass does a vital part, where they execute an important part in the cycling of nutrients and decay, thus sustaining the fertility of soil and accumulation of nutrients such as microbial P, N, and C. These are nutrients which are discharged following the microorganism's death. Total N and organic C have also been seen to get condensed due to the consequences of pesticides and chemical fertilizers like triadimefon and ammonium bicarbonate by a considerable amount of 58.5, 54.8, and 55.0% as compared to the soil without chemical pollution (Yang et al. 2000). Thus, it may be said that an alteration in the population dynamics of microbes under the effect of pesticide application disturbs the nutrient stability and availability in soil.

12.9.3 Use of Microbes in Phytoremediation

The contaminants including metals are altered due to microbes associated with plant roots. For instance, the H^+ particles secreted by root-associated microbes intensify the bioavailability of metals. In the natural habitats, the role of Arbuscular mychorrhizal fungi, P solubilizing bacteria is crucial for plant growth. Compared to the rural soils, their role is minimized as agricultural practices heavily rely on pesticides which limit the role played by soil microorganisms. Their symbiotic association produces proteins including glycoprotein and glomalin which are essential for toxic metal removal. Conventional methods can be replaced by phytoextraction which is purely based on the use of microbes like rhizobial microflora, arbuscular mychorrhizal fungi using accumulator or nonaccumulator plant species (Khan 2005). Arbuscular mycorrhizal fungi have proven to be effective in metal uptake in various contaminated soils. The difference in tolerance levels has been observed in a wide range of species. Different natural fungal strains were reported from contaminated *soils. Rhizophagusirregularis and Funneliformismosseae have been reported to sequester* cadmium and zinc in contaminated soils (Hassan et al. 2013). Among the heterotrophic fungi, *Aspergillusniger* was tolerant to most contaminants. Three basidiomycetes species including *Pleurotusostreatus*, *Ganodermalucidum*, and *Schizophyllum commune* have been used for removing metals including nickel, copper, zinc, and chromium, in the case of the electroplating industry. In a study conducted by (Javaid and Bajwa 2008) on removal of toxic constituents, *S. Commune* has shown positive results by removing copper, zinc, cadmium from the leachate of a landfill site. The plant growth-promoting bacteria like *Enterobacter cloacae*, *Rhizobium leguminosarum*, and *Pseudomonas* have shown the highest remediation potential against copper toxicity (Fatnassi et al. 2015). The bacteria that are resistant toward lead can tolerate concentrations of lead up to 1000 μg/ml and also act as plant growth-promoting bacteria with good heavy metal removal efficiency (Meryem and Yasmin 2013). Two mechanisms are associated with remediation based on the effect of interaction among plant and microbe, physiology of leaf, and growth of the plant (Cocozza et al. 2014).

1) Cadmium accumulation and phytostabilization in roots.
2) Translocation of metals with high cadmium transfer to shoots.

This is the representation of phytoremediation by microbial mediated. Fungi in particular have a significant role in the bioremediation of contaminated soils. As a matter of fact, that cycling of elements like copper, chromium is actually stored in fungal fruiting bodies. These have a role in bioprospecting and mycoremediation being the least expensive and safer for nature and mankind. However this technology is in the early stage and needs to be refined for its use in agricultural systems, and its potential to build a clean green technology.

References

Abreu, C.A., Coscione, A.R., and Pires, A.M.M. (2012). Organic matter and barium absorption by plant species grown in an area polluted with scrap metal residue. *Appl. Environ. Soil Sci.* 2012: 1–7.

Ali, H., Khan, E., and Sajad, M.A. (2013). Phytoremediation of heavy metals concepts and applications. *Chemosphere* 91: 869–881.

Alloway, B.J. (2008). *Zinc in Soils and Crop Nutrition*, 2e. Brussels; Paris: International Zinc Association (IZA) and International Fertilizer Industry Association (IFA).

Anderson, T.A., Kruger, E.L., and Coats, J.R. (1994). Enhanced degradation of a mixture of three herbicides in the rhizosphere of an herbicide tolerant plant. *Chemosphere* 28: 1551–1557.

Andrea, M.M., Papini, S., and Nakagawa, L.E. (2001). Optimizing microwave-assisted solvent extraction (MASE) of pesticides from soil. *J. Environ. Sci. Health* B36: 87–93.

Anon (1993). *The Environmental Effects of Pesticide Drift*, 9–17. Peterborough: English Nature.

Arora, S., Sahini, D., Sehgal, M. et al. (2019). Pesticides use and its effect on soil bacteria and fungal populations, microbial biomass carbon and enzymatic activity. *Curr. Sci.* 116: 643–649.

Balderacchi, M., Filippini, M., Gemitzi, A. et al. (2014). Does groundwater protection in Europe require new EU-wide environmental quality standards. *Front. Chem.* 2: 32.

Banerjee, A. and Banerjee, A.K. (1987). Influence of captan on some microorganisms and microbial processes related to the nitrogen cycle. *Plant Soil* 102: 239–246.

Bharagava, R.N. and Mishra, S. (2018). Hexavalent chromium reduction potential of *Cellulosimicrobium sp.* isolated from common effluent treatment industries. *Ecotoxicol. Environ. Saf.* 147: 102–109.

Breitburg, D.L., Sanders, J.G., and Gilmour, C.C. (1999). Variability in responses to nutrients and trace elements, and transmission of stressor effects through an estuarine food web. *Limnol. Oceanogr.* 44: 837–863.

Brunetti, G., Soler-Rovira, P., Farrag, K., and Senesi, N. (2009). Tolerance and accumulation of heavy metals by wild plant species grown in contaminated soils in Apulia region, Southern Italy. *Plant Soil* 318: 285–298.

Bulak, P., Walkiewicz, M., and Brzezinka, M. (2014). Plant growth regulators-assisted phytoextraction. *Biol. Plant.* 58: 1–8.

Burken, J.G. and Schnoor, J.L. (1996). Phytoremediation: plant uptake of atrazine and role of root exudates. *J. Environ. Eng.* 122: 958–963.

Burton, A.G. (1992). *Assessing Sediment Quality in Sediment Toxicity Assessment*. Lewis Publishers.

Casabé, N., Piola, L., Fuchs, J. et al. (2007). Ecotoxicological assessment of the effects of glyphosate and chlorpyrifos in an Argentine soya field. *J. Soils Sediments* 7: 232–239.

Chakravarty, P. and Sidhu, S.S. (1987). Effects of glyphosate, hexazinone and triclopyr on in vitro growth of five species of ectomycorrhizal fungi. *Eur. J. For. Pathol.* 17: 204–210.

Chen, M., Tao, L., McLean, J., and Lu, C. (2014). Quantitative analysis of neonicotinoid insecticide residues in foods: implication for dietary exposures. *J. Agric. Food Chem.* 62 (26): 6082–6090.

Cheng, S. (2003). Heavy metal pollution in China: origin, pattern and control. *Environ. Sci. Pollut. Res.* 10: 192–198.

Chhotu, D.J. and Fulekar, M.H. (2009). Phytoremediation of heavy metals: recent techniques. *Afr. J. Biotechnol.* 8: 921–928.

Clemens, S., Palmgren, M.G., and Kraemer, U. (2002). A long way ahead: understanding and engineering plant metal accumulation. *Trends Plant Sci.* 7: 309–315.

Cocozza, C., Vitullo, D., Lima, G. et al. (2014). Enhancing phytoextraction of Cd by combining poplar (clone "I-214") with *Pseudomonas fluorescens* and microbial consortia. *Environ. Sci. Pollut. Res.* 2: 1796–1808.

Cotter-Howells, J.D. and Capom, S. (1996). Remediation of contaminated land by formation of heavymetal phosphates. *Appl. Geochem.* 11: 335–342.

Cunningham, S.D. and Ow, D.W. (1996). Promises of phytoremediation. *Plant Physiol.* 110: 715–719.

Cycon, M. and Piotrowska, S.Z. (2007). Effect of selected pesticides on soil microflora involved in organic matter and nitrogen transformations: pot experiment. *Pol. J. Ecol.* 55: 207–220.

Damalas, C.A. and Eleftherohorinos, I.G. (2011). Pesticide exposure, safety issues, and risk assessment indicators. *Int. J. Environ. Res. Public Health* 8: 1402–1419.

Dec, J. and Bollag, J.M. (1994). Use of plant material for the decontamination of water polluted with phenols. *Biotechnol. Bioeng.* 44: 1132–1139.

Delhaize, A. (1996). A metal accumulator mutant of *Arabidopsis thaliana*. *Plant Physiol.* 111: 849–855.

Dobos, L. and Puia, C. (2009). The most important methods for depollution of hydrocarbons polluted soils. *Bull. UASMV Agric.* 66: 67–72.

Doneche, B., Seguin, G., and Ribereau-Gayon, P. (1993). Mancozeb effect on soil microorganisms and its degradation in soils. *Soil Sci.* 135: 361–366.

Eapen, S., Suseelan, K.N., Tivarekar, S. et al. (2003). Potential for rhizofiltration of uranium using hairy root cultures of *Brassica juncea* and *Chenopodium amaranticolor*. *Environ. Res.* 91: 127–133.

Edwards, C.A. (2002). Assessing the effects of environmental pollutants on soil organisms, communities, processes and ecosystems. *Eur. J. Soil Biol.* 38: 225–231.

Erakhrumen, A. and Agbontalor, A. (2007). Review phytoremediation: an environmentally sound technology for pollution prevention, control and remediation in developing countries. *Educ. Res. Rev.* 2: 151–156.

Erdei, L., Mezosi, G., Mecs, I. et al. (2005). Phytoremediation as a program for decontamination of heavy-metal polluted environment. *Acta. Biol. Szeged.* 49: 75–76.

Ernst, W.H.O. (2005). Phytoextraction of mine wastes-options and impossibilities. *Chem. Erde Geochem.* 65: 29–42.

Fatnassi, I.C., Chiboub, M., Saadani, O. et al. (2015). Phytostabilization of moderate copper contaminated soils using co-inoculation of *Vicia faba* with plant growth promoting bacteria. *J. Basic Microbiol.* 55: 303–311.

Frankenberger, W.T., Tabatabai, M.A. Jr., and Tabatabai, M.A. (1991). Factor's affecting *L-asparaginase* activity in soils. *Biol. Fertil. Soils* 11: 1–5.

Fritioff, A. and Greger, M. (2003). Aquatic and terrestrial plant species with potential to remove heavy metals from storm water. *Int. J. Phytoremed.* 5 (3): 211–224.

Fu, S., Wei, C., Xiao, Y. et al. (2019). Heavy metals uptake and transport by native wild plants: implications for phytoremediation and restoration. *Environ. Earth Sci.* 78: 103.

Funke, B.R. and Schulz, J.T. (1971). Effects of pesticides on soil fertility. *Farm Res.* 28 (4): 38–39.

Gallegos-Saliner, A., Poater, A., Jeliazkova, N. et al. (2008). Toxmatch - a new software tool to aid in the development and evaluation of chemically similar groups. *Regul. Toxicol. Pharmacol.* 52: 77–84.

Gavrilescu, M., Demnerova, K., Aamand, J. et al. (2015). Emerging pollutants in the environment: present and future challenges in biomonitoring, ecological risks and bioremediation. *New Biotechnol.* 32: 147–156.

Ghosh, M. and Singh, S.P. (2005). A review on phytoremediation of heavy metals and utilization of its byproducts. *Appl. Ecol. Environ. Res.* 3: 1–18.

Gill, M. (2014). Phytoremediation: green technology to clean the environment. *Int. J.* 2: 879–886.

Ginneken, V.L., Meers, E., Guisson, R. et al. (2007). Phytoremediation for heavy metal-contaminated soils combined with bioenergy production. *J. Environ. Eng. Landsc. Manag.* 15 (4): 227–236.

Goulson, D.J. (2013). An overview of the environmental risks posed by neonicotinoid insecticides. *J. Appl. Ecol.* 50: 977.

Hardy, I.A.J., Carter, A.D., and Leeds-Harrison, P.B. (1998). *Water Quality from Contrasting Drained Clay Soils: The Relative Importance of Sorbed and Aqueous Phase Transport Mechanisms*. London: The Food-Environment Challenge. Royal Society of Chemistry.

Hassan, S.E., Hijri, M., and St-Arnaud, M. (2013). Effect of arbuscular mycorrhizal fungi on trace metal uptake by sunflower plants grown on cadmium contaminated soil. *New Biotechnol.* 30 (6): 780–787.

Hayo, M.G. and Werf, V.D. (1996). Assessing the impact of pesticides on the environment. *Agric. Ecosyst. Environ.* 60: 81–96.

Helfrich, L.A., Weigmann, D.L., Hipkins, P., and Stinson, E.R. (2009). *Pesticides and Aquatic Animals: A Guide to Reducing Impacts on Aquatic Systems*. Virginia Polytechnic Institute and State University.

Helweg, A. (1994). Threats to water quality from pesticides – case histories from Denmark. *Pesticide Outlook* 5 (5): 12–18.

Hirata, K., Tsuji, N., and Miyamoto, K. (2005). Biosynthetic regulation of phytochelatins, heavy metalbinding peptides. *J. Biosci. Bioeng.* 100: 593–599.

Horne, A.J. and Goldman, C.R. (1994). *Limnology*, 2e. Singapore: McGraw-Hill.

Ibanez, S., Talano, M., Ontanon, O. et al. (2015). Transgenic plants and hairy roots: exploiting the potential of plant species to remediate contaminants. *New Biotechnol.* S1871–6784: 00267–00268.

Javaid, A. and Bajwa, R. (2008). Biosorption of electroplating heavy metals by some basidiomycetes. *Mycopathology* 6 (182): 1–6.

Khan, A.G. (2005). Role of soil microbes in the rhizospheres of plants growing on trace metal contaminated soils in phytoremediation. *J. Trace Elem. Med. Biol.* 18 (4): 355–364.

Kirk, J.L., Beaudette, L.A., Hart, M. et al. (2004). Methods of studying soil microbial diversity. *J. Microbiol. Methods* 58: 169–188.

Kozminska, A., Alina, W., Ewa, H.F., and Ewa, M. (2018). Recent strategies of increasing metal tolerance and phytoremediation potential using genetic transformation of plants. *Plant Biotechnol. Rep.* 12 (1): 1–14.

Kramer, U. (2005). Phytoremediation: novel approaches to cleaning up polluted soils. *Curr. Opin. Biotechnol.* 1 (16): 133–141.

Lang, M. and Cai, Z. (2009). Effects of chlorothalonil and carbendazim on nitrification and denitrification in soils. *J. Environ. Sci.* 21: 458–467.

Lee, J.H. (2013). An overview of phytoremediation as a potentially promising technology for environmental pollution control. *Biotechnol. Bioprocess Eng.* 18 (3): 431–439.

Li, J.T., Liao, B., Lan, C.Y. et al. (2010). Cadmium tolerance and accumulation in cultivars of a high-biomass tropical tree (Averrhoacarambola) and its potential for phytoextraction. *J. Environ. Qual.* 39: 1262–1268.

Lo, C.C. (2010). Effect of pesticides on soil microbial community. *J. Environ. Sci. Health* B45: 348–359.

Lu, J., Fu, Z., and Yin, Z. (2008). Performance of a water hyacinth (*Eichhornia crassipes*) system in the treatment of wastewater from a duck farm and the effects of using water hyacinth as duck feed. *J. Environ. Sci.* 20 (5): 513–519.

Magan, N., Fragoeiro, S., and Bastos, C. (2010). Environmental factors and bioremediation of xenobiotics using white rot fungi. *Mycobiol* 38 (4): 238–248.

Mahmood, I., Imadi, S.R., Shahzadi, K. et al. (2016). Effects of pesticides on environment. In: *Plant, Soil and Microbes* (ed. K.R. Hakeem et al.), 253–269. Switzerland: Springer International Publishing.

Maria, A.C.G., Rachel, A.H.D., de Souza, A.N., and Angela, P.V. (2016). Metal phytoremediation: general strategies, genetically modified plants and applications in metal nanoparticle contamination. *Ecotoxicol. Environ. Saf.* 134: 133–147.

Martinez Toledo, M.V., Rubia, T., Moreno, J., and Gonzalez Lopez, J. (1988). Effect of diflubenzuron on *Azotobacter* nitrogen fixation in soil. *Chemosphere* 17: 829–834.

Martinez Toledo, M.V., Salmerón, V., and Gonzalez-Lopez, J. (1992a). Effect of an organophosphorus insecticide, profenofos, on agricultural soil microflora. *Chemosphere* 24: 71–80.

Martinez Toledo, M.V., Salmeron, V., and Gonzalez Lopez, J. (1992b). Effect of the insecticides pyrimifos methyl and chlorpyrifos on soil microflora in an agricultural loam. *Plant Soil* 147: 25–30.

McCutcheon, S.C. and Schnoor, J.L. (2003). *Phytoremediation Transformation and Control of Contaminants*. Hoboken, NJ: Wiley.

Memon, A.R. and Schroder, P. (2009). Implications of metal accumulation mechanisms to phytoremediation. *Environ. Sci. Pollut. Res.* 16: 162–175.

Meryem, S.S. and Yasmin, A. (2013). Effects of lead resistant bacteria on the early growth of *Vigna Mungo* L. (Hepper) under lead stress. *Pak. J. Bot.* 45 (S1): 481–485.

Mesjasz-Przybyłowicz, J., Nakonieczny, M., Migula, P. et al. (2004). Uptake of cadmium, lead, nickel and zinc from soil and water solutions by the nickel hyperaccumulator Berkheya coddii. *Acta Biol. Cracoviensia Ser. Botanica.* 46: 75–85.

Miller, R. (1996). Phytoremediation. Technology Overview Report, Ground-Water Remediation Technologies Analysis Center, Series Os, 3.

Milner, M.J. and Kocian, L.V. (2008). Investigating heavy-metal hyperaccumulation using Thlaspi caerulescens as a model system. *Ann. Bot. (Lond.)* 102: 3–13. https://doi.org/10.1093/aob/mcn063.

Muszynska, E., Hanus-Fajerska, E., and Ciarkowska, K. (2015). Studies on *Gypsophila fastigiata* parameters verifying its suitability to reclamation of post-flotation Zn-Pb wastes. *Geol. Geophys. Environ.* 41 (1): 17–24.

Nakamura, M., Ochiai, T., Noji, M. et al. (2014). An improved tolerance to cadmium by overexpression of two genes for cysteine synthesis in tobacco. *Plant Biotechnol.* 31: 141–147.

Neha, K.D., Shukla, P., Kumar, S. et al. (2017). Metal distribution in the sediments, water and naturally occurring macrophytes in the river Gomti, Lucknow, Uttar Pradesh. *India. Curr. Sci.* 113 (8): 1578–1585.

Notenboom, J., Verschoor, A., van der Linden, A. et al. (1998). Pesticides in groundwater: occurrence and ecological impacts. RIVM Report 601506002, Bilthoven.

Nouri, J., Khorasani, N., Lorestani, B. et al. (2009). Accumulation of heavy metals in soil and uptake by plant species with phytoremediation potential. *Environ. Earth Sci.* 59: 315–323.

Novakova, M., Mackova, M., Chrastilova, Z. et al. (2010). Cloning the bacterial bph C gene into *Nicotiana tabacum* to improve the efficiency of phytoremediation of polychlorinated biphenyls. *Biotechnol. Bioeng.* 1: 419–423.

Nugroho, B.Y.H., Wulandari, S.Y., and Ridlo, A. (2014). Analisis Residu Pestisida Organofosfat Di Perairan Mlonggo Kabupaten Jepara. *J. Oseanogr.* 4 (3): 541–544.

Olgun, E.J. and Sanchez-Galvan, G. (2012). Heavy metal removal in phytofiltration and phyco remediation: the need to differentiate between bio adsorption and bioaccumulation. *Nat. Biotechnol.* 30 (1): 3–8.

O'Toole, C. and Irvine, K. (2006). *Response of Aquatic Biota to Combined Pressure from Nutrients and Priority Substances. Indicators and Methods for the Ecological Status Assessment under the Water Framework Directive: Linkages between Chemical and Biological Quality of Surface Waters*. Luxembourg: Institute for Environment and Sustainability.

Padmavathiamma, P.K. and Li, L.Y. (2007). Phytoremediation technology: hyper-accumulation metals in plants. *Water Air Soil Pollut.* 184: 105–126.

Papendick, R.I., Elliot, L.F., and Dahlgren, R.B. (1986). Environmental consequences of modern production agriculture: how can alternative agriculture address these issues and concerns? *Am. J. Altern. Agric.* 1: 3–10.

Pell, M., Stenberg, B., and Torstensson, L. (1998). Potential denitrification and nitrification tests for evaluation of pesticide effects in soil. *Ambio* 27: 24–28.

Peters, K., Bundschuh, M., and Schäfer, R.B. (2013). Review on the effects of toxicants on freshwater ecosystem functions. *Environ. Pollut.* 180: 324–329.

Power, E.A., Chapman, P.M., and Burton, G.A. (1992). Assessing sediment quality. In: *Sediment Toxicity Assessment* (ed. G.A. Burton Jr.), 1–16. Lewis Publishers.

Prasad, M.N.V. and De Oliveira Freitas, H.M. (2003). Metal hyperaccumulation in plants—biodiversity prospectings for phytoremediation technology. *Electron. J. Biotechnol.* 6 (3): 110–146.

Racke, K.D., Skidmore, M.W., Hamilton, D.J. et al. (1997). Pesticide fate in tropical soils. *Pure Appl. Chem.* 69: 1349–1372.

Rahman, M.S., Belal Hossain, M.B., Babu, S.M.O.F. et al. (2019). Source of metal contamination in sediment, their ecological risk, and phytoremediation ability of the studied mangrove plants in ship breaking area, Bangladesh. *Mar. Pollut. Bull.* 141: 137–146.

Ramos, M.L.G. and Ribeiro, W.O.J. (1993). Effect of fungicides on survival on rhizobium on seeds and the nodulation of bean (*Phaseolus vulgaris* L.). *Plant Soil* 152: 145–150.

Rawat, K., Fulekar, M.H., and Pathak, B. (2012). Rhizofiltration: a green technology for remediation of heavy metals. *Int. J. Innovation Biosci.* 2 (4): 193–199.

Reinecke, S.A. and Reinecke, A.J. (2007). The impact of organophosphate pesticides in orchards on earthworms in the Western Cape, South Africa. *Ecotoxicol. Environ. Saf.* 66 (2): 244–251.

Relyea, R.A. and Hoverman, J. (2006). Assessing the ecology in ecotoxicology: a review and synthesis in freshwater systems. *Ecol. Lett.* 9: 1157–1171.

Revellin, C., Canson, B., and Catroux, G. (1992). Effect of a mixture of chlorpyrifos and lindane on the symbiosis of *Bradyrhizobium japonicum* and soybean. *Pestic. Sci.* 36: 69–74.

Rodriguez, L., Lopez-Bellido, F.J., Carnicer, A. et al. (2005). Mercury recovery from soils by phytoremediation. In: *Book of Environmental Chemistry* (ed. E. Lichtfouse, J. Schwarzbauer, and D. Robert), 197–204. Berlin, Germany: Springer.

Ruppert, JLW., Travers, MJ., Smith, LL., Fortin, M-J., Meekan, MG. (2013). Caught in the middle: combined impacts of shark removal and coral loss on the fish communities of coral reefs. *PLoS One* 8(9): e74648. https://doi.org/10.1371/journal.pone.0074648

Salt, D.E., Blaylock, M., Kumar, N.P.B.A. et al. (1995). Phytoremediation: a novel strategy for the removal of toxic metals from the environment using plants. *Biotechnology* 13: 468–474.

Salt, D.E., Smith, R.D., and Raskin, I. (1998). Phytoremediation. *Annu. Rev. Plant Physiol. Plant Mol. Biol.* 49: 643–668.

Santos, A. and Flores, M. (1995). Effects of glyphosate on nitrogen fixation of free-living heterotrophic bacteria. *Lett. Appl. Microbiol.* 20: 349–352.

Sarma, H. (2011). Metal hyperaccumulation in plants: are views focusing on phytoremediation technology. *J. Environ. Sci. Technol.* 4: 118–138.

Scholz, N.L., Fleishman, E., Brown, L. et al. (2012). A perspective on modern pesticides, pelagic fish declines, and unknown ecological resilience in highly managed ecosystems. *Bioscience* 62 (4): 428–434.

Schreck, E., Geret, F., Gontier, L., and Treilhou, M. (2008). Neurotoxic effect and metabolic responses induced by a mixture of six pesticides on the earthworm *Aporrectodea caliginosa* nocturna. *Chemosphere* 71 (10): 1832–1839.

Schroeder, J.I., Delhaize, E., Frommer, W.B. et al. (2013). Using membrane transporters to improve crops for sustainable food production. *Nature* 497: 60–66.

Senesi, N. (1992). Metal-humic substances complexes in the environment. Molecular and mechanistic aspects by multiple spectroscopic approaches. In: *Biogeochemistry of Trace Metals* (ed. C. Adriano Domy), 425–491. Boca Raton, FL: CRC Press.

Shanker, A.K., Cervates, C., Loza-Tavera, H., and Avudainayagam, S. (2005). Chromium toxicity in plants. *Environ. Int.* 31: 739–753.

Sharma, S., Singh, B., and Manchanda, V.K. (2015). Phytoremediation: role of terrestrial plants and aquatic macrophytes in the remediation of radionuclides and heavy metal contaminated soil and water. *Environ. Sci. Pollut. Res.* 22: 946–962.

Shirzadi, A., Simpson, M.J., Xu, Y., and Simpson, A.J. (2008). Application of saturated transfer double difference NMR to elucidate the mechanistic interactions of pesticides with humic acid. *Environ. Sci. Technol.* 42: 1084–1090.

Sinha, R.K., Herat, S., and Tandon, P.K. (2004). Phytoremediation: role of plants in contaminated site management. In: *Environmental Bioremediation Technologies* (ed. S.N. Singh and R.D. Tripathi), 315–330. Berlin, Germany: Springer 10.1007/978-3-540-34793-4_14.

Sivarajasekar, N., Mohanraj, N., Sivamani, S., et al. (2018). Comparative modeling of fluoride biosorption onto waste *Gossypium hirsutum* seed microwave-bichar using response surface methodology and artificial neural networks. *IEEE Xplore*, 1631–1635.

Song, W.Y., Park, J., Eisenach, C. et al. (2014). ABC transporters and heavy metals. In: *Plant ABC transporters*, 1–17. New York: Springer International Publishing.

Spaczynski, M., Aleksandra, S.K., Paweł, P. et al. (2012). Phytodegradation and biodegradation in rhizosphere as efficient methods of reclamation of soil contaminated by organic chemicals (a review). *Acta Agrophysica* 19: 155–169.

Suthersan, S.S. and McDonough, J. (1996). *Remediation Engineering: Design Concepts*. Boca Raton, FL: CRC Press, Kj l4k4klllugnlkjUEJ X.

Tao, L. and Yang, H. (2011). Fluroxypyr biodegradation in soils by multiple factors. *Environ. Monit. Assess.* 175: 227–238.

Taylor, A.W. (1995). The volatilization of pesticide residues. In: *Environmental Behaviour of Agrochemicals* (ed. T.R. Roberts and P.C. Kearney), 257–306. Chichester: J. Wiley & Sons Ltd.

Traunfeld, J.H. and Clement, D.L. (2001). *Lead in Garden Soils*. Home and Garden: Maryland Cooperative Extention, University of Maryland.

Tullberg, J. (2010). Reduce soil damage for more sustainable crop production. *Nature* 466: 920–920.

U. S. Department of Energy, Plume Focus Area (1994). Mechanisms of plant uptake, translocation, and storage of toxic elements. Summary Report of a workshop on phytoremediation research needs.

Van Dijk, T.C., Van Staalduinen, M.A., and Van der Sluijs, J.P. (2013). Macro invertebrate decline in surface water polluted with imidacloprid. *PLoS One* 8: e62374.

Vangronsveld, J., Herzig, R., Weyens, N. et al. (2009). Phytoremediation of contaminated soils and ground water: lessons from the field. *Environ. Sci. Pollut. Res.* 16: 765–794.

Vazquez, S., Agha, R., Sarro, M. et al. (2006). Use of white lupin plant for phytostabilization of Cd and As polluted acid soil. *Water Air Soil Pollut.* 177: 349–365. https://doi.org/10.1007/s11270-006-9178-y.

Wang, M., Webber, M., Finlayson, B., and Barnett, J. (2008). Rural industries and water pollution in China. *J. Environ. Manage.* 86 (4): 648–659. https://doi.org/10.1016/j.jenvman.2006.12.019.

Wasim, A., Dwaipayan, S., and Ashim, C. (2009). Impact of pesticides use in agriculture, their benefits and hazards. *Interdiscip. Toxicol.* 2: 1–12.

Windham, L., Weis, J.S., and Weis, P. (2003). Uptake and distribution of metals in two dominant salt marsh macrophytes, *Spartina alterniflora* (cordgrass) and *Phragmites australis* (common reed). *Estuar. Coast. Shelf Sci.* 56: 63–72.

Wright, D.J. and Otte, M.L. (1999). Wetland plant effects on the biogeochemistry of metals beyond the rhizosphere. *Biol. Environ.* 99: 3–10.

Wuana, R.A. and Okieimen, F.E. (2011). Heavy metals in contaminated soils: a review of sources, chemistry, risks and best available strategies for remediation. *ISRN Ecol.* 2011: 402–647.

Xia, H. and Ma, X. (2006). Phytoremediation of ethion by water hyacinth (*Eichhorniacrassipes*) from water. *Bioresour. Technol.* 97: 1050–1054.

Yang, Y.H., Yao, J., Hu, S., and Qi, Y. (2000). Effects of agricultural chemicals on DNA sequence diversity of soil microbial community: a study with RAPD marker. *Microb. Ecol.* 39: 72–79.

Yasmin, S. and D'Souza, D. (2010). Effects of pesticides on the growth and reproduction of earthworm: a review. *Appl. Environ. Soil Sci.* 2010: 678360.

Ye, Z.H., Whiting, S.N., Lin, Z.Q. et al. (2001). Removal and distribution of iron, manganese, cobalt and nickel within a Pennsylvania constructed wetland treating coal combustion by-product leachate. *J. Environ.* Q30: 1464–1473.

13

Biotechnological Interventions for Removal of Heavy Metals and Metalloids from Water Resources

Munir Ozturk[1], *Bengu Turkyilmaz Unal*[2], *and Huseyin Turker*[2]

[1] *Botany Department and Centre for Environmental Studies, Ege University, Izmir, Turkiye*
[2] *Art and Sciences Faculty, Biotechnology Department, Nigde Omer Halisdemir University, Nigde, Turkiye*

13.1 Introduction

Environmental pollution, which poses a great threat to our future, is one of the most serious problems of our century (Gündüz 2008). The environment has been exposed to many pollutants of natural or anthropogenic origin that adversely affect biological systems.

Global environmental pollution continues to increase with industrial developments, rapid population growth, urbanization, and especially the release of chemical wastes into the environment.

One of the areas where environmental pollution is felt most is water resources. Large quantities of heavy metal and metalloids (HMMs) are released into nature due to mining, iron and steel factories, automotive industry, paint industry, textile industry, petroleum chemical industry, and fertilizer industry. The increase in the accumulation of heavy metals in water causes many problems (Galiatsatou et al. 2002). Toxic heavy metal accumulation causes serious damage to both the environment and living things (Ozturk et al. 2021b).

People can take heavy metals directly through drinking water or indirectly through plants exposed to sewage water, fertilizers, or industrial wastes. These have very significant harmful effects on humans (Vural and Güvendik 1983; Shamsul et al. 2021; Ozturk et al. 2021a). Removal of heavy metals in water resources is of vital importance (Joseph et al. 2019).

Chemical and physical methods such as adsorption, electrodialysis, ion exchange, precipitation, and reverse osmosis can be applied for heavy metal-containing water removal (Saxena et al. 2017). The biotechnological methods are less costly, easier, and environmentally friendly (Mani and Kumar 2014; Haq et al. 2020).

In this chapter, biotechnological methods used in the purification of water from heavy metals will be discussed.

13.2 Water Pollution

Water is very important for being the best solvent, being used in biochemical reactions, sustaining life, and being the habitat of many organisms. It also plays a role in maintaining biodiversity, human existence, and socioeconomic status (Aldieri et al. 2020).

Anthropogenic factors such as agricultural activities, industrial and domestic wastes, mining, refinery wastes, using of fossil fuels, have been the main causes of increasing pollution in the world. Water pollution mostly occurs in the form of acidification, heavy metal, and eutrophication (Kurt 2019).

Heavy metals and their salts are among the most important water pollutants. Heavy metals can be mixed with water from various sources (Figure 13.1). Although many harmful substances can be removed from the environment or biodegraded, they accumulate at a high rate and adversely affect many biological and chemical processes (Artan and Keskinkan 2008).

Aquatic Contamination: Tolerance and Bioremediation, First Edition. Edited by Rouf Ahmad Bhat, Gowhar Hamid Dar, Fernanda Maria Policarpo Tonelli, and Saima Hamid.

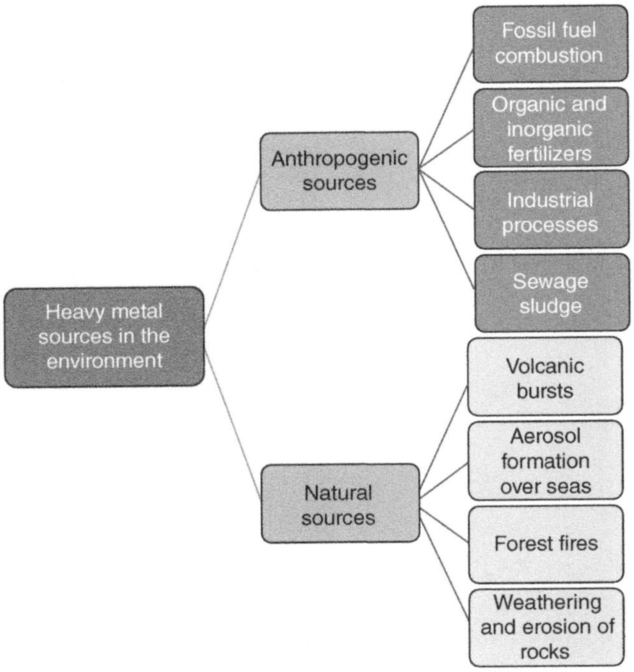

Figure 13.1 Heavy metal sources causing environmental pollution.

13.3 Heavy Metals and Metalloids

The heavy metal term is used for metals and metalloids with a density higher than 5 g/cm^3 and has toxic effects even at low concentrations (Järup 2003). This name is given to metals and metalloids that pollute the environment biologically and have a high toxicity effect (Duffus 2001). The US Environmental Protection Agency has listed heavy metals among 129 priority environmental pollutants (Batir 2014).

Some of the heavy metals (such as Zn and Fe) are essential metals required for the continuity of life activities. Some heavy metals and metalloids such as cadmium, lead, and mercury do not have any function in living things and show toxic effects even at very low doses. These are called nonessential metals. However, essential metals can also have toxic effects at high concentrations (Järup 2003; Özbolat and Abdullah 2016; Ergönül and Atasagun 2017; Ghori et al. 2019).

Heavy metals have long half-lives and are not biodegradable (Tripathi and Ranjan 2015). It can accumulate in the water, soil, sediment, or living tissues (Fisher 1995) and threatens human health through the food chain (Taylan and Özkoç 2007; Jaishankar et al. 2014).

Heavy metals not only pose a great threat to aquatic organisms but also pose a significant risk because they limit the use of water resources. For this reason, many studies have been carried out recently on the removal of heavy metals from aquatic environments (Raghav and Shrivastava 2016) (Table 13.1).

13.4 Effects of Heavy Metals and Metalloids on Water Pollution

HMMs enter the waters through agricultural, domestic, and industrial wastes. In addition to causing great damage to the environment, they accumulate in living things by being included in the food chain and cause damage to biological systems (Gavrilescu 2004).

Heavy metals are emitted into the environment as a result of industrial activities such as cement production, glass production, iron and steel industry, thermal power plants, garbage and waste sludge incineration plants (Horsfall Jnr and Spiff 2005), mining activities (Kurt 2019) and agricultural activities such as pesticide usage.

Aquatic organisms constitute an important place in the human diet. Humans live a large part of their heavy metal intake as a result of long-term exposure of consumed aquatic organisms to these pollutants. The risk is very high as aquatic organisms are in direct contact with heavy metal-contaminated media (Ay 2019).

Table 13.1 Ecological status of some heavy metals and metalloids.

Elements	Pollution status	The necessity for living things	Specific gravity (g/cm^3)
As	✓	X	2.3
B	✓	✓	2.3
Cd	✓	X	13.5
Co	✓	✓	13.9
Cr	✓	✓	7.2
Cu	✓	✓	13.9
Fe	✓	✓	7.9
Hg	✓	X	13.6
Mn	X	✓	7.4
Mo	✓	✓	10.2
Ni	✓	✓	13.9
Pb	✓	X	11.3
Zn	✓	✓	7.1

Since most of the bioaccumulative substances are soluble in fat, they are retained in animal fat tissues, most of them are not biodegradable or have a long half-life, and their excretion from the body is insufficient and causes undesirable accumulations in living tissues. It is also known that accumulation increases as you go up the food chain (Uysal and Taner 2010).

Heavy metals and metalloids can cause toxic effects and even death on aquatic organisms (Bhat et al. 2022). Toxicity in fish and many aquatic organisms that people prefer in their diets varies according to the heavy metal type, density, and physicochemical effects in water (Kurt 2019).

13.5 Heavy Metal and Metalloids Removal

Heavy metals and metalloids are resistant to decomposition and cannot be easily removed from the environment (Tripathi and Ranjan 2015). Their removal is generally achieved by using physical, chemical, and biological methods (Fu and Wang 2011). Precipitation and filtration, which are among the most important physical methods, are used alone or together with chemical methods (adsorption, ion exchange, and reverse osmosis). However, these systems have high operating and maintenance costs. They are also specific for the treatment of certain heavy metals (Mishra and Tripathi 2008) and are rather used for the purification of highly polluted waters (Fu and Wang 2011). In addition, substances used in chemical methods have the potential to pose a risk to living things.

Biological methods such as phytoremediation, in which plants are used as an alternative and promising technology for HMMs removal, are environmentally friendly and low-cost methods (Öztürk et al. 2015). It has been frequently used for heavy metal removal in moderately polluted waters (Miretzky et al. 2004). The method, which was used for the first time in the United States in 1972 for the removal of pollution caused by fuel pipeline leakage, became the focus of attention when it succeeded (Wang et al. 2012). Bioremediation refers to the pollutants removal from the medium using microorganisms, macrophytes, algae or other organisms (Kulshreshtha et al. 2014). Although bacteria, fungi and archaea are frequently used in this process, the potential of various algae and macrophytes in the removal of heavy metals in water has been investigated recently (Zeraatkar et al. 2016).

13.6 Bioremediation in Pollution Management

Thanks to biotechnological methods, renewable resources can be used and the amount of industrial waste can be reduced (Stottmeister et al. 2005). The bioremediation method has great advantages over traditional ones (Boopathy 2000). It is presented as a low-cost, efficient, and environmentally friendly solution (Deshmukh et al. 2016).

Bioremediation is the method of using micro (bacteria, fungi, etc.) and macro (plants, earthworms) organisms for the removal of pollution of polluted water in an environmentally friendly manner (Anonymous 2015). This method includes long-term treatment processes in which these organisms are used in order to break down dangerous and harmful substances into harmless or less harmful substances (Dua et al. 2002). Appropriate microorganisms break down harmful substances by performing various physical and chemical reactions (Mishra et al. 2014).

Bioremediation has many advantages as well as disadvantages. Bacteria and fungi play significant roles in bioremediation (Strong and Burgess 2008), although only a few have a wide range of influence. In addition, long periods are required for bioremediation to take effect. It takes several years or more to improve an area (Chowdhury et al. 2012). The process can be speeded up by using genetic manipulation techniques to obtain advanced new breeds, strengthen them biochemically or increase the bioremediant population.

For bioremediation to take place effectively, the organisms that provide bioremediation must enter into enzymatic reactions with pollutants. Therefore, these creatures should be in close contact with pollutants (Dindar et al. 2010). During bioremediation, nutrients, electron acceptors, pH and temperature should be regulated (Baker and Herson 1994). In addition, this method should encourage the growth of living things that are effective in bioremediation (Ceyhan and Esmeray 2012). If the population of living things is insufficient, some engineering methods should be used to bring them into contact with pollutants or to increase their population (Singh and Ward 2004). For example, the nutrient transfer can be made to the area where the pollutants are located and thus the population can be supported (Ceyhan and Esmeray 2012). Environmental conditions can be controlled and conditions can be changed to optimize the growth of living things.

Many cities in the world use microorganisms with bioremediation methods in hazardous waste management, and the number of cities using bioremediation has been increasing in recent years. For example, in the United States, bioremediation is used in more than 100 sites by EPA projects. In particular, bioremediation is being tried to be developed with bioengineering applications and it is expected that the techniques applied will increase in parallel with the increasing number of toxic substances (Alexander 1999).

Biodegradation, biostimulation, blastofiltration, cyanoremediation, biosparging, and mycoremediation, etc. to improve polluted waters. Biotechnological interventions can be used. In order to prevent the passage of pollutants into the aquatic environment, biotechnological tools such as biological fuel cells and biosensors are important in ensuring preferable sustainability (Dervash et al. 2020).

Below are some techniques used in bioremediation.

13.6.1 Biodegradation

It is the process of microorganisms breaking down organic compounds with the help of metabolic or enzymatic processes (Gaur et al. 2018; Bhatt et al. 2019; Touliabah et al. 2022). Organic materials can be decomposed aerobically in the O_2 presence and anaerobically in the O_2 absence (Bhat et al. 2017a; Jain et al. 2022). Factors such as humidity, nutrients, oxygen, pH, soil properties and temperature can limit the biodegradation of organic materials in the environment (Haripriyan et al. 2022).

13.6.2 Biosorption

Biosorption; It is a biotechnological method that enables the removal of metal or metalloid species from polluted water using various biomaterials (Mustapha and Halimoon 2015). Bacteria, algae, plants and fungi can be used as biosorbents for the removal of HMMs (Kumar et al. 2014; Mustapha and Halimoon 2015). Biosorption advantages include ease and cheapness, high efficiency, and recovery of metals (Azouaou et al. 2008; Hamutoglu et al. 2012). One of its disadvantages is that the cell surface is quickly saturated with metal and the valency of the metal cannot be (Hamutoglu et al. 2012).

Bacterial biosorption is basically an effective method for removing pollutants from wastewater that has been exposed to strong pollutants such as metal ions and dyes. Different bacterial strains are used (Mustapha and Halimoon 2015).

Algae are inexpensive and competent biosorbents with low nutritional requirements and high absorption compared to other biosorbents (Mustapha and Halimoon 2015). Biosorption of metals takes place on the cell surface in the form of ion exchange. For example, Brown algae use amino, carboxyl, sulfhydryl, and sulfonates on their surfaces for the absorption of Cd, Ni, and Pb (Mustapha and Halimoon 2015). Various fungi (Li et al. 2015) and lichens are also known to have very effective biosorption capacity (Ates et al. 2007; Tay et al. 2009).

The amount of metal ions bound to the biomass during biosorption depends not only on the type of biomass, but also on the amount of biomass, pH, temperature, etc. It has been stated that it depends on physicochemical factors (Volesky 2007).

13.6.3 Mycoremediation

It is basically the healing of polluted areas through fungi. It refers to fungi and their enzymes, which have the ability to decompose numerous environmentally persistent pollutants and convert agro-industrial and industrial wastes into products (Rhodes 2014). *Aspergillus*, *Trichoderma*, *Fusarium*, and *Penicillium* species have been reported to have the potential for remediation in contaminated wastewater by many heavy metals (Chen et al. 2017; Dhami et al. 2017; Pundir et al. 2018; Kumar and Dwivedi 2021a). The most commonly used fungal species are *Aspergillus niger*, *Penicillium* spp., *Funalia trogii*, *Cladosporium resinae*, *Aureobasidium pullulans*, and *Ganoderma lucidum* (Loukidou et al. 2003; Say et al. 2003). *Trametes versicolor* and *Rhizopus arrhizus* are well-known polluted water heavy metal recoverers (Dervash et al. 2020).

Fungi can be used in two ways (growing/living form or dead/dry biomass form) in mycoremediation. In the living form, the fungus has the ability to tolerate heavy metals, while the surface functional groups of the biomass are effective in the removal of heavy metals in the form of dead or dry biomass (Kumar and Dwivedi 2021b).

In addition, many researchers have emphasized that Arbuscular mycorrhizal fungi have the potential to reduce heavy metal pollution in wastewater-irrigated soils (Bhat et al. 2017b; Elhindi et al. 2018; Xu et al. 2018; Kumar and Saxena 2019; Wahid et al. 2022). However, few data on mycorrhizal species associated with heavy metal accumulator plants are available and more detailed studies are needed (Kumar and Saxena 2019).

13.6.4 Cyanoremediation

The use of cyanobacteria and some algae species to remove heavy metals from polluted waters and wetlands is expressed as cyanoremediation (Kumar and Singh 2017; Yaashikaa et al. 2022).

Anabaena, *Chlorella*, *Hydrodictylon*, *Nostoc*, *Oedogonium*, *Rhizoclonium*, *Spirulina*, and *Spirogyra* can efficiently remove heavy metals from wastewater (Saunders et al. 2012; Yaashikaa et al. 2022).

Cyanoremediation consists of two stages (Singh et al. 2019): (i) adsorption by extracellular components involving physical adsorption, and (ii) adsorption and deposition inside cells. The outer layer of cyanobacteria consists of an extra polymeric structure with various binding sites. They can also synthesize metal-binding proteins (metallothioneins) that can bind to cysteine ligands with heavy metals (Bind et al. 2019).

The fact that cyanobacteria can live in extreme conditions and clean the environment provides a great advantage. In addition, their phototropicity, nitrogen fixation, and bioenergy-generating abilities are important for a sustainable environment (Yaashikaa et al. 2022). Cyanobacteria-mediated heavy metal removal, which is an in situ, easy-to-apply, inexpensive, and green chemistry approach has a high potential for use (Al-Amin et al. 2021).

13.6.5 Phytoremediation

The term phytoremediation was first used in 1991 and it is the technology that enables to improve the environment using plants. According to another definition, it is an important chelation technique that is effective in removing heavy metal pollution in both terrestrial and aquatic ecosystems (Vangronsveld et al. 2009). Different names are used, such as "botanical remediation," "phytoremediation," or "green remediation" (US-ERPA 2000).

Phytoremediation is the removal of pollutants such as HMMs from the environment by using mako algae or plants or the elimination of their toxic effects (Ashraf et al. 2010; Caparrós et al. 2022; Parnian et al. 2022). The use of phytoremediation techniques is increasing due to their low cost, being usable in areas with more than one pollutant, and being environmentally friendly. It can also be used as an ore to obtain precious heavy metals. However, it is a technique that has disadvantages such as the fact that the process can be done in a limited area (due to the depth of the roots and the area it covers), the plants are degradable, and the importance of plant tolerance used (Sood et al. 2012).

In recent years, aquatic macrophytes have started to be used more frequently in the remediation of wastewater contaminated with heavy metals and natural waters. It was stated by Vasavi et al. (2010) that the removal of Cd, Cu, Hg, Ni, Zn, and many other heavy metals can be achieved with techniques of phytoremediation (Nassouhi et al. 2018).

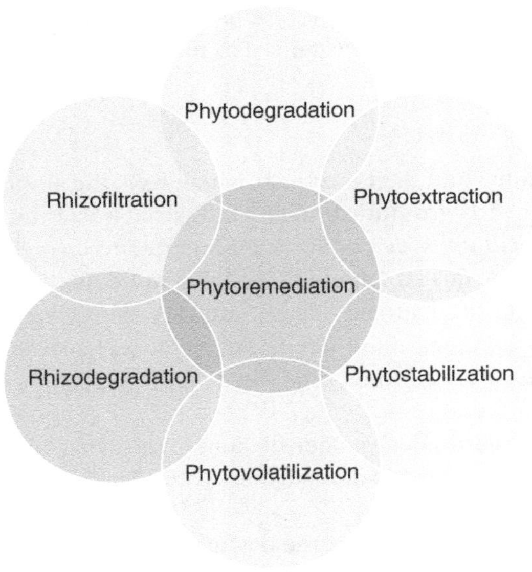

Figure 13.2 Various mechanisms used in phytoremediation.

Macrophytes used in phytoremediation use various mechanisms such as phytodegradation, phytostabilization, phytoextraction, phytovolatilization, rhizodegradation, and rhizofiltration (Tangahu et al. 2011; Tiryaki and Potur 2017) (Figure 13.2). Phytoextraction is the removal of pollutants from the environment by the harvesting of these tissues by being taken by the roots and transmitted to the stem and leaves; Phytostabilization; Preventing the transport of pollutants by adsorbing them by the roots is rhizofiltration. Phytodegradation has limited use and complex pollutants are converted into simple compounds and transported to tissues. Phytovolatilization, on the other hand, is the release of chemicals from the leaves to the atmosphere, which are in gaseous form or that the plants convert into gaseous form by metabolic processes. In rhizodegradation, transformation processes are carried out by microorganisms in the plant rhizosphere (Nassouhi et al. 2018).

In phytoremediation, the plant's tolerance to heavy metals in the environment in which it will be used, its ability to accumulate, transport and remove metals, root structure and depth, biomass yield, growth rate, and resistance to extreme conditions should be considered in plant selection (Ay 2019).

13.7 Biosensors

Biosensors consist of a biological sensing element, a transducer, and a signal processing system (Figure 13.3). The analyte is identified by a bioreceptor and then converted into a measurable signal by the transducer (Nigam and Shukla 2015).

In most of the bacterial biosensors used for the detection of HMMs, genes resistant to these metals act as bioreceptors. Bacterial bioreceptors are preferred for the detection of cobalt, copper, mercury, silver, tin, zinc, etc. Some biosensors, on the other hand, consist of a combination of genes expressing bioluminescent proteins such as luciferin and genes for the resistance mechanism. In addition, heavy metals can be detected by enzyme-catalyzed reactions since most of the ions inhibit the enzyme even at low concentrations (Nigam and Shukla 2015).

Conventional detection methods, which vary according to different times and sources, are not effective in monitoring a great number of pollutants in wastewater. Biosensors, on the other hand, are in greater demand owing to their high sensitivity, low cost, real-time response, reliability, selectivity, and simplicity (Ejeian et al. 2018).

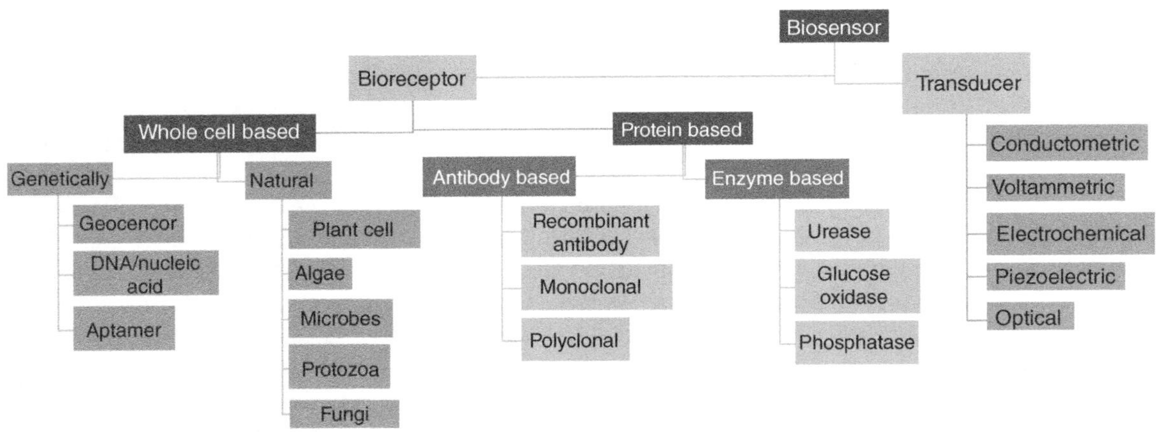

Figure 13.3 Some types of biosensors.

13.8 Biotechnological Methods Used in the Removal of HMMs

Bioremediation is a cost-effective and environmentally friendly method, but there is little information on the factors that regulate the metabolism and growth of microorganisms in a polluted environment, resulting in limited use of this technique.

In order to eliminate these limitations, the number of biotechnological, genetic, and proteomic studies has increased in recent years. In this way, the genes, promoters, etc., of microorganisms. Genetically modified microorganisms were obtained and used in bioremediation.

For example, *Geobacter metallireducens* has been used to remove uranium from contaminated groundwater and drainage waters (Verma and Kuila 2019). For heavy metals, the *Arabidopsis thaliana* gene for phytochelatin synthesis was transferred to *Mesorhizobium huakuii* subsp. Rengei strain B3 (Sriprang et al. 2003). When used against Cd toxicity from the genetically modified organisms *Chlamydomonas reinhardtii* (Ibuot et al. 2017). *Corynebacterium glutamicum* has been genetically modified for the decontamination of arsenic (Mateos et al. 2017). It has been used against *Escherichia coli* to arsenic III (Kostal et al. 2004) and *Saccharomyces cerevisiae* to cadmium and zinc (Vinopal et al. 2007).

The use of genetically modified algae, other aquatic macrophytes, and plants can increase heavy metal removal efficiency. Genetically modified organisms can reduce heavy metal contamination in wastewater by using genes related to the cation diffusion facilitator family (Mani and Kumar 2014) or by using a triphosphate-binding cassette (Song et al. 2014). It is thought that the integration of bacterial symbiosis, genetic engineering, and chelate-assisted approach will provide more effective phytoremediation (Das et al. 2022).

13.9 Conclusion

Heavy metal accumulation and toxicity in the waters, which are increasing day by day, both affect natural ecosystems and cause significant damage to living things.

Biotechnological interventions that are environmentally friendly, easy to implement, and attract attention with their low cost are promising. Biodegradation, biostimulation, blastofiltration, cyanoremediation, biosparging, mycoremediation, and biotechnological interventions can be used to improve heavy metal waters.

The use of biosensors can be encouraged to determine the metal presence and concentration and to take necessary precautions.

References

Al-Amin, A., Parvin, F., Chakraborty, J. et al. (2021). Cyanobacteria mediated heavy metal removal: a review on mechanism, biosynthesis, and removal capability. *Environ. Technol. Rev.* 10: 44–57.

Aldieri, L., Makkonen, T., and Vinci, C.P. (2020). Environmental knowledge spillovers and productivity: a patent analysis for large international firms in the energy, water and land resources fields. *Res. Policy* 69: 101877.

Alexander, M. (1999). *Biodegradation and Bioremediation*. Gulf Professional Publishing.

Anonymous (2015). https://www.teachengineering.org/view_lesson.php?url=collection/cub_/lessons/cub_bio/cub_bio_lesson06.xml. Accessed 15 March 2018.

Artan, R. and Keskinkan, O. (2008). Use of duckweed (Lemna minor) plant in advanced treatment of cadmium-containing wastewater. *Cukurova Univ. J. Grad. Sch. Nat. Appl. Sci.* 19 (2): 11–19.

Ashraf, M., Ozturk, M., and Ahmad, M.S.A. (2010). *Plant Adaptation and Phytoremediation*. Springer.

Ates, A., Yildiz, A., Yildiz, N. et al. (2007). Heavy metal removal from aqueous solution by *Pseudevernia furfuracea* (L.) *Zopf. Ann. Chim.* 97: 385–393.

Ay, İ. (2019). Phytoremediation technique in heavy metal chelation and effective macrophytes in practice. *Mediterr. Fish. Aqua. Res.* 2: 77–82.

Azouaou, N., Sadaoui, Z., and Mokaddem, H. (2008). Removal of cadmium from aqueous solution by adsorption on vegetable wastes. *J. Appl. Sci.* 8: 4638–4643.

Baker, K. and Herson, D. (1994). Microbiology and biodegradation. In: *Bioremediation* (ed. K. Baker and D. Herson), 9–60. New York: McGraw Hill Inc.

Batir, M.B. (2014). *Determination of DNA Changes in Seedlings of Artichoke (Cynara scolymus L.) Seeds Subjected to Heavy Metal Stress of Lead (Pb) and Copper (Cu)*. ESOGU, Graduate School of Natural and Applied Sciences.

Bhat, R., Dervash, M., Mehmood, M. et al. (2017a). Municipal solid waste generation and its management, a growing threat to fragile ecosystem in Kashmir Himalaya. *Am. J. Environ. Sci.* 13 (6): 388–397. https://doi.org/10.31344/ajessp.

Bhat, R.A., Dervash, M.A., Mehmood, M.A. et al. (2017b). Mycorrhizae: a sustainable industry for plant and soil environment. In: *Mycorrhiza-Nutrient Uptake, Biocontrol, Ecorestoration* (ed. A. Varma, R. Prasad, and N. Tuteja), 473–502. Cham: Springer.

Bhat, R.A., Singh, D.V., Qadri, H. et al. (2022). Vulnerability of municipal solid waste: an emerging threat to aquatic ecosystems. *Chemosphere* 287: 132223.

Bhatt, P., Pathak, V.M., Joshi, S. et al. (2019). Major metabolites after degradation of xenobiotics and enzymes involved in these pathways. In: *Smart Bioremediation Technologies* (ed. P. Bhatt), 205–215. Elsevier.

Bind, A., Kushwaha, A., Devi, G. et al. (2019). Biosorption valorization of floating and submerged macrophytes for heavy-metal removal in a multi-component system. *Appl Water Sci* 9: 1–9.

Boopathy, R. (2000). Factors limiting bioremediation technologies. *Bioresour. Technol.* 74: 63–67.

Caparrós, P.G., Ozturk, M., Gul, A. et al. (2022). Halophytes have potential as heavy metal phytoremediators: a comprehensive review. *Environ. Exp. Bot.* 193: 104666.

Ceyhan, N. and Esmeray, E. (2012). Oil pollution and bioremediation. *Turk. J. Sci. Rev.* 1: 95–101.

Chen, S.H., Ng, S.L., Cheow, Y.L. et al. (2017). A novel study based on adaptive metal tolerance behavior in fungi and SEM-EDX analysis. *J. Hazard. Mater.* 334: 132–141.

Chowdhury, S., Bala, N., and Dhauria, P. (2012). Bioremediation–a natural way for cleaner environment. *Int. J. Pharm. Chem. Biol. Sci.* 2: 600–611.

Das, S., Das, S., and Ghangrekar, M.M. (2022). Efficacious bioremediation of heavy metals and radionuclides from wastewater employing aquatic macro-and microphytes. *J. Basic Microbiol.* 62: 260–278.

Dervash, M.A., Bhat, R.A., Shafiq, S. et al. (2020). Biotechnological intervention as an aquatic clean up tool. In: *Fresh Water Pollution Dynamics and Remediation* (ed. H. Qadri, R. Bhat, M. Mehmood, and G. Dar), 183–196. Singapore: Springer.

Deshmukh, R., Khardenavis, A.A., and Purohit, H.J. (2016). Diverse metabolic capacities of fungi for bioremediation. *Indian J. Microbiol.* 56: 247–264.

Dhami, N.K., Quirin, M.E.C., and Mukherjee, A. (2017). Carbonate biomineralization and heavy metal remediation by calcifying fungi isolated from karstic caves. *Ecol. Eng.* 103: 106–117.

Dindar, E., Topac, S.F., and Başkaya, H. (2010). Bioremediation of contaminated soil. *Uludağ Univ. J. Fac. Engin.* 15 (2): 123–137.

Dua, M., Singh, A., Sethunathan, N. et al. (2002). Biotechnology and bioremediation: successes and limitations. *Appl. Microbiol. Biotechnol.* 59: 143–152.

Duffus, J.H. (2001). "Heavy metals"–a meaningless term. *Chem. Int.* 23: 163–167.

Ejeian, F., Etedali, P., Mansouri-Tehrani, H.A. et al. (2018). Biosensors for wastewater monitoring: a review. *Biosens. Bioelectron.* 118: 66–79.

Elhindi, K.M., Al-Mana, F.A., El-Hendawy, S. et al. (2018). Arbuscular mycorrhizal fungi mitigates heavy metal toxicity adverse effects in sewage water contaminated soil on *Tagetes erecta* L. *Soil Sci. Plant Nutr.* 64: 662–668.

Ergönül, M.B. and Atasagun, S. (2017). The effects of chronic low level zinc (Zn) exposure on the hematological profile of tench, *Tinca tinca* L., 1758. In: *Trends in Fisheries and Aquatic Animal Health* (ed. P. Berillis), 324–333. Sharjah: Bentham Science Publisher.

Fisher, S. (1995). Mechanisms of bioaccumulation in aquatic systems. *Rev. Environ. Contam. Toxicol.* 142: 87–117.

Fu, F. and Wang, Q. (2011). Removal of heavy metal ions from wastewaters: a review. *J. Environ. Manage.* 92: 407–418.

Galiatsatou, P., Metaxas, M., and Kasselouri-Rigopoulou, V. (2002). Adsorption of zinc by activated carbons prepared from solvent extracted olive pulp. *J. Hazard. Mater.* 91: 187–203.

Gaur, N., Narasimhulu, K., and Pydisetty, Y. (2018). Recent advances in the bio-remediation of persistent organic pollutants and its effect on environment. *J. Clean. Prod.* 198: 1602–1631.

Gavrilescu, M. (2004). Removal of heavy metals from the environment by biosorption. *Eng. Life Sci.* 4: 219–232.

Ghori, N.H., Ghori, T., Hayat, M. et al. (2019). Heavy metal stress and responses in plants. *Int. J. Environ. Sci. Technol.* 16: 1807–1828.

Gündüz, T. (2008). *Environmental Chemistry*. Ankara, Turkey: Gazi Bookstore.

Hamutoglu, R., Dincsoy, A.B., Cansaran-Duman, D. et al. (2012). Biosorption, adsorption, phytoremediation methods and applications. *Turk. J. Hyg. Exp. Biol.* 69: 235–253.

Haq, S., Bhatti, A.A., Dar, Z.A. et al. (2020). Phytoremediation of heavy metals: an eco-friendly and sustainable approach. In: *Bioremediation and Biotechnology* (ed. K. Hakeem, R. Bhat, and H. Qadri), 215–231. Cham: Springer.
Haripriyan, U., Gopinath, K., Arun, J. et al. (2022). Bioremediation of organic pollutants: a mini review on current and critical strategies for wastewater treatment. *Arch. Microbiol.* 204: 1–9.
Horsfall, M. Jnr and Spiff, A.I. (2005). Effects of temperature on the sorption of Pb^{2+} and Cd^{2+} from aqueous solution by Caladium bicolor (wild cocoyam) biomass. *Electron. J. Biotechnol.* 8: 43–50.
Ibuot, A., Dean, A.P., Mcintosh, O.A. et al. (2017). Metal bioremediation by CrMTP4 over-expressing chlamydomonas reinhardtii in comparison to natural wastewater-tolerant microalgae strains. *Algal Res.* 24: 89–96.
Jain, P., Bhavani, A.G., Singh, P. et al. (2022). Fundamental of aerobic and anaerobic processes in dye wastewater. In: *Biological Approaches in Dye-Containing Wastewater* (ed. A. Khadir and S.S. Muthu), 39–55. Singapore: Springer.
Jaishankar, M., Tseten, T., Anbalagan, N. et al. (2014). Toxicity, mechanism and health effects of some heavy metals. *Interdiscip. Toxicol.* 7: 60.
Järup, L. (2003). Hazards of heavy metal contamination. *Br. Med. Bull.* 68: 167–182.
Joseph, L., Jun, B.M., Flora, J.R. et al. (2019). Removal of heavy metals from water sources in the developing world using low-cost materials: a review. *Chemosphere* 229: 142–159.
Kostal, J., Yang, R., Wu, C.H. et al. (2004). Enhanced arsenic accumulation in engineered bacterial cells expressing ArsR. *Appl. Environ. Microbiol.* 70: 4582–4587.
Kulshreshtha, A., Agrawal, R., Barar, M. et al. (2014). A review on bioremediation of heavy metals in contaminated water. *IOSR-JESTFT* 8: 44–50.
Kumar, V. and Dwivedi, S.K. (2021a). Bioremediation mechanism and potential of copper by actively growing fungus Trichoderma lixii CR700 isolated from electroplating wastewater. *J. Environ. Manage.* 277: 111370.
Kumar, V. and Dwivedi, S.K. (2021b). Mycoremediation of heavy metals: processes, mechanisms, and affecting factors. *Environ. Sci. Pollut. Res.* 28: 10375–10412.
Kumar, S. and Saxena, S. (2019). Arbuscular mycorrhizal fungi (AMF) from heavy metal-contaminated soils: molecular approach and application in phytoremediation. In: *Biofertilizers for Sustainable Agriculture and Environment* (ed. B. Giri, R. Prasad, Q.S. Wu, and A. Varma), 489–500. Cham: Springer.
Kumar, A. and Singh, J.S. (2017). Cyanoremediation: a green-clean tool for decontamination of synthetic pesticides from agro-and aquatic ecosystems. In: *Agro-Environmental Sustainability* (ed. J. Singh and G. Seneviratne), 59–83. Cham: Springer.
Kumar, R., Sharma, A.K., Singh, P. et al. (2014). Potential of some fungal and bacterial species in bioremediation of heavy metals. *J. Nucl. Phys. Mater. Sci. Radi. Appl.* 1 (2): 213–223.
Kurt, B. (2019). Bioremediation of some heavy metals (tin, cobalt, antimony, vanadium) in different concentrations using *Lemna minor* L. and *Lemna gibba* L. (duckweed) and determination of possible stress effects of heavy metals. Master's thesis. Institute of Science and Technology.
Li, X., Li, A., Long, M. et al. (2015). Equilibrium and kinetic studies of copper biosorption by dead *Ceriporia lacerata* biomass isolated from the litter of an invasive plant in China. *J. Environ. Health Sci. Eng.* 13: 1–8.
Loukidou, M.X., Matis, K.A., Zouboulis, A.I. et al. (2003). Removal of As (V) from wastewaters by chemically modified fungal biomass. *Water Res.* 37: 4544–4552.
Mani, D. and Kumar, C. (2014). Biotechnological advances in bioremediation of heavy metals contaminated ecosystems: an overview with special reference to phytoremediation. *Int. J. Environ. Sci. Technol.* 11: 843–872.
Mateos, L.M., Villadangos, A.F., Alfonso, G. et al. (2017). The arsenic detoxification system in corynebacteria: basis and application for bioremediation and redox control: *Adv. Appl. Microbiol.* 99: 103–137.
Miretzky, P., Saralegui, A., and Cirelli, A.F. (2004). Aquatic macrophytes potential for the simultaneous removal of heavy metals (Buenos Aires, Argentina). *Chemosphere* 57: 997–1005.
Mishra, V.K. and Tripathi, B. (2008). Concurrent removal and accumulation of heavy metals by the three aquatic macrophytes. *Bioresour. Technol.* 99: 7091–7097.
Mishra, A., Kumari, M., Pandey, S. et al. (2014). Biocatalytic and antimicrobial activities of gold nanoparticles synthesized by *Trichoderma* sp. *Bioresour. Technol.* 166: 235–242.
Mustapha, M.U. and Halimoon, N. (2015). Microorganisms and biosorption of heavy metals in the environment: a review paper. *J. Microb. Biochem. Technol.* 7: 253–256.
Nassouhi, D., Ergönül, M.B., Fikirdeşici, Ş. et al. (2018). Use of aquatic macrophytes in bioremediation of heavy metal pollution. *SDU-JEFF* 14: 148–165.

Nigam, V.K. and Shukla, P. (2015). Enzyme based biosensors for detection of environmental pollutants-a review. *J. Microbiol. Biotechnol.* 25: 1773–1781.

Özbolat, G. and Abdullah, T. (2016). Effects of heavy metal toxicity on human health. *Arch. Source Rev. J.* 25: 502–521.

Öztürk, M., Ashraf, M., Aksoy, A. et al. (2015). *Plants, Pollutants and Remediation*. Springer.

Ozturk, M., Metin, M., Altay, V. et al. (2021a). Arsenic and human health: genotoxicity, epigenomic effects, and cancer signaling. *Biol. Trace Elem. Res.* 200: 1–14.

Ozturk, M., Metin, M., Altay, V. et al. (2021b). Molecular biology of cadmium toxicity in *Saccharomyces cerevisiae*. *Biol. Trace Elem. Res.* 199: 4832–4846.

Parnian, A., Chorom, M., Jaafarzadeh, N. et al. (2022). Bioremediation of cadmium and nickel from a saline aquatic environment using *Ceratophyllum demersum*. *Chiang Mai J. Sci.* 49: 339–347.

Pundir, R., Chary, G., and Dastidar, M. (2018). Application of Taguchi method for optimizing the process parameters for the removal of copper and nickel by growing *Aspergillus* sp. *Water Resour. Ind.* 20: 83–92.

Raghav, N. and Shrivastava, J. (2016). Toxic pollution in river water and bacterial remediation: an overview. *Int. J. Curr. Microbiol. App. Sci.* 5: 244–266.

Rhodes, C.J. (2014). Mycoremediation (bioremediation with fungi)–growing mushrooms to clean the earth. *Chem. Speciat. Bioavailab.* 26: 196–198.

Saunders, R.J., Paul, N.A., Hu, Y. et al. (2012). Sustainable sources of biomass for bioremediation of heavy metals in waste water derived from coal-fired power generation. *PLoS One* 7: e36470.

Saxena, A., Bhardwaj, M., Allen, T. et al. (2017). Adsorption of heavy metals from wastewater using agricultural–industrial wastes as biosorbents. *Water Sci.* 31: 189–197.

Say, R., Yilmaz, N., and Denizli, A. (2003). Removal of heavy metal ions using the fungus *Penicillium canescens*. *Adsorpt. Sci. Technol.* 21: 643–650.

Shamsul, H., Bharose, R., Ozturk, M. et al. (2021). Impact of treated sewage water on nutrient status of alfisols and vegetable crops. *Not. Bot. Horti Agrobot. Cluj Napoca* 49: 12255.

Singh, A. and Ward, O.P. (2004). *Applied Bioremediation and Phytoremediation*. Springer.

Singh, J.S., Kumar, A., and Singh, M. (2019). Cyanobacteria: a sustainable and commercial bio-resource in production of bio-fertilizer and bio-fuel from waste waters. *Environ. Sustainability Indic.* 3: 100008.

Song, W.Y., Park, J., Eisenach, C. et al. (2014). ABC transporters and heavy metals. In: *Plant ABC Transporters* (ed. M. Geisler), 1–17. Cham: Springer.

Sood, A., Uniyal, P.L., Prasanna, R. et al. (2012). Phytoremediation potential of aquatic macrophyte, *Azolla*. *Ambio* 41: 122–137.

Sriprang, R., Hayashi, M., Ono, H. et al. (2003). Enhanced accumulation of Cd^{2+} by a *Mesorhizobium* sp. transformed with a gene from Arabidopsis thaliana coding for phytochelatin synthase. *Appl. Environ. Microbiol.* 69: 1791–1796.

Stottmeister, U., Aurich, A., Wilde, H. et al. (2005). White biotechnology for green chemistry: fermentative 2-oxocarboxylic acids as novel building blocks for subsequent chemical syntheses. *J. Ind. Microbiol. Biotechnol.* 32: 651–664.

Strong, P.J. and Burgess, J.E. (2008). Treatment methods for wine-related and distillery wastewaters: a review. *Biorem. J.* 12: 70–87.

Tangahu, B.V., Sheikh Abdullah, S.R., Basri, H. et al. (2011). A review on heavy metals (As, Pb, and Hg) uptake by plants through phytoremediation. *Int. J. Chem. Eng.* 2011: 939161.

Tay, T., Candan, M., Erdem, M. et al. (2009). Biosorption of cadmium ions from aqueous solution onto non-living lichen *Ramalina fraxinea* biomass. *CLEAN-Soil Air Water* 37: 249–255.

Taylan, Z.S. and Özkoç, H.B. (2007). Bioavailability of aquatic organisms in the determination of potential heavy metal pollution. *Nat. Appl. Sci. J.* 9: 17–33.

Tiryaki, O. and Potur, T. (2017). An important tool for removal of pesticides from soil: phytoremediation. *Nat. Appl. Sci. J.* 33: 59–68.

Touliabah, H.E.S., El-Sheekh, M.M., Ismail, M.M. et al. (2022). A review of microalgae-and cyanobacteria-based biodegradation of organic pollutants. *Molecules* 27: 1141.

Tripathi, A. and Ranjan, M.R. (2015). Heavy metal removal from wastewater using low cost adsorbents. *J. Bioremed. Biodeg.* 6: 315.

US-ERPA (2000). Environmental protection agency. National Risk Management Authority. Introduction to Phytoremediation, Environmental Protection Agency.

Uysal, Y. and Taner, F. (2010). Bioremoval of cadmium by *Lemna minor* in different aquatic conditions. *CLEAN–Soil Air Water* 38: 370–377.

Vangronsveld, J., Herzig, R., Weyens, N. et al. (2009). Phytoremediation of contaminated soils and groundwater: lessons from the field. *Environ. Sci. Pollut. Res.* 16: 765–794.

Vasavi, A., Usha, R., and Swamy, P. (2010). Phytoremediation–an overview review. *J. Ind. Pollut. Control.* 26: 83–88.

Verma, S. and Kuila, A. (2019). Bioremediation of heavy metals by microbial process. *Environ. Technol. Innov.* 14: 100369.

Vinopal, S., Ruml, T., and Kotrba, P. (2007). Biosorption of Cd^{2+} and Zn^{2+} by cell surface-engineered *Saccharomyces cerevisiae*. *Int. Biodeterior. Biodegradation* 60: 96–102.

Volesky, B. (2007). Biosorption and me. *Water Res.* 41: 4017–4029.

Vural, N. and Güvendik, G. (1983). Investigation of air and human blood lead level in Ankara. *J. Nat. Sci. Med.* 7: 191–200.

Wahid, F., Sharif, M., and Turkyilmaz Unal, B. (2022). Role of arbuscular mycorrhizal fungi and biochar in remediation. In: *Introduction and Application of Organic Fertilizers as Protectors of our Environment* (ed. M. Ozturk, N.A. Akram, B. Turkyilmaz Unal, and M. Ashraf), 143–181. Cambridge Scholars Publishing.

Wang, J., Feng, X., Anderson, C. et al. (2012). Remediation of mercury contaminated sites–a review. *J. Hazard. Mater.* 221: 1–18.

Xu, Z., Wu, Y., Jiang, Y. et al. (2018). Arbuscular mycorrhizal fungi in two vertical-flow wetlands constructed for heavy metal-contaminated wastewater bioremediation. *Environ. Sci. Pollut. Res.* 25: 12830–12840.

Yaashikaa, P., Kumar, P.S., Jeevanantham, S. et al. (2022). A review on bioremediation approach for heavy metal detoxification and accumulation in plants. *Environ. Pollut.* 301: 119035.

Zeraatkar, A.K., Ahmadzadeh, H., Talebi, A.F. et al. (2016). Potential use of algae for heavy metal bioremediation, a critical review. *J. Environ. Manage.* 181: 817–831.

14

Microbial Biofilms – Pollutant Load Suppressor

Tanaji V. Latha[1], Uzma Sultana[2], Podduturi Vanamala[2], and Mir Z. Gul[3]

[1] *Telangana Social Welfare Residential Degree College for Women, Nizamabad, Telangana State, India*
[2] *Telangana Social Welfare Residential Degree College for Women, Kamareddy, Telangana State, India*
[3] *Department of Biochemistry, University College of Science, Osmania University, Hyderabad, Telangana State, India*

14.1 Introduction

The environment is getting polluted day by day by the activities of humans, which are posing a serious concern to future generations. The environment started getting polluted with the advent of the industrial and agricultural revolutions. Though these revolutions had benefited mankind in various ways but at the same time have resulted in the introduction of myriad contaminants into the environment. Now it has become the biggest challenge to reduce the load of these pollutants from the environment. Air, water, and soil are not only the ones that are affected but underground water is also showing the footprints of contamination. Hence, serious efforts are needed to reduce these contaminants from nature. It was found that the native microbial population has the ability to immobilize and biodegrade organic and heavy metal pollutants. Though indigenous biofilm communities can reduce pollutant load, certain factors such as less abundance of biofilms, non-availability, or non-accessibility of pollutants to the biomass are the major hindrance. Despite these biofilms, usage for bioremediation is of more interest because they can form aggregates and get fixed to the substratum which confers protection from the surroundings and increased the opportunity to exchange genetic material. Moreover, among all remediation processes, bioremediation using biofilms is considered eco-friendly and economical.

14.2 Characteristic Features of Biofilms that are Exploited for Bioremediation

Biofilms can be formed symbiotically by the same species or by different kingdoms too, namely bacteria, fungi, algae, and archaebacterial. Hence, they rely on different nutritional requirements, electron donors or acceptors, and sources of energy (Ferrera et al. 2004; Baker et al. 2009). Biofilms are mainly used as aggregates in a fixed form (sessile) or free-floating form (activated sludge). Unlike solitary forms, these aggregates exhibit huge genetic heterogenicity which makes them susceptible to different types of environmental stress such as changes in pH, temperature, salt concentration, predators, and dehydration (Heipieper et al. 1991; Beveridge et al. 1997; Boon et al. 2003; Boles et al. 2004; Hall-Stoodley et al. 2004; Vu et al. 2009; Accinelli et al. 2012). In addition, they are also resistant to high concentrations of toxic pollutants which otherwise would be inimical to indigenous planktons and this forms the basis for exploiting microorganisms for bioremediation. Biofilms also confer increased resistance to antibiotics as they possess the ability to transfer genetic material horizontally (Beaudoin et al. 1998; Ghigo 2001; Wolcott et al. 2013). The concentration of antibiotics required to kill the planktonic bacteria was observed to be much lower than that required for biofilms (Mah et al. 2003). Microorganisms in the biofilms exist at different growth rates which induces the persistence of biofilms (Wentland et al. 1996; Davies et al. 1998; Costerton 1999; Woo et al. 2012). Biofilms are found to secrete extracellular polymeric substances (EPS) around them that enable them to exchange or pass on the nutrients, electron donors/acceptors among them including water (Latch et al. 2003; Hall-Stoodley et al. 2004; Wijeyekoon et al. 2004; Seo et al. 2009b; Wang et al. 2011; Xiao et al. 2012). The cells on the

Aquatic Contamination: Tolerance and Bioremediation, First Edition. Edited by Rouf Ahmad Bhat, Gowhar Hamid Dar, Fernanda Maria Policarpo Tonelli, and Saima Hamid.

periphery procure more nutrients, etc., compared to the ones in the center of the film (Picioreanu et al. 2000; Chen et al. 2013). Additionally, the EPS also increases the availability of hydrophobic or inaccessible substances by solubilizing them (Latch et al. 2003; Seo et al. 2009b; Wang et al. 2011). Certain gram-negative bacteria such as *Pseudomonas aeruginosa* release membrane vesicles (MVs) that feeds on other bacteria in the surroundings, thus eliminating competition and utilizing the lysed cells as nutrients (Beveridge et al. 1997) MVs also aid in altering the cell surface making them resistant to antibiotics, and change their surface charge and hydrophobicity (Baumgarten et al. 2012). Hence, the versatility of biofilms makes them most effective in bioremediating groundwater and other aquatic ecosystems.

Biofilm organisms are found to be resistant to various antibiotics. The dose of vancomycin antibiotic needed to kill *Staphylococcus aureus* biofilm is 10 times greater to reduce three exponential phases (Williams et al. 1997). Resistance may be either acquired or intrinsic. Acquired susceptibility will be mainly due to the transfer of resistance plasmids horizontally among the biofilm organisms. The resistance plasmids get exchanged in the biofilms by a large group of species (Joklik et al. 1988). Resistance genes encoding β lactams, aminoglycosides, erythromycin, tetracyclin, glycopeptides, trimethoprim, and sulfa drugs were getting transferred mostly by conjugation (Tenover et al. 1998). This is because of the proximity of the bacterial cells with no predominant external forces to disrupt the cell-to-cell contact or impair the pilli required for conjugation. The intrinsic effect of antimicrobial susceptibility is mainly because of EPS which retards the movement of antibiotics across the biofilm to kill the organism (Hoyle et al. 1992). Biofilm organisms also have very low growth rates which reduce the uptake of the antimicrobial agent into the cell making them resistant intrinsically. If the growth rate of *Staphylococcus epidermidis* in the biofilm increased, susceptibility was also found to increase (Otto 2018). Internal environment of *E. coli* in the biofilm is deprived of oxygen, which, in turn, decreases the possibility of uptake of aminoglycoside antibiotics by the cells in the biofilm (Tresse et al. 1995).

14.3 Environmental Pollutants

Biofilms can effectively degrade pollutants such as phenantherene and pyrene (PHAHs), naphthalene, polychlorinated biphenyls (PCBs), mixed PCB contamination, natural organic acids, and humic acids, perchlorate and nitrate, nitrate, trichloroethene, trichloroethane, chloroform, toluene, parachloronitrobenzene (*p*-CNB), acenaphthene, pyrenes, phenols, 2-chlorophenols, pentachlorophenol (PCP), cadmium (Cd), copper (Cu), zinc (Zn), chromium (Cr), manganese (Mn), Cobalt (Co), Lead (Pb), Mercury (Hg), Nickle (Ni), cadmium (Cd), cobalt (Co), Strontium (Sr), and landfill leachate.

14.4 Microbial Biofilms

Antonie van Leeuwenhoek was the first person to identify biofilm bacteria in his and others' teeth scrapings. But he termed them "animalcules". The term biofilm was coined by Bill Costerton in 1978. Biofilms are known to be ubiquitous. They may be formed by single species or by mixed species. Biofilms formed by multispecies have synergistic interactions among them.

14.4.1 Structure of Biofilm

The composition of biofilms is completely based on the type of bacteria, environmental conditions,, and the nutrients available. However, biofilms are mainly composed of 97% of water and 65–95% of extra polymeric substances (EPS) and microbes that form the remaining components. Water plays a pivotal role in circulating nutrients to all the microbes in it (Flemming et al. 2016). EPS is a self-synthesized matrix that protects the microbes from varied contaminants, predators, and other environmental stress such as desiccation, osmotic lysis, etc. (Mortensen 2014). Fourier-transform infrared (FTIR) spectroscopic observations elucidated that EPS is composed majorly of polysaccharides and proteins (Mosharaf et al. 2018).

14.4.1.1 Polysaccharides

It is primarily composed of neutral or polyanionic polysaccharides. Neutral polysaccharides include hexoses, pentoses, and uronic acids. Uronic acids such as mannuronic acids, galacturonic acids, and D-glucuronic acids impart an anionic charge to EPS. This charge is of high significance as it provides a high binding force for the development of EPS. On the contrary, gram-positive staphylococci contain cationic EPS instead of anionic EPS (Hussain et al. 1993). EPS is also characterized by the existence of organic or inorganic functional groups like acetate ester, formats, succinates, phosphates, and sulfates. EPS maintains the structure of biofilms after fixing onto the surface. PsI is secreted or cell-associated neutral polysaccharide whose role in the EPS formation and adhesion in *P. aeruginosa* has been well documented (Ma et al. 2012). Other polysaccharides such as xanthan, alginate, and cellulose also determine the biofilm structure. Cellulose is the main component in biofilms of algae, amoeba, and bacteria. *Desulfovibrio vulgaris* contain mannose, fucose, and N-acetyl galactosamine along with cellulose in its biofilm (Poosarla et al. 2017). Polysaccharides of biofilms are of various types.

14.4.1.1.1 Aggregative Polysaccharides Aggregative polysaccharides play a crucial role in adhesion, development of complex structures, and detachment of biofilm organisms. Based on the type of function performed, aggregative polysaccharides are again classified into different types.

a) Polysaccharide Intercellular Adhesion
   In *S. aureus* biofilm, EPS exists as extracellular polymeric substance intercellular adhesion (PIA), which mainly mediates the formation of micro and macro colonies by forming aggregates. PIA also mediates the detachment process of biofilm by producing detergent-like peptides that cleave the noncovalent interactions among PIA and bacterial cells in the biofilm (Arciola et al. 2015).
b) *P. aeruginosa* Pel
   Pel is another such EPS that forms a thick pellicle (a small protrusion formed at the air-liquid interface) in *P. aeruginosa* and hence the name. Pel plays a pivotal role in the adherence of biofilm to the substratum.
c) *P. aeruginosa* Psl
   It helps in the initiation of attachment, and the formation of thick and stable biofilms with increased biomass and microcolony height (Ma et al. 2009). Psl also acts as a signaling molecule that can trigger biofilm formation (Irie et al. 2012). Psl-producing cells were found to be highly resistant to a wide range of antibiotics. This characteristic feature is transferable to the cultures when grown together including other species (Billings et al. 2013).

14.4.1.1.2 Protective Polysaccharides Initially, PS was the protective layer that retards the entry of antibiotics into the colony but altered studies revealed that PS just delays the diffusion of antibiotics so that resistance genes get activated and neutralizing enzymes are produced (Dibdin et al. 1996; Høiby et al. 2010). PS also protects from phagocytosis. PS retards the diffusion of antibodies and inflammatory cells into the biofilm (Lam et al. 1980; Leid et al. 2002; Jesaitis et al. 2003; Kocianova et al. 2005; Rohde et al. 2007; Boles et al. 2010). As a result, the biofilm organisms can maintain a stable chronic infectious phase in the host.

a) Alginate
   Alginate usually helps in promoting cell–cell adhesion in mucoid biofilms by acetylation (Hentzer et al. 2001; Nivens et al. 2001; Stapper et al. 2004; Tielen et al. 2005). In mucoid colony-forming biofilms, particularly P. aeruginosa, it establishes a microenvironment and protects the cells from antimicrobial agents (Hentzer et al. 2001; Alkawash et al. 2006; Hengzhuang et al. 2011). Alginate is also responsible for providing resistance against host innate immunity by inhibiting phagocytosis (Mai et al. 1993; Pier et al. 2001) and protection against ROS (Learn et al. 1987; Simpson et al. 1989) and prevents the effect of cationic antibacterial peptides (Chan et al. 2004; Limoli et al. 2014).
b) Capsular Polysaccharide (CPS)
   Bacterial capsular polysaccharide (CPS) protects the biofilm from desiccation, opsonin-mediated phagocytosis, complement-mediated, and antibacterial peptide-mediated neutralization (Roberts 1996). During biofilm formation, CPS expression is suppressed to gain attachment (Deghmane et al. 2002; Hall-Stoodley et al. 2008; Sanchez et al. 2011). Decreased CPS can establish an acute infection in the host (Tuchscherr et al. 2005; Sanchez et al. 2011).
c) Levan
   Levan is a high molecular weight homopolymer that contains β-D-Fructans with irregular branching which confers mucoid nature to the biofilm (Velázquez-Hernández et al. 2009). It is truly not necessary in biofilm formation but levan with prolonged exposure to sucrose can form dental caries (Kiska and Macrina 1994). This property can be used in stable biofilm formation (Kolenbrander and London 1993; Sutherland 2001; Dogsa et al. 2013).

14.4.1.1.3 Architectural Polysaccharides These PSs play a significant role in regulating the structure of biofilms. It is also crucial in gene regulation, second messenger signaling, and matrix assembly (Krasteva et al. 2010; Berk et al. 2012; Shikuma et al. 2012).

a) Colanic Acid
   It is a branched PS containing glucose, galactose, and glucuronic acid (Stevenson et al. 1996). It is the first studied PS which is also called M antigen. It forms a mature and three-dimensional biofilm (Danese et al. 2000). Mutants of this are unable to produce stable three-dimensional structures indicating its significance in pathogenesis.
b) Vibrio Polysaccharides (VPS)
   Vibrio polysaccharide (VPS) is important in forming biofilms and providing resistance to chlorine. It forms corrugated colonies. Mutants of this produce smooth colonies. Rough colonies have increased attachment ability (Yildiz and Schoolnik 1999; Yildiz et al. 2001). VPS together with proteins Bap1, and RbmC forms an envelope around the biofilm which plays a crucial role in pathogenesis and also in the environment.
c) *Bacillus subtilis* Polysaccharide
   B. subtilis forms a widespread architectural polysaccharide called extracellular polysaccharide (EPS) (Branda et al. 2001; Branda et al. 2004; Marvasi et al. 2010). It is important in forming pellicles and other surface structures (Branda et al. 2004). This improves the stability of biofilm. Overexpression of this could result in increased cell aggregation, pellicle formation, and wrinkled colonies with no swarming motility (Kearns et al. 2005).
d) Cellulose
   It furnishes structure and shields the cells in the biofilm In *Glucnoacetobacter xylinus*, cellulose is important in forming cell aggregates, and pellicles and provides buoyancy to the biofilm aiding in oxygen intake by the bacterium (Cook and Colvin 1980.). It is even involved in the protection of the cells from nonionizing UV rays (Williams and Cannon 1989; Ross et al. 1991). In *E. coli,* Nissle 1917 (EcN) cellulose establishes a commensal relationship with the host epithelial cells by enhancing attachment with the host epithelial cells and protecting it from host immunity (Monteiro et al. 2009).

14.4.1.2 Proteins

EPS is also found to be associated with proteins whose composition depends on the changing conditions in the biofilms. Biofilms derive it from the proteins secreted by cell debris and the proteins that are associated with extracellular membrane proteins (Toyofuku et al. 2012), cell surface adhesins, and protein subunits of flagella or pili. In addition to these, the matrix also contains proteins of periplasmic space, cytoplasm, inner membrane, and outer membrane. These proteins aid in attachment and colonization. Without matrix proteins, the stability and integrity of the biofilm may not be possible (Cucarella et al. 2001; Toledo-Arana et al. 2001; Latasa et al. 2005; Tielker et al. 2005; Diggle et al. 2006; Fong et al. 2006; Fong and Yildiz 2007; Borlee et al. 2010; Ostrowski et al. 2011; Kobayashi and Iwano 2012). A detailed study of biofilm structure and localization revealed that proteins can facilitate matrix organization or disintegration. For instance, the enzymatic activity of glycosyl hydrolase dispersion B hydrolyses polysaccharides (Kaplan et al. 2003), proteases hydrolyze proteins of the matrix (Martí et al. 2010), and DNases degrade nucleic acids (Mann et al. 2009; Nijland et al. 2010).

Vibrio cholerae can survive in the natural aquatic environment through the formation of biofilms. The three major proteins that facilitate biofilm formation in *V cholerae* are RbmA, Bap 1, and RbmC (Kirn et al. 2005). Mutants of these proteins have very low efficiency in mature biofilm formation. Proteomic studies of *P. aeruginosa* elucidated the presence of 13.3% secreted proteins, 28.9% cytoplasmic proteins, 11.1% periplasmic proteins, 2.2% cytoplasmic membrane proteins, and 35.6% outer membrane proteins (OMPs). The presence of cytoplasmic proteins and OMPs in the matrix is attributed to the lysis of cells and also due to the presence of outer MVs (OMVs). In addition, a few enzymes are also found in the matrix such as alkaline protease, chitinase, protease IV, and putative magnesium-dependent components, viz. exopolysaccharides or proteins whose functions are yet to be determined.

S. aureus, *S. epidermidis*, and *Salmonella enterica* are found to have a special type of protein called biofilm-associated proteins (Bap) in their matrix. These proteins do not just aid in adhesion and biofilm formation but also act as virulence factors. Mutants of Bap are found to be devoid of colony corrugation ability and decreased cell adhesion, which, in turn, affects biofilm formation (Cucarella et al. 2002; Cucarella et al. 2004; Valle et al. 2012).

B. subtilis is found to be associated with an amphiphilic biofilm surface layer protein (BslA), which is necessary for the formation of colonies, pellicles, and biofilms, and bestow the biofilm with surface-repelling ability. Mutants of BslA form colonies and pellicles but with no colony corrugation, changing surface structure, and surface repellency (Verhamme et al. 2009; Hobley et al. 2013). Surprisingly, matrix proteins also include nucleoid-binding proteins that play a crucial role in the formation of biofilm (Goodman et al. 2011; Gustave et al. 2013; Novotny et al. 2013). Usually, HU (histone-like protein) (Toyofuku et al. 2012) and IHF (integration host factor) proteins are localized extracellularly probably by cell lysis and confer stability to the biofilm. On the contrary, other nucleoid-binding proteins such as H-NS and DPS were found to play no role in biofilm formation (Goodman et al. 2011).

14.4.1.3 Membrane Vesicles

Certain gram-negative bacteria such as *P. aeruginosa* (Schooling and Beveridge 2006; Toyofuku et al. 2012), *Helicobacter pylori* (Yonezawa et al. 2009), *V. cholerae* (Kondo et al. 1993; Duperthuy et al. 2013), and *Vibrio fischeri* (Aschtgen et al. 2016) produce dynamic and versatile OMVs by bulging out from outer membrane in the form of a bleb. It is composed of proteins of the outer membrane, enzymes such as proteases, alkaline phosphatases and lipases, toxins, and lipopolysaccharide. OMVs facilitate the formation of biofilms, virulence, quorum sensing (QS), procuring nutrients, and horizontal gene transfer (Beveridge et al. 1997; Bielig et al. 2011; Li and Tian 2012). They also improve the resistance of bacteria in the biofilm by neutralizing the effect of some harmful chemicals.

14.4.1.4 Extracellular DNA (eDNA)

Extracellular DNA is found in the biofilm as a result of cell lysis. It does not simply carry the genetic information but can even provide structural stability, enhances attachment (Montanaro et al. 2011; Das et al. 2017; Panlilio and Rice 2021), and protects antimicrobial agents by chelating cations (Mulcahy et al. 2008). The concentration of eDNA increases during the late exponential phase and peaks at the stationary phase and decline phase.

14.4.2 Biofilm Formation

The formation of biofilm is a complex process that requires moisture nutrients and a suitable substratum. Biofilms can be found in various natural environments. Specifically, biofilms found in the liquid/liquid interface are exploited in the degradation of hydrocarbons such as petroleum oils, etc. (Grimaud 2010; Omarova et al. 2019). It begins with a single free-floating microorganism which undergoes the following steps to form a mature biofilm.

14.4.3 Attachment

The first step in biofilm formation is the attachment of the bacterium to the suitable substratum. Initially, the attachment could be reversible because of weak van der Waals forces and hydrophobic interactions. If remains undisturbed it gets converted to irreversible attachment using various cell adhesins. For successful attachment the properties of both the substratum and the surface of the cell are critical. The substratum could be rough, smooth, hydrophobic, or charged hydrophilic materials. Generally, rough hydrophobic substratum supported the rapid development of biofilm (Donlan 2002; Silyn-Roberts and Lewis 2003; Trevor Roger Garrett et al. 2008). However, if the substratum is found in moisture-rich environments such as an ocean, blood, urinary tract, etc., biofilm forms a proteinaceous layer acquired from the environment around itself making it more complex. As a result, the significance of the substratum becomes obscure. The characteristics of the cell surface are equally critical. Bacterial appendages such as flagell, pili, fimbriae, or glycocalyx aid in the initial attachment of microbes on the substratum until the actual attachment process initiates (Lappin-Scott and Bass 2001). Hydrophobic cell surfaces also play an important role in the attachment (Lappin-Scott and Bass 2001).

14.4.4 Formation of Microcolony

The cell that gains successful attachment will now start dividing to form a microcolony (Lappin-Scott and Bass 2001; Vu et al. 2009). These cells secrete EPS that are primarily composed of polysaccharides. EPS gives a proper configuration to the biofilm. Biofilm is a highly heterogeneous layer consisting of different microbial cells which change with time. This heterogenicity facilitates the cells with proper nutrients, exchange of metabolic products, etc. (Boelee et al. 2011). Water channels keep the biofilm hydrated, and help in the uptake of nutrients and oxygen for the developing cells (Lewandowski and Evans 2000). Biofilm organisms exhibit syntrophic association among themselves encouraging the growth of varied microbial communities. Microcolony now with time changes to a mature macrocolony. Because of nutrient and oxygen depletion, biofilms tend to grow more slowly than planktonic organisms (Donlan 2000).

14.4.5 Formation of Macrocolony (Maturation)

Mature biofilm organisms communicate with each other via signaling molecules and trigger quorum-sensing mechanisms within them. This extensive communication leads to the activation gene that plays a pivotal role in EPS formation and stabilization. With this, a mature three-dimensional biofilm structure is formed which can be several inches thick in some instances. EPS forms a base for the circulatory system in the biofilm that helps in the dissemination of nutrients, excretion of cell waste supply of oxygen, etc. (Karatan and Watnick 2009). Mature biofilms tend to fix firmly on the substratum till dispersal initiates (Randrianjatovo-Gbalou et al. 2017).

14.4.6 Detachment (Dispersal)

Detachment or dispersal of microbial cells from biofilm is crucial as it promotes the spread and colonization of cells on the new surface. With dispersion, sessile cells transform into motile ones (Flemming and Wingender 2010). But the dispersion process is not a compulsion in the life cycle of biofilm. It may initiate because of environmental stress, nutrient depletion, etc. (Lappin-Scott and Bass 2001; Kaplan 2010). Dispersal is made possible with the release of enzymes such as dispersin B, deoxyribonucleases, etc. Nitric oxide was also found to initiate dispersal. During the process of dispersal, bacterial cells turn on the genes responsible for the synthesis of flagella, which helps the released bacterium reach the new substratum. Biofilms have different roles to play in the process of bioremediation.

14.5 Pesticide Degradation

Pesticides are widely used to prevent and control crop diseases and pests. Residues from pesticides have caused considerable harm to human health and the environment. In the field of worldwide environmental (aquatic and soil) restoration, studying the breakdown of pesticides by microbial biofilms is an important subject. Bacteria can thrive in biofilms at locations with high concentrations of toxic chemicals. Constituents of EPS, such as polysaccharides and proteins, have various functional groups. They provide adsorption sites for various pesticides via hydrophobic, electrostatic, and hydrogen-bonding interactions. Hence, biofilm-forming microbes can be applied to various bioremediation processes.

In aquatic environments subjected to agrochemical contamination, microbiomes exhibit biofilm production. With the help of extracellular enzymes and organized gene expression mediated by cell–cell communication or QS, microbiomes can capture organic compounds for metabolism (Zhao et al. 2014; Lerch et al. 2017; Anutrakunchai et al. 2018).

In vitro biostimulation of the mixture with low concentrations (0.1% and 0.5%) of inorganic fertilizers (N, P, and K) led to a noticeable increase in the breakdown of chlorpyrifos (Proia et al. 2013). According to several reports, suspension cultures are more sensitive to environmental changes than biofilms produced on substrates or supports.

A three-phase mechanism may be involved in the metabolism of pesticides. In phase I metabolism, a parent compound's initial characteristics are changed through hydrolysis or oxidation–reduction reactions to generate a more water-soluble and less dangerous derivative than its original parent compound. In phase II, a pesticide metabolite undergoes conjugation to a sugar or amino acid, which decreases its toxicity and increases its water solubility in comparison, to the original pesticide. Metabolites from phase II are converted into secondary conjugates, which are also nontoxic, in phase III. These processes are carried out by bacteria and fungi, by secreting various extracellular or intracellular enzymes including oxygenases, peroxidases, hydrolytic enzymes, etc. (Van Eerd et al. 2003; Laura et al. 2013).

There are several reports of the immobilization of biofilms for biodegradation of pesticides. A study with bacterial consortia immobilized in a biofilm upon a volcanic rock known as tezontle exhibited a tremendous capacity for the removal of organophosphorus pesticides (Yañez-Ocampo et al. 2009, 2011). In a study, epilithic biofilms sampled from the river bioaccumulated the active ingredients of two insecticides (chlorpyrifos and imidacloprid), two herbicides (chlorpyrifos and atrazine), seven fungicides (boscalid, carbendazim, epoxiconazole, tebuconazole, metconazole, prochloraz, and prothioconazole), and 10 herbicides (2,4 (desethyl-atrazine). Following their application in the summer cultures, the principal pesticides under observation saw a sharp decline in concentration in epilithic biofilm samples taken several months later, demonstrating their excellent detoxifying capacities (Rheinheimer Dos Santos et al. 2020). There have been reports of the genera *Pseudomonas* and *Arthrobacter* degrading atrazine. Specifically, one strain of *Arthrobacter aurescens* may get nitrogen from the methylthiotriazines mercaptoatrazine and ametryn (Strong et al. 2002) According to reports, biofilms quickly mineralize the herbicides bentazone, metazachlor, and metribuzin rather than storing them for extended periods, demonstrating their significance in the self-purification of surface waters (Bighiu and Goedkoop 2021).

14.6 Wastewater Treatment

Industrial wastewater treatment methods mediated by biofilms are extremely effective strategies because of their discrete advantages over traditional approaches. Industrial wastewaters are frequently identified by the basic criteria, which include pH, turbidity, color, pH, suspended solids (SS), ammonium nitrogen, chemical oxygen demand (COD), biochemical oxygen demand (BOD), heavy metal concentration, and biological parameters. Depending on the industrial activity, these qualities fluctuate greatly. Several microbial species, including algae, fungi (Badia-Fabregat et al. 2017), yeast (Nhe Cong et al. 2014), and bacteria (Abzazou et al. 2016), have been successfully used as biofilm forms for the treatment of wastewater (Hoh et al. 2016). The microbial consortia have the potential to remove or change the problematic contaminant. The complex contaminant should be presented to the consortia of microbes in the best possible manner in terms of matrix preparation to successfully treat wastewater using biofilm-based systems. In biofilm reactors, microorganisms may be supported on various materials used as matrices.

The fixed bed is a typical biofilm-based reactor, also known as the packed bed reactor. In this reactor, solid supports (media) are closely packed where biofilms are colonized and offer good contact between the mass of the biofilm and the liquid. The bioremediation of mercury has been successfully carried out using a biofilm of strains that are resistant to mercury in a packed bed reactor.

One of the earliest and most widely used types of biofilm reactors for wastewater treatment is the trickle-bed biofilm bioreactor (TBR), a special kind of fixed-bed reactor. Typically, plastic, stone, ceramic, and other materials are utilized as TBR media. In TBR, wastewater trickles across the biofilm surface held on a fixed media as it is distributed from the top of the system. Pollutants are broken down as the water diffuses through the biofilms. The biofilm can be reached by oxygen that is delivered either upwards or downward and diffuses through the water. When there is a net generation of SS in the TBR, a clarifier must separate the liquid from the solid (Asri et al. 2019).

The use of membrane biofilm reactors (MBRs) to remove and/or recover pollutants from aqueous solutions has long been used as promising biotechnology. It primarily refers to a mix of physical separation carried out by a membrane unit in place of the secondary settler and biological degradation of waste compounds utilizing a biofilm-based system. It is used to treat a variety of industrial and urban wastewaters, demonstrating excellent removal of both organic and inorganic materials, and offers specific advantages over other treatment systems (Di Fabio et al. 2013).

14.7 Microbial Fuel Cells (MFCs)

A recently created biofilm-based system is the microbial fuel cell (MFC). By converting the chemical energy held in organic matter into electricity, this technique enables the simultaneous solution of two real-world problems, namely wastewater decontamination and the production of green energy. MFCs use a certain class of microbial species that can generate current by trading an electron network with an electrode. On the electrodes' surface, these microorganisms create a biofilm known as an "electroactive biofilm," or "EAB." Heavy metal removal from contaminated soil and groundwater can be done using an EAB (Jasu and Ray 2021). There is a vast variety of microorganisms that could be present in MFCs, according to community analysis. However, the two most prevalent are types of bacteria and algae (Saba et al. 2017). MFC technology has demonstrated its efficacy in the removal of numerous contaminants. Numerous industrial wastewaters, such as those from refineries and food processing, have been employed as inoculums in anode chambers to create electroactive biofilms, or "EAB." MFCs are available in a variety of configurations.

A typical MFC is made up of two chambers – anodic and cathodic – that are connected by a conductive substance that contains a resistor. These chambers host separate electrochemical and biological reactions. The organic material in the anodic chamber (negative terminal) oxidizes as a result of the microbial cells in the biofilm reacting with it, by releasing electrons and protons in the process. Following the transfer to the anode, the electrons go over an external circuit to the other MFC component, which houses the cathode (positive terminal). The direct membrane-associated electron transfer method, so-called bacterial nanowires, electron mediators, shuttles, or any combination of these methods can be used to transfer electrons from the anode to the cathode (Logan et al. 2006). Oxygen serves as an electron acceptor in the cathodic chamber. When these electrons and protons diffuse from the anode through a separator with the help of oxygen from aerobic conditions, water molecules are created (Min and Logan 2004). It has been proven that some bacteria, most significantly metal-reducing bacteria, may transfer electrons to the anode directly in MFC. Sediments contain metal-reducing bacteria like *Shewanella putrefaciens*, *Geobacter sulfurreducens*, and *Rhodoferax ferrireducens* that utilize insoluble electron acceptors like manganese (IV) oxide and iron (III). These bacteria become electrochemically active by certain specific cytochromes on the outer surface of the cell membrane. The MFC anode is the final electron acceptor (Marsili et al. 2008). These bacteria become electrochemically active thanks to certain cytochromes that are located on the cell membrane's outer surface. The MFC anode is capable of serving as the final electron acceptor (Marsili et al. 2008). Some Geobacter and Shewanella strains can produce nanowires, which are electronically conducting molecular peels that enable the usage of more distant solid electronic acceptors by the microbe.

MFCs have excellent denitrification capabilities without the requirement for an external electron donor since organic matter oxidation produces electrons that are released into an external circuit and transmitted to a cathodic chamber, which facilitates nitrate reduction (Clauwaert et al. 2007). Due to the anaerobic conditions in the anode chamber additionally, the MFC technique is appropriate for the treatment of high-strength wastewaters (COD exceeding 8000 mg/L).

14.8 Bioremediation of Organic Pollutants

Polycyclic aromatic hydrocarbons (PAHs) are environmental contaminants that are commonly produced by anthropogenic activities like incomplete combustion of fossil fuels and petroleum refining. PAHs, polychlorinated biphenyls, and polychlorinated ethenes are examples of persistent organic pollutants. Due to their hydrophobicity, persistent organic pollutants (POPs) can become poisonous when they enter the food chain. POPs, which are found in air, water, and sediments, are the most persistent of all pollutants with long half-lives due to their hydrophobicity. Many bacterial biofilm-forming microorganisms that can be further modified and employed in bioremediation have been identified from habitats that break down persistent organic pollutants. The circulation of metabolic substrate and product is heterogeneous (or gradient) within a mature biofilm (Mishra et al. 2022). POP detoxification using bacterial biofilms has been developed. According to studies, biofilm development and cometabolism within biofilms are crucial elements in the removal of PAHs.

In biological wastewater treatment processes, QS and quorum quenching (QQ) are increasingly reported due to their innate roles in the production of biofilms granulation, bacterial aggregation, biotransformation of pollutants, and colonization. PAH degradation is positively influenced by QS. QS systems control the expression of enzyme synthesis during biofilm-mediated degradation. *P. aeruginosa*'s RhlI/RQS system participates in the denitrification as well as biodegradation of organic pollutants found in urban and industrial wastewater (Yong and Zhong 2013). The expression of the QS genes lasI and rhlI dramatically boosted PAH degradation by the marine bacteria *P. aeruginosa* N6P6. An increase in QS gene expression promotes the development of biofilms and the creation of EPS, which increases the rate of PAH breakdown (Mangwani et al. 2015).

Rapidly growing hydrocarbon degradation bacteria can be utilized to clean up oil spills in marine environments. Gamma proteobacteria species such as *Pseudomonas*, *Arthrobacter*, *Rhodococcus*, *Bacillus*, *Alcanivorax*, and *Cycloclasticus* have been used. New microbial strains that can degrade oil via biofilms are being continuously screened. The genomes of several strains of *Pseudomonas stutzeri*, including the AN10 naphthalene-degrading strain, have been sequenced (Brunet-Galmés et al. 2012). *P. stutzeri* produces biofilms and exhibits a wide range of phenotypic and genotypic variations. Additionally, *P. stutzeri* strain T102 accelerated naphthalene breakdown in biofilms. It has been shown that microbial consortiums containing a surfactant-producing strain plus *B. subtilis* and *Acinetobacter radioresistens* break down petroleum more efficiently than single strains. It has been demonstrated that biofilm-forming bacteria can change or degrade substances including phenol, hexadecane, phenanthrene, pyrene, nitrogen, and total organic carbon (Zhang et al. 2015). The degradation of PAHs can be enhanced by sorbing PAHs or other organic contaminants to a matrix before treatment. The proximity of cells to a poorly soluble substrate may enhance degradation kinetics (Bastiaens et al. 2000). The main bacteria enriched by high molecular weight PAHs were *Pseudomonas*, *Bacillus*, and *Microbacterium* species with the capacity to degrade PAHs by the production of biofilms.

14.9 Bioremediation of Heavy Metals

Given public health concerns, heavy metals are the major environmental pollutants due to their persistence and bioaccumulative nature in contaminating aquatic and terrestrial ecosystems. The toxic heavy metals are released into the ecosystem through natural sources such as volcanic eruptions, weathering of rocks, anthropogenic sources, and also from industrial operations which include mining, distilleries, textiles, plastics, leather, manufacturing units of electronic devices, foodstuff processing, and pharmaceuticals production (Shukla et al. 2014; Mohapatra et al. 2017c), and so on. Once they accumulate in the ecosystem, they consequently initiate the contamination of the food chain at different levels.

As they are biorecalcitrant and exist in the environment in enormous concentrations, may become carcinogenic or teratogenic (Shukla et al. 2017). These pollutants affect the ecosystem adversely and induce serious threats to human beings and animals. They are the major factors responsible for alarming cases related to birth defects, and physical and mental retardation these days (Edwards and Kjellerup 2013; Shukla et al. 2014, 2017).

Employing bacterial biofilm in bioremediation is a quite promising approach in comparison to other attainable methods of reducing pollutants. Bioremediation mediated by the activities of biofilms is a sustainable alternative in virtue of their surpassing microbial biomass, adhesive properties, and immobilizing potential of microorganisms (Edwards and Kjellerup 2013; Gupta and Diwan 2016; Das et al. 2017; Shukla et al. 2017). Biodegradation is accelerated by the accumulation of bacteria in response to chemical stimulus, which elevates the digestibility of toxic pollutants and the transfer of genes within biofilm species also contributes to intensified biodegradation (Singh et al. 2006).

14.10 Toxicity of Heavy Metals

Metals with a molecular density of more than 5 g/cm^3 impose harmful effects on the biosphere. Such metallic elements have relatively high density and are reported to affect the biological functions of a cell. These metals if accumulated in higher concentrations have a toxic impact on biota, although few metals are inevitable for cellular functions in minute quantities. Deficiency of these metals overt symptoms of physiological stress and reduction of growth. Despite this fact, they cause serious health hazards if available in slightly higher concentrations as they form nonspecific compounds with cellular organelles (Shukla et al. 2014). The recalcitrant properties of various heavy metals induce biomagnifications which may lead to detrimental health impacts. Several mechanisms are involved in exerting the toxic effect such as free radical imbalance, oxidative stress, structural damage of DNA or protein, and metals including chromium, arsenic, mercury, cadmium, and aluminum may replace integral cofactors in vital proteins. Cr (VI) alters the permeability of ion channels. Excess iron levels may induce organ incursion and oxidation of lipids (Jaishankar et al. 2014; Gupta and Diwan 2016).

14.10.1 Remediation of Hazardous Metals

Several physiochemical approaches are effective and applied for the reduction of toxic heavy metal concentrations such as reverse osmosis (RO), nanofiltration, ion exchange, and many new developments in bioengineering (Shukla et al. 2017; Sinha et al. 2018). These techniques have certain limitations as they require excess energy and sophisticated equipment, the cost of the process is huge, and release a vast quantity of sludge. These methods may form recalcitrant secondary intermediate substrates (Shukla et al. 2014; Gupta and Diwan 2016). Necessarily, the practical utilization of biofilm for the detoxification of heavy metals, and metalloids offers a preferable alternate mechanism that is effective, economical, and eco-friendly. Such biological treatment is more advantageous than any other approach in use.

The application of microorganisms in the degradation of heavy metals is the preferred approach that also transforms various organic pollutants into nontoxic forms. Alleviation of environmental pollutants by using site-specific native flora is a characteristic feature of the bioremediation technique which is further classified into three parts: bioattenuation, biostimulation, and bioaugmentation (Shukla et al. 2014). Bioattenuation is a natural transformation of pollutants using microbial flora whereas biostimulation includes the addition of electron acceptors or donors for enhancement of biodegradation. Bioaugmentation is the practice of adding specific microbes with pollutant degradation ability, as the indigenous microflora in the contaminant site may not possess suitable mechanisms to make use of the pollutants. This method will reduce the contaminant load by either mineralization or immobilization. The bacterial population use metallothionein-mediated entrapment. Various intracellular and extracellular enzymes encoded by pah, nid, nah, and bph are involved in the degradation of organic pollutants (Seo et al. 2009a). There are several factors such as the physiochemical status of the site, contamination levels of toxic pollutants, and the density of indigenous bacteria analyzed to choose a suitable bioremediation technique.

Bacterial biofilm acts as a proficient tool in the removal of heavy metals as they have a high tolerance against toxic concentrations of inorganic metals. The bacterial biofilms are more suitable with the advent of their high microbial mass, sustainability, metabolic diversification of signaling mechanisms, and the presence of surfactant or emulsifier properties (Singh et al. 2006; Shukla et al. 2017). Bacteria adopt different methods to remove metals and dyes, for instance, biosorption mechanism. The competent properties of the biofilm EPS are used in binding with metal ions by chemical bonding (Shukla et al. 2017).

Bioremediation mediated by biofilms can be enhanced by intracellular genetic exchange among mixed population biofilms, and the chemotaxis, escalates the bioavailability of the pollutants to the biofilm.

Bioremediation of hazardous metals is greatly alleviated by several factors, such as suitable temperature, pH, aeration, contact time, chemotaxis, and QS (Pearce et al. 2003; Shukla et al. 2014; Mohapatra et al. 2017c). Effective degradation of organic metal pollutants is achieved by optimization of the above physiochemical and biological operational procedures. Optimal temperature is an essential factor that enhances the bioavailability and activities of the biofilm matrix (Aryal and Liakopoulou-Kyriakides 2014; Shukla et al. 2014). The duration of contact between the pollutants and the active surface of biofilms is also a crucial factor that regulates the rate of biosorption.

14.10.2 Microplastic Degradation

Microplastic (MP) pollution is another challenging environmental issue affecting the ecological balance. Most of the research studies describe the serious threats associated with the persistence of MPs and the development of methods for effective microbial remediation. MPs come in contact with various organic and inorganic particles and microorganisms once they accumulate in the environment (Parrish and Fahrenfeld 2019). The microbes colonize the surfaces of MPs and form the biofilm that is clusters of microorganisms associated with a complex of secreted polymers and other components of the ecosystem. The biofilms initiate damaging the primary structure and characteristic features of MP in different ways which include the secretion of enzymes that modify or degrade the MP and masking surface components (Miao et al. 2019).

As per certain research studies, the physiochemical properties of MPs are crucial for bacterial colonization. The four freshly obtained and two beached MPs with microcosms were assayed by Hossain et al. to observe the MP degradation process. It was evident that eroded PP disks promote the colonization of bacteria abundantly which promotes the degradation of MP (Hossain et al. 2019). Another study by Miao comparing the adhesion activity of biofilms on MPs and other natural contaminants like cobblestone and wood has uncloaked the diversity and species richness of the microbial community (Miao et al. 2019). Degradation of MPs mediated by biofilms has shown the accumulation of Pirellulaceae, Phycisphaerales, Roseococcus, and Cyclobacteriaceae than those found associated with natural substrates degradation, which implies the fact that few microorganisms selectively colonize the MPs. The biodegradation of MPs mediated by biofilms is quite more complex than observed with natural bacterial or fungal pure cultures involved mechanisms. The biofilm-mediated biodegradation is divided into four stages. Stage one begins with the adherence of microbes to the external surface of Micro plastic that modifies its surface properties such as hydrophilicity. In stage two, the leaching of monomeric components out of the MPs occurs, and the reduction of compounds by secreted enzymes increases. The third stage involves the enzymes or free radicals, originating from microbes which leads to a loss of conformational stability. The MPs are accumulated with water and penetration of microbial filaments occurs in the fourth stage, in which MPs are further utilized by microorganisms (Flemming 1998). With these observations, it is evident that MPs act as prominent habitats that provide a favorable environment for the proliferation of microorganisms. Biofilms use MPs as adhesion media from which they absorb the nutrients and degrade them through the attack of microbial enzymes. MPs are a good source to immobilize the biofilm community, involved in the effective utilization of MPs as a carbon source and energy source (Galloway et al. 2017).

The pathways involved in the biodegradation of MPs are almost similar to those of other polymers. They are initially converted to simple monomers before they undergo mineralization. The MPs particle size is big compared to the pore size of the cell membrane. Hence, MP particles must be depolymerized to pass through the cell membranes and degraded within microbial cells. The enzymes secreted by biofilms play a major role in hydrolytic cleavage which is catalyzed by the action of either intracellular enzymes or extracellular enzymes. Intracellular degradation proceeds through the hydrolysis of an endogenous carbon reservoir by the accumulated indigenous microflora themselves, whereas the extracellular enzymes utilize exogenous carbon sources and break down the MPs into smaller molecules (monomers, dimers, and oligomers) which can easily pass through the semipermeable membrane. The mineralization of these short chains results in the formation of end products such as CO_2, H_2O, or CH_4 which are further utilized as carbon and energy sources (Yoon et al. 1996; Gu 2003).

Several studies in recent years describe the separation and identification of microbial enzymes (Shosuke Yoshida et al. 2016). According to the studies on a novel bacterium, Ideonella sakaiensis 201-F6, allowed to grow on PET, produced two enzymes (ISF6–4831 AND ISF6–0224) that are efficient in hydrolyzing PET into environmentally benign simple units, terephthalic acid, and ethylene glycol monomers. As per the research observations, microbial enzymes are important in the MP degradation process, but the current findings about the association of biofilms and MP are obtained through routine data analysis. and there is a need for extensive research to understand the pathways and structural features of intermediate compounds. Comprehensive research about microbial interactions and degradation properties in controlled environments is yet to be reviewed.

Biodegradation is mainly influenced by two main factors one is associated with the diversity of the microbial community and another one is related to the distinct properties of MPs. The characteristic features of MPs include the basic physical and chemical properties like density, crystallinity, molecular weight, and functional groups present as part of the structure (Shah et al. 2008). Certain environmental factors, particularly light (UV), and humidity also have an impact on MPs degradation, as the external environment can oxidize and cause the aging of MPs, which facilitates the degradation process and utilization of MPs by the microorganisms. Sepu et al. (Volke-Sepúlveda et al. 2002) analyzed the effect of chemical agents on the degradation of TO-LDPE by fungi. In the presence of ethanol, the mineralization rate of TO-LDPE is enhanced after incubation with *A.niger* or *P. Pinophlum*. In another study, the biodegradation of HDPE and LDPE was investigated using indigenously established microbial associations (Satlewal et al. 2008). The potential increase of biodegradation is observed in microbial consortium treated HDPE than LDPE, which states the impact of density that induces the biodegradation of MPs to a greater extent.

14.10.3 Degradation of Pharmaceuticals

Biofilm provides certain advantages over indigenous microflora, as microbes in biofilm provide structure and possess potential self-protection in an assembled complex polymeric matrix (Vu et al. 2009). The concentration of pollutants increases in the water bodies usually when the flow of water reduces especially in the winter months. Several pharmaceutical compounds such as trimethoprim, metoprolol, diclofenac, and gemfibrozil are effectively degraded by employing membrane film bioreactors (Sui et al. 2011).

Certain recalcitrant chemicals fall under the category of pharmaceutical and personal care products (PPCPs). Particularly triclosan is the most commonly employed antimicrobial agent at an industrial scale, which consequently phototransformed into a dioxin analog in water columns (Latch et al. 2003; Buth et al. 2009). In water-logged areas the concentration of the pollutant increases. It was evident that microbial biofilms exposed to triclosan or diuran have a competent capacity to convert them into simple degradable homeostatic compounds in a few weeks (Proia et al. 2011). As per a few reviews, the microbial population is responsible for the degradation of triclosan and plays a key role in wastewater treatment (Lolas et al. 2012). The MBRs are potentially useful in removing several PPCPs, including diclofenac, trimethoprim, gemfibrozil, and metoprolol than traditional activated sludge (Sui et al. 2011). The levels of these PPCPs increase in wastewater effluent especially in winter when the bacterial biofilm communities are less abundant. Notable efficacy was found in the reduction of PPCPs when wastewater sludge is added as inoculums to develop a biofilm community on sand column filters (Onesios and Bouwer 2012). As per the observations of this experiment, 10 out of 14 PPCPs (ketoprofen, gemfibrozil, ibuprofen, naproxen, biphenylol, p-chlorom-cresol, chlorophene, 5-fluorouracil, troclosan, and valproic acid) were degraded exhaustively, and the detrimental effects of PPCPs in the effluent water are significantly reduced.

Despite these emerging new approaches, the significance of biofilm-mediated bioremediation is to accelerate detoxification. And the longevity of microbial diversity and proliferation has just recently been actively appreciated as a proficient and safer mechanism for reducing the pollutant load (Boon et al. 2003; Accinelli et al. 2012). The screening and identification of bacteria possessing similar characteristics to that of the phylum Chloroflexi (Dehalococcoides, Dehalobium chlorocoercia, and o-17) capable of halogenated compounds' respiration have established the sustainable use of microbial processes for the degradation of halogenated hydrocarbons (Chaudhry and Chapalamadugu, 1991; Margesin and Schinner 2001). In addition to this, the members of the genus Dehalococcoides have been recognized as efficient valuable residents of a biofilm reactor community for the detoxification of trichloroethene (Saiyari et al. 2018).

14.11 Conclusion

Pollution is an extremely long-standing issue concerning the ecological balance and the most widely studied subject to formulate novel technologies to resolve the problems associated with it. Biofilms act as a natural habitat where microbes exchange gene segments, signaling molecules, and metabolites. Bacterial cells use QS for communication between different species. Metagenomic approaches such as 16srRNA sequencing are an essential technique to analyze bacterial diversity in biofilms for designing better strategies to bioremediate environmentally contaminated sites such as wastewaters. A crucial strategy that emerged in the current global scenario is the microbiome-based bioremediation process. The formation of biofilms via genetic engineering promotes the bioremediation of contaminants. The emerging technologies using genetically engineered biofilms enhance the bioremediation of contaminants. To improve the effectiveness of microbial enzymes in bioremediation, synthetic biology methods are used. Biological procedures for the degradation of toxic pollutants in industrial wastewater treatment plants are improved by QS technology. QS regulates the production of biofilm, EPS, and biosurfactants, which contribute significantly to the removal of heavy metals and organic pollutants. However, biofilm-mediated degradation of the recalcitrant compounds has several advantages as discussed in several reviews about the degradation behavior, characteristics of MPs, and other pollutants. There is a need for intense research for a better understanding of interactions among the biofilms and environmental pollutants in aquatic and soil atmosphere. Understanding the function of AHL-based QS and QQ systems in wastewater treatment is crucial. Metabolic engineering, genome editing, and omic-based techniques, as well as in silico databases and bioinformatics tools, may pave the way for novel biofilm-based bioremediation research directions.

References

Abzazou, T., Araujo, R.M., Auset, M., and Salvadó, H. (2016). Tracking and quantification of nitrifying bacteria in biofilm and mixed liquor of a partial nitrification MBBR pilot plant using fluorescence in situ hybridization. *Sci. Total Environ.* 541: 1115–1123. https://doi.org/10.1016/j.scitotenv.2015.10.007.

Accinelli, C., Saccà, M.L., Mencarelli, M., and Vicari, A. (2012). Application of bioplastic moving bed biofilm carriers for the removal of synthetic pollutants from wastewater. *Bioresour. Technol.* 120: 180–186. https://doi.org/10.1016/j.biortech.2012.06.056.

Alkawash, M.A., Soothill, J.S., and Schiller, N.L. (2006). Alginate lyase enhances antibiotic killing of mucoid *Pseudomonas aeruginosa* in biofilms. *APMIS* 114 (2): 131–138. https://doi.org/10.1111/j.1600-0463.2006.apm_356.x.

Anutrakunchai, C., Bolscher, J., Krom, B.P. et al. (2018). Impact of nutritional stress on drug susceptibility and biofilm structures of *Burkholderia pseudomallei* and *Burkholderia thailandensis* grown in static and microfluidic systems. *PLoS One* 13 (3): e0194946. https://doi.org/10.1371/journal.pone.0194946.

Arciola, C.R., Campoccia, D., Ravaioli, S., and Montanaro, L. (2015). Polysaccharide intercellular adhesin in biofilm: structural and regulatory aspects. *Front Cell Infect. Microbiol.* 10 (5): 7.

Aryal, M. and Liakopoulou-Kyriakides, M. (2014). Characterization of *Mycobacterium* sp. strain Spyr1 biomass and its biosorption behavior towards Cr (III) and Cr (VI) in single, binary and multi-ion aqueous systems. *J. Chem. Technol. Biotechnol.* 89 (4): 559–568.

Aschtgen, M.S., Wetzel, K., Goldman, W. et al. (2016). *Vibrio fischeri*-derived outer membrane vesicles trigger host development. *Cell Microbiol.* 18 (4): 488–499.

Asri, M., Elabed, S., Ibnsouda Koraichi, S., and El Ghachtouli, N. (2019). Biofilm-based systems for industrial wastewater treatment. Handbook of Environmental Materials Management Hussain, C. Springer, Cham 1767–1787. https://doi.org/10.1007/978-3-319-73645-7_137

Badia-Fabregat, M., Lucas, D., Tuomivirta, T. et al. (2017). Study of the effect of the bacterial and fungal communities present in real wastewater effluents on the performance of fungal treatments. *Sci. Total Environ.* 579: 366–377. https://doi.org/10.1016/j.scitotenv.2016.11.088.

Baker, B.J., Tyson, G.W., Goosherst, L., and Banfield, J.F. (2009). Insights into the diversity of eukaryotes in acid mine drainage biofilm communities. *Appl. Environ. Microbiol.* 75 (7): 2192–2199. https://doi.org/10.1128/AEM.02500-08.

Bastiaens, L., Springael, D., Wattiau, P. et al. (2000). Isolation of adherent polycyclic aromatic hydrocarbon (PAH)-degrading bacteria using PAH-sorbing carriers. *Appl. Environ. Microbiol.* 66 (5): 1834–1843. https://doi.org/10.1128/AEM.66.5.1834-1843.2000.

Baumgarten, T., Sperling, S., Seifert, J. et al. (2012). Membrane vesicle formation as a multiple-stress response mechanism enhances *Pseudomonas putida* DOT-T1E cell surface hydrophobicity and biofilm formation. *Appl. Environ. Microbiol. 78* (17): 6217–6224.

Beaudoin, D.L., Bryers, J.D., Cunningham, A.B., and Peretti, S.W. (1998). Mobilization of broad host range plasmid from *Pseudomonas putida* to established biofilm of *Bacillus azotoformans*. I. Experiments. *Biotechnol. Bioeng.* 57 (3): 272–279.

Berk, V., Fong, J.C., Dempsey, G.T. et al. (2012). Molecular architecture and assembly principles of *Vibrio cholerae* biofilms. *Science* 337 (6091): 236–239. https://doi.org/10.1126/science.1222981.

Beveridge, T.J., Makin, S.A., Kadurugamuwa, J.L., and Li, Z. (1997). Interactions between biofilms and the environment. *FEMS Microbiol. Rev.* 20 (3–4): 291–303. https://doi.org/10.1111/j.1574-6976.1997.tb00315.x.

Bielig, H., Dongre, M., Zurek, B. et al. (2011). A role for quorum sensing in regulating innate immune responses mediated by *Vibrio cholerae* outer membrane vesicles (OMVs). *Gut Microbes* 2: 274–279.

Bighiu, M.A. and Goedkoop, W. (2021). Interactions with freshwater biofilms cause rapid removal of common herbicides through degradation – evidence from microcosm studies. *Environ Sci Process Impacts.* 23 (1): 66–72.

Billings, N., Millan, M., Caldara, M. et al. (2013). The extracellular matrix component Psl provides fast-acting antibiotic defense in *Pseudomonas aeruginosa* biofilms. *PLoS Pathog.* 9 (8): e1003526. https://doi.org/10.1371/journal.ppat.1003526.

Boelee, N.C., Temmink, H., Janssen, M. et al. (2011). Nitrogen and phosphorus removal from municipal wastewater effluent using microalgal biofilms. *Water Res.* 45 (18): 5925–5933. https://doi.org/10.1016/j.watres.2011.08.044.

Boles, B.R., Thoendel, M., and Singh, P.K. (2004). Self-generated diversity produces "insurance effects" in biofilm communities. *Proc. Natl. Acad. Sci. U. S. A.* 101 (47): 16630–16635. https://doi.org/10.1073/pnas.0407460101.

Boles, B.R., Thoendel, M., Roth, A.J., and Horswill, A.R. (2010). Identification of genes involved in polysaccharide-independent *Staphylococcus aureus* biofilm formation. *PLoS One* 5 (4): e10146. https://doi.org/10.1371/journal.pone.0010146.

Boon, N., Top, E.M., Verstraete, W., and Siciliano, S.D. (2003). Bioaugmentation as a tool to protect the structure and function of an activated-sludge microbial community against a 3-chloroaniline shock load. *Appl. Environ. Microbiol.* 69 (3): 1511–1520. https://doi.org/10.1128/AEM.69.3.1511-1520.2003.

Borlee, B.R., Goldman, A.D., Murakami, K. et al. (2010). *Pseudomonas aeruginosa* uses a cyclic-di-GMP-regulated adhesin to reinforce the biofilm extracellular matrix. *Mol. Microbiol.* 75 (4): 827–842. https://doi.org/10.1111/j.1365-2958.2009.06991.x.

Branda, S.S., González-Pastor, J.E., Ben-Yehuda, S. et al. (2001). Fruiting body formation by *Bacillus subtilis*. *Proc. Natl. Acad. Sci. U. S. A.* 98 (20): 11621–11626. https://doi.org/10.1073/pnas.191384198.

Branda, S.S., González-Pastor, J.E., Dervyn, E. et al. (2004). Genes involved in formation of structured multicellular communities by *Bacillus subtilis*. *J. Bacteriol.* 186: 3970–3979.

Brunet-Galmés, I., Busquets, A., Peña, A. et al. (2012). Complete genome sequence of the naphthalene-degrading bacterium *Pseudomonas stutzeri* AN10 (CCUG 29243). *J. Bacteriol.* 194 (23): 6642–6643. https://doi.org/10.1128/JB.01753-12.

Buth, J.M., Grandbois, M., Vikesland, P.J. et al. (2009). Aquatic photochemistry of chlorinated triclosan derivatives: potential source of polychlorodibenzo-*P*-dioxins. *Environ. Toxicol. Chem.* 28 (12): 2555–2563. https://doi.org/10.1897/08-490.1.

Chan, C., Burrows, L.L., and Deber, C.M. (2004). Helix induction in antimicrobial peptides by alginate in biofilms. *J. Biol. Chem.* 279 (37): 38749–38754. https://doi.org/10.1074/jbc.M406044200.

Chaudhry, G.R. and Chapalamadugu, S. (1991). Biodegradation of halogenated organic compounds. *Microbiol. Rev.* 55 (1): 59–79.

Chen, X., Suwarno, S.R., Chong, T.H. et al. (2013). Dynamics of biofilm formation under different nutrient levels and the effect on biofouling of a reverse osmosis membrane system. *Biofouling* 29 (3): 319–330. https://doi.org/10.1080/08927014.2013.772141.

Clauwaert, P., Rabaey, K., Aelterman, P. et al. (2007). Biological denitrification in microbial fuel cells. *Environ. Sci. Technol.* 41 (9): 3354–3360. https://doi.org/10.1021/es062580r.

Cook, K.E. and Colvin, J.R. (1980). Evidence for a beneficial influence of cellulose production on growth of *Acetobacter xylinum* in liquid medium. *Curr. Microbiol.* 3 (4): 203–205. https://doi.org/10.1007/BF02602449.

Costerton, J.W. (1999). Introduction to biofilm. *Int. J. Antimicrob. Agents* 11 (3–4): 217–239. https://doi.org/10.1016/s0924-8579(99)00018-7.

Cucarella, C., Solano, C., Valle, J. et al. (2001). Bap, a *Staphylococcus aureus* surface protein involved in biofilm formation. *J. Bacteriol.* 183 (9): 2888–2896. https://doi.org/10.1128/JB.183.9.2888-2896.2001.

Cucarella, C., Tormo, M.A., Knecht, E. et al. (2002). Expression of the biofilm-associated protein interferes with host protein receptors of *Staphylococcus aureus* and alters the infective process. *Infect. Immun.* 70 (6): 3180–3186. https://doi.org/10.1128/IAI.70.6.3180-3186.2002.

Cucarella, C., Tormo, M.A., Ubeda, C. et al. (2004). Role of biofilm-associated protein bap in the pathogenesis of bovine *Staphylococcus aureus*. *Infect. Immun.* 72 (4): 2177–2185. https://doi.org/10.1128/IAI.72.4.2177-2185.2004.

Danese, P.N., Pratt, L.A., and Kolter, R. (2000). Exopolysaccharide production is required for development of *Escherichia coli* K-12 biofilm architecture. *J. Bacteriol.* 182 (12): 3593–3596. https://doi.org/10.1128/JB.182.12.3593-3596.2000.

Das, N., Basak, L.V.G., and Salam, J.A. (2017). Application of biofilms on remediation of pollutants—an overview. *J. Microbiol. Biotechnol. Res.* 2 (5): 783–790.

Davies, D.G., Parsek, M.R., Pearson, J.P. et al. (1998). The involvement of cell-to-cell signals in the development of a bacterial biofilm. *Science* 280 (5361): 295–298. https://doi.org/10.1126/science.280.5361.295.

Deghmane, A.E., Giorgini, D., Larribe, M. et al. (2002). Down-regulation of pili and capsule of *Neisseria meningitidis* upon contact with epithelial cells is mediated by CrgA regulatory protein. *Mol. Microbiol.* 43 (6): 1555–1564. https://doi.org/10.1046/j.1365-2958.2002.02838.x.

Di Fabio, S., Lampis, S., Zanetti, L. et al. (2013). Role and characteristics of problematic biofilms within the removal and mobility of trace metals in a pilot-scale membrane bioreactor. *Process Biochem.* 48: 1757–1766.

Dibdin, G.H., Assinder, S.J., Nichols, W.W., and Lambert, P.A. (1996). Mathematical model of beta-lactam penetration into a biofilm of *Pseudomonas aeruginosa* while undergoing simultaneous inactivation by released beta-lactamases. *J. Antimicrob. Chemother.* 38 (5): 757–769. https://doi.org/10.1093/jac/38.5.757.

Diggle, S.P., Stacey, R.E., Dodd, C. et al. (2006). The galactophilic lectin, LecA, contributes to biofilm development in *Pseudomonas aeruginosa*. *Environ. Microbiol.* 8 (6): 1095–1104. https://doi.org/10.1111/j.1462-2920.2006.001001.x.

Dogsa, I., Brloznik, M., Stopar, D., and Mandic-Mulec, I. (2013). Exopolymer diversity and the role of Levan in *Bacillus subtilis* biofilms. *PLoS One* 8 (4): e62044. https://doi.org/10.1371/journal.pone.0062044.

Donlan, R.M. (2000). Role of biofilms in antimicrobial resistance. *ASAIO J.* 46 (6): S47–S52. https://doi.org/10.1097/00002480-200011000-00037.

Donlan, R.M. (2002). Biofilms: microbial life on surfaces. *Emerg. Infect Dis.* 8 (9): 881–890. 10.3201/eid0809.020063. PMID: 12194761; PMCID: PMC2732559.

Duperthuy, M., Sjöström, A.E., Sabharwal, D. et al. (2013). Role of the *Vibrio cholerae* matrix protein Bap1 in cross-resistance to antimicrobial peptides. *PLoS Pathog.* 9 (10): e1003620.

Edwards, S.J. and Kjellerup, B.V. (2013). Applications of biofilms in bioremediation and biotransformation of persistent organic pollutants, pharmaceuticals/personal care products, and heavy metals. *Appl. Microbiol. Biotechnol.* 97 (23): 9909–9921. https://doi.org/10.1007/s00253-013-5216-z.

Ferrera, I., Massana, R., Casamayor, E.O. et al. (2004). High-diversity biofilm for the oxidation of sulfide-containing effluents. *Appl. Microbiol. Biotechnol.* 64 (5): 726–734. https://doi.org/10.1007/s00253-004-1582-x.

Flemming, H.C. (1998). Relevance of biofilms for the biodeterioration of surfaces of polymeric materials. *Polym. Degrad. Stab.* 59: 309–315.

Flemming, H.C. and Wingender, J. (2010). The biofilm matrix. *Nat. Rev. Microbiol.* 8 (9): 623–633. https://doi.org/10.1038/nrmicro2415.

Flemming, H.C., Wingender, J., Szewzyk, U. et al. (2016). Biofilms: an emergent form of bacterial life. *Nat. Rev. Microbiol.* 14 (9): 563–575. https://doi.org/10.1038/nrmicro.2016.94.

Fong, J.C. and Yildiz, F.H. (2007). The rbmBCDEF gene cluster modulates development of rugose colony morphology and biofilm formation in *Vibrio cholerae*. *J. Bacteriol.* 189 (6): 2319–2330. https://doi.org/10.1128/JB.01569-06.

Fong, J.C., Karplus, K., Schoolnik, G.K., and Yildiz, F.H. (2006). Identification and characterization of RbmA, a novel protein required for the development of rugose colony morphology and biofilm structure in *Vibrio cholerae*. *J. Bacteriol.* 188 (3): 1049–1059. https://doi.org/10.1128/JB.188.3.1049-1059.2006.

Galloway, T.S., Cole, M., and Lewis, C. (2017). Interactions of microplastic debris throughout themarine ecosystem. *Nature Ecol. Evol.* 1 (5): 116. https://doi.org/10.1038/s41559-017-0116.

Garrett, T.R., Bhakoo, M., and Zhang, Z. (2008). Bacterial adhesion and biofilms on surfaces. *Prog. Nat. Sci.* 18 (9): 1049–1056.

Ghigo, J.M. (2001). Natural conjugative plasmids induce bacterial biofilm development. *Nature* 412 (6845): 442–445. https://doi.org/10.1038/35086581.

Goodman, S.D., Obergfell, K.P., Jurcisek, J.A. et al. (2011). Biofilms can be dispersed by focusing the immune system on a common family of bacterial nucleoid-associated proteins. *Mucosal Immunol.* 2011 (4): 625–637.

Grimaud, R. (2010). Biofilm development at interfaces between hydrophobic organic compounds and water. In: *Handbook of Hydrocarbon and Lipid Microbiology* (ed. K.N. Timmis). Berlin, Heidelberg: Springer https://doi.org/10.1007/978-3-540-77587-4_102.

Gu, J.D. (2003). Microbiological deterioration and degradation of synthetic polymeric materials: recent research advances. *Int. Biodeterior. Biodegrad.* 52: 69–91.

Gupta, P. and Diwan, B. (2016). Bacterial exopolysaccharide mediated heavy metal removal: a review on biosynthesis, mechanism and remediation strategies. *Biotechnol. Rep.* 13: 58–71. https://doi.org/10.1016/j.btre.2016.12.006.

Gustave, J.E., Jurcisek, J.A., McCoy, K.S. et al. (2013). Targeting bacterial integration host factor to disrupt biofilms associated with cystic fibrosis. *J. Cyst. Fibros.* 12 (4): 384–389. https://doi.org/10.1016/j.jcf.2012.10.011.

Hall-Stoodley, L., Costerton, J.W., and Stoodley, P. (2004). Bacterial biofilms: from the natural environment to infectious diseases. *Nat. Rev. Microbiol.* 2 (2): 95–108. https://doi.org/10.1038/nrmicro821.

Hall-Stoodley, L., Nistico, L., Sambanthamoorthy, K. et al. (2008). Characterization of biofilm matrix, degradation by DNase treatment and evidence of capsule downregulation in *Streptococcus pneumoniae* clinical isolates. *BMC Microbiol.* 8: 173. https://doi.org/10.1186/1471-2180-8-173.

Heipieper, H.J., Keweloh, H., and Rehm, H.J. (1991). Influence of phenols on growth and membrane permeability of free and immobilized *Escherichia coli*. *Appl. Environ. Microbiol.* 57 (4): 1213–1217. https://doi.org/10.1128/aem.57.4.1213-1217.1991.

Hengzhuang, W., Wu, H., Ciofu, O. et al. (2011). Pharmacokinetics/pharmacodynamics of colistin and imipenem on mucoid and nonmucoid *Pseudomonas aeruginosa* biofilms. *Antimicrob. Agents Chemother.* 55 (9): 4469–4474. https://doi.org/10.1128/AAC.00126-11.

Hentzer, M., Teitzel, G.M., Balzer, G.J. et al. (2001). Alginate overproduction affects *Pseudomonas aeruginosa* biofilm structure and function. *J. Bacteriol.* 183 (18): 5395–5401. https://doi.org/10.1128/JB.183.18.5395-5401.2001.

Hobley, L., Ostrowski, A., Rao, F.V. et al. (2013). BslA is a self-assembling bacterial hydrophobin that coats the *Bacillus subtilis* biofilm. *Proc. Natl. Acad. Sci. U. S. A.* 110 (33): 13600–13605. https://doi.org/10.1073/pnas.1306390110.

Hoh, D., Watson, S., and Kan, E. (2016). Algal biofilm reactors for integrated wastewater treatment and biofuel production a review. *Chem. Eng. J.* 287: 466–473.

Høiby, N., Bjarnsholt, T., Givskov, M. et al. (2010). Antibiotic resistance of bacterial biofilms. *Int. J. Antimicrob. Agents* 35 (4): 322–332. https://doi.org/10.1016/j.ijantimicag.2009.12.011.

Hossain, M.R., Jiang, M., Wei, Q., and Leff, L.G. (2019). Microplastic surface properties affect bacterial colonization in freshwater. *J. Basic Microbiol.* 59 (1): 54–61. https://doi.org/10.1002/jobm.201800174.

Hoyle, B.D., Wong, C.K., and Costerton, J.W. (1992). Disparate efficacy of tobramycin on Ca(2+)-, Mg(2+)-, and HEPES-treated *Pseudomonas aeruginosa* biofilms. *Can. J. Microbiol.* 38 (11): 1214–1218. https://doi.org/10.1139/m92-201.

Hussain, M., Wilcox, M.H., and White, P.J. (1993). The slime of coagulase-negative staphylococci: biochemistry and relation to adherence. *FEMS Microbiol. Rev.* 10 (3–4): 191–207. https://doi.org/10.1111/j.1574-6968.1993.tb05867.x.

Irie, Y., Borlee, B.R., O'Connor, J.R. et al. (2012). Self-produced exopolysaccharide is a signal that stimulates biofilm formation in *Pseudomonas aeruginosa*. *Proc. Natl. Acad. Sci. U. S. A.* 109 (50): 20632–20636. https://doi.org/10.1073/pnas.1217993109.

Jaishankar, M., Tseten, T., Anbalagan, N. et al. (2014). Toxicity, mechanism and health effects of some heavy metals. *Interdiscip. Toxicol.* 7 (2): 60–72. https://doi.org/10.2478/intox-2014-0009.

Jasu, A. and Ray, R.R. (2021). Biofilm mediated strategies to mitigate heavy metal pollution: a critical review in metal bioremediation. *Biocatal. Agric. Biotechnol.* 37: 102183, ISSN 1878-8181, https://doi.org/10.1016/j.bcab.2021.102183.

Jesaitis, A.J., Franklin, M.J., Berglund, D. et al. (2003). Compromised host defense on *Pseudomonas aeruginosa* biofilms: characterization of neutrophil and biofilm interactions. *J. Immunol.* 171 (8): 4329–4339. https://doi.org/10.4049/jimmunol.171.8.4329.

Joklik, W.K., Willett, H.P., Amos, D.B., and Wilfert, C.M. (1988). *Zinsser Microbiology*, vol. 19. Norwalk, CT: Appleton and Lange.

Kaplan, J.B. (2010). Biofilm dispersal: mechanisms, clinical implications, and potential therapeutic uses. *J. Dent. Res.* 89 (3): 205–218.

Kaplan, J.B., Ragunath, C., Ramasubbu, N., and Fine, D.H. (2003). Detachment of *Actinobacillus actinomycetemcomitans* biofilm cells by an endogenous beta-hexosaminidase activity. *J. Bacteriol.* 185 (16): 4693–4698. https://doi.org/10.1128/JB.185.16.4693-4698.2003.

Karatan, E. and Watnick, P. (2009). Signals, regulatory networks, and materials that build and break bacterial biofilms. *Microbiol. Mol. Biol. Rev.* 73 (2): 310–347. https://doi.org/10.1128/MMBR.00041-08.

Kearns, D.B., Chu, F., Branda, S.S. et al. (2005). A master regulator for biofilm formation by *Bacillus subtilis*. *Mol. Microbiol.* 55 (3): 739–749. https://doi.org/10.1111/j.1365-2958.2004.04440.x.

Kirn, T.J., Jude, B.A., and Taylor, R.K. (2005). A colonization factor links *Vibrio cholerae* environmental survival and human infection. *Nature* 438 (7069): 863–866. https://doi.org/10.1038/nature04249.

Kiska, D.L. and Macrina, F.L. (1994). Genetic analysis of fructan-hyperproducing strains of Streptococcus mutans. *Infect. Immun.* 62 (7): 2679–2686. https://doi.org/10.1128/iai.62.7.2679-2686.1994.

Kobayashi, K. and Iwano, M. (2012). BslA(YuaB) forms a hydrophobic layer on the surface of *Bacillus subtilis* biofilms. *Mol. Microbiol.* 85 (1): 51–66. https://doi.org/10.1111/j.1365-2958.2012.08094.x.

Kocianova, S., Vuong, C., Yao, Y. et al. (2005). Key role of poly-gamma-DL-glutamic acid in immune evasion and virulence of *Staphylococcus epidermidis*. *J. Clin. Invest.* 115 (3): 688–694. https://doi.org/10.1172/JCI23523.

Kolenbrander, P.E. and London, J. (1993). Adhere today, here tomorrow: oral bacterial adherence. *J. Bacteriol.* 175 (11): 3247–3252. https://doi.org/10.1128/jb.175.11.3247-3252.1993.

Kondo, K., Takade, A., and Amako, K. (1993). Release of the outer membrane vesicles from *Vibrio cholerae* and *Vibrio parahaemolyticus*. *Microbiol. Immunol.* 37: 149–152.

Krasteva, P.V., Fong, J.C., Shikuma, N.J. et al. (2010). *Vibrio cholerae* VpsT regulates matrix production and motility by directly sensing cyclic di-GMP. *Science* 327 (5967): 866–868. https://doi.org/10.1126/science.1181185.

Lam, J., Chan, R., Lam, K., and Costerton, J.W. (1980). Production of mucoid microcolonies by *Pseudomonas aeruginosa* within infected lungs in cystic fibrosis. *Infect. Immun.* 28 (2): 546–556. https://doi.org/10.1128/iai.28.2.546-556.1980.

Lappin-Scott, H.M. and Bass, C. (2001). Biofilm formation: attachment, growth, and detachment of microbes from surfaces. *Am. J. Infect Control.* 29 (4): 250–251. https://doi.org/10.1067/mic.2001.115674. PMID: 11486266.

Latasa, C., Roux, A., Toledo-Arana, A. et al. (2005). BapA, a large secreted protein required for biofilm formation and host colonization of *Salmonella enterica* serovar Enteritidis. *Mol. Microbiol.* 58 (5): 1322–1339. https://doi.org/10.1111/j.1365-2958.2005.04907.x.

Latch, D.E., Packer, J.L., Arnold, W.A., and Mc. Neill, K. (2003). Photochemical conversion of triclosan to 2,8-dichlorodibenzo-p-dioxin in aqueous solution. *J. Photochem. Photobiol. A Chem.* 158: 63–66.

Laura, M., Snchez-Salinas, E., Dantn Gonzlez, E., and Luisa, M. (2013). *Pesticide Biodegradation: Mechanisms, Genetics and Strategies to Enhance the Process*. InTech https://doi.org/10.5772/56098.

Learn, D.B., Brestel, E.P., and Seetharama, S. (1987). Hypochlorite scavenging by *Pseudomonas aeruginosa* alginate. *Infect. Immun.* 55 (8): 1813–1818. https://doi.org/10.1128/iai.55.8.1813-1818.1987.

Leid, J.G., Shirtliff, M.E., Costerton, J.W., and Stoodley, P. (2002). Human leukocytes adhere to, penetrate, and respond to *Staphylococcus aureus* biofilms. *Infect. Immun.* 70 (11): 6339–6345. https://doi.org/10.1128/IAI.70.11.6339-6345.2002.

Lerch, T.Z., Chenu, C., Dignac, M.F. et al. (2017). Biofilm vs. planktonic lifestyle: consequences for pesticide 2,4-D metabolism by *Cupriavidus necator* JMP134. *Front. Microbiol.* 8: 904.

Lewandowski, Z. and Evans, L.V. (2000). Structure and function of biofilms. In: *Biofilms: Recent Advances in their Study and Control* (ed. L.V. Evans), 1–17. London: CRC Press.

Li, Y.-H. and Tian, X. (2012). Quorum sensing and bacterial social interactions in biofilms. *Sensors.* 12 (3): 2519–2538.

Limoli, D.H., Rockel, A.B., Host, K.M. et al. (2014). Cationic antimicrobial peptides promote microbial mutagenesis and pathoadaptation in chronic infections. *PLoS Pathog.* 10 (4): e1004083. https://doi.org/10.1371/journal.ppat.1004083.

Logan, B.E., Hamelers, B., Rozendal, R. et al. (2006). Microbial fuel cells: methodology and technology. *Environ. Sci. Technol.* 40 5181–5192. https://doi.org/10.1021/es0605016.

Lolas, I.B., Chen, X., Bester, K., and Nielsen, J.L. (2012). Identification of triclosan-degrading bacteria using stable isotope probing, fluorescence in situ hybridization and microautoradiography. *Microbiology* 158 (11): 2796–2804. https://doi.org/10.1099/mic.0.061077-0.

Ma, L., Conover, M., Lu, H. et al. (2009). Assembly and development of the *Pseudomonas aeruginosa* biofilm matrix. *PLoS Pathog.* 5 (3): e1000354.

Ma, L., Wang, S., Wang, D. et al. (2012). The roles of biofilm matrix polysaccharide Psl in mucoid *Pseudomonas aeruginosa* biofilms. *FEMS Immunol. Med. Microbiol.* 65 (2): 377–380. https://doi.org/10.1111/j.1574-695X.2012.00934.x.

Mah, T.F., Pitts, B., Pellock, B. et al. (2003). A genetic basis for *Pseudomonas aeruginosa* biofilm antibiotic resistance. *Nature* 426 (6964): 306–310. https://doi.org/10.1038/nature02122.

Mai, G.T., Seow, W.K., Pier, G.B. et al. (1993). Suppression of lymphocyte and neutrophil functions by *Pseudomonas aeruginosa* mucoid exopolysaccharide (alginate): reversal by physicochemical, alginase, and specific monoclonal antibody treatments. *Infect. Immun.* 61 (2): 559–564. https://doi.org/10.1128/iai.61.2.559-564.1993.

Mangwani, N., Kumari, S., and Das, S. (2015). Involvement of quorum sensing genes in biofilm development and degradation of polycyclic aromatic hydrocarbons by a marine bacterium *Pseudomonas aeruginosa* N6P6. *Appl. Microbiol. Biotechnol.* 99 (23): 10283–10297.

Mann, E.E., Rice, K.C., Boles, B.R. et al. (2009). Modulation of eDNA release and degradation affects *Staphylococcus aureus* biofilm maturation. *PLoS One* 4 (6): e5822. https://doi.org/10.1371/journal.pone.0005822.

Margesin, R. and Schinner, F. (2001). Biodegradation and bioremediation of hydrocarbons in extreme environments. *Appl. Microbiol. Biotechnol.* 56 (5–6): 650–663.

Marsili, E., Baron, D.B., Shikhare, I.D. et al. (2008). *Shewanella* secretes flavins that mediate extracellular electron transfer. *Proc. Natl. Acad. Sci. U. S. A.* 105 (10): 3968–3973. https://doi.org/10.1073/pnas.0710525105.

Martí, M., Trotonda, M.P., Tormo-Más, M.A. et al. (2010). Extracellular proteases inhibit protein-dependent biofilm formation in *Staphylococcus aureus*. *Microbes Infect.* 12 (1): 55–64. https://doi.org/10.1016/j.micinf.2009.10.005.

Marvasi, M., Visscher, P.T., and Casillas Martinez, L. (2010). Exopolymeric substances (EPS) from *Bacillus subtilis*: polymers and genes encoding their synthesis. *FEMS Microbiol. Lett.* 313 (1): 1–9. https://doi.org/10.1111/j.1574-6968.2010.02085.x.

Miao, L., Wang, P., Hou, J. et al. (2019). Distinct community structure and microbial functions of biofilms colonizing microplastics. *Sci. Total Environ.* 650: 2395–2402.

Min, B. and Logan, B.E. (2004). Continuous electricity generation from domestic wastewater and organic substrates in a flat plate microbial fuel cell. *Environ. Sci. Technol.* 38 (21): 5809–5814. https://doi.org/10.1021/es0491026.

Mishra, S., Huang, Y., Li, J. et al. (2022). Biofilm-mediated bioremediation is a powerful tool for the removal of environmental pollutants. *Chemosphere.* 294: 133609.

Mohapatra, R.K., Parhi, P.K., Patra, J.K. et al. (2017c). Biodetoxification of toxic heavy metals by marine metal resistant bacteria—a novel approach for bioremediation of the polluted saline environment. In: *Microbial Biotechnology* (ed. J. Patra, C. Vishnuprasad, and G. Das), 343–376. Singapore: Springer.

Montanaro, L., Poggi, A., Visai, L. et al. (2011). Extracellular DNA in biofilms. *Int. J. Artif Organs.* 34 (9): 824–831.

Monteiro, C., Saxena, I., Wang, X. et al. (2009). Characterization of cellulose production in *Escherichia coli* Nissle 1917 and its biological consequences. *Environ. Microbiol.* 11 (5): 1105–1116. https://doi.org/10.1111/j.1462-2920.2008.01840.x.

Mortensen, B.K. (2014). *Formation and Detection of Biofilms*. Bactoforce International.

Mosharaf, M.K., Tanvir, M., Haque, M.M. et al. (2018). Metal-adapted bacteria isolated from wastewaters produce biofilms by expressing proteinaceous curli fimbriae and cellulose nanofibers. *Front. Microbiol.* 9: 1334. https://doi.org/10.3389/fmicb.2018.01334.

Mulcahy, H., Charron-Mazenod, L., and Lewenza, S. (2008). Extracellular DNA chelates cations and induces antibiotic resistance in *Pseudomonas aeruginosa* biofilms. *PLOS Pathog.* 4 (11): e1000213.

le Nhi Cong, T., Ngoc Mai, C.T., Thanh, V.T. et al. (2014). Application of a biofilm formed by a mixture of yeasts isolated in Vietnam to degrade aromatic hydrocarbon polluted wastewater collected from petroleum storage. *Water Sci Technol.* 70 (2): 329–336.

Nijland, R., Hall, M.J., and Burgess, J.G. (2010). Dispersal of biofilms by secreted, matrix degrading, bacterial DNase. *PLoS One* 5 (12): e15668. https://doi.org/10.1371/journal.pone.0015668.

Nivens, D.E., Ohman, D.E., Williams, J., and Franklin, M.J. (2001). Role of alginate and its O acetylation in formation of *Pseudomonas aeruginosa* microcolonies and biofilms. *J. Bacteriol.* 183 (3): 1047–1057. https://doi.org/10.1128/JB.183.3.1047-1057.2001.

Novotny, L.A., Amer, A.O., Brockson, M.E. et al. (2013). Structural stability of *Burkholderia cenocepacia* biofilms is reliant on eDNA structure and presence of a bacterial nucleic acid binding protein. *PLoS One* 8 (6): e67629. https://doi.org/10.1371/journal.pone.0067629.

Omarova, M., Swientoniewski, L.T., Tsengam, I.K.M. et al. (2019). Biofilm formation by hydrocarbon-degrading marine bacteria and its effects on oil dispersion. *ACS Sustain. Chem. Eng.* 7 (17): 14490–14499.

Onesios, K.M. and Bouwer, E.J. (2012). Biological removal of pharmaceuticals and personal care products during laboratory soil aquifer treatment simulation with different primary substrate concentrations. *Water Res.* 46 (7): 2365–2375. https://doi.org/10.1016/j.watres.2012.02.001.

Ostrowski, A., Mehert, A., Prescott, A. et al. (2011). YuaB functions synergistically with the exopolysaccharide and TasA amyloid fibers to allow biofilm formation by *Bacillus subtilis*. *J. Bacteriol.* 193 (18): 4821–4831. https://doi.org/10.1128/JB.00223-11.

Otto, M. (2018). Staphylococcal biofilms. Microbiology spectrum. *ASM J.* 6 (4): 10.

Panlilio, H. and Rice, C.V. (2021). The role of extracellular DNA in the formation, architecture, stability, and treatment of bacterial biofilms. *Biotechnol. Bioeng.* 118 (6): 2129–2141.

Parrish, K. and Fahrenfeld, N.L. (2019). Microplastic biofilm in fresh- and wastewater as a function of microparticle type and size class. *Environ. Sci. Water Res. Technol.* 5: 495–505.

Pearce, C.I., Lloyd, J.R., and Guthrie, J.T. (2003). The removal of colour from textile wastewater using whole bacterial cells: a review. *Dyes Pigm.* 58 (3): 179–196.

Picioreanu, C., Van Loosdrecht, M.C., and Heijnen, J.J. (2000). Effect of diffusive and convective substrate transport on biofilm structure formation: a two-dimensional modeling study. *Biotechnol. Bioeng.* 69 (5): 504–515. https://doi.org/10.1002/1097-0290(20000905)69:5<504::aid-bit5>3.0.co;2-s.

Pier, G.B., Coleman, F., Grout, M. et al. (2001). Role of alginate O acetylation in resistance of mucoid *Pseudomonas aeruginosa* to opsonic phagocytosis. *Infect. Immun.* 69 (3): 1895–1901. https://doi.org/10.1128/IAI.69.3.1895-1901.2001.

Poosarla, V.G., Wood, T.L., Zhu, L. et al. (2017). Dispersal and inhibitory roles of mannose, 2-deoxy-d-glucose and *N*-acetylgalactosaminidase on the biofilm of *Desulfovibrio vulgaris*. *Environ. Microbiol. Rep.* 9 (6): 779–787. https://doi.org/10.1111/1758-2229.12595.

Proia, L., Morin, S., Peipoch, M. et al. (2011). Resistance and recovery of river biofilms receiving short pulses of Triclosan and Diuron. *Sci. Total Environ.* 409: 3129–3137.

Proia, L., Osorio, V., Soley, S. et al. (2013). Effects of pesticides and pharmaceuticals on biofilms in a highly impacted river. *Environ. Pollut.* 178: 220–228. https://doi.org/10.1016/j.envpol.2013.02.022.

Randrianjatovo-Gbalou, I., Rouquette, P., Lefebvre, D. et al. (2017). In situ analysis of *Bacillus licheniformis* biofilms: amyloid-like polymers and eDNA are involved in the adherence and aggregation of the extracellular matrix. *J. Appl. Microbiol.* 122 (5): 1262–1274. https://doi.org/10.1111/jam.13423.

Rheinheimer Dos Santos, D., de Castro, M., Lima, J.A. et al. (2020). Pesticide bioaccumulation in epilithic biofilms as a biomarker of agricultural activities in a representative watershed. *Environ. Monit. Assess.* 192 (6): 381. https://doi.org/10.1007/s10661-020-08264-8.

Roberts, I.S. (1996). The biochemistry and genetics of capsular polysaccharide production in bacteria. *Annu. Rev. Microbiol.* 50: 285–315. https://doi.org/10.1146/annurev.micro.50.1.285.

Rohde, H., Burandt, E.C., Siemssen, N. et al. (2007). Polysaccharide intercellular adhesin or protein factors in biofilm accumulation of *Staphylococcus epidermidis* and *Staphylococcus aureus* isolated from prosthetic hip and knee joint infections. *Biomaterials* 28 (9): 1711–1720. https://doi.org/10.1016/j.biomaterials.2006.11.046.

Ross, P., Mayer, R., and Benziman, M. (1991). Cellulose biosynthesis and function in bacteria. *Microbiol. Rev.* 55 (1): 35–58. https://doi.org/10.1128/mr.55.1.35-58.1991.

Saba, B., Christy, A.D., Yu, Z. et al. (2017). Characterization and performance of anodic mixed culture biofilms in submersed microbial fuelcells. *Bioelectrochemistry* 113: 79–84. https://doi.org/10.1016/j.bioelechem.2016.10.003.

Saiyari, D.M., Chuang, H.-P., Senoro, D.B. et al. (2018). A review in the current developments of genus Dehalococcoides, its consortia and kinetics for bioremediation options of contaminated groundwater. *Sustain. Environ. Res.* 28 (4): 149–157, ISSN 2468-2039. https://doi.org/10.1016/j.serj.2018.01.006.

Sanchez, C.J., Kumar, N., Lizcano, A. et al. (2011). *Streptococcus pneumoniae* in biofilms are unable to cause invasive disease due to altered virulence determinant production. *PLoS One* 6 (12): e28738. https://doi.org/10.1371/journal.pone.0028738.

Satlewal, A., Soni, R., Zaidi, M. et al. (2008). Comparative biodegradation of HDPE and LDPE using an indigenously developed microbial consortium. *J. Microbiol. Biotechnol.* 18 (3): 477–482.

Schooling, S.R. and Beveridge, T.J. (2006). Membrane vesicles: an overlooked component of the matrices of biofilms. *J. Bacteriol.* 188 (16): 5945–5957. https://doi.org/10.1128/JB.00257-06.

Seo, J.S., Keum, Y.S., and Li, Q.X. (2009a). Bacterial degradation of aromatic compounds. *Int. J. Environ. Res. Public Health* 6 (1): 278–309. https://doi.org/10.3390/ijerph6010278.

Seo, Y., Lee, W.H., Sorial, G., and Bishop, P.L. (2009b). The application of a mulch biofilm barrier for surfactant enhanced polycyclic aromatic hydrocarbon bioremediation. *Environ. Pollut.* 157 (1): 95–101. https://doi.org/10.1016/j.envpol.2008.07.022.

Shah, A.A., Hasan, F., Hameed, A., and Ahmed, S. (2008). Biological degradation of plastics: a comprehensive review. *Biotechnol. Adv.* 26 (3): 246–265. https://doi.org/10.1016/j.biotechadv.2007.12.005.

Shikuma, N.J., Fong, J.C., and Yildiz, F.H. (2012). Cellular levels and binding of c-di-GMP control subcellular localization and activity of the *Vibrio cholerae* transcriptional regulator VpsT. *PLoS Pathog.* 8 (5): e1002719. https://doi.org/10.1371/journal.ppat.1002719.

Shukla, S.K., Mangwani, N., Rao, T.S., and Das, S. (2014). Biofilm-mediated bioremediation of polycyclic aromatic hydrocarbons. In: *Microbial Biodegradation and Bioremediation* (ed. S. Das and H.R. Dash), 203–232. Elsevier.

Shukla, S.K., Mangwani, N., Karley, D., and Rao, T.S. (2017). Bacterial biofilms and genetic regulation for metal detoxification. In: *Handbook of Metal-Microbe Interactions and Bioremediation*, vol. 19 (ed. S. Das and H.R. Dash), 317. CRC Press, Taylor & Francis.

Silyn-Roberts, G. and Lewis, G. (2003). Substrata effects on bacterial biofilm development in a subsurface flow dairy waste treatment wetland. *Water Sci. Technol.* 48 (8): 261–269. PMID: 14682595.

Simpson, J.A., Smith, S.E., and Dean, R.T. (1989). Scavenging by alginate of free radicals released by macrophages. *Free Radic. Biol. Med.* 6 (4): 347–353. https://doi.org/10.1016/0891-5849(89)90078-6.

Singh, R., Paul, D., and Jain, R.K. (2006). Biofilms: implications in bioremediation. *Trends Microbiol.* 14 (9): 389–397. https://doi.org/10.1016/j.tim.2006.07.001.

Sinha, S., Behera, S.S., Das, S. et al. (2018). Removal of Congo red dye from aqueous solution using Amberlite IRA-400 in batch and fixed bed reactors. *Chem. Eng. Commun.* 205 (4): 432–444.

Stapper, A.P., Narasimhan, G., Ohman, D.E. et al. (2004). Alginate production affects *Pseudomonas aeruginosa* biofilm development and architecture, but is not essential for biofilm formation. *J. Med. Microbiol.* 53 (Pt 7): 679–690. https://doi.org/10.1099/jmm.0.45539-0.

Stevenson, G., Andrianopoulos, K., Hobbs, M., and Reeves, P.R. (1996). Organization of the *Escherichia coli* K-12 gene cluster responsible for production of the extracellular polysaccharide colanic acid. *J. Bacteriol.* 178 (16): 4885–4893. https://doi.org/10.1128/jb.178.16.4885-4893.1996.

Strong, L.C., Rosendahl, C., Johnson, G. et al. (2002). *Arthrobacter aurescens* TC1 metabolizes diverse s-triazine ring compounds. *Appl. Environ. Microbiol.* 68 (12): 5973–5980. https://doi.org/10.1128/AEM.68.12.5973-5980.2002.

Sui, Q., Huang, J., Deng, S. et al. (2011). Seasonal variation in the occurrence and removal of pharmaceuticals and personal care products in different biological wastewater treatment processes. *Environ. Sci. Technol.* 45 (8): 3341–3348. https://doi.org/10.1021/es200248d.

Sutherland, I. (2001). Biofilm exopolysaccharides: a strong and sticky framework. *Microbiology* 147 (Pt 1): 3–9. https://doi.org/10.1099/00221287-147-1-3.

Tenover, F.C., Schaberg, D.R., Bennett, J.V., and Brachman, P.S. (1998). *Molecular Biology of Resistance, Hospital Infections*, 4e, 237–247. Philadelphia: Lippincott-Raven.

Tielen, P., Strathmann, M., Jaeger, K.E. et al. (2005). Alginate acetylation influences initial surface colonization by mucoid *Pseudomonas aeruginosa*. *Microbiol. Res.* 160 (2): 165–176. https://doi.org/10.1016/j.micres.2004.11.003.

Tielker, D., Hacker, S., Loris, R. et al. (2005). *Pseudomonas aeruginosa* lectin LecB is located in the outer membrane and is involved in biofilm formation. *Microbiology* 151 (Pt 5): 1313–1323. https://doi.org/10.1099/mic.0.27701-0.

Toledo-Arana, A., Valle, J., Solano, C. et al. (2001). The enterococcal surface protein, Esp, is involved in *Enterococcus faecalis* biofilm formation. *Appl. Environ. Microbiol.* 67 (10): 4538–4545. https://doi.org/10.1128/AEM.67.10.4538-4545.2001.

Toyofuku, M., Roschitzki, B., Riedel, K., and Eberl, L. (2012). Identification of proteins associated with the *Pseudomonas aeruginosa* biofilm extracellular matrix. *J. Proteome Res.* 11 (10): 4906–4915. https://doi.org/10.1021/pr300395j.

Tresse, O., Jouenne, T., and Junter, G.A. (1995). The role of oxygen limitation in the resistance of agar-entrapped, sessile-like *Escherichia coli* to aminoglycoside and beta-lactam antibiotics. *J. Antimicrob. Chemother.* 36 (3): 521–526. https://doi.org/10.1093/jac/36.3.521.

Tuchscherr, L.P., Buzzola, F.R., Alvarez, L.P. et al. (2005). Capsule-negative *Staphylococcus aureus* induces chronic experimental mastitis in mice. *Infect. Immun.* 73 (12): 7932–7937. https://doi.org/10.1128/IAI.73.12.7932-7937.2005.

Valle, J., Latasa, C., Gil, C. et al. (2012). Bap, a biofilm matrix protein of *Staphylococcus aureus* prevents cellular internalization through binding to GP96 host receptor. *PLoS Pathog.* 8 (8): e1002843. https://doi.org/10.1371/journal.ppat.1002843.

Van Eerd, L.L., Hoagland, R.E., Zablotowicz, R.M., and Hall, J.C. (2003). Pesticide metabolism in plants and microorganisms. *Weed Sci.* 51: 472–495.

Velázquez-Hernández, M.L., Baizabal-Aguirre, V.M., Bravo-Patiño, A. et al. (2009). Microbial fructosyltransferases and the role of fructans. *J. Appl. Microbiol.* 106 (6): 1763–1778. https://doi.org/10.1111/j.1365-2672.2008.04120.x.

Verhamme, D.T., Murray, E.J., and Stanley-Wall, N.R. (2009). DegU and Spo0A jointly control transcription of two loci required for complex colony development by *Bacillus subtilis*. *J. Bacteriol.* 191 (1): 100–108. https://doi.org/10.1128/JB.01236-08.

Volke-Sepulveda, T., Saucedo-Castaneda, G., Gutierrez-Rojas, M. et al. (2002). Thermally treated low density polyethylene biodegradation by penicillium pinophilum and aspergillus niger. *J. Appl. Polym. Sci.* 83: 305–314.

Vu, B., Chen, M., Crawford, R.J., and Ivanova, E.P. (2009). Bacterial extracellular polysaccharides involved in biofilm formation. *Molecules* 14 (7): 2535–2554. https://doi.org/10.3390/molecules14072535.

Wang, R., Khan, B.A., Cheung, G.Y. et al. (2011). *Staphylococcus epidermidis* surfactant peptides promote biofilm maturation and dissemination of biofilm-associated infection in mice. *J. Clin. Invest.* 121 (1): 238–248. https://doi.org/10.1172/JCI42520.

Wentland, E.J., Stewart, P.S., Huang, C.T., and McFeters, G.A. (1996). Spatial variations in growth rate within *Klebsiella pneumoniae colonies* and biofilm. *Biotechnol. Prog.* 12 (3): 316–321. https://doi.org/10.1021/bp9600243.

Wijeyekoon, S., Mino, T., Satoh, H., and Matsuo, T. (2004). Effects of substrate loading rate on biofilm structure. *Water Res.* 38 (10): 2479–2488. https://doi.org/10.1016/j.watres.2004.03.005.

Williams, W.S. and Cannon, R.E. (1989). Alternative environmental roles for cellulose produced by *Acetobacter xylinum*. *Appl. Environ. Microbiol.* 55 (10): 2448–2452. https://doi.org/10.1128/aem.55.10.2448-2452.1989.

Williams, I., Venables, W.A., Lloyd, D. et al. (1997). The effects of adherence to silicone surfaces on antibiotic susceptibility in *Staphylococcus aureus*. *Microbiology* 143 (7): 2407–2413. https://doi.org/10.1099/00221287-143-7-2407.

Wolcott, R., Costerton, J.W., Raoult, D., and Cutler, S.J. (2013). The polymicrobial nature of biofilm infection. *Clin. Microbiol. Infect.* 19 (2): 107–112. https://doi.org/10.1111/j.1469-0691.2012.04001.x.

Woo, J.K., Webb, J.S., Kirov, S.M. et al. (2012). Biofilm dispersal cells of a cystic fibrosis *Pseudomonas aeruginosa* isolate exhibit variability in functional traits likely to contribute to persistent infection. *FEMS Immunol. Med. Microbiol.* 66 (2): 251–264. https://doi.org/10.1111/j.1574-695X.2012.01006.x.

Xiao, J., Klein, M.I., Falsetta, M.L. et al. (2012). The exopolysaccharide matrix modulates the interaction between 3D architecture and virulence of a mixed-species oral biofilm. *PLoS Pathog.* 8 (4): e1002623. https://doi.org/10.1371/journal.ppat.1002623.

Yañez-Ocampo, G., Sanchez-Salinas, E., Jimenez-Tobon, G.A. et al. (2009). Removal of two organophosphate pesticides by a bacterial consortium immobilized in alginate or tezontle. *J. Hazard Mater.* 168 (2–3): 1554–1561.

Yañez-Ocampo, G., Sánchez-Salinas, E., and Ortiz-Hernández, M.L. (2011). Removal of methyl parathion and tetrachlorvinphos by a bacterial consortium immobilized on tezontle-packed up-flow reactor. *Biodegradation* 22: 1203–1213.

Yildiz, F.H. and Schoolnik, G.K. (1999). *Vibrio cholerae* O1 El Tor: identification of a gene cluster required for the rugose colony type, exopolysaccharide production, chlorine resistance, and biofilm formation. *Proc. Natl. Acad. Sci. U. S. A.* 96 (7): 4028–4033. https://doi.org/10.1073/pnas.96.7.4028.

Yildiz, F.H., Dolganov, N.A., and Schoolnik, G.K. (2001). VpsR, a member of the response regulators of the two-component regulatory systems, is required for expression of vps biosynthesis genes and EPS(ETr)-associated phenotypes in *vibrio cholerae* O1 El Tor. *J. Bacteriol.* 183 (5): 1716–1726. https://doi.org/10.1128/JB.183.5.1716-1726.2001.

Yonezawa, H., Osaki, T., Kurata, S. et al. (2009). Outer membrane vesicles of Helicobacter pylori TK1402 are involved in biofilm formation. *BMC Microbiol.* 9: 197.

Yong, Y.C. and Zhong, J.J. (2013). Regulation of aromatics biodegradation by rhl quorum sensing system through induction of catechol meta-cleavage pathway. *Bioresour. Technol.* 136: 761–765. https://doi.org/10.1016/j.biortech.2013.03.134.

Yoon, J.S., Chin, I.J., Kim, M.N., and Kim, C. (1996). Degradation of microbial polyesters: a theoretical prediction of molecular weight and polydispersity. *Macromolecules* 29: 3303–3307.

Yoshida, S., Hiraga, K., Takehana, T. et al. (2016). A bacterium that degrades and assimilates poly(ethylene terephthalate). *Science* 351 (6278): 1196–1199. https://doi.org/10.1126/science.aad6359.

Zhang, K., Zheng, X., Shen, D.S. et al. (2015). Evidence for existence of quorum sensing in a bioaugmented system by acylated homoserine lactone-dependent quorum quenching. *Environ. Sci. Pollut. Res. Int.* 22 (8): 6050–6056. https://doi.org/10.1007/s11356-014-3795-6.

Zhao, K., Li, Y., Yue, B., and Wu, M. (2014). Genes as early responders regulate quorum-sensing and control bacterial cooperation in *Pseudomonas aeruginosa*. *PLoS One* 9 (7): e101887. https://doi.org/10.1371/journal.pone.0101887.

15

Recent Advances in the Biodegradation of Petroleum Hydrocarbons: Insights from Whole Genome Sequencing

Yahaya Y. Riko[1] *and Zubairu U. Darma*[2]

[1] *Department of Microbiology, Faculty of Natural and Applied Sciences, School of Postgraduate Studies, Umaru Musa 'Yar'adua University, Katsina, Nigeria*
[2] *Department of Microbiology, Umaru Musa 'Yar'adua University, Katsina, Nigeria*

15.1 Introduction: Aquatic Contamination Through Petroleum Hydrocarbons – Sources, Statistics, Impact, and Solution

Among the major problems in the current Anthropocene is the contamination of soil and water environments (Nikhil et al. 2013). This can result in the secondary contamination of water supplies within the underlying basement, thereby threatening the water supply. About 600,000 metric tons of oil that are spilled onshore every year eventually leak and seep into underground water aquifers, according to Kvenvolden and Cooper (2003). Spillage of petroleum products in aquatic ecosystems is further worsened by wave action and evaporation, leading to continuous widespread contamination of large areas of water (Jagtap et al. 2014). In certain water bodies, the spread of these petrochemical pollutants creates compressed mounds that later spread extensively due to the upward pressure of the water flow. Seasonal variations and fluctuations in water behavior can lead to further spread of the oil into the subsoil, upon the adsorption and stabilization of the petrochemicals. Ultimately, they seep, accumulate on the water table, and penetrate deeper into the saturated zone, causing long-term effects (Dorn and Salanitro 2000).

Numerous incidences of the spillage of petroleum hydrocarbons into the marine environment from crude oil sources, and/or its derivatives, such as engine oil, had been reported before, and such debilitating incidences continue to occur (Darma and Riko 2022). For instance, according to a 2001 estimate by Zhu et al. (2001), the amount of petroleum deposits into marine ecosystems ranges between 1.7 and 8.8 million metric tons, annually. Furthermore, the 2004 Gulf of Mexico spill incidence, which involved the Taylor Energy Well, initiated by a hurricane, leaked 0.03–0.05 t of oil per day into water bodies. Another spillage event occurred in Mumbai, India, and involved the leakage of 55 t of oil into the Arabian Sea (Nikhil et al. 2013). Additionally, another recent estimate (Chowdhury 2015) approximates that 0.08–0.4% of all produced crude oil worldwide is spilled into the water. By and large, every year witnesses the onshore and offshore spillage of a huge amount of crude oil and its derivatives, including engine oil (Fingas 2015). These spillage events that contaminate the aquatic environment (including marine and subsurface environments) extensively are typically due to accidents involving ships, failure of underground storage tanks, bursts, and leaks in buried pipes, and illicit activities such as improper disposal, oil bunkering, and pipeline vandalism (Darma and Riko 2022).

The impact of these water pollution events is detrimental to human health, directly or indirectly such as bioaccumulation/biomagnification of toxic, carcinogenic, or mutagenic substances across the food chain (Darma and Riko 2022), from low concentrations to overwhelming levels at the quaternary trophic level, usually occupied by humans (Das and Chandran 2011).

Aquatic Contamination: Tolerance and Bioremediation, First Edition. Edited by Rouf Ahmad Bhat, Gowhar Hamid Dar, Fernanda Maria Policarpo Tonelli, and Saima Hamid.

15.1.1 Need for Remediating Contaminated Aquatic Environments

Sustainable development has the overall goal of creating improved socioeconomic conditions and healthier environments for current and future generations. The concept was birthed as an ideology to build a modern world, and, by 2030, take concrete steps toward solving the major current global challenges (Ortiz et al. 2009; UN 2015).

Numerous issues threaten the achievement of many SDGs, one of which is the persistent pollution of aquatic environments with priority pollutants, primary among which are petroleum hydrocarbons, due to their toxicity (Curtis et al. 2021). Therefore, the achievement of many SDGs rests on the remediation of contaminated aquatic environments. Specifically, these SDGs include SDG 3 (good health and wellbeing, due to the impact of petroleum hydrocarbons on water bodies, and consequently on humans, through biomagnification, as explained before, or through contamination of contaminated waters), 6 (clean water and sanitation, due to contamination of water bodies/sources), 12 (responsible production and consumption, since the majority of spills/contaminations are during the process of producing/consuming crude oil), 13 (climate action, due to the impact of the pollutants on climate), 14 (life below water, because of the effects of petroleum contaminants on marine life), among others (Radelyuk et al. 2021).

In particular, goal 6, "Clean Water and Sanitation" encompasses the elimination of potential hazards of disposal of effluents and the protection of water-related ecosystems (rivers, lakes, aquifers, etc). (Radelyuk et al. 2021). Aquatic contamination by petroleum hydrocarbons threatens the achievement of these goals, in that it leads to the direct discharge of hazardous materials into the oceans, and it contaminates water environments and makes them hazardous.

Additionally, the 14th goal, "Life below Water", examines the impacts of water pollution. Failure to remediate aquatic ecosystems can lead to the deterioration of ecosystems and health. This, therefore, necessitates global endeavors to eradicate threats to the successful implementation of the SDGs, through innovate techno-scientific approaches, such as the bioremediation of contaminated aquatic ecosystems (UNDP 2016).

15.1.2 Bioremediation Using Hydrocarbonoclastic Bacteria as the Effective Solution to Contamination of Aquatic Environments by Petroleum Hydrocarbons

Numerous negative consequences of spillage of petrochemicals in water in particular and ultimately humans create an increased sense of environmental hazard and impeding catastrophe; subsequently making bioremediation indispensable and imperative (Umar 2017) that requires serious commitments from both agents behind the pollution and the regulatory agencies (Stroud et al. 2007). For example, the commercial, large-scale removal of pollutants from a contaminated site requires an estimated \$200,000 (Arthur et al. 2005).

The major methods for effectively remediating contaminated water environments in use today include the physicomechanical, chemical, and biological methods (Ameen et al. 2014; Darma and Riko 2022). Nonetheless, most of the remediation projects currently in use employ the conventional (physicochemical) approaches (Li et al. 2008; Hu et al. 2013; Yunusa and Umar 2021). Even though many physicochemical strategies for removing petrochemical contaminants were reported to have scaled laboratory tests, and have been applied in on-the-field experiments, nonetheless, physiochemical methods of remediation are associated with multiple inherent drawbacks (Darma and Riko 2022).

For instance, physical remediation methods, such as vapor extraction, are often ex situ in terms of the approach to remediation. Even though these methods are associated with the advantages of fast, permanent removal of pollutants and being ideal for high levels of pollution. However, their disadvantages include high cost (it costs approximately \$405–1485 to remediate just one cubic meter of the polluted environment), its destructive nature, the fact that it is prone to secondary pollution, etc. Chemical methods, on the other hand (such as thermal decomposition), also involve remediation away from the site of contamination, and have the advantage of fast removal of contaminants; likewise, they do not generate a large number of byproducts, and hence are recommended for use when the amount of pollutants is considerably high; nevertheless, their disadvantages also include cost (\$80–440/m^3), its destructive nature, and being prone to secondary pollution (Koshlaf and Ball 2017).

Bioremediation/biological remediation approaches (such as biostimulation), on the other hand, have been put forward as a cheap alternative to other methods of remediating contaminated environments (it costs just \$30–100 to bioremediate a cubic meter of polluted area) which involves minimal site disruption and environmental damage; occurs at the contaminated area, and is particularly useful for low levels of pollution; however, it often takes longer time to achieve, has low predictability, and is dependent on environmental conditions (Chowdhury 2015; Koshlaf and Ball 2017; Umar et al. 2018a,b).

Therefore, the need for bioremediation stems from the yearning to tackle many environmental challenges faced by physicochemical removal methods (Yunusa and Umar 2021), which also destroy ecosystems, and produce hazardous byproducts (Ghazali et al. 2004; Umar et al. 2019). Moreover, physical methods, such as photo-oxidation, burying, washing, incineration, thermal conversion, and pyrolysis are susceptible to releasing incomplete decomposition products (Li et al. 2008). Other methods, such as chemical oxidation and vitalization fail to remove aromatic hydrocarbons (Hu et al. 2013). However, bioremediation either involves the complete removal of the pollutants or at least making them become less harmful, through the activities of biological cells (Vijay 2012).

Bioremediation involves the use of living organisms (both macro- and microorganisms) as an alternative to physicochemical means for the removal of contaminants from spoilt areas, thus restoring the area to its pristine state (Yunusa and Umar 2021).

The advantages of bioremediation over conventional remediation technologies are numerous. Bioremediation using microorganisms is efficient and effective, labor- and cost-wise; likewise, it has minimal personnel and expenditure requirements; it is easy, simple, and cheap, and has no undesired repercussions on the other environmental flora and fauna (Agarry and Latinwo 2015; Dixit et al. 2018; Umar et al. 2020a,b), and therefore should be at the forefront during negotiations on the selection of appropriate cleansing techniques to be used in the event of spillage (Vijay 2012).

Bioremediation is especially important in the case of aquatic contamination because it ensures cleaner groundwater supplies, as it does not contaminate underground aquifers (Patowary et al. 2016; Darma and Riko 2022). Additionally, it can be used in the remediation of contaminated groundwater without excavation, and can perfectly be implemented below or within existing buildings, piping and paved surfaces (Vidali 2001; Das 2010; Darma and Riko 2022). Microbial bioremediation is more advantageous over other bioremediation techniques such as phytoremediation, because microorganisms have the ability to inhabit virtually all environmental niches, including extreme environments, where plants cannot survive, furthermore, they possess a wide metabolic repertoire (Desai et al. 2009; Umar et al. 2018a,b).

The use of microorganisms, especially bacteria, to remediate aquatic environments contaminated with petroleum hydrocarbons hinges on the principle of biodegradation, where various microorganisms are employed to degrade petroleum hydrocarbons using aerobic, anaerobic, or fermentation metabolisms (Vidali 2001). Through this alternative, living cells are used to completely mineralize the petroleum hydrocarbon contaminants to H_2O and CO_2 (Gallego et al. 2001).

The hydrocarbon-degrading bacteria (HDB), or hydrocarbonoclastic bacteria, comprise a taxonomically diverse group (~ 79 genera), and are metabolically versatile, with rich enzymatic repertoires and diverse pathways for hydrocarbonoclasticity (Mahjoubi et al. 2021). The process of biodegradation of an HC involves its breakdown either to harmless mineralization products (CO_2, H_2O, NH_3, etc.) or to simpler nontoxic forms, through actions of microbes or their enzymes (Kumari and Kumar 2021). This technique is successfully applicable in bioremediating polluted aquatic environments (Yunusa and Umar 2021).

Specific examples of some of these key bacterial genera shown to biodegrade petroleum are *Pseudomonas*, *Burkholderia*, *Alcaligenes*, *Flavobacterium*, *Xanthomonas*, *Norcadia*, *Stenotrophomonas*, *Cycloasticus*, *Franconibacter*, *Planococcus*, *Brevibacillus*, *Alcaligenes*, *Sphingomonas*, *Citrobacter*, *Mycobacterium*, *Corynebacterium*, *Arthrobacter*, *Bacillus*, *Staphylococcus*, *Halomonas*, *Klebsiella*, *Salmonella*, *Enterobacter*, *Chromium*, *Micrococcus*, and *Rhodococcus* (Darma and Riko 2022; Yunusa and Umar 2021; Musa 2019; Ukaegbu-Obi et al. 2017; Mwamura 2017; Naila 2015; Rehman et al. 2015; Karthika et al. 2014; Ganesh and Lin 2009; Uzoamaka et al. 2009; Fritsche and Hofrichter 2008; Van Hamme and Ward 2001; Solano-Serena et al. 2000).

15.2 Whole Genome Sequencing (WGS): History, Concepts, Methodology, Analyses, and Relevance to Biodegradation of Petroleum Hydrocarbons

15.2.1 A Historical Prelude to (WGS) of Petroleum Hydrocarbon-Degrading Bacteria

According to Darma and Riko (2022), the history of studies in the biodegradation of hydrocarbons can be said to have evolved in the following four eras:

1) The first era, characterized by the use of selective enrichment techniques to isolate HDB from diverse environmental samples. These bacteria depend on the particular petroleum hydrocarbon as their only source of carbon/energy (Bushnell and Haas 1941; Umar et al. 2019);

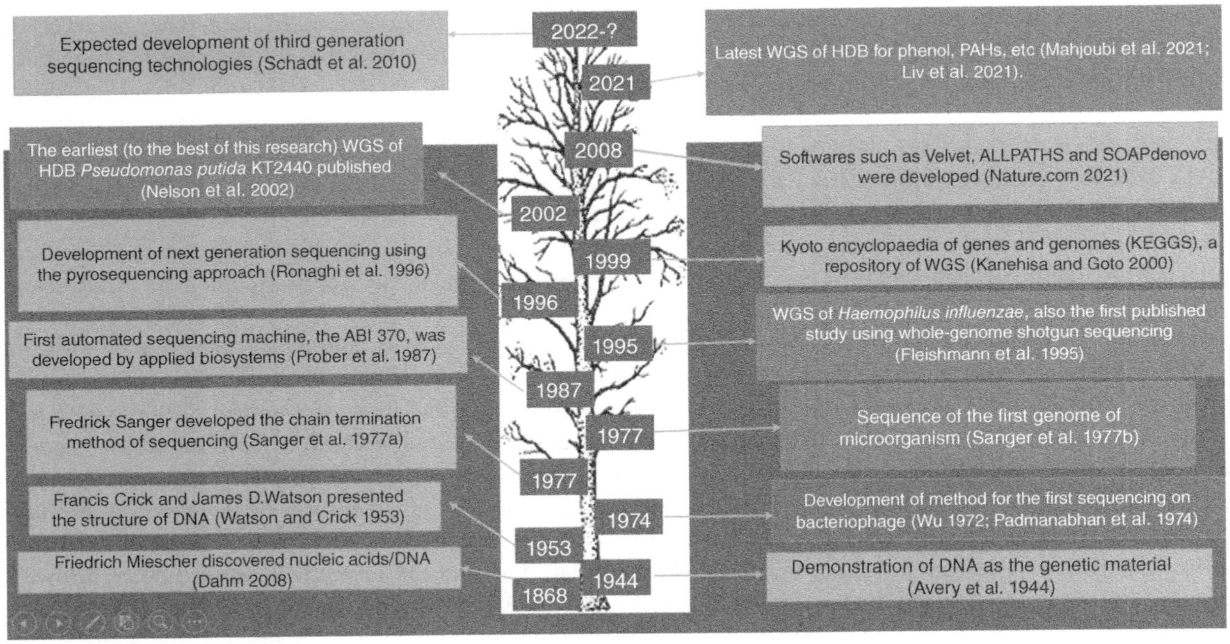

Figure 15.1 Key milestones in research on the WGS of hydrocarbonoclastic bacteria.

2) The second era, exemplified by the quantification of the rate/percentage of biodegradation using spectrometric/GC–MS techniques, etc. (Payne et al. 1970; Ruiz et al. 2021);
3) The third era, which witnessed the application of genetic/molecular techniques for the identification of hydrocarbonoclastic bacteria, the characterization of genes involved in the degradation, 16S rRNA-enabled identification of bacteria, and comparison of existing strains with previous strains, establishing phylogenetic relationships, etc. The era also witnessed the beginning of metagenomics/proteomics studies, etc. (Gauthier et al. 1992; Henckel et al. 2000; Yunusa and Umar 2021).
4) The fourth and current era, which started around the beginning of the new century, and involved the whole genome sequencing (WGS) of HDB, construction of genome sequences and "cell metabolism diagrams" (Nelson et al. 2002; Schneiker et al. 2006; Joye and Kostka 2020).

WGS itself can be traced to the discovery of nucleic acids by Freidrich Miescher in the year 1868 (Dahm 2008); down to the 2002 publication of the earliest (to the best of this research) whole genome sequence of an HDB (Nelson et al. 2002), and ultimately to recent (2021) developments, including the introduction of third-generation sequencing technologies (characterized by high throughput, decreased time to get results and lower costs) and the continual publishing of whole genome sequences. The important milestones in the journey are summarized in Figure 15.1 overleaf.

15.2.2 WGS: Definition and Basic Methodological Protocols Involved in a Typical WGS Pipeline

WGS, also known as complete/entire/full genome sequencing, refers to the method for identifying the entire sequence of DNA in a microorganism, including coding and noncoding sequences (NCI – National Cancer Institute 2021). It is capable of providing us with a comprehensive snapshot of the "life" of the microbe, including its degradation capability, production of metabolites (including useful extracellular metabolites), nutrition, survival mechanisms during stress, etc. (Schneiker et al. 2006).

It remains the echelon of current molecular techniques applicable in microbial studies in the field of biodegradation and other fields, as it is capable of painting the microbial cell at levels of sophisticated detail. Currently, greater than 130,000 whole genomes of bacteria exist, including complete and near-complete genomes, from more than 50 different bacterial phyla. Sequences for other microbes (archaea, fungi, and viruses) also exist (Arriola et al. 2020).

A typical experimental pipeline for a WGS experiment is highlighted in Figure 15.2.

The process begins by culturing the bacteria whose genome is to be sequenced in pure culture. Hydrocarbonoclastic bacteria can be isolated from a variety of environments: soil, water, extreme environments, etc. (see sunburst chart here) (Figure 15.3).

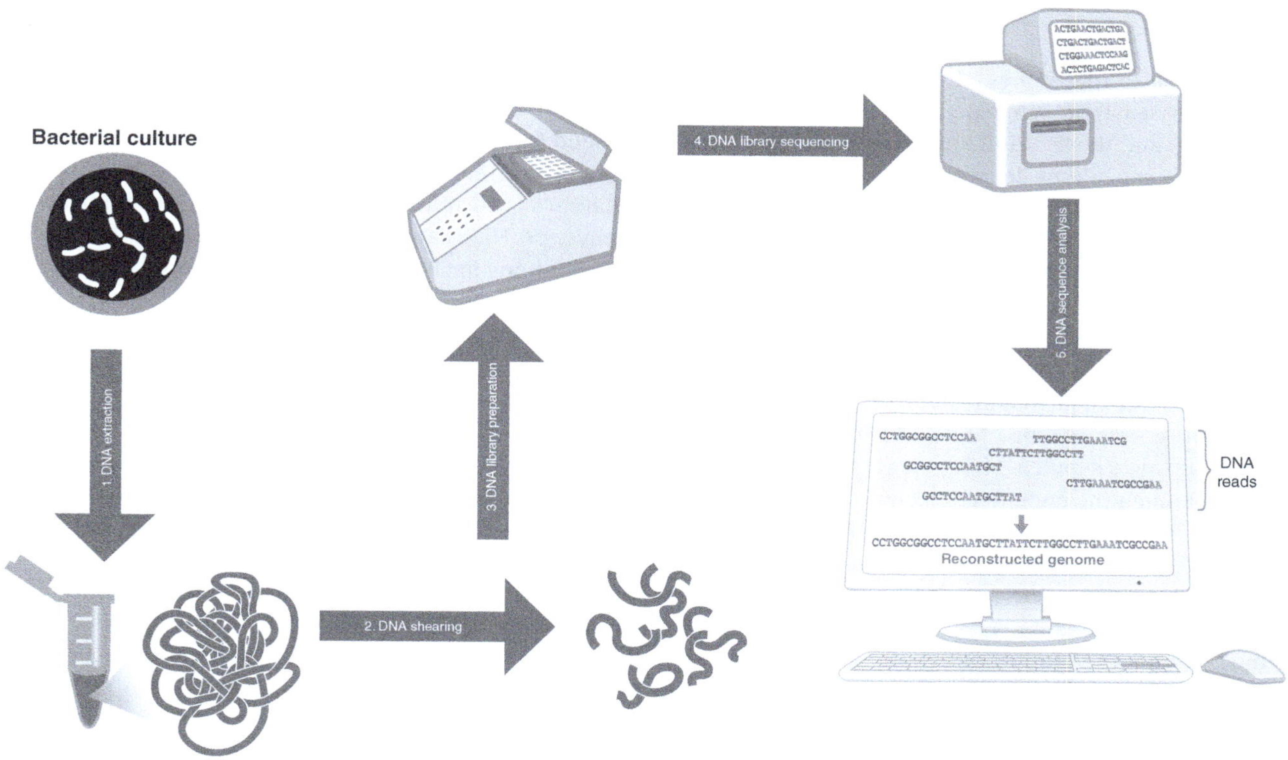

Figure 15.2 Stages in a typical WGS pipeline. *Source:* Adapted from CDC – Centres for Disease Prevention and Control, USA (2021).

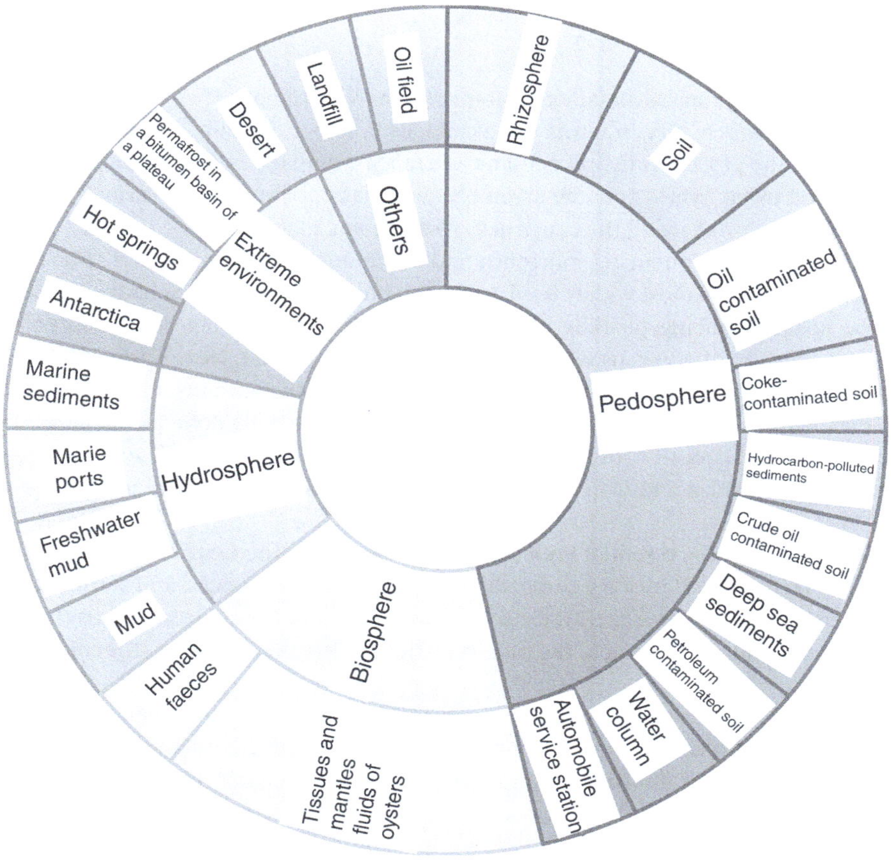

Figure 15.3 Sunburst chart showing the diversity of environments from where HDB whose WGSs were published were isolated. *Source:* Data considered was within the years (2012–2021) across six search engines.

The two basic approaches for the isolation of microorganisms in all environmental microbiology approaches are the objective-based approach, e.g. the isolation of bacteria that can degrade diesel from diesel-contaminated soils (Umar et al. 2019) and the shotgun approach, where bacteria are isolated from a generalist source, for instance, the search for bacteria capable of degrading bisphenol from human feces (López-Moreno et al. 2021); or identifying phenol-degrading bacteria from nonhuman impacted portions of Antarctica (Lee et al. 2020). This showed the applicability of the shotgun method and its advantage in serendipitous discoveries. HDBs are also capable of surviving in extreme environments since they possess physiologic and metabolic adaptations for life at the extreme ends of the environment/earth (Rampelotto 2013).

After the collection of samples, culturing/isolation of the bacteria in the laboratory can be done using various media. These include Brain-Heart Infusion agar, Bushnell-Haas agar, De Man, Ragosa and Sharpe medium, minimal media, mineral salt media, phenol medium, peptone-yeast extract-glucose-vitamins medium, reasoner's 2A agar, stenoVIA agar, tetrathionate agar, among others. This illustrates the diversity of media used in isolating HDB. The culturing process typically involves supplementing the various media with the hydrocarbon to be degraded as the only source of energy and carbon, through the enrichment technique. Screening is subsequently carried out, typically on a Luria-Bertani medium, and with increasing concentrations of the substrate, until one pure strain was obtained (Mahto and Das 2020; Xiang et al. 2020). Degradation studies are also carried out in the lab, using both OFAT (single-factor optimization) and sometimes RSM (response surface methodology) optimizations, to understand the best conditions for biodegradation (Lee et al. 2020; Zhao et al. 2021).

Inocula from the pure culture plates of the selected bacteria are then enzymatically treated with DNA to release the DNA. DNA extraction is carried out on the pure strain of the bacterium to obtain a sufficient amount of DNA. Extraction is usually done via a variety of extraction kits (Zhao et al. 2021). The released DNA is then purified, sheared into fragments of known sizes using restriction enzymes or mechanical disruption, and then used to make a DNA library through PCR amplification. The library is then sequenced to obtain reads. Finally, the obtained sequence is then analyzed using specialized software, and used to reconstruct the genome, for further analyses (CDC – Centres for Disease Prevention and Control, USA 2021).

15.2.3 Analysis of WGSs

When the sequence is obtained, the next step is analyzing the obtained genome. As shown in Figure 15.4, WGS data can be obtained using various sequencing approaches, which vary in terms of advantages, number of reads, speed of analysis, quality of obtained reads, etc. The next step of the process is hybrid genome assembly, where the reads obtained from the various sequencing approaches are assembled using hybrid genome assembly, and then the function of various parts of the genome is determined using software such as Prokka, and the construction of genome maps.

Sequencing itself is carried out using a variety of sequencing machines and techniques, including Oxford Nanopore MinION Sequencing and Illumina Sequencing, the two most widely used sequencing techniques today (Schadt et al. 2010; http://Nature.com 2021). Thereafter, the post-sequencing pipeline begins, which starts with the determination of the quality of the generated sequences, and the closure of any gaps which might have been obtained. Next, various kinds of software are utilized for the determination of protein-coding sequences, the classification of the proteins into their respective subsystems and the identification of functions of particular genes. This comparison is usually done using databases, the two major ones among which are the COG (cluster of orthologous groups) database (Galperin et al. 2015) and The Kyoto Encyclopedia of Genes and Genomes (Kanehisa and Goto 2000). This process is called functional annotation of the genome (Xiang et al. 2020).

The potential of WGS is very vast and diverse, and the essential premise that enables this is the Central Dogma Theory, i.e. the nucleotide sequence of the bases can be translated into the corresponding amino acid sequences, and subsequently proteins. This process is made possible using software such as the Molecular Evolutionary and Genomics Analysis (MEGA 10) software, together with other software and databases. As such, the totality of the information regarding the microbe is easily revealed, as exemplified in Figure 15.5.

Upon completion of analyses, the whole sequence is deposited in accessible online databases, and usually, a whole genome announcement is first made, and later, (a) complete paper(s) on the study is published (Chauhan et al. 2013; Das et al. 2015; Ruiz et al. 2015; Ruiz et al. 2021). However, the information provided must be adherent to MIGS – The Minimum Information about a Genome Sequence Specification, developed by the Genetic Standards Consortium, which stipulated the necessary requirements all papers detailing WGS studies must comply with, including details on the sampling,

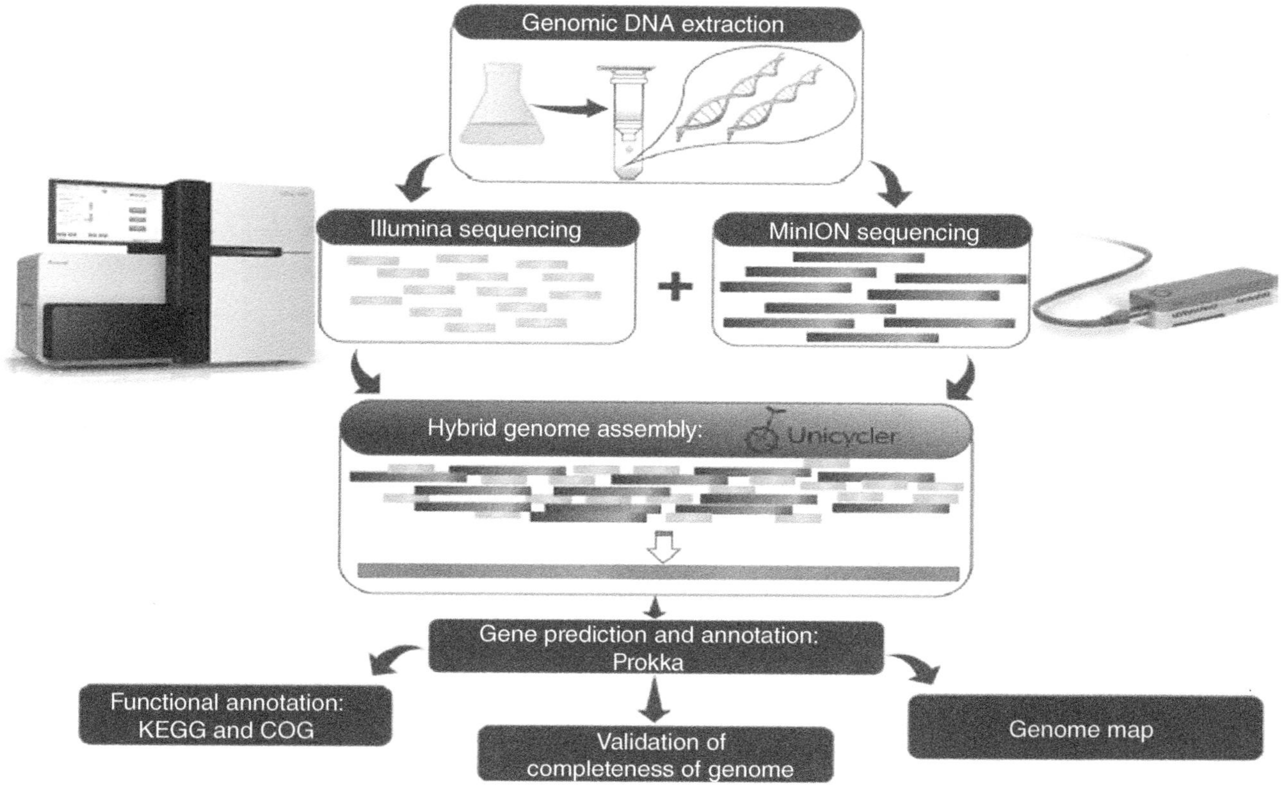

Figure 15.4 A typical workflow for analyses of WGSs. *Source:* Ravintheran et al. (2019) / Springer Nature / CC BY 4.0.

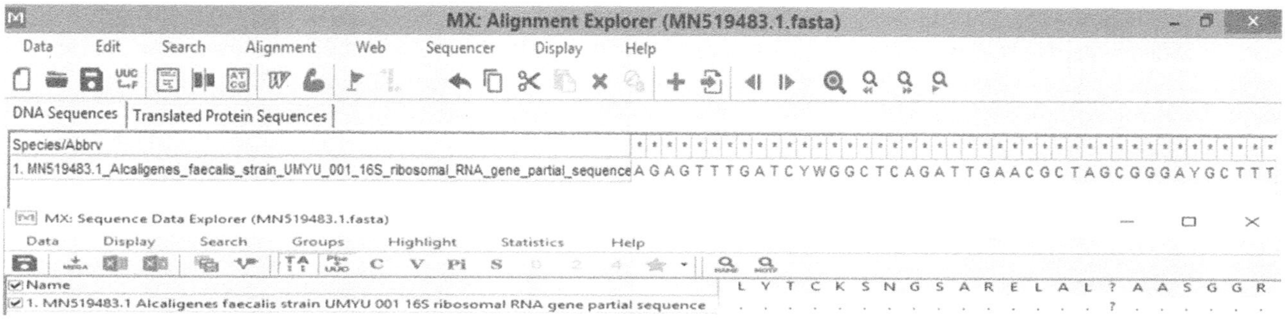

Figure 15.5 Translated amino acid sequences from the partial 16S rRNA gene sequence of a diesel degrading bacterium, *Alcaligenes faecalis* UMYU001, from Daura, Katsina State, Nigeria. *Source:* Sequence obtained from Umar et al. (2020a, 2020b).

exhaustive characterization of the microbe, and details on the protocols employed in carrying out the sequencing, and results obtained (Field et al. 2008).

15.2.4 Relevance of WGS in Studies of Hydrocarbonoclastic Bacteria

Advancement in current strategies of using microorganisms in bioremediating hydrocarbon-contaminated environments requires an impetus from knowledge of the genetic background behind the biodegradation of hydrocarbons. This is critical especially in the current fourth era of biodegradation studies, catalyzed by the continual emergence of molecular techniques for characterizing a single microbial cell at levels of sophisticated detail (Arriola et al. 2020).

WGS is capable of illuminating the microbial cell's metabolic circuitry, physiology, and metabolism of the degradation processes, the presence of novel/extent pathways for the biodegradation, elucidating the roles of nutrient sources in biodegradation and highlighting the enzymes secreted and involved in the process. It can also highlight the presence or absence of known and novel genes responsible for both the biodegradation itself and other associated processes. Therefore, it remains the ultimate technique for characterizing HDB. Such studies also bring about perpetual advancements in the field, replacing methodology trends and unveiling new techniques, which can be ultimately applicable in future WGS studies on hydrocarbon degradation and ultimately in bioremediation of contaminated environments (Yunusa and Umar 2021).

15.3 Key Insights and Recent Advances from Studies on the WGS of Petroleum Hydrocarbon-Degrading (Hydrocarbonoclastic) Bacteria in the Past Decade (2012–2021)

A systematic review approach (adopted from Dehkordi et al. 2021 and Tawfik et al. 2021, with major modifications), using data sourced from six search engines (Google, GoogleScholar, PubMed, SCIELO, Z-Library [Articles], and ResearchGate) yielded a pool of 26 journal articles involving published WGSs of hydrocarbonoclastic bacteria. These articles were as follows:

1 = Liu et al. (2012); 2 = Chauhan et al. (2013); 3 = Patel et al. (2014); 4 = Paliwal et al. (2014); 5 = Lai et al. (2012); 6 = Miyazaki et al. (2014); 7 = Ruiz et al. (2015); 8 = Das et al. (2015); 9 = Kwak et al. (2016); 10 = Pal et al. (2017); 11 = Li et al. (2017); 12 = Yang et al. (2018); 13 = Khalil et al. (2018); 14 = Mahjoubi et al. (2019); 15 = Ravintheran et al. (2019); 16 = Mahto and Das (2020); 17 = Lee et al. (2020); 18 = Garrido-Sanz et al. (2020); 19 = Filonov et al. (2020); 20 = Elufisan et al. (2020); 21 = Xiang et al. (2020); 22 = Zhao et al. (2021); 23 = Weiten et al. (2021); 24 = Liu et al. (2021); 25 = López-Moreno et al. (2021); and 26 = Mahjoubi et al. (2021).

15.3.1 Key Bacterial Species Involved, the Most Commonly Degraded or Potentially Degradable Substrates/Classes of Petroleum Hydrocarbons, etc.

During the past decade, the commonest hydrocarbonoclastic bacteria whose whole genome sequences were published were *Pseudomonas* sp. (13 WGSs), followed by *Stenotrophomonas* and *Rhodococcus* (2 WGSs each), and then *Alcaligenes*, *Brevibacillus*, *Cyclocalsticus*, *Franconibacter*, *Microbacterium*, *Mycobacterium*, *Planococcus*, and *Sphingomonas* (1 WGS respectively) (Figure 15.6). This illustrates the diversity of bacteria involved in the biodegradation of hydrocarbons. The reason *Pseudomonas* species were the most frequently associated with the biodegradation of hydrocarbons, followed by *Rhodococci* and *Stenotrophomonas* is probably that they possess a vast array of genes/enzymes that enable the isolates to metabolize the hydrocarbons (Elufisan et al. 2020; Zampolli et al. 2019; He et al. 2018).

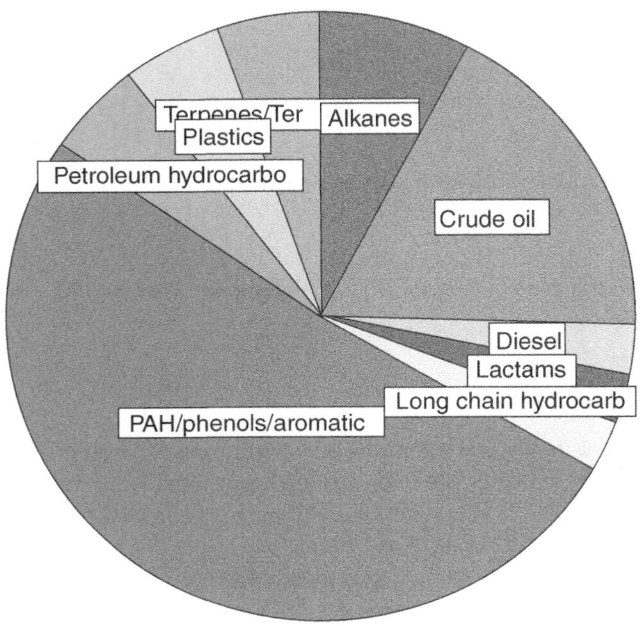

Figure 15.6 Principal substrates/classes of hydrocarbons degraded by HDB whose whole genome sequences are available across six search engines (2012–2021). Petroleum hydrocarbons = assorted or unspecified petroleum hydrocarbon(s). *Source:* Adapted from Darma and Riko (2022).

On the basis of substrates degraded, the majority of the strains metabolized polycyclic aromatic hydrocarbons, phenols, and aromatic compounds, followed by crude oils, alkanes, and other petroleum hydrocarbons. Interest in these compounds may be due to the fact that research is being focused on PAHs as they are classified as priority pollutants having potential carcinogenic, teratogenic, mutagenic, and toxigenic tendencies (Umar et al. 2019, 2020b). Interest in crude oils, alkanes, and other petroleum hydrocarbons may be due to its position in the global economy, or due to the frequency with which its spills occur (Elufisan et al. 2020; Mahjoubi et al. 2021).

By expanding the scope of the definition of degradation to encompass either: (i) growth using the hydrocarbon substrate as the only carbon/energy source; (ii) complete mineralization of substrate(s) to end product(s); (iii) metabolizing the substrate in the presence of other energy sources, e.g. glucose/yeast extract; or (iv) partial breakdown of the substrate to give assayable metabolic products (Yunusa and Umar 2021; Darma and Riko 2022); it can be inferred that the scope of degradable petroleum hydrocarbons goes beyond those described earlier (Table 15.1).

Table 15.1 Petroleum hydrocarbons degraded or potentially degradable by HDB in the reviewed studies.

S. no.	Petroleum hydrocarbon(s)	Degradation status	Bacteria involved	References
1	*(Petroleum) Hydrocarbons*[a]	Potential	*Pseudomonas aeruginosa* Strain PK6	Patel et al. (2014)
2	1-butanol	Complete	*Rhodococcus* sp. WAY2	Garrido-Sanz et al. (2020)
3	2,6-dimethylnaphthalene	Complete	*Cycloclasticus* sp. Strain P1	Lai et al. (2012)
4	2-methylnaphthalene	Complete	*Cycloclasticus* sp. Strain P1	Lai et al. (2012)
5	Acenaphthene	Complete	*Cycloclasticus* sp. Strain P1	Lai et al. (2012)
6	Aliphatic hydrocarbons (C_6–C_{35})	Complete	*Pseudomonas stutzeri* MF28, *Pseudomonas alcaligenes* OT69, *Pseudomonas aeruginosa* WC55, *Stenotrophomonas maltophilia* MF89 and *Marinobacterium maritypicum* MF109	Chauhan et al. (2013)
7	*Anthanilate*[a]	Potential	*Pseudomonas knackmussii* B13	Miyazaki et al. (2014)
8	Anthracene	Complete	*Cycloclasticus* sp. Strain P1	Lai et al. (2012)
9	Anthraquinone	Complete	*Pseudomonas putida* BS3701	Elufisan et al. (2020)
10	*Aromatic compounds*[a]	Potential	*P. aeruginosa* Strain PK6	Patel et al. (2014)
11	Benzene	Complete	*Franconibacter pulveris* strain DJ34	Pal et al. (2017)
12	*Benzoate*[a]	Partial	*P. aeruginosa* Strain PK6	Paliwal et al. (2014)
13	Crude oil	Complete	*P. aeruginosa* N002, *Franconibacter pulveris* strain DJ34, *Planococcus maritimus* Y42, *Alcaligenes aquatilis* strain BU33N, *P. putida* BS3701, *Pseudomonas* sp. strain BUN14	Das et al. (2015), Pal et al. (2017), Yang et al. (2018), Mahjoubi et al. (2019), Filonov et al. (2020), Mahjoubi et al. (2021)
14	*Cyclohexane carboxylate*[a]	Partial	*Aromatoleum petrolei* ToN1[T]	Weiten et al. (2021)
15	Dibenzofuran	Complete	*Cycloclasticus* sp. Strain P1, *Pseudomonas brassicacearum* strain MPDS, *Pseudomonas* sp. strain BUN14	Lai et al. (2012), Liu et al. (2021), Mahjoubi et al. (2021)
16	*Dibenzofuran*[b]	Partial	*Rhodococcus* sp. WAY2	Garrido-Sanz et al. (2020)
17	Dibenzothiophene	Complete	*Cycloclasticus* sp. Strain P1, *P. brassicacearum* strain MPDS, *Pseudomonas* sp. strain BUN14	Lai et al. (2012), Liu et al. (2021), Mahjoubi et al. (2021)
18	Diesel	Complete	*Sphingomonas paucimobilis* AIMST S2	Ravintheran et al. (2019)
19	Ethylbenzene	Complete	*Pseudomonas frederiksbergensis* SI8, *Franconibacter pulveris* strain DJ34	Ruiz et al. (2015), Pal et al. (2017)
20	*Ethylbenzene*[b]	Partial	*Rhodococcus* sp. WAY2	Garrido-Sanz et al. (2020)

(*Continued*)

Table 15.1 (Continued)

S. no.	Petroleum hydrocarbon(s)	Degradation status	Bacteria involved	References
21	Fluoranthene	Complete	*Mycobacterium rufum* JS14T, *Pseudomonas* sp. strain BUN14	Kwak et al. (2016), Mahjoubi et al. (2021)
22	Fluorene	Complete	*Cycloclasticus* sp. Strain P1, *P. brassicacearum* strain MPDS	Lai et al. (2012), Liu et al. (2021)
23	Heptadecane	Complete	*Rhodococcus* sp. WAY2	Garrido-Sanz et al. (2020)
24	*Hydroxyphenol*[a]	Partial	*P. knackmussii* B13	Miyazaki et al. (2014)
25	*n*-alkanes (C_{12}-C_{36})	Complete	*A. aquatilis* strain BU33N	Mahjoubi et al. (2019)
26	Naphthalene	Complete	*Cycloclasticus* sp. Strain P1, *P. aeruginosa* PFL-P1, *Rhodococcus* sp. WAY2, *P. putida* BS3701, *P. brassicacearum* strain MPDS, *Pseudomonas* sp. strain BUN14	Lai et al. (2012), Mahto and Das (2020), Garrido-Sanz et al. (2020), Elufisan et al. (2020), Liu et al. (2021), Mahjoubi et al. (2021), Paliwal et al. (2014)
27	*Naphthalene*[a]	Partial	*P. aeruginosa* Strain PK6	Paliwal et al. (2014)
28	*n*-docosane	Complete	*P. aeruginosa* Strain SJTD-1	Liu et al. (2012)
29	*n*-hexadecane/hexadecane	Complete	*P. aeruginosa* Strain SJTD1, *Franconibacter pulveris* strain DJ34	Liu et al. (2012), Pal et al. (2017)
30	*n*-octadecane	Complete	*P. aeruginosa* Strain SJTD-1, *Pseudomonas* sp. strain BUN14	Liu et al. (2012), Mahjoubi et al. (2021)
31	*n*-tetracosane	Complete	*P. aeruginosa* Strain SJTD-1, *Rhodococcus* sp. WAY2	Liu et al. (2012), Garrido-Sanz et al. (2020)
32	Pentadecane	Complete	*Rhodococcus* sp. WAY2	Garrido-Sanz et al. (2020)
33	Phenanthredine	Complete	*Stenotrophomonas* sp. Pemsol	Elufisan et al. (2020)
34	Phenanthrene	Complete	*Cycloclasticus* sp. Strain P1, *A. aquatilis* strain BU33N, *P. aeruginosa* PFL-P1, *P. aeruginosa* PFL-P1, *Pseudomonas* sp. strain BUN14	Lai et al. (2012), Mahjoubi et al. (2019), Mahto and Das (2020), Elufisan et al. (2020), Mahjoubi et al. (2021)
35	*Phenanthrene*[b]	Partial	*Rhodococcus* sp. WAY2	Garrido-Sanz et al. (2020)
36	Phenol	Complete	*Rhodococcus* sp. Strain AQ5–07, *Rhodococcus ruber* C1	Lee et al. (2020), Zhao et al. (2021)
37	Polycyclic aromatic hydrocarbons, including those with chain length C7–C35	Complete	*P. stutzeri* MF28, *P. alcaligenes* OT69, *P. aeruginosa* WC55, *S. maltophilia* MF89 and *Marinobacterium maritypicum* MF109, *P. putida* BS3701	Chauhan et al. (2013), Filonov et al. (2020)
38	Pristine	Complete	*Pseudomonas* sp. strain BUN14	Mahjoubi et al. (2021)
39	Pyrene	Complete	*Cycloclasticus* sp. Strain P1, *A. aquatilis* strain BU33N, *P. aeruginosa* PFL-P1, *Pseudomonas* sp. strain BUN14	Lai et al. (2012), Mahjoubi et al. (2019), Mahto and Das (2020), Mahjoubi et al. (2021)
40	Tetradecane	Complete	*Pseudomonas* sp. strain BUN14	Mahjoubi et al. (2021)
41	Toluene	Complete	*P. frederiksbergensis* SI8, *Franconibacter pulveris* strain DJ34, *Rhodococcus* sp. WAY2, *Aromatoleum petrolei* ToN1^T	Ruiz et al. (2015), Pal et al. (2017), Garrido-Sanz et al. (2020), Weiten et al. (2021)
42	*Toluene*[a]	Potential	*Aromatoleum petrolei* ToN1^T	Weiten et al. (2021)
43	*Toluene*[b]	Partial	*Rhodococcus* sp. WAY2	Garrido-Sanz et al. (2020)
44	Xylene (incluing m-, p- and o-xylene)	Complete	*Franconibacter pulveris* strain DJ34, *Rhodococcus* sp. WAY2	Pal et al. (2017), Garrido-Sanz et al. (2020)

[a] potential; indicates that the isolate was not shown to degrade the named hydrocarbon in the lab, rather, it was shown to potentially degrade the hydrocarbon through genome mining (i.e. demonstrating the gene/genes responsible for that hydrocarbon's degradation in the bacteria's genome).
[b] partial; indicates that the substrate was only partially transformed.

15.3.2 Highlights from a Review of Key Contents of Bacterial Whole Genomes from Published Studies Involving WGS of Petroleum Hydrocarbon-Degrading Bacteria (2012–2021)

Table 15.2 overleaf shows a summary highlighting key genome features of the WGS of some petroleum hydrocarbon-degrading bacteria whose whole genomes were published (2012–2021). Specifically, the summary of the main features of the genome (including the genome size, the G + C content, the number of genes/coding sequences in the genome, presence of plasmid and gene aggregates) was stated, together with gene annotation results (number/function of specific genes that function specifically in petroleum hydrocarbon biodegradation). The relationship of this information to biodegradation is also outlined, from the context of application in biodegradation, with regards to the strain of bacterium involved, the techniques followed in its isolation and the substrates it was initially shown to tolerate during the enrichment/isolation process. Subsequently, the range of the physiochemical tolerance limits of the bacteria was explored, followed by results from laboratory-scale biodegradation experiments (substrate(s) degraded, their concentrations, kinetics of the biodegradation, the pathway/mechanism of the biodegradation, a summary of optimization studies, etc.

15.3.3 Pathways for Petroleum Hydrocarbon Biodegradation

The reviewed studies highlighted many pathways for the biodegradation of numerous petroleum hydrocarbons (some 20 pathways in total, see Figure 15.7). These pathways involve both previously established and novel pathways.

Summarily, the pathways involved in petroleum hydrocarbon biodegradation of petroleum hydrocarbons follow a recognizable pattern. Initially, the petroleum pollutants are intracellularly attacked via an oxidative process, where an oxygen atom is incorporated into the compound, in a process enzymatically catalyzed by oxygenases and peroxidases (Chen et al. 2011). Thereafter, the oxygenated product is subjected to transformation via peripheral degradation pathways, where they end up in the central intermediate pathways for metabolism, e.g. tricarboxylic acid cycle. The compound is eventually transformed into precursor metabolites used for the biosynthesis of cell biomass. These include acetyl-CoA, succinate, and pyruvate (Darma and Riko 2022).

Specifically, the aerobic degradation of alkanes follows the following sequence: the methyl group is first oxidized to primary alcohol, using terminal hydroxylase enzymes specific to the substrate (terminal oxidation, even though oxidation can also occur subterminally, independently, or together with terminal oxidation); and thereafter to an aldehyde, and then a fatty acid (through β-oxidation), and then to acetyl CoA, upon conjugation with CoA molecules through β-oxidation. These are then utilized for numerous cellular processes (Rojo 2009; Mwamura 2017). Another approach is subterminal oxidation. Here, after oxidation, the alkane transforms into a secondary alcohol, and later a ketone, then an ester. This last step is mediated by an enzyme, the Baeyer-Villinger monooxygenase. Furthermore, this is subsequently hydrolyzed to generate an alcohol and a fatty acid, in a reaction catalyzed by an esterase (Thorne-Holst et al. 2007; Rojo 2009).

Similarly, in the case of aromatic compounds, their degradation process starts with the generation of a diol, which is then cleaved to form a diacid, e.g. cis, cismuconic acid (Olajire and Essien 2014). In the case of simple aromatic hydrocarbons with alkyl substituents on the aromatic ring, the first step is the attack of the alkyl substituent or directly on the aromatic ring, mediated by enzymes (Rehman et al. 2015). Degradation of complex aromatic hydrocarbons follows the same pattern as monocyclic aromatic hydrocarbons (Olajire and Essien 2014), even though highly complex structures are often difficult to be biodegraded, as they are resistant to enzymatic action (Boonchan et al. 2000).

As a representative, the pathway for the degradation of phenol is shown in Figure 15.8 overleaf, after the table.

15.3.4 Operons and Gene Families Involved in the Biodegradation of the Petroleum Hydrocarbons

Gene clusters or operons are found located at specific portions of the genome and are under the control of the same promoter. They can sometimes be part of genetic islands involved in biodegradation, and they are often regulated in the same manner and are responsible for the biodegradation of the same or similar substrates. Some gene clusters were hypothesized before and confirmed later (Miyazaki et al. 2014; Paliwal et al. 2014; Elufisan et al. 2020).

Table 15.3 shows the gene clusters and their function in biodegradation.

15.3.5 Major Genes Involved in Biodegradation

A wide variety of genes are involved in the biodegradation process, either in the synthesis of enzymes or proteins, that are subsequently useful in the biodegradation of petroleum hydrocarbons, as shown in the table here. These enzymes are often

Table 15.2 Summary of highlights from WGS and biodegradation potential of petroleum HDB (2012–2021).

S. no.	Key results from whole genome analysis	Biodegradation-related results from the study	Reference
1	General Genome Features: The strain possessed a single circular chromosome 6,213,959 bases in size. It was found to have G + C content of 60.25%, with 5534 protein-coding sequences. Additionally, there are five rRNA operons and 65 tRNA genes in the circular chromosome, with the percentage of Coding SequenceS (CDSs) being 87.3% Gene Annotation and Transcriptomics Studies: Genes for PAHs and heterocyclic aromatic hydrocarbons were predicted on a 27,933 kb region. The authors performed a transcriptome analysis, based on the genes identified using WGS. The strain grown on a glycerol medium (MSM + 1% glycerol) served as the control, while the strain grown on DBF medium (MSM + 5 mg DBF) was the experimental group. The results highlighted the differential expression levels of 870 genes, with 487 genes being up-regulated and 383 genes down-regulated. The authors reported a hypothetical protein (gene 4284) downstream of the naphthalene degradation gene cluster, which was upregulated 7.8 times. Seven other up-regulated genes were also found nearby. The genes *nahA* and *nahB*, synthesizers of naphthalene dioxygenase and dehydrogenase, were found to be highly up-regulated (37.8 times and 42.2 times). The *nahH* gene, which encodes catechol 2,3-dioxygenase, was found to be up-regulated 36.1 times, all in relation to the control	Strain, Isolation, and Substrates Degraded: The bacterium, *Pseudomonas brassicacearum* strain MPDS, was originally isolated using naphthalene enrichment culture, however, it can also survive on the expense of fluorene, dibenzofuran (DBF) and dibenzothiophene (DBT) as sole sources of carbon/energy. It could effectively degrade these PAHs/heterocyclic derivatives mentioned Biodegradation Pathway/Mechanism: Fluorene, DBF and DBT were found to be uniquely degraded via a lateral dioxygenation pathway, not the angular deoxygenation pathway usually reported. They are also solely degraded, not co-metabolized, as reported before Physiochemical Parameters and Optimization Studies: The optimal growth temperature was 25 °C, with the strain hardly growing beyond 37 °C. *Optimum* concentration of the substrate (naphthalene) was 50 mg, with decreases obtained at 100 mg concentrations. In the case of fluorene, DBT and DBF, growth is not directly linked to substrate addition, however, 5 mg was found to yield better growths. Optimum flask shaking speed was 200 rpm. Studies subsequent to optimization showed that the isolate completely degraded 50 mg naphthalene (in 50 mL medium) in 84 h, with OD_{600} reaching 1.0–1.1. In the case of 5 mg fluorene or DBF or DBT, the OD_{600} value plateaued at 0.5–0.6; with 65.7% DBF/32.1% DBT and 40.3% fluorene being degraded in 96 h and 72 h, respectively	Liu et al. (2021)
2	General Genome Features: The WGS showed a total of 22 contigs with a size of 4,898,053 bp. The mean G + C content was of 62.4%, and the protein-coding sequences (CDSs) were 4551, and approximately 87% of the proteins could be assigned to EggNOG orthologous genes (OGs) Gene Annotation Studies: Out of these genes, 154 code for proteins related to xenobiotic biodegradation and metabolism. When analyzed using the KEGG database, 2564 (56.34%) of the genes could be assigned to 2062 KEGG Orthologues (KOs), including 154 linked to xenobiotic biodegradation and metabolism. Out of these, 44 proteins were linked to the pathways for the complete degradation of benzoate and naphthalene	Strain, Isolation and Substrates Degraded: The *Pseudomonas* sp. strain BUN14 was a halotolerant strain originally isolated from polluted sediment recovered from the refinery harbor on the Bizerte coast, north Tunisia, via enrichment techniques on mineral medium ONR7a supplemented with 1% crude oil as the sole carbon source, for 21 days, at 30 °C. The organism was found to degrade crude oil, pristane, dibenzofuran, dibenzothiophene and naphthalene, as a sole carbon source Biodegradation Experiments: The laboratory experiments on the total degradation of petroleum Total Extracted and Resolved Hydrocarbons and their derivatives (TERHCs) by strain BUN14 was estimated at 90% after 21 days, with rapid degradation of almost all alkanes C_{12}–C_{36}, Pristane, naphthalene, DBT and DBF achieved degradation percentages of 22%, 40%, 30%, and 14%, respectively, by the strain (BUN14), over a period of 21 days	Mahjoubi et al. (2021)

3	General Genome Features: The manually annotated genomes of *A. petrolei* ToN1T and *A. bremense* PbN1T, two microbes at the proximal and distal points of the phylogenetic tree of the genus *Aromatoleum* display diverse features. The size of the former is 5.41 Mbp, while the latter is 4.38 Mbp in size, respectively. Both display richness in terms of horizontal genes transferred and the presence of mobile genetic elements. Their composite network of catabolic genes is 88-genes in strength, occupying 1.59% and 1.75% of their genomes, respectively. The G + C content of the former was 65.72%, while that of the latter was 65.59%. Likewise, *A. petrolei* ToN1T had one incomplete phage in its genome, while *A. bremense* PbN1T had 7. The latter also exhibited Class 1, type I-E transposons CRISPR sequences, which were absent in the former. The coding genes in the former were 4964 (90.5%), with an average size of 986 bp; while in the latter, they were 4198 (90.5% still), with average size of 944 bp. Out of these, 1484 and 1499 were assigned functions in the former and latter, with 257 and 457 being uncharacterized, respectively	Strain, Isolation and Substrates Degraded: The microbes (*Aromatoleum petrolei* ToN1T and *Aromatoleum bremense*) PbN1T were originally isolated from toluene (*A. petrolei* ToN1T) and ethylbenzene (*A. bremense* PbN1T). Specifically, they possess the genetic capability to degrade ethyl/*n*-propylbenzene (strain PbN1T) and toluene/monoterpenes (strain ToN1T), anaerobically. Both can degrade benzoate/benzoyl-COA, and *A. petrolei* ToN1T can potentially degrade cyclohexane carboxylate	Weiten et al. (2021)
4	General Genome Features: The genome of the isolate has a size of 5.66 Mb, and comprises a single circular chromosome (5.47 Mb), and two plasmids A and B, of size, 0.15 Mb and 0.03 Mb, respectively. The isolate's G + C content ranged from 68.81% to 70.52%. The genome of the isolate was compared to that of 36 other *Rhodococci*. High variations in COGs were observed in genes involved in transcription (1433 gains, 354 losses) and those of unknown function (8196 gains, 2061 losses), during the course of evolution of the *Rhodococcus* genus Gene Annotation Studies: Genes for benzoate, aminobenzoate, steroid, xylene, caprolactam, dioxin and chloroalkane/chloroalkene metabolism were the xenobiotic degrading genes most frequently identified in the strain	Strain, Isolation and Substrates Degraded: The *Rhodococcus ruber* C1 strain was first isolated from a landfill in Chongqing, China, through selection by enrichment in M9 medium, supplemented with 500 mg/L of phenol as the sole carbon source Biodegradation Experiments: They were found to degrade up to 1806 mg/L of phenol in 38 hours. Isolates can generally perform degradation at 20–45 °C within 72 hours. Mineralization rates of up to 88% obtained. Extended time required for 10, 15, and 50 °C Biodegradation Pathway/Mechanism: Three pathways elucidated for the phenol degradation (two via the ortho, one via the meta position)	Zhao et al. (2021)
5	General Genome Features: The genome size is 4,373,402 bp and it contains 4037 genes, out of which 3905 are protein-coding, with 4,370,061 bases. A total of 145 genes involved in the degradation of PAHs were found in its genome. Its G + C content is 66.59%. There are 3 rRNA genes and 67 tRNA genes in the genome Gene Annotation Studies: The bacterium was found to possess 154 unique genes, mostly present in genomic islands, and some of which are associated with PAH degradation	Strain, Isolation and Substrates Degraded: The *Stenotrophomonas* sp. Pemsol strain was isolated from a crude oil contaminated soil, and was found to tolerate and degrade polycyclic aromatic hydrocarbons PAHs such as anthraquinone, biphenyl, naphthalene, phenanthrene and phenanthridine Biodegradation Experiments: The ability of the strain to degrade those substrates was evaluated in Bushnell Hass medium containing PAHs as the sole carbon source. The assays were performed in BH minimal media with PAHs as a sole carbon source at a concentration ranging from 1 to 5 mg/ml. The bacterium showed a good capacity to grow at these concentrations in the 8-day study, with the concentration of 1 mg/mL being the optimum. The results of the assay showed that anthraquinone, biphenyl, naphthalene, phenanthrene, and phenanthridine, and not xylene, can be used as sole carbon sources for Pemsol's growth in the Bushnell Hass medium. The degradation of naphthalene at a concentration of 1 mg/mL within 30 days was also assessed. The strain can also grow in a mix of the five PAHs at the final concentration of 5 mg/mL (1 mg/mL for each)	Elufisan et al. (2020)

(*Continued*)

Table 15.2 (Continued)

S. no.	Key results from whole genome analysis	Biodegradation-related results from the study	Reference
6	General Genome Features: The whole genome sequence of the bacterium comprises a chromosome of size 6,337,358-bp, with a G + C content of 61.6%; and two circular plasmids: pBS1141 (107,388 bp in size, with a G + C content of 57.8%), and pBS1142 (54,501 bp in size, G + C content = 57.4%). The genome comprises 5756 coding sequences, 7 rRNA clusters (5S, 16S, 23S), and 79 tRNAs. Results from the mating experiment indicated that pBS1142 is a helper plasmid, which could assist the transfer of plasmid pBS1141 during conjugation	Strain, Isolation and Substrates Degraded: The bacterium *Pseudomonas putida* BS3701 was originally isolated from soil contaminated with coke byproduct waste (sewage from the Moscow Coke and Gas Plant, Vidnoye, Moscow Region, Russian Federation). It was found to grow using naphthalene as the sole carbon and energy source	Filonov et al. (2020)
7	General Genome Features: The draft genome sequence contained 6,749221 bp (6.75 Mb), with a G + C content of 62.4%. The bacterium contained 34 contigs, with the longest contig being 1,78,314 bp. Genome annotation using Rapid Annotation Subsystems Technology (RAST) server showed that the bacterium contained 6545 coding sequences, 4 RNAs and 75 tRNAs	Strain, Isolation and Substrates Degraded: The bacterium *Rhodococcus* sp. Strain AQ5-07 was originally isolated from soil from King George Island (South Shetland Islands, maritime Antarctic), through enrichment with phenol as the sole carbon and energy source	Lee et al. (2020)
	Gene Annotation Studies: Among these genes, 88 were found to be associated with the metabolism of aromatic compounds, including complete gene clusters containing catA, catB, catC, catR, pheR, pheA2, and pheA1 genes, involved in catechol, phenol and phenanthrene metabolism	Physiochemical Parameters and Optimization Studies: Initial degradation studies using Plackett–Burman experimental design matrix yielded degradation levels ranging from 44.12% to 61.08%, before Central Compartmentalized Design (CCD) optimization predicted degradation values of 3.43–93.45%. These CCD values were experimentally validated, with degradation levels of 2.76–94.10% observed. Maximal degradation was obtained at a temperature of 12.5 °C, pH of 7.0 and $(NH4)_2SO_4$ concentration of 0.40. However, optimization using response surface methodology (RSM) predicted that the bacterium can achieve optimum degradation of up to 94.61%, within 84 h, at a temperature of 14.8 °C, pH of 7.05 and an ammonium sulphate concentration of 0.41 g/L. Experimental validation showed that the bacterium achieved 92.91% degradation of 0.5 g/L phenol under the conditions predicted by the model	
		Biodegradation Pathway/Mechanism: In addition to housing complete enzyme systems required for phenol and catechol degradation, the genome suggests that phenol degradation occurs via the β-ketoadipate pathway	

8	General Genome Features: The total genome sequence of the bacterium measures 6,333,060 bp in sze, with a G + C content of 66.51%. The genome contains 5857 CDSs, comprising 90.09% of the total genome. The average gene length was found to be 974 bp. The prediction of non-coding RNA in the genome revealed 60 tRNAs which included 59 tRNAs and 1 pseudogene, along with 4 rRNAs. Annotation using the COG database revealed 4747 genes, and gene ontology distribution analysis revealed 2862 genes. Out of the 4747 genes, 318–383 were found to be horizontally transferred, upon comparison of this genome with the reference genomes of other *Pseudomonads* Gene Annotation Studies: Analyses involving gene annotation and critical studies of pathways highlighted that 145 genes from this bacterium are involved in xenobiotic biodegradation and metabolism	Strain, Isolation and Substrates Degraded: *The Pseudomonas aeruginosa* PFL-P1 bacterium was initially isolated using phenanthrene enrichment Biodegradation Experiments: The biofilm cultures of *P. aeruginosa* PFL-P1 could degrade ~74% phenanthrene (100pppm concentration) within 120 h while degradation increased up to ~76% in co-culture condition. Analysis of degradation bypoducts through GC–MS analyses revealed the presence of diverse metabolites, pointing to the existence of multiple pathways for PAHs (phenanthrene) degradation. Additionally, the strain was found to possess the genetic machinery for the utilization of diverse toxic aromatic compounds such as naphthalene, benzoate, aminobenzoate, fluorobenzoate, toluene, xylene, styrene, atrazine, caprolactam etc., which is aided by the presence of familiar catabolic gene clusters, including *benABCD*, *xylXYZ* and *catAB* in the genome, known to play major roles in the degradation of these and other toxic aromatic compounds Biodegradation Pathway/Mechanism: Upon comparison of the genome of this bacterium (*P. aeruginosa* PFL-P1), with that of *P. putida*, containing TOL plasmids, similarities were observed in regards to the meta cleavage pathway employed for degradation of aromatic compounds like xylene and toluene	Mahto and Das (2020)
9	General Genome Features: The genome of the strain comprises of 3,838,299 bp, with a G + C content of 56.1%. Annotation using RAST subsystems successfully covered 56% of the genes. There are 3506 protein-coding genes and 56 rRNA genes Gene Annotation Studies: Among the CDSs, multiple genes related to the osmotic stress (23 genes), oxidative stress (63 genes) and heat and cold stress (21 genes) were observed, likewise, genes related to the metabolism of aromatic compounds (97 genes). The genome also possesses all the enzymes necessary for the aerobic degradation of benzoate	Strain, Isolation and Substrates Degraded: The *Alcaligenes aquatilis* strain BU33N was isolated from hydrocarbon polluted sediments from the Bizerte coast (northern Tunisia), through enrichment techniques using mineral medium ONR7a supplemented with 1% crude oil as the sole carbon source, at 30 °C; and thereafter, the strain was tested for its ability to utilize various hydrocarbons (pristane, phenanthrene, pyrene, naphtalene, carbazole, octadecane, fluoranthene, dibenzothiophen, dibenzofuran, squalene, anthracene and xylene) as a sole carbon and energy sources, by inoculation on ONR7a agar media containing specified amounts of hydrocarbons and incubation at 30 ± 1 °C for seven days. It successfully grew on crude oil, phenanthrene, naphthalene, fluoranthene, pristane, pyrene, carbazole, octadecane and xylene, and but not on dibenzothiophen, dibenzofuran, squalene and anthracene Biodegradation Experiments: It successfully degraded ~86%, ~56%, and 70% of TERHc (Total Extracted and Recovered Hydrocarbons), *n*-alkanes and phenanthrene, respectively, when grown on crude oil and phenanthrene, for 21 days. It can survive the temperature range of 20–40 °C, with an optimum temperature of 28 °C, and a pH of 5.5–10.0, with an optimum growth pH of 7.5 Biodegradation Pathway/Mechanism: Based on the identified genes, BU33N could predictably catabolize phenanthrene via the ortho-cleavage pathway	Mahjoubi et al. (2019)

(*Continued*)

Table 15.2 (Continued)

S. no.	Key results from whole genome analysis	Biodegradation-related results from the study	Reference
10	General Genome Features: The whole genome comprises 4,005,505 bases, 3612 CDSs and 56 tRNAs. The G + C content was 65.73%, and an array of genes involved in xenobiotic degradation (approximately 32 genes) were elucidated from the genome	Strain, Isolation and Substrates Degraded: The *Sphingomonas paucimobilis* AIMST S2 bacterium was originally isolated from oil-polluted soil at Kedah, Malaysia, and enriched in LB broth. Subsequently, it was acclimatized in M9 minimal medium supplemented with diesel and chlorpyrifos in increasing concentrations, with maximum concentrations of 1% v/v and 100 mg/L, respectively	Ravintheran et al. (2019)
11	General Genome Features: The genome has a total length of 5,155,092 bp., with the G + C bases totalling 2,680,425, giving a G + C content of 52.00%. The coding region encompasses 4,471,011 bases (86.73%). In total, there are 5090 genes in the genome, of which 4951 are protein coding (97.27%), and 3946 code for proteins with known functions (77.52%). Out of the genes, 3231 are assigned to COGs (63.48%). Among these, 1186 (23.30%) code for enzymes, while 1347 (21.84%) have unknown functions Gene Annotation Studies: Genes that can potentially be involved in the degradation of polyethylene and hydrocarbons were found, including 1 cutinase, 67 lipases, 99 hydroxylases, 2 laccases, 1 polyphenol oxidase, and 116 proteases. Additionally, 159 monooxygenases, 136 dioxygenases, and 118 lyases are identified. The RNA genes observed were 139	Strain, Isolation and Substrates Degraded: The *Brevibacillus borstelensis* AK1 bacterium was first isolated from a thermophilic hot spring, using 250 ml of tetrathionate (TT) broth as the culture medium, which was incubated at 55 °C for 24 to 48 h, at a shaking speed of 300 rpm Physiochemical Parameters: The temperature range for the growth of the bacterium was between 45 °C and 70 °C. The optimum temperature was 45–50 °C. The bacterium can also survive the pH range of 5.5–8.5, with an optimum pH of 7.5	Khalil et al. (2018)
12	General Genome Features: The whole genome analysis of the bacterium revealed the presence of one circular DNA chromosome, with a size of 3,718,896 bp and the G + C bp are 2,005,184, for a G + C content of 48.32%. a G + C content of 48.8%. There are three plasmids of size 329,482 bp; 89,073 bp; and 12,282 bp, respectively. The CDSs are of 3,541,381 bp size (85.34%). From a total of 4283 genes, 4172 were protein coding (97.41%), of which 3947 were annotated as predicted protein-coding sequences (CDS). From these, 3162 have function prediction (75.83%), and 2696 (62.95%) were assigned function prediction, based on COG analysis. In addition, the genome included 111 RNA genes (70 tRNA genes, 27 rRNA genes, and 4 ncRNA genes), and 108 pseudogenes. The genes having an unknown function were 180 (5.87%) Gene Annotation Studies: The presence of nine genes putatively encoding for crude oil metabolites in the genome confirm the ability of this bacterium to degrade crude oil-related products, likewise 238 genes were predicted to be involved in transport systems for aromatic compounds, amino acids, carbohydrates, lipids and inorganic ions	Strain, Isolation and Substrates Degraded: The *Planococcus maritimus* Y42 bacterium was initially isolated from petroleum-contaminated soil of the Qaidam Basin, through culture using strain Y42, with MM medium supplemented with crude oil as sole source of carbon and energy at 20 °C, for 10 d Physiochemical Parameters: The temperature range for the survival of the bacterium was 4–30 °C, with an optimum of 25 °C. Likewise, it can survive a pH of 6–9; with an optimum of 7.5. The strain tolerates salinity <15% NaCl (w/v) Biodegradation Experiments: Biodegradation studies showed that upon 10 days' incubation at 20 °C, short-chain *n*-alkanes (12–18 carbons in length) were degraded more than other longer, straight chain alkanes and aromatic hydrocarbons (50% vs 20–30%, respectively). Approximately, the rate of degradation follows the pattern of dodecane > tridecane > tetradecane > pentadecane > octadecane > heneicosane > heptadecane > hexadecane > nonadecane > eicosane > docosane > tricosane > pentacosane > tetracosane > octacosane > hexacosane > heptacosane > trimethyl pentadecane > trimethyloctane > indene	Yang et al. (2018)

13	General Genome Features: The genome consists of 4,856,096 bp, with a G + C content of 56.5%. There were 4382 protein coding sequences (CDSs), of which 3478 were designated to COGs. Among these, 23 genes code for rRNA and 76 for Trna; whilst 60 were pseudogenes and 14 were noncoding RNA. There are 24 different groups into which the orthologous groups were assigned	Strain, Isolation and Substrates Degraded: The *Franconibacter pulveris* strain DJ34 bacterium was isolated from Duliajan oil fields, Assam, India, on Reasoner's 2A (R2A) agar plate Physiochemical Parameters: The strain can survive a temperature range of 4–45 °C, and can grow at a pH range of 3–9. The bacterium can also tolerate up to 10% NaCl Biodegradation Experiments: Using colony counts as indexes of biodegradation activity, from day 0 to day 7, there was an increase in the counts during growth in presence of each alkane. Thereafter, it gradually decreased, until the 28th day, with high rates of degradation of hydrocarbons observed during the period. The shorter hydrocarbon, pentadecane, was degraded more than hexadecane, the longer one (99% removal vs 86.33% removal, respectively), after 28 days. When BTEX was used as sole carbon source, increase in growth was observed for three days, and on petroleum sludge, the growth (bacterial cell count) increased from the first to the 14th day. The reduction in Total Petroleum Hydrocarbons (TPH) occurred by more than 50% during the period. With crude oil as the sole carbon source, a 10-fold increase in the cell count was observed within seven days of inoculation, with concomitant increase up to the 28th day (10^6–10^7 cells/mL), with a decrease observed thereafter. Approximately 26% of the TPH was reduced during this period	Pal et al. (2017)
14	General Genome Features: The size of the whole genome is 6,176,413 bp, with a G + C content of 69.25% G + C. There were a total of 5864 genes of which 5810 are protein-coding, and 54 are RNA genes, comprising of 6 rRNAs, 47 tRNAs, and 1 ncRNA Gene Annotation Studies: Among the genes, 4498 genes have their putative functions assigned, and 3669 genes (approximately 62.57%) have functional categories based on COG annotation. There were 367 pseudogenes (6.26%). In the genome, genes encoding proteins for the aromatic compound degradation were found, which belonged to six (6) major groups	Strain, Isolation and Substrates Degraded: *Mycobacterium rufum* JS14T, The bacterium was originally isolated from polycyclic aromatic hydrocarbon (PAH)-contaminated soil in Hilo. Hawaii, USA Physiochemical Parameters: It was found to have an optimum temperature of 28 °C, and it can survive the pH range of 7.0–8.0, with the optimum pH being 7.5. The bacterium was non-salt tolerant, it cannot survive beyond 2.5–5.0% NaCl, w/v) Biodegradation Experiments: The bacterium was capable of using the four-fused PAH fluoranthene as a sole carbon/energy source, at a concentration of 40 mg/L, and the strain utilized the added compound completely within 10 days	Kwak et al. (2016)
15	General Genome Features: The whole genome was comprised of a single circular chromosome with a total sequence length of 6,537,648 bases. The G + C content was 62.36%. The genome has 11,038 open reading frames (ORFs) and 5764 annotated genes, out of which 5629 were protein coding genes (CDS), and 135 code for RNA (including 63 tRNA genes and 12 rRNA genes) Gene Annotation Studies: Genes for crude oil components' degradation (alkanes, alkenes and aromatic compounds) were observed in the genome	Strain, Isolation and Substrates Degraded: The *P. aeruginosa* N002 strain was originally isolated from crude oil contaminated soil of Geleky, Assam, India, through enrichment via M1 mineral medium supplemented with 2% crude oil as the sole carbon source	Das et al. (2015)

(*Continued*)

Table 15.2 (Continued)

S. no.	Key results from whole genome analysis	Biodegradation-related results from the study	Reference
16	General Genome Features: The draft genome was found to comprise 6,569,267 bases (6.57 Mb in size), with a G + C content of 60.5%. There were 5904 coding sequences detected upon RAST annotation, and the genome consists of 64 RNAs	Strain, Isolation and Substrates Degraded: The *Pseudomonas frederiksbergensis* SI8 bacterium was first isolated from a desert soil sample from under a fuel storage bladder in Kuwait, and has a range of temperature tolerance of 4–30 °C. It can metabolize toluene, ethylbenzene, and propylbenzene Biodegradation Pathway/Mechanism: The isopropylbenzene (cumene) degradation pathway and central protocatechuate catabolic pathway for aromatic degradation were predicted/observed from the whole genome	Ruiz et al. (2015)
17	General Genome Features: The whole genome consisted of 6,469,780 bp (6.4 Mb), with a G + C content of 61.85%. Annotation using the Prokaryotic Genome Automatic Annotation Pipeline (PGAAP) showed that there 5836 coding sequences (CDSs), and 1689 hypothetical proteins. There are also 60 genes that code for tRNA Gene Annotation Studies: The genome analysis likewise reveals genes for naphthalene, benzoate, aromatic alcohol, phenylacetic, p- hydroxyphenylacetic acid, p-Hydroxybenzoate and phenylpropanoid degradation pathways	Strain, Isolation and Substrates Degraded: The *P. putida* CSV86 strain is capable of degrading naphthalene, as evidenced by the presence of the *nah* operon, and this supports an earlier hypothesis of the existence of a genomic island (GI) for naphthalene degradation. Even though it is plasmid-free, the bacterium possesses the ability to transfer the naphthalene degradation property to other strains. The bacterium also has the genetic capability for the degradation or metabolism of naphthalene, salicylate, benzoate, benzylalcohol, phydroxybenzoate, phenylacetate, p-hydroxyphenylacetate, phenylpropanoid and homogentisate (as buttressed by the detection of inducible homogentisate dioxygenase within its genome). It can also grow on vanillin, veratraldehyde and ferulic acid	Paliwal et al. (2014)
18	General Genome Features: The whole genome consists of 187 contigs of size 6,045,648 bp (6.04 Mb), with a G + C content of 66.9%. Annotations using the Rapid Annotations using Subsystems Technology (RAST) server highlighted that 5954 protein-coding sequences (CDSs) are present in the genome, 3027 of which were categorized into one of the 553 RAST subsystems. There were 60 tRNAs	Strain, Isolation and Substrates Degraded: The *P. aeruginosa* Strain PK6 bacterium was isolated from a site contaminated by a petroleum hydrocarbon spill from an automobile service station in Junagadh, Saurashtra Region, Gujarat, India, and analysis of its genome revealed the presence of 124 CDSs related to the degradation of aromatic compounds	Patel et al. (2014)
19	General Genome Features: Five whole genomes were analyzed, and they possess different properties. *Pseudomonas stutzeri* MF28 has a genome 4,943,564 bp in size, consisting of 91 contigs. It's G + C content is 62.29%, and it has 4630 putative genes. The next bacterium is *Pseudomonas alcaligenes* OT69, which possesses 223 contigs, with a total size of 7,029,758 bp. Its G + C content is 65.96%, and it has 6543 putative genes. The next bacterium is *P. aeruginosa* WC55, which had 342 contigs, and a genome size of 6,844,176 bp. It had a G + C content of 66.14. The bacterium has 6650 putative genes. *Stenotrophomonas maltophilia* MF89 is the next bacterium, having 209 contigs, and a size of 4,649,035 bp. Its G + C content is 66.18%, and it has 4376 putative genes. The final bacterium was *Marinobacterium maritypicum* MF109, whose genome consists of 260 contigs, totalling 3,996,920 bp in size. It's G + C content is 68.24%, and it has 4094 putative genes Gene Annotation Studies: The *P. stutzeri* MF28 strain was isolated from mantle fluid, and it possesses 813 genes for xenobiotic degradation/metabolism; while *P. alcaligenes* OT69; *P. aeruginosa* WC55, *S. maltophilia* MF89 and *Marinobacterium maritypicum* MF109 were isolated from oyster tissue, water column, mantle fluid and mantle fluid, respectively; and they possessed 976, 958, 619, and 608 genes for xenobiotic degradation/metabolism, correspondingly	Strains, Isolation and Substrates Degraded: All the bacteria were oyster-associated, and possess hydrocarbon degradation abilities. There was a fairly constant distribution of the putative genes for xenobiotic degradation and metabolism relative to the total number of genes possessed by the five isolates (they ranged from 14.1% to 17.6%). Among these, genes for the degradative pathways of toluene, xylene, and PAHs, as well as those for the degradation of chloroalkane, chloroalkene, and nitrotoluene, were common among all the strains	Chauhan et al. (2013)

20	General Genome Features: The genome was assembled into 23 contigs, having a size of 2,363,215 bp. It has an average G + C content of 42.0%. There are 2249 putative open reading frames, giving a coding intensity of 91.2%. There were 35 tRNA genes for all 20 amino acids and one 16S–23S-5S rRNA operon. There were 18 PAH-degrading dioxygenase alpha (or large) subunits found in the genome	Strain, Isolation and Substrates Degraded: The *Cycloclasticus* sp. Strain P1. Strain was isolated from deep-sea sediments of the Pacific Ocean, and it can degrade pyrene, a four-ring polycyclic aromatic hydrocarbon (PAH), additionally, it can also use naphthalene, 2-methylnaphthalene, 2,6-dimethylnaphthalene, biphenyl, fluorene, acenaphthene, dibenzofuran, dibenzothiophene, phenanthrene and anthracene as the only carbon sources for growth. It was reported to be the only Strain in the genus that could use pyrene as the only carbon source for growth	Lai et al. (2012)
21	General Genome Features: The genome sequence of this bacterium revealed that it has a size of 6,074,058 bp, with a GC content of 66.83%. There were 5647 predicted coding sequences (CDSs), one 16S–23S-5S operon, and 50 tRNAs. The majority of the predicted proteins (88.9%) were classified into 22 COG categories using the COG database, with 557 subsystems	Strain, Isolation and Substrates Degraded: The *P. aeruginosa* Strain SJTD-1 bacterium was isolated from oil-contaminated soil, and was found to be capable of efficiently utilizing long-chain alkanes (*n*-hexadecane, *n*-octadecane, *n*-docosane, *n*-tetracosane, and *n*-triacontane), diesel oil and crude oil as its sole carbon and energy source	Liu et al. (2012)

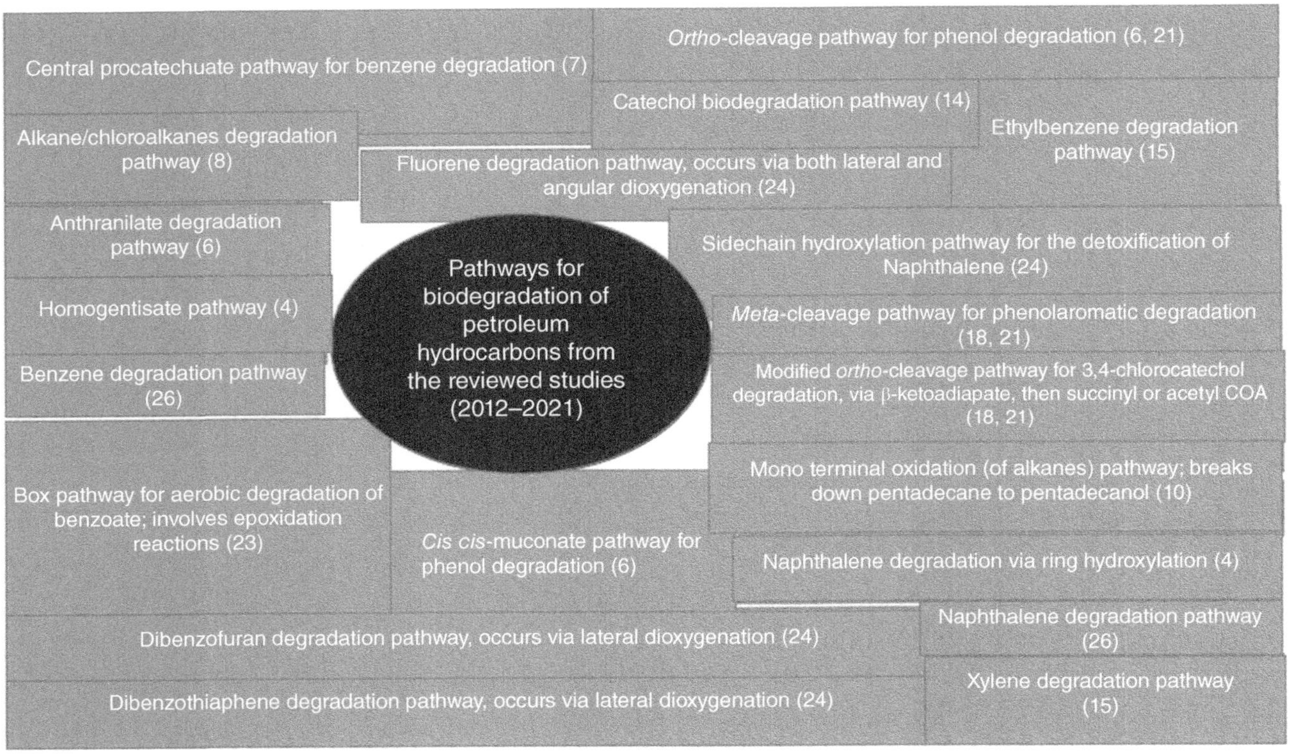

Figure 15.7 Literature map of the pathways involved in the biodegradation of petroleum hydrocarbons from the reviewed papers. Key: References are stated in brackets, and are the paper IDs stated before, namely: 4 = Paliwal et al. (2014); 6 = Miyazaki et al. (2014); 7 = Ruiz et al. (2015); 8 = Das et al. (2015); 10 = Pal et al. (2017); 14 = Mahjoubi et al. (2019); 15 = Ravintheran et al. (2019); 18 = Garrido-Sanz et al. (2020); 19 = Filonov et al. (2020); 20 = Elufisan et al. (2020); 21 = Xiang et al. (2020); 23 = Weiten et al. (2021); 24 = Liu et al. (2021) and 26 = Mahjoubi et al. (2021).

abbreviated, using their genes, and most have EC numbers. The same gene may have more than one function, and some genes have not yet had their functions fully characterized. Often, the same gene is reported in more than one function, i.e. possessed by different microbes, degrading the same, similar, or even different substrates (Table 15.4).

15.3.6 Whole Genome Sequence Maps

Whole genome maps summarize the whole genome sequence of the bacterium, in the form of rings. The meaning of the rings may differ, but generally, they follow a sequence akin to this: the first ring represents the genomic position/size markers; the second ring represents the genes that were said to belong to categories in the COG; the third ring represents the genes found in the KEGG database pathways; the fourth ring represents the genes found assigned to GO (genome online) database terms; the fifth ring represents the genes assigned to ncRNA (noncoding RNA); the sixth ring shows the GC content; and the seventh ring indicates the GC skew (Xiang et al. 2020). However, these may slightly differ depending on the databases used and the software involved in generating the whole genome maps (Figure 15.9).

15.3.7 Cell Metabolism Diagrams

These diagrams are constructed using insights from the WGS. Basically, after the arbitrary shape of the cell is drawn, transporters and/or transport channels/proteins/efflux pumps are constructed, and the paths for the entry and subsequent metabolism of the particular hydrocarbons are then drawn. Other Central Metabolic Pathways and other basic cellular metabolism processes can then be outlined. Such diagrams are clear manifestations of models of cell metabolism and energy utilization (Yasemi and Jolicoeur 2021) (Figures 15.10 and 15.11).

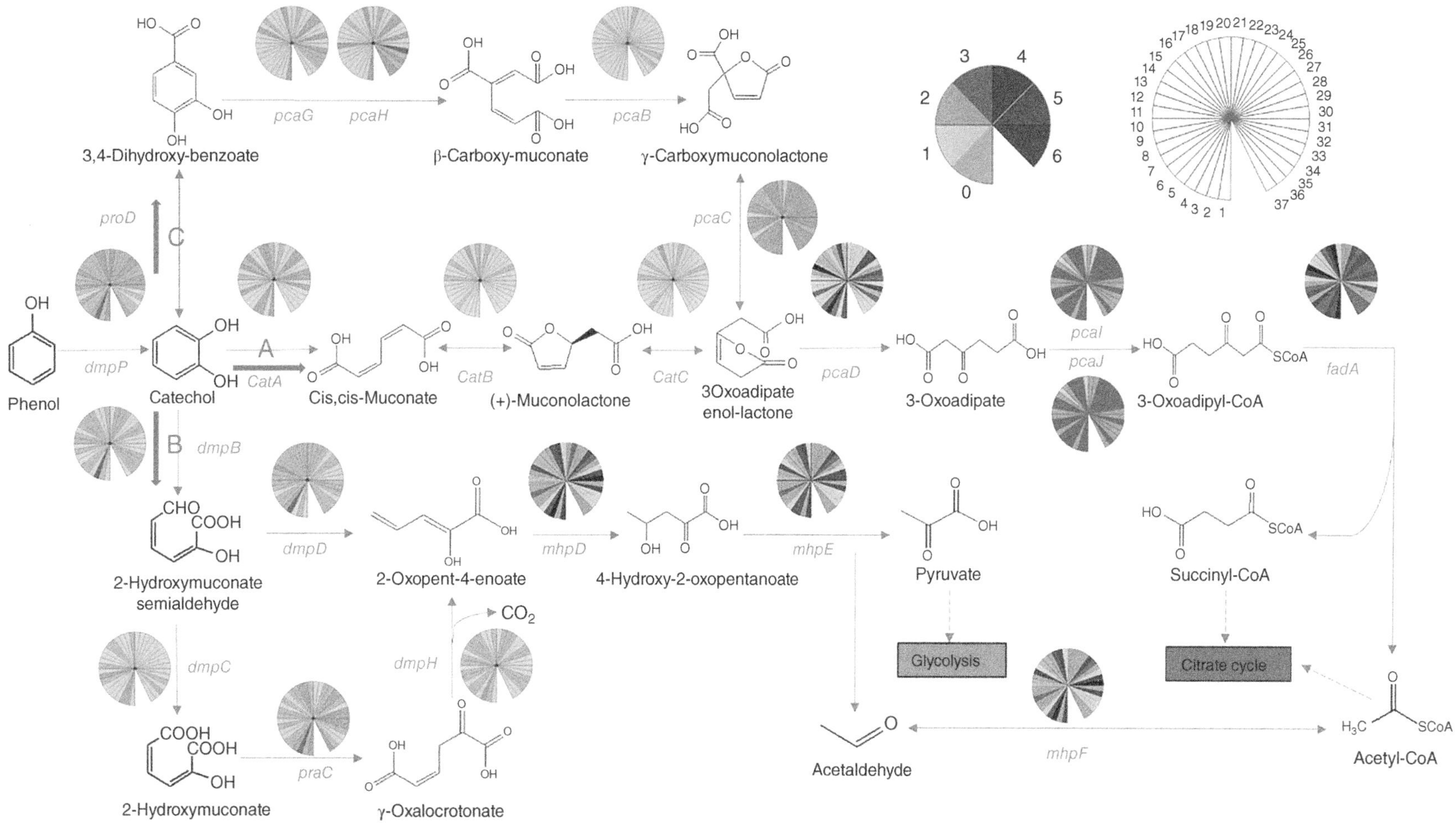

Figure 15.8 An integrated pathway for phenol metabolism in *Rhodococcus ruber* C1. A and C are the ortho-cleavage pathways, while B is the meta-cleavage pathway. The circle numbered 0–6 represents the copy number of the genes. Circle number 1–37 shows the number of *Rhodococcus* genomes used for reconstructing the pathway. The genes shown include phenol hydroxylase (*dmpP*), catechol 1,2-dioxygenase (*CatA*), catechol 2,3-dioxygenase (*dmpB*), and protocatechuate 3,4-dioxygenase (*pcaG, pcaH*). *Source:* Zhao et al. (2021).

Table 15.3 Putative gene clusters/operons involved in the biodegradation of various petroleum hydrocarbons.

S. no.	Gene clusters/operons	Genes encoded and their function in biodegradation	Reference
1	*nahAaAbAcAd cluster*	Produces naphthalene dioxygenase, which is involved in the naphthalene to naphthalene dihydrodiol	Liu et al. (2021)
2	*nahBFCED*	Converts cis naphthalene dihydrodiol to salicylic acid	Liu et al. (2021)
3	*nahHIJKL* cluster	Involved in the conversion of salicylic acid to catechol	Liu et al. (2021)
4	*nahDECFBAdAcAbAa* cluster	Involved in the complete biodegradation of naphthalene to salicylate	Mahjoubi et al. (2021)
5	*bphA1aA2aA3A4BCDE*	Involved in the metabolism of dibenzofuran	Garrido-Sanz et al. (2020)
6	*bphA1bA2bBD*	Involved in the metabolism of dibenzofuran, naphthalene and phenanthrene	Garrido-Sanz et al. (2020)
7	*bhpE1–3F1–3G1–3,bphE, bph G1–2* cluster	Involved in the metabolism of 2-hydroxypentadienoate	Garrido-Sanz et al. (2020)
8	*etbA1Aa2aA4BCDEFST*	Involved in the biodegradation of naphthalene, ethylbenzene and o-xylene	Garrido-Sanz et al. (2020)
9	*etbA4betbA2netbA3d*	Incomplete etb/bph cluster; involved in the biodegradation of ethylbenzene	Garrido-Sanz et al. (2020)
10	*etbA4RetbA1Aa2Aa2c + bphDEF*	Complete etb/bph cluster for metabolism of ethylebenzene	Garrido-Sanz et al. (2020)
11	*nah* gene cluster	Involved in the biodegradation of naphthalene, ethylbenzene and o-xylene	Garrido-Sanz et al. (2020)
12	*pcaGHBCD(D1-D5)I-J* and *catABC*	Degradation of β-ketoadiapate	Garrido-Sanz et al. (2020)
13	*tmo ABCDEF* gene cluster	Metabolizes toluene to p-cresol	Garrido-Sanz et al. (2020)
14	*X2Y2B2C2Z2*	Encodes for methane monooxygenases	Garrido-Sanz et al. (2020)
15	*ttg 2* operon	Encodes for ABC transporters involved in the uptake of petroleum hydrocarbons from the environment	Ruiz et al. (2015)
16	*α subunit,* 18 PAH degrading genes, including the naphthalene gene cluster	Involved in the metabolism of naphthalene	Lai et al. (2012)

15.3.8 Comparative Genomics

Comparative genomics is interested in making comparisons between genomes of similar, related, or even different bacteria, with regard to their characteristic features. This is capable of elucidating evolution, relationships, convergence/divergence, etc. (Zhao et al. 2021) (Table 15.5).

15.3.9 Gene Gain and Loss Events

Comparative genomics had been utilized in outlining the gain and loss of genes across species, with the passage of time. This is illustrated overleaf. The phenomenon becomes widespread when specified portions of the genomes of bacterial species belonging to the same genera are compared. This also clearly demonstrates the adaptation of the species to varying environmental conditions, and, possibly, the effect of mutation (Zhao et al. 2021).

15.3.10 Synteny Maps

Synteny maps are used for the tracking of location/shifts in genes position across different species in the same genus, or, generally, between different bacteria. They indicate locus/loci of genes, and possibly their variation between species.

Table 15.4 Summary of the genes involved in biodegradation of selected hydrocarbons based on the study area, and their respective function.

S. no.	Enzyme/Gene name (abbreviation and enzymes commission/EC number, if any)	Function(s)	References
1	1, 2-dihydroxycyclohexa-3, 5-diene-1-carboxylate dehydrogenase (EC 1.3.1.25)	Converts cis-1,6-dihydroxy-2,3-cyclohexadiene-1-carboxyclic acid to catechol, involved in the biodegradation of phenanthrene (reaks down cis-3,4-Dihydroxy-3,4-Dihydrophenanthrene to 3,4-Dihydroxyphenanthrene)	Garrido-Sanz et al. (2020), Xiang et al. (2020)
2	1,2-dihydroxynaphthalene deoxygenase	Forms salicylaldehyde from 1,2-dihydroxynaphthalene	Paliwal et al. (2014)
3	1-phenylethanol/−propanol dehydrogenase (*Ped*)	Involved in ethylbenzene and *n*-propylbenzene degradation	Weiten et al. (2021)
4	2,4-dihydroxyacetophenone dioxygenase	Facilitates the cleavage of carbon–carbon bond in a substituent aromatic ring	Elufisan et al. (2020)
5	2-hydroxychromene-2-carboxylate isomerase (*nahD*)	Naphthalene biodegradation	Liu et al. (2021)
6	2-hydroxymuconate semialdehyde hydrolase (EC 3.7.1.9)	Converts 4-oxalocrotonate to 2-oxo-pent-4-enoate	López-Moreno et al. (2021)
7	2-hydroxymuconic semialdehyde hydrolase	Naphthalene degradation	Filonov et al. (2020)
8	2-polyprenylphenol hydroxylase	Xylene degradation	Das et al. (2015)
9	3- oxoadipyl-CoA thiolase (*pcaF*)	Involved in phenanthrene degradation	Mahto and Das (2020)
10	3,4-dihydroxyphenylacetate 2,3-dioxygenase (*hpcB*)	Involved in phenanthrene degradation	Mahto and Das (2020)
11	3,4-dihydroxyphthalate decarboxylase	*NA*	Kwak et al. (2016)
12	3-carboxy cis, cis muconate cycloisomerase (*pcaB*; EC 5.5.1.2)	Produces 4-carboxymuconolactone from 3-carboxy cis, cis muconate; involved in phenanthrene biodegradation	Pal et al. (2017), Mahto and Das (2020)
13	3-carboxymuconate cycloisomerase	*NA*	Ruiz et al. (2015)
14	3-hydroxybutyryl-CoA dehydrogenase (*paaH/ fadB/mmgB)*	Xenobiotic degradation; also, it converts (S)-3-hydroxybutanoyl COA to acetoacetyl CoA	Mahto and Das (2020), López-Moreno et al. (2021)
15	3-oxoadipate CoA-transferase, including alpha and β-subunits (*pcaI/ pcaJ*; EC 2.8.3.6)	Breaks down 3-oxoadipate to 3-oxoadipyl COA; involved in phenol degradation	Mahjoubi et al. (2019), Lee et al. (2020), López-Moreno et al. (2021)
16	3-oxoadipate enol-lactonase (*pcaD/ praD*)	Involved in phenol and phenanthrene degradation (breaks down 3-oxoadipate enol lactone to 3-oxoadipate)	Mahto and Das (2020) Lee et al. (2020) Xiang et al. (2020)
17	3-phenylpropionate dioxygenase (*hcaE*; EC 1.1.4.1.-)	Breaks down phenanthrene into cis-3,4-dihydroxy-3,4-dihydrophenanthrene, involved in the degradation of monobenzene ring compounds	Mahjoubi et al. (2019), Liu et al. (2021)
18	4-carboxymuconolactone decarboxylase (*pcaC*; EC 4.1.1.44)	Breaks down 3,4-dihydroxyphthalate to 3,4-Dihydroxybenzoate; also, it transforms 4-carboxymuconolactone into β-ketoadipate enol lactone; it is likewise involved in phenanthrene biodegradation	Pal et al. (2017), Mahjoubi et al. (2019), Mahto and Das (2020)
19	4-hydroxy-2-oxovalerate aldolase (*PobA*; EC 4.1.3.39)	Breaks down trans-2'-Carboxybenzalpyruvate to 2-Carboxybenzaldehyde; converts 4-hydroxy 2-ocopentenoate to acetaldehyde; involved in the biodegradation of phenanthrene	Mahjoubi et al. (2019), López-Moreno et al. (2021)
20	4-hydroxyphenylacetate isomerase	Degradation of monobenzene ring compounds	Liu et al. (2021)

(Continued)

Table 15.4 (Continued)

S. no.	Enzyme/Gene name (abbreviation and enzymes commission/EC number, if any)	Function(s)	References
21	4-oxalocrotonate tautomerase (*praC, xylH*; EC 5.3.2.6)	Crude oil degradation; breaks down 2-hydroxymuconate to 4-oxalocrotonate; also catalyzes the degradation of toluene, o-xylene, 3-ethyltoluene, and 1,2,4-trimethylbenzene	Yang et al. (2018), Elufisan et al. (2020), López-Moreno et al. (2021)
22	5-carboxymethyl-2- hydroxymuconic semialdehyde dehydrogenase (*hpcD*)	Involved in phenanthrene degradation	Mahto and Das (2020)
23	Acetaldehyde dehydrogenase	Biodegradation of phenanthrene	Liu et al. (2021)
24	Acetylating Acetaldehyde dehydrogenase (EC 1.2.1.10)	Involved in naphthalene degradation (breaks down 1-Hydroxy-2-naphthaldehyde to 1-Hydroxy-2-naphthoate)	Mahjoubi et al. (2021)
25	Aceto−/propiophenone carboxylase (*ApcA-E*)	Involved in ethylbenzene and *n*-propylbenzene degradation	Weiten et al. (2021)
26	Acetyl-CoA acetyl transferase (*fadA*; *fadI*)	Phenol degradation	Das et al. (2015), Lee et al. (2020)
27	Acetyl-CoA C-acetyltransferase (*atoB*)	Xenobiotic degradation	Mahto and Das (2020)
28	Acyl-CoA desaturase	Degradation of *n*-alkanes	Mahjoubi et al. (2021)
29	Acylphosphatase (*acyP*)	Xenobiotic degradation	Mahto and Das (2020)
30	Acyltransferase	Produces biosurfactants, which reduce the interfacial surface tension of hydrocarbons, hence, enhancing biodegradation	Pal et al. (2017)
31	Alcohol dehydrogenase (*frmA, adhC*; EC 1.1.1.1)	Involved in the oxidation of alcohols (the second step of alkanes mineralization)	Das et al. (2015), Pal et al. (2017), Yang et al. (2018)
32	Aldehyde dehydrogenase (NAD+) (ALDH; EC: 1.2.1.3)	Alkane degradation; breaks down 2-Carboxybenzaldehyde to phthalate; degrades 4-hydroxymuconic semialdehyde to 2-maleyl acetate; oxidizes aldehydes (the third step of alkanes mineralization); converts 2-hydroxymuconate semialdehyde to 2-hydroxymuconate	Liu et al. (2012), Pal et al. (2017), Yang et al. (2018), Mahjoubi et al. (2019), López-Moreno et al. (2021)
33	Alkane-1-monooxygenase (*alkB, alkM*; EC 1.14.15.3)	Hydrocarbons degradation, i.e. degradation of linear alkanes (converts alkanes to alkyl alcohols)	Patel et al. (2014), Pal et al. (2017), Mahto and Das (2020), Garrido-Sanz et al. (2020)
34	Alkane hydroxylase	Alkane degradation	Liu et al. (2012)
35	Alkane sulfonate monoo xygenase (EC 1.14.14.5)	Hydrocarbons degradation/degradation of alkanes	Patel et al. (2014), Ruiz et al. (2015), Pal et al. (2017)
36	Aminobenzoate deoxygenase	*NA*	Ruiz et al. (2015)
37	Anthranilate 1,2-dioxygenase, including large subunit (*AntA*), small subunit (*AntB*) and reductase component (*AntC*)	Xenobiotic degradation	Mahto and Das (2020)
38	Aromatic-ring-hydroxylating/opening/cleaving dioxygenase (*ligB*; EC 1.14.12.13),	Converts phenanthrene into cis-3,4-Dihydroxy-3,4-dihydrophenanthrene; converts catechol into 2-hydroxymuconate semialdehyde; converts benzoyl COA to crotonyl CoA	Kwak et al. (2016), Pal et al. (2017), Mahjoubi et al. (2019), Mahto and Das (2020), López-Moreno et al. (2021)
39	Aryl aldehyde dehydrogenase	Removes -H moiety from aryl alcohols	Paliwal et al. (2014)
40	Benzaldehyde dehydrogenase (*bzdh*; EC:1.2.1.28)	Converts 1- and 2-hydroxymethylnaphthalene to respective naphthoic acids	Paliwal et al. (2014)

Table 15.4 (Continued)

S. no.	Enzyme/Gene name (abbreviation and enzymes commission/EC number, if any)	Function(s)	References
41	Benzene 1,2-dioxygenase, including reductase component, alpha and β-subunits (*benA-xylX, benB-xylY, benC-xylZ*; EC 1.14.12.10)	Degradation of xylene, it breaks down phthalate to phthalate 3,4-cis-dihydrodiol	Ruiz et al. (2015), Das et al. (2015), Li et al. (2017), Mahjoubi et al. (2019)
42	Benzoylacetate/−propionate CoA-ligase (*bal*)	Involved in ethylbenzene and *n*-propylbenzene degradation	Weiten et al. (2021)
43	Benzoylformate decarboxylase (*mdlC*)	Xenobiotic degradation	Mahto and Das (2020)
44	Benzylalcohol dehydrogenase (*badh*)	Converts 1- and 2-hydroxymethylnaphthalene to respective naphthoic acids	Paliwal et al. (2014)
45	Biphenyl-2,3-diol 1,2-dioxygenase (EC 1.13.11.39)	Breaks down 3,4-Dihydroxyphenanthrene to 2-Hydroxy-2H-benzo[h]chromene-2-carboxylate, catalyzes the degradation of biphenyl and gamma-hexachlorocyclohexane	Mahjoubi et al. (2019), Elufisan et al. (2020)
46	Carboxymethylenebutenolidase	Involved in toluene degradation (facilitates the conversion of methyl catechol to 4-carboxymethyl-4-methylbut-2-en-4-olide, to 4-oxohex-2-enedioate)	Elufisan et al. (2020)
47	Carboxymuconolactone decarboxylase (EC 4.1.1.44)e	Breaks down trans-2'-Carboxybenzalpyruvate to 2-Carboxybenzaldehyde; degrades 4-carboxymuconolactone to ß-ketoadipate enol-lactone; also breaks down trans-4-(1'-Hydroxynaphth-2'-yl)-2-oxobut-3-enoate to 1-Hydroxy-2-naphthaldehyde	Mahjoubi et al. (2019)
48	Catechol 1,2-dioxygenase (*catA*; EC 1.13.11.1)	Toluene, phenol and phenanthrene, degradation (breaks down catechol into cis,cis muconate)	Ruiz et al. (2015), Das et al. (2015), Yang et al. (2018), Mahjoubi et al. (2019), Mahto and Das (2020), Lee et al. (2020), Xiang et al. (2020), Zhao et al. (2021)
49	Catechol 2,3-dioxygenase (*catE, dmpB; nahH;* EC: 1.13.11.2)	Degradation of aromatic rings, degradation of naphthalene (conversion of salicylate aldehyde to catechol) and Bisphenol A	Pal et al. (2017), Li et al. (2017), Yang et al. (2018), Elufisan et al. (2020), Zhao et al. (2021), López-Moreno et al. (2021)
50	Catechol dioxygenase	Aromatic compounds degradation	Patel et al. (2014)
51	Cytochrome P450 (*CYP120A*)	Degradation of small, medium or long chain *n*-alkanes by oxidizing them to primary alcohols, monocarboxylic fatty acids, secondary alcohols and ketones	Kwak et al. (2016), Pal et al. (2017), Yang et al. (2018)
52	Dienelactone hydrolase	Toluene degradation	Das et al. (2015)
53	Dihydrodiol dehydrogenase (*benD, xylL*)	Transforms 3- or 4-chloro-3,5-cyclohexadiene-1,2-diol-1-carboxylic acid to 3- and 4-chlorocatechol	Miyazaki et al. (2014)
54	Enol-CoA hydratase (*echA/paaF*)	Xenobiotic degradation	Mahto and Das (2020)
55	Enoyl COA hydratase (EC 4.2.1.17)	Converts crotonyl COA to (S)-3-hydroxybutanoyl COA	López-Moreno et al. (2021)
56	Ethyl−/*n*-propylbenzene dehydrogenase (*EbdA – D*)	Involved in ethylbenzene and *n*-propylbenzene degradation	Weiten et al. (2021)
57	Extradiol ring-cleavage dioxygenase	Initializes PAH degradation	Kwak et al. (2016)
58	Flavin-dependent oxidoreductase	*NA*	Kwak et al. (2016)

(*Continued*)

Table 15.4 (Continued)

S. no.	Enzyme/Gene name (abbreviation and enzymes commission/EC number, if any)	Function(s)	References
59	FMNH2-dependent monooxygenase	Converts 3-dehydroxybenzoate to catechol	Lopez-Moreno et al. (2021)
62	Fumaryl acetoacetase (*HmgB*)	Hydrolyses fumarylacetoacetate to fumarate and acetoacetate	Paliwal et al. (2014)
61	Gentisate 1,2-dioxygenase	Involved in phenanthrene degradation	Mahto and Das (2020)
62	Glycosyl transferase (*rhlB*)	Produces biosurfactants (rhamnolipids) which reduce the interfacial surface tension of hydrocarbons, hence, enhancing biodegradation	Pal et al. (2017), Das et al. (2015)
63	Homogentisate 1,2-dioxygenase/ Inducible homogentisate dioxygenase (*HmgA*; EC 1.13.11.5)	Catalyzes the transformation of homogentisate to 4-maleylacetoacetate; breaks down 3,4-Dihydroxybenzoate to -Carboxy-cis,cis-muconate; it is involved in phenanthrene biodegradation and the catabolism of aromatic rings	Patel et al. (2014), Paliwal et al. (2014), Ruiz et al. (2015), Das et al. (2015), Majoubi et al. (2019), Mahto and Das (2020), Elufisan et al. (2020)
64	Long-chain alkane monooxygenases (*ladA*)	Biodegradation of long-chain alkanes	Garrido-Sanz et al. (2020)
65	L-rhamnosyl transferase (RhlC)	Synthesis of rhamnolipids	Das et al. (2015)
66	Maleylacetoacetate isomerase (EC 5.2.1.2)	Breaks down 2-Hydroxy-2H-benzo[h] chromene-2-carboxylate to trans-4-(1'-Hydroxynaphth-2'-yl)-2-oxobut-3-enoate	Mahjoubi et al. (2019)
67	Maleylacetoacetate isomerase (*HmgC*)	Isomerises 4-maleylacetoacetate to fumarylacetoacetate	Paliwal et al. (2014)
68	Methyltransferase	*NA*	Kwak et al. (2016)
69	m-phenylacetyl-COA ligase	Transforms phenylacetic acid to its CoA-thioester (phenylacetyl CoA)	Paliwal et al. (2014)
70	Muconate cycloisomerase (*catB*; EC 5.5.1.1)	Benzoic acid, phenanthrene, phenol and crude oil degradation; breaks down 3,4-Dihydroxybenzoate to 3-Carboxy-cis,cis-muconate; can also degrade 3-Carboxy-cis,cis-muconate to 4-carboxymuconolactone	Li et al. (2017), Yang et al. (2018), Mahjoubi et al. (2019), Mahto and Das (2020), Lee et al. (2020), Xiang et al. (2020)
71	Muconolactone delta-isomerase/D--isomerase (*catC*)	Phenanthrene degradation (breaks down muconolactone to 3-oxoadipate enol-lactone)	Li et al. (2017), Mahto and Das (2020), Xiang et al. (2020)
72	Muconolactone isomerase (*catC*; EC 5.3.3.4),	Breaks down muconolactone to enol lactone; involved in phenol degradation	Mahjoubi et al. (2019), Lee et al. (2020)
73	NAD(P)H-dependent oxidoreductase (EC 1.3.1.32)	Converts 2-malelyl acetate to 3-oxoadipate	López-Moreno et al. (2021)
74	Naphthalene 1,2-dioxygenase (including alpha and β-subunit)	Catalyzes the hydroxylation of the aromatic ring to yield 1,2-dihydroxynaphthalene	Paliwal et al. (2014), Filonov et al. (2020)
75	Naphthalene dehydrogenase (*nahB*)	Degrades naphthalene into 1,2- dihydroxynaphthalene	Liu et al. (2021)
76	Naphthalene dioxygenase (*ndoB; nahAaAbAcAd*)	Involved in aromatic hydrocarbon degradation, breaks down naphthalene giving cis-naphthalene dihydrodiol	Pal et al. (2017), Liu et al. (2021)
77	Ortho-halobenzoate 1, 2-dioxygenase, including alpha and β-subunit	Breaks down phthalate to Phthalate 3,4-cis-dihydrodiol	Mahjoubi et al. (2019)
78	Phenol hydroxylase (*dmpP*), including large (*PheA1*) and small (*PheA2*) subunits	Biodegradation of phenol (hydroxylates phenol to catechol)	Lee et al. (2020), Zhao et al. (2021)

Table 15.4 (Continued)

S. no.	Enzyme/Gene name (abbreviation and enzymes commission/EC number, if any)	Function(s)	References
79	Phenylacetate hydroxylase	Metabolism of phenylacetate	Lai et al. (2012)
80	Phosphomannomutase (*AlgC*)	Produces/synthesizes biosurfactants, which reduce interfacial surface tension of hydrocarbons, hence, enhancing biodegradation	Das et al. (2015), Pal et al. (2017)
81	Phthalate 3,4-dioxygenase ferredoxin/ferredoxin reductase subunit (*phtAc, phtAd*; EC:1.18.1.3)	PAH degradation	Kwak et al. (2016)
82	p-hydroxybenzoate 3-monooxygenase (EC 1.14.13.2)	Plays an essential role in the β-ketoadipate pathway of aromatic compound degradation	Pal et al. (2017)
83	p-hydroxyl benzoyl-COA thioesterase	Catalyzes the conversion of p-hydroxybenzoyl-CoA to p-hydroxybenzoate	Paliwal et al. (2014)
84	Protocatechuate dioxygenase/ protocatechuate 3,4-dioxygenase alpha and β-subunits (*ligA/ligB*, *PcaG*, and *PcaH*; EC 1.13.11.3)	Cleaves the ring of protocatechuate, yielding 3-carboxy cis, cis muconate; converts 3,4-dihydroxybenzoate to carboxy-cis,cis-muconate via *ortho*-cleavage; involved in phenanthrene biodegradation	Paliwal et al. (2014), Ruiz et al. (2015), Kwak et al. (2016), Pal et al. (2017), Mahto and Das (2020), Zhao et al. (2021)
85	Rubredoxin reductase	Alkane degradation	Liu et al. (2012)
86	Rubredoxin-NAD(+) reductase	Degradation of *n*-alkanes	Mahjoubi et al. (2021)
87	S-(hydroxymethyl) glutathione Dehydrogenase (EC 1.1.1.284)	Naphthalene degradation	Pal et al. (2017)
88	S-(hydroxymethyl) glutathione dehydrogenase/class III alcohol dehydrogenase	Naphthalene degradation	Das et al. (2015)
89	Salicylaldehyde dehydrogenase (EC 1.2.1.65)	Naphthalene degradation	Pal et al. (2017)
90	Salicylaldehyde dehydrogenase (*nahF*)	Naphthalene biodegradation	Liu et al. (2021)
91	Salicylate 1-monooxygenase	Naphthalene degradation	Filonov et al. (2020)
92	Salicylate hydroxylase	Catalyzes the removal of the carboxyl group at position 1 of salicylic acid, replacing it with a hydroxyl group in the same position, thus converting the salicylic acid to catechol	Elufisan et al. (2020)
93	Soluble methane monooxygenases (*mmoXCYB*)	Biodegradation of *n*-alkanes 3–5 carbons in length; also utilize methane	Garrido-Sanz et al. (2020)
94	Succinate dehydrogenase	Toluene degradation	Das et al. (2015)
95	β-keto adipyl CoA thiolase (EC 2.3.1.174)	Transforms β-keto adipyl CoA to succinyl COA and acetyl COA/ Breaks down 3-oxoadipyl COA to acetyl and succinyl-COA	Pal et al. (2017), Mahjoubi et al. (2019)
96	β-ketoadipate enol lactone hydrolase (EC 3.1.1.24)	Hydrolyzes 4-carboxymuconolactone decarboxylase into β-keto adipate; breaks down enol lactone to 3-oxoadipate	Pal et al. (2017), Mahjoubi et al. (2019)
97	β-ketoadipate succinyl CoA transferase (EC 2.8.3.6)	Conversion of β-keto adipate to β-keto adipyl CoA	Pal et al. (2017)

NB: Genes involved in the biodegradation of the substrates from pristine states to intermediaries of Central Metabolic Pathways were shown/considered. Genes involved in the breakdown/transformation of these intermediaries, downwards, were not considered in the study. Xenobiotic degradation means that the gene is confirmed to be functional in petroleum hydrocarbons biodegradation of one or more xenobiotic(s), but had not been determined experimentally.
NA = Function not yet known.

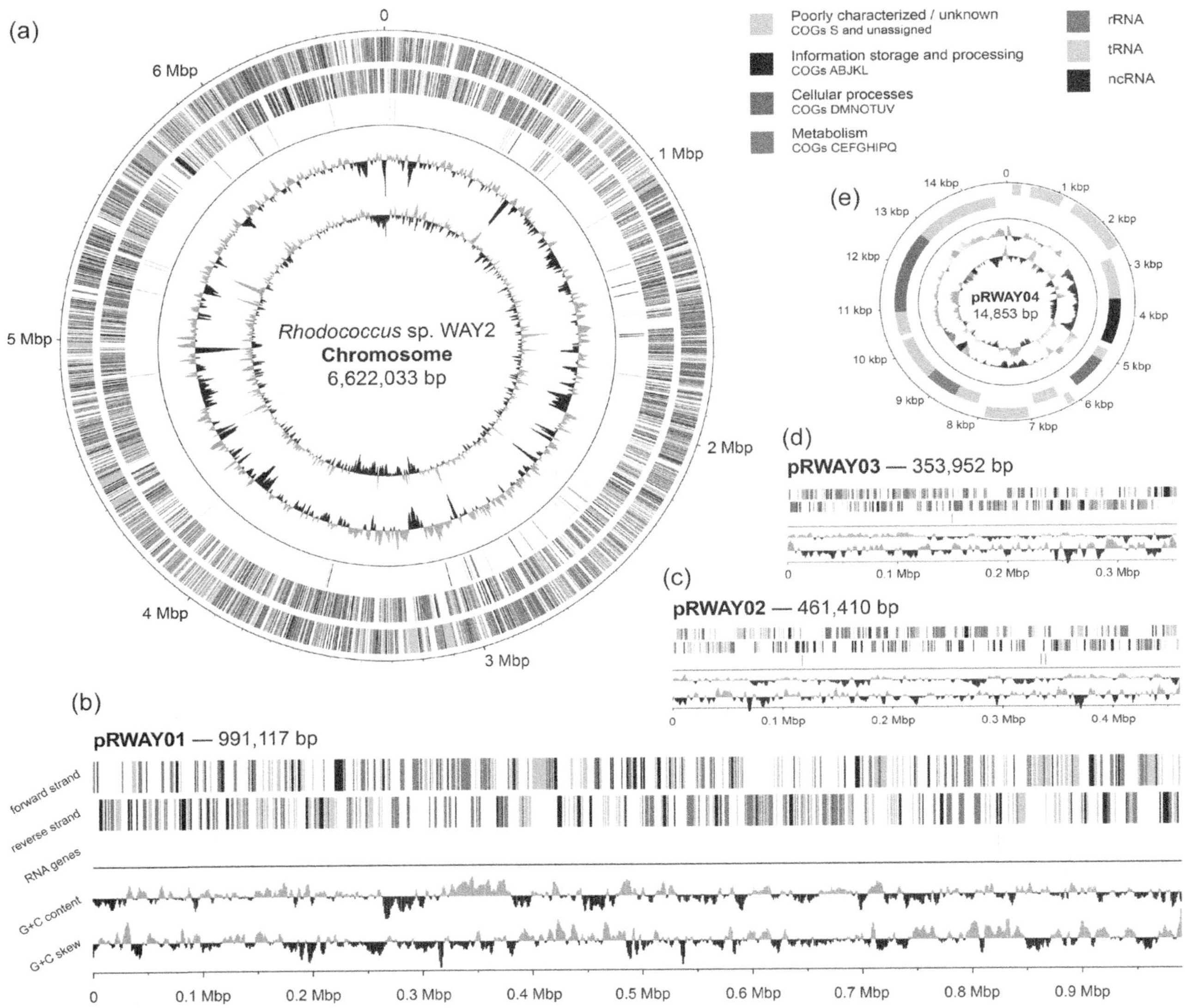

Figure 15.9 An example of a whole genome sequence map of the bacterium: *Rhodococcus* sp. WAY2 showing both chromosome a, the linear mega plasmids: pRWAY01 b, pRWAY02 c, and pRWAY03 d; and the circular plasmid pRWAY04 e. The rings are as follows: Rings 1 and 2 = coding strands in the forward and reverse strands, respectively. Ring 3 = RNA genes (. Ring 4 = deviation in G+C content. Ring 5 = G+C skew. *Source:* Detailed interpretation of the figure can be found in Garrido-Sanz et al. (2020) / Microbiology society / CC BY.

This can be harnessed in understanding relationships between genes and/or the bacteria expressing the genes. A model synteny map is the one highlighted by Mahjoubi et al. (2021), which showed the loci and relationship of genes responsible for benzoate, dioxin, and naphthalene degradation across selected *Pseudomonads*.

15.3.11 Other Physiological Characteristics of the Bacterium

The WGS data includes both genes related to biodegradation and those involved in other life processes of the organism, including infectious disease causation. Upon annotation of the genome, the comparison of various genes within the same subfamily (e.g. those involved in biodegradation), or the whole genes in total, can be carried out, as shown in Figures 15.12 and 15.13.

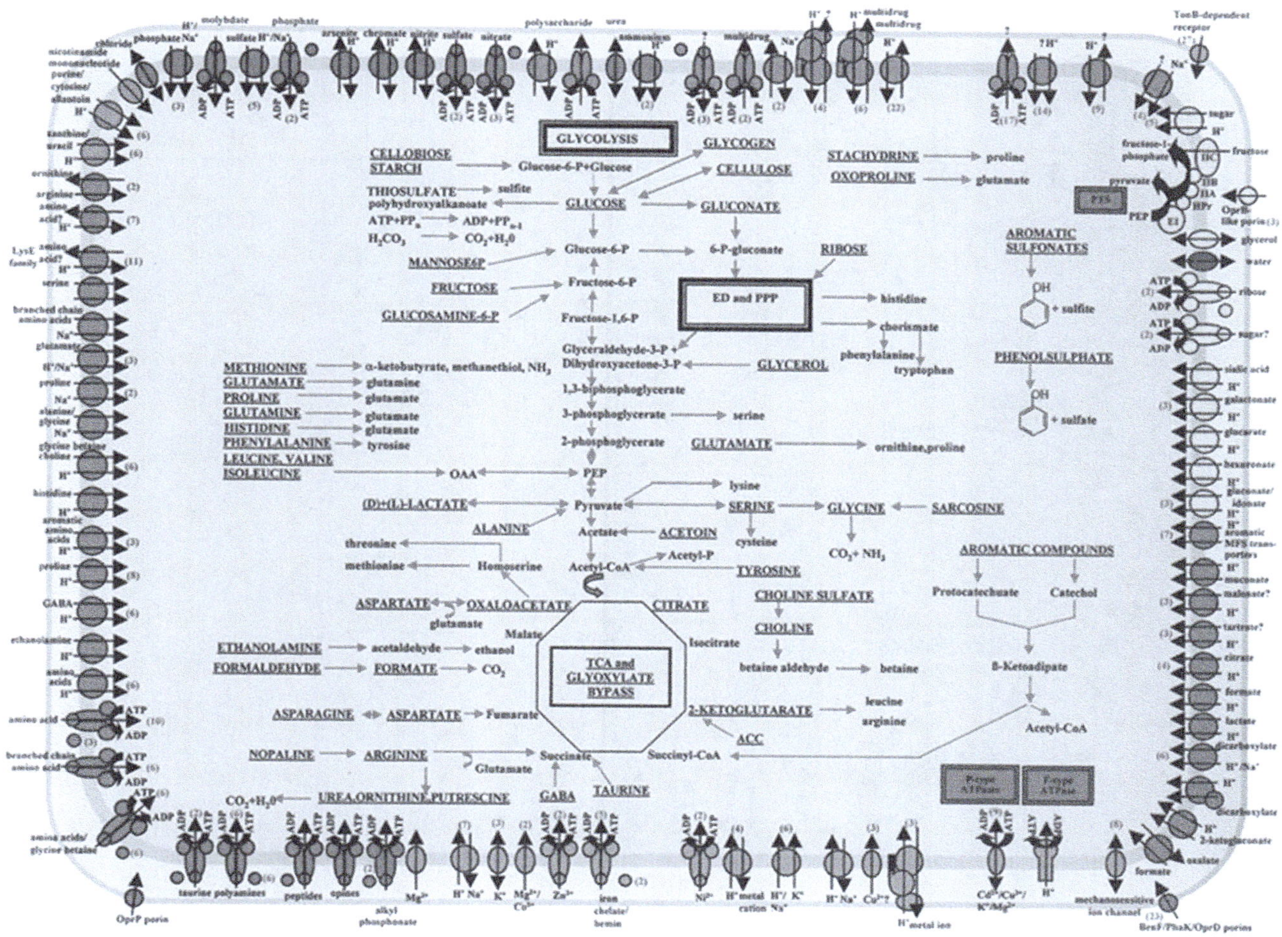

Figure 15.10 Cell metabolism diagram of the first bacterium with potential HC degradation ability constructed using WGS data. Key highlights include the outline of the cell's membranes and cytosol, pathways for the metabolism of aromatic HCs, energy metabolism, transport proteins, efflux pumps, and heavy metal regulators. *Source:* Nelson et al. (2002) / Blackwell Science Ltd.

15.4 Future Research Directions in WGS Studies of Petroleum Hydrocarbon-Degrading Bacteria

There is still the need for more WGS studies, and from more diverse regions/countries. Among the reviewed articles throughout the period (2012–2021) that this chapter considers, only 21 WGS sequences were published, from just 12 countries. There is indeed the need for further research, especially from oil-producing, developing countries whose land and water environments are bedeviled by petroleum spill events, such as Nigeria, not only to gain new perspectives and insights into the issue but also to unleash the tremendous potential of these bacteria in remediating the spoiled land/aquatic environments. Future studies should also involve the WGS of hydrocarbonoclastic bacteria that showed great results based on the evaluation of biodegradation studies in vitro/in situ, to further elucidate the metabolic bases behind biodegradation, and reveal other features of the strains which may not be readily observable or assayable using in vitro techniques alone.

Another priority is the need to establish the function of novel/unknown genes from the WGS diagrams/data, considering the fact that some genes were not assigned any function and that many coding sequences (hundreds, in fact) belong to genes having "unknown functions". Moreover, gene regulation/transcriptomic studies are necessary to establish the true roles of individual genes in an operon/gene cluster involved in biodegradation, including upregulation, downregulation, and constitutive expression, to determine leading genes in pathways for degradative metabolism of certain substrates.

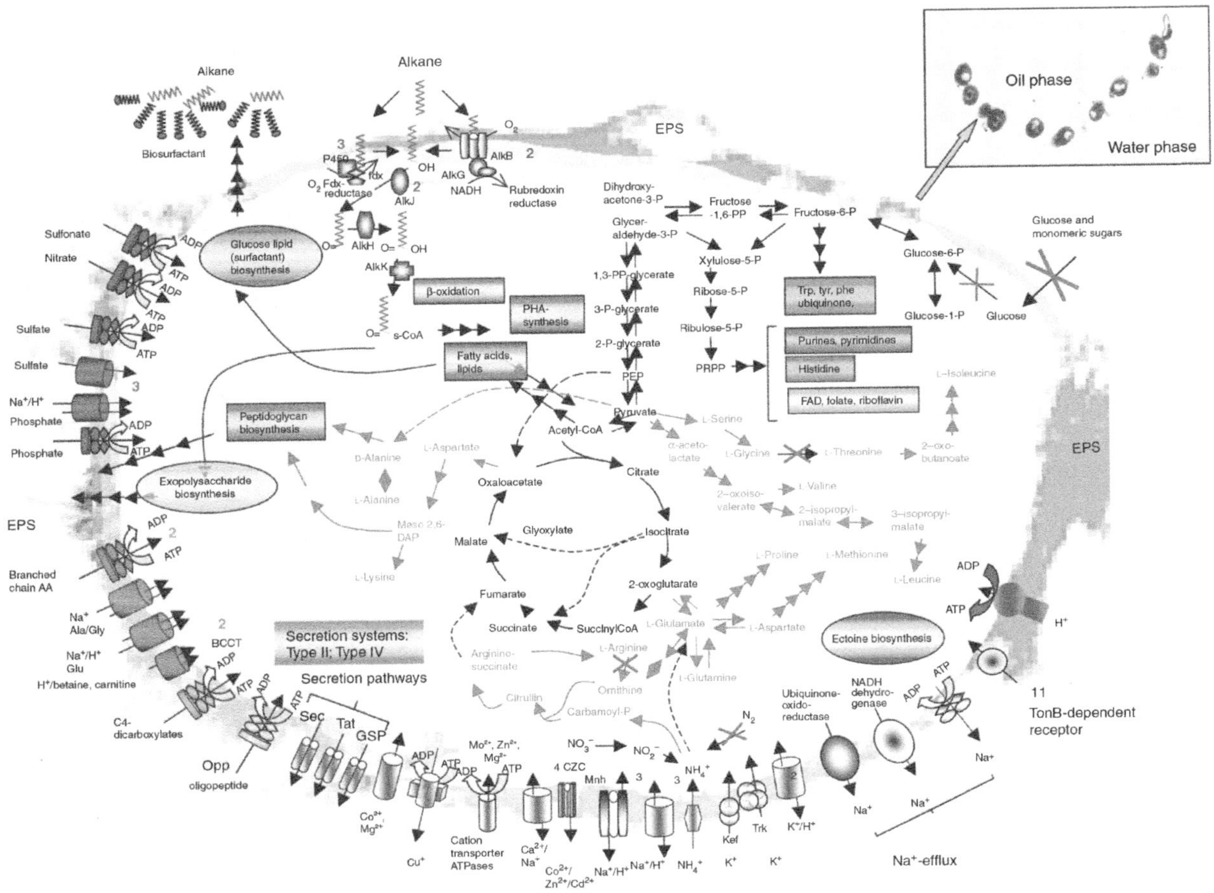

Figure 15.11 Cell metabolism diagram for *Alcanivorax borkumensis* SK2. Cell shape chosen based on transmission electron micrograph (TEM) of the bacterium's cell grown on hexadecane. Highlighted are pathways for the degradation of alkanes. *Source:* Transporters are also shown Schneiker et al. (2006) / Springer Nature / CC BY.

Results from WGS studies still need the application, i.e. the utilization of strains having novel/robust/rapid degradation abilities, and those capable of performing other functions apart from the degradation, e.g. plant growth promotion, production of bioplastics, etc., in experimental, on-the-field settings, to verify and ascertain the true ability of the strains to perform biodegradation in contaminated sites, toward bioremediation of polluted land and/or aquatic environments.

15.5 Conclusions

An in-depth analysis of the WGS of petroleum HDB revealed to us a number of insights. Firstly, a great number of diverse genes are involved in petroleum hydrocarbon biodegradation. These genes may be distributed across different species of microorganisms within the same genus, e.g. the distribution of genes for the anaerobic dearomatization of aromatics and terpenoids in *Aromatoleum* sp. (vide Weiten et al. 2021). Moreover, one gene can have multiple functions in biodegradation, depending on the strain, the type of substrate on which the bacterium possessing the gene is acting, and other factors. This also occurs in the case of general/unspecified genes/enzymes, e.g. monooxygenase, laccase, etc. (vide López-Moreno et al. 2021). Thirdly, microbes adapted to the degradation of a specific metabolite often possess additional metabolic capabilities that can be utilized in the biodegradation of various xenobiotics (vide Zhao et al. 2021). Such xenobiotics include heavy metals, such as arsenic, cadmium, cobalt, copper, lead, manganese, mercury, tellurium, zinc, etc. (vide Paliwal et al. 2014; Li et al. 2017; Mahjoubi et al. 2019; Garrido-Sanz et al. 2020). Sometimes, the degradation ability is coupled with other important properties, such as plant growth promotion (including auxin biosynthesis, alkaloid biosynthesis, and nitrogen metabolism) and/or biosurfactant production (vide Das et al. 2015; Pal et al. 2017; Ravintheran et al. 2019; Xiang

Table 15.5 Results of comparative genomics analysis of bacteria belonging to the genera: *Aromatoleum* and *Azoarcus*, based on typical genome features.

	Aromatoleum				Azoarcus
Genome feature	***petrolei* ToN1**[T]	***bremense* PbN1**[T]	***aromaticum* EbN1**[Ta]	**sp. CIB**[b]	***olearius* BH72**[c]
Chromosome size, bp	5,407,396	4376,837	4,296,230	5,257,030	4376,040
Plasmid(s) [size, bp]	0	0	2 [207355233670]	0	0
G + C-content, %	65.72	65.59	65.12[d]	65.8	67.92
Phages (intact/questionable/incomplete)	1 (0/0/1)	7 (0/0/7)	3 (1/2/0)[d,f]	5 (0/2/3)	1 (0/0/1)
CRISPR	none	class 1 type I-E	class 1 type I-C	none	none
Transposase-related genes	54	170	237	101	19
Coding regions	4964	4198	4133[d]	4739	3992
Coding, %	90.5	90.5	90.9	90.3	91.2
Average size, bp	986	944	945[d]	1002	999
Assigned function[e]	1484	1499	2560[d]	N.A.[e]	3418
Predicted function[e]	2446	1734	N.A.[e]	2347	N.A.[e]
Uncharacterized, conserved[e]	777	508	628[d]	1427[e]	517
Uncharacterized[e]	257	457	945[d]		57
rRNA operons	4	4	4	4	4
tRNAs	55[g]	58[g]	58	57	56

Source: Weiten et al. (2021).
[a] Rabus et al. (2005).
[b] Martin-Moldes et al. (2015).
[c] Krause et al. (2006).
[d] Referring to chromosome only.
[e] Different categorizing used for strains EbN1T/BH72/CIB. Categories were taken from the annotation of strains ToN1T/PbN1T, where uncharacterized refers to not conserved uncharacterized proteins.
[f] phage search was redone in strain EbN1T to enable comparison.
[g] tRNAscan-SE 2.0 was used.

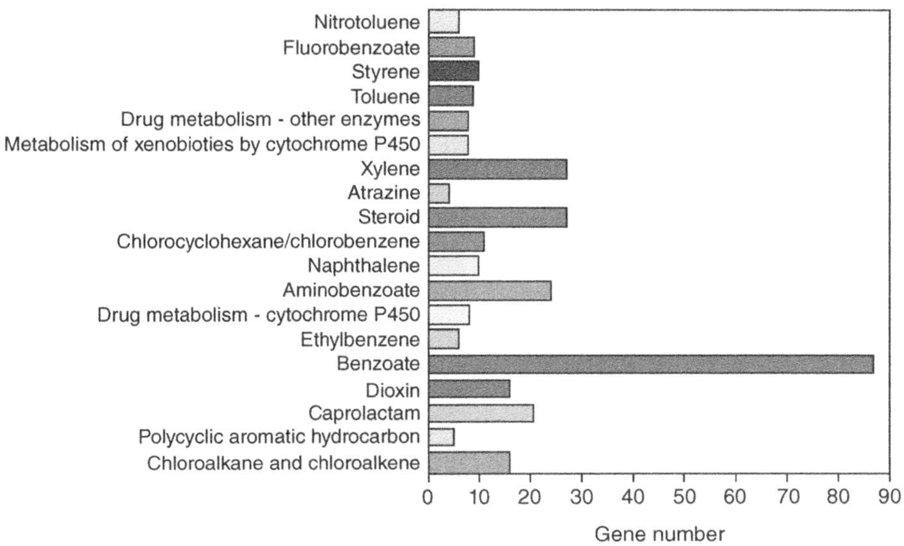

Figure 15.12 Distribution of genes responsible for xenobiotics degradation from *Rhodococcus ruber* C1, a phenol-degrading bacterium. *Source:* Zhao et al. (2021) / with permission of Elsevier.

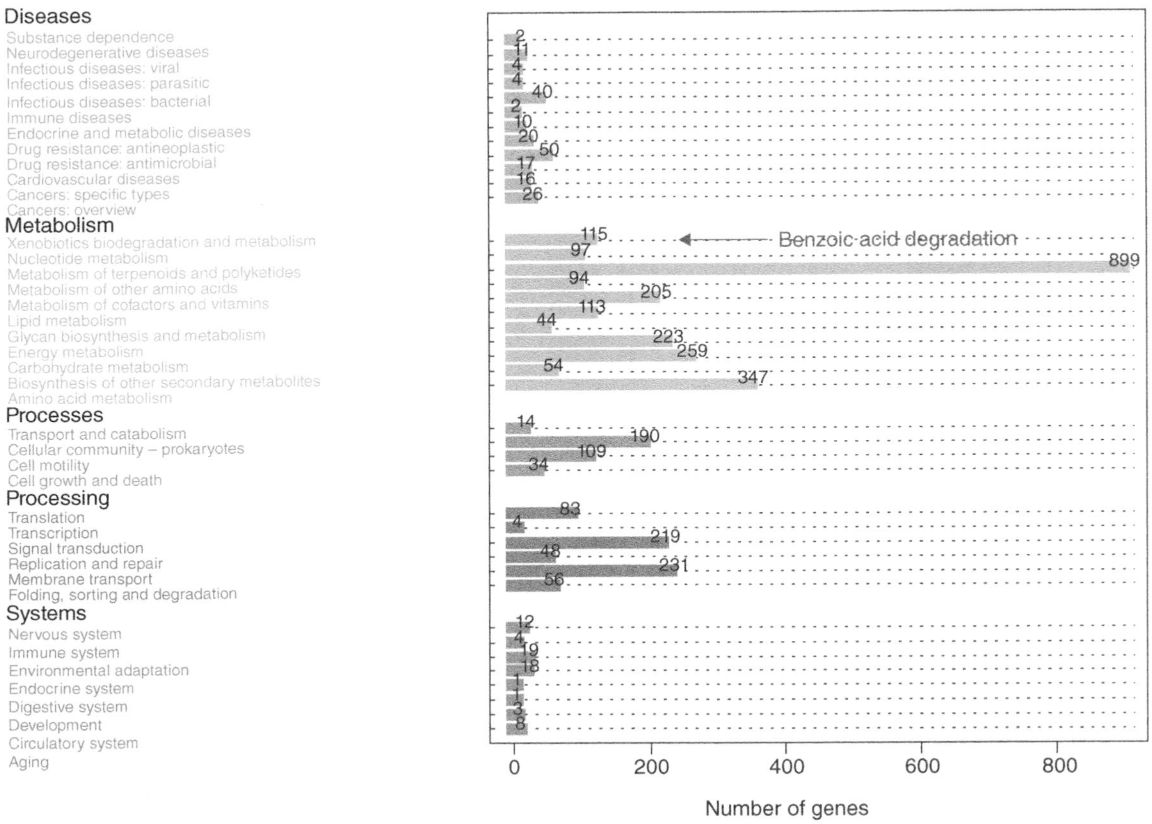

Figure 15.13 Relative distribution of genes responsible for xenobiotics degradation compared to other genes, from the rhizobium *Pseudomonas* sp. SCB32, in comparison to other genes related to other life processes. *Source:* Zhao et al. (2021) / with permission of Elsevier.

et al. 2020; Mahjoubi et al. 2021); production of antimicrobials, such as fengycin (vide Filonov et al. 2020), etc. Some may also possess efflux pumps that can be used both for drug resistance and extrusion of xenobiotics from the environment (vide Das et al. 2015). This can rationalize coselection, and be exploited for the bioremediation of environments polluted with a diverse array of pollutants, and can also delineate the use of these microbes for other useful applications.

These bacteria can also possess other capabilities outside the sphere of biodegradation, e.g. synthesis of compounds beneficial to plants, such as indole acetic acid (IAA), fixing of atmospheric nitrogen and endophytic colonization, e.g. the aromatic degraders *Aromatoleum petrolei* ToN1T and *Aromatoleum bremense* PbN1T (vide Weiten et al. 2021). The comparison of genome profiles from different species within the same genus reveals significant similarities and differences in copy number of genes involved in biodegradation, presence of novel genes, gene gain, and loss across a species over time and evolution of individual genes/gene families, which can further illuminate adaptation of specified species to particular environments. Substrates and/or niches, with potential for application in biodegradation. An instance is Zhao et al. (2021). This tends to point to the specific adaptation of various microbial species to particular biodegradation niches/conditions, making each species' repertoire of genes, pathways, etc., adapted to particular tasks. Additionally, the diversity of the sources from which the HDB were isolated means that the "shotgun method" of bioprospecting is still important (vide López-Moreno et al. 2021). Also, WGS studies illuminate to us the full range of HCs degrading capacity of individual bacteria plus other astonishing metabolic features; more than any other method today (vide Paliwal et al. 2014). Moreover, the presence of varying numbers of uncharacterized genes ("functions unknown") in the WGS of the various HDB (vide Li et al. 2017; Khalil et al. 2018; Yang et al. 2018; Zhao et al. 2021) means that the filed is still wide open for researches and potentially exciting discoveries, and the full extent of the metabolic/biodegradation capability of the HDB had not been wholly grasped yet.

Similarly, the WGS results highlight how the bacteria are adapted to the extreme environments they are isolated from, such as deserts, Antarctica, hot springs, etc. This adaptation can occur through the elaboration of stress-related genes and altered physiology and metabolism; and these properties are exploitable in the bioremediation of specified niches (vide Ruiz et al. 2015; Khalil et al. 2018; Lee et al. 2020). Furthermore, genes involved in hydrocarbon degradation show both wide and narrow substrate specificities. The same gene may have two or more names and may be involved in the degradation of one or more, related or unrelated substrates. The function of some genes is rather general and simply delineated as "xenobiotic/hydrocarbon degradation", etc. (vide Tables 15.2–15.4). In the same vein, bacteria belonging to the same genus may show substrate(s) specificity/preference, which indicates the particular adaptation of that genus to the specified hydrocarbon(s) (vide Table 15.4).

Finally, WGS analyses alone do not provide all the answers to questions in biodegradation. However, they provide the majority of the answers, and, when coupled with other studies, such as transcriptomics and other OMICs studies (vide Yunusa and Umar 2021), laboratory and on-the-field degradation studies (vide Table 15.2), efficient information about biodegradation can be effectively gathered, for further application in bioremediation of hydrocarbon-contaminated environments.

References

Agarry, S. and Latinwo, G.K. (2015). Biodegradation of diesel oil in soil and its enhancement by application of bioventing and amendment with brewery waste effluents as biostimulation-bioaugmentation agents. *J. Ecol. Eng.* 16: 82–91.

Ameen, F., Moslem, M.A., Hadi, S. et al. (2014). Biodegradation of cellulosic materials by marine fungi isolated from south Coniche of Jedda Saudi Arabia. *J. Pure Appl. Microbiol.* 8: 3617–3626.

Arriola, L.A., Cooper, A., and Weyrich, L.S. (2020). Palaeomicrobiology: application of ancient DNA sequencing to better understand bacterial genome evolution and adaptation. *Front. Ecol. Evol.* 8: 40.

Arthur, E.L., Rice, P.J., and Rice, P.J. (2005). Phytoremediation—an overview. *CRC Crit. Rev. Plant Sci.* 24: 109–122.

Boonchan, S., Britz, M.L., and Stanley, G.A. (2000). Degradation and mineralization of high-molecular weight polycyclic aromatic hydrocarbons by defined fungal bacterial cocultures. *Appl. Environ. Microbiol.* 66: 10–16.

Bushnell, L.D. and Haas, H.F. (1941). The utilisation of certain hydrocarbons by microorganisms. *J. Bacteriol.* 41 (5): 653–673.

CDC – Centres for Disease Prevention and Control, USA (2021). What is whole genome sequencing? CDC PulseNet. https://www.cdc.gov/pulsenet/pathogens/wgs.html. Accessed 5 September 2021.

Chauhan, A., Green, S., Pathak, A. et al. (2013). Whole-genome sequences of five oyster-associated bacteria show potential for crude oil hydrocarbon degradation. *Genome Announc.* 1 (5): e00802–e00813.

Chen, J., Yang, Q., Huang, T. et al. (2011). Enhanced bioremediation of soil contaminated with viscous oil through microbial consortium construction and ultraviolet mutation. *World J. Microbiol. Biotechnol.* 27: 1381–1389.

Chowdhury, M.N. (2015). Isolation of kerosene degrading bacteria from soil samples and determination of their optimum growth conditions. Unpublished BSc. Dissertation, BRAC University, Dhaka, Bangladesh. https://core.ac.uk/download/pdf/61805969. Accessed 1 December 2018.

Curtis, J., Metheny, E., and Sergent, S.R. (2021). Hydrocarbon toxicity. StatPearls Publishing. https://www.ncbi.nlm.nih.gov/books/NBK499883/#_NBK499883_pubdet_. Accessed 5 September 2021.

Dahm, R. (2008). Discovering DNA: Friedrich Miescher and the early years of nucleic acid research. *Hum. Genet.* 122 (6): 565–581.

Darma, Z.U. and Riko, Y.Y. (2022). *Biodegradation of Petroleum Hydrocarbons: Understanding Basic Concepts*. Lambert Academic Publishing.

Das, S. (2010). *Microbial Biodegradation and Bioremediation*. Elsevier.

Das, N. and Chandran, P. (2011). Microbial degradation of petroleum hydrocarbon contaminants: an overview. *Biotechnol. Res. Int.* 2: 1–13.

Das, D., Baruah, R., and Roy, A.S. (2015). Complete genome sequence analysis of *Pseudomonas aeruginosa* N002 reveals its genetic adaptation for crude oil degradation. *Genomics* 105: 182–190.

Dehkordi, H.A., Mazaheri, A., Ibrahim, H.A. et al. (2021). How to write a systematic review: a narrative review. *Int. J. Prev. Med.* 12 (3): 27.

Desai, C., Pathak, H., and Madamwar, D. (2009). Advances in molecular and "-Omics" technologies to gauge microbial communities and bioremediation at xenobiotic/anthropogen contaminated sites. *Bioresour. Technol.* 101 (6): 1558–1569.

Dixit, H., Lowry, M., Mohsin, U. et al. (2018). Screening and identification of diesel oil degrading bacterial isolates from petroleum contaminated soil of Barmer. *J. Pharm. Chem. Biol. Sci.* 6: 34–40.

Dorn, P.B. and Salanitro, J.P. (2000). Temporal ecological assessment of oil contaminated soils before and after bioremediation. *Chemosphere* 40: 419–426.

Elufisan, T.O., Rodríguez-Luna, I.C., Oyedara, O.O. et al. (2020). The polycyclic aromatic hydrocarbon (PAH) degradation activities and genome analysis of a novel strain *Stenotrophomonas* sp. Pemsol isolated from Mexico. *PeerJ* 8: e8102.

Field, D., Garrity, G., Gray, T. et al. (2008). The minimum information about a genome sequence specification. *Nat. Biotechnol.* 26 (5): 541–547.

Filonov, A., Delegan, Y., Puntus, I. et al. (2020). Complete genome sequence of *Pseudomonas putida* BS3701, a promising polycyclic aromatic hydrocarbon-degrading strain for bioremediation technologies. *Microbiol. Resour. Announc.* 9 (40): e00892–e00820.

Fingas, M. (2015). *Handbook of Oil Spill Science and Technology*. Wiley.

Fritsche, W. and Hofrichter, M. (2008). Chapter 6: aerobic degradation by microorganisms. In: *Biotechnology: A Comprehensive Treatise, Set 5*, 2e (ed. H.-J. Rehm and G. Reed), 144–167. Wiley.

Gallego, J.L.R., Loredo, J., Llamas, J.F. et al. (2001). Bioremediation of diesel contaminated soils; evaluation of potential *in situ* techniques by study of bacterial degradation. *Biodegradation* 12: 325–335.

Galperin, M.Y., Makarova, K.S., Wolf, Y.I. et al. (2015). Expanded microbial genome coverage and improved protein family annotation in the COG database. *Nucleic Acids Res.* 43 (D1): D261–D269.

Ganesh, A. and Lin, J. (2009). Diesel degradation and biosurfactant production by grampositive isolates. *Afr. J. Biotechnol.* 8: 5847–5854.

Garrido-Sanz, D., Sansegundo-Lobato, P., Redondo-Nieto, M. et al. (2020). Analysis of the biodegradative and adaptive potential of the novel polychlorinated biphenyl degrader *Rhodococcus* sp. WAY2 revealed by its complete genome sequence. *Microb. Genom.* 6: 323.

Gauthier, M.J., Lafay, B., Christen, R. et al. (1992). Marinobacter hydrocarbonoclasticus gen. Nov., sp. nov., a new, extremely halotolerant, hydrocarbon-degrading marine bacterium. *Int. J. Syst. Evol. Microbiol.* 42 (4): 568–576.

Ghazali, F., Rahman, R., Salleh, A. et al. (2004). Biodegradation of hydrocarbons in soil by microbial consortium. *Int. J. Biodeterior. Biodegrad.* 54: 61–67.

He, C., Li, Y., Huang, C. et al. (2018). Genome sequence and metabolic analysis of a fluoranthene-degrading strain *Pseudomonas aeruginosa* DN1. *Front. Microbiol.* 9, Article 2595.

Henckel, T., Jackal, U., Schnell, S. et al. (2000). Molecular analyses of novel methanotrophic communities in forest soil that oxidise atmospheric methane. *Appl. Environ. Microbiol.* 66: 1801–1808.

Hu, G., Li, J., and Zeng, G. (2013). Recent development in the treatment of oily sludge from petroleum industry: a review. *J. Hazard. Mater.* 261: 470–490.

Jagtap, S.S., Woo, S.M., Kim, T.S. et al. (2014). Phytoremediation of diesel-contaminated soil and saccharification of the resulting biomass. *Fuel* 116: 292–298.

Joye, S., and Kostka, J. (2020). Microbial genomics of the global ocean system. American Society for Microbiology. https://www.ncbi.nlm.nih.gov/books/NBK556286. Accessed 5 September 2021.

Kanehisa, M. and Goto, S. (2000). KEGG: Kyoto Encyclopedia of Genes And Genomes. *Nucleic Acids Res.* 28 (1): 27–30.

Karthika, R., Gopinath, L.R., Archaya, S. et al. (2014). Isolation of diesel degrading bacteria, identification of catechol gene and its biogas production. *IOSR J. Environ. Sci. Toxicol. Food Technol.* 8: 76–82.

Khalil, A.B., Sivakumar, N., Arslan, M. et al. (2018). Insights into *Brevibacillus borstelensis* AK1 through whole genome sequencing: a thermophilic bacterium isolated from a hot spring in Saudi Arabia. *Biomed. Res. Int.* 2018: 5862437.

Koshlaf, E. and Ball, A.S. (2017). Soil bioremediation approaches for petroleum hydrocarbon polluted environments. *AIMS Microbiol.* 3 (1): 25–49.

Krause, A., Ramakumar, A., Bartels, D. et al. (2006). Complete genome of the mutualistic, N2-fixing grass endophyte *Azoarcus* sp. strain BH72. *Nat. Biotechnol.Nature Biotechnology* 24 (11): 1385–1391.

Kumari, P. and Kumar, Y. (2021). Chapter 19: bioinformatics and computational tools in bioremediation and biodegradation of environmental pollutants. In: *Bioremediation for Environmental Sustainability: Approaches to Tackle Pollution for Cleaner and Greener Society* (ed. V. Kumar, G. Sxena, and M.P. Shah), 421–444. Elsevier.

Kvenvolden, K.A. and Cooper, C.K. (2003). Natural seepage of crude oil into the marine environment. *Geo-Mar. Lett.* 23: 140–146.

Kwak, Y., Li, Q.X., and Shin, J.-H. (2016). Draft genome sequence of *Mycobacterium rufum* JS14T, a polycyclic-aromatic hydrocarbon-degrading bacterium from petroleum-contaminated soil in Hawaii. *Stand. Genomic Sci.* 11: 47.

Lai, Q., Li, W., Wang, B. et al. (2012). Complete genome sequence of the pyrene-degrading bacterium *Cycloclasticus* sp. strain P1. *J. Bacteriol.* 194 (23): 6677.

Lee, G.L.Y., Zakaria, N.N., Convey, P. et al. (2020). Statistical optimisation of phenol degradation and pathway identification through whole genome sequencing of the cold-adapted antarctic bacterium, *Rhodococcus* sp. strain AQ5-07. *Int. J. Mol. Sci.* 21: 9363.

Li, Y., Liu, H., Tian, Z. et al. (2008). Diesel pollution biodegradation: synergistic effect of *Mycobacterium* and filamentous fungi. *J. Biomed. Environ. Sci.* 21: 181–187.

Li, Y., Ren, Y., and Jiang, N. (2017). Analysis of draft genome sequence of *Pseudomonas* sp. QTF5 reveals its benzoic acid degradation ability and heavy metal tolerance. *Biomed. Res. Int.* 2017: 4565960.

Liu, H., Liang, R., Tao, F. et al. (2012). Genome sequence of *Pseudomonas aeruginosa* strain SJTD-1, a bacterium capable of degrading long-chain alkanes and crude oil. *J. Bacteriol.* 194 (17): 4783–4784.

Liu, Y., Hu, H., Zanaroli, G. et al. (2021). A *Pseudomonas* sp. strain uniquely degrades PAHs and heterocyclic derivatives via lateral dioxygenation pathways. *J. Hazard. Mater.* 403: 123956.

López-Moreno, A., Torres-Sánchez, A., Acuña, I. et al. (2021). Representative *Bacillus* sp. AM1 from gut microbiota harbor versatile molecular pathways for Bisphenol a biodegradation. *Int. J. Mol. Sci.* 22: 4952.

Mahjoubi, M., Aliyu, H., Cappello, S. et al. (2019). The genome of *Alcaligenes aquatilis* strain BU33N: insights into hydrocarbon degradation capacity. *PLoS One* 14 (9): e0221574.

Mahjoubi, M., Aliyu, H., Neifar, M. et al. (2021). Genomic characterization of a polyvalent hydrocarbonoclastic bacterium *Pseudomonas* sp. strain BUN14. *Sci. Rep.* 11: 8124.

Mahto, K.U. and Das, S. (2020). Whole genome characterization and phenanthrene catabolic pathway of a biofilm forming marine bacterium *Pseudomonas aeruginosa* PFL-P1. *Ecotoxicol. Environ. Saf.* 206: 111087.

Martin-Moldes, Z., Zamarro, M.T., del Cerro, C. et al. (2015). Wholegenome analysis of *Azoarcus* sp. strain CIB provides genetic insights to its different lifestyles and predicts novel metabolic features. *Syst. Appl. Microbiol.* 38 (7): 462–471.

Miyazaki, R., Bertelli, C., Benaglio, P. et al. (2014). Comparative genome analysis of *pseudomonas knackmussii* B13, the first bacterium known to degrade chloroaromatic compounds. *Environ. Microbiol.* 17 (1): 91–104.

Musa, S.I. (2019). Isolation and identification of diesel oil-degrading bacteria in used engine oil contaminated soil. *J. Appl. Sci. Environ. Manage.* 23: 431–435.

Mwamura, A. (2017). Screening, isolation and characterization of hydrocarbonoclastic bacteria from oil contaminated soils. Unpublished MSc. Thesis. University of Nairobi, Kenya. http://erepository.uonbi.ac.kebitstream/handle/11295/101451/Mwamura_Screening%252C%2520Isolation%2520And%2520Characterization%2520Of%2520Hydrocarbonoclastic%2520Bacteria%2520From%2520Oil%2520Contaminated%2520Soils.pdf. Accessed 1 December 2018.

Naila, T. (2015). Isolation and study of optimum growth conditions of diesel degrading bacteria from soil. Unpublished BSc. (Biotechnology) Project, BRAC University, Dhaka, Bangladesh. http://dspace.brac.ac.bd/bitsream/handle/10361/4163/Isolation%20of%20diesel%20derading%20bacteria.pdf. Accessed 1 December 2018.

Nature.com (2021). Milestones in genomic sequencing. https://www.nature.com/immersive/d42859-020-00099-0/index.html. Accessed 5 September 2021.

NCI – National Cancer Institute (2021). Definition of whole genome sequencing – NCI dictionary of genetics terms. https://www.cancer.gov/publications/dictionaries/genetics-dictionary/def/740456?cdrid=74056. Accessed 5 September 2021.

Nelson, K.E., Weinel, C., Paulsen, I.T. et al. (2002). Complete genome sequence and comparative analysis of the metabolically versatilse *Pseudomonas putida* KT2440. *Environ. Microbiol.* 4 (12): 798–808.

Nikhil, T., Deepa, V., Rohan, G. et al. (2013). Isolation, characterisation and identification of bacteria from garage soil and comparison of their bioremediation potential. *Int. Res. J. Environ. Sci.* 2: 48–50.

Olajire, A.A. and Essien, J.P. (2014). Aerobic degradation of petroleum components by microbial consortia. *Pet. Environ. Biotechnol.* 5: 1–22.

Ortiz, O., Castells, F., Sonnemann, G. et al. (2009). Sustainability in the construction industry: a review of recent developments based on LCA. *Constr. Build. Mater.* 23 (1): 28e39.

Pal, S., Kundu, A., Banerjee, T.D. et al. (2017). Genome analysis of crude oil degrading *Franconibacter pulveris* strain DJ34 revealed its genetic basis for hydrocarbon degradation and survival in oil contaminated environment. *Genomics 109*: 374–382.

Paliwal, V., Raju, S.C., Modak, A. et al. (2014). *Pseudomonas putida* CSV86: a candidate genome for genetic bioaugmentation. *PLoS One* 9 (1): e84000.

Patel, P.A., Kothari, V.V., Kothari, C.R. et al. (2014). Draft genome sequence of petroleum hydrocarbon-degrading *Pseudomonas aeruginosa* strain PK6, isolated from the Saurashtra region of Gujarat, India. *Genome Announc.* 2 (1): e00002–e00014.

Patowary, K., Patowary, R., Kalita, M.C. et al. (2016). Development of an efficient bacterial consortium for the potential remediation of hydrocarbons from contaminated sites. *Front. Microbiol.* 7: 1–14.

Payne, W.J., Wiebe, W.J., and Christian, R.R. (1970). Assays for biodegradability essential to unrestricted usage of organic compounds. *Bioscience* 20: 862–865.

Rabus, R., Kube, M., Heider, J. et al. (2005). The genome sequence of an anaerobic aromatic-degrading denitrifying bacterium, strain EbN1. *Arch. Microbiol.* 183 (1): 27–36.

Radelyuk, I., Tussupova, K., Klemeš, J.J. et al. (2021). Oil refinery and water pollution in the context of sustainable development: developing and developed countries. *J. Clean. Prod.* 302 (2021): 126987.

Rampelotto, P.H. (2013). Extremophiles and extreme environments. *Life (Basel)* 3 (3): 482–485.

Ravintheran, S.K., Sivaprakasam, S., Loke, S. et al. (2019). Complete genome sequence of *Sphingomonas paucimobilis* AIMST S2, a xenobiotic-degrading bacterium. *Sci. Data* 6: 280.

Rehman, Z.U., Khan, K., Faisal, S. et al. (2015). Isolation and identification of diesel degrading bacteria from oil contaminated soil in Mansehra, Pakistan. *Int. J. Sci. Eng. Res.* 6: 1438–1661.

Rojo, F. (2009). Degradation of alkanes by bacteria. *Environ. Microbiol.* 11 (10): 2477–2490.

Ruiz, O.N., Brown, L.M., Striebich, R.C. et al. (2015). Draft genome sequence of *Pseudomonas frederiksbergensis* SI8, a psychrotrophic aromatic degrading bacterium. *Genome Announc.* 3 (4): e00811–e00815.

Ruiz, O.N., Radwan, O., and Striebich, R.C. (2021). GC–MS hydrocarbon degradation profile data of *Pseudomonas frederiksbergensis* SI8, a bacterium capable of degrading aromatics at low temperatures. *Data Br.* 35: 106864.

Schadt, E.E., Turner, S., and Kasarskis, A. (2010). A window into third generation sequencing. *Hum. Mol. Genet.* 19 (R2): R227–R240.

Schneiker, S., dos Santos, V.A.P.M., Bartels, D. et al. (2006). Genome sequence of the ubiquitous hydrocarbon-degrading marine bacterium *Alcanivorax borkumensis*. *Nat. Biotechnol.* 24 (8): 997–1004.

Solano-Serena, F., Marchal, R., Casaregola, S. et al. (2000). A new *Mycobacterium* strain with extended degradation capacities for gasoline hydrocarbons. *Appl. Environ. Biotechnol.* 66: 2392–2399.

Stroud, J.L., Paton, G.I., and Semple, K.T. (2007). Microbe-aliphatic hydrocarbon interactions in soil implications for biodegradation and bioremediation. *J. Appl. Microbiol.* 102: 1239–1253.

Tawfik, G.M., Dila, K.A.S., and Mohamed, M.Y.F. (2021). A step by step guide for conducting a systematic review and meta-analysis with simulation data. *Trop. Med. Health* 47: 46.

Thorne-Holst, M., Wentzel, A., Ellingsen, T.E. et al. (2007). Identification of novel genes involved in long-chain n-alkane degradation by *Acinetobacter* sp. strain DSM 17874. *Appl. Environ. Microbiol.* 73 (10): 3327–3332.

Ukaegbu-Obi, K.M., Iweha, O.N., and Okonkwo, E.C. (2017). Bacterial community dynamics of diesel-polluted soil and their bioremediation potential. *FUTO J. Ser.* 3: 161–165.

Umar, Z.D. (2017). Biodegradation of phenanthrene and pyrene using bacteria isolated from used vehicle lubricant-contaminated soil. Doctoral Thesis (PhD). Published by Universiti Putra Malaysia Institutional Repository (UPMIR), 1–172. http://psasir.upm.edu.my/id/eprint/71027. Accessed 5 September 2021.

Umar, Z.D., Aziz, N.A.A., Zulkifli, S.Z. et al. (2018a). Efficiency of polycyclic aromatic hydrocarbons (PAHs) degrading consortium in resisting heavy metals during PAHs degradation. *Int. J. Environ.* 7 (1): 14–27.

Umar, Z.D., Azwady, A.A.N., Zulkifli, S.Z. et al. (2018b). Effective phenanthrene and pyrene biodegradation using *Enterobacter* sp.MM087 (KT933254) isolated from used engine oil contaminated soil. *Egypt. J. Pet.* 27 (3): 349–359.

Umar, Z.D., Mansir, A.Z., and Riko, Y.Y. (2019). Compatibility and formulation of diesel degrading consortia using bacteria isolated from contaminated soil. *Bayero J. Pure Appl. Sci.* 12 (1): 199–208.

Umar, Z.D., Aminu, M., and Yahaya, Y.R. (2020a). Survival response of consortium isolates from diesel contaminated soil within Katsina state, Nigeria. *Int. J. Environ.* 9: 51–66.

Umar, Z.D., Musa, A., and Yunusa, Y.R. (2020b). Optimization of diesel biodegrading conditions using response surface methodology based on central composite design. *Polycycl. Aromat. Compd.* 40 (4): 1–11.

UN – United Nations (2015). Transforming our world: the 2030 Agenda for sustainable development. https://sustainabledevelopment.un.org/post2015/transformingourworld. Accessed 14 December 2020.

UNDP – United Nations Development Programme (2016). Parliaments and the sustainable development goals. A Self-Assessment Toolkit. ICA. Inter-Parliamentary Union, France. https://info.undp.org/docs/pdc/Documents/H05/Parliaments%20for%202030%20Agenda%20Pilot%20Phase.pdf. Accessed 14 December 2020.

Uzoamaka, G.O., Tasie, F., and Muote-Okafor, F. (2009). Hydrocarbon degradation potentials of indigenous fungal isolates from petroleum contaminated soils. *Phys. Nat. Sci.* 3: 1–6.

Van Hamme, J.D. and Ward, O.P. (2001). Physical and metabolic interactions of *Pseudomonas* sp. strain JA5–B45 and *Rhodococcus* sp. strain F9–D79 during growth on crude oil and effect of a chemical surfactant on them. *Appl. Environ. Microbiol.* 67: 4874–4879.

Vidali, M. (2001). Bioremediation: an overview. *Pure Appl. Chem.* 73: 1163–1172.
Vijay, A. (2012). Biodegradation of petroleum hydrocarbons. Unpublished M.Tech. Thesis, IIT Bombay, Mumbai, India. https://core.ac.uk/download/pdf/53189038. Accessed 1 December 2018.
Weiten, A., Kalvelage, K., Becker, P. et al. (2021). Complete genomes of the anaerobic degradation specialists *Aromatoleum petrolei* ToN1T and *Aromatoleum bremense* PbN1T. *Microb. Physiol.* 31: 16–35.
Xiang, W., Wei, X., Tang, H. et al. (2020). Complete genome sequence and biodegradation characteristics of benzoic acid-degrading bacterium *Pseudomonas* sp. SCB32. *Biomed. Res. Int.* 2020: 6146104.
Yang, R., Liu, G., Chen, T. et al. (2018). The complete genomic sequence of a novel cold-adapted bacterium, *Planococcus maritimus* Y42, isolated from crude oil-contaminated soil. *Standards Genom. Sci. 13*, Article 23.
Yasemi, M. and Jolicoeur, M. (2021). Modelling cell metabolism: a review on constrain-based steady-state and kinetic approaches. *Processes* 9 (2): 322.
Yunusa, Y.R. and Umar, Z.D. (2021). Effective microbial bioremediation via the multiomics approach: an overview of trends, problems and prospects. *UMYU J. Microbiol. Res.* 6 (1): 127–145.
Zampolli, J., Zeaiter, Z., Di Canito, A., and Di Gennaro, P. (2019). Genome analysis and -omics approaches provide new insights into the biodegradation potential of *Rhodococcus*. *Appl. Microbiol. Biotechnol. 103* (3): 1069–1080.
Zhao, T., Gao, Y., Yu, T. et al. (2021). Biodegradation of phenol by a highly tolerant strain *Rhodococcus ruber* C1: biochemical characterization and comparative genome analysis. *Ecotoxicol. Environ. Saf.* 208: 111709.
Zhu X., Venosa, A.D., Suidan, M.T. et al. (2001). Guidelines for the bioremediation of marine shorelines and freshwaters waterlands. US Environmental Protection Agency Office of Research and Development, National Risk Management Research Laboratory.

16

Green Synthesized Nanomaterials as Tools to Remediate Aquatic Pollution

Charu Gupta[1], Mahendra K. Gupta[2], Mir S. Rabani[2], Shivani Tripathi[2], and Anjali Pathak[2]

[1] *School of Studies in Microbiology, Jiwaji University, Gwalior, Madhya Pradesh, India*
[2] *Microbiology Research Laboratory, School of Studies in Botany, Jiwaji University, Gwalior, Madhya Pradesh, India*

16.1 Introduction

Water is considered a vital resource for the existence of all life forms. Along with the expanding population, urbanization and rapid advancement of industrialization, freshwater scarcity, and water pollution are becoming critical across the globe. Various types of nonbiodegradable pollutants including heavy metal ions, textile dyes, and organic and inorganic metals are discharged from different industries. These toxic pollutants can lead to mutagenicity, carcinogenicity, respiratory toxicity, chromosomal fractures, and teratogenicity and can also cause mutation and various cancers (Ahsan et al. 2020). Another significant issue is the presence of harmful pathogenic bacteria in water such as *Vibrio cholera*, *Salmonella typhosa*, and *Escherichia coli* that cause life-threatening diseases like cholera, typhoid, and diarrhea, respectively. Consequently, there is a dire need to look for advanced remediation technologies for the rapid and efficient removal of toxins and impurities from aquatic environments. In recent times, various materials like low-cost adsorbents, activated carbon and biochar developed from different resources have been developed for the remediation of aquatic environments (Gupta and Suhas 2009). However, these modes of treatment are less compatible, time-consuming, and not much effective for the degradation of toxic waste from aquatic ecosystem. Consequently, there is an immense need for a technology for the degradation and scavenging of noxious waste from aquatic ecosystem.

Owing to its exclusive properties, nanotechnology has become a broad horizon in the various fields of the food industry, agriculture, environmental remediation, textiles, drug delivery, medicines etc. (Gupta et al. 2022). Recently, the advancement of nanoclusters for the improvement of the quality of water bodies is gaining the attention of researchers. NPs are ultrafine particles with less than 100 nm and possess unique properties such as a large reactive surface area and small size. Nanomaterials are widely applied in the remediation of toxins from aquatic ecosystem as these toxins get involved in the food chain and cause various life-threatening diseases.

Various approaches like chemical, physical, and biological processes are employed in the fabrication of different nanomaterials, for instance, nanofibers, nanoparticles, nanowires, nanoribbons, nanotubes, nanorods, etc. (Tharmavaram et al. 2018; Rawtani et al. 2019). Physical methods involve irradiation, condensation, sputtering, deposition, and ball milling, etc. (Dhand et al. 2015), and chemical methods involve pyrolysis, polyol synthesis, sol-gel method, hydrothermal synthesis, etc. (Darroudi et al. 2013). However, these approaches are not suitable and eco-friendly and produce harmful by products that perturb the ecosystem quality. In the recent times, the green method of **nanoparticles synthesis** has been used extensively. It is an environmentally benign, sustainable, and reliable approach than other conventional approaches. However, it is also desirable to mitigate the generation of toxic residues. In biological methods, plants, bacteria, fungi, algae, and yeast are exploited for the production of natural extracts. These biological entities produce flavonoids, alkaloids, carboxylic acid, ascorbic acid, ketones, proteins, polysaccharides, and teichoic acid and the extracts are used as reducing, capping, and stabilizing agents. However, during the synthesis of **nanoparticles**, a conglomeration of particles occurs between the nanoparticles due to attractive forces. Therefore, these stabilizing or reducing agents are required to avoid conglomeration and obtain uniform products (Pandey et al. 2020).

Aquatic Contamination: Tolerance and Bioremediation, First Edition. Edited by Rouf Ahmad Bhat, Gowhar Hamid Dar, Fernanda Maria Policarpo Tonelli, and Saima Hamid.

In this regard, researchers are focusing on the green synthesis of nanomaterial through plants and microbial extract. This recent technology has now become a fascinating area for the efficient remediation of detrimental waste from water bodies.

16.2 Approaches of Nanoparticle Synthesis

Two traditional approaches on the basis of their assemblies are used to fabricate nanoparticles with the desired shape and size: "top-down" and "bottom-up" approaches (Gwo et al. 2016). Both of these methods use different synthesis principles that result in NPs (nanoparticles) with the desired properties.

16.2.1 Top-Down Approach

In order to create fine generations of NPs, bulk materials are broken up into tiny pieces. Photolithographic methods, milling, sputtering, and grinding are used to produce this type of NPs (Meyers et al. 2006). Each technique has its own drawbacks and levels of NP production. This approach is very attainable that produces a significant amount of NPs. However, the drawbacks of the top-down method for NP synthesis are that they may have imperfect surfaces and occasionally suffer damage (Thakkar et al. 2010).

16.2.2 Bottom-Up Approach

This approach involves the coalescence of atoms by atoms, clusters by clusters, and molecules by molecules to generate a variety of NPs (Decarolis et al. 2019) by adding reducing agents. For the synthesis of NPs, processes like bio-assisted synthesis, chemical vapor deposition, sol–gel processing, electrochemical synthesis, flame-spraying synthesis, and laser pyrolysis are used (Dhand et al. 2015). In order to create nanoclusters for application in a wide variety of applications, bottom-up is a viable technique as shown in Figure 16.1.

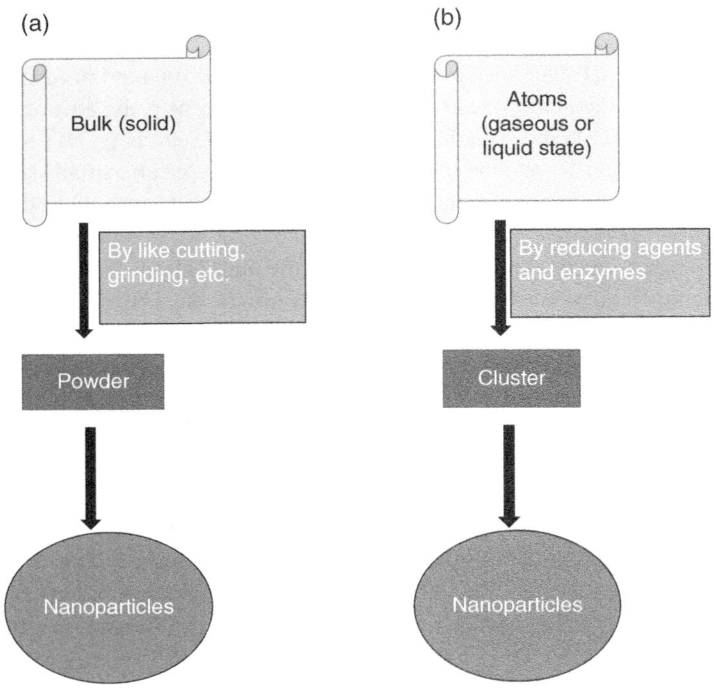

Figure 16.1 Two approaches for synthesizing nanoparticles (a) Top-down approach; (b) Bottom-up approach.

16.3 Routes of Metal Nanoparticle Synthesis

16.3.1 Physical Synthesis of Nanoparticles

This method involves sputtering, deposition, plasma-based techniques, and ball milling for the synthesis of metal NPs (Dhand et al. 2015). In most of these methods, the synthesis rate of metal nanoparticles is extremely slow. For instance, the ball milling technique has a yield of 50% or less of NPs (Yadav et al. 2012). In the sputtering method, a large particle size distribution is yielded with only 6–8% of the sputtered material being less than 100 nm. Plasma techniques and laser ablation have some drawbacks like consumption of high energy, wide size distribution, waste byproducts, and slow production rates. Because of these demerits, most physical methods are prohibitively expensive and cannot be affordable for commercial applications (Seetharaman et al. 2018).

16.3.2 Chemical Synthesis of Nanoparticles

Chemical methods involving pyrolysis, hydrothermal synthesis, sol–gel, chemical reduction, chemical vapor deposition, and microemulsion are now widely used to create nanostructured materials (Darroudi et al. 2013). Furthermore, using these hazardous reagents and chemicals during the synthesis process yields byproducts, which are harmful to the whole ecosystem (Zhang et al. 2019). Hence, the use of such NPs in biological applications is restricted.

16.3.3 Biosynthesis of Metal Nanoparticles

The purpose of this chapter is to emphasize the biological route of synthesizing metal nanoparticles because of its ability to produce nanoparticles quickly, in controlled size, with biocompatibility, nontoxicity and in an environmentally friendly manner. Plants and a variety of microbes are involved in the assembly and synthesis of nanomaterials by biological routes. These biological entities are used as a nontoxic and environmentally benign, stabilizing, and reducing agent to combat chemical toxicity. However, all the synthetic methods have an adverse effect on the environment (Parashar et al. 2009). Therefore, researchers have shifted toward the use of biocompatible and environmentally friendly methods for nanoparticle synthesis, as they are safe for various applications. As a result, biological extracts could be exploited as organic nanofactories for the fabrication of nanomaterials (Sheoran and Kaur 2018).

16.3.3.1 From Plants

Green synthesis with plants is a novel technology. In the development of green chemistry, the distinct parts of plants such as flowers, seeds, shoots, roots, leaves, stems, and bark have been used to reduce metal ions into metal NPs (Rauwel and Rauwel 2017). This process has gained the attention of researchers as the process is easy, biocompatible, nonbiodegradable, cheapest, and environmentally friendly over chemical and physical methods (Rather and Sundarapandian 2020). Plant extract contains secondary metabolites, amino acids, coenzymes, carbohydrates, enzymes, and polysaccharides. These biomolecules function as capping and reducing agents and facilitate the stabilization and formation of nanoparticles (Rai et al. 2008; Rajeshkumar and Bharath 2017).

In a recent study, ex-vivo synthesis of metal nanoparticles has been employed with a variety of plants. These plants like oat (*Avena sativa*) (Armendariz et al. 2004), *aloe vera* (*Aloe barbadensis Miller*) (Chandran et al. 2006), alfalfa (*Medicago sativa*) (Herrera-Becerra et al. 2008), lemon (*Citrus limon*) (Prathna et al. 2011), tulsi (*Ocimum sanctum*) (Philip and Unni 2011), neem (*Azadirachta indica*) (Shankar et al. 2004a), mustard (*Brassica juncea*) (Haverkamp and Marshall 2009), lemon grass (*Cymbopogon fexuosus*) (Shankar et al. 2005), and coriander (*Coriandrum sativum*) (Narayanan and Sakthivel 2008) have been exploited for silver and gold nanoparticles. Whereas, the in vivo metallic nanoparticle synthesis includes nickel, copper, zinc, and cobalt has been observed in alfalfa (*M. sativa*), mustard (*B. juncea*), and sunflower (*Helianthus annuus*) (Marchiol 2012). In this synthesis process, these metal NPs can be synthesized from plants by reducing metal salt ions.

A study found that polycrystalline iron nanorods synthesized from mango leaves were efficiently applied for the remediation of heavy oil viscosity (Al-Ruqeishi et al. 2016). Moreover, Xiao et al. (2017) demonstrated that Fe nanoparticles fabricated by *S. jambos* (powdered biomass) have a remarkable effect against Cr(VI) in the water by the adsorption method. In this way, plant-mediated nanoparticles could be feasible for the treatment of aquatic ecosystems.

16.3.3.2 From Bacteria

Bacterial strains have been extensively used in various biotechnological applications such as genetic engineering, bioleaching, and bioremediation (Gericke and Pinches 2006). However, the use of bacteria in nanotechnology has bloomed. It is an environmentally benign method in the field of life sciences. The first evidence came from the production of AgNPs by the bacterial strain *Pseudomonas stutzeri* AG259 that was isolated from a silver mine (Prabhu and Poulose 2012).

Several bacterial strains such as *P. stutzeri*, *E. coli*, *Plectonema boryanum*, *Staphylococcus currens*, *Pseudomonas aeruginosa*, *Vibrio cholerae*, *Salmonella typhus*, etc., have been reported for fabrication of nanomaterials by both intracellular and extracellular methods (Klaus et al. 1999; Konishi et al. 2004; Pal et al. 2019). However, *Bacillus licheniformis* produced AgNPs with about 50 nm particles size. It was isolated from sewage collected from municipal sources (Kalimuthu et al. 2008). Earlier investigations have proposed that ecologically and environmentally important bacterial species are also responsible for Au (Gold) nanoparticles. *E. coli* (Du et al. 2007), *Rhodopseudomonas capsulata* (He et al. 2007), *Klebsiella pneumoniae* (Srinath and Rai 2015), *Brevibacillus formosus* (Srinath et al. 2017), and *Arthrobacter nitroguajacolicus* (Dehnad et al. 2015) were reported for biologically reducing Au^{3+} ions in a single-step process (Mukherjee et al. 2002).

In a study, Srivastava et al. (2013) found that the extracellular fabrication of gold nanoparticles (Au NPs) from *E. coli* K12 strain was remarkably effective for the catalytic reduction of 4-nitrophenol (nitroaromatic pollutant) from water. Furthermore, Hennebel et al. (2011) demonstrated that palladium nanoparticles (PdNPs) synthesized by using mixed strains of *B. vulgates*, *C. butyricum*, *K. pneumoniae*, *C. braakii*, and *E. faecium* were found efficient for the biodegradation of diatrizoate in wastewater.

16.3.3.3 From Fungi

Myconanotechnology has flourished with a broad range of applications in various fields, especially in biomedicine (Rai et al. 2009). Fungi consist of diverse range of intracellular enzymes so that they play a better role as reducing or stabilizing agents in the fabrication of metal nanoparticles (Chen et al. 2009). Furthermore, their cell wall surfaces contain protein, enzymes, and other biological reducing components. Therefore, fungi have numerous advantages over other microorganisms. Fungi-mediated synthesis process is relatively simple to isolate, having a wide range of diversity and having much simpler downstream processing than bacterial fermentation.

In a study, Duran et al. (2007) stated that extracellularly synthesized AgNPs with *F. oxysporum* were found promising for mitigating textile effluents. In another study, rod-shaped iron nanoparticles were synthesized with *Itajahya* sp. (Basidiomycetous fungi). It was remarkably used as a stabilizing polymer for the elimination of lindane from wastewater (Paknikar et al. 2005). Within eight hours, 94% of the toxic component lindane could be removed, as demonstrated by a chromatographic study. In a further study, gold nanoparticles of quasi-spherical-shaped and 72.32 ± 21.80 nm size were synthesized with *C. oxysporum* AJP03 (a soil fungus) (Bhargava et al. 2016). It was applied for $NaBH_4$-mediated remediation of toxic dye rhodamine B-contaminated water. These synthesized particles were found to be highly efficient, with complete rhodamine B degradation occurring after seven minutes of treatment (Table 16.1).

16.4 Applications of Green Nanomaterials in the Remediation of Aquatic Pollution

Trade and globalization lead to a negative impact on the quality of water. Since the disturbance in the quality of water causes deterioration of the water ecosystem and human health, thus, there is an urgent necessity to achieve such technology that helps in remediation of toxic compounds from contaminated water bodies. Green nanotechnology is now emerging with the distinct applications in environmental remediation, drug delivery, food technology, agricultural engineering, etc. However, biosynthesized nanoparticles have been remarkable efficiency in degradation and scavenging of detrimental nonbiodegradable pollutants like dyes, heavy metals, oil spills, pharmaceuticals, radioactive, etc., discharged from different industries. Furthermore, studies have also revealed the application of biologically fabricated nanomaterials in the removal of aquatic pollution in Table 16.2 (Figure 16.2).

16.4.1 Removal of Dyes

Textile and other chemical industries' discharges harmful dyes in the aquatic ecosystem. These dyes are pernicious and frequently found to be mutagenic and carcinogenic. Structures of dyes are complex, stable, and nonbiodegradable,

Table 16.1 Biosynthesis of metal nanoparticles.

Source	Metal NPs	Size (nm)	Shape	References
Plants				
Berberis aristata (Indian burberry)	Zinc oxide (ZnO)	2–5	Spherical	Shankar et al. (2004b)
Carica papaya (Papaya)	Silver (Ag)	60–80	Spherical	Mude et al. (2009)
Cassia fistula (Golden shower tTree)	Silver (Ag)	50–60	Face-centered cubic	Kumari et al. (2012)
Eucalyptus citriodora (lemon scented gum)	Silver (Ag)	20	Spherical	Ravindra et al. (2020)
Ficus carica (Fig)	Silver (Ag)	2–15	Spherical	Naseem and Farrukh (2015)
Garcinia mangostana (Mangosteen)	Silver (Ag)	35	Spherical	Veerasamy et al. (2010)
Jatropha curcas (Jatropha)	Silver (Ag)	15–50	Spherical	Bar et al. (2009)
Medicago sativa (alfalfa)	Fe_2O_3 (Ironoxide)	2–10	Crystalline	Herrera-Becerra et al. (2008)
Bacteria				
Bacillus athrophaeus	Silver (Ag)	5–30	Polydispersed	Hosseini-Abari et al. (2013)
Bacillus cereus	Silver (Ag)	20–40	Spherical	Sunkar and Nachiyar (2012)
Bacillus subtilis 168	Gold (Au)	5–10	Hexagonal-octahedral	Southam and Beveridge (1994)
E. coli	Cadmium sulfide (CdS)	2–5	Fluorescent labels	Chen et al. (2009)
Klebsiella pneumoniae	Silver (Ag)	15–37	Spherical	Duraisamy and Yang (2013)
Lactobacillus casei	Silver (Ag)	20–50	Spherical	Korbekandi et al. (2012)
Nostoc commune	Silver (Ag)	15–54	Spherical	Morsy et al. (2014)
Pseudomonas aeruginosa BS-161R	Silver (Ag)	15.1 ± 5.8	Spherical	Kumar et al. (2010)
Streptomyces coelicolor	Silver (Ag)	28–50	Irregular	Manikprabhu and Lingappa (2013)
Fungi				
Aspergillus flavus TFR7	Titanium dioxide (TiO_2)	12–15	Spherical	Binupriya et al. (2010)
Alternata alternate	Silver (Ag)	20–60	Spherical	Gajbhiye et al. (2009)
Penicillium fellutanum	Silver (Ag)	5–25	Spherical	Kathiresan et al. (2009)
Phoma glomerata	Silver (Ag)	60–80	Spherical	Birla et al. (2009)
Rhizopus nigricans	Silver (Ag)	35–40	Round	Ravindra and Rajasab (2014)
Rhizopus stolonifer	Silver (Ag), Gold (Au)	25–30	Spherical	Binupriya et al. (2010)
Trichoderma asperellum	Silver (Ag)	13–18	Nanocrystalline	Mukherjee et al. (2008)

and they persist in ecosystems for a longer period of time. However, it has become a challenge for researchers to attain a reliable technique for the complete degradation of dyes because traditional methods were not very effective and affordable.

- Physical method for decolorization of dyes – reverse osmosis, sedimentation, and filtration adsorption
- Chemical method for decolorization of dyes – chemical oxidation, neutralization, ion exchange, and recovery methods
- Biological method for decolorization of dyes – bacteria, fungi, and actinomyces are some conventional treatment techniques used to remove dyes from wastewater (Morshedi et al. 2013) (Pandey et al. 2020).

These chemical, physical, and biological methods have some demerits in the elimination of dyes as they are high operational costs, produce toxic byproducts, and are time consuming. Green synthesis of nanobiotechnology has emerged to apply in the remediation of wastewater. Some studies have proved that biosynthesized nanomaterials have high efficiency for the treatment of wastewater.

Table 16.2 Application of green synthesized nanoparticles in wastewater treatment.

Nanomaterials	Biological source	Size and Shape	Application	References
Mn	*Pseudomonas putida* MnB1	8–9 nm, stick like	Removal of Pb(II), Cd(II), and Zn(II)	Zhou et al. (2015)
Fe	*C. reticulate*	50 nm and spherical	Cd(II)	Ehrampoush et al. (2015)
Fe	*E. globules*	80 nm and spherical	Removal of Nitrate	Wang et al. (2014a)
Fe	*Aloe vera*	5.5 nm and rod like	Removal of As (V)	Mukherjee et al. (2016)
Fe	*Sorghum bran*	40–50 nm and Amorphous	Removal of Bromo-thymol blue	Njagi et al. (2011)
Ag	*Terminalia chebula*	25 nm and spherical	Elimination of methylene blue	Edison and Sethuraman (2012)
Fe	Oolong tea	40–50 nm and spherical	Removal of Malachite green	Huang et al. (2014)
Fe	Grape	15–100 and Quasi spherical	Removal of acid orange (Azo dye)	Luo et al. (2014)
TiO_2 NPs	*Jatropha curcas* L.	10–20 nm and Spherical	Photocatalytic treatment of tannery wastewater	Goutam et al. (2018)
Iron oxide NPs	Tangerine	50 nm and spherical	Remediation of contaminated solution (Cd removal)	Ehrampoush et al. (2015)
FeNPs	*Eucalyptus* sp.	20–80 nm and spherical	Remediation of eutrophic wastewater	Wang et al. (2014a)
ZnO NPs	Lemon fruits	21.5 nm and spherical	Photocatalytic degradation of methyl red and methylene blue	Davar et al. (2015)

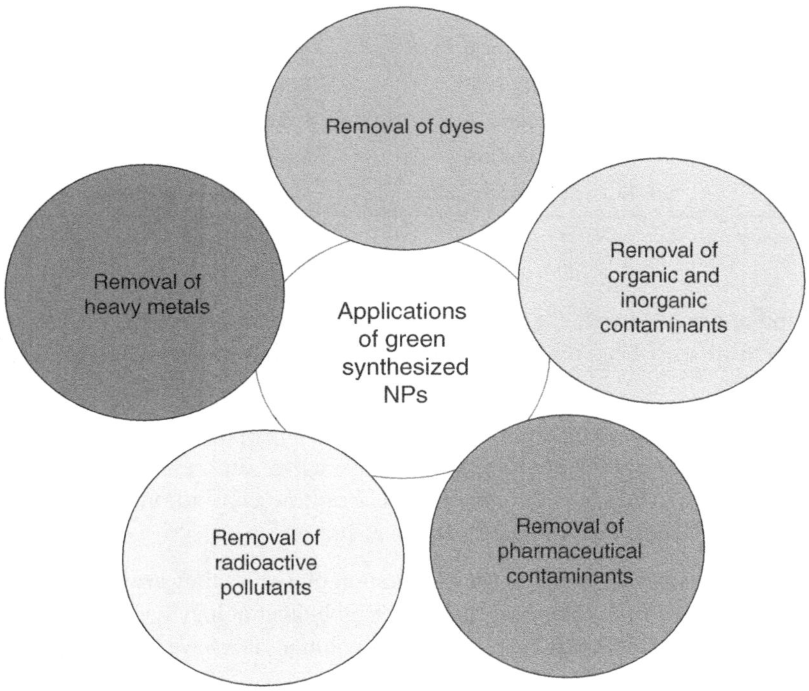

Figure 16.2 Application of green synthesized nanoparticles (NPs) in wastewater treatments.

Types of Dyes

- **Natural dyes:** Natural dyes are produced from minerals, plants, and animals. Lajerd, sindur, sajeda, and sangraj are some minerals, different parts of plants such as stem, wood, bark, and root, whereas microorganisms such as fungi and lichens are used to produce natural dyes.
- **Synthetic dyes:** Azo dyes, basic dyes, acid dyes, and mordant dyes are some examples of synthetic dyes. These dyes are widely used as they are bright in color. However, the production of synthetic dyes are simple, cheap, and time-consuming but produces toxic byproducts.

According to Kouhbanani et al. (2019), iron oxide NPs were synthesized with the leaf extract of *Teucrium polium* with the size of about 5.68–30.29 nm. These green synthesized nanomaterials have 73.6% efficiency to degrade the methyl orange dye in six hours' reaction. In further study, de Barizaoa et al. 2020 reported that iron oxide nanoparticles synthesized by *Beta vulgaris* stalks and *Cucurbita moschata* leaves (agroindustrial residues) have efficiency to degrade Bordeaux and tartrazine red dye.

16.4.2 Removal of Organic and Inorganic Contaminants

Environmental pollution is increasing globally due to discharge of wide variety of organic and inorganic pollutants. These pollutants are released into water systems as a result of extensive industrial, agricultural, urban, and anthropogenic developmental processes. This leads to detrimental effects on the environment.

- **Organic pollutants:** These refer to a wide range of chemical compounds that contain hydrogen and carbon as key elements, as well as trace amounts of other elements like halogens, sulfur, nitrogen, and phosphorus. Pesticides, insecticides, halogenated hydrocarbons, pharmaceuticals, phenols, synthetic dyes, and aromatic hydrocarbons are examples of organic pollutants (Lapworth et al. 2012). POPs (persistent organic pollutants) are highly harmful and noxious compounds as it introduces to the water systems via domestic sewage, agricultural runoff, and industrial discharge (Yadav et al. 2015). Their accumulation in animal and plant tissues causes biomagnifications, which can enter our body through the food chain and cause a variety of health issues, including cancer and immune system disruption (Zhang et al. 2003).

  For the elimination of organic pollutants, nanotechnology-based methods are employed to use nanocatalysts, nanoadsorbents, and nanofilms. In this context, Mishra et al. (2014) created gold nanoparticles (AuNPs) from the extract of two fungal species, *Trichoderma viride* and *Hypocrea lixii*. These AuNPs have potentially converted 4-nitrophenol to 4-aminophenol for the treatment of toxic 4-nitrophenol from water. In a further study, Nadaf and Kanase (2016) proposed that AuNPs fabricated with secondary metabolites of *Bacillus marisflavi* exhibited catalytic activity against methylene blue and Congo red dyes.
- **Inorganic pollutants:** Inorganic contaminants such as phosphate and nitrate have been found responsible for polluting ground and surface water for a long time. Agriculture runoff and municipal wastewater are the primary sources of these pollutants. The presence of excessive nutrients such as nitrate and phosphate causes eutrophication in aquatic bodies. These nutrients stimulate algal bloom and promote the production of an algal biofilm on the water surface. Consequently, this causes anaerobic conditions in the water systems. Nitrates are extremely persistent and highly soluble in aqueous medium, thus resulting in a variety of health diseases like methemoglobinemia (Fewtrell 2004). Although many researchers have conducted studies regarding the elimination of nitrates with various types of NPs from water bodies, still very little research has been done with organic NPs.

In a study, Wang et al. (2014b) revealed the fabrication of Fe-mediated NPs using the liquid extract from green tea and eucalyptus leaves to remove nitrates via a batch adsorption method. These NPs have nitrate-eliminating potentials of approximately 59% and 41.4%, respectively. Furthermore, Yong et al. (2004) created a biogenic nanoadsorbent by using *Serratia* sp. and hydroxyapatite crystallite nanopowder in order to remediate phosphate-contaminated water.

16.4.3 Removal of Pharmaceutical Contaminants

Currently, the availability of pharmaceutical products in aquatic environments is recognized as an evolving environmental problem. The extensive use of pharmaceutical compounds and their improper release is an important environmental concern. Pharmaceutical compounds are a vast and distinct category of chemical substances such as antibiotics, steroids, hormones, caffeine, anti-inflammatory products, and other endocrine disturbing compounds (Kim et al. 2007) used to treat, prevent, and cure diseases. Due to the growth of the population and other factors, their use and consumption are consistently rising (Husain Khan et al. 2020; Khan et al. 2021). Different sources that contribute to the contamination in the aquatic ecosystems are surface runoff (Pedersen et al. 2005), sewer leaking (Wolf et al. 2012), and sewer overflows (Launay et al. 2016). This contaminated water enters the human body and causes diseases like cancer, lung diseases, kidney failure, etc. Furthermore, these pharmaceutical pollutants also cause adverse effects on aquatic animals such as organ damage, reduced hatching success, disruption with hormones causing reproduction toxicity, and disruption in growth and feeding rate, etc. These hazardous effects of pharmaceutical pollutants have received immense attention from the scientists toward remediation of wastewater.

Nanotechnology has found applications as a potential technology to combat these issues (Malik et al. 2017). In this regard, Martins et al. (2017) reported that bio-Pd and bio-Pt nanocatalysts were integrated with *D. vulgaris* and performed efficient removal of about 94.85% sulfamethoxazole and 17b-estradiol and 70% ciprofloxacin. Furthermore, De Gusseme et al. (2011) stated that PdNPs synthesized with *S. oneidensis* as a cathode rod in microbial electrolysis cells (MECs). These PdNPs are employed to achieve the complete removal of diatrizoate, a highly persistent and nonbiodegradable pollutant.

16.4.4 Removal of Radioactive Pollutants

Radioactive pollutants are the residues of nuclear power generation. These pollutants are generated by the testing of nuclear weapons, medical research, and the mining industry and are responsible for contamination in the aquatic environment. Radioisotopes such as ^{90}Sr, ^{241}Am, ^{99}Mo, ^{137}Cs, ^{75}Se, ^{152}Eu, ^{131}I, ^{125}I, ^{129}I, ^{144}Ce, ^{106}Ru, ^{99m}Tc, and ^{60}Co are frequently responsible for chronic and acute diseases in human beings and other animals. It is considered that the majority of radionuclides in the aquatic environment have extremely long half-lives. Their long-term radiological toxicities can pose a serious threat to the environment, even in trace amounts (Kinly III 2006). It has been reported that dissolved radionuclides are nonbiodegradable and pernicious for the environment and have the potential to bioaccumulate through the food chain.

However, investigators have a serious concern about the detrimental effects of these radioactive isotopes and have found an efficient remediation technology. Biosynthesized nanoclusters are considered an effective solution for the degradation of radioactive pollutants. In this regard, Bio-HAP (biogenic nanometered hydroxyapatite material) was developed by using *Serratia* sp. NCIMB 40259 strain, applied for the efficient dissociation of strontium (Sr^{2+}) from simulated groundwater. The adsorption capacity of Bio-HAP for the separation of Sr^{2+} (5.35 mg/g) was much higher than commercial HAP (Handley-Sidhu et al. 2011a). Further studies reported that biosynthesized Bio-HAP nanomaterials successfully degraded and scavenged harmful radionuclides like europium, uranium, cobalt, and strontium from groundwater (Handley-Sidhu et al. 2014).

16.4.5 Removal of Heavy Metals

Heavy metals are known to be poisonous and unfavorable to human health and environment. Even a low concentration of heavy metal ions such as cadmium (Cd), lead (Pb), nickel (Ni), mercury (Hg), chromium (Cr), and arsenic (As) causes hazardous effects on the terrestrial and aquatic environments. When metal-contaminated water enters a source of drinking water, it has a negative impact on human health. Humans are affected by a wide variety of harmful diseases, including prostate cancer, lung cancer, breast cancer, kidney failure, and cardiovascular diseases, as well as other harmful diseases. Aquatic animals also experienced organ damage, decreased hatching success, mutagenicity, and disruption with hormones. These detrimental effects of heavy metals have now become an important concern for researchers. However, several investigators have focused on using nanobiotechnology to address the issues of environmental pollution.

In this regard, Ehrampoush et al. (2015) reported that FeNPs are synthesized with tangerine-peel extract and applied for 91% elimination of Cd(II) from wastewater within 90 minutes at pH 4. In a further experiment, Raj et al. (2016) demonstrated that cadmium sulfide (CdS) NPs synthesized from extracellular polymeric substances (EPSs) of a marine bacterium, *P. aeruginosa* JP-11 found remarkable potential for the elimination of 88% cadmium (Cd) from wastewater within 48 hours. Furthermore, Mukherjee et al. (2016) stated that Fe_2O_3 NPs, synthesized with the leaf extract of *A. vera*, were successfully applied for the degradation of As (Arsenic) from the water ecosystem. All these studies revealed that nanobiotechnology has potential in the treatment of aquatic pollution.

16.5 Conclusion

Environmental pollution has become a major concern globally with the rapid rise in industrialization and urbanization. Industries frequently release various types of pollutants into aquatic bodies, including textile dyes, pharmaceutical waste, heavy metal ions, radioactive pollutants, organic contaminants, and inorganic contaminants. These actions have a negative impact on the aquatic ecosystem. However, contaminated water reaches humans and animals through drinking sources, leading to serious diseases. In humans, this contaminated water leads to diseases like cancers, genetic disorders, respiratory conditions, hormonal imbalance, kidney failure, etc. In aquatic animals, it causes diseases like reproduction toxicity, hormone disruption, mutagenicity, carcinogenicity, toxicity to the fetus, development delay, etc. Thus, there is a dire need to provide a technology that could offer a fruitful solution to this problem. Hence, nanotechnology has been addressed to eliminate or mitigate wastewater pollution. Metal nanoparticles have been created using chemical, physical, and biological processes. However, the chemical and physical methods for creating metal nanomaterials require a lot of energy, are expensive, produce a low yield, and are hazardous. Hence, green synthesis of metal nanomaterials is found to be more effective, requiring less energy, being nontoxic, and having a high yield as well. Numerous microorganisms and plants are involved in the assembly of metal nanoparticles. These natural entities contain carbohydrates, proteins, flavonoids, polymers, teichoic acid, and alkaloids and act as better stabilizing and reducing agents. Further, the chapter elucidates how to use environmentally friendly metal nanomaterials to degrade and scavenge pollutants from water bodies in order to make them suitable for human and animal use.

References

Ahsan, M.A., Jabbari, V., Imam, M.A. et al. (2020). Nanoscale nickel metal organic framework decorated over graphene oxide and carbon nanotubes for water remediation. *Sci. Total Environ.* 698: 134214.

Al-Ruqeishi, M.S., Mohiuddin, T., and Al-Saadi, L.K. (2016). Green synthesis of iron oxide nanorods from deciduous Omani mango tree leaves for heavy oil viscosity treatment. *Arabian J. Chem.* 12: 4084–4090.

Armendariz, V., Herrera, I., Peralta-Videa, J.R. et al. (2004). Size controlled gold nanoparticle formation by *Avena sativa* biomass: use of plants in nanobiotechnology. *J. Nanopart. Res.* 6: 377–382.

Bar, H., Bhui, D.K., Sahoo, G.P. et al. (2009). Green synthesis of silver nanoparticles using latex of *Jatropha curcas*. *Colloids Surf., A* 339: 134–139.

de Barizaoa, A.C.L., Silvab, M.F., Andradeb, M. et al. (2020). Green synthesis of iron oxide nanoparticles for tartrazine and Bordeaux red dye removal. *J. Environ. Chem. Eng.* 8: 103618.

Bhargava, A., Jain, N., Khan, M.A. et al. (2016). Utilizing metal tolerance potential of soil fungus for efficient synthesis of gold nanoparticles with superior catalytic activity for degradation of rhodamine B. *J. Environ. Manage.* 183: 22–32.

Binupriya, A.R., Sathishkumar, M., and Yun, S.I. (2010). Biocrystallization of silver and gold ions by inactive cell filtrate of *Rhizopus stolonifer*. *Colloids Surf., B* 79: 531–534.

Birla, S.S., Tiwari, V.V., Gade, A.K. et al. (2009). Fabrication of silver nanoparticles by *Phoma glomerata* and its combined effect against *Escherichia coli*, *Pseudomonas aeruginosa* and *Staphylococcus aureus*. *Lett. Appl. Microbiol.* 48: 173–179.

Chandran, S.P., Chaudhary, M., Pasricha, R. et al. (2006). Synthesis of gold nanotriangles and silver nanoparticles using *Aloe vera* plant extract. *Biotechnol. Prog.* 22: 577–583.

Chen, Y.L., Tuan, H.Y., Tien, C.W. et al. (2009). Augmented biosynthesis of cadmium sulfide nanoparticles by genetically engineered *Escherichia coli*. *Biotechnol. Prog.* 25: 1260–1266.

Darroudi, M., Sabouri, Z., Kazemi, R. et al. (2013). Sol–gel synthesis, characterization, and neurotoxicity effect of zinc oxide nanoparticles using gum tragacanth. *Ceram. Int.* 39: 9195–9199.

Davar, F., Majedi, A., and Mirzaei, A. (2015). Green synthesis of ZnO nanoparticles and its application in the degradation of some dyes. *J. Am. Ceram. Soc.* 98: 1739–1746.

De Gusseme, B., Hennebel, T., Vanhaecke, L. et al. (2011). Biogenic palladium enhances diatrizoate removal from hospital wastewater in a microbial electrolysis cell. *Environ. Sci. Technol.* 45 (13): 5737–5745.

Decarolis, D., Odarchenko, Y., Herbert, J.J. et al. (2019). Identification of the key steps in the self-assembly of homogeneous gold metal nanoparticles produced using inverse micelles. *Phys. Chem. Chem. Phys.* 22 (34): 18824–18834.

Dehnad, A., Hamedi, J., Derakhshan-Khadivi, F., and Abusov, R. (2015). Green synthesis of gold nanoparticles by a metal resistant *Arthrobacter nitroguajacolicus* isolated from gold mine. *IEEE Trans. Nanobiosci.* 14: 393–396.

Dhand, C., Dwivedi, N., Loh, X.J. et al. (2015). Methods and strategies for the synthesis of diverse nanoparticles and their applications: a comprehensive overview. *RSC Adv.* 5: 105003–105037.

Du, L., Jiang, H., Liu, X., and Wang, E. (2007). Biosynthesis of gold nanoparticles assisted by *Escherichia coli* DH5 and its application on direct electrochemistry of hemoglobin. *Electrochem. Commun.* 9: 1165–1170.

Duraisamy, K. and Yang, S.L. (2013). Synthesis and characterization of bactericidal silver nanoparticles using cultural filtrate of simulated microgravity grown *Klebsiella pneumoniae*. *Enzyme Microb. Technol.* 52: 151–156.

Duran, N., Marcato, P.D., De Souza, G.I.H. et al. (2007). Antibacterial effect of silver nanoparticles produced by fungal process on textile fabrics and their effluent treatment. *J. Biomed. Nanotechnol.* 3: 203–208.

Edison, T.J.I. and Sethuraman, M.G. (2012). Instant green synthesis of silver nanoparticles using *Terminalia chebula* fruit extract and evaluation of their catalytic activity on reduction of methylene blue. *Process Biochem.* 47 (9): 1351–1357.

Ehrampoush, M.H., Miria, M., Salmani, M.H., and Mahvi, A.H. (2015). Cadmium removal from aqueous solution by green synthesis iron oxide nanoparticles with tangerine peel extract. *J. Environ. Health Sci. Eng.* 13: 84.

Fewtrell, L. (2004). Drinking-water nitrate, methemoglobinemia, and global burden of disease: a discussion. *Environ. Health Perspect.* 112 (14): 1371–1374.

Gajbhiye, M., Kesharwani, J., Ingle, A. et al. (2009). Fungus-mediated synthesis of silver nanoparticles and their activity against pathogenic fungi in combination with fuconazole. *Nanomed. Nanotechnol. Biol. Med.* 5: 382–386.

Gericke, M. and Pinches, A. (2006). Microbial production of gold nanoparticles. *Gold Bull.* 39: 22–28.

Goutam, S.P., Saxena, G., Singh, V. et al. (2018). Green synthesis of TiO_2 nanoparticles using leaf extract of *Jatropha curcas* L. for photocatalytic degradation of tannery wastewater. *Chem. Eng. J.* 336: 386–396.

Gupta, V.K. and Suhas (2009). Application of low-cost adsorbents for dye removal–a review. *J. Environ. Manage.* 90: 2313–2342.

Gupta, C., Rabani, M.S., Gupta, M.K. et al. (2022). Nanoscience in biotechnology. In: *Diverse Applications of Nanotechnology in the Biotechnology Sciences* (ed. K.R. Hakeem, M. Kamli, J.S.M. Sabir, and H.F. Alharby), 241–267. Taylor &Francis Group.

Gwo, S., Chen, H.Y., Lin, M.H. et al. (2016). Nanomanipulation and controlled self-assembly of metal nanoparticles and nanocrystals for plasmonics. *Chem. Soc. Rev.* 45: 5672–5716.

Handley-Sidhu, S., Hriljac, J., Cuthbert, M.O. et al. (2014). Bacterially Produced Calcium Phosphate Nano-Biominerals: Sorption Capacity, Site Preferences and Stability of Captured Radionuclides. *Environmental Science and Technology* 48(12): 6891–6898.

Handley-Sidhu, S., Renshaw, J.C., Yong, P. et al. (2011a). Nano-crystalline hydroxyapatite bio-mineral for the treatment of strontium from aqueous solutions. *Biotechnology Lett.* 33(1): 79–87.

Haverkamp, R.G. and Marshall, A.T. (2009). The mechanism of metal nanoparticle formation in plants: limits on accumulation. *J. Nanopart. Res.* 11: 1453–1463.

He, S., Guo, Z., Zhang, Y. et al. (2007). Biosynthesis of gold nanoparticles using the bacteria Rhodopseudomonas capsulata. *Mater. Lett.* 61 (18): 3984–3987.

Hennebel, T., Van Nevel, S., Verschuere, S. et al. (2011). Palladium nanoparticles produced by fermentatively cultivated bacteria as catalyst for diatrizoate removal with biogenic hydrogen. *Appl. Microbiol. Biotechnol.* 91: 1435–1445.

Herrera-Becerra, R., Zorrilla, C., Rius, J.L., and Ascencio, J.A. (2008). Electron microscopy characterization of biosynthesized iron oxide nanoparticles. *Appl. Phys. A* 91: 241–246.

Hosseini-Abari, A., Emtiazi, G., and Ghasemi, S.M. (2013). Development of an ecofriendly approach for biogenesis of silver nanoparticles using spores of *Bacillus athrophaeus*. *World J. Microbiol. Biotechnol.* 29: 2359–2364.

Huang, L., Weng, X., Chen, Z. et al. (2014). Synthesis of iron-based nanoparticles using oolong tea extract for the degradation of malachite green. *Spectrochim. Acta Part A* 117: 801–804.

Husain Khan, A., Abdul Aziz, H., Khan, N.A. et al. (2020). Pharmaceuticals of emerging concern in hospital wastewater: removal of ibuprofen and ofloxacin drugs using MBBR method. *Int. J. Environ. Anal. Chem.* 19: 1–5.

Kalimuthu, K., Babu, S.R., Venkataraman, D. et al. (2008). Biosynthesis of silver nanoparticles by *Bacillus licheniformis*. *Colloids Surf., B* 65: 150–153.

Kathiresan, K., Manivannan, S., Nabeel, M.A., and Dhivya, B. (2009). Studies on silver nanoparticles synthesized by a marine fungus, *Penicillium fellutanum* isolated from coastal mangrove sediment. *Colloids Surf., B* 71: 133–137.

Khan, A.H., Aziz, H.A., Khan, N.A. et al. (2021). Impact, disease outbreak and the eco-hazards associated with pharmaceutical residues: a critical review. *Int. J. Environ. Sci. Technol.* 10: 1–2.

Kim, S.D., Cho, J., Kim, I.S. et al. (2007). Occurrence and removal of pharmaceuticals and endocrine disruptors in South Korean surface, drinking, and waste waters. *Water Res.* 41: 1013–1021.

Kinly, D III. (2006). Chernobyl's legacy: Health, environmental and socio-economic impacts and recommendations to the Governments of Belarus, the Russian Federation and Ukraine. *The Chernobyl Forum 2003–2005*. Second revised version.

Klaus, T., Joerger, R., Olsson, E., and Granqvist, C.G. (1999). Silver based crystalline nanoparticles, microbially fabricated. *Proc. Natl. Acad. Sci. U.S.A.* 96: 13611–13614.

Konishi, Y., Nomura, T., Tsukiyama, T., and Saitoh, N. (2004). Microbial preparation of gold nanoparticles by an anaerobic bacterium. *Trans. Mater. Res. Soc. Jpn* 29: 2341–2343.

Korbekandi, H., Iravani, S., and Abbasi, S. (2012). Optimization of biological synthesis of silver nanoparticles using *Lactobacillus casei* subsp. *casei*. *J. Chem. Technol. Biotechnol.* 87: 932–937.

Kouhbanani, M.A.J., Beheshtkhoo, N., Taghizadeh, S. et al. (2019). One-step green synthesis and characterization of iron oxide nanoparticles using aqueous leaf extract of *Teucrium polium* and their catalytic application in dye degradation. *Adv. Nat. Sci.: Nanosci. Nanotechnol.* 10: 015007.

Kumar, C.G., Mamidyala, S.K., Das, B. et al. (2010). Synthesis of biosurfactant-based silver nanoparticles with purified rhamnolipids isolated from *Pseudomonas aeruginosa* BS-161R. *J. Microbiol. Biotechnol.* 20: 1061–1068.

Kumari, A., Kumar, V., and Yadav, S.K. (2012). Plant extracts synthesized PLA nanoparticles for controlled and sustained release of quercetin: a green approach. *PLoS One* 7 (7): e41230.

Lapworth, D.J., Baran, N., Stuart, M.E., and Ward, R.S. (2012). Emerging organic contaminants in groundwater: a review of sources, fate and occurrence. *Environ. Pollut.* 163: 287–303.

Launay, M.A., Dittmer, U., and Steinmetz, H. (2016). Organic micropollutants discharged by combined sewer overflows – characterisation of pollutant sources and storm water-related processes. *Water Res.* 2016 (104): 82–92.

Luo, F., Chen, Z., Megharaj, M., and Naidu, R. (2014). Biomolecules in grape leaf extract involved in one-step synthesis of iron-based nanoparticles. *RSC Adv.* 4 (96): 53467–53474.

Malik, B., Pirzadah, T.B., Kumar, M., and Rehman, R.U. (2017). Biosynthesis of nanoparticles and their application in pharmaceutical industry. In: *Metabolic Engineering for Bioactive Compounds*, 331–349. Singapore: Springer.

Manikprabhu, D. and Lingappa, K. (2013). Antibacterial activity of silver nanoparticles against methicillin-resistant *Staphylococcus aureus* synthesized using model *Streptomyces* sp. pigment by photo-irradiation method. *J. Pharm. Res.* 6: 255–260.

Marchiol, L. (2012). Synthesis of metal nanoparticles in living plants. *Ital. J. Agron.* 7: 274–282.

Martins, M., Mourato, C., Sanches, S. et al. (2017). Biogenic platinum and palladium nanoparticles as new catalysts for the removal of pharmaceutical compounds. *Water Res.* 108: 160–168.

Meyers, M.A., Mishra, A., and Benson, D.J. (2006). Mechanical properties of nanocrystalline materials. *Prog. Mater. Sci.* 51: 427–556.

Mishra, A., Kumari, M., Pandey, S. et al. (2014). Biocatalytic and antimicrobial activities of gold nanoparticles synthesized by *Trichoderma* sp. *Bioresour. Technol.* 166: 235–242.

Morshedi, D., Mohammadia, Z., Mashhadi, M. et al. (2013). Using protein nanofibrils to remove azo dyes from aqueous solution by the coagulation process. *Colloids Surf., B* 112: 245–254.

Morsy, F.M., Nafady, N.A., Abd-Alla, M.H., and Elhady, D.A. (2014). Green synthesis of silver nanoparticles by water soluble fraction of the extracellular polysaccharides/matrix of the cyanobacterium *Nostoc commune* and its application as a potent fungal surface sterilizing agent of seed crops. *Univ. J. Microbiol. Res.* 2: 36–43.

Mude, N., Ingle, A., Gade, A., and Rai, M. (2009). Synthesis of silver nanoparticles using callus extract of *Carica papaya* – a first report. *J. Plant Biochem. Biotech.* 18: 83–86.

Mukherjee, P., Senapati, S., Mandal, D. et al. (2002). Extracellular synthesis of gold nanoparticles by the fungus *Fusarium oxysporum*. *ChemBioChem* 3: 461–463.

Mukherjee, P., Roy, M., Mandal, B.P. et al. (2008). Green synthesis of highly stabilized nanocrystalline silver particles by a non-pathogenic and agriculturally important fungus *T. asperellum*. *Nanotechnology* 19: 1–7.

Mukherjee, D., Ghosh, S., Majumdar, S., and Annapurna, K. (2016). Green synthesis of α-Fe_2O_3 nanoparticles for arsenic (V) remediation with a novel aspect for sludge management. *J. Environ. Chem. Eng.* 4: 639–650.

Nadaf, N.Y. and Kanase, S.S. (2016). Biosynthesis of gold nanoparticles by *Bacillus marisflavi* and its potential in catalytic dye degradation. *Arabian J. Chem.* 12: 4806–4814.

Narayanan, K.B. and Sakthivel, N. (2008). Coriander leaf mediated biosynthesis of gold nanoparticles. *Mater. Lett.* 62: 4588–4590.

Naseem, T. and Farrukh, M.A. (2015). Antibacterial activity of green synthesis of iron nanoparticles using *Lawsonia inermis* and *Gardenia jasminoides* leaves extract. *J. Chem.* 2015 (2015): 1–7.

Njagi, E.C., Huang, H., Stafford, L. et al. (2011). Biosynthesis of iron and silver nanoparticles at room temperature using aqueous sorghum bran extracts. *Langmuir* 27 (1): 264–271.

Paknikar, K., Nagpal, V., Pethkar, A., and Rajwade, J. (2005). Degradation of lindane from aqueous solutions using iron sulfide nanoparticles stabilized by biopolymers. *Sci. Technol. Adv. Mater.* 6: 370–374.

Pal, G., Rai, P., and Pandey, A. (2019). Green synthesis of nanoparticles: a greener approach for a cleaner future. In: *Green synthesis, Characterization and Applications of Nanoparticles* (ed. A.C. Garcia), 1–26. Elsevier.

Pandey, A., Shukla, P., and Srivastava, P.K. (2020). Remediation of dyes in water using green synthesized nanoparticles. *Int. J. Plant Environ.* 6 (1): 68–84.

Parashar, V., Parashar, R., Sharma, B., and Pandey, A.C. (2009). Parthenium leaf extract mediated synthesis of silver nanoparticles: a novel approach towards weed utilization. *Dig. J. Nanomater. Biostruct.* 4: 45–50.

Pedersen, J.A., Soliman, M., and Suffet, I.H. (2005). Human pharmaceuticals, hormones, and personal care product ingredients in runoff from agricultural fields irrigated with treated wastewater. *J. Agric. Food Chem.* 53 (5): 1625–1632.

Philip, D. and Unni, C. (2011). Extracellular biosynthesis of gold and silver nanoparticles using Krishna tulsi (*Ocimum sanctum*) leaf. *Phys. E Low Dimens. Syst. Nanostruct.* 43: 1318–1322.

Prabhu, S. and Poulose, E.K. (2012). Silver nanoparticles: mechanism of antimicrobial action, synthesis, medical applications, and toxicity effects. *Int. Nano Lett.* 2 (1): 32.

Prathna, T.C., Chandrasekaran, N., Raichur, A.M., and Mukherjee, A. (2011). Biomimetic synthesis of silver nanoparticles by *Citrus limon* (lemon) aqueous extract and theoretical prediction of particle size. *Colloids Surf., B* 82: 152–159.

Rai, M., Yadav, A., and Gade, A. (2008). CRC 675–current trends in phytosynthesis of metal nanoparticles. *Crit. Rev. Biotechnol.* 28: 277–284.

Rai, M., Yadav, A., and Gade, A. (2009). Silver nanoparticles as a new generation of antimicrobials. *Biotechnol. Adv.* 27: 76–83.

Raj, R., Dalei, K., Chakraborty, J., and Das, S. (2016). Extracellular polymeric substances of a marine bacterium mediated synthesis of CdS nanoparticles for removal of cadmium from aqueous solution. *J. Colloid Interface Sci.* 462: 166–175.

Rajeshkumar, S. and Bharath, L. (2017). Mechanism of plant-mediated synthesis of silver nanoparticles–a review on biomolecules involved, characterisation and antibacterial activity. *Chem. Biol. Interact.* 273: 219–227.

Rather, M.Y. and Sundarapandian, S. (2020). Magnetic iron oxide nanorod synthesis by *Wedelia urticifolia* (Blume) DC. Leaf extract for methylene blue dye degradation. *Appl. Nanosci.* 10: 2219–2227.

Rauwel, P. and Rauwel, E. (2017). Emerging trends in nanoparticle synthesis using plant extracts for biomedical applications. *Global J. Nanomedicine* 1 (3): 1–3.

Ravindra, B.K. and Rajasab, A.H. (2014). A comparative study on biosynthesis of silver nanoparticles using four different fungal species. *Int. J. Pharm. Pharm. Sci.* 6 (1): 372–376.

Ravindra, S., Murali, M.Y., Reddy, N.N., and Raju, M.K. (2020). Fabrication of antibacterial cotton fibres loaded with silver nanoparticles via "green approach". *Colloids Surf., A* 367: 31–40.

Rawtani, D., Tharmavaram, M., Pandey, G., and Hussain, C.M. (2019). Functionalized nanomaterial for forensic sample analysis. *TrAC, Trends Anal. Chem.* 120: 115661.

Seetharaman, P.K., Chandrasekaran, R., Gnanasekar, S. et al. (2018). Antimicrobial and larvicidal activity of eco-friendly silver nanoparticles synthesized from endophytic fungi *Phomopsis liquidambaris*. *Biocatal. Agric. Biotechnol.* 16: 22–30.

Shankar, S.S., Rai, A., Ahmad, A., and Sastry, M. (2004a). Rapid synthesis of Au, Ag, and bimetallic Au core Ag shell nanoparticles using neem (*Azadirachta indica*) leaf broth. *J. Colloid Interface Sci.* 1: 1.

Shankar, S.S., Rai, A., Ankamwar, B. et al. (2004b). Biological synthesis of triangular gold nanoprisms. *Nat. Mater.* 3: 482–488.

Shankar, S.S., Rai, A., Ahmad, A., and Sastry, M. (2005). Controlling the optical properties of lemongrass extract synthesized gold nanotriangles and potential application in infrared-absorbing optical coatings. *Chem. Mater.* 17: 566–572.

Sheoran, N. and Kaur, P. (2018). Biosynthesis of nanoparticles using eco-friendly factories and their role in plant pathogenicity: a review. *Biotechnol. Res. Innovation.*

Southam, G. and Beveridge, T.J. (1994). The in vitro formation of placer gold by bacteria. *Geochim. Cosmochim. Acta* 58: 4527–4530.

Srinath, B.S. and Rai, V.R. (2015). Biosynthesis of highly monodispersed, spherical gold nanoparticles of size 4–10 nm from spent cultures of *Klebsiella pneumoniae*. *3 Biotech* 5: 671–676.

Srinath, B.S., Namratha, K., and Byrappa, K. (2017). Eco-friendly synthesis of gold nanoparticles by gold mine bacteria *Brevibacillus formosus* and their antibacterial and biocompatible studies. *IOSR J. Pharm.* 7: 53–60.

Srivastava, S.K., Yamada, R., Ogino, C., and Kondo, A. (2013). Biogenic synthesis and characterization of gold nanoparticles by Escherichia coli K12 and its heterogeneous catalysis in degradation of 4-nitrophenol. *Nanoscale Res. Lett.* 8 (1): 1–9.

Sunkar, S. and Nachiyar, C.V. (2012). Biogenesis of antibacterial silver nanoparticles using the endophytic bacterium *Bacillus cereus* isolated from *Garcinia xanthochymus*. *Asian Pac. J. Trop. Biomed.* 2: 953–959.

Thakkar, K.N., Mhatre, S.S., and Parikh, R.Y. (2010). Biological synthesis of metallic nanoparticles. *Nanomedicine* 6: 257–262.

Tharmavaram, M., Pandey, G., and Rawtani, D. (2018). Surface modified halloysite nanotubes: a flexible interface for biological, environmental and catalytic applications. *Adv. Colloid Interface Sci.* 261: 82–101.

Veerasamy, R., Xin, T.Z., Gunasagaran, S. et al. (2010). Biosynthesis of silver nano-particles using mangosteen leaf extract and evaluation of their anti-microbial activities. *J. Saudi Chem. Soc.* 15: 113–120.

Wang, T., Jin, X., Chen, Z. et al. (2014a). Green synthesis of Fe nanoparticles using eucalyptus leaf extracts for treatment of eutrophic wastewater. *Sci. Total Environ.* 466, 467: 210–213.

Wang, T., Lin, J., Chen, Z. et al. (2014b). Green synthesized iron nanoparticles by green tea and eucalyptus leaves extracts used for removal of nitrate in aqueous solution. *J. Clean. Prod.* 83: 413–419.

Wolf, L., Zwiener, C., and Zemann, M. (2012). Tracking artificial sweeteners and pharmaceuticals introduced into urban groundwater by leaking sewer networks. *Sci. Total Environ.* 430: 8–19.

Xiao, Z., Zhang, H., Xu, Y. et al. (2017). Ultraefficient removal of chromium from aqueous medium by biogenic iron based nanoparticles. *Sep. Purif. Technol.* 174: 466–473.

Yadav, P.T., Yadav, M.R., and Singh, P.D. (2012). Mechanical milling: a top down approach for the synthesis of nanomaterials and nanocomposites. *Nanosci. Nanotechnol.* 2: 22–48.

Yadav, I.C., Devi, N.L., Syed, J.H. et al. (2015). Current status of persistent organic pesticides residues in air, water, and soil, and their possible effect on neighboring countries: a comprehensive review of India. *Sci. Total Environ.* 511: 123–137.

Yong, P., Macaskie, L.E., Sammons, R.L., and Marquis, P.M. (2004). Synthesis of nanophase hydroxyapatite by a *Serratia* sp. from waste-water containing inorganic phosphate. *Biotechnol. Lett.* 26 (22): 1723–1730.

Zhang, Z.L., Hong, H.S., Zhou, J.L. et al. (2003). Fate and assessment of persistent organic pollutants in water and sediment from Minjiang River Estuary, Southeast China. *Chemosphere* 52 (9): 1423–1430.

Zhang, M., Yang, J., Cai, Z. et al. (2019). Detection of engineered nanoparticles in aquatic environments: current status and challenges in enrichment, separation, and analysis. *Environ. Sci. Nano* 6: 709–735.

Zhou, D., Kim, D.G., and Ko, S.O. (2015). Heavy metal adsorption with biogenic manganese oxide generated by *Pseudomonas putida* strain MnB1. *J. Ind. Eng. Chem.* 24: 132–139.

17

Nanotechnology-Based Applications: A Valuable Tool for Wastewater Clean-up

Mir Z. Gul, Beedu S. Rao and Karuna Rupula

Department of Biochemistry, University College of Sciences, Osmania University, Hyderabad, 5000 07, Telangana, India

17.1 Introduction

Water quality is vital for all human activities from basic domestic needs to the very intricate agricultural and industrial processes. While three-quarters of the Earth's surface is covered by water, still past reports published by the World Health Organization and other agencies concerned put forth alarming facts about fresh water availability as well as its affordability. Demand for water has increased manifolds (more than twice the rate of population) and since the 1980s, its usage has been growing around the world by 1% annually, due to several factors such as climate change, rising population, socioeconomic evolution and fluctuating consumption patterns (UNESCO 2019). Worldwide water requirement is anticipated to rise at 20–30% higher than the existing consumption levels, owing to the massive growth in the domestic and industrial setup. More than two billion inhabitants living across the world and approximately four billion inhabitants for at least one month a year experience acute water shortages. The stress on water resources will persist as water requirement intensify and the impact of the changes in climate escalate (UNESCO 2019). Therefore, about 52% of the world's projected 9.7 billion populace will be dwelling in water-stressed regions by the middle of the century, i.e. by 2050 (World Health Organization 2016).

The scarcity of water supplies has been attributed to the depletion of water reserves for industrial, irrigation, and domestic needs (Gleick and Iceland 2018). The stress on freshwater supplies is due to ever-increasing global food demands, energy consumption, etc., and more importantly because of the exponential population boom and the aggressions of climatic changes (Vörösmarty et al. 2000). Another reason for the decline of freshwater supplies is the pollution of surface/groundwater supplies (Smith 2003; Singhal and Gupta 2010; Walker et al. 2019). Aquifers all over the world are depleting and damaged due to industrial waste discharges (heavy metals, PAHs), animal wastes from agricultural farms, fertilizer and pesticide runoff, algal blooms, untreated sewage, etc. (Smith 2003; Vörösmarty et al. 2010; Qu et al. 2013). The disposal of such pollutants into the water ecosystems has adversely impacted all forms of life. Nature has its mechanism to recycle water and provide an adequate quantity of clean water to us. However, unregulated population growth, lifestyle patterns and unplanned industrialization have progressively degraded the water resources and consequently disrupted the natural purification mechanism. The presence of emerging micropollutants like endocrine-disrupting compounds (EDCs) in wastewater has made the current conventional treatment facilities unsuitable in achieving environmental regulations (Westerhoff et al. 2005). The existing materials and treatment practices, such as activated carbon, and filtration membranes, are not efficient in controlling diverse and highly polluted waters that are infested with a myriad of pollutants. Moreover, the traditional approaches for water remediation cannot sufficiently manage the disposal of a wide variety of hazardous toxins and pathogenic microorganisms from wastewater. Biological treatment methods like activated sludge and biological filtration are not efficient to eliminate a broad range of toxic elements and the majority of such pollutants endure dissolved in the wastewater (Lakatos 2018). Physicochemical approaches such as coagulation, flocculation, or bleaching that have been extensively investigated over the past decade have proved futile in removing various EDCs and industrial pollutants (Westerhoff et al. 2005; Qu et al. 2013). While chlorination ensures residual protection against the regrowth of

Aquatic Contamination: Tolerance and Bioremediation, First Edition. Edited by Rouf Ahmad Bhat, Gowhar Hamid Dar, Fernanda Maria Policarpo Tonelli, and Saima Hamid.

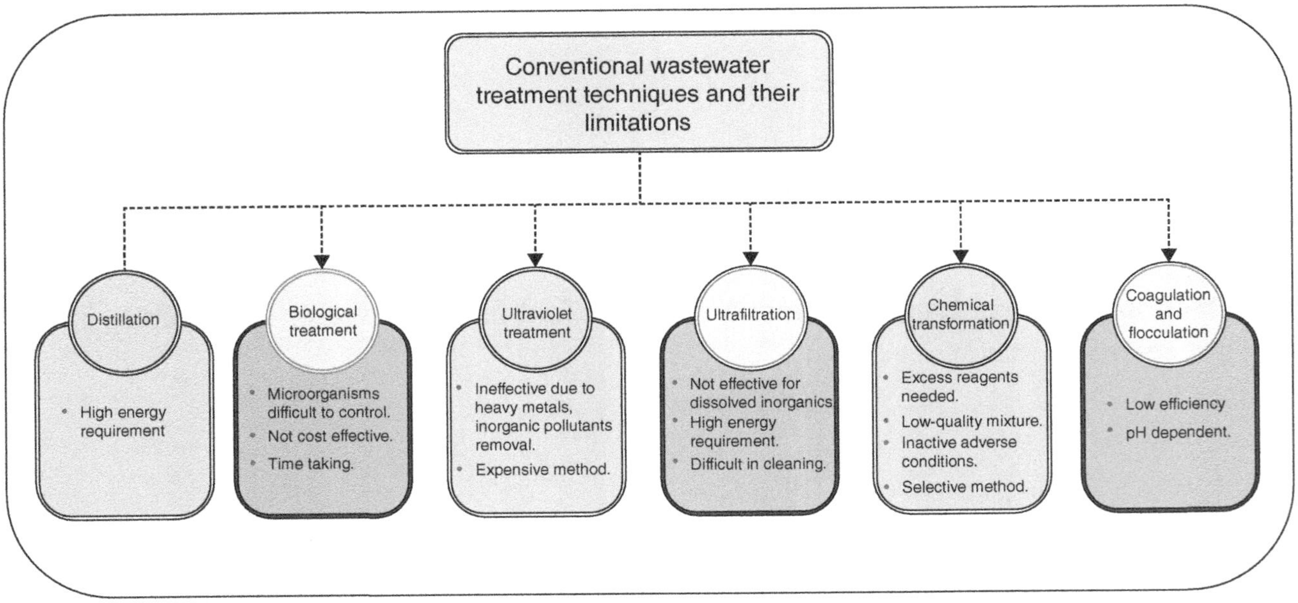

Figure 17.1 A drawing depicting the disadvantages of some of the conventional methods employed in water purification and treatment processes.

pathogenic microbes, it leads to unpleasant flavors and odor as well as the formation of various disinfection byproducts (DBPs) in the water supply (Boorman et al. 1999). Ozonation is now a less desirable choice due to its high costs and limited lifespan. Ultraviolet photolysis and ion-exchange methods though available are not feasible solutions for the removal of micropollutants, although they are an advanced class of treatment methods (Beltrán et al. 2008). RO is comparatively more efficient than nanofiltration (NF). However, high power demand in RO renders it less desirable as compared to NF, where pollutant elimination is influenced by various processes including convections, diffusion, and charging effects (Chung et al. 2012). To summarize, all these existing treatment methods involve high capital investment, high operational and maintenance expenditure, higher demands of energy, and huge facility areas. Some of these conventional methods employed in water treatment processes and their disadvantages are listed in Figure 17.1. The industries and other sectors are finding it hard to implement such costly methods because of less profitability. Therefore, compliance with environmental laws and regulations with these industrial setups remains quite low. To address these concerns, the research and development institutions need to establish efficient, cost-effective innovative and sustainable wastewater treatment technologies. A promising approach was found to be the application of advanced treatment technology to improve wastewater treatment plants, which seek to convert or even mineralize contaminants into less harmful substances.

17.2 Nanotechnology: A Reliable Tool

The twenty-first century ushered new and exciting opportunities associated with nanotechnology to address the prevailing water-related issues (Qu et al. 2013). Recent advances in nanotechnology enable us to develop technologically feasible strategies for conventional water remediation. Nanotechnology not only has been able to transcend the limitations of traditional methods but also provide sustainable and economic utilization of available sources (Kumar et al. 2014; Gehrke et al. 2015). The involvement of nanotechnology in the treatment has envisioned us with high performance, inexpensive wastewater treatment, thereby ending the reliability of infrastructure (Qu et al. 2013). Nanotechnology is the solution for improving costs and effectiveness and has a significant potential to boost long term water quality, accessibility and the feasibility of uninterrupted water supplies with the progressive filtration methods that improve overall water reuse, reclamation, and desalination of water. Most of the applications of nanotechnology fall into three main categories: (i) environmentally safe and/or sustainable materials (viz. green synthesis approach), (ii) treatment and mitigation of hazardous compounds, and (iii) environmental sensors and detectors (Rickerby and Morrison 2007; Vaseashta et al. 2007). New and

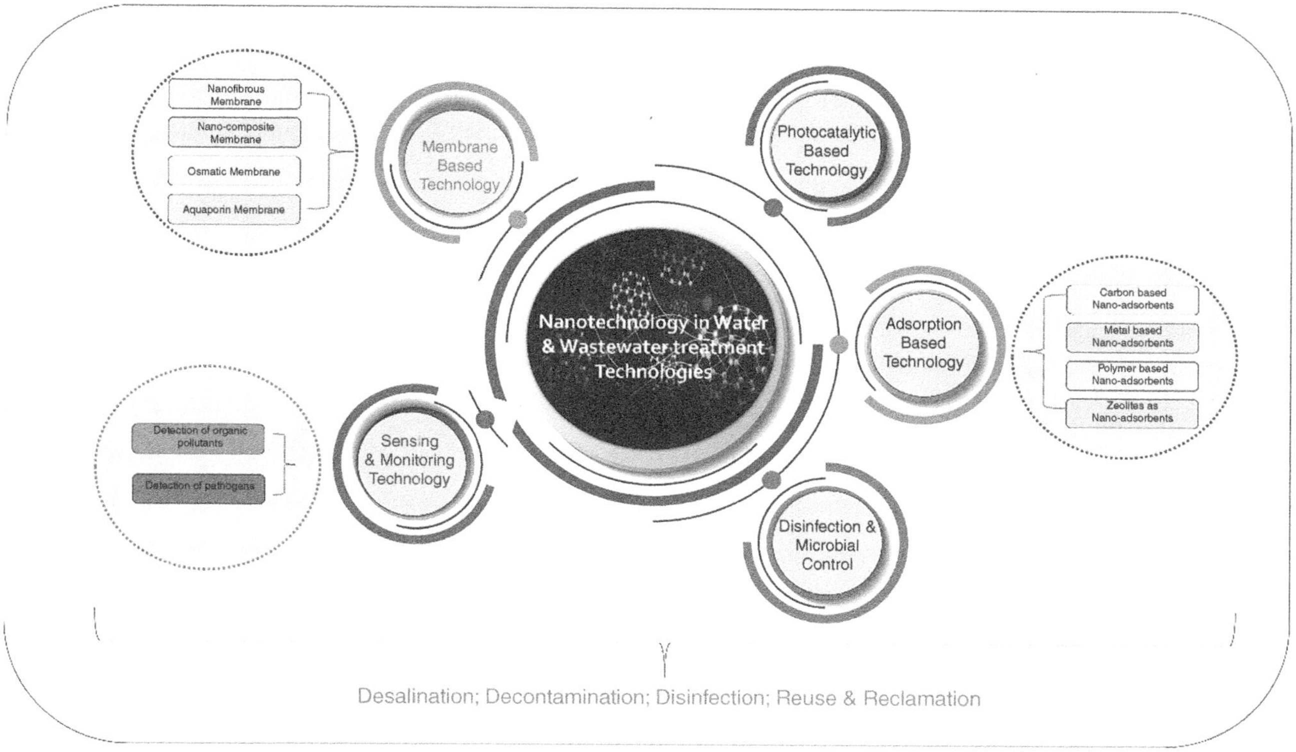

Figure 17.2 A generalized schematic representation of nanotechnology and its applications in water purification and treatment processes.

improved sensors are being developed to identify and monitor the biological, chemical, and other emerging pollutants at minimal levels in the environment including water resources. Furthermore, nanotechnology tends to enhance the implementation of constant monitoring systems to provide low-cost, real-time insights with improved specificity (Vaseashta et al. 2007; Zhang and Fang 2010).

Among the wide array of applications of nanotechnology, the application of nanocomposite materials for water remediation can be deemed as a novel application of nanotechnology. Because of the larger volume-to-surface ratio, these nanosized materials offer improved reactivity compared to their larger equivalents and thus greater effectiveness. Compared to traditional methods, nanomaterials provide the opportunity of exploiting unique chemical properties so that they will be functionalized or grafted with specific functional groups that will in turn target pollutants for an efficient remedy. Additionally, deliberate refinement of physical properties of nanomaterials like scale, shape, permeability, and chemical composition may impart added valuable attributes that significantly affect the material's efficiency for contaminant remediation. In contrast, the processes are more precise, potent and reliable as a combination of a different material (composites/hybrids), integrating similar characteristics from its components than the processes that are based on a single nanoplatform. The functionalization of materials with specific molecules primarily responsible for striking the pollutants would significantly improve selectiveness and competency (Bora and Dutta 2014; Guerra et al. 2018). The role of nanotechnology and its applications in purification and treatment processes is illustrated in Figure 17.2.

17.3 Main Nanotechnological Processes for Water Purification and Wastewater Treatment

So far, several distinct nanotechnology-based pathways have been explored extensively for their efficient use in water purification and wastewater treatment. Among them, most explored technologies may be classified into the following specific categories:

17.3.1 Adsorption-Based Technology

Adsorption is a process by which particles, atoms, or molecules of a liquid or gas or a dissolved substance physically adhere to the surface of solid materials. Adsorption is one of the realistic strategies for filtering out pollutants from wastewater by the accumulation of pollutants at the interface of nanosorbents. Nanomaterials possess two fundamental properties that render them incredibly effective adsorbents. These are high surface area and functionalized structure or binding tendency toward different adjacent molecules or atoms (Santhosh et al. 2016).

These features not only end up making nanomaterials efficient adsorbents in wastewater for various contaminants, but also ensure long-term stability in their degradation (in the presence of catalytic properties of nanomaterials) and improve adsorption efficiency. Additionally, these nanoadsorbents have distinctive optical, electrical, and mechanical along with many additional features which enable them to have much stronger adsorption qualities toward certain pollutants (Hua et al. 2012). The process of adsorption is primarily focused on the electrostatic activity between metal ions in water and the nanoadsorbent surfaces. The alteration in the pH of the solution may substantially affect the rate of interaction. Hence, the nanoadsorbent surface can be:

- acidic with positive anion attracting charges
- basic with a negative attracting charge (Qu et al. 2013).

Present research efforts concentrate mainly on the following types of nanoadsorbents:

17.3.2 Carbon-Based Nanoadsorbents

Carbon nanocomposite materials have been used primarily for the adsorption of diverse organic and inorganic pollutants in water. With the discovery of carbon nanotubes (CNTs) by Japanese physicist, Sumio Iijima in 1991, a unique carbon-based adsorption material was introduced to the world (Iijima 1991). Those CNTs were portrayed as the "material of the twenty-first century" (Ren et al. 2011). Due to their relatively low cost, nontoxicity and versatile physicochemical properties, this class of nanoadsorbents hold a great ability for application in wastewater treatment and remediations (Mauter and Elimelech 2008). There are several reports in the scientific literature on the use of surface-functionalized carbon materials, CNTs, graphene, and fullerenes for the efficacious degradation of organic contaminants as well as other heavy metal ions such as Fe, Cu, Ni, Pb, and Zn, so on and so forth (Sankararamakrishnan et al. 2014; Burakov et al. 2018). Activated carbon has also been demonstrated to be efficient in removing heavy metal ions like Ni(II), Cu(II), Pb(II), Cd(II), Hg(II), Co(II), Cr(II), and Cr(VI) (Kadirvelu et al. 2001; Kobya et al. 2005; Al-Omair and El-Sharkawy 2007). Permeable activated carbon sheets and fibers, obtained from the organic residues (mainly from agricultural wastes), also were reported to be significantly efficient in the remediation of wastewaters that contains dyes, and chlorinated toxic chemicals apart from hazardous metal ions like As, Cd, Cd, and Hg. They are desirable over other nanoabsorbents because of their excellent adsorption ability, low cost, and easy recyclability (Ali et al. 2012). In combination with activated carbon, CNTs often attract enormous attention from the scientific community as an impressive adsorbent product owing to their greater surface area and reasonable chemical and thermal stability. Moreover, the CNT's reactivity can be precisely modified easily by functionalizing the CNT's surfaces (Tamburri et al. 2005; Ren et al. 2011; Yang et al. 2013). Acid or alkaline treatment is highly effective in further optimizing the CNT's adsorption efficiency. The treatment with potent oxidizing agents such as nitric acid (HNO_3), sulfuric acid (H_2SO_4), and potassium permanganate ($KMnO_4$) ultimately leads to surface modification of CNTs with functional groups like hydroxyl (–OH), carboxyl (–COOH), or ketonic (–C=O). The functionalization ensures the nanoadsorbents effectively attach metal ions through electrostatic binding forces (Fu and Wang 2011). An icosahedral fullerene (C_{60}) has been extensively used in the adsorption of hydrophobic organic moieties. Owing to its unique mechanical, physical, and chemical properties, graphene and its composites with metal and metal oxides have attracted enormous attention all over the world. The extremely large surface area, enhanced binding sites, large delocalized π electron systems and excellent chemical stability make it suitable for use as an adsorbent for wastewaters (Theron et al. 2008; Amin et al. 2014).

17.3.3 Metal-Based Nanoadsorbents

Metals and their composites with a large surface-to-volume ratio are also impressive adsorbents for various organic pollutants and simultaneously have been explored extensively for wastewater treatment with great precision and accuracy (Hua et al. 2012). The metal and metal oxides have several outstanding features, such as fast adsorption kinetics, strong energy efficiency, and swift extraction post-adsorption process, attributed to their unique sensitivity toward harmful metal ions, magnetic behavior and variable morphology or size (Xu et al. 2012). It has been shown that pure metallic iron neutralizes wastewater, which includes wastes from agriculture (fertilizers, pesticides, herbicides, etc.), industrial wastes (polychlorinated biphenyls, etc.), and other organic contaminants (Wang and Zhang 1997). Further iron oxides, hydroxides, and oxy-hydroxides are extremely effective in the adsorption of As, Pb, and Cd water species at a very low pH. The larger surface-to-volume ratio of iron nanocomposites and their innate propensity to bind As(III) and As(V) ions via electrostatic attraction enables these magnetically separable nanoadsorbents very attractive for the removal of arsenic from the wastewaters. Besides, they were also employed to remove Ni, Pb, Cr, Co, Cu, and As from wastewaters (Andjelkovic et al. 2015; Ghosh et al. 2015; Kumari et al. 2015; Khan et al. 2016; Alswata et al. 2017; Lee et al. 2017; Ullah Khan et al. 2017; Verma et al. 2017; Zhou et al. 2020). Metal oxide nanoadsorbents have added a feature of magnetism that gives them greater efficacy in removing heavy metals. It has been observed that small-sized nanosorbents have higher magnetism characteristics, therefore, present relatively higher adsorption potential. Since most of the pollutants are not magnetic, filtration aids are needed to absorb which is normally followed by magnetic separation (Campos et al. 2011). Magnetic nanorods and nanotubes of γ-Fe_2O_3 are known to be highly effective in binding Fe, Ni, Cu, Pb, Zn, and Cd. Nanorods have also been observed to show better adsorption efficiency compared to nanotubes. Superparamagnetic nanoparticles have exhibited faster and more selective adsorption toward Hg^{+2} (Shipley et al. 2013). Graphene oxide-deposited hematite nanocomposites have been effectively employed for the precise binding of highly toxic Hg(II) ions from wastewaters (Jung et al. 2011; Diagboya et al. 2015).

In addition to iron-based nanomaterials, nanocomposites like TiO_2, MgO, and ZnO have also been highly effective in eradicating several unwanted hazardous compounds from the wastewaters because they provide specific affinity and higher surface-to-volume ratio to absorb these heavy metals and also possess the merits of being economically cheap materials (Hua et al. 2012; Rafiq et al. 2014; Ghosh et al. 2015; Hu et al. 2015; Ray and Shipley 2015; Srivastava et al. 2015). Moist amorphous titanium is also reported to be significantly competent in the elimination of different pollutants like Fe, Cu, As, Pb, Cd, and Mn (Hu et al. 2015). Magnesium oxide nanoflowers with a quite high surface area are described as successful adsorbents for As, Pb, and Cd (Srivastava et al. 2015). Few other economically cheap nanoadsorbents, viz. ZnO and Al_2O_3 are often employed for the adsorption as well as the separation of Cr, Cd, and Hg ions from wastewaters (Sankararamakrishnan et al. 2014; Ray and Shipley 2015). Moreover, the application of nonmetallic oxides, like SiO_2 as nano-adsorbents has also exhibited promising outcomes in degrading organic contaminants as well as other toxic metals from wastewaters (Zhao et al. 2005; Li et al. 2007a; Yantasee et al. 2010; Anbia and Salehi 2012; El-Safty et al. 2012).

17.4 Polymer-Based Nanoabsorbents

Polymeric nanoadsorbents have gained tremendous attention over the past decade as an emerging class of materials for the remediation of water resources (Yin and Deng 2015). Multiple composite polymers of organic and organic–inorganic origin were developed over the years and subsequently used for the adsorptive elimination of different pollutants (heavy metals, etc.) from different contaminated waters. Metal-doped polymeric materials with porosity, high surface-to-volume ratio, and ideal functional groups are validated to effectively bind azo dyes and other toxic metal ions (Pb, Cd, and Zn) from the wastewaters (Lofrano et al. 2016). Because of the presence of the hydroxyl group at the outer surface on these polymers, heavy metals are absorbed, while organic pollutants are eliminated by the hydrophobic activity at its inner surface due to sorption. A polymeric nanoabsorbent system such as the dendrimer-UF system has been reported to significantly remove organic pollutants as well as other heavy metals from the aqueous medium (Diallo et al. 1999). An internal covering of these polymeric nanoabsorbents is hydrophobic, adsorbing organic pollutants, while the outer hydrophilic shells absorb toxic metals through their tailor-made branch. The polymer's adsorption mechanism is driven by complexion, electrostatic interactions, hydrophobic properties, and hydrogen bonding. Furthermore, composite materials based on molecularly imprinted polymers (MIPs) have emerged in the recent past as one of the attractive wastewater treatment strategies by mimicking natural detection and identification, showing high affinity and selectivity for its structural analogs (Advincula 2011). The good selectivity of the polymers is because of their synthetic methods where a model is connected to suitable monomer(s) having functional groups by covalent, semicovalent, or noncovalent bonds offering corresponding unique active sites to MIPs. The use of composites such as quantum dots (QDs) and magnetic nanoparticles allows MIPs to preserve their prominent selectivity and high adsorption capacity.

17.4.1 Zeolites

Zeolites, also known as "molecular sieves," are crystalline aluminosilicate materials with the rigid anionic framework of compact dispersed micropores (Camblor et al. 1998). Besides, a higher specific surface area, the main reason for their widespread use in catalysis, isolation, and ion exchange is physicochemical features. Moreover, zeolite's high porosity offers it an elevated adsorption potential, and photocatalytic reduction capability helps reduce high valence to lower valence metal ions, thus reducing their toxic levels (Sakaguchi et al. 2005; Song et al. 2005; Tavolaro et al. 2007). Owing to these unique features, these zeolites present excellent and low-budget options for the degradation of organics from the aqueous solution. Both natural and synthetic zeolites were extensively employed for the adsorption-based treatment of wastewaters. Zeolites of natural origin namely, clinoptilolite, chabazite, heulandite, etc., are 3D microporous structures of silicate alumina in which cations like Na, K, Ca, and Mg are loosely bound which enables them to integrate with other pollutant cations in wastewater quite rapidly (Yuna 2016). Synthetic zeolites have efficient cation exchange and sorption performance for the removal of heavy metal ions. Due to their designed anionic framework, regulated pore size, shape as well as deliberately adjusted surfaces, these synthetic zeolites are much superior to their natural counterparts in terms of specificity, thermal and structural stability, and adsorption capacity. They are typically functionalized with various moieties of organosilane (Yuna 2016). Cationic surface-functionalized zeolites have been widely used to remove toxic anions as well as other organics from wastewaters (Li and Bowman 1998). Silver and lead-filled zeolites were used to eliminate pathogens and harmful anions (Cr, Ni, Cu, Cd, and Zn) from the acidified wastewaters because of their outstanding metal ion exchange properties (Camblor et al. 1998; Kwakye-Awuah et al. 2008).

17.5 Membrane-Based Technology

Membrane-based techniques are recognized as the major advancements in modern water and wastewater treatments and desalination strategies due to their high separation efficiency, and easy operational process where there is no need for chemical addition, thermal input, or regeneration of spent media and more importantly, this does not lead to secondary pollution (Balamurugan et al. 2011; Pendergast and Hoek 2011; Buonomenna 2013). An evolving strategy in membrane technology is the incorporation of nanomaterials in the membrane framework for surface manipulation and optimization of membrane performances. In recent years, relatively efficient membranes have been introduced to segregate particulates, micropollutants, pathogens, and organic matter that impart color, taste, and odors to water (Amin et al. 2014; Ezugbe and Rathilal 2020). Membrane process has been classified based on pore sizes and filtration application as (i) microfiltration (MF) for segregating suspended solids, the removal of bacteria and protozoa; (ii) ultrafiltration (UF) for removal of viruses and colloids; (iii) NF for removal of water hardness, toxic metal ions, total dissolved solids; and (iv) RO and forward osmosis (FO) for the desalination, water reclamation and ultrapure water processes (Ulbricht 2006; Balamurugan et al. 2011; Amin et al. 2014; Ezugbe and Rathilal 2020). In the last decade, nanotechnology advances led to a variety of membranes for water purification and treatment by incorporating nanomaterials into these membranes either by surface grafting and/or through blending. The incorporation of functional nanomaterials into these membranes provides an excellent opportunity to enhance membrane permeability, resistance to fouling, mechanical and thermal stabilization, and render additional functionality for degradation of contaminants and self-cleaning (Pendergast and Hoek 2011; Vatanpour et al. 2012; Ezugbe and Rathilal 2020).

17.5.1 Nanofibrous Membranes

Nanofibrous membranes, an important class of membranes, are obtained through electro-spinning, a flexible and effective technique for the composition of ultrafine fibers using different materials that include ceramics, polymers, and also metals of between 20 and 200 nm in diameter (Li and Xia 2004; Yang et al. 2009; Li et al. 2013b; Qu et al. 2013). The large specific area, fine-tuning pore size and high water-flux of electro-spun nanofibrous membranes have attracted researchers to investigate their role in MF and UF (Gopal et al. 2006, 2007; Ramakrishna et al. 2006; Balamurugan et al. 2011; Gopakumar et al. 2017). Such electrospun nanofibers can be simply modified as per the requirement and subsequently used as affinity membranes to filter out the toxic heavy metal as well as other organic pollutants throughout the process of purification by introducing specified functional groups (Li and Xia 2004; Li et al. 2013b; Qu et al. 2013). Without major fouling (bad odour), these nanofiber membranes can filter out micro-pollutants from the aqueous medium at a relatively higher exclusion rate (Ramakrishna et al. 2006). Consequently, it was suggested that they should be employed as a pre-treatment before the RO or UF. Specialized nanomaterials are being conveniently conjugated into the spinning solutions to construct nanoparticle-impregnated nanofibers (Li and Xia 2004). Moreover, the bio-nanofiber membranes based on polyurethane, polylactic acid, and polyethylene oxide would effectively remove particles of micron sizes from the aqueous medium at a higher exclusion pace in the absence of any significant fouling. Hence, the membrane can be effectively recovered after cleansing (Ramakrishna et al. 2006; Rana and Matsuura 2010). It is also established that bio-mimetic membranes are more selective and have relatively high water flux. The vesicle-integrated polymer-supported aquaporin NF membrane is an example of a biomimetic membrane and is highly efficient in the desalination of marine waters (Tang et al. 2013).

17.5.2 Nanocomposite Membranes

Though membrane filtration technique plays a prominent role in water purification and treatment, fouling caused by various contaminants like organic and inorganic matter, colloids and microorganisms is the most highlighted impediment to the efficient utilization of membranes since it results in flux degradation during the process and restricting its application due to increased energy consumption, operational costs as well as difficulty in the operational process (Meng et al. 2009; Rahimpour 2011; Qu et al. 2013; Goh et al. 2015; Werber et al. 2016; Zhang et al. 2016a). Organic fouling (caused by the accumulation of organics in water), biological fouling (due to the high concentration of bacterial load on the membrane surfaces) and inorganic scaling by sporadical water-soluble salts are the major categories of membrane fouling identified in membrane systems (Wang et al. 2005; Ulbricht 2006; Herzberg and Elimelech 2007; Bjørkøy and Fiksdal 2009; Ciston et al. 2009; Sawada et al. 2012). Concerning the extreme implications of membrane fouling, augmenting the antifouling and antibacterial properties of the membranes is imperative. The attributes of nanocomposite membranes to be applied for water treatments are influenced by the development of customized membranes with high sensitivity and selectivity, self-cleaning ability, competitive water flux, and sustainability standards to minimize environmental impacts, flexibility, and ease of use (Le and Nunes 2016). Several efforts were made to evaluate several modification methods for improving hydrophilicity and minimizing membrane fouling, including blending, grafting, and coating with hydrophilic metallic nanocomposites that are demonstrated as an efficient approach to developing nanocomposite membranes without any complex operating procedure (Rahimpour 2011; Razmjou et al. 2011; Yu et al. 2013). The application of metallic nanoparticles is shown to have no adverse impact on the structure of these membranes, although it improves their thermal stability and overall performance (Ebert et al. 2004; Pendergast et al. 2010; Shen et al. 2011). A plethora of studies have been performed using a wide variety of nanoparticles, such as copper oxide (CuO), silver (Ag), titanium (TiO_2), zinc (ZnO), CNTs, graphene oxide (GOs), aluminum (Al_2O_3), cobalt (Co), zirconium (ZrO_2), silicon (SiO_2), iron (Fe_3O_4), clay nanoparticles, and zeolite (NaX) in the synthesis of innovative nanocomposite membranes for a variety of organic pollutants as well as the strong antibacterial ability (Shen et al. 2011; Celik et al. 2011; Pendergast and Hoek 2011; Rahimpour 2011; Razmjou et al. 2011; Balta et al. 2012; Shi et al. 2013; Ben-Sasson et al. 2014, 2016; Prince et al. 2014; Zhang et al. 2016a; Pi et al. 2016; García et al. 2018). The functionalized MWCNTs were incorporated with polymer membranes with great success as they significantly enhanced the membrane permeability, and hydrophilicity, and, therefore, presented fouling resistance (Choi et al. 2006b; Daraei et al. 2013). (Photo)catalytic nanomaterial integrated membranes incorporate their physical segregations ability and catalytic reaction to pollutant degradations. Significant efforts have been made to establish inorganic photocatalytic membranes composed of nanophotocatalysts such as nano-TiO_2 (Choi et al. 2006a). In addition, nZVI, a nanozerovalent iron, that serves as both catalyst and electron donor was also integrated into membranes for reductive dichlorination especially chlorinated organic compounds (COCs) (Wu et al. 2005; Wang et al. 2008; Wu and Ritchie 2008; Qu et al. 2013).

17.5.3 Osmatic Membranes

Both RO and FO utilize semipermeable membranes for water filtration and desalination process and their performance is characterized by salt repudiation and power usage apart from antifouling properties. Elevated consumption of energy and irreversible membrane fouling are the main limitations of these RO membranes. Moreover, the degradation of polyamide in the presence of chemical oxidants that are used to inhibit microbial growth also hampers the use of conventional RO membranes (Liu et al. 2011; Tiraferri et al. 2011; Chung et al. 2012). Accordingly, several attempts were made to exploit nanomaterials to functionalize the active layer to improve the performance of thin-film composite membranes. Modification approaches include the integration of nanomaterials within the functional TFC layer to create better polyamide nanoparticle membranes, referred to as thin-film nanocomposite (TFN) membranes having improvised fouling protection, better porosity as well as excellent salt rejection ability (Fathizadeh et al. 2011; Lee et al. 2011). Polyamide is a predominant substance for this activated layer and works on the principle of diffusion to segregate contaminants from the water (Paul 2004; Tiraferri et al. 2011). The introduction of hydrophilic nanomaterials into the thin-film composite FO membranes can contribute to better permeability, increased hydrophilicity, and low tortuosity that mitigates internal concentration polarization (ICP) in a concerted manner (Sakai et al. 2019). Primarily, metallic as well as carbon-based nanocomposites integrated with hydrophilic moieties were commonly used for this purpose (Amini et al. 2013; Wan Azelee et al. 2017; Bi et al. 2018). SiO_2/MWNTs derived from tetraethyl orthosilicate hydrolysis onto the aminated MWCNTs were integrated into polyvinylidene fluoride substrate for TFC-FO membrane fabrication (Zhang et al. 2018). The synergistic attributes of SiO_2 MWNTs blended nanostructured material increased the permeability and hydrophilicity of the membrane so that an augmented configuration with an appropriate distribution of pores was established to promote the development of defect-free polyamide coating (Zhang et al. 2018).

The reduction of water flux induced by the ICP in the FO process is influenced mainly by the supporting layer's permeability, tortuosity, and diameter. Thus, the structure of the TFC FO membrane microporous substrate is one of the important aspects to minimize ICP and achieve FO flux. An ideal FO membrane substrate should be quite small, highly porous with low tortuosity and have sufficient stability when certain applications need hydraulic pressure (Yang et al. 2018). Electrospun nanofibers have been explored as thin-film composite FO membrane substrates and selective layers through ICP with success. The relatively higher permeability and minimal thickness make it a potential candidate for the development of high porosity substrate structure (Huang et al. 2016). Compared to previously synthesized membranes, a notable increase in the water flux was reported. Owing to the three-dimensional interlinked porous configuration of the electrospun nanofibers support sheet, the diffusive resistance has been significantly minimized. Subsequently, the highly permeable morphology and low tortuosity impeded ICP. Cross-linked electrospun polyvinyl alcohol nanofiber is being recommended mostly as a prospective support layer for TFC FO applications since it meets the requirement of low porosity, significantly higher permeability and excellent hydrophilic qualities (Puguan et al. 2014).

17.5.4 Aquaporin Membranes

Aquaporins represent protein channels that regulate the water flow across the biological membranes. These channels are commonly present in mammalian tissues for the fast, passive transportation of water molecules through the cell membrane (Pendergast and Hoek 2011). Aquaporin can preserve its compact geometric configuration with a high turnover rate of water molecules (10^9/second) in a single channel, due to selective, rapid diffusion induced by osmatic gradients (Jensen and Mouritsen 2006). Biomimetic FO membrane development is now at the centerstage of current research endeavors. Influenced by the rapid transport potential of aquaporin channels, the pore-farming protein is integrated into the activated polyamide surface of FO membranes to facilitate gradient-driven water diffusion, thereby enhancing permeability. The selective layer of FO membrane with symmetric triblock copolymer was designed and created by mimicking the natural aquaporins to achieve high specificity and rapid permeability of water (Kumar et al. 2007). The resultant protein-polymer membrane showed a significant increment in water flux over the conventional polymer membrane. Aquaporin-based biomimetic FO membranes have also shown excellent tolerance for most of the chemicals including NaOCl and Alconox that are used in membrane clean-up, thus facilitating better flux recovery and at the same time maintaining salt rejection after the cleaning process (Li et al. 2017). Aquaporin-1, with specific extracellular vestibules at each side, enables the movement of water swiftly through the single-file sheet, whereas, the electrostatic tuning mechanism excludes proteins and ions (Sui et al. 2001; Agre 2004; Bowen 2006). These AQP1 membranes also spurned salts, urea, glucose and glycerol to the limits of detection and even further promoted the permeation of water (Zhu et al. 2004). Aquaporin-based membranes offer incredible potential for desalination of seawater where the operating method is an osmotic pressure gradient instead of gradient pressure imposed mechanically as in the case of RO (Pendergast and Hoek 2011). For an aquaporin distribution of around 75% on a membrane, hydraulic permeability has been predicted to increase in the order of magnitude (Kaufman et al. 2010). Given the lack of pressure applied, the energy demand is foreseen to be considerably lesser as compared to RO-based membranes. However, aquaporin-based membranes are yet to be commercially accessible because of the difficulties in collecting a higher amount of proteins and processing wide areas of membrane substrate (Pendergast and Hoek 2011).

17.6 Nanomaterials for Microbial Control and Disinfection

Water that has been infested with microbial pathogens is fatal to people's well-being. The disinfection of water is an essential step in the purification and treatment of wastewater (Zhang et al. 2016b). The widely used chemical disinfectants in the water treatment process include chlorine, chloramines, and ozone. However, these create undesirable chemical hazards because of the formation of carcinogenic – disinfection by-products (DBPs), e.g. haloacetic acids, trihalomethanes, and aldehydes during the disinfection process (Savage and Diallo 2005; Li et al. 2008; Hossain et al. 2014; Kunduru et al. 2017). Furthermore, certain pathogen tolerance to such conventional chemical disinfectants needs extremely higher dosages of disinfectants, leading to increasing incidents of waterborne diseases. Such limitations prompt an urgent need to manage the risks associated with pathogenic microbes as well as the emergence of carcinogenic DBPs. It is, therefore, critical to choose an advanced substitute approach which can successfully prevent the formation of DBPs and increase the performance of the disinfection process by using safe, chemically inert and water-soluble disinfectants (Amin et al. 2014; Samanta et al. 2016). The consequent development of nanoscience has created considerable interest in exploring and incorporating the antimicrobial effects of several nanostructured materials for water disinfection and purification processes. These nanomaterials have demonstrated an enticing direction to be considered replacements for traditional disinfectants and simultaneously coupled with several other proven techniques such as photoexcitation, to increase the effectiveness of decontamination (Li et al. 2008; Hossain et al. 2014). Accordingly, it has been proposed that a wide range of nanomaterials/nanocomposites, namely polymer films, metal and metal oxide nanoparticles (titanium, magnesium, zinc, iron, aluminum, copper, etc.), nanosilver (nAg), CNTs, chitosan, fullerene NPs (nC_{60}) and antibacterial regulate microbial infestation and neutralize microbes in wastewaters (Lyon and Alvarez 2008; Vecitis et al. 2011; Ahmed et al. 2013; Dizaj et al. 2015; Khan and Al-Khedhairy 2017; Wang et al. 2017). Using gum kondagogu, a natural biopolymer silver nanocomposites were synthesised exhibited significant antimicrobial properties (Kora et al. 2010; Malkapur et al. 2017). These nanocomposite materials have displayed superior antimicrobial potential with strong oxidation through multiple pathways, and thus do not produce harmful DBPs. The most extensively studied nanostructured antimicrobial agents include nAg, TiO_2, ZnO, and carbon-based nanomaterials (Malik et al. 2018). They are designed and developed to enhance their antimicrobial effects and synergism with carriers or other substances to augment disinfection efficiency. Nanomaterials can be employed for disinfection process by several approaches (Li et al. 2008) like:

- directly on bacterial cells to avoid the passage of electrons across the membrane,
- disruption of the cell membrane,
- cellular oxidation,
- generation of hydroxyl radicals (nanoparticles as photocatalysts), and
- metal ion release.

This section highlighted the overview of nanomaterials against antifouling caused by microbial growth. Nanosized particle consistently exhibits broad-spectrum antimicrobial activity because of its strong interaction with the cell membrane. By establishing strong bonds, these materials change the structure and permeability of the cell membrane, hence causing the rupture of the membrane (Karlsson et al. 2013). Earlier scientific reports suggested that nanomaterials such as nAg, nZnO, $nTiO_2$, nCe_2O_4, CNTs, and fullerenes, etc., inhibit the growth of microorganisms by inducing ROS ($^{\cdot}OH$, $^{\cdot-}O_2$) and H_2O_2 that results in oxidation of some cellular components of bacterial cells (Li et al. 2008; Krishnamoorthy et al. 2012; Dizaj et al. 2015). Industrial devices utilizing nAg, viz. Marathon® and Aquapure® systems are now available in the market. Another novel disinfection method incorporates the integrated use of light energy (photons) and nanostructured particles. Nanophotocatalysts are good disinfection candidates. The nanophotocatalysts utilize both UV as well as visible light through a simple catalytic method to eliminate microbes including viruses (Li et al. 2007b). The commonly used photocatalysts are TiO_2, ZnO, TiO_2 doped with nitrogen (TiON), or a transition metal like palladium. The algal blooms in the freshwater resources and the resulting water contaminants, microcystin toxins have been shown to degrade on immobilized TiO_2 catalysts (Shephard et al. 2002). The best part of employing photocatalysts in membranes is that they eradicate all micro-organisms including bacteria, viruses, protozoans, and their oocytes without producing any kind of harmful substances (DBPs) and provide environmentally friendly alternatives (Gill and McLoughlin 2007). In brief, the nanostructured materials emerged as an effective disinfectant for decontamination processes by using various antimicrobial approaches and subsequently overcoming the constraints that impede the feasibility of traditional disinfection methods (Mahendra et al. 2014).

17.7 Photocatalytic-Based Technology

Photocatalysis, a familiar innovative oxidation method, has been described as a beneficial robust, approach to the breakdown of recalcitrant organic compounds and hazardous nonbiodegradable pollutants in water into innocuous byproducts that uses light active nanocatalysts to transform pollutants in water (Wang et al. 2015; Kunduru et al. 2017). Water pollutants can also be degraded by a direct transition of the photogenerated electrons to the pollutant molecules from the catalyst surface. During the polishing operations, as in tertiary stages in sewage treatment plants, persistent compounds (micropollutants) can be photo-catalytically removed. Lately, nanomaterials have been versatile photocatalysts, attracting much of the latest scientific research interests. The nanocomposites offer diverse morphologies, high specific surface area and viable photocatalytic modeling devices (Tong et al. 2012). Metal oxide semiconductor nanomaterials like TiO_2, Ta_2O_5, ZnO, Cu_2O, CeO_2, MoS_2, O_3W, SnO, ZnS, CdS, and their composites with other substrates have gained significant attention as photocatalysts for the degradation of various contaminants including dyes from the wastewaters (Hagfeldt and Grätzel 1995; Amin et al. 2014; Kumar et al. 2016; Malik et al. 2016; Massey et al. 2016; Wu et al. 2016; Ganguly et al. 2018). Besides these metallic oxides, organic semiconductors such as graphene, carbon nitride, etc., have also gained interest as futuristic photocatalysts (Wang et al. 2009; Xiong et al. 2010). It is also important to highlight that the higher surface area to volume ratio of these nanomaterials offers many surface adsorption sites and better light absorbance, which facilitate color aggregation and improves photocatalytic degradation efficiency. It is the fundamental principle of these nanostructured photocatalysts' enhanced photocatalytic function. The organic pollutants undergo oxidation in the presence of these photo-catalysts and transform into stable and nontoxic products. (Yunus et al. 2012). Nanocomposite photocatalytic capabilities can also greatly reduce groundwater contaminants toluene, 2-dichloroethane, xylene, 1-dichloroethane and cis-1, and 3 chloromethanes, etc. (Yunus et al. 2012). These properties of nanostructures may help to mitigate dyes and thus provide resistance to biological and chemical destabilization (Demming 2011). The application of the photocatalysis technique has effectively degraded different forms of alcohols, phenolic compounds, carboxylic acids, and chlorinated aromatic pollutants (Bhatkhande et al. 2002; Kunduru et al. 2017).

Photocatalysis with TiO_2 is an efficient strategy among several semiconductors due to its high photoactivity, stability, conductivity, perfect bandgap of 3.6 eV, and ease of synthesis (Kaur and Singh 2007). TiO_2 has generated a great deal of interest in water and wastewater treatment approaches, involving photocatalytic degradation of various organic compounds, photoreduction of inorganic pollutants and inhibition of microbial pathogens due to their eco-friendly merits like reduced toxicity, high photoconductivity, chemical stability, competitiveness, and commercially viability (Chong et al. 2010; Kurniawan et al. 2012; Lazar et al. 2012). The photocatalytic mechanism of TiO_2 which relied heavily on the initiation of ROS like $^{\cdot}OH$, $^{\cdot-}O_2$, H_2O_2, etc., could actively participate in the photodegradation or disinfection process of organic pollutants (Henderson 2011). Moreover, TiO_2-nanotubes are far more efficient than TiO_2 nanomaterials owing to their improved photocatalytic activity and better deterioration levels of organic pollutants guided by small diffusion routes within the tubular wall. Additional functionalization together with unique metals, semi-conductors, anions, and semiconductors has been found to accelerate the intensity of photocatalytic activity by lowering the band-gap that permits even visible light to excite. Apart from TiO_2, O_3W and certain derivatives of fullerene also have great promise to be explored in water treatment through photocatalysis The smaller band-gap of O_3W than TiO_2 nanomaterials enables its activation by visible light at < 450 nm (Kominami et al. 2001). The platinum doping also enhances O_3W catalytic activity by allowing multielectron O_2 reduction and e^-/h^+ differentiation (Henderson 2011). The catalytic amino-fullerenes generate $^1O_2$ under the influence of visible light and are reported to have successfully degraded pharmaceutical-derived pollutants and inhibition of viruses (Lof et al. 1995; Lee et al. 2009). Poly(*N*-vinylpyrrolidone) encapsulated fullerol and C_{60} can generate $^1O_2$ and $^{\cdot-}O_2$ under the influence of UV light conditions (Al-Jumaili et al. 2017). Amino-fullerenes are quite suitable for surface modifications than fullerol and are efficacious in the process of decontamination because of their positive charge. Hence, nanomaterials with photocatalytic properties, therefore, are effective, safe, inexpensive environmentally friendly and efficient water disinfection strategies on an industrial scale.

17.7.1 Sensing- and Monitoring-Based Technology

The abundance of pollutants in contaminated wasters is critical for selection, setup as well as the functionality of processing methods and waste recycling. The dynamic proportion of water pollutants across times and places has attracted widespread interest in the advent of new waste monitoring technologies employing affordable and real-time approaches (Dewan et al. 2016). Multiple quality indicators are routinely used to analyze the various attributes of discarded or recycled waste fluid. Most of them are compared with laboratory techniques that involve regular sampling of data, pretreatment of collected samples and thus, are time-consuming and high-priced (Su et al. 2012; Chong et al. 2013; Yang et al. 2015). Such factors endorse the need to explore alternative approaches that are even more flexible, compact and effective for on-site cross-pollution detection (Dewan et al. 2016). The growing advancements in nanotechnologies have a tremendous impact on environmental surveillance and detection, wherein numerous nanomaterials have been developed to detect and remediate wide-ranging aqueous pollutants (Theron et al. 2008, 2010; Dewan et al. 2016). In recent years, considerable attention has been drawn toward the application of biosensors for the detection of pollution in wastewaters. Biosensors are also vital for the effective management of wastewater quality, particularly in the process of treatment of biological wastewaters. It is because this type of treatment depends on the biological entities for the decomposition of the organic material and the elimination of nutrient load (Xiao et al. 2015). These biosensors provide a significant advantage in detecting the minimum contaminant level in complex matrices like wastewaters. Moreover, these sensors are micro-electromechanical devices which facilitate the development of compact and lightweight sensors capable of monitoring onsite effluents (Tsopela et al. 2016). Over the last few years, a fast-expanding area of research has helped promote versatile nano-biosensors that display unique characteristics for monitoring hazardous chemical pollutants as well as a biological phenomenon (Zhang et al. 2009; Li et al. 2013a; Tsopela et al. 2016). Nanobiosensors have indicated enormous potential in chemical and biological monitoring and detecting investigations owing to their physicochemical, electromagnetic, and catalytic properties as well as their high sensitivity and selectivity (Zhang et al. 2009; Wang et al. 2010; Qu et al. 2013). Lab-on-chip and other portable monitoring systems based on electrochemical sensing tend to be very well designed to reinforce existing methodological approaches for a variety of applications for water management, in particular, water quality assessment (Jang et al. 2011; Tsopela et al. 2016). The sophisticated lab-on-chip electrochemical biosensor is a comprehensive tool that employs microfluidics to conduct analysis and manages very small quantities of fluids in microchannel deals (Jang et al. 2011). Integrating microfluidic and biosensor technology offers the fusion of biological and chemical components into a single platform and, therefore, provides new capabilities for biosensing applications. This technology in water analysis significantly decreases sampling time as well as resources as it can deliver instant onsite results. Additionally, because of the small volumes evaluated and the prospect of minimizing the sampling process, this model system is much less costly and provides higher precision, which, in turn, greatly reduces human errors. QDs are widely used nanomaterials in the development of sensing devices for the detection of various organic contaminants, cyanotoxins as well as harmful microorganisms by utilizing their fluorescence properties (Wang et al. 2010). Nanocomposite coupled sensors such as CuO/AgO, based on multiple inorganic electrocatalysts, exhibited a positive correlation with sensor responses (Gutiérrez-Capitán et al. 2015).

17.7.2 Detection of Organic Pollutants

In addition to water treatment approaches and techniques, it is essential to utilize analytical procedures such as detection methods that are capable of specifically detecting the presence of varied forms of water pollutants, while enabling constant onsite tracking. These techniques not only can be used as primary detection tools but they could also be applied involved in verifying the effectiveness of wastewater treatments, simply by analyzing the water quality following subsequent sanitation and disinfection. Nanosensors-based CNTs, graphene, metal and metal oxides, dye-doped nanomaterials, QDs, etc., have been widely used for the specific and sensitive detection of various hazardous organic pollutants in various matrices because of their high surface-to-volume ratio, tuneable characteristics, better adsorption potential and electrical conductivity (Mauter and Elimelech 2008; Cialla-May et al. 2020). The resonance-free transmitting abilities and good conductive behavior of the CNTs are attributed to the widespread application in detection sensors (Mauter and Elimelech 2008; Cialla-May et al. 2020). Recently, CNTs were coated with surface-enhanced Raman spectroscopy (SERS)-active AuNPs for the detection of azo dyes from the aqueous solution due to their large surface area (Cialla-May et al. 2020). Copper(III) oxide NPs have been used to alter MWCNTs/epoxy nanocomposite sensors to ensure sensitive detection of low levels of free chlorine in water amperometrically (Muñoz et al. 2015). In addition to the use of SERS-based strategies for the detection of pollutants like polycyclic aromatic hydrocarbons (PAHs), a fluorometric QD-graphene oxide nanocomposite sensor was designed and subsequently used for the detection of various pollutants such as As(III) in wastewaters with high sensitivity (Vaishanav et al. 2017; Cialla-May et al. 2020). The CdTe QDs have been embedded in TiO_2 nanotubes for the detection of low levels of PAHs including organic dyes in water samples utilizing the process of fluorescence resonance energy (Yang et al. 2010; Cialla-May et al. 2020). The application of AuNPs and complexes were used for the sensitive detection of low levels of multiple pollutants such as pesticide residues, As, Hg, Pb, and Cd, due to their stability, ease of surface control, increased surface area, miniaturization and compatibility to the aqueous phase (Lin and Tseng 2010; Hayat and Marty 2014; Priyadarshini and Pradhan 2017). Furthermore, the development of microfluidic paper-based electrochemical sensing devices (μPED) conjugated with bio-nanocomposites has also allowed the precise detection of various types of target analytes (Pb^{+2}, Cd^{+2}, As(III), etc.) in water samples, particularly in limited-resource environments. On-site water monitoring using microfluidics is a thriving technology as it leads to fast, sensitive affordable and user-friendly methods of detection of contaminants (Jaywant and Arif 2019).

17.7.3 Detection of Harmful Pathogens

Accurate pathogen detection is extremely important since it is closely related to public health. Moreover, pathogen detection is the main component of the diagnosis regimen-based water disinfection strategy, in which the treatment is triggered by the detection of target organisms. The detection of pathogenic microorganisms is the main feature of the water disinfection process in which the treatment is initiated by the presence and identification of target organisms. Conventional methods of microbial growth indicators are not effective, time-consuming, labor extensive, and most of the time fail to detect certain harmful bacteria, viruses, and protozoans (Gu et al. 2006). This facilitates the development of precise diagnostic procedures that can resolve these lacunae in detection methods. Thus, in the past decade, novel nanomaterial-based sensing and tracking technologies have evolved for detecting the microorganism in different environments including wastewaters. Nanomaterials-based sensors offer the ability to design and development of a sensitive method for the rapid detection of pathogens in wastewaters due to their specific physicochemical properties. The significant characteristics of these nanostructures facilitate these materials to be integrated with biomolecules such as antibodies, DNA, enzymes, carbohydrates, aptamers, and peptides to specifically detect the targeting entities (Theron et al. 2010; Vikesland and Wigginton 2010; Mustafa et al. 2017). In addition to their extensive use in the production of biosensors, many types of nanocomposites/nanostructured materials such as silver, gold, iron, copper, CNTs, nanowires, and QDs have been shown to possess significant antimicrobial potential (Vikesland and Wigginton 2010; Mustafa et al. 2017). Hence, the design and development of smart biosensors as toxicity indicators and functionalized systems combining capture, detection, and inhibition attributes have lately emerged as an interesting research area. The high surface-to-volume ratio and the nano-size attributes have been optimized to make them appealing for the design of multipurpose detecting systems in response to a target. For example, QDs hold great promise for the sensing operation due to their highly tunable optical as well as photo-electrochemical properties. QDs such as TiO_2, SnO_2, CdS, CdSe, and ZnS, etc., are widely employed in wastewater for the detection and differentiation of several bacterial pathogens (Hahn et al. 2005; Slowing et al. 2007; McCreery 2008). Due to high conductivity, simple surface alteration, and good adsorption capability, CNTs and nanowires provide enough scope for the electrochemical sensing of pathogens in the water medium. They have been widely used in the detection process as they provide strong and detectable signals from trace quantities of targeted pathogens or small molecules. Moreover, with CdTe QDs fluorescence detection was successfully achieved as fluorescent markers integrated with aptamers for specific recognition and molecular validation (Zelada-Guillén et al. 2009; García-Aljaro et al. 2010; Petryayeva and Krull 2011; Duan et al. 2015). Functionalized nanocomposites of silver and gold have been used to avert bacterial manifestations, while at the same time enabling bacteria to be detected as well as inactivated (Raj et al. 2015; Mustafa et al. 2017). Aptamer-functionalized Au-nanocomposites have been used to detect bacterial pathogens calorimetrically by measuring aggregation and deaggregation at target binding (Zhao et al. 2018). The integration of nanomaterials into the sensing platforms has ensured major improvements in the detection capacity and functionality of these electrochemical sensors to track and manage wastewaters.

17.8 Conclusions and Future Outlook

While water performs a significant role in every facet of our daily life, its shortage and deteriorating quality have raised the alarm as a present and future challenge to mankind and its existence on earth. Existing practices and methods of handling can regulate the water from organic and inorganic waste. Yet due to the apparent lack of ability to cleanse water completely and reuse the retentates, these approaches are energy-consuming and economically unviable. Accordingly, new technologies of water treatment are going global recognition. In the coming future, nanotechnology and applications thereof can substantially influence the wastewater domain. Nanotechnology aims to improve the existing tools by upgrading the performance process and enhancing the reusability of nanocomposite materials, therefore reducing the running cost of plants or processes. The nanostructured materials are associated with remarkable attributes like higher surface volume, strong sensitivity, and reactivity, possessing the ability of self-assembly on substances to create films, high adsorption, shape-dependent catalytic properties, etc., making them ideal for water treatment processes. Nanomaterials are effective against various environmental pollutants, and multiple pathogenic organisms found in polluted water because of all these important characteristics. The new developments in research endeavors including those discussed earlier have made it apparent that nanotechnology has enormous potential that seems to be proven into a very valuable wastewater management strategy of the twenty-first century. However, the path to wide-ranging nanotechnological applications is still in its infancy. Most of the wastewater treatment methods and strategies involving nanotechnology reviewed so far have been mostly investigated at the laboratories, fewer are studied at a pilot scale or even commercialized. Although the future looks extremely progressive, it needs concerted efforts from diverse fields to get the speedy, economical, and complete elimination of target pollutants from these wastewaters. Nanotechnology can occupy a vital position by offering a wide variety of unparalleled ways to develop certain water treatment systems, especially resilient treatment technologies and processes which can be customized to strategic goals. However, future studies must aim to address the following issues of compliance with current treatment strategies from the safety, reliability, feasibility, and viability perspectives:

i) Long-term efficiencies and less expensive
ii) Development of hybrid materials with enhanced stability, dispersion, and low toxicity
iii) Optimized nanomaterial designs for enhanced activities
iv) Large-scale production and reproducibility
v) Identification of potential adverse effects on humans and the environment
vi) Validation of nanomaterial-enabled sensing techniques
vii) Legal framework for assessment, management, and research risks associated with nanomaterials.

Notwithstanding these knowledge gaps, it can be anticipated that emerging innovation waves with careful management designed to avoid undesirable consequences may ultimately empower advanced water purification and treatment processes that substantially enhance the safety and resilience of water supplies to present as well as future generations.

Acknowledgments

The author, Mir Zahoor Gul, gratefully acknowledges the Council of Scientific and Industrial Research (CSIR), New Delhi, for their financial support through CSIR-Research Associateship (File No.: 09/132(0883)/2019-EMR-I).

References

Advincula, R.C. (2011). Engineering molecularly imprinted polymer (MIP) materials: developments and challenges for sensing and separation technologies. *Korean J. Chem. Eng.* 28: 1313–1321.

Agre, P. (2004). Aquaporin water channels (Nobel lecture). *Angew. Chem. Int. Ed.* 43: 4278–4290.

Ahmed, F., Santos, C.M., Mangadlao, J. et al. (2013). Antimicrobial PVK: SWNT nanocomposite coated membrane for water purification: performance and toxicity testing. *Water Res.* 47 (12): 3966–3975. https://doi.org/10.1016/j.watres.2012.10.055.

Ali, I., Asim, M., and Khan, T.A. (2012). Low cost adsorbents for the removal of organic pollutants from wastewater. *J. Environ. Manage.* 113: 170–183. https://doi.org/10.1016/j.jenvman.2012.08.028.

Al-Jumaili, A., Alancherry, S., Bazaka, K., and Jacob, M.V. (2017). Review on the antimicrobial properties of carbon nanostructures. *Materials* 10 (9): 1066. https://doi.org/10.3390/ma10091066.

Al-Omair, M.A. and El-Sharkawy, E.A. (2007). Removal of heavy metals via adsorption on activated carbon synthesized from solid wastes. *Environ. Technol.* 28 (4): 443–451. https://doi.org/10.1080/09593332808618808.

Alswata, A.A., Ahmad, M.B., Al-Hada, N.M. et al. (2017). Preparation of zeolite/zinc oxide nanocomposites for toxic metals removal from water. *Results Physics* 7: 723–731. https://doi.org/10.1016/j.rinp.2017.01.036.

Amin, M.T., Alazba, A.A., and Manzoor, U. (2014). A review of removal of pollutants from water/wastewater using different types of nanomaterials. *Adv. Mater. Sci. Eng.* 1: 125–146.

Amini, M., Jahanshahi, M., and Rahimpour, A. (2013). Synthesis of novel thin film nanocomposite (TFN) forward osmosis membranes using functionalized multi-walled carbon nanotubes. *J. Membr. Sci.* 435: 233–241. https://doi.org/10.1016/j.memsci.2013.01.041.

Anbia, M. and Salehi, S. (2012). Removal of acid dyes from aqueous media by adsorption onto amino-functionalized nanoporous silica SBA-3. *Dyes Pigm.* 94 (1): 1–9. https://doi.org/10.1016/j.dyepig.2011.10.016.

Andjelkovic, I., Tran, D.N.H., Kabiri, S. et al. (2015). Graphene aerogels decorated with α-FeOOH nanoparticles for efficient adsorption of arsenic from contaminated waters. *ACS Appl. Mater. Interfaces* 7 (18): 9758–9766. https://doi.org/10.1021/acsami.5b01624.

Balamurugan, R., Sundarrajan, S., and Ramakrishna, S. (2011). Recent trends in nanofibrous membranes and their suitability for air and water filtrations. *Membranes* 1 (3): 232–248. https://doi.org/10.3390/membranes1030232.

Balta, S., Sotto, A., Luis, P. et al. (2012). A new outlook on membrane enhancement with nanoparticles: the alternative of ZnO. *J. Membr. Sci.* 389: 155–161. https://doi.org/10.1016/j.memsci.2011.10.025.

Beltrán, F.J., Aguinaco, A., García-Araya, J.F., and Oropesa, A. (2008). Ozone and photocatalytic processes to remove the antibiotic sulfamethoxazole from water. *Water Res.* 42 (14): 3799–3808. https://doi.org/10.1016/j.watres.2008.07.019.

Ben-Sasson, M., Lu, X., Bar-Zeev, E. et al. (2014). In situ formation of silver nanoparticles on thin-film composite reverse osmosis membranes for biofouling mitigation. *Water Res.* 62: 260–270. https://doi.org/10.1016/j.watres.2014.05.049.

Ben-Sasson, M., Lu, X., Nejati, S. et al. (2016). In situ surface functionalization of reverse osmosis membranes with biocidal copper nanoparticles. *Desalination* 388: 1–8. https://doi.org/10.1016/j.desal.2016.03.005.

Bhatkhande, D.S., Pangarkar, V.G., and Beenackers, A.A.C.M. (2002). Photocatalytic degradation for environmental applications – a review. *J. Chem. Technol. Biotechnol.* 77: 102–116.

Bi, R., Zhang, Q., Zhang, R. et al. (2018). Thin film nanocomposite membranes incorporated with graphene quantum dots for high flux and antifouling property. *J. Membr. Sci.* 553: 17–24. https://doi.org/10.1016/j.memsci.2018.02.010.

Bjørkøy, A. and Fiksdal, L. (2009). Characterization of biofouling on hollow fiber membranes using confocal laser scanning microscopy and image analysis. *Desalination* 245: 474–484. https://doi.org/10.1016/j.desal.2009.02.011.

Boorman, G.A., Dellarco, V., Dunnick, J.K. et al. (1999). Drinking water disinfection byproducts: review and approach to toxicity evaluation. *Environ. Health Perspect.* 107: 207–217.

Bora, T. and Dutta, J. (2014). Applications of nanotechnology in wastewater treatment-a review. *J. Nanosci. Nanotechnol.* 14: 613–626.

Bowen, W.R. (2006). Biomimetic separations - learning from the early development of biological membranes. *Desalination* 199: 225–227. https://doi.org/10.1016/j.desal.2006.03.053.

Buonomenna, M.G. (2013). Membrane processes for a sustainable industrial growth. *RSC Adv.* 3: 5694–5740.

Burakov, A.E., Galunin, E.V., Burakova, I.V. et al. (2018). Adsorption of heavy metals on conventional and nanostructured materials for wastewater treatment purposes: a review. *Ecotoxicol. Environ. Saf.* 148: 702–712.

Camblor, M.A., Corma, A., and Valencia, S. (1998). Synthesis in fluoride media and characterisation of aluminosilicate zeolite beta. *J. Mater. Chem.* 8: 2137–2145. https://doi.org/10.1039/a804457k.

Campos, A.F.C., Aquino, R., Cotta, T.A.P.G. et al. (2011). Using speciation diagrams to improve synthesis of magnetic nanosorbents for environmental applications. *Bull. Mater. Sci.* 34: 1357–1361. https://doi.org/10.1007/s12034-011-0328-5.

Celik, E., Park, H., Choi, H., and Choi, H. (2011). Carbon nanotube blended polyethersulfone membranes for fouling control in water treatment. *Water Res.* 45: 274–282. https://doi.org/10.1016/j.watres.2010.07.060.

Choi, H., Stathatos, E., and Dionysiou, D.D. (2006a). Sol-gel preparation of mesoporous photocatalytic TiO_2 films and TiO_2/Al_2O_3 composite membranes for environmental applications. *Appl. Catal., B* 63: 60–67. https://doi.org/10.1016/j.apcatb.2005.09.012.

Choi, J.H., Jegal, J., and Kim, W.N. (2006b). Fabrication and characterization of multi-walled carbon nanotubes/polymer blend membranes. *J. Membr. Sci.* 284: 406–415. https://doi.org/10.1016/j.memsci.2006.08.013.

Chong, M.N., Jin, B., Chow, C.W.K., and Saint, C. (2010). Recent developments in photocatalytic water treatment technology: a review. *Water Res.* 44: 2997–3027.

Chong, S.S., Aziz, A.R., and Harun, S.W. (2013). Fibre optic sensors for selected wastewater characteristics. *Sensors* 13 (7): 8640–8668. https://doi.org/10.3390/s130708640.

Chung, T.S., Li, X., Ong, R.C. et al. (2012). Emerging forward osmosis (FO) technologies and challenges ahead for clean water and clean energy applications. *Curr. Opin. Chem. Eng.* 1 (3): 246–257.

Cialla-May, D., Weber, K., Popp, J. (2020) Sensitive detection of organic pollutants by advanced nanostructures. In: In Micro and Nano Technologies, Advanced Nanostructures for Environmental Health (eds. L.Baia, Z. Pap, K. Hernadi, M. Baia), 35–74. Elsevier.

Ciston, S., Lueptow, R.M., and Gray, K.A. (2009). Controlling biofilm growth using reactive ceramic ultrafiltration membranes. *J. Membr. Sci.* 342: 263–268. https://doi.org/10.1016/j.memsci.2009.06.049.

Daraei, P., Madaeni, S.S., Ghaemi, N. et al. (2013). Enhancing antifouling capability of PES membrane via mixing with various types of polymer modified multi-walled carbon nanotube. *J. Membr. Sci.* 444: 184–191. https://doi.org/10.1016/j.memsci.2013.05.020.

Demming, A. (2011). Big challenges and nanosolutions. *Nanotechnology* 22 (29): 290201. https://doi.org/10.1088/0957-4484/22/29/290201.

Dewan, N., Ahmed, P., Chowdhury, G. et al. (2016). Nanotechnology based biosensors and its application. *Pharm. Innovation J.* 5 (6): 18–25.

Diagboya, P.N., Olu-Owolabi, B.I., and Adebowale, K.O. (2015). Synthesis of covalently bonded graphene oxide-iron magnetic nanoparticles and the kinetics of mercury removal. *RSC Adv.* 5: 2536–2542. https://doi.org/10.1039/c4ra13126f.

Diallo, M.S., Balogh, L., Shafagati, A. et al. (1999). Poly(amidoamine) dendrimers: a new class of high capacity chelating agents for Cu(II) ions. *Environ. Sci. Technol.* 33: 820–824. https://doi.org/10.1021/es980521a.

Dizaj, S.M., Mennati, A., Jafari, S. et al. (2015). Antimicrobial activity of carbon-based nanoparticles. *Adv. Pharm. Bull.* 5 (1): 19–23. https://doi.org/10.5681/apb.2015.003.

Duan, N., Wu, S., Dai, S. et al. (2015). Simultaneous detection of pathogenic bacteria using an aptamer based biosensor and dual fluorescence resonance energy transfer from quantum dots to carbon nanoparticles. *Microchim. Acta* 182: 917–923. https://doi.org/10.1007/s00604-014-1406-3.

Ebert, K., Fritsch, D., Koll, J., and Tjahjawiguna, C. (2004). Influence of inorganic fillers on the compaction behaviour of porous polymer based membranes. *J. Membr. Sci.* 233: 71–78. https://doi.org/10.1016/j.memsci.2003.12.012.

El-Safty, S.A., Shahat, A., and Ismael, M. (2012). Mesoporous aluminosilica monoliths for the adsorptive removal of small organic pollutants. *J. Hazard. Mater.* 201, 202: 23–32. https://doi.org/10.1016/j.jhazmat.2011.10.088.

Ezugbe, E.O. and Rathilal, S. (2020). Membrane technologies in wastewater treatment: a review. *Membranes* 10 (5): 89. https://doi.org/10.3390/membranes10050089.

Fathizadeh, M., Aroujalian, A., and Raisi, A. (2011). Effect of added NaX nano-zeolite into polyamide as a top thin layer of membrane on water flux and salt rejection in a reverse osmosis process. *J. Membr. Sci.* 375: 88–95. https://doi.org/10.1016/j.memsci.2011.03.017.

Fu, F. and Wang, Q. (2011). Removal of heavy metal ions from wastewaters: a review. *J. Environ. Manage.* 92 (3): 407–418. https://doi.org/10.1016/j.jenvman.2010.11.011.

Ganguly, P., Byrne, C., Breen, A., and Pillai, S.C. (2018). Antimicrobial activity of photocatalysts: fundamentals, mechanisms, kinetics and recent advances. *Appl. Catal., B* 225: 51–75.

García, A., Rodríguez, B., Oztürk, D. et al. (2018). Incorporation of CuO nanoparticles into thin-film composite reverse osmosis membranes (TFC-RO) for antibiofouling properties. *Polym. Bull.* 75: 2053–2069. https://doi.org/10.1007/s00289-017-2146-4.

García-Aljaro, C., Cella, L.N., Shirale, D.J. et al. (2010). Carbon nanotubes-based chemiresistive biosensors for detection of microorganisms. *Biosens. Bioelectron.* 26 (4): 1437–1441. https://doi.org/10.1016/j.bios.2010.07.077.

Gehrke, I., Geiser, A., and Somborn-Schulz, A. (2015). Innovations in nanotechnology for water treatment. *Nanotechnol. Sci. Appl.* 8: 1–17. https://doi.org/10.2147/NSA.S43773.

Ghosh, A., Pal, M., Biswas, K. et al. (2015). Manganese oxide incorporated ferric oxide nanocomposites (MIFN): a novel adsorbent for effective removal of Cr(VI) from contaminated water. *J. Water Process Eng.* 7: 176–186. https://doi.org/10.1016/j.jwpe.2015.06.008.

Gill, L.W. and McLoughlin, O.A. (2007). Solar disinfection kinetic design parameters for continuous flow reactors. *J. Sol. Energy Eng.* 129 (1): 111–118. https://doi.org/10.1115/1.2391316.

Gleick, P. and Iceland, C. (2018). *Water, Security, and Conflict*, 16. World Resour Institute.

Goh, P.S., Ng, B.C., Lau, W.J., and Ismail, A.F. (2015). Inorganic nanomaterials in polymeric ultrafiltration membranes for water treatment. *Sep. Purif. Rev.* 44 (3): 216–249. https://doi.org/10.1080/15422119.2014.926274.

Gopakumar, D.A., Pasquini, D., Henrique, M.A. et al. (2017). Meldrum's acid modified cellulose nanofiber-based polyvinylidene fluoride microfiltration membrane for dye water treatment and nanoparticle removal. *ACS Sustainable Chem. Eng.* 5 (2): 2026–2033. https://doi.org/10.1021/acssuschemeng.6b02952.

Gopal, R., Kaur, S., Ma, Z. et al. (2006). Electrospun nanofibrous filtration membrane. *J. Membr. Sci.* 281: 581–586. https://doi.org/10.1016/j.memsci.2006.04.026.

Gopal, R., Kaur, S., Feng, C.Y. et al. (2007). Electrospun nanofibrous polysulfone membranes as pre-filters: particulate removal. *J. Membr. Sci.* 289: 210–219. https://doi.org/10.1016/j.memsci.2006.11.056.

Gu, H., Xu, K., Xu, C., and Xu, B. (2006). Biofunctional magnetic nanoparticles for protein separation and pathogen detection. *Chem. Commun.* 9: 941–949. https://doi.org/10.1039/b514130c.

Guerra, F.D., Attia, M.F., Whitehead, D.C., and Alexis, F. (2018). Nanotechnology for environmental remediation: materials and applications. *Molecules* 23: 1760.

Gutiérrez-Capitán, M., Baldi, A., Gómez, R. et al. (2015). Electrochemical nanocomposite-derived sensor for the analysis of chemical oxygen demand in urban wastewaters. *Anal. Chem.* 87 (4): 2152–2160. https://doi.org/10.1021/ac503329a.

Hagfeldt, A. and Grätzel, M. (1995). Light-induced redox reactions in nanocrystalline systems. *Chem. Rev.* 95 (1): 49–68. https://doi.org/10.1021/cr00033a003.

Hahn, M.A., Tabb, J.S., and Krauss, T.D. (2005). Detection of single bacterial pathogens with semiconductor quantum dots. *Anal. Chem.* 77: 4861–4869. https://doi.org/10.1021/ac050641i.

Hayat, A. and Marty, J.L. (2014). Disposable screen printed electrochemical sensors: tools for environmental monitoring. *Sensors* 14: 10432–10453.

Henderson, M.A. (2011). A surface science perspective on TiO_2 photocatalysis. *Surf. Sci. Rep.* 66 (6, 7): 185–297.

Herzberg, M. and Elimelech, M. (2007). Biofouling of reverse osmosis membranes: role of biofilm-enhanced osmotic pressure. *J. Membr. Sci.* 295: 11–20. https://doi.org/10.1016/j.memsci.2007.02.024.

Hossain, F., Perales-Perez, O.J., Hwang, S., and Román, F. (2014). Antimicrobial nanomaterials as water disinfectant: applications, limitations and future perspectives. *Sci. Total Environ.* 466, 467: 1047–1059.

Hu, S., Shi, Q., and Jing, C. (2015). Groundwater arsenic adsorption on granular TiO_2: integrating atomic structure, filtration, and health impact. *Environ. Sci. Technol.* 49 (16): 9707–9713. https://doi.org/10.1021/acs.est.5b01520.

Hua, M., Zhang, S., Pan, B. et al. (2012). Heavy metal removal from water/wastewater by nanosized metal oxides: a review. *J. Hazard. Mater.* 211, 212: 317–331.

Huang, L., Arena, J.T., and McCutcheon, J.R. (2016). Surface modified PVDF nanofiber supported thin film composite membranes for forward osmosis. *J. Membr. Sci.* 499: 352–360. https://doi.org/10.1016/j.memsci.2015.10.030.

Iijima, S. (1991). Helical microtubules of graphitic carbon. *Nature* 354: 56–58. https://doi.org/10.1038/354056a0.

Jang, A., Zou, Z., Lee, K.K. et al. (2011). State-of-the-art lab chip sensors for environmental water monitoring. *Meas. Sci. Technol.* 22 (03): 032001.

Jaywant, S.A. and Arif, K.M. (2019). A comprehensive review of micro fluidic water quality monitoring sensors. *Sensors* 19 (21): 4781. https://doi.org/10.3390/s19214781.

Jensen, M.Ø. and Mouritsen, O.G. (2006). Single-channel water permeabilities of *Escherichia coli* aquaporins AqpZ and GlpF. *Biophys. J.* 90 (7): 2270–2284. https://doi.org/10.1529/biophysj.105.073965.

Jung, J.H., Lee, J.H., and Shinkai, S. (2011). Functionalized magnetic nanoparticles as chemosensors and adsorbents for toxic metal ions in environmental and biological fields. *Chem. Soc. Rev.* 40: 4464–4474. https://doi.org/10.1039/c1cs15051k.

Kadirvelu, K., Thamaraiselvi, K., and Namasivayam, C. (2001). Removal of heavy metals from industrial wastewaters by adsorption onto activated carbon prepared from an agricultural solid waste. *Bioresour. Technol.* 76: 63–65. https://doi.org/10.1016/S0960-8524(00)00072-9.

Karlsson, H.L., Cronholm, P., Hedberg, Y. et al. (2013). Cell membrane damage and protein interaction induced by copper containing nanoparticles-importance of the metal release process. *Forensic Toxicol.* 313: 59–69. https://doi.org/10.1016/j.tox.2013.07.012.

Kaufman, Y., Berman, A., and Freger, V. (2010). Supported lipid bilayer membranes for water purification by reverse osmosis. *Langmuir* 26: 7388–7395. https://doi.org/10.1021/la904411b.

Kaur, S. and Singh, V. (2007). Visible light induced sonophotocatalytic degradation of reactive red dye 198 using dye sensitized TiO_2. *Ultrason. Sonochem.* 14: 531–537. https://doi.org/10.1016/j.ultsonch.2006.09.015.

Khan, S.T. and Al-Khedhairy, A.A. (2017). Metals and metal oxides: important nanomaterials with antimicrobial activity. In: *Antimicrobial Nanoarchitectonics: From Synthesis to Applications* (ed. A.M. Grumezescu), 195–222. Elsevier.

Khan, S.U., Zaidi, R., Hassan, S.Z. et al. (2016). Application of Fe-Cu binary oxide nanoparticles for the removal of hexavalent chromium from aqueous solution. *Water Sci. Technol.* 74: 165–175. https://doi.org/10.2166/wst.2016.172.

Kobya, M., Demirbas, E., Senturk, E., and Ince, M. (2005). Adsorption of heavy metal ions from aqueous solutions by activated carbon prepared from apricot stone. *Bioresour. Technol.* 96: 1518–1521. https://doi.org/10.1016/j.biortech.2004.12.005.

Kominami, H., Yabutani, K.I., Yamamoto, T. et al. (2001). Synthesis of highly active tungsten(VI) oxide photocatalysts for oxygen evolution by hydrothermal treatment of aqueous tungstic acid solutions. *J. Mater. Chem.* 11: 3222–3227. https://doi.org/10.1039/b104223h.

Kora A.J., Sashidhar, R.B., and Arunachalam, J. (2010). Gum kondagogu (Cochlospermum gossypium): A template for the green synthesis and stabilization of silver nanoparticles with antibacterial application. *Carbohydr. Polym.* 82 (3): 670–679. https://doi.org/10.1016/j.carbpol.2010.05.034.

Krishnamoorthy, K., Veerapandian, M., Zhang, L.H. et al. (2012). Antibacterial efficiency of graphene nanosheets against pathogenic bacteria via lipid peroxidation. *J. Phys. Chem. C* 116 (32): 17280–17287. https://doi.org/10.1021/jp3047054.

Kumar, M., Grzelakowski, M., Zilles, J. et al. (2007). Highly permeable polymeric membranes based on the incorporation of the functional water channel protein Aquaporin Z. *Proc. Natl. Acad. Sci. U.S.A.* 104: 20719–20724. https://doi.org/10.1073/pnas.0708762104.

Kumar, S., Ahlawat, W., Bhanjana, G. et al. (2014). Nanotechnology-based water treatment strategies. *J. Nanosci. Nanotechnol.* 14 (2): 1838–1858. https://doi.org/10.1166/jnn.2014.9050.

Kumar, S., Parlett, C.M.A., Isaacs, M.A. et al. (2016). Facile synthesis of hierarchical Cu_2O nanocubes as visible light photocatalysts. *Appl. Catal., B* 189: 226–232. https://doi.org/10.1016/j.apcatb.2016.02.038.

Kumari, M., Pittman, C.U., and Mohan, D. (2015). Heavy metals [chromium (VI) and lead (II)] removal from water using mesoporous magnetite (Fe_3O_4) nanospheres. *J. Colloid Interface Sci.* 442: 120–132. https://doi.org/10.1016/j.jcis.2014.09.012.

Kunduru, K.R., Nazarkovsky, M., Farah, S. et al. (2017). Nanotechnology for water purification: applications of nanotechnology methods in wastewater treatment. In: *Water Purification* (ed. A.M. Grumezescu), 33–74. Academic Press.

Kurniawan, T.A., Sillanpää, M.E.T., and Sillanpää, M. (2012). Nanoadsorbents for remediation of aquatic environment: local and practical solutions for global water pollution problems. *Crit. Rev. Env. Sci. Technol.* 42 (12): 1233–1295.

Kwakye-Awuah, B., Williams, C., Kenward, M.A., and Radecka, I. (2008). Antimicrobial action and efficiency of silver-loaded zeolite X. *J. Appl. Microbiol.* 104 (5): 1516–1524. https://doi.org/10.1111/j.1365-2672.2007.03673.x.

Lakatos, G. (2018). Biological wastewater treatment. In: *Wastewater and Water Contamination: Sources, Assessment and Remediation* (ed. M.D. Holm and M.C. Lund), 105–128. Nova Science publishers.

Lazar, M.A., Varghese, S., and Nair, S.S. (2012). Photocatalytic water treatment by titanium dioxide: recent updates. *Catalysts* 2: 572–601.

Le, N.L. and Nunes, S.P. (2016). Materials and membrane technologies for water and energy sustainability. *Sustainable Mater. Technol.* 7: 1–28.

Lee, J., Mackeyev, Y., Cho, M. et al. (2009). Photochemical and antimicrobial properties of novel C_{60} derivatives in aqueous systems. *Environ. Sci. Technol.* 43: 6604–6610. https://doi.org/10.1021/es901501k.

Lee, K.P., Arnot, T.C., and Mattia, D. (2011). A review of reverse osmosis membrane materials for desalination-development to date and future potential. *J. Membr. Sci.* 370: 1–22.

Lee, C.G., Lee, S., Park, J.A. et al. (2017). Removal of copper, nickel and chromium mixtures from metal plating wastewater by adsorption with modified carbon foam. *Chemosphere* 166: 203–211. https://doi.org/10.1016/j.chemosphere.2016.09.093.

Li, Z., Bowman, R.S. (1998). Sorption of perchloroethylene by surfactant-modified zeolite as controlled by surfactant loading. *Environ. Sci. Technol.* 32(15), 2278–2282. https://doi.org/10.1021/es971118r

Li, D. and Xia, Y. (2004). Electrospinning of nanofibers: reinventing the wheel? *Adv. Mater.* 16: 1151–1170.

Li, F., Du, P., Chen, W., and Zhang, S. (2007a). Preparation of silica-supported porous sorbent for heavy metal ions removal in wastewater treatment by organic-inorganic hybridization combined with sucrose and polyethylene glycol imprinting. *Anal. Chim. Acta* 585: 211–218. https://doi.org/10.1016/j.aca.2006.12.047.

Li, Q., Liang, W., and Shang, J.K. (2007b). Enhanced visible-light absorption from PdO nanoparticles in nitrogen-doped titanium oxide thin films. *Appl. Phys. Lett.* 90: https://doi.org/10.1063/1.2454344.

Li, Q., Mahendra, S., Lyon, D.Y. et al. (2008). Antimicrobial nanomaterials for water disinfection and microbial control: potential applications and implications. *Water Res.* 42: 4591–4602.

Li, M., Gou, H., Al-Ogaidi, I., and Wu, N. (2013a). Nanostructured sensors for detection of heavy metals: a review. *ACS Sustainable Chem. Eng.* 1: 713–723. https://doi.org/10.1021/sc400019a.

Li, X., Zhang, C., Zhao, R. et al. (2013b). Efficient adsorption of gold ions from aqueous systems with thioamide-group chelating nanofiber membranes. *Chem. Eng. J.* 229: 420–428. https://doi.org/10.1016/j.cej.2013.06.022.

Li, Z., Valladares Linares, R., Bucs, S. et al. (2017). Aquaporin based biomimetic membrane in forward osmosis: chemical cleaning resistance and practical operation. *Desalination* 420: 208–215. https://doi.org/10.1016/j.desal.2017.07.015.

Lin, Y.H. and Tseng, W.L. (2010). Ultrasensitive sensing of Hg2+ and CH3Hg+ based on the fluorescence quenching of lysozyme type VI-stabilized gold nanoclusters. *Anal. Chem.* 82: 9194–9200. https://doi.org/10.1021/ac101427y.

Liu, Z., Bai, H., Lee, J., and Sun, D.D. (2011). A low-energy forward osmosis process to produce drinking water. *Energy Environ. Sci.* 4: 2582–2585. https://doi.org/10.1039/c1ee01186c.

Lof, R.W., van Veenendaal, M.A., Jonkman, H.T., and Sawatzky, G.A. (1995). Band gap, excitons and Coulomb interactions of solid C_{60}. *J. Electron. Spectrosc. Relat. Phenom.* 72: 83–87. https://doi.org/10.1016/0368-2048(94)02308-5.

Lofrano, G., Carotenuto, M., Libralato, G. et al. (2016). Polymer functionalized nanocomposites for metals removal from water and wastewater: an overview. *Water Res.* 92: 22–37.

Lyon, D.Y. and Alvarez, P.J.J. (2008). Fullerene water suspension (nC_{60}) exerts antibacterial effects via ROS-independent protein oxidation. *Environ. Sci. Technol.* 42: 8127–8132. https://doi.org/10.1021/es801869m.

Mahendra, S., Li, Q., Lyon, D.Y. et al. (2014). Nanotechnology-enabled water disinfection and microbial control: merits and limitations. In: *Nanotechnology Applications for Clean Water*, 2e (ed. A. Street, R. Sustich, J. Duncan, and N. Savage), 319–327. William Andrew Publishing.

Malik, R., Rana, P.S., Chaudhary, V. et al. (2016). Nanostructured WO_3/SnO_2 and TiO_2/SnO_2 heterojunction with enhanced photocatalytic performance. *Energy Environ. Focus* 5 (2): 108–115. https://doi.org/10.1166/eef.2016.1200.

Malkapur, D., Devi, M.S., Rupula, K., and Sashidhar, R.B. (2017). Biogenic synthesis, characterisation and antifungal activity of gum kondagogu-silver nano bio composite construct: assessment of its mode of action. *IET Nanobiotechnol.* 11 (7): 866–873. https://doi.org/10.1049/iet-nbt.2017.0043.

Malik, R., Tomer, V.K., Kienle, L. et al. (2018). Ordered mesoporous Ag-ZnO@g-CN nanohybrid as highly efficient bifunctional sensing material. *Adv. Mater. Interfaces* 5: 1701357. https://doi.org/10.1002/admi.201701357.

Massey, A.T., Gusain, R., Kumari, S., and Khatri, O.P. (2016). Hierarchical microspheres of MoS_2 nanosheets: efficient and regenerative adsorbent for removal of water-soluble dyes. *Ind. Eng. Chem. Res.* 55 (26): 7124–7131. https://doi.org/10.1021/acs.iecr.6b01115.

Mauter, M.S. and Elimelech, M. (2008). Environmental applications of carbon-based nanomaterials. *Environ. Sci. Technol.* 42: 5843–5859.

McCreery, R.L. (2008). Advanced carbon electrode materials for molecular electrochemistry. *Chem. Rev.* 108: 2646–2687.

Meng, F., Chae, S.R., Drews, A. et al. (2009). Recent advances in membrane bioreactors (MBRs): membrane fouling and membrane material. *Water Res.* 43: 1489–1512.

Muñoz, J., Céspedes, F., and Baeza, M. (2015). Modified multiwalled carbon nanotube/epoxy amperometric nanocomposite sensors with CuO nanoparticles for electrocatalytic detection of free chlorine. *Microchem. J.* 122: 189–196. https://doi.org/10.1016/j.microc.2015.05.001.

Mustafa, F., Hassan, R.Y.A., and Andreescu, S. (2017). Multifunctional nanotechnology-enabled sensors for rapid capture and detection of pathogens. *Sensors* 17: 2121. https://doi.org/10.3390/s17092121.

Paul, D.R. (2004). Reformulation of the solution-diffusion theory of reverse osmosis. *J. Membr. Sci.* 241: 371–386. https://doi.org/10.1016/j.memsci.2004.05.026.

Pendergast, M.M. and Hoek, E.M.V. (2011). A review of water treatment membrane nanotechnologies. *Energy Environ. Sci.* 4: 1946–1971.

Pendergast, M.T.M., Nygaard, J.M., Ghosh, A.K., and Hoek, E.M.V. (2010). Using nanocomposite materials technology to understand and control reverse osmosis membrane compaction. *Desalination* 261: 255–263. https://doi.org/10.1016/j.desal.2010.06.008.

Petryayeva, E. and Krull, U.J. (2011). Localized surface plasmon resonance: nanostructures, bioassays and biosensing-a review. *Anal. Chim. Acta* 706: 8–24.

Pi, J.K., Yang, H.C., Wan, L.S. et al. (2016). Polypropylene microfiltration membranes modified with TiO_2 nanoparticles for surface wettability and antifouling property. *J. Membr. Sci.* 500: 8–15. https://doi.org/10.1016/j.memsci.2015.11.014.

Prince, J.A., Bhuvana, S., Boodhoo, K.V.K. et al. (2014). Synthesis and characterization of PEG-ag immobilized PES hollow fiber ultrafiltration membranes with long lasting antifouling properties. *J. Membr. Sci.* 454: 538–548. https://doi.org/10.1016/j.memsci.2013.12.050.

Priyadarshini, E. and Pradhan, N. (2017). Gold nanoparticles as efficient sensors in colorimetric detection of toxic metal ions: a review. *Sens. Actuators, B* 238: 888–902.

Puguan, J.M.C., Kim, H.S., Lee, K.J., and Kim, H. (2014). Low internal concentration polarization in forward osmosis membranes with hydrophilic crosslinked PVA nanofibers as porous support layer. *Desalination* 336: 24–31. https://doi.org/10.1016/j.desal.2013.12.031.

Qu, X., Alvarez, P.J.J., and Li, Q. (2013). Applications of nanotechnology in water and wastewater treatment. *Water Res.* 47: 3931–3946. https://doi.org/10.1016/j.watres.2012.09.058.

Rafiq, Z., Nazir, R., Durr-E-Shahwar et al. (2014). Utilization of magnesium and zinc oxide nano-adsorbents as potential materials for treatment of copper electroplating industry wastewater. *J. Environ. Chem. Eng.* 2: 642–651. https://doi.org/10.1016/j.jece.2013.11.004.

Rahimpour, A. (2011). UV photo-grafting of hydrophilic monomers onto the surface of nano-porous PES membranes for improving surface properties. *Desalination* 265: 93–101. https://doi.org/10.1016/j.desal.2010.07.037.

Raj, V., Vijayan, A.N., and Joseph, K. (2015). Cysteine capped gold nanoparticles for naked eye detection of *E. coli* bacteria in UTI patients. *Sens. Bio-Sens. Res.* 5: 33–36. https://doi.org/10.1016/j.sbsr.2015.05.004.

Ramakrishna, S., Fujihara, K., Teo, W.E. et al. (2006). Electrospun nanofibers: solving global issues. *Mater. Today* 9: 40–50. https://doi.org/10.1016/S1369-7021(06)71389-X.

Rana, D. and Matsuura, T. (2010). Surface modifications for antifouling membranes. *Chem. Rev.* 110: 2448–2471. https://doi.org/10.1021/cr800208y.

Ray, P.Z. and Shipley, H.J. (2015). Inorganic nano-adsorbents for the removal of heavy metals and arsenic: a review. *RSC Adv.* 5: 29885–29907. https://doi.org/10.1039/c5ra02714d.

Razmjou, A., Mansouri, J., Chen, V. et al. (2011). Titania nanocomposite polyethersulfone ultrafiltration membranes fabricated using a low temperature hydrothermal coating process. *J. Membr. Sci.* 380: 98–113. https://doi.org/10.1016/j.memsci.2011.06.035.

Ren, X., Chen, C., Nagatsu, M., and Wang, X. (2011). Carbon nanotubes as adsorbents in environmental pollution management: a review. *Chem. Eng. J.* 170: 395–410.

Rickerby, D.G. and Morrison, M. (2007). Nanotechnology and the environment: a European perspective. *Sci. Technol. Adv. Mater.* 8: 19–24. https://doi.org/10.1016/j.stam.2006.10.002.

Sakaguchi, K., Matsui, M., and Mizukami, F. (2005). Applications of zeolite inorganic composites in biotechnology: current state and perspectives. *Appl. Microbiol. Biotechnol.* 67: 306–311.

Sakai, M., Seshimo, M., and Matsukata, M. (2019). Hydrophilic ZSM-5 membrane for forward osmosis operation. *J. Water Process Eng.* 32: https://doi.org/10.1016/j.jwpe.2019.100864.

Samanta, H., Das, R., and Bhattachajee, C. (2016). Influence of nanoparticles for wastewater treatment – a short review. *Austin Chem. Eng.* 3: 1–6.

Sankararamakrishnan, N., Jaiswal, M., and Verma, N. (2014). Composite nanofloral clusters of carbon nanotubes and activated alumina: an efficient sorbent for heavy metal removal. *Chem. Eng. J.* 235: 1–9. https://doi.org/10.1016/j.cej.2013.08.070.

Santhosh, C., Velmurugan, V., Jacob, G. et al. (2016). Role of nanomaterials in water treatment applications: a review. *Chem. Eng. J.* 306: 1116–1137.

Savage, N. and Diallo, M.S. (2005). Nanomaterials and water purification: opportunities and challenges. *J. Nanopart. Res.* 7: 331–342.

Sawada, I., Fachrul, R., Ito, T. et al. (2012). Development of a hydrophilic polymer membrane containing silver nanoparticles with both organic antifouling and antibacterial properties. *J. Membr. Sci.* 387–388: 1–6. https://doi.org/10.1016/j.memsci.2011.06.020.

Shen, J., Ruan, H., Wu, L., Gao, C. (2011). Preparation and characterization of PES-SiO_2 organic-inorganic composite ultrafiltration membrane for raw water pretreatment. *Chem. Eng. J.* 168: 1272–1278. doi: https://doi.org/10.1016/j.cej.2011.02.039

Shephard, G.S., Stockenström, S., De Villiers, D., et al. (2002). Degradation of microcystin toxins in a falling film photocatalytic reactor with immobilized titanium dioxide catalyst. *Water Res.* 36: 140–146. doi: https://doi.org/10.1016/S0043-1354(01)00213-5

Shi, F., Ma, Y., Ma, J. et al. (2013). Preparation and characterization of PVDF/TiO_2 hybrid membranes with ionic liquid modified nano-TiO_2 particles. *J. Membr. Sci.* 427: 259–269. https://doi.org/10.1016/j.memsci.2012.10.007.

Shipley, H.J., Engates, K.E., and Grover, V.A. (2013). Removal of Pb(II), Cd(II), Cu(II), and Zn(II) by hematite nanoparticles: effect of sorbent concentration, pH, temperature, and exhaustion. *Environ. Sci. Pollut. Res.* 20: 1727–1736. https://doi.org/10.1007/s11356-012-0984-z.

Singhal, B.B.S. and Gupta, R.P. (2010). Groundwater contamination. In: *Applied Hydrogeology of Fractured Rocks* (ed. B.B.S. Singhal and R.P. Gupta), 221–236. Springer.

Slowing, I., Trewyn, B., Giri, S., and Lin, V.-Y. (2007). Mesoporous silica nanoparticles for drug delivery and biosensing applications. *Adv. Funct. Mater.* 17: 1225–1236. https://doi.org/10.1002/adfm.200601191.

Smith, V.H. (2003). Eutrophication of freshwater and coastal marine ecosystems: a global problem. *Environ. Sci. Pollut. Res.* 10: 126–139.

Song, W., Grassian, V.H., and Larsen, S.C. (2005). High yield method for nanocrystalline zeolite synthesis. *Chem. Commun.* 2951–2953. https://doi.org/10.1039/b501768h.

Srivastava, V., Sharma, Y.C., and Sillanpää, M. (2015). Green synthesis of magnesium oxide nanoflower and its application for the removal of divalent metallic species from synthetic wastewater. *Ceram. Int.* 41: 6702–6709. https://doi.org/10.1016/j.ceramint.2015.01.112.

Su, S., Wu, W., Gao, J. et al. (2012). Nanomaterials-based sensors for applications in environmental monitoring. *J. Mater. Chem.* 22: 18101–18110.

Sui, H., Han, B.G., Lee, J.K. et al. (2001). Structural basis of water-specific transport through the AQP1 water channel. *Nature* 414: 872–878. https://doi.org/10.1038/414872a.

Tamburri, E., Orlanducci, S., Terranova, M.L. et al. (2005). Modulation of electrical properties in single-walled carbon nanotube/conducting polymer composites. *Carbon* 43: 1213–1221. https://doi.org/10.1016/j.carbon.2004.12.014.

Tang, C.Y., Zhao, Y., Wang, R. et al. (2013). Desalination by biomimetic aquaporin membranes: review of status and prospects. *Desalination* 308: 34–40.

Tavolaro, A., Tavolaro, P., and Drioli, E. (2007). Zeolite inorganic supports for BSA immobilization: comparative study of several zeolite crystals and composite membranes. *Colloids Surf., B* 55: 67–76. https://doi.org/10.1016/j.colsurfb.2006.11.010.

Theron, J., Walker, J.A., and Cloete, T.E. (2008). Nanotechnology and water treatment: applications and emerging opportunities. *Crit. Rev. Microbiol.* 34: 43–69.

Theron, J., Eugene, C., and T, De Kwaadsteniet, M. (2010). Current molecular and emerging nanobiotechnology approaches for the detection of microbial pathogens. *Crit. Rev. Microbiol.* 36: 318–339.

Tiraferri, A., Vecitis, C.D., Elimelech, M. (2011). Covalent binding of single-walled carbon nanotubes to polyamide membranes for antimicrobial surface properties. *ACS Appl. Mater. Interfaces* 3: 2869–2877. doi: https://doi.org/10.1021/am200536p

Tong, H., Ouyang, S., Bi, Y., et al. (2012). Nano-photocatalytic materials: possibilities and challenges. *Adv. Mater.* 24: 229–251. doi: https://doi.org/10.1002/adma.201102752.

Tsopela, A., Laborde, A., Salvagnac, L., et al. (2016). Development of a lab-on-chip electrochemical biosensor for water quality analysis based on microalgal photosynthesis. *Biosens. Bioelectron.* 79: 568–573. doi: https://doi.org/10.1016/j.bios.2015.12.050

Ulbricht, M. (2006). Advanced functional polymer membranes. *Polymer* 47: 2217–2262.

Ullah Khan, S., Farooqi, I.H., and Ayub, S. (2017). Studies on application of Fe based binary oxide nanoparticles for treatment of lead (Pb^{2+}) contaminated water – a batch study. *Mater. Today Proc.* 4 (9): 9650–9655.

UNESCO (2019). UN World Water Development Report 2019: Leaving No One Behind. UNESCO, Paris, 188 p. https://unesdoc.unesco.org/ark:/48223/pf0000367306.

Vaishanav, S.K., Korram, J., Pradhan, P., et al. (2017). Green luminescent CdTe quantum dot based fluorescence nano-sensor for sensitive detection of arsenic(III). *J. Fluoresc.*, 27(3), 781–789. https://doi.org/10.1007/s10895-016-2011-0

Vaseashta, A., Vaclavikova, M., Vaseashta, S. et al. (2007). Nanostructures in environmental pollution detection, monitoring, and remediation. *Sci. Technol. Adv. Mater.* 8: 47–59. https://doi.org/10.1016/j.stam.2006.11.003.

Vatanpour, V., Madaeni, S.S., Khataee, A.R. et al. (2012). TiO_2 embedded mixed matrix PES nanocomposite membranes: influence of different sizes and types of nanoparticles on antifouling and performance. *Desalination* 292: 19–29. https://doi.org/10.1016/j.desal.2012.02.006.

Vecitis, C.D., Schnoor, M.H., Rahaman, M.S. et al. (2011). Electrochemical multiwalled carbon nanotube filter for viral and bacterial removal and inactivation. *Environ. Sci. Technol.* 45: 3672–3679. https://doi.org/10.1021/es2000062.

Verma, M., Tyagi, I., Chandra, R., and Gupta, V.K. (2017). Adsorptive removal of Pb(II) ions from aqueous solution using CuO nanoparticles synthesized by sputtering method. *J. Mol. Liq.* 225: 936–944. https://doi.org/10.1016/j.molliq.2016.04.045.

Vikesland, P.J. and Wigginton, K.R. (2010). Nanomaterial enabled biosensors for pathogen monitoring – a review. *Environ. Sci. Technol.* 44: 3656–3669.

Vörösmarty, C.J., Green, P., Salisbury, J., and Lammers, R.B. (2000). Global water resources: Vulnerability from climate change and population growth. *Science 289* (5477): 284–288. https://doi.org/10.1126/science.289.5477.284.

Vörösmarty, C.J., McIntyre, P.B., Gessner, M.O. et al. (2010). Global threats to human water security and river biodiversity. *Nature* 467: 555–561. https://doi.org/10.1038/nature09440.

Walker, D.B., Baumgartner, D.J., Gerba, C.P., and Fitzsimmons, K. (2019). Surface water pollution. *Environ. Pollut. Sci.* 261–292.

Wan Azelee, I., Goh, P.S., Lau, W.J. et al. (2017). Enhanced desalination of polyamide thin film nanocomposite incorporated with acid treated multiwalled carbon nanotube-titania nanotube hybrid. *Desalination* 409: 163–170. https://doi.org/10.1016/j.desal.2017.01.029.

Wang, C.B., Zhang, W.X. (1997). Synthesizing nanoscale iron particles for rapid and complete dechlorination of TCE and PCBs. *Environ. Sci. Technol.* 31(7): 2154–2156. https://doi.org/10.1021/es970039c

Wang, S., Guillen, G., and Hoek, E.M.V. (2005). Direct observation of microbial adhesion to membranes. *Environ. Sci. Technol.* 39: 6461–6469. https://doi.org/10.1021/es050188s.

Wang, X., Chen, C., Liu, H., and Ma, J. (2008). Preparation and characterization of PAA/PVDF membrane-immobilized Pd/Fe nanoparticles for dechlorination of trichloroacetic acid. *Water Res.* 42: 4656–4664. https://doi.org/10.1016/j.watres.2008.08.005.

Wang, X., Maeda, K., Thomas, A. et al. (2009). A metal-free polymeric photocatalyst for hydrogen production from water under visible light. *Nat. Mater.* 8 (1): 76–80. https://doi.org/10.1038/nmat2317.

Wang, L., Ma, W., Xu, L. et al. (2010). Nanoparticle-based environmental sensors. *Mater. Sci. Eng. R Rep.* 70 (3–6): 265–274.

Wang, W., Huang, G., Yu, J.C., and Wong, P.K. (2015). Advances in photocatalytic disinfection of bacteria: development of photocatalysts and mechanisms. *J. Environ. Sci.* 34: 232–247.

Wang, L., Hu, C., and Shao, L. (2017). The antimicrobial activity of nanoparticles: present situation and prospects for the future. *Int. J. Nanomed.* 12: 1227–1249.

Werber, J.R., Osuji, C.O., and Elimelech, M. (2016). Materials for next-generation desalination and water purification membranes. *Nat. Rev. Mater.* 1: 16018.

Westerhoff, P., Yoon, Y., Snyder, S., and Wert, E. (2005). Fate of endocrine-disruptor, pharmaceutical, and personal care product chemicals during simulated drinking water treatment processes. *Environ. Sci. Technol.* 39: 6649–6663. https://doi.org/10.1021/es0484799.

World Health Organization (2016). WHO | Drinking-water Fact Sheet.

Wu, L. and Ritchie, S.M.C. (2008). Enhanced dechlorination of trichloroethylene by membrane-supported Pd-coated iron nanoparticles. *Environ. Prog.* 27: 218–224. https://doi.org/10.1002/ep.10277.

Wu, L., Shamsuzzoha, M., and Ritchie, S. (2005). Preparation of cellulose acetate supported zero-valent iron nanoparticles for the dechlorination of trichloroethylene in water. *J. Nanopart. Res.* 7: 469–476. https://doi.org/10.1007/s11051-005-4271-5.

Wu, X., Cai, J., Li, S. et al. (2016). Au@Cu_2O stellated polytope with core-shelled nanostructure for high-performance adsorption and visible-light-driven photodegradation of cationic and anionic dyes. *J. Colloid Interface Sci.* 469: 138–146. https://doi.org/10.1016/j.jcis.2016.01.064.

Xiao, Y., De Araujo, C., Sze, C.C., and Stuckey, D.C. (2015). Toxicity measurement in biological wastewater treatment processes: a review. *J. Hazard. Mater.* 286: 15–29.

Xiong, Z., Zhang, L.L., Ma, J., and Zhao, X.S. (2010). Photocatalytic degradation of dyes over graphene-gold nanocomposites under visible light irradiation. *Chem. Commun.* 46: 6099–6101. https://doi.org/10.1039/c0cc01259a.

Xu, P., Zeng, G.M., Huang, D.L. et al. (2012). Use of iron oxide nanomaterials in wastewater treatment: a review. *Sci. Total Environ.* 424: 1–10.

Yang, H.L., Lin, J.C., and Huang, C. (2009). Application of nanosilver surface modification to RO membrane and spacer for mitigating biofouling in seawater desalination. *Water Res.* 43 (15): 3777–3786. https://doi.org/10.1016/j.watres.2009.06.002.

Yang, L., Chen, B., Luo, S. et al. (2010). Sensitive detection of polycyclic aromatic hydrocarbons using CdTe quantum dot-modified TiO_2 nanotube array through fluorescence resonance energy transfer. *Environ. Sci. Technol.* 44: 7884–7889. https://doi.org/10.1021/es101760c.

Yang, Z., Li, Z., Xu, M. et al. (2013). Controllable synthesis of fluorescent carbon dots and their detection application as nanoprobes. *Nano-Micro Lett.* 5: 247–259.

Yang, L., Han, D.H., Lee, B.M., and Hur, J. (2015). Characterizing treated wastewaters of different industries using clustered fluorescence EEM-PARAFAC and FT-IR spectroscopy: implications for downstream impact and source identification. *Chemosphere* 127: 222–228. https://doi.org/10.1016/j.chemosphere.2015.02.028.

Yang, Y., Gao, X., Li, Z. et al. (2018). Porous membranes in pressure-assisted forward osmosis: flux behavior and potential applications. *J. Ind. Eng. Chem.* 60: 160–168. https://doi.org/10.1016/j.jiec.2017.10.054.

Yantasee, W., Rutledge, R.D., Chouyyok, W. et al. (2010). Functionalized nanoporous silica for the removal of heavy metals from biological systems: adsorption and application. *ACS Appl. Mater. Interfaces* 2: 2749–2758. https://doi.org/10.1021/am100616b.

Yin, J. and Deng, B. (2015). Polymer-matrix nanocomposite membranes for water treatment. *J. Membr. Sci.* 479: 256–275.

Yu, H., Zhang, X., Zhang, Y. et al. (2013). Development of a hydrophilic PES ultrafiltration membrane containing SiO_2@N-Halamine nanoparticles with both organic antifouling and antibacterial properties. *Desalination* 326: 69–76. https://doi.org/10.1016/j.desal.2013.07.018.

Yuna, Z. (2016). Review of the natural, modified, and synthetic zeolites for heavy metals removal from wastewater. *Environ. Eng. Sci.* 33: 443–454.

Yunus, I.S., Harwin, Kurniawan, A. et al. (2012). Nanotechnologies in water and air pollution treatment. *Environ. Technol. Rev.* 1 (1): 136–148. https://doi.org/10.1080/21622515.2012.733966.

Zelada-Guillén, G., Riu, J., Düzgün, A., and Rius, F. (2009). Immediate detection of living bacteria at ultralow concentrations using a carbon nanotube based potentiometrie aptasensor. *Angew. Chem. Int. Ed.* 48: 7334–7337. https://doi.org/10.1002/anie.200902090.

Zhang, L. and Fang, M. (2010). Nanomaterials in pollution trace detection and environmental improvement. *Nano Today* 5: 128–142.

Zhang, X., Guo, Q., and Cui, D. (2009). Recent advances in nanotechnology applied to biosensors. *Sensors* 9 (2): 1033–1053. https://doi.org/10.3390/s90201033.

Zhang, R., Liu, Y., He, M. et al. (2016a). Antifouling membranes for sustainable water purification: strategies and mechanisms. *Chem. Soc. Rev.* 45: 5888–5924.

Zhang, Y., Wu, B., Xu, H. et al. (2016b). Nanomaterials-enabled water and wastewater treatment. *NanoImpact* 3, 4: 22–39.

Zhang, X., Shen, L., Guan, C.Y. et al. (2018). Construction of SiO_2@MWNTs incorporated PVDF substrate for reducing internal concentration polarization in forward osmosis. *J. Membr. Sci.* 564: 328–341. https://doi.org/10.1016/j.memsci.2018.07.043.

Zhao, Y.X., Ding, M.Y., and Chen, D.P. (2005). Adsorption properties of mesoporous silicas for organic pollutants in water. *Anal. Chim. Acta* 542 (2): 193–198. https://doi.org/10.1016/j.aca.2005.04.005.

Zhao, Y.W., Wang, H.X., Jia, G.C., and Li, Z. (2018). Application of aptamer-based biosensor for rapid detection of pathogenic *Escherichia coli*. *Sensors* 18 (8): 2518. https://doi.org/10.3390/s18082518.

Zhou, J., Zhou, X., Yang, K., et al. (2020). Adsorption behavior and mechanism of arsenic on mesoporous silica modified by iron-manganese binary oxide (FeMnOx/SBA-15) from aqueous systems. *J. Hazard. Mater.* 384:. https://doi.org/10.1016/j.jhazmat.2019.121229

Zhu, F., Tajkhorshid, E., and Schulten, K. (2004). Theory and simulation of water permeation in Aquaporin-1. *Biophys. J.* 86 (1 Pt 1): 50–57. https://doi.org/10.1016/S0006-3495(04)74082-5.

18

Reliability on Nanoscience: A Valuable Cleaning Tool for Wastewaters

Fernanda M. P. Tonelli[1], Helon G. Cordeiro[2], Danilo R. C. Ferreira[3], and Flávia C. P. Tonelli[4]

[1] *Department of Biochemistry, CCO/UFSJ, Divinópolis, Brazil*
[2] *Department of Biochemistry and Tissue Biology, IB/UNICAMP, Campinas, Brazil*
[3] *Department of Materials, CDTN/CNEN, Belo Horizonte, Brazil*
[4] *Department of Pharmacy, UFSJ/CCO, Divinópolis, Brazil*

18.1 Introduction

The rapid urbanization and industrialization that the world has been experiencing severely impact in a negative way the ecosystems once sustainability is not a major concern for human beings (Chaoua et al. 2019; Zheng et al. 2020). Environmental pollution, as a consequence of anthropic action, can cause serious damage specially to living forms that may accumulate contaminants (such as heavy metals, pesticides, fertilizers, oil, explosives, and dyes) (Guerra et al. 2018; Zhang et al. 2019) that can induce from acute intoxication to complex diseases like cancer or even death (Gavrilescu et al. 2015; Fatima et al. 2018; Singh et al. 2018; Bagazgoïtia et al. 2018; Fulekar and Pathak 2020). Some of these pollutants can persist on polluted sites for a long period of time and accumulate through the food chain increasing their harmful potential (Varjani et al. 2019). Furthermore, domestic wastewater has a huge concentration of organic material (like feces, urine, and the rest of food) and it can cause a deregulation in the ecosystem when the sewer does not have a treatment of wastewater (Boutin and Eme 2016). This occurs because of the favoring of decomposers species in relation to the others, which results in an unregulated population growth that can drastically decrease the natural resources of this ecosystem (Moretti et al. 2015). Moreover, many exotic microorganisms coming from human intestinal microbiota can be inserted in this ecosystem (Newton et al. 2015), and they can be pathogenic for resident fauna and flora or compete for the natural resources (Moretti et al. 2015). Another problem that can be observed in mostly sewage of great urban centers is the high concentration of human estrogenic hormones and the presence of drugs, both licit and illicit, that are wrongly discarded or naturally eliminated by the human organism (Manickum et al. 2011; Gimiliani et al. 2016; Deng et al. 2020).

Water pollution deserves special attention once water is a vital resource with a limited self-recovery capacity and a critically increasing challenge of scarcity due to purification strategies not sustainable and a continuous enhancement in demand and contamination that result from human activities (Schwarzenbach et al. 2006; Ying et al. 2017).

18.2 Wastewater's Pollution

When it comes to water pollution, wastewaters are a relevant source of pollution. Pollutants not eliminated by common treatments (for example, the waste sludge produced by chemically enhanced primary sedimentation [CEPS] commonly possesses phosphorus and organics) or thrown in water bodies without any treatment can affect not only the chemical composition but also the biological equilibrium initially of the environ they are thrown in Kazlauskienė et al. (2012) and Zhou et al. (2020). However, the pollution may not be restricted to the spot in which wastewater is thrown. It is commonly discharged into water (for example, river) and can end up contaminating a large region of superficial water and also the groundwater (Kim et al. 2017; Sposito et al. 2018; Ziadi et al. 2019).

Aquatic Contamination: Tolerance and Bioremediation, First Edition. Edited by Rouf Ahmad Bhat, Gowhar Hamid Dar, Fernanda Maria Policarpo Tonelli, and Saima Hamid.

Industrialization brought to attention the pollution present in industrial wastewater, mainly persistent organic contaminants and heavy metals (Tayeb et al. 2015). The urbanization that accompanied the process required water treatments' development. However, not only this was able to avoid biodiversity reduction and other negative impacts of pollutants (Melián 2020).

Wastewater is generally contaminated by mixed pollutants and as the years pass and urbanization/industrialization increases, wastewater composition changes; wastewater composition reflects the population's lifestyle (Gray 2004). Recently emerging trace contaminants such as retinoids and/or oestrogenic endocrine-disrupting chemicals (such as atrazine, bisphenol A, dichlorodiphenyltrichloroethane, dioxin, endosulfan, polybrominated biphenyls, polychlorinated biphenyls, phthalate esters, and zeranol (Roy et al. 2009)) are becoming a more serious threat (Zhou et al. 2019) with deleterious effects already studied in aquatic living forms (Yeung et al. 2020)). Pharmaceuticals are also concerning emerging contaminants that can be present in wastewater and even in treated wastewater (Klatte et al. 2017; Yang et al. 2017). These xenobiotics are more present as water pollutants due to the worldwide increase of medicines available for human and veterinary use; by being bioactive compounds that can act in small amounts they can be harmful to not only aquatic species but also to species that will consume the contaminated water (Bebianno and Gonzalez-Rey 2015; Sangion and Gramatica 2016; Mezzelani et al. 2018). Synthetic dyes are also concerning xenobiotics that are generally present in industrial's wastewaters being the textile dyes (for example, azo dyes), the major pollutants among them (El-Sikaily et al. 2012). The carcinogenic and/or mutagenic potential of these substances can be enhanced by water treatment that can convert them into aromatic amines even more harmful (Bhaskar et al. 2003; Pinheiro et al. 2004; Khan and Malik 2018).

Thus, remediation of wastewater and change in human behavior in order to avoid new pollution cycles are urgent actions. It is urgent to take good care of this vital natural resource that is water, and wastewater treatment plants (WWTPs) cannot be the only solution being proposed to do so. WWTPs require the input of energy and chemicals (for example, polyaluminum chloride or ferric chloride is applied in CEPS) and are a concerning source of air pollution (Hao et al. 2019). It is relevant to dedicate attention to remediation and change in behavior to reduce human dependence on WWTPs.

18.3 Nanotechnology and Nanomaterials

Nanomaterials are products of nanoscience field development since the late 1950s. These are very versatile materials that possess sizes ranging from 1 to 100 nm at least in one dimension, which results in a high surface-to-volume ratio and the process of quantum confinement (Werkneh et al. 2018). Both of them ensure the nanomaterial's new impressive proprieties in relation to the bulk materials (Bethi and Sonawane 2018). The nanomaterials can have a chemical constitution based on carbon atoms (such as carbon nanotubes, fullerene, carbon quantum dots, graphene, and its oxide) or not (such as gold nanorods, magnetic nanoparticles, synthetic or organic polymers) (Drexler 2004; Weiss et al. 2006; Zhang et al. 2008; Wicki et al. 2015; Werkneh and Rene 2019; Trivedi et al. 2020).

The nanomaterials have contributed to improvements in several fields such as vaccine development, drug delivery, imaging and/or diagnosis, tissue engineering, energy storage, and the development of some research methodologies (Tonelli et al. 2015, 2016, 2020).

The large array of functions that nanomaterials can perform can be amplified by their surface chemical modification known as functionalization. Specific protocols can be developed to transform these materials into tools to perform specific functions with high efficiency like being precise sensors to detect pollution in aqueous samples (Priyadarshni et al. 2018; Das et al. 2019; Rao et al. 2019; Xie et al. 2019) or promoting water treatment removing undesirable pollutants – that is the focus of this chapter (Santhosh et al. 2016; Khan and Malik 2019).

18.3.1 Carbon-Based Nanomaterials

A carbon atom is considered the most versatile element possessing a number of allotropes (naturally occurring ones such as diamond, amorphous carbon and graphite, and synthetic ones like carbon-based nanomaterials). This tetravalent atom has the capacity to hybridize in sp, sp^2, and sp^3 configurations that allow the capacity of forming bonds with other carbon atoms or with many other elements (Silva 2008; Rauti et al. 2019).

Carbon quantum dots or simply carbon dots are small nanomaterials (considered zero-dimensional) that possess strong fluorescence characteristics, superconductivity, and enhanced catalytic activity (Semeniuk et al. 2019). They also present good dispersibility and high crystallization and are easy to be functionalized by a large array of groups. Although their main application is in the chemical sensor/bioimaging, field, they can also be designed to deal with pollutant present in wastewater (Wang et al. 2019a).

The allotrope family of fullerene is usually represented by its most famous member, the Fullerene C_{60}. The C_{60} is an sp^2-hybridized equivalent carbon atoms-based hollow spherical nanomaterial (containing 12 five-membered and 20 six-membered rings) maintained by bonds located on the borders of two hexagons (considered double bonds – 1.38 Å) and bonds located on the borders of hexagons and pentagons (1.45 Å) (Hirsch 1995); its discovery was recognized in 1996 by the Nobel Prize in Chemistry (Biglova and Mustafina 2019). It can be functionalized to generate systems to remediate wastewater.

A carbon nanotube is a tube-like nanomaterial generated by carbon atoms in hexagonal structure originated from graphite sheet rolled up resulting in various interesting characteristics like high elasticity and thermal conductivity (Dresselhaus et al. 2000). If only one carbon layer is present, the structure is known as single-walled carbon nanotube if two layers are present: double-walled carbon nanotubes, and if more layers are present: multiwalled carbon nanotubes. A carbon nanotube can be generated by three main methods: electric arc, laser deposition, or chemical vapor deposition method (Anzar et al. 2020). This carbon nanomaterial besides being very useful in electrical, electronics, sensors, and thermal devices can be used to remediate wastewater (Wang et al. 2019b). It can be used, for example, to promote the adsorption of pollutants and after being functionalized. Its capacity can also be optimized by the attachment of other materials like activated carbon (Sousa-Moura et al. 2019).

Graphene is a 2D carbon nanomaterial that possesses a high thermal conductivity and excellent optical and mechanical properties. It is built by sp^2 carbon atoms in hexagonal rings in a unique electronic cloud once they are tightly bonded to their neighbors (Novoselov et al. 2004). Its oxide can be generated by the oxidation of expanded graphite flakes and posterior exfoliation (Aliyev et al. 2019). Graphene oxide (GO) has the advantageous proprieties of being easily dispersed in water without using surfactants. Furthermore, GO has been shown to be more biocompatible and less toxic in relation to graphene (Pinto et al. 2013; Kiew et al. 2016). Graphene and its oxide can be functionalized to perform, efficiently, desirable tasks (Tiwari et al. 2020) like cleaning polluted samples (Wang et al. 2019b).

Activated carbon is the name of a group of carbonized materials that have been used to remove macropollutants from water since antiquity. Although, activated carbon is not a nanomaterial itself, nowadays is possible to design their pores to have a nanometric dimension, ensuring great new proprieties (Tanaka et al. 2010). The nanometric pores (nanopores) afford a large superficial area to the activated carbon, allowing a higher capacity of adsorption (Kaneko et al. 1999; Guo et al. 2012). Furthermore, the nanometric diameter of nanopores can guarantee retention of small molecules (like alcohol, hormones, drugs, dyes, and molecular hydrogen), pathogens and even metallic complex and ions (Efremenko and Sheintuch 2006; Cabria et al. 2007; Tolmachev et al. 2009; Li et al. 2011; Contescu et al. 2012; Plerdsranoy and Utke 2015).

18.3.2 Non-Carbon-Based Nanomaterials

Metal-based nanomaterials have been extensively studied to remediate a variety of heavy metals' pollution such as arsenic, lead, mercury, copper, cadmium, chromium, and nickel. Furthermore, they have shown the potential to outcompete activated carbon (Sharma et al. 2009). Noteworthy, many metal oxide nanomaterials such as nanosized magnetite and TiO_2 showed arsenic adsorption performance superior to activated carbon (Deliyanni et al. 2003; Mayo et al. 2007). Interestingly, metal (hydr)oxide nanoparticles also can be used on the skeleton of activated carbon or other porous materials to remove concomitantly arsenic and organic co-contaminants (Hristovski et al. 2009a, 2009b).

Nanoscale zero-valent iron (nZVI) is a nanomaterial that possesses high reactivity and can be used to remediate, at low cost and with low toxicity, organic pollutants, and heavy metals from water especially in coated versions or bimetallic versions (O'Carroll et al. 2013; Jiang et al. 2018). Zero-valent iron (Fe^0) nanoparticles (NPs) have shown excellent ability to remediate hexavalent chromium (Gheju 2011; Guan et al. 2015). Interestingly, these NPs with 4% bentonite were more efficient for the treatment of Cr^{6+}-containing wastewaters (Wang et al. 2018).

Iron oxide nanoparticles present magnetic behavior and can exist as ferrimagnetic maghemite (γ-Fe_2O_3) or as ferrimagnetic magnetite ($Fe_3O_4 \equiv FeO{\cdot}Fe_2O_3$) being both biodegradable and biocompatible and presenting interesting electronic properties. They can be designed for specific functions by functionalization (Lesiak et al. 2019). The development of polyvinylpyrrolidone (PVP)-coated magnetite nanoparticles made it possible to remove oil from oil–water mixture (Mirshahghassemi and Lead 2015).

Quantum dots are not only generated from carbon. They can be generated from metals that can also perform water remediation like silver quantum dots that can promote photocatalytic reactions to degrade organic contaminants (Kumar et al. 2019).

18.4 Nanoscience and Wastewater Remediation

The development of new technologies in the nanotechnology field has offered new strategies to efficiently deal with water pollution and they also offer advantages when compared to classic physicochemical technologies that were basically the only available options, and even when compared to new biotechnological remediation strategies. The main disadvantage related to standard physicochemical technologies (such as plasma photocatalysis, oxidation, and ozonation (Baldissarelli et al. 2019)) and that can also make unfeasible even some new biotechnologies (such as remediation assisted by enzymes) is the high cost to perform remediation in large scale (Ye et al. 2017). However, nanomaterials can help to make these strategies more applicable by overcoming this problem. For example, inexpensive and corrosion resistant nanoparticles such as nano-TiO_2 can offer a large photocatalytic activity to degrade pollutants like methylene blue (Jain and Vaya 2017); modified Fe_3O_4 nanoparticles could immobilize the enzyme Protocatechuate 3,4-dioxygenase from *Rhizobium* sp. LMB-1. The nanocomplex increased its stability and allowed reusability to remediate samples contaminated with aromatic compounds, reducing the process costs (Zhang et al. 2017a).

This topic will be dedicated to discussing nanomembranes, nanosorbents, nanophotocatalysts, and nanomotors (Figure 18.1) regarding their potential to be used to remediated wastewater (Table 18.1).

The main concern regarding nanomaterials use to treat water is related to how risky they can be to the quality of drinking water at the end of the process. Although in the current scenario, the risks are considered low, it is necessary to carry out studies specifically for each new proposed strategy, considering the possibility of its scale increase and the biocompatibility of the nanomaterial to be used (Pulizzi and Sun 2018).

18.4.1 Nanomembranes

The use of nanomembranes (the first mechanism present in Figure 18.1) to remediate wastewater makes it possible to separate the solid pollutants by nanofiltration, ultrafiltration, or reverse osmosis (Singh et al. 2020). The presence of nanopores among the nanofibers in the nanomembranes can retain dispersed contaminants with small size, like biomolecules and metallic ions. Furthermore, with a specific functionalization it is possible to the nanomembrane to offer selective

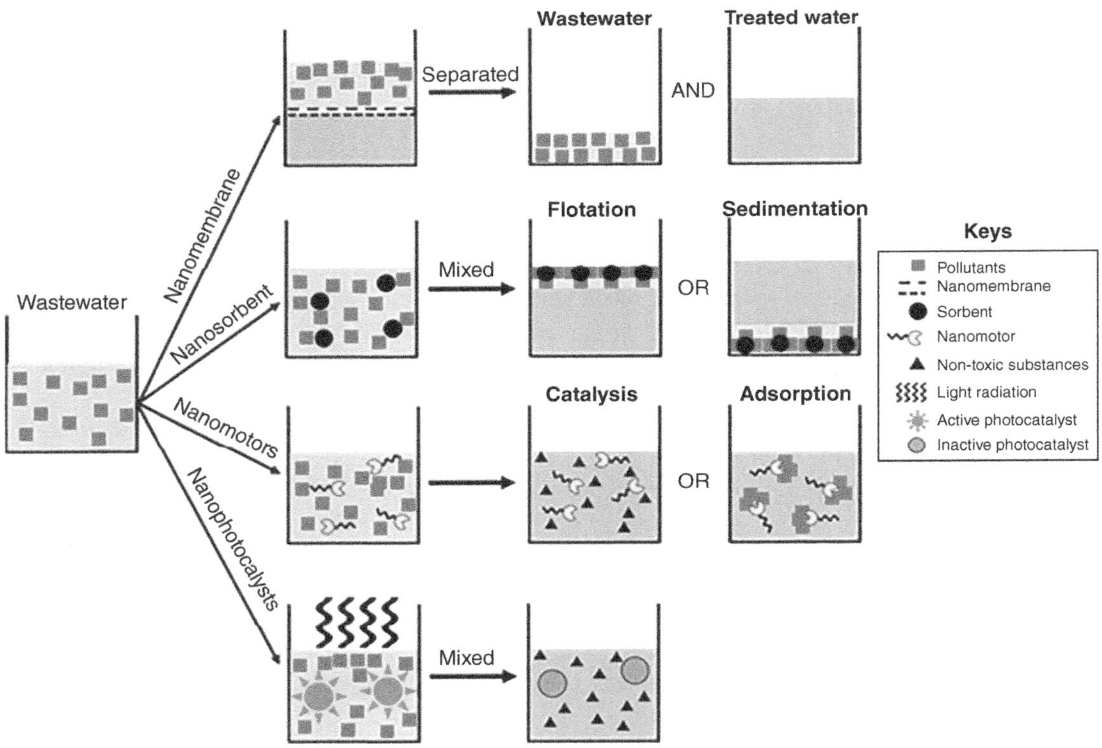

Figure 18.1 Mechanisms of treating wastewater to promote remediation using nanomaterials.

Table 18.1 Some examples of nanomaterials that can be used to remediate aqueous samples.

Nanomaterial	Strategy classification	Pollutant remediated	Reference
Fullerene functionalized with zinc porphyrin and TiO_2	Nanophotocatalyst	Phenol and methylene blue	Regulska et al. (2019)
Green-emissive carbon quantum dots	Nanophotocatalyst	Methylene blue	Das et al. (2019)
Carbon nanotubes functionalized with TiO_2/SiO_2	Nanophotocatalyst	Heavy metals and organic compounds	Rasheed et al. (2019)
Ag-graphene-Bi_2WO_6 nanosheets	Nanophotocatalyst	Rhodamine B	Low et al. (2014)
Silver quantum dots with 2D SnO_2 nanoflakes	Nanophotocatalyst	Rhodamine B	Kumar et al. (2019)
Prussian blue/TiO_2 nanocomposites	Nanophotocatalyst	Organic pollutants	Li et al. (2015b)
ZnO-graphene composites	Nanophotocatalyst	Rhodamine B	Li and Cao (2011)
Nanomembranes containing TiO_2	Nanomembrane/ Nanophotocatalyst	Organic pollutants	Leong et al. (2014)
γ-Al_2O_3 spheres with Prussian blue nanoparticles	Nanophotocatalyst/ Nanosorbent	Orange G dye	Doumic et al. (2015)
Carbon nanotube combined with activated carbon	Nanosorbent	Fluoxetine hydrochloride	Sousa-Moura et al. (2019)
Sodium functionalized graphene oxide fibers	Nanosorbent	Cs^+	Lee et al. (2019)
Date syrup-based graphene sand hybrid	Nanosorbent	Cationic and anionic dyes and also Pb^{2+} and Cd^{2+}	Khan et al. (2019)
nZVI (Nanofer25)	Nanosorbent	Metals	Oprčkal et al. (2017)
Magnetic Fe_3O_4 nanoparticles functionalized with humic acid attached to the oyster shell surface	Nanosorbent	Hg^{2+}	He et al. (2019)
Magnetic Fe_3O_4 nanoparticles	Nanosorbent	As^{3+} and As^{5+}	Lunge et al. (2014)
Microspheres of silica and Fe_3O_4	Nanosorbent	Hg^{2+} and Pb^{2+}	Hu et al. (2010)
Magnetite nanoparticles (Fe_3O_4) coated with citric acid	Nanosorbent	Cd^{2+}	Singh et al. (2014)
Cupric oxide nanoparticles	Nanosorbent	Methylene blue dye	Mustafa et al. (2013)
Oval CuO nanostructures	Nanosorbent	Pb^{2+}	Farghali et al. (2013)
Polyacrylonitrile nanofibers with Prussian blue nanoparticles	Nanosorbet	$^{137}CS+$	Kim et al. (2018)
Zeolite nanoparticles in a polysulfone support	Nanomembrane	Negatively charged pharmaceuticals	Dong et al. (2016)
Silica nanospheres in trimesoyl chloride/ piperazine over polysulfone support	Nanomembrane	Negatively charged pollutants	Li et al. (2015a)
ZnO nanospheres in inner-skinned hollow fiber composite	Nanomembrane	Negatively charged pollutants	Li et al. (2016b)
Bubble-propelled Fe^0 Janus nanomotor	Nanomotor	Azo-dyes	Teo et al. (2016)
Mesoporous magnetic nanorods CoNi-Pt	Nanomotor	4-Nitrophenol, methylene blue and rhodamine B	García-Torres et al. (2017)

permeability, removing target pollutants (Yin et al. 2013; Turchanin and Gölzhäuser 2016; Yaqoob et al. 2020). At the end of the process, it is possible to distinguish two liquid portions, one with a bigger volume of treated water and the other with a small volume of wastewater containing the pollutants in a more concentrated way.

The performance of membrane systems is directly associated with the membrane material. Besides, contributing to new functions for contaminant degradation and self-cleaning, the incorporation of functional nanomaterials into membranes offers a great opportunity to improve membrane peculiarities such as permeability, and mechanical and thermal stability.

In addition, much research in membrane nanotechnology has focused on creating synergism or multifunction, adding nanomaterials in polymeric or inorganic membranes (Qu et al. 2013).

It is common to use carbon nanotubes or nanoparticles containing silver atoms to develop a composite thin layer that will serve as a support to a thin film of nanomaterials to generate the nanomembranes. These nanomaterials can prevent biofilm formation by possessing anti-microbial properties to avoid the attachment of bacteria (Liu et al. 2018; Qing et al. 2018). For example, polysulfone membranes can present antibacterial activity when attached to a silver nanoparticles layer (Andrade et al. 2015). However, graphene and its oxide are also versatile materials for the construction of nanomembranes to treat wastewater (Bodzek et al. 2020) and can be modified to also present antimicrobial effects (Li et al. 2016a; Akhavan and Ghaderi 2009).

Metallic/bimetallic catalyst nanoparticles such as nano zero-valent iron (nZVI) and noble metals supported on nZVI have been incorporated into polymeric membranes for reductive degradation of contaminants, particularly chlorinated compounds (Wu et al. 2005; Wu and Ritchie 2008).

During the creation of protocols to synthesize nanomembranes, special attention should be dedicated to developing ways to prevent membrane clogging and membrane fouling. Surface functionalization with layers of nanosilver and also TiO_2 nanoparticles can help prevent clogging once these nanomaterials can oxidize organic contaminants and prevent fouling layers to build up (Gehrke et al. 2015). Functionalized titanate nanotubes combined with polyethyleneimine as polymeric material made possible the production of nanomembranes with optimized performance in salt rejection and also in antifouling (Sumisha et al. 2015). Antifouling and permeability of nanomembranes could also be improved by TiO_2 nanoparticles (Mollahosseini and Rahimpour 2014) and zirconia nanoparticles grown over polydopamine/polyethyleneimine could optimize the structure stability of TFC membranes offering high water flux high salt retention in long-term nanofiltration (Lv et al. 2016). GO-Ag nanoparticles sheets could efficiently improve the anti-biofouling properties of polyvinylidene difluoride membranes also offering antibacterial properties (Li et al. 2016a).

Thin film composites (TFCs) from diverse characteristics have been developed to serve as nanomembranes in water treatment. For example, a polysulfone support could be in situ embedded with zeolite nanoparticles, then scientists promoted interfacial polymerization to generate a polyamide layer, offering a TFC with improved water permeability and improved rejection to negatively charged pharmaceuticals possessing small molecular weights (Dong et al. 2016). The negatively charged substance rejection potential of TFC composed of trimesoyl chloride and piperazine over polysulfone support was also higher, as same as the salt rejection capacity, due to the addition of silica nanospheres (Li et al. 2015a). Low concentration of ZnO nanospheres could also offer increased performance and enhance rejection of negative charge to inner-skinned hollow fiber composite nanofiltration membranes (Li et al. 2016b).

Nanomembranes can also present activity depending on the nanomaterials that constitute them. The presence of TiO_2 in nanomembranes offers, besides the physical separation possibility of filtration, the capacity of degrading organic pollutants through photocatalysis (Leong et al. 2014).

18.4.2 Nanosorbents

Nanosorbents (the second mechanism represented in Figure 18.1) can be used to interact with the pollutants (by polar or van der Walls interactions) and remove them from the wastewater (Yaqoob et al. 2020). After the initial interaction between the nanosorbent and the pollutant, a complex is formed (a process called coagulation) and it can have different densities. When the nanosorbent-pollutant complex is denser than water, the complex will precipitate and accumulate in the bottom of the recipient through a process called sedimentation (Boutilier et al. 2009; Medvedeva et al. 2015). However, when the density is lesser than the one from the water, the complex will float, by a process called flotation (Kyzas and Matis 2018). The same nanosorbent can do both of these processes, depending on the pollutants present in wastewater.

Carbon nanotubes can be combined with activated carbon to generate a nontoxic complex to remove, by adsorption, pharmaceuticals such as fluoxetine hydrochloride from contaminated wastewater (Sousa-Moura et al. 2019).

The sodium functionalization of the GO fibers can remove from aqueous samples 220 mg/g of Cs^+. NaOH can be used to wash the nanomaterial from the polluted and prepare it for a new removal cycle (Lee et al. 2019). Cationic and anionic dyes and also Pb^{2+} and Cd^{2+} can be removed from contaminated water by a sustainable date syrup-based graphene sand hybrid (Khan et al. 2019).

nZVI could remove metals from polluted water from a small biological WWTP. Nanofer25 at optimized conditions could perform an efficient removal (Oprčkal et al. 2017).

Magnetic Fe_3O_4 nanoparticles functionalized with humic acid were able to remove Hg^{2+} from water samples (at a maximum of 141.57 mg/g). They were attached to oyster shells to generate adsorbent surfaces (He et al. 2019). These magnetic nanoparticles, synthesized using a tea waste template, could also promote the adsorption of As^{3+} and As^{5+} in up to five adsorption cycles (Lunge et al. 2014). Magnetite nanoparticles (Fe_3O_4) when coated with citric acid could remove Cd^{2+} ions from aqueous samples (Singh et al. 2014). Microspheres of silica and Fe_3O_4 could remove heavy metal ions (Hg^{2+} and Pb^{2+}) from industrial wastewater, and the most interesting feature is that they could be reused, easily removed with an external magnet, and the pollutants adsorbed removed from SiO_2 with medium acidification (Hu et al. 2010).

The changing solution pH can regenerate metal oxide nanoadsorbents (Sharma et al. 2009). In many circumstances, the adsorption capacity of metal oxide nanoadsorbents is well maintained after several regeneration and reuse cycles (Hu et al. 2006). However, adsorption capacity can be decreased after regeneration (Deliyanni et al. 2003). The high adsorption capacity, low cost, easy separation and regeneration make metal-based nanoadsorbents technologically and economically advantageous.

Oval CuO nanostructures possess the potential to promote the adsorption of Pb^{2+} removing it from aqueous samples up to 125 mg/g (Farghali et al. 2013). Cupric oxide nanoparticles can also remove methylene blue dye at a rate of 88.93% (Mustafa et al. 2013).

Dendrimers are tailored adsorbents that are capable of removing both organics and heavy metals. Their interior shells can be hydrophobic for the sorption of organic compounds while the exterior branches can be tailored (e.g. hydroxyl- or amine-terminated) for the adsorption of heavy metals. The sorption can be based on complexation, electrostatic interactions, hydrophobic effect, and hydrogen bonding (Crooks et al. 2001). A dendrimer-ultrafiltration system could be designed to recover metal ions from aqueous solutions (Diallo et al. 2005).

The ferric hexacyanoferrate, most commonly called Prussian blue, is an insoluble dye with a great capacity to act like a heavy metal adsorbent. Its crystal has face-centered cubic Bravais lattices, in which the lattice spacing has a similar size of hydration radius of Cs ions (Kim et al. 2017). Because of this, Prussian blue is a great adsorbent of Cs, used especially in the treatment of patients with contamination of Cs radionuclides, as in the worst radiological accident in the world, which happened in Brazil in 1987 (IAEA 1988). However, the Prussian blue synthesized as nanoparticles and immobilized in a platform has been used for wastewater treatment of contamination by cesium. Another interesting characteristic of Prussian blue nanoparticles is their magnetism, favoring the process of separating the nanosorbent-pollutant complex from the water by application of a magnetic field (Wang and Huang 2011). Some examples of Prussian blue nanoparticles in nanocomposites are shown in Table 18.1 where it is possible to notice another quality of this nanoparticle: photocatalysis. This mechanism, present in Figure 18.1 will be discussed in Section 18.4.4.

Another useful adsorbent for radionuclides is GO. Because its hydroxyl, carboxyl, phenol, carbonyl, and epoxy groups, the GO is greatly dispersed in water (Aliyev et al. 2019). On the other hand, in the presence of radionuclides such as U^{6+}, Pu^{2+}, Th^{4+}, the GO will adsorb them and coagulate, forming a complex easily removed by forced-sedimentation by centrifugation (Li et al. 2012; Romanchuk et al. 2013; Laver 2020).

18.4.3 Nanomotors

Nanomotors (the third mechanism represented in Figure 18.1) are a complex type of synthetic nanomaterials (that can be generated using inorganic, organic and biological materials) able to interact with pollutants and move by themselves by converting different sources of energy into motion (Gao and Wang 2014).

These characteristics allow treatment of a big volume of wastewater without a step of mixing, to increase the encounter of nanomotors and pollutants (Sánchez et al. 2015). These nanomotors can perform two different methods of pollutant removal: catalysis or adsorption (Ying and Pumera 2019). The adsorption adopts the same mechanism as nanosorbents and can result in the precipitation or flotation of a nanomotor-pollutant complex. Trough the second method, catalysis of the pollutant, the nanomotors can degrade the pollutant into non-toxical subproducts.

Biocompatible and low-cost bubble-propelled Fe^0 Janus nanomotors stabilized by an ultrathin iron oxide shell could be propelled by their decomposition in citric acid and degraded azo-dyes acting as reducing agents (Teo et al. 2016).

Mesoporous magnetic nanorods containing a core of CoNi and a Pt shell could be propelled by a rotating magnetic field and in the presence of $NaBH_4$ could degrade the organic pollutants 4-nitrophenol, methylene blue and rhodamine B (García-Torres et al. 2017).

Nanostructures can also be assemble to generate micromotors (Yaqoob et al. 2020); for example, a light-driven Au-WO_3-C Janus micromotor was generated by colloidal carbon WO_3 nanoparticle composite spheres and presented high sensitivity toward sodium-2,6-dichloroindophenol and Rhodamine B in low concentration, which promoted their photocatalytic degradation (Zhang et al. 2017b).

18.4.4 Nanophotocatalysts

Nanophotocatalysts (the fourth mechanism represented in Figure 18.1) are nanomaterials with the capacity to convert the energy of light radiation (usually the sunlight) into chemical energy to stimulate the production of ROS (reactive oxygen species) (Gómez-Pastora et al. 2017). Photocatalysis can act in two different ways: the first one is the advanced oxidative processes (AOPs), wherein the redox of photocatalyst can produce ROS in the surrounding (Werkneh and Rene 2019) and the second method is the formation of electron–hole pair in a semiconductor making two redox polos that can act directly in degradation of pollutants or catalyze the formation of ROS (Li et al. 2016c).

The AOPs are made mainly from metallic oxides, and iron and zinc oxides are the most widely studied. These oxides, after absorbing the energy by a luminous photon (in the spectrum of visible, UV and infrared), can react to the water or hydrogen peroxide to produce hydroxyl ROS (OH^{-} and $OH^{\cdot}$) by a reaction called photo-Fenton (Vaiano et al. 2020). Thereby, the ROS will react with the pollutants and degrade him. In POA, the nanomaterial will act indirectly in the degradation of pollutants, having the function of producing the ROS, directly responsible for it (Singh and Pandey 2020).

The process of the formation of an electron–hole pair in nanomaterial is the most common semiconductor due to its capacity to have an electron jumping from the valence band to the conduction band when stimulated (Lam et al. 2016; Li et al. 2016c). In the case of photocatalysts, the stimulus to the band gap is the energy absorbed by an electron from the photon of light. After the excitation of an electron, the conduction band has an electron and soon a negative charge. On the other hand, the valence band will stay without the electron, and this absence is called a hole and has a positive electric charge (Horikoshi and Serpone 2020). Because of the high ratio of the superficial area to the volume of the nanomaterials, the electron in the conductive band will accumulate in a negative polo with the potential to reduce substances surrounding the nanomaterial and the hole in the valence band will form a positive polo able to oxidize (Kuang et al. 2020). Therefore, photocatalysts semiconductors nanomaterials are able to act directly in the degradation of pollutants by the redox process or indirectly by the production of ROS (Lam et al. 2016).

Some research groups dedicate their efforts to study the photocatalytic properties of one single nanomaterial, such as iron oxide nanoparticles (Pastrana-Martinez et al. 2015), TiO_2 (Das et al. 2014), ZnO (El-Kemary et al. 2010), and Prussian blue nanoparticles (Singh and Pandey 2020). However, the strategy that is the most promising for the treatment of wastewaters is the formulation of nanocomposites with two or more photocatalysts nanomaterials, which can enhance the efficiency of catalysis (Bhanvase et al. 2017). For example, carbon nanotubes possess a hollow structure with a high surface area that favors its functionalization, for example, with TiO_2/SiO_2 to remediate heavy metals and organic compounds by exhibiting photocatalytic properties (Rasheed et al. 2019).

Graphene and GO have been studied in the formulation of photocatalysts nanocomposite because of their unique electric, thermal, and mechanical proprieties. This makes the graphene and GO a high-quality support to photocatalyst, enhancing the efficiency of redox sites through its large surface area and fast electrical conductivity (An and Jimmy 2011; Li et al. 2016c). Because of this, there is one of the most common supports utilized in photocatalyst nanocomposites like that shown in Table 18.1.

In Section 18.4.3, the Prussian blue nanoparticle is described as a great absorbent of Cs, but beyond that, it has an excellent photocatalysis capacity too. The Prussian blue possesses a peroxidase-like and ascorbic acid oxidase-like activity and, in some cases, can exceed the efficiency of a natural enzyme, being called a "nanozyme" (Zhang et al. 2018; Komkova et al. 2018). This capacity is a promising AOP strategy for sewage and wastewater treatment when polluted by organic pollutants (Wang and Huang 2011; Mai et al. 2017). The version of fullerene functionalized with zinc porphyrin and attached to TiO_2 could remediate phenol and methylene blue in wastewater through its photocatalytic activity (Regulska et al. 2019).

Green-emissive carbon quantum dots could present photocatalytic activity to degrade Methylene blue (Das et al. 2019).

Silver quantum dots could promote the photocatalytic degradation of rhodamine B; the nanomaterials were used with 2D SnO_2 nanoflakes and could remediate water samples (Kumar et al. 2019).

Vanadium (V^{5+}) and chromium (Cr^{6+}) could be photocatalytically reduced to less-toxic V^{6+} and Cr^{3+} through $ZnWO_4$ nanoparticles (Zhao et al. 2016). This new procedure may be promisingly implemented in contaminated wastewaters, combining the remediation and possible recovery of V and Cr.

Synthesized mixed-phase bismuth ferrite nanoparticles, prepared by co-precipitation method using potassium hydroxide as the precipitant, were presented as promising photocatalytic material to promote the remediation of dye-contaminated industrial wastewater (Kalikeri and Kodialbail 2018).

18.5 Conclusions

Human actions to promote development, especially the industrial ones, ended up threatening living forms on Earth through environmental pollution. Water is a vital resource and its pollution is a serious increasing problem worldwide. Diverse remediation strategies have been developed, for example, the use of nanotechnology. Various nanomaterials, functionalized or not, can be used to deal with pollutants present in wastewaters mainly through nanomembranes, nanosorbents, nanophotocatalysts, and nanomotors.

18.6 Future Perspectives

It is expected that there will be further advances in the nanotechnology field especially to design new biodegradable materials with high capacity to remediate wastewater and avoid environmental pollution to increase. However, it is also important to highlight that technology development is not sufficient to substitute the necessity of environmental education to transform human behavior into a sustainable one. It is also expected that attention continue to be dedicated to studying the biocompatibility of the nanomaterials to be used in water treatment and the potential risks they may offer to living beings. As same as that it is necessary that experiments regarding increase in scale are performed for potential new strategies in order to guarantee their viability.

References

Akhavan, O. and Ghaderi, E. (2009). Photocatalytic reduction of graphene oxide nanosheets on TiO_2 thin film for photoinactivation of bacteria in solar light irradiation. *J. Phys. Chem. C* 113 (47): 20214–20220.

Aliyev, E., Filiz, V., Khan, M.M. et al. (2019). Structural characterization of graphene oxide: surface functional groups and fractionated oxidative debris. *Nanomaterials* 9 (8): 1180.

An, X. and Jimmy, C.Y. (2011). Graphene-based photocatalytic composites. *RSC Adv.* 1 (8): 1426–1434.

Andrade, P.F., de Faria, A.F., Oliveira, S.R. et al. (2015). Improved antibacterial activity of nanofiltration polysulfone membranes modified with silver nanoparticles. *Water Res.* 81: 333–342.

Anzar, N., Hasan, R., Tyagi, M. et al. (2020). Carbon nanotube – a review on synthesis, properties and plethora of applications in the field of biomedical science. *Sens. Int.* 1: 100003.

Bagazgoïtia, N.V.E., Bailey, H.D., Orsi, L. et al. (2018). Maternal residential pesticide use during pregnancy and risk of malignant childhood brain tumors: a pooled analysis of the ESCALE and ESTELLE studies (SFCE). *Int. J. Cancer* 142: 489–497.

Baldissarelli, D.P., Vargas, G.D.L.P., Korf, E.P. et al. (2019). Remediation of soils contaminated by pesticides using physicochemical processes: a brief review. *Planta Daninha* 37: e019184975.

Bebianno, M.J. and Gonzalez-Rey, M. (2015). Ecotoxicological risk of personal care products and pharmaceuticals. In: *Aquatic Ecotoxicology: Advancing Tools for Dealing with Emerging Risks*, vol. 16 (ed. C. Amiard-Triquet, J.-C. Amiard, and C. Mouneyrac). Cambridge: Academic Press, 518 pp.

Bethi, B. and Sonawane, S.H. (2018). Nanomaterials and its application for clean environment. In: *Nanomaterials for Green Energy* (ed. B.A. Bhanvase, V.B. Pawade, S.J. Dhoble, et al.), 385–409. Elsevier.

Bhanvase, B.A., Shende, T.P., and Sonawane, S.H. (2017). A review on graphene–TiO_2 and doped graphene–TiO_2 nanocomposite photocatalyst for water and wastewater treatment. *Environ. Technol. Rev.* 6 (1): 1–14.

Bhaskar, M., Gnanamani, A., Ganeshjeevan, R.J. et al. (2003). Analyses of carcinogenic aromatic amines released from harmful azo colorants by *Streptomyces* sp. SS07. *J. Chromatogr. A* 1018: 117–123.

Biglova, Y.N. and Mustafina, A.G. (2019). Nucleophilic cyclopropanation of [60] fullerene by the addition–elimination mechanism. *RSC Adv.* 9: 22428–22498.

Bodzek, M., Konieczny, K., and Kwiecińska-Mydlak, A. (2020). Nanotechnology in water and wastewater treatment. Graphene – the nanomaterial for next generation of semipermeable membranes. *Crit. Rev. Env. Sci. Technol.* 50 (15): 1515–1579.

Boutilier, L., Jamieson, R., Gordon, R. et al. (2009). Adsorption, sedimentation, and inactivation of *E. coli* within wastewater treatment wetlands. *Water Res.* 43 (17): 4370–4380.

Boutin, C. and Eme, C. (2016). Domestic wastewater characterization by emission source. *13eme Congres Spécialisé IWA on Small Water and Wastewater Systems*, Athènes, Greece. 8 p. https://hal.archives-ouvertes.fr/hal-01469077/document

Cabria, I., López, M.J., and Alonso, J.Á. (2007). The optimum average nanopore size for hydrogen storage in carbon nanoporous materials. *Carbon* 45 (13): 2649–2658.

Chaoua, S., Boussaa, S., El Gharmali, A. et al. (2019). Impact of irrigation with wastewater on accumulation of heavy metals in soil and crops in the region of Marrakech in Morocco. *J. Saudi Soc. Agric. Sci.* 18 (4): 429–436.

Contescu, C.I., Saha, D., Gallego, N.C. et al. (2012). Restricted dynamics of molecular hydrogen confined in activated carbon nanopores. *Carbon* 50 (3): 1071–1082.

Crooks, R.M., Zhao, M., Sun, L. et al. (2001). Dendrimer-encapsulated metal nanoparticles: synthesis, characterization, and applications to catalysis. *Acc. Chem. Res.* 31: 181–190.

Das, R., Sarkar, S., Chakraborty, S. et al. (2014). Remediation of antiseptic components in wastewater by photocatalysis using TiO2 nanoparticles. *Ind. Eng. Chem. Res.* 53 (8): 3012–3020.

Das, G.S., Shim, J.P., Bhatnagar, A. et al. (2019). Biomass-derived carbon quantum dots for visible-light-induced photocatalysis and label-free detection of Fe(III) and ascorbic acid. *Sci. Rep.* 9 (1): 15084.

Deliyanni, E.A., Bakoyannakis, D.N., Zouboulis, A.I. et al. (2003). Sorption of As(V) ions by akaganeite-type nanocrystals. *Chem. Tech.* 50 (1): 155–163.

Deng, Y., Guo, C., Zhang, H. et al. (2020). Occurrence and removal of illicit drugs in different wastewater treatment plants with different treatment techniques. *Environ. Sci. Eur.* 32: 28.

Diallo, M.S., Christie, S., Swaminathan, P. et al. (2005). Dendrimer enhanced ultrafiltration – recovery of Cu(II) from aqueous solutions using PA-MAM dendrimers with ethylene diamine core and termi-nal NH2 groups. *Environ. Sci. Technol.* 39: 1366–1377.

Dong, L.X., Huang, X.C., Wang, Z. et al. (2016). A thin-film nanocomposite nanofiltration membrane prepared on a support with in situ embedded zeolite nanoparticles. *Sep. Purif. Technol.* 166: 230–239.

Doumic, L., Salierno, G., Cassanello, M. et al. (2015). Efficient removal of Orange G using Prussian blue nanoparticles supported over alumina. *Catal. Today* 240: 67–72.

Dresselhaus, M.S., Dresselhaus, G., Eklund, P.C. et al. (2000). Carbon nanotubes. In: *The Physics of Fullerene-Based and Fullerene-Related Materials* (ed. W. Andreoni), 331–379. Dordrecht: Springer.

Drexler, K.E. (2004). Nanotechnology: from Feynman to funding. *Bull. Sci., Technol. Soc.* 24: 21–27.

Efremenko, I. and Sheintuch, M. (2006). Predicting solute adsorption on activated carbon: phenol. *Langmuir* 22 (8): 3614–3621.

El-Kemary, M., El-Shamy, H., and El-Mehasseb, I. (2010). Photocatalytic degradation of ciprofloxacin drug in water using ZnO nanoparticles. *J. Lumin.* 130 (12): 2327–2331.

El-Sikaily, A., Khaled, A., and El Nemr, A. (2012). Textile dyes xenobiotic and their harmful effect. In: *Non-Conventional Textile Waste Water Treatment* (ed. A. El-Nemr), 31–64. Nova Science Publishers.

Farghali, A.A., Bahgat, M., Allah, A.E. et al. (2013). Adsorption of Pb(II) ions from aqueous solutions using copper oxide nanostructures. *Beni-Seuf Univ. J. Appl. Sci.* 2: 61–71.

Fatima, S.A., Hamid, A., Yaqub, G. et al. (2018). Detection of volatile organic compounds in blood of farmers and their general health and safety profile. *Nat. Environ. Pollut. Technol.* 17: 657–660.

Fulekar, M.H. and Pathak, B. (2020). *Bioremediation Technology: Hazardous Waste Management.* CRC Press, 366 pp.

Gao, W. and Wang, J. (2014). The environmental impact of micro/nanomachines: a review. *ACS Nano* 8: 3170–3180.

García-Torres, J., Serra, A., Tierno, P. et al. (2017). Magnetic propulsion of recyclable catalytic nanocleaners for pollutant degradation. *ACS Appl. Mater. Interfaces* 9: 23859–23868.

Gavrilescu, M., Demnerová, K., Aamand, J. et al. (2015). Emerging pollutants in the environment: present and future challenges in biomonitoring, ecological risks and bioremediation. *New Biotechnol.* 32 (1): 147–156.

Gehrke, I., Geiser, A., and Somborn-Schulz, A. (2015). Innovations in nanotechnology for water treatment. *Nanotechnol. Sci. Appl.* 8: 1–17.

Gheju, M. (2011). Hexavalent chromium reduction with zero-valent iron (ZVI) in aquatic systems. *Water Air Soil Pollut.* 222: 103–148.

Gimiliani, G.T., Fontes, R.F.C., and Abessa, D.M.S. (2016). Modeling the dispersion of endocrine disruptors in the Santos Estuarine System (Sao Paulo State, Brazil). *Braz. J. Oceanogr.* 64 (1): 1–8.

Gómez-Pastora, J., Dominguez, S., Bringas, E. et al. (2017). Review and perspectives on the use of magnetic nanophotocatalysts (MNPCs) in water treatment. *Chem. Eng. J* 310: 407–427.

Gray, N.F. (2004). *Biology of Wastewater Treatment*, 2e. London: Imperial College, 1444 pp.

Guan, X., Sun, Y., Qin, H. et al. (2015). The limitations of applying zero-valent iron technology in contaminants sequestration and the corresponding countermeasures: the development in zero-valent iron technology in the last two decades (1994–2014). *Water Res.* 75: 224–248.

Guerra, F.D., Attia, M.F., Whitehead, D.C. et al. (2018). Nanotechnology for environmental remediation: materials and applications. *Molecules* 23 (7): 1760.

Guo, J., Morris, J.R., Ihm, Y. et al. (2012). Topological defects: origin of nanopores and enhanced adsorption performance in nanoporous carbon. *Small* 8 (21): 3283–3288.

Hao, X., Wang, X., Liu, R. et al. (2019). Environmental impacts of resource recovery from wastewater treatment plants. *Water Res.* 160: 268–277.

He, C., Qu, J., Yu, Z. et al. (2019). Preparation of micro-nano material composed of oyster shell/Fe_3O_4 nanoparticles/humic acid and its application in selective removal of Hg(II). *Nanomaterials* 9 (7): 953.

Hirsch, A. (1995). Addition reactions of buckminsterfullerene (C_{60}). *Synthesis* 1995 (8): 895–913.

Horikoshi, S. and Serpone, N. (2020). Can the photocatalyst TiO_2 be incorporated into a wastewater treatment method? Background and prospects. *Catal. Today* 340: 334–346.

Hristovski, K.D., Nguyen, H., and Westerhoff, P.K. (2009a). Removal of arsenate and 17-ethinyl estradiol (EE2) by iron (hydr) oxide modified activated carbon fibers. *J. Environ. Sci. Health., Part A* 44 (4): 354–361.

Hristovski, K.D., Westerhoff, P.K., Moller, T. et al. (2009b). Effect of synthesis conditions on nano-iron (hydr)oxide impregnated granulated activated carbon. *Chem. Eng. J* 146 (2): 237–243.

Hu, Q.H., Qiao, S., Haghseresht, F. et al. (2006). Adsorption study for removal of basic red dye using bentonite. *Ind. Eng. Chem. Res.* 45 (2): 050889.

Hu, H., Wang, Z., and Pan, L. (2010). Synthesis of monodisperse Fe_3O_4@silica core–shell microspheres and their application for removal of heavy metal ions from water. *J. Alloys Compd.* 492: 656–661.

International Atomic Energy Agency (1988). *The Radiological Accident in Goiânia*. Vienna: IAEA.

Jain, A. and Vaya, D. (2017). Photocatalytic activity of TiO_2 nanomaterial. *J. Chil. Chem. Soc.* 62 (4): 3683–3690.

Jiang, D., Zeng, G., Huang, D. et al. (2018). Remediation of contaminated soils by enhanced nanoscale zero valent iron. *Environ. Res.* 163: 217–227.

Kalikeri, S. and Kodialbail, V. (2018). Solar light-driven photocatalysis using mixed-phase bismuth ferrite ($BiFeO_3$/$Bi_{25}FeO_{40}$) nanoparticles for remediation of dye-contaminated water: kinetics and comparison with artificial UV and visible light-mediated photocatalysis. *Environ. Sci. Pollut. Res. Int.* 25 (14): 13881–13893.

Kaneko, K., Hanzawa, Y., Iiyama, T. et al. (1999). Cluster-mediated water adsorption on carbon nanopores. *Adsorption* 5 (1): 7–13.

Kazlauskienė, N., Svecevičius, G., Marčiulionienė, D. et al. (2012). The effect of persistent pollutants on aquatic ecosystem: a complex study. *2012 IEEE/OES Baltic International Symposium* (8–10 May 2012). IEEE, 1–6. http://doi.org/10.1109/BALTIC.2012.6249198.

Khan, S. and Malik, A. (2018). Toxicity evaluation of textile effluents and role of native soil bacterium in biodegradation of a textile dye. *Environ. Sci. Pollut. Res. Int.* 25 (5): 4446–4458.

Khan, S.T. and Malik, A. (2019). Engineered nanomaterials for water decontamination and purification: from lab to products. *J. Hazard. Mater.* 363: 295–308.

Khan, S., Edathil, A.A., and Banat, F. (2019). Sustainable synthesis of graphene-based adsorbent using date syrup. *Sci. Rep.* 9: 18106.

Kiew, S.F., Kiew, L.V., Lee, H.B. et al. (2016). Assessing biocompatibility of graphene oxide-based nanocarriers: a review. *J. Controlled Release* 226: 217–228.

Kim, K.H., Kabir, E., and Jahan, S.A. (2017). Exposure to pesticides and the associated human health effects. *Sci. Total Environ.* 575: 525–529.

Kim, H., Kim, M., Lee, W. et al. (2018). Rapid removal of radioactive cesium by polyacrylonitrile nanofibers containing Prussian blue. *J. Hazard. Mater.* 347: 106–113.

Klatte, S., Schaefer, H.C., and Hempel, M. (2017). Pharmaceuticals in the environment: a short review on options to minimize the exposure of humans, animals and ecosystems. *Sustainable Chem. Pharm.* 5: 61–66.

Komkova, M.A., Karyakina, E.E., and Karyakin, A.A. (2018). Catalytically synthesized Prussian blue nanoparticles defeating natural enzyme peroxidase. *J. Am. Chem. Soc.* 140 (36): 11302–11307.

Kuang, P., Sayed, M., Fan, J. et al. (2020). 3D graphene-based H_2-production photocatalyst and electrocatalyst. *Adv. Energy Mater.* 10 (14): 1903802.

Kumar, N.S., Asif, M., Reddy, T.R.K. et al. (2019). Silver quantum dot decorated 2D-SnO_2 nanoflakes for photocatalytic degradation of the water pollutant rhodamine B. *Nanomaterials* 9 (11): 1536.

Kyzas, G.Z. and Matis, K.A. (2018). Flotation in water and wastewater treatment. *Processes* 6 (8): 116.

Lam, S.M., Sin, J.C., and Mohamed, A.R. (2016). A review on photocatalytic application of g-C_3N_4/semiconductor (CNS) nanocomposites towards the erasure of dyeing wastewater. *Mater. Sci. Semicond. Process.* 47: 62–84.

Laver, N. R. (2020). The removal of radionuclides from contaminated water samples using graphene oxide nano-flakes. Doctoral dissertation. UCL (University College London).

Lee, H., Lee, K., Kim, S.O. et al. (2019). Effective and sustainable Cs^+ remediation via exchangeable sodium-ion sites in graphene oxide fibers. *J. Mater. Chem. A* 7: 17754–17760.

Leong, S., Razmjou, A., Wang, K. et al. (2014). TiO_2 based photocatalytic membranes: a review. *J. Membr. Sci.* 472 (15): 167–184.

Lesiak, B., Rangam, N., Jiricek, P. et al. (2019). Surface study of Fe_3O_4 nanoparticles functionalized with biocompatible adsorbed molecules. *Front. Chem.* 7: 642.

Li, B. and Cao, H. (2011). ZnO@graphene composite with enhanced performance for the removal of dye from water. *J. Mater. Chem.* 21 (10): 3346–3349.

Li, J., Xia, J.T., Liu, H.B. et al. (2011). Nanoporous carbon synthesized from sol–gel template for adsorbing gibberellic acid in aqueous solution. *Mater. Sci. Eng., A* 31 (7): 1313–1319.

Li, Z., Chen, F., Yuan, L. et al. (2012). Uranium (VI) adsorption on graphene oxide nanosheets from aqueous solutions. *Chem. Eng. J* 210: 539–546.

Li, Q., Wang, Y., Song, J. et al. (2015a). Influence of silica nanospheres on the separation performance of thinfilm composite poly(piperazineamide) nanofiltration membranes. *Appl. Surf. Sci.* 324: 757–764.

Li, X., Wang, J., Rykov, A.I. et al. (2015b). Prussian blue/TiO_2 nanocomposites as a heterogeneous photo-Fenton catalyst for degradation of organic pollutants in water. *Catal. Sci. Technol.* 5 (1): 504–514.

Li, J., Liu, X., Lu, J. et al. (2016a). Anti-bacterial properties of ultrafiltration membrane modified by graphene oxide with nano-silver particles. *J. Colloid Interface Sci.* 484: 107–115.

Li, H., Shi, W., Zhu, H. et al. (2016b). Effects of zinc oxide nanospheres on the separation performance of hollow fiber poly(piperazine-amide) composite nanofiltration membranes. *Fibers Polym.* 17 (6): 836–846.

Li, X., Yu, J., Wageh, S. et al. (2016c). Graphene in photocatalysis: a review. *Small* 12 (48): 6640–6696.

Liu, D., Mao, Y., and Ding, L. (2018). Carbon nanotubes as antimicrobial agents for water disinfection and pathogen control. *J. Water Health* 16 (2): 171–180.

Low, J., Yu, J., and Li, Q. (2014). Enhanced visible-light photocatalytic activity of plasmonic Ag and graphene co-modified Bi_2WO_6 nanosheets. *Phys. Chem. Chem. Phys.* 16 (3): 1111–1120.

Lunge, S., Singh, S., and Sinha, A. (2014). Magnetic iron oxide (Fe_3O_4) nanoparticles from tea waste for arsenic removal. *J. Magn. Magn. Mater.* 356: 21–31.

Lv, Y., Yang, H.C., Liang, H.Q. et al. (2016). Novel nanofiltration membrane with ultrathin zirconia film as selective layer. *J. Membr. Sci.* 500: 265–271.

Mai, N.X.D., Yoon, J., Kim, J.H. et al. (2017). Hybrid hydrogel and aerogel membranes based on chitosan/Prussian blue for photo-Fenton-based wastewater treatment using sunlight. *Sci. Adv. Mater.* 9 (9): 1484–1487.

Manickum, T., John, W., and Terry, S. (2011). Determination of selected steroid estrogens in treated sewage effluent in the Umsunduzi (Duzi) river water catchment area. *Hydrol. Curr. Res.* 2 (3): 117.

Mayo, J.T., Yavuz, C., Yean, S. et al. (2007). The effect of nanocrystalline magnetite size on arsenic removal. *Sci. Technol. Adv. Mater.* 8 (1e2): 71–75.

Medvedeva, I., Bakhteeva, I., Zhakov, S. et al. (2015). Separation of Fe_3O_4 nanoparticles from water by sedimentation in a gradient magnetic field. *J. Water Resour. Prot.* 7 (02): 111.

Melián, J.A.H. (2020). Sustainable wastewater treatment systems (2018–2019). *Sustainability* 12 (5): 1940.

Mezzelani, M., Gorbi, S., and Regoli, F. (2018). Pharmaceuticals in the aquatic environments: evidence of emerged threat and future challenges for marine organisms. *Mar. Environ. Res.* 140: 41–60.

Mirshahghassemi, S. and Lead, J.R. (2015). Oil recovery from water under environmentally relevant conditions using magnetic nanoparticles. *Environ. Sci. Technol.* 49 (19): 11729–11736.

Mollahosseini, A. and Rahimpour, A. (2014). Interfacially polymerized thin film nanofiltration membranes on TiO_2 coated polysulfone substrate. *J. Ind. Eng. Chem.* 20 (4): 1261–1268.

Moretti, S.M.L., Bertoncini, E.I., and Abreu-Junior, C.H. (2015). Composting sewage sludge with green waste from tree pruning. *Sci. Agric.* 72 (5): 432–439.

Mustafa, G., Tahir, H., Sultan, M. et al. (2013). Synthesis and characterization of cupric oxide (CuO) nanoparticles and their application for the removal of dyes. *Afr. J. Biotechnol.* 12: 6650–6660.

Newton, R.J., McLellan, S.L., Dila, D.K. et al. (2015). Sewage reflects the microbiomes of human populations. *mBio* 6 (2): e02574.

Novoselov, K.S., Geim, A.K., Morozov, S.V. et al. (2004). Electric field effect in atomically thin carbon films. *Science* 306 (5696): 666–669.

O'Carroll, D., Sleep, B., Krol, M. et al. (2013). Nanoscale zero valent iron and bimetallic particles for contaminated site remediation. *Adv. Water Resour.* 51: 104–122.

Oprčkal, P., Mladenovič, A., Vidmar, J. et al. (2017). Critical evaluation of the use of different nanoscale zero-valent iron particles for the treatment of effluent water from a small biological wastewater treatment plant. *Chem. Eng. J* 321: 20–30.

Pastrana-Martinez, L.M., Pereira, N., Lima, R. et al. (2015). Degradation of diphenhydramine by photo-Fenton using magnetically recoverable iron oxide nanoparticles as catalyst. *Chem. Eng. J* 261: 45–52.

Pinheiro, H.M., Thomas, O., and Touraud, E. (2004). Aromatic amines from azo dye reduction: status review with emphasis on direct UV spectrophotometric detection in textile industry wastewaters. *Dyes Pigm.* 61: 121–139.

Pinto, A.M., Goncalves, I.C., and Magalhães, F.D. (2013). Graphene-based materials biocompatibility: a review. *Colloids Surf., B* 111: 188–202.

Plerdsranoy, P. and Utke, R. (2015). Confined $LiBH_4$–$LiAlH_4$ in nanopores of activated carbon nanofibers. *Int. J. Hydrogen Energy* 40 (22): 7083–7092.

Priyadarshni, N., Nath, P., Nagahanumaiah et al. (2018). DMSA-functionalized gold nanorod on paper for colorimetric detection and estimation of arsenic (III and V) contamination in groundwater. *ACS Sustainable Chem. Eng.* 6 (5): 6264–6272.

Pulizzi, F. and Sun, W. (2018). Treating water with nano. *Nat. Nanotechnol.* 13 (8): 633.

Qing, Y., Cheng, L., Li, R. et al. (2018). Potential antibacterial mechanism of silver nanoparticles and the optimization of orthopedic implants by advanced modification technologies. *Int. J. Nanomed.* 13: 3311–3327.

Qu, X., Alvarez, P.J., and Li, Q. (2013). Applications of nanotechnology in water and wastewater treatment. *Water Res.* 47 (12): 3931–3946.

Rao, H., Xue, X., Wang, H. et al. (2019). Gold nanorod etching-based multicolorimetric sensors: strategies and applications. *J. Mater. Chem. C* 7: 4610–4621.

Rasheed, T., Adeel, M., Nabeel, F. et al. (2019). TiO_2/SiO_2 decorated carbon nanostructured materials as a multifunctional platform for emerging pollutants removal. *Sci. Total Environ.* 688: 299–311.

Rauti, R., Musto, M., Bosi, S. et al. (2019). Carbon review article properties and behavior of carbon nanomaterials when interfacing neuronal cells: how far have we come? *Carbon* 143: 430–446.

Regulska, E., Rivera-Nazario, D.M., Karpinska, J. et al. (2019). Zinc porphyrin-functionalized fullerenes for the sensitization of titania as a visible-light active photocatalyst: river waters and wastewaters remediation. *Molecules* 24 (6): 1118.

Romanchuk, A.Y., Slesarev, A.S., Kalmykov, S.N. et al. (2013). Graphene oxide for effective radionuclide removal. *Phys. Chem. Chem. Phys.* 15 (7): 2321–2327.

Roy, J.R., Chakraborty, S., and Chakraborty, T.R. (2009). Estrogen-like endocrine disrupting chemicals affecting puberty in humans – a review. *Med. Sci. Monit.* 15 (6): 137–145.

Sánchez, S., Soler, L., and Katuri, J. (2015). Chemically powered micro-and nanomotors. *Angew. Chem. Int. Ed.* 54 (5): 1414–1444.

Sangion, A. and Gramatica, P. (2016). PBT assessment and prioritization of contaminants of emerging concern: pharmaceuticals. *Environ. Res.* 147: 297–306.

Santhosh, C., Velmurugan, V., Jacob, G. et al. (2016). Role of nanomaterials in water treatment applications: a review. *Chem. Eng. J.* 306: 1116–1137.

Schwarzenbach, R.P., Escher, B.I., Fenner, K. et al. (2006). The challenge of micropollutants in aquatic systems. *Science* 313 (5790): 1072–1077.

Semeniuk, M., Yi, Z., Poursorkhabi, V. et al. (2019). Future perspectives and review on organic carbon dots in electronic applications. *ACS Nano* 13 (6): 6224–6255.

Sharma, Y.C., Srivastava, V., Singh, V.K. et al. (2009). Nano-adsorbents for the removal of metallic pollutants from water and wastewater. *Environ. Technol.* 30 (6): 583–609.

Silva, G.A. (2008). Nanotechnology approaches to crossing the blood-brain barrier and drug delivery to the CNS. *BMC Neurosci.* 9 (Suppl 3): S4.

Singh, S. and Pandey, P.C. (2020). Synthesis and application of functional Prussian blue nanoparticles for toxic dye degradation. *J. Environ. Chem. Eng.* 8 (3): 103753.

Singh, D., Gautam, R.K., Kumar, R. et al. (2014). Citric acid coated magnetic nanoparticles: synthesis, characterization and application in removal of Cd(II) ions from aqueous solution. *J. Water Process Eng.* 4: 233–241.

Singh, N. S., Sharma, R., Parween, T. et al. (2018). "Pesticide contamination and human health risk factor," in Modern Age Environmental Problems and Their Remediation, eds O. Mohammad, Z. K. Mohammad, and M. I. I. Iqbal (Cham: Springer), 49–68. doi: https://doi.org/10.1007/978-3-319-64501-8_3.

Singh, R., Bhadouria, R., Singh, P. et al. (2020). Nanofiltration technology for removal of pathogens present in drinking water. *Waterborne Pathogens* 2020: 463–489.

Sousa-Moura, D., Matsubara, E.Y., Machado Ferraz, I.B. et al. (2019). CNTs coated charcoal as a hybrid composite material: adsorption of fluoxetine probed by zebrafish embryos and its potential for environmental remediation. *Chem. Tech.* 230: 369–376.

Sposito, J.C.V., Montagner, C.C., Casado, M. et al. (2018). Emerging contaminants in Brazilian rivers: occurrence and effects on gene expression in zebrafish (*Danio rerio*) embryos. *Chem. Tech.* 209: 696–705.

Sumisha, A., Arthanareeswaran, G., Ismail, A.F. et al. (2015). Functionalized titanate nanotube–polyetherimide nanocomposite membrane for improved salt rejection under low pressure nanofiltration. *RSC Adv.* 5 (49): 39464–39473.

Tanaka, A., Iiyama, T., Ohba, T. et al. (2010). Effect of a quaternary ammonium salt on propylene carbonate structure in slit-shape carbon nanopores. *J. Am. Chem. Soc.* 132 (7): 2112–2113.

Tayeb, A., Chellali, M.R., and Hamou, A. (2015). Impact of urban and industrial effluents on the coastal marine environment in Oran, Algeria. *Mar. Pollut. Bull.* 98 (1, 2): 281–288.

Teo, W.Z., Zboril, R., Medrik, I. et al. (2016). Fe^0 nanomotors in ton quantities (10^{20} units) for environmental remediation. *Chem. Eur. J.* 22 (14): 4789–4793.

Tiwari, S.K., Sahoo, S., Wang, N. et al. (2020). Graphene research and their outputs: status and prospect. *J. Sci. Adv. Mat. Dev.* 5 (1): 10–29.

Tolmachev, A.M., Firsov, D.A., Kuznetsova, T.A. et al. (2009). DFT modeling of the adsorption of benzene, methanol, and ethanol molecules in activated carbon nanopores. *Prot. Met. Phys. Chem.* 45 (2): 163–168.

Tonelli, F.M.P., Lacerda, S.M.S.N., Paiva, N.C.O. et al. (2015). Functionalized nanomaterials: are they effective to perform gene delivery to difficult-to-transfect cells with no cytotoxicity? *Nanoscale* 7 (43): 18036–18043.

Tonelli, F.M.P., Lacerda, S.M.S.N., Paiva, N.C.O. et al. (2016). Efficiently and safely in gene transfection in fish spermatonial stem cells using nanomaterials. *RSC Adv.* 58: 1058.

Tonelli, F.M.P., Tonelli, F.C.P., Ferreira, D.R.C. et al. (2020). Biocompatibility and functionalization of nanomaterials. Intelligent nanomaterials for drug delivery applications. In: *Intelligent Nanomaterials for Drug Delivery Applications*, 212 pp (ed. N. Ahmad and P. Gopinath). New York: Elsevier.

Trivedi, M., Johri, P., Singh, A. et al. (2020). Latest tools in fight against cancer: nanomedicines. In: *NanoBioMedicine* (ed. S.K. Saxena and S.M.P. Khurana), 139–164. Singapore: Springer.

Turchanin, A. and Gölzhäuser, A. (2016). Carbon nanomembranes. *Adv. Mater.* 28 (29): 6075–6103.

Vaiano, V., Sannino, D., and Sacco, O. (2020). The use of nanocatalysts (and nanoparticles) for water and wastewater treatment by means of advanced oxidation processes. In: *Nanotechnology in the Beverage Industry* (ed. A. Amrane, S. Rajendran, T.A. Nguyen, et al.), 241–264. Elsevier.

Varjani, S., Kumar, G., and Rene, E.R. (2019). Developments in biochar application for pesticide remediation: current knowledge and future research directions. *J. Environ. Manage.* 232: 505–513.

Wang, H. and Huang, Y. (2011). Prussian-blue-modified iron oxide magnetic nanoparticles as effective peroxidase-like catalysts to degrade methylene blue with H_2O_2. *J. Hazard. Mater.* 191 (1–3): 163–169.

Wang, F., Yang, W., Zheng, F. et al. (2018). Removal of Cr(VI) from simulated and leachate wastewaters by bentonite-supported zero-valent iron nanoparticles. *Int. J. Environ. Res. Public Health* 15 (10): 2162.

Wang, X., Feng, Y., Dong, P. et al. (2019a). A mini review on carbon quantum dots: preparation, properties, and electrocatalytic application. *Front. Chem.* 7: 671.

Wang, Y., Pan, C., Chu, W. et al. (2019b). Environmental remediation applications of carbon nanotubes and graphene oxide: adsorption and catalysis. *Nanomaterials* 9 (3): 439.

Weiss, J., Takhistov, P., and McClements, D.J. (2006). Functional materials in food nanotechnology. *J. Food Sci.* 71: R107–R116.

Werkneh, A.A. and Rene, E.R. (2019). Applications of nanotechnology and biotechnology for sustainable water and wastewater treatment. In: *Water and Wastewater Treatment Technologies* (ed. X.T. Bui, C. Chiemchaisri, T. Fujioka, and S. Varjani), 405–430. Singapore: Springer.

Werkneh, A.A., Rene, R.R., and Lens, P.N.L. (2018). Application of nanomaterials in food, cosmetics and other process industries. In: *Nanotoxicology: Toxicity Evaluation, Risk Assessment and Management* (ed. V. Kumar, N. Dasgupta, and S. Ranjan), 63–79. CRC Press Taylor and Francis Group.

Wicki, A., Witzigmann, D., Balasubramanian, V. et al. (2015). Nanomedicine in cancer therapy: challenges, opportunities, and clinical applications. *J. Controlled Release* 200: 138–157.

Wu, L. and Ritchie, S.M.C. (2008). Enhanced dechlorination of trichloroethylene by membrane-supported Pd-coated iron nanoparticles. *Environ. Prog.* 27 (2): 218–224.

Wu, L., Shamsuzzoha, M., and Ritchie, S.M.C. (2005). Preparation of cellulose acetate supported zero-valent iron nanoparticlesfor the dechlorination of trichloroethylene in water. *J. Nanopart. Res.* 7: 469–476.

Xie, H., Li, P., Shao, J. et al. (2019). Electrostatic self-assembly of Ti_3C_2Tx MXene and gold nanorods as an efficient surface-enhanced Raman scattering platform for reliable and high-sensitivity determination of organic pollutants. *ACS Sens.* 4 (9): 2303–2310.

Yang, Y., Ok, Y.S., Kim, K.H. et al. (2017). Occurrences and removal of pharmaceuticals and personal care products (PPCPs) in drinking water and water/ sewage treatment plants: a review. *Sci. Total Environ.* 596: 303–320.

Yaqoob, A.A., Parveen, T., Umar, K. et al. (2020). Role of nanomaterials in the treatment of wastewater: a review. *Water* 12 (2): 495.

Ye, S., Zeng, G., Wu, H. et al. (2017). Biological technologies for the remediation of co-contaminated soil. *Crit. Rev. Biotechnol.* 37 (8): 1062–1076.

Yeung, K.W.Y., Zhou, G.J., Hilscherová, K. et al. (2020). Current understanding of potential ecological risks of retinoic acids and their metabolites in aquatic environments. *Environ. Int.* 136: 105464.

Yin, J., Yang, Y., Hu, Z. et al. (2013). Attachment of silver nanoparticles (AgNPs) onto thin-film composite (TFC) membranes through covalent bonding to reduce membrane biofouling. *J. Membr. Sci.* 441: 73–82.

Ying, Y. and Pumera, M. (2019). Micro/nanomotors for water purification. *Chemistry* 25 (1): 106–121.

Ying, Y.L., Ying, W., Li, Q.C. et al. (2017). Recent advances of nanomaterial-based membrane for water purification. *Appl. Mater. Today* 7: 144–158.

Zhang, Y., Yang, M., Portney, N.G. et al. (2008). Zeta potential: a surface electrical characteristic to probe the interaction of nanoparticles with normal and cancer human breast epithelial cells. *Biomed. Microdevices* 10 (2): 321–328.

Zhang, L.S., Fang, Y., Zhou, Y. et al. (2017a). Improvement of the stabilization and activity of protocatechuate 3,4-dioxygenase isolated from *Rhizobium* sp. LMB-1 and immobilized on Fe_3O_4 nanoparticles. *Appl. Biochem. Biotechnol.* 183 (3): 1035–1048.

Zhang, Q., Dong, R., Wu, Y. et al. (2017b). Light-driven Au-WO_3@C Janus micromotors for rapid photodegradation of dye pollutants. *ACS Appl. Mater. Interfaces* 9 (5): 4674–4683.

Zhang, W., Wu, Y., Dong, H.J. et al. (2018). Sparks fly between ascorbic acid and iron-based nanozymes: a study on Prussian blue nanoparticles. *Colloid Surf., B* 163: 379–384.

Zhang, Z., Zhu, Z., Shen, B. et al. (2019). Insights into biochar and hydrochar production and applications: a review. *Energy* 171 (C): 581–598.

Zhao, Z., Zhang, B., Chen, D. et al. (2016). Simultaneous reduction of Vanadium(V) and Chromium(VI) in wastewater by nanosized $ZnWO_4$ photocatalysis. *J. Nanosci. Nanotechnol.* 16 (3): 2847–2852.

Zheng, H., Zhang, C., Liu, B. et al. (2020). Biochar for water and soil remediation: production, characterization, and application. In: *A New Paradigm for Environmental Chemistry and Toxicology* (ed. G. Jiang and X. Li), 153–196. Singapore: Springer.

Zhou, G.J., Li, X.Y., and Leung, K.M.Y. (2019). Retinoids and oestrogenic endocrine disrupting chemicals in saline sewage treatment plants: removal efficiencies cological risks to marine organisms. *Environ. Int.* 127: 103–113.

Zhou, G.J., Lin, L., Li, X.Y. et al. (2020). Removal of emerging contaminants from wastewater during chemically enhanced primary sedimentation and acidogenic sludge fermentation. *Water Res.* 175: 115646.

Ziadi, A., Hariga, N.T., and Tarhouni, J. (2019). Mineralization and pollution sources in the costal aquifer of Lebna, Cap Bom, Tunisia. *J. Afr. Earth. Sci.* 151: 391–402.

19

Transgenic Plant Technology and its Role in Bioremediation

Gulzar A. Rathar[1,2], Romica Verma[1], and Bhavana Sharma[1]

[1] *Department of Botany, University of Jammu, Jammu, India*
[2] *The Institute of Plant Sciences, Department of Ornamental Plants and Agricultural Biotechnology, The Volcani Center, ARO, Rishon LeZion, Israel*

19.1 Introduction

Environmental stress, population growth, and food scarcity have all generated severe issues for humanity across the world. The global population is rapidly expanding and is likely to reach 8.5 billion by 2025. With fewer natural resources available, it is difficult to supply everyone's needs for food (Ahmad et al. 2012). This factor has contributed to food insecurity, which has caused malnutrition, which is a stern health concern in the modern world. For sustainable agriculture to meet the growing need for food, it is essential to produce crops of higher quantity and quality. This can be done via genetic engineering or traditional selection and breeding (Ashraf and Akram 2009). The use and advancement of biotechnology resulted in new openings and fresh opportunities for improving the quantitative and qualitative characteristics of biological organisms (Yamaguchi and Blumwald 2005; Sun 2008). For enhancing the composition of proteins, carbohydrates, fats, vitamins, and micronutrients, biotechnology for crop development has become a viable method for addressing nutritional deficiencies (Zimmermann and Hurrell 2002; Sun 2008). Insect and herbicide tolerance, quality nutritional components, viral resistance, shelf life, and the generation of biofuel have all been areas of emphasis for agricultural biotechnology since the 1990s. Crop enhancement via genetic engineering is a complex process because many of these qualities involve many genes. Being unfamiliar with the fundamentals of plant species' molecular biology and genetics makes this more challenging. The creation of transgenic plants using various genetic engineering methods has raised a number of ethical, political, and sociological questions (Ashraf and Akram 2009). According to the World Health Organization (WHO), a few key issues with genetically improved crops, are first leading to the production of allergenic foods and also with the altered genes of food crops are consumed by human bodies which may be harmful, and the mating of transgenic plant species with nontransgenic plants. The safety of food may be compromised by all of these factors. Despite these challenges, certain countries, including China, Canada, the United States, Argentina, and Brazil, are currently legalizing transgenic agricultural agriculture (James 2006). The development of high-yielding and high-quality transgenic crops is still urgently needed despite the fact that transgenic methods have significantly increased the productivity of many crops, including tobacco, wheat, brassica, rice, and soybean, among others. The world's food supply, environmental protection, and energy production seem to be at odds, and genetically modified crops seem to be the only way to resolve it (Sexton and Zilberman 2011). At a time when conventional means of yield augmentation have mostly run their course, increased global demand for biofuels, for instance, is significantly straining agricultural systems (Sexton and Zilberman 2011). Enhancing the capacities of fuels made from biomass using biotechnology is a practical option (Rosegrant and Cavalieri 2008). It is vital to examine the effect of biofuel manufacture on food security, nevertheless, because of a considerable alteration in the use of land and crop growing. The economic and environmental effects of this are substantial. The energy catastrophe and climate variation must remove restrictions on biotechnology development, allowing the field to flourish and investing to improve biofuel production technologies. There is an example that these constraints hinder the potential advancement of agriculture biotechnology (Wolt 2009). According to estimates, biofuels contributed 2.7% of total global transportation fuel in

Aquatic Contamination: Tolerance and Bioremediation, First Edition. Edited by Rouf Ahmad Bhat, Gowhar Hamid Dar, Fernanda Maria Policarpo Tonelli, and Saima Hamid.

2010 (REN21 2001). Genetically modified strains of algae, especially blue-green algae, are being created by scientists to be used in the production of fuels (cyanobacteria). Since they simply need water and sunlight for growth, algae can be grown in nonarable locations without competing with food crops and have the potential to generate considerable quantities of fatty acids. In relation to production per hectare, food source, and the need for heavy processing, scientists think about algal biofuel that can perform better than grass and maize biofuels (The Biology Refugia 2011).

New expression methods were being discovered for the production of high-value therapeutic proteins. This has been investigated since the advent of recombinant protein technology (Schillberg et al. 2005). Instead of using standard expression techniques based on the mass cultivation of microbial or animal cells, or the generation of transgenic animals, plants have emerged as a useful, secure, and affordable substitute (Fischer et al. 1999; Giddings 2001; Larrick and Thomas 2001; Houdebine 2002). The effective expression of a wide number of valuable proteins by plants has been shown by various research studies over the last few decades. Antibodies, vaccines, industrial enzymes, biopolymers, serum proteins, human growth regulators, and molecular biology reagents are a few examples (Hood et al. 2002; Twyman et al. 2003). The accomplishment of these experimental demonstrations is that these plant-based platforms may be employed to produce desired recombinant-proteins in a viable setting. This practice is known as "molecular farming," and despite early skepticism, the initial generation of the desired recombinant vaccines/proteins being engendered in genetically modified transgenic plant species is currently marketed.

Transgenic crops are now generally acknowledged around the world and have become a crucial component of modern agriculture (James 2011). The likelihood of transgenes unintentionally entering the environment increases as the number of transgenic plants increases. Low-level gene transfer from transgenic plants to nontransgenics through pollen and seeds has occurred for a few species, but it has had no discernible effects (Warwick et al. 2009). Although crops regularly interchange DNA with their wild and weedy cousins, authorities have concentrated explicitly on the risks associated with transgenic crops' gene flow. However, gene introgression and hybridization are conceivable but it does not inherently entail that transgenic give their weedy cousins a discerning advantage (Gressel and Al-Ahmad 2012). In fact, endogenous crop gene introgression to wild families is occasional for the majority of crops in the majority of locations (Ellstrand et al. 2013). There is an evident benefit to completely avoiding gene flow, even if the ecological risk is frequently linked to a crop and wild relative's characteristics, environment, and mating system (Stewart Jr et al. 2003). This seems to be particularly true for transgenic perennial crops that the transgene-linked traits confer resistance like salt, drought, or cold tolerance. But these are not entirely characterized (Kausch et al. 2010; Stewart et al. 2003; Moon et al. 2010a; Warwick et al. 2009). Before therapeutic crops may be commercialized, regulators may need biocontainment (Moon et al. 2010a, b; Stewart Jr. 2008). Technologies that diminution or stop gene flow would therefore be advantageous from both an ecological and regulatory point of view.

The biological arrest of transgenes, sometimes known as "bioconfinement," has been a contentious topic since the beginning of the plant biotechnology. There aren't any commercially viable confined transgenic plants on the market that have completely controlled pollen and seed-mediated gene transfer. There have been several procedures proposed (Gressel 1999; Daniell 2002; Moon et al. 2011), such as methods for male bareness or sterility, maternal inheritance (Mariani et al. 1990; Daniell et al. 1998; Iamtham and Day 2000; Ruf et al. 2001), transgenic extenuation (Al-Ahmad et al. 2006), transgene editing (Gressel 2010a, b; Oliver et al. 1999a, b). Moreover, even though naturally occurring confinement systems like ploidy hindrances, apomixis, and the genomic incompatibility are useful in a small number of plant species that are more difficult to manage for genetic manipulation (Daniell 2002; Kausch et al. 2010), they have caught the attention of researchers with thrilling advancements (Hanna 1981; Johnson et al. 2006; Sandhu et al. 2010; Kannan et al. 2012). These approaches have compensations as well as disadvantages. For instance, a seed-based system originally known as the "Technology Protection System" that later changed its name to "GeneSafe Technologies" and was dubbed the "Terminator" was one of the first technologies presented and successfully proven (Oliver et al. 1998). To prevent the spread of transgenic organisms, the GeneSafe technique produced seeds that could not germinate (Oliver and Hake 2012). The "Terminator" moniker was joyfully adopted by GeneSafe's detractors, who claimed that it would harm farmers in developing countries by prohibiting them from conserving healthy seeds for the upcoming season. GeneSafe critics claim that cross-pollination with GeneSafe plants will result in the loss of viable seed in regions nearby that are growing nontransgenic crop plants of the same species. Despite having the potential to stop gene flow, this method has never been used commercially because of public outrage and other issues.

The term "gene use restriction technologies" (GURTs) was introduced in a 2001 report to the United Nations Commission of Genetic Organization to refer to a collection of this biocontained technologies (Oliver and Hake 2012). As the primary goal is to limit the usage of features or rational property controlled by the agricultural business, we believe the word

"GURTs" may not be the most appropriate one (Oliver and Hake 2012). GURTs can, of course, be employed to prevent the rational and intellectual property, but they could be also a useful tool in the confinement of transgenics, which we have faith in, will become a more valuable function in the future.

19.2 Transgenic Plant Technology

Techniques for protein expression have been developed for both transiently transformed plant tissues as well as permanently altered plants. For this strategy, the gene has been incorporated into the plant genome. Effects from transient gene expression appear within days, making it faster than continuous plant transformation. Temporary gene expression, which can only be expressed in small amounts, is frequently used to test protein expression plans before steady transformation. Stable transformation in plant tissues is the genomic insertion of a new transgene from the same or other species. The plastid and nuclear genomes can both be altered using a number of transgenesis techniques (Horsch et al. 1985; Staub et al. 2000). The two main transformation techniques now being used in plant transformation are as; one is biolistic gene transfer to the monocot plants like wheat, rice, and maize, or agrobacterium-mediated gene delivery to dicots like pea, potato, tomato, tobacco, ashwagandha (Horsch et al. 1985; Christou 1995). However, in plant biotechnology, the ease of tissue transformation is a major bottleneck, and it still impacts the choice of the crop in molecular farming. Although the basis for a stable transition is well acknowledged, it takes time. We need three to nine months, dependent on the plant species, to create transgenic varieties that can be further evaluated for the functions of transgene genes and phenological characteristics. Although transient expression of the desired genes can be evaluated in an expression system like *N. benthamiana*, to check the construct's functionality before switching to the large-scale generation of stable transgenic lines. Moreover, transient expression systems can be possibly used for protein synthesis. Using recombinant *Agrobacterium tumefaciens* to enter plant tissue is known as agroinfiltration (Kapila et al. 1997). During transformation, T-DNA is carried to the nucleus and the expression of transgene causes the production of that protein in milligram amounts (Vaquero et al. 1999). Since viral infection is swift and systemic and the infected cells generate many copies of viral DNA and a significant amount of viral gene products, viral vectors have also caught the attention of researchers (Verch et al. 1998; Gleba et al. 2004). Additionally, plant viruses' genetic material does not integrate into infected plant genomes. Therefore, no long-lasting change takes place and the foreign genes are not transferred to the germ line. Instead, plant viruses typically infect a wide variety of hosts, and quickly spread by means of mechanical injuries and can even travel from plant to plant, allowing the rapid infection of a large number of plant species. For the biotechnological purpose, viral vectors have been utilized to generate a number of antibodies (Verch et al. 1998; Hendy et al. 1999; McCormick et al. 1999). For example, tobacco mosaic virus vectors bearing the antibody's heavy and light chains were produced. Tobacco plants were transformed with these vectors the transgenes or any other marker genes were expressed. The implant antibody's production was verified.

19.3 Transgenic Plants in Bioremediation

Large-scale mining, agricultural, industrial, and military operations have all caused the release of enormous amounts of poisonous compounds into the environment. Plants, animals, and people are more vulnerable to toxic metals or metalloids as well as organic pollutants. It is therefore imperative to get rid of these toxins from polluted areas as soon as feasible (Cherian and Oliveira 2005). Traditional remediation treatments for these locations include soil mining, washing soil, filling of land and immobilization or extraction utilizing physicochemical techniques, which are ineffectual and expensive. Furthermore, they typically destroy natural ecosystems and leave unsightly marks on the ground. As a result, practical alternatives like bioremediation have gained popularity. In the bioremediation process, microorganisms are used to convert impurities/pollutants into safe byproducts. When bioreactors are utilized, this process necessitates a high microbial density and a steady supply of nutrients (French et al. 1999). The practice of using plants to remove metal or organic toxins has gained popularity recently, giving rise to the use of the concept of phytoremediation (Raskin 1996; Garbisu et al. 2002). A brand-new, low-cost technique called phytoremediation makes use of plants to eliminate, or soothe contaminants like organic ones found in soils, water, and sediments. The advantageous effects of phytoremediation over conventional bioremediation by microbes include the fact that plants, being autophytic systems with huge biomass, only need a little amount of nutrients and stop poisons from spreading by water and wind (Pulford and Watson 2003). In order to create and maintain a microbial population for additional pollution detoxification, plants additionally feed nutrients to rhizosphere bacteria.

The ability of plants to heal themselves can be pointedly increased via genetic alteration and plant transformation strategies (Pilon-Smits and Pilon 2002). The recent developments in transgenic plants with enhanced phytoremediation potential have demonstrated how very advantageous it is to find novel genes in hyperaccumulators and transfer them to fast-growing species (Ellis et al. 2004; LeDuc et al. 2004).

The extraordinary potential of various plant species for the accumulation of high concentrations of some dangerous metals in their tissues or organs gave rise to the concept of phytoremediation in the 1980s. The word "phytoremediation" initially emerged in the scientific literature in 1993, and numerous related technologies that allowed practical usage of higher plant species to purify soil and water were created throughout time (Cunningham et al. 1995; Salt et al. 1995). "Phytotechnologies" refer to a broad range of technologies which can be utilized to remove pollutants through four different processes: accretion and sequestration, metabolism, volatilization, and metabolism, including rhizosphere degradation. A thorough discussion of phytotechnologies is provided by McCutcheon and Schnoor (2003). Marmiroli and McCutcheon (2003) provided evidence of how phytoremediation evolved from the identification of characteristics in wild species found in particular settings (such as endemic hyperaccumulators) to the exploitation of agricultural plants and the creation of genetically altered genotypes. The objective of this research was to develop plants that would produce a lot of biomass, be able to use a wide range of environmental and climatic conditions, be simple to grow and care for, and potentially be able to remove a sizable amount of pollutants. For this, researchers looked for plant species with characteristics that facilitate phytoremediation, such as deep root systems, quick development, and contamination tolerance. Transgenic techniques have been utilized to impart specialized characteristics for detoxification, accumulation, or resistance. This approach is now applicable to a large number of plant species due to improvements in plant genetic transformation (Meyers et al. 2010).

In recent years, a rise in interest in the concept of phytoremediation has been observed, which involves using plants to remove or deactivate contaminants, toxic compounds from soil, garbage, and surface waters (Salt et al. 1995, Cunningham et al. 1995; Raskin et al. 1997; Chaney et al. 1997; Meagher 2000). Plants are more environmentally friendly and inexpensive in comparison to the use of traditional physiochemical methods of soil repair like excavation, digging, and burial (Cunningham and Berti 2000). Over the next five years, it is anticipated that the global phytoremediation market, which was estimated to be worth about $35 million in 1999, will increase more than 10-fold (Kramer and Chardonnens 2001). It has a good chance of growing its share of the global cleanup business, which totaled more than US$18–19 billion in 1998 (Glass 1999). In order to gather ions from soil and further concentrate these in their biomass, plants use photosynthetic energy. Pollutants can penetrate higher plants when present in greater quantities due to their chemical resemblance to other nutrient ions. Cd^{2+} or AsO_4, for instance, can be absorbed for roots either for PO_4, Fe^{2+} or Ca^{2+} (Meharg and Macnair 1992; Clemens et al. 1998; Cohen et al. 1998). Utilizing plants' mechanisms for acquiring nutrients, phytoextraction aims to maximize the accumulation of pollution trace elements in tissues above ground. Then, over a brief period of time, the contamination is cleared from the area by collecting above-ground biomass. To concentrate the pollution even further, plant debris can be recycled and disposed of in specific landfills. When grownup in a polluted environment, a plant used for phytoextraction must: (i) accumulate large amounts of trace elements and compounds in their shoots; (ii) have a high rate of photosynthesis in these conditions and produce high biomass; (iii) establish a large root system. Additionally, the chemical properties of a few dangerous trace elements, such as selenium and mercury, permit the use of phytovolatilization technology. The trace compounds/element is enzymatically changed into less dangerous, volatile molecule, and then released into the environment rather than building up inside the plant (Rugh et al. 1996; Meagher 2000; Pilon-Smits and Pilon 2000). Polluted streams can be cleaned up using phytovolatilization as well as phytoextraction because streams frequently have better trace element bioavailability than soils do. In this case, rhizofiltration – the accumulation of toxic and unwanted trace elements into root biomass and further followed its removal is sufficient for the restoration of the contaminated soil quality (Dushenkov et al. 1995).

19.4 Metal Accumulators

Metal hyperaccumulator plants growing well in polluted or contaminated soils with elevated levels of metals and concentrate metals in their cells and tissues at high concentrations up to high orders of magnitude higher than "normal" plants grown nearby (Baker and Brooks 1989). All four metals have been found to be hyperaccumulating: cobalt up to 1.2% Co in shoot dry biomass (SDB), cadmium (0.2% Cd in SDB), nickel (up to 3.8% Ni in SDB), and zinc (up to 4% Zn in SDB) (Baker and Brooks 1989; Lombi et al. 2000). More recently, a hyperaccumulator of the metalloid arsenic (up to 0.75% as in SDB) and plants that accumulate extremely high levels of the nonmetal selenium (0.4% Se in SDB) have also been discovered (Shrift 1969). Metal hyperaccumulators are particularly fascinating model organisms having overwhelmed substantial physiological restrictions by limiting the accumulation of metals in shoots (Salt and Krämer 2000; Ma et al. 2001).

19.4.1 Metal Hyperaccumulation in Relation to Molecular and Genetic Basis

Thlaspi caerulescens is a hyperaccumulator, while *T. arvense*, a similar nonaccumulator, has different transcriptional control of ZNT1, transporters in the membrane belong to the ZIP family (Saier 2000). According to Pence et al. (2000) and Lasat et al. (2000), ZNT1 expression in roots of plants enhances the zinc uptake rate in *T. caerulescens*. Moreover, *T. arvense*. ZNT1 is also capable of facilitating Cd^{2+} and Zn^{2+} absorption when over-expressed in *Saccharomyces cerevisiae*. Approximately 50 times more external metal is present in the hyperaccumulator than the nonhyperaccumulator, which causes zinc to mediate the downregulation of ZNT1 mRNA levels. In nickel hyperaccumulators plants, metal acquaintance results in a significant enhancement in concentrations of the free amino acid histidine, a specific chelator capable of detoxifying Ni^{2+} and enhancing the rate of nickel absorption and translocation to the shoot via the transpiration stream (Kramer et al. 1996). Metal chelation and the subcellular compart-mentalization into the apoplast and vacuole is well sequestration inside specialized tissues, like the epidermis or trichomes, are ways that metal detoxification occurs in hyperaccumulator shoots (Vázquez et al. 1992). These detoxification processes have not yet been explored at genetic and molecular levels. The closely related, nontolerant, nonaccumulator *Arabidopsis lyrata* and cadmium, zinc-tolerant zinc hyperaccumulator *Aphyosemion halleri* were crossed to produce a number of F2 offspring. Only a few numbers of crucial *A. halleri* genes are associated with both zinc hyperaccumulation and tolerance (Macnair et al. (1999)). The validity of this conclusion will require additional research on the F2 offspring. The production and study of crosses between hyperaccumulators and other related nonhyperaccumulators will be a key strategy in exposing the genes which could be responsible for metal hyperaccumulation.

19.5 Need for Transgenic Plants

19.5.1 Improve Plant Performance

Brassicaceae crop species including *Brassica oleracea* and *R. sativus*, metal hyperaccumulators like *T. caerulescens* and *Silene vulgaris* were used in the initial phytoremediation research. These plants were also known to concentrate heavy metals or have a high tolerance to metal accumulation (Baker et al. 1991; Brown et al. 1994). Metal hyperaccumulators demonstrated the best metal removal performance in these field tests. According to the authors, cleaning up heavily polluted soil would require 6 and 130 croppings, respectively. Brassica plants with high biomass that are non-accumulators were more successful at removing zinc from polluted soil than *T. caerulescens*, which had a low rate of biomass development, which is typical of most hyperaccumulators (Ebbs et al. 1997). According to the results of the pilot field testing, significant increases in either hyper-accumulator biomass output or nonaccumulator metal accumulation would be required for phytoextraction to become a practical approach (Felix 1997; Chaney et al. 2000; McGrath et al. 2000). These results provide evidence for one potential benefit of using transgenic techniques for phytoremediation.

19.5.2 Exploring Genetic Diversity beyond the Plant Kingdom

Polluted soils frequently contain elevated levels of multiple contaminants and trace elements, indicating that phytoremediation will almost certainly require engineered plants to accumulate several toxic and trace elements. In contrast, naturally occurring hyper-accumulation plant species are largely specific to trace elements (Cunningham and Ow 1996; Chaney et al. 1997). Cleanup efforts should focus on soils that have arsenic, cadmium, lead, or mercury contamination. There have not been any plants found yet that can continuously hyper-accumulate mercury or lead. In most naturally existing hyperaccumulator plants thus far identified, tolerance to lead, selenium, or arsenic seems to be on elimination from the plant (Meharg and Macnair 1992). It is almost clear that the gene transfer among species boundaries will be necessary for the development of improved phytoremediation technique for toxic and trace elements. Despite the lack of knowledge regarding the detoxification and hyperaccumulation in plants, the decontamination systems from yeast and bacteria have yet to be explored (Ehrlich 1997), for example, for the decontamination of arsenic (Mukhopadhyay et al. 1998; Ghosh et al. 1999), cadmium (Nies et al. 1989), lead (R (Gupta et al. 1999). The positive insertion of an improved bacterial mercuric ion reductase gene into *Nicotiana tabacum* and yellow poplar (*Liriodendron tulipifera*) showed that the transfer of bacterial genes can be incredibly helpful in phytoremediation, despite the exertion of envisaging the effects of the microbial gene in a multifaceted multicellular organism like a plant (Heaton et al. 1998; Rugh et al. 1998; Nies 1999). Cosuppression can stop implanted plant genes from being overexpressed in plants since they are identical to native genes. On the other hand, plants have successfully expressed microbial genes in a number of instances. Some microbial genes must have their GC contents altered since they cannot be expressed in plants due to their unfavorable expression (Rugh et al. 1996).

19.6 Phytoremediation Via Chelation

The only economically viable plant-based method for removing lead contamination from soils is chelator-assisted phytoremediation. Lead bioremediation requires high biomass growing Brassicaceae crop plant *B. juncea*, followed by adding the EDTA chelator to the soil to cause lead mobilization into the soil solution. Because of this, plant shoots quickly accumulate very high levels of lead. Therefore, a few days after EDTA administration, plant material can be harvested.

19.6.1 Uptake of Toxic Elements by Transgenic Plants

To sustain elevated levels of the controlled absorption of toxic elements at the low concentration in the soil solution, it is ideal for there to be an increased inflow into the root symplasm system (Marschner 1995). Additionally, there are so many evidences of apoplastic channels for solute transfer into the xylem, which appear to be minor and only occur at sites where lateral roots enter the endodermis or in immature root regions (Marschner et al. 1987; Yeo et al. 1987).

19.6.2 Metal Transport across the Plasma Membrane

There are a number of transporters that are known to move divalent nutritional cations like Ca^{2+}, Fe^{2+}, and Zn^{2+} as well as other divalent cations. For instance, the iron-repressed metal transporter IRT1 from the ZIP family of plants, when expressed heterologously in yeast, exhibits a wide range of transport selectivity for Cd^{2+}, Mn^{2+}, Fe^{2+}, Zn^{2+}, and maybe other divalent cations (Eide et al. 1996; Korshunova et al. 1999). When the iron levels are low, plants' IRT1 expression is strongly promoted, whereas when iron levels are high, plants' IRT1 expression is inhibited. This is in line with the finding that iron-deficient pea seedlings take in more cadmium (Cohen et al. 1998). There have not been any instances of plants bred to overexpress ZIP family transporters, albeit these transporters are being regulated at the molecular or transcriptional and posttranscriptional levels. NtCBP4 protein, a potential cyclic-nucleotide and calmodulin regulated cation channel in cell membrane was bred into tobacco plants, and these plants displayed improved lead sensitivity, 1.5–2.0 fold lead accumulation in the shoot, and higher nickel tolerance (Arazi et al. 1999). The wheat LCT1 gene which is a low-affinity cation transporter causes hypersensitivity to Cd^{2+} and increased cadmium accumulation in baker's yeast cells (Clemens et al. 1998). Despite no evidence of increased cadmium accumulation, plants overexpressing *At*Nramp3 a metal transporter from the Nramp family, exhibited hypersensitivity to Cd^{2+} (Thomine et al. 2000). A transporter that can move a particular contaminated metal cation in a natural soil environment may also move other competing cations, such Ca^{2+} or Zn^{2+}, if those ions are present in significant excess. In order to create mutant transporters with various specificities, it is vital to comprehend better what determines the specificity of membrane transporters (Rogers et al. 2000). The analysis of Arabidopsis mutants with various metal responses and knowledge of how the ZIP family members are regulated in *T. caerulescens* can both help in the search for new target genes and methods for cultivating plants with elevated metal absorption.

19.6.3 Metal Speciation's Impact on Metal Uptake

A change in metal speciation may result in higher transport rates through the root plasma membrane's existing transporters. Grasses may also take up zinc phytosiderophore complexes in addition to iron phytosiderophore complexes (von Wirén et al. 1996; Curie et al. 2001). Transportability tests for further trace metal phytosiderophore complexes are still pending.

The source of Ni(II) as a compound containing the amino acid histidine in its place of the free metal-cation results in a 40-fold elevation in nickel translocation in the nonhyperaccumulator *Arabidopsis montanum* (Kramer et al. 1996). This might be due to the chelation of the nickel ion with amino acid histidine which facilitates their transport via the plasma membrane with the involvement of specific transporters. After the transported trace element species are removed from the cytoplasmic chemical equilibrium through chelation, chemical transformation, or transit, such as into the vacuole, increased net absorption rates may take place at the tissue or overall plant level.

19.6.4 Trace Element Detoxification

In the phytoremediation process, tolerance of toxic elements to higher concentrations in the roots or other plant tissues is highly needed. The level of a potential toxin, such as a free metal cation, was decreased using a variety of transgenic approaches. Three main strategies are available for use: (i) efflux from cytoplasm and transport into the vacuole; (ii) volatilization or chemical transformation; (iii) overproduction of metal chelators or other toxic compounds that bind the trace element.

19.7 Phytovolatilization

Chemical alteration of trace elements into anoxic, volatile molecules is a particularly good detoxification strategy since the potentially hazardous element is eliminated from the tissues. As a result of microbial activity in mercury-contaminated soils and sediments, deadly Hg(II) is transformed into extremely hazardous organomercurials called methylmercury (CH_3Hg^+) and other toxic mercury compounds. These areas include mercury-resistant bacteria that can break down Hg(II) and organomercurials into elemental mercury, which is far less harmful. The merB gene produces organomercurial lyase, which changes methylmercury into the less dangerous mercury (Hg) (II). A second enzyme that uses NADPH as an electron source to catalyze the reduction of mercury to elemental mercury is encoded by MerA. By volatilization under highly toxic environment, elemental mercury passes into the biogeochemical cycle (Hamlett et al. 1992; Bizili et al. 2000). The rate of bacterial mercury volatilization is slow for application in bioremediation, Rugh et al. (1996) changed the pathway of the volatilization of mercury into plants. A bacterial merA gene's nucleotide sequence must be altered to allow for high level of expression in plants. On agarose plates containing 50 M and 100 M $HgCl_2$, concentrations that completely blocked wild-type seed germination, *Arabidopsis thaliana* seedlings expressing merAPe under the control of CaMV-35S promoter sprouted and thrived. Improved Au^{3+} tolerance has also been shown in transgenics. Then, altered merA genes were transformed into plants to produce greater biomass. Mercury removal from soil substrates is unknown, although tobacco transformants carrying a merA gene were able to grow and bloom in soil having up to 500 ppm Hg(II) (Heaton et al. 1998). Trees may be helpful in phytovolatilization because of their extensive root systems, long lifespans, and large litter creation because they may increase the availability of metals in the soil (Rugh et al. 1998). Volatilization of mercury was found to be 10 times high in merA-expressing transgenic poplar plants than in non-transgenic control plantlets. The transgenics usually volatilize Hg at a rate of about 1 g/day when grown on agarose media with 10 M $HgCl_2$ (Rugh et al. 1998). Sub micromolar quantities of very toxic organo-mercurials hinder germination in both wild as control and merAP-expressing *A. thaliana* (Bizili et al. 1999). Seedlings over-expressing a bacterial merB encoding organo-mercurial lyase under the control of a CaMV-35S promoter, remain unaffected by 2 M organomercurials (Bizili et al. 1999). *A. thaliana* seedlings that express merA as well as merB can grow in methylmercury concentrations up to 10 M, which is more than 40 times higher than the level that wild type seedlings can endure (Bizili et al. 2000). Moreover, roots of the transgenic having merA merB were administered on 25 M organomercurials in solution and there was an estimated rate of 14.4 to 85.0 g Hg g-1 fresh biomass day-1 of elemental mercury volatilization. As expected, under these conditions, mercury volatilization from *A. thaliana* expressing either merB or merA was less or negligible. In order to create a superior mercury phytoremediator plant, combining the expression of merA and merB in a high-biomass plant may be a viable first step. The researchers suggest that a readily saturable transport step could become rate-limiting in the presence of several MerA and MerB genes (Bizili et al. 2000). The accessibility of mercurial elements for root absorption is expected to become rate-limiting on soil substrates, though. Some plants have been seen to just slightly volatilize selenium, either as dimethyl diselenide or dimethyl selenide. The relevant enzymes have not yet been found. Selenate is transformed into selenium organic forms before selenium volatilizes (De Souza et al. 1998). Any additional plant trace elements that are to be volatilized appear to necessitate the insertion of important microbial genes (Cheng and Focht 1979).

19.8 Chemical Modification

Higher plants can absorb and digest selenate, the selenium oxoanion, because of its chemical resemblance to sulphate. Overproduction of selenocysteine and selenomethionine was produced as a result of growth on selenate-contaminated soils. Instead of cysteine and methionine, they are absorbed into sensitive plant proteins, rendering affected inactive proteins (Neuhierl and Böck 1996). *A. thaliana* gene APS1, which encodes ATP sulfurylase, was overexpressed in *B. juncea* under the direction of a CaMV-35S promoter in an effort to improve selenium absorption by plants. Transgenic plants had somewhat higher selenium tolerance than wild-type controls and roughly doubled selenium accumulation in their shoots. These transgenic plants' enhanced sulfur uptake resulted in higher glutathione levels by approximately 97% and 31% in the shoots and underground roots, respectively. This suggests that ATP sulfurylase may be a promising target for phytoremediation of other toxic metals, such as cadmium.

19.8.1 Trace Element Binding

Low molecular weight proteins called metallothioneins (MTs) have high cysteine and strong affinity for binding metal cations like cadmium, zinc, and copper (Hamer 1986). Metallothionein-like proteins have been found associated with metal homeostasis in humans, yeasts, cyanobacteria, and also in plants (Huckle et al. 1993; Murphy et al. 1997). A number of scientists and researcher groups have introduced metallothioneins from animals to plants via a transgenic approach, mostly to prevent metal buildup in shoots by capturing the metal in the roots. The production of a mammalian MT in *N. tabacum* under constitutive promoter decreased cadmium translocation into the shoot (Yeargan et al. 1992, Elmayan and Tepfer 1994). After being exposed to modest (0.02 M) cadmium levels in the rooting media, transgenic plants' leaf cadmium accumulation was 22% lower than those of control plants. But in the real world, there was no discernible variance between transgenic and control plants in terms of leaf cadmium concentrations or plant development (Yeargan et al. 1992). These results showed that trace element absorption data obtained in a glasshouse or growth chamber on nonsoil substrates cannot be used to predict transgenic plant performance on soil substrates or in the field. Other research is focused on the overexpression of MT as a means of improving cadmium tolerance (Misra and Gedamu 1989; Pan et al. 1994a, b). Wild-type plants were incapable to live in a hydroponic media at dosages greater than 25 M cadmium, but transgenics of *B. oleracea* expressing CUP1 metallothionein gene of yeast survived 400 M cadmium (Hasegawa et al. 1997). The top leaves of transformants growing at 50 M cadmium contained 10–70% more cadmium than untransformed plants growing at 25 M cadmium. This shows that cadmium removal from the leaves was not the cause of the elevated tolerance which was observed in transgenic plants. Seedlings of *A. thaliana* that were altered to express pea *Ps*MTA under constitutive promoter has accumulated up to eight times as much copper in comparison to control (Evans et al. 1992). Finally, increased plant tolerance to particular metals like cadmium and copper can be achieved by overexpressing MTs. However, this needs to be demonstrated in practice. Only a few successful instances of MT overexpression led to increased accumulation of metals in the shoots. This severely limits the application of MTs in phytoremediation. It may be possible to adapt MT overexpression to the requirements of a certain application by directing the production of MTs through root or shoot-specific promoters. Numerous trace elements can be scavenged and subsequently detoxified by organic compounds that contain low molecular weight sulfur (Grill et al. 1985). Cad1 mutants of *A. thaliana* are highly copper- and cadmium-sensitive (Howden et al. 1995). PCs may bind metals like arsenic and cadmium that have a strong affinity for sulfhydryl groups (Grill et al. 1987; Schmoger et al. 2000). In response to exposure to metals, metal-sensitive plants manufacture more PCs than metal-tolerant plants (Harmens et al. 1993). This shows that while all plants have some level of metal tolerance, those acclimated to soils rich in metalliferous elements have additional tolerance mechanisms (Cobbett 2000). Numerous initiatives have been made to hasten the production of glutathione, a precursor to PC, which is known to occur in some plants following exposure to cadmium (Xiang and Oliver 1998). The ATP-driven transfer of glycine onto -glutamylcysteine to produce -glutamylcysteinylglycine (-Glu-Cys-Gly; glutathione) is catalyzed by the enzyme glutathione synthase. A fivefold increase in glutathione content was seen in the roots of *B. juncea* after the expression of the *Escherichia coli* glutathione synthase gene (gsh2), but only after exposure to cadmium (Zhu et al. 1999a). Additionally, plants exposed to cadmium had around twice as much (-Glu-Cys)2-Gly (PC2). Overall, transgenic plants' non-protein thiol content increased by less than 10% in the shoots and by almost twice as much in the roots. The researchers also found that plants that produce gsh2 had 26% more cadmium in their shoots (Zhu et al. 1999a). 150 M cadmium on an agarose medium, which reduced wild-type seedlings' root growth by 50%, had no effect on the transgenic seedlings' root development. However, the synthesis of glutamyl-cysteine from glutamate and cysteine, which is fueled by ATP and is known to be negatively regulated by glutathione, is a more important point of metabolic control in glutathione formation (Noctor et al. 1998). Glutathione levels were three times greater in *B. juncea* transformed with a chimeric *E. coli* gsh1 gene that produces glutamyl-cysteine synthetase coupled to a 35S promoter and a chloroplast transit peptide than in nontransgenic control plants (Zhu et al. 1999a, b). Gsh1 expression showed higher cadmium tolerance on par with gsh2 overexpressors. Compared to controls, PC levels increased by around 30% in the roots and shoots of cadmium-exposed gsh1 transformants, whereas Cd concentrations surged by 40–90% in the shoots. During a 14-day period, metal-uptake studies were carried out in the hydroponic culture system, and the metal concentration lessen the generation of new biomass by at least 70% and 40% in control and transgenic plants, respectively (Zhu et al. 1999a, b). Experiments must be conducted at lower harmful cadmium concentrations and on a cadmium-containing soil substrate to determine these plants' eligibility for phytoremediation. Cadmium exposure alone caused nonprotein thiol levels in shoots and roots of untransformed *B. juncea* to increase by 400% and around 2000%, respectively (Zhu et al. 1999a, b). The overexpression of gsh1 only caused a 50–70% additional increase when exposed to cadmium, as shown by a similar rise in cadmium concentrations (Zhu et al. 1999a, b). According to Dominguez-Solis et al. (2000), simultaneous expression of the sulfur assimilation pathway genes PC synthetase and -glutamylcysteine synthetase may raise tissue concentrations of nonprotein thiols and boost cadmium tolerance and probably cadmium accumulation in the shoot. There has not been much research done on other ligands that may improve plants' tolerance to metals. Exogenous histidine applied topically can reduce or remove the nickel sensitivity of *A. montanum* (Krämer et al. 1996). According to primary data, plants created to produce excessive amounts of histidine are more nickel-resistant (Krämer et al. 2000).

19.8.2 Membrane Transporters

Transformant tobacco seedlings had increased nickel tolerance due to the overexpression of *Nt*CBP4 (Arazi et al. 1999). The biomass production of wild-type and transgenic lines was 50% suppressed at 90 M nickel and 200 M nickel, respectively. Reduced nickel buildup in shoots was associated with increased transgenic nickel tolerance, suggesting that removing nickel from shoots may increase NtCBP4 overexpressors' tolerance. Numerous genes have been identified that could be involved in the compartmentalization of intracellular structures like the vacuole or cellular efflux of toxic and trace elements. These include ATPases, cation diffusion facilitator (CDF) transporters (Nies et al. 1989), ATP-binding cassettes and divalent cation/proton antiporters (Shaul et al. 1999). However, scientists are only now starting to explain how these transporters work in plants and how their overexpression affects metal homeostasis. Using *in vitro* absorption tests to identify the substrates of these transporters requires caution because heavy metals are likely to be chelated in the cytoplasm (Williams et al. 2000). In comparison to controls, plants overexpressing ZAT1, which encodes a potential membrane transporter for zinc from the CDF family, accumulated more zinc in their roots. Approximately 85% of the root development of control plants was reduced at 200 M zinc in a hydroponic solution, but only 15% of the root growth of transgenic plants was affected (Van der Zaal et al. 1999). Hirschi (1999) found that the similar CAX2 was able to transport the chemically identical ion, Cd^{2+}, whereas the *A. thaliana* $Ca2^{+}/H^{+}$ antiporter CAX1 (calcium exchanger 1) appeared to be Ca^{2+} specific (Hirschi et al. 2000). Roots of tobacco plants expressing the *A. thaliana* CAX2 expressed under CaMV-35S promoter accumulated amounts of calcium, manganese, and cadmium that were almost twice as high as those of empty vector transformants and showed somewhat improved tolerance to Mn^{2+} (Hirschi et al. 2000). Only a small percentage of primary transformants that expressed CAX2 showed symptoms of disturbed nutritional cation homeostasis. Additionally, it showed that Mg^{2+} and Zn^{2+} hypersensitivity was present in transgenic tobacco plants over-expressing the *A. thaliana* vacuolar Mg^{2+}/H^{+} antiporter AtMHX1 under the control of the 35S promoter (Shaul et al. 1999). This emphasizes the importance of comprehending the function of these transporters better. Mutagenesis of specific residues in these Ca^{2+} and Mg^{2+} transporters may be one method of preventing disruption of the nutritional cation balance by altering cation selectivity. Inhibiting vacuolar compartmentation in the roots while enhancing it in the shoots is the aim of phytoextraction. Therefore, it is envisaged that the use of tissue-specific promoters will be essential in cutting-edge transgenic phytoremediation techniques.

19.9 Risk Assessment

Transgenic plants with advanced and modified metal tolerance, efficient accumulation, or transformation/modification properties are advantageous for a variety of reasons. They provide new details on the underlying biological processes that underlie these actions, such as the pathways that are activated and the rate-limiting enzymes (Pilon-Smits and Pilon 2002).

In addition to enhancing agricultural production in areas with low soil metal levels, plants with modified metal accumulation abilities may be beneficial as "fortified diets" for humans or cattle or for phytoremediation (Guerinot and Salt 2001). When genetically modified plant species are being used for any of these goals, exhaustive risk analysis assessment studies should be carried out in each instance (Wolfenbarger and Phifer 2000). The biological conversion of metals into more bioavailable forms, increased exposure of wildlife and people to metals (due to enhanced metal accumulations in edible plant tissues/parts or being volatilized), unchecked spread of modified plants with transgenic nature due to good fitness (e.g. enhanced tolerance of metals) or generally weedy nature or unchecked transfer of the transgene through inter-breeding with the populations of the wild plant relative are some of the potential risks, for an inclusive report on (Glass 1997). The rewards, as well as the hazards of doing nothing or using alternative technology, must be balanced against these risks on a case-by-case basis. As far as we are aware, there are currently no transgenic plants being utilized for phytoremediation. As a result, no studies have ever been done to determine the true risks associated with using transgenic plants for phytoremediation. On the other hand, Meagher and colleagues did theoretical analyses of the risks associated with the use of mercury volatilizing plants (Meagher et al. 2000; Rugh et al. 2000). They estimate that the mercury released by these facilities offers no significant environmental risk and is negligible in comparison to other mercury sources, such as the burning of fossil fuels and the dumping of medical waste. Although the phytoremediation site's volatile mercury level is 400 times elevated than background levels, it is still 25 times below the recommended regulatory limits. Moreover, elemental mercury is retained in the atmosphere for one to two years before precipitation occurs, during this time of exposure, it is diluted/thinned to safe levels. In a related theoretical assessment of the risk of volatile selenium emitted by plants (Lin et al. 2000), Norman Terry and associates came to the conclusion that the volatile Selenium will likely be useful rather than harmful since it will precipitate/accumulate in Selenium-deficient areas. Any increase in metal buildup caused by biotechnology would boost this risk correspondingly because metal deposition in plant shoots raises the probability of animal ingestion. The length of exposure time may be lessened when transgenic plants are being employed, though, if a place can be cleansed more quickly. By railing off the land area, using toxic deterrents such as periodic noise, and using less desirable plant species, metal consumption by wildlife can be decreased. According to Meagher et al. (2000), transgenic plants often provide little to no benefit over untransformed plants, whether in a clean or dirty environment. Hence, the possibility of transgenics or their inserted genes "escaping" is not a major worry. However, a greenhouse experiment measuring transgenic gene frequency across a number of generations on the polluted and uncontaminated soil might be done before using specific transgenics for phytoremediation in the field. Transgenic plant species with no acceptable wild relatives may be chosen, male sterile transgenics may be produced, and/or the plants may be removed before blooming to lessen the likelihood of outcrossing to wild relatives. The use of transgenic plant species in phytoremediation in the United States should not be significantly hampered by government restrictions (Glass 1997). Although the US Department of Agriculture recommends submitting permit requests around four months before the planned start date, clearance for research field experiments should typically be standard in most places.

19.10 Future Perspectives

The genetic alteration in plants has been shown in numerous investigations to produce plants with modified metal tolerance, elevated accumulation and high biotransformation capabilities. Metal accumulation per plant rose by a factor of two to three when genetic engineering sped up natural plant processes. This may diminish the cost of phyto-remediation to a comparable level if the same or repeated results are produced in the field. Additionally, the development of a unique method has led to the emergence of plants that can detoxify contaminated soils in many ways that other plants cannot, perhaps making them even more beneficial. In the next years, it is conceivable that some of the newly available transgenics will undergo testing in a setting more akin to actual phytoremediation. The discovery of additional metal-related genes as a result of genome sequencing activities may open up a variety of new possibilities for the creation of novel transgenics that have characteristics that are advantageous for phytoremediation. Numerous genes can be overexpressed simultaneously, in addition to constitutive overexpression of a single gene, and the overexpression can be tailored in certain tissues, conditions, or cellular compartments. An intriguing alternative method for affecting metal absorption or exclusion involves manipulating the type and quantity of chemicals emitted by the roots. Together, these new findings are projected to quickly produce a plethora of new knowledge on metal metabolism in plants, possibly paving the way for useful applications in food production, environmental cleansing, and agriculture productivity.

References

Ahmad, P., Ashraf, M., Younis, M. et al. (2012). Role of transgenic plants in agriculture and biopharming. *Biotechnol. Adv.* 30 (3): 524–540.

Al-Ahmad, H., Dwyer, J., Moloney, M., and Gressel, J. (2006). Mitigation of establishment of *Brassica napus* transgenes in volunteers using a tandem construct containing a selectively unfit gene. *J. Plant Biotechnol.* 4: 7–21.

Arazi, T., Sunkar, R., Kaplan, B., and Fromm, H. (1999). A tobacco plasma membrane calmodulin-binding transporter confers Ni^{2+} tolerance and Pb^{2+} hypersensitivity in transgenic plants. *Plant J.* 20: 171–182.

Ashraf, M. and Akram, N.A. (2009). Improving salinity tolerance of plants through conventional breeding and genetic engineering: an analytical comparison. *Biotechnol. Adv.* 27: 744–752.

Baker, A.J.M. and Brooks, R.R. (1989). Terrestrial higher plants which hyperaccumulate metallic elements – a review of their distribution, ecology and phytochemistry. *Biorecovery* 1: 81–126.

Baker, A.J.M., Reeves, R.D., and McGrath, S.P. (1991). In situ decontamination of heavy metal polluted soils using crops of metalaccumulating plants – a feasibility study. In: *In situ bioreclamation* (ed. R.E. Hinchee and R.F. Olfenbuttel), 539–544. Stoneham, MA: Butterworth-Heinemann.

Bizili, S.P., Rugh, C.L., Summers, A.O., and Meagher, R.B. (1999). Phytoremediation of methylmercury pollution: *MerB* expression in *Arabidopsis thaliana* confers resistance to organomercurials. *Proc. Natl. Acad. Sci. U. S. A.* 96: 6808–6813.

Bizili, S.P., Rugh, C.L., and Meagher, R.B. (2000). Phytodetoxification of hazardous organomercurials by genetically engineered plants. *Nat. Biotechnol.* 18: 213–217.

Brown, S.L., Chaney, R.L., Angle, J.S., and Baker, A.J.M. (1994). Phytoremediation potential of *Thlaspi caerulescens* and bladder campion for zinc- and cadmium-contaminated soil. *J. Environ. Qual.* 23: 1151–1157.

Chaney, R.L., Malik, M., Li, Y.M. et al. (1997). Phytoremediation of soil metals. *Curr. Opin. Biotechnol.* 8: 279–284.

Chaney, R.L., Li, Y.M., Brown, S.L. et al. (2000). Improving metal hyperaccumulator wild plants to develop commercial phytoextraction systems: approaches and progress. In: *Phytoremediation of Contaminated Soil and Water* (ed. N. Terry and G. Bañuelos), 129–158. Boca Raton, FL: CRC Press.

Cheng, C.N. and Focht, D.D. (1979). Production of arsine and methylarsines in soil and in culture. *Appl. Environ. Microbiol.* 38: 494–498.

Cherian, S. and Oliveira, M. (2005). Transgenic plants in phytoremediation: recent advances and new possibilities. *Crit. Rev.* 39: 24.

Christou, P. (1995). Particle bombardment. *Methods Cell Biol.* 50: 375–382.

Clemens, S., Antosiewicz, D.M., Ward, J.M. et al. (1998). The plant cDNA LCT1 mediates the uptake of calcium and cadmium in yeast. *Proc. Natl. Acad. Sci. U. S. A.* 95: 12043–12048.

Cobbett, B. (2000). Phytochelatin biosynthesis and function in heavy-metal detoxification. *Curr. Opin. Plant Biol.* 3: 211–216.

Cohen, C.K., Fox, T.C., Garvin, D.F., and Kochian, L.V. (1998). The role of iron-deficiency stress responses in stimulating heavy-metal transport in plants. *Plant Physiol.* 116: 1063–1072.

Cunningham, S.D. and Berti, W.R. (2000). Phytoextraction and phytostabilization: technical, economic, and regulatory considerations of the soil-lead issue. In: *Phytoremediation of Contaminated Soil and Water* (ed. N. Terry and G. Bañuelos), 359–376. Boca Raton, FL: CRC Press.

Cunningham, S.D. and Ow, D.W. (1996). Promises and prospects of phytoremediation. *Plant Physiol.* 110: 715–719.

Cunningham, S.D., Berti, W.R., and Huang, J.W. (1995). Phytoremediation of contaminated soils. *Trends Biotechnol.* 13: 393–397.

Curie, C., Panaviene, Z., Loulergue, C. et al. (2001). Maize yellow stripe1 encodes a membrane protein directly involved in Fe(III) uptake. *Nature* 409: 346–349.

Daniell, H. (2002). Molecular strategies for gene containment in transgenic crops. *Nat. Biotechnol.* 20: 581–586.

Daniell, H., Datta, R., Varma, S. et al. (1998). Containment of herbicide resistance through genetic engineering of the chloroplast genome. *Nat. Biotechnol.* 16: 345–348.

De Souza, M.P., Pilon-Smits, E.A.H., Lytle, C.M. et al. (1998). Rate-limiting steps in selenium assimilation and volatilization by Indian mustard. *Plant Physiol.* 117: 1487–1494.

Dominguez-Solis, J.R., Gutierrez-Alcala, G., Romero, L.C., and Gotor, C. (2000). Cytosolic o-acetylserine(thiol)lyase gene is regulated by heavy metals and can function in cadmium tolerance. *J. Biol. Chem.* 276: 8623–9582. https://doi.org/10.1074/jbc.M009574200.

Dushenkov, V., Kumar, P.B.A.N., Motto, H., and Raskin, I. (1995). Rhizofiltration: the use of plants to remove heavy metals from aqueous streams. *Environ. Sci. Technol.* 29: 1239–1245.

Ebbs, S.D., Lasat, M.M., Brady, D.J. et al. (1997). Phytoextraction of cadmium and zinc from a contaminated soil. *J. Environ. Qual.* 26: 1424–1430.

Ehrlich, H.L. (1997). Microbes and metals. *Appl. Microbiol. Biotechnol.* 48: 687–692.

Eide, D., Broderius, M., Fett, J., and Guerinot, M.L. (1996). A novel iron regulated metal transporter from plants identified by functional expression in yeast. *Proc. Natl. Acad. Sci. U. S. A.* 93: 5624–5628.

Ellis, D.R., Sors, T.G., Brunk, D.G. et al. (2004). Production of Se-methylselenocysteine in transgenic plants expressing selenocysteine methyltransferase. *BMC Plant Biol.* 4: 1–1.

Ellstrand, N.C., Meirmans, P., Rong, J. et al. (2013). Introgression of crop alleles into wild or weedy populations. *Annu. Rev. Ecol. Evol. Syst.* 44: 325–345.

Elmayan, T. and Tepfer, M. (1994). Synthesis of a bifunctional metallothionein/beta-glucuronidase fusion protein in transgenic tobacco plants as a means of reducing leaf cadmium levels. *Plant J.* 6: 433–440.

Evans, K.M., Gatehouse, J.A., Lindsay, W.P. et al. (1992). Expression of metallothionein-like gene PsMT-ALPHA in *Escherichia coli* and *Arabidopsis thaliana* and analysis of trace metal ion accumulation: implications for PSMT-ALPHA function. *Plant Mol. Biol.* 20: 1019–1028.

Felix, H. (1997). Fields trials for in situ decontamination of heavy metal polluted soils using crops of metal-accumulating plants. *Z. Pflanzenernaehr. Bodenk.* 160: 525–529.

Fischer, R., Drossard, J., Emans, N. et al. (1999). Towards molecular farming in the future: *Pichia pastoris*-based production of single-chain antibody fragments. *Biotechnol. Appl. Biochem.* 30: 117–120.

French, C.E., Rosser, S.J., Davies, G.J. et al. (1999). *Nat. Biotechnol.* 17: 491–494.

Garbisu, C., Hernandez, A.J., Barrutia, O. et al. (2002). Phytoremediation: a technology using green plants to remove contaminants from polluted areas. *Rev. Environ. Health* 17 (3): 173–188.

Ghosh, M., Shen, J., and Rosen, B.P. (1999). Pathways of As(III) detoxification in *Saccharomyces cerevisiae*. *Proc. Natl. Acad. Sci. U. S. A.* 96: 5001–5006.

Giddings, G. (2001). Transgenic plants as protein factories. *Curr. Opin. Biotechnol.* 12: 450–454.

Glass, D.J. (1997). Prospects for use and regulation of transgenic plants in phytoremediation. In: *In Situ and On-Site Bioremediation*, vol. 4 (ed. B.C. Alleman and A. Leeson), 51–56. Columbus, OH: Battelle Press.

Glass, D.J. (1999). *Markets for Phytoremediation, 1999–2000*. Needham, MA: Glass Associates. http://www.channel1.com/dglassassoc/index.htm.

Gleba, Y., Marillonnet, S., and Klimyuk, V. (2004). Engineering viral expression vectors for plants: the 'full virus' and the 'deconstructed virus' strategies. *Curr. Opin. Plant Biol.* 7: 182–188.

Gressel, J. (1999). Tandem constructs: preventing the rise of superweeds. *Trends Biotechnol.* 17: 361–366.

Gressel, J. (2010a). Gene flow of transgenic seed-expressed traits: biosafety considerations. *Plant Sci.* 179: 630–634.

Gressel, J. (2010b). Needs for and environmental risks from transgenic crops in the developing world. *New Biotechnol.* 27: 522–527.

Gressel, J. and Al-Ahmad, H. (2012). Transgenic mitigation of transgene dispersal by pollen and seed. In: *Plant Gene Containment* (ed. M.J. Oliver and Y. Li), 125–146. Oxford, UK: Blackwell Publishing Ltd.

Grill, E., Winnacker, E.L., and Zenk, M.H. (1985). Phytochelatins, the principal heavy-metal complexing peptides of higher plants. *Science* 230: 674–676.

Grill, E., Winnacker, E.-L., and Zenk, M.H. (1987). Phytochelatins, a class of heavy-metal-binding peptides from plants, are functionally analogous to metallothioneins. *Proc. Natl. Acad. Sci. U. S. A.* 84: 439–443.

Guerinot, M.L. and Salt, D.E. (2001). Fortified foods and phytoremediation. Two sides of the same coin. *Plant Physiol.* 125: 164–167.

Gupta, A., Matsui, K., Lo, J.F., and Silver, S. (1999). Molecular basis for resistance to silver cations in *Salmonella*. *Nat. Med.* 5: 183–188.

Hamer, D.H. (1986). Metallothionein. *Annu. Rev. Biochem.* 55: 913–951.

Hamlett, N.V., Landale, E.C., Davis, B.H., and Summers, A.O. (1992). Roles of the Tn21 merT, merP, and merC gene products in mercury resistance and mercury binding. *J. Bacteriol.* 174: 6377–6385.

Hanna, W.W. (1981). Method of reproduction in Napiergrass and in the 3x and 6x alloploid hybrids with pearl millet. *Crop Sci.* 21: 122–126.

Harmens, H., Den Hartog, P.R., Ten Ookum, W.M., and Verkleij, J.A.C. (1993). Increased zinc tolerance in *Silene vulgaris* (Moench) Garcke is not due to increased production of phytochelatins. *Plant Physiol.* 103: 1305–1309.

Hasegawa, I., Terada, E., Sunairi, M. et al. (1997). Genetic improvement of heavy metal tolerance in plants by transfer of the yeast metallothionein gene (CUP1). *Plant Soil* 196 (277–281): 670.

Heaton, A.C.P., Rugh, C.L., Wang, N.-J., and Meagher, R.B. (1998). Phytoremediation of mercury- and methylmercury-polluted soils using genetically engineered plants. *J. Soil Contam.* 7: 497–509.

Hendy, S., Chen, Z.C., Barker, H. et al. (1999). Rapid production of single-chain Fv fragments in plants using a potato virus X episomal vector. *J. Immunol. Methods* 231: 137–146.

Hirschi, K.D. (1999). Expression of *Arabidopsis* CAX1 in tobacco: altered calcium homeostasis and increased stress sensitivity. *Plant Cell* 11: 2113–2122.

Hirschi, K.D., Korenkov, V.D., Wilganowski, N.L., and Wagner, G.J. (2000). Expression of Arabidopsis *CAX2* in tobacco. Altered metal accumulation and increased manganese tolerance. *Plant Physiol.* 124: 125–134.

Hood, E.E., Woodard, S.L., and Horn, M.E. (2002). Monoclonal antibody manufacturing in transgenic plants—myths and realities. *Curr. Opin. Biotechnol.* 13: 630–635.

Horsch, R.B., Fry, J.E., Hoffmann, N.L. et al. (1985). A simple and general method for transferring genes into plants. *Science* 227: 1229–1231.

Houdebine, L.M. (2002). Antibody manufacture in transgenic animals and comparisons with other systems. *Curr. Opin. Biotechnol.* 13: 625–629.

Howden, R., Goldsbrough, P.B., Andersen, C.R., and Cobbett, C.S. (1995). Cadmium-sensitive, cad1, mutants of *Arabidopsis thaliana* are phytochelatin deficient. *Plant Physiol.* 107: 1059–1066.

Huckle, J.W., Morby, A.P., Turner, J.S., and Robinson, N.J. (1993). Isolation of a prokaryotic metallothionein locus and analysis of transcriptional control by trace metal ions. *Mol. Microbiol.* 7: 177–187.

Iamtham, S. and Day, A. (2000). Removal of antibiotic resistance genes from transgenic tobacco plastids. *Nat. Biotechnol.* 18: 1172–1176.

James, C. (2006). Global status of commercialized biotech/GM crops. ISAAA publication; Brief No. 35-2006.

James, C. (2011). *Global Status of Commercialized Biotech/GM Crops: 2011*. ISAAA Brief. No. 43. Ithaca, NY: ISAAA.

Johnson, P.G., Larson, S.R., Anderton, A.L. et al. (2006). Pollen-mediated gene flow from Kentucky bluegrass under cultivated field conditions. *Crop Sci.* 46: 1990–1997.

Kannan, B., Sinche, M., Corsato, C.E. et al. (2012). Developing seedless genotypes of the biofuel crop elephantgrass for improved biosafety. *12th International Symposium on Biosafety of Genetically Modified Organisms (ISBGMO)*, 2–49.

Kapila, J., DeRycke, R., VanMontagu, M., and Angenon, G. (1997). An agrobacterium-mediated transient gene expression system for intact leaves. *Plant Sci.* 122: 101–108.

Kausch, A.P., Hague, J., Oliver, M. et al. (2010). Transgenic perennial biofuel feedstocks and strategies for bioconfinement. *Biofuels* 1: 163–176.

Korshunova, Y.O., Eide, D., Clark, W.G. et al. (1999). The IRT1 protein from *Arabidopsis thaliana* is a metal transporter with a broad substrate range. *Plant Mol. Biol.* 40: 37–44.

Kramer, U. and Chardonnens, A.N. (2001). The use of transgenic plants in the bioremediation of soils contaminated with trace elements. *Appl. Microbiol. Biotechnol.* 55: 661–672.

Kramer, U., Cotter-Howells, J.D., Charnock, J.M. et al. (1996). Free histidine as a metal chelator in plants that accumulate nickel. *Nature* 379: 635–638.

Kramer, U., Pickering, I.J., Prince, R.C. et al. (2000). Subcellular localization and speciation of Ni in hyperaccumulator and non-accumulator Thlaspi species. *Plant Physiol.* 122: 1343–1353.

Larrick, J.W. and Thomas, D.W. (2001). Producing proteins in transgenic plants and animals. *Curr. Opin. Biotechnol.* 12: 411–418.

Lasat, M.M., Pence, N.S., Garvin, D.F. et al. (2000). Molecular physiology of zinc transport in the Zn hyperaccumulator *Thlaspi caerulescens*. *J. Exp. Bot.* 51: 71–79.

LeDuc, D.L., Tarun, A.S., Montes-Bayon, M. et al. (2004). Overexpression of selenocysteine methyltransferase in Arabidopsis and Indian mustard increases selenium tolerance and accumulation. *Plant Physiol.* 135 (1): 377–383.

Lin, Z.-Q., Schemenauer, R.S., Cervinka, V. et al. (2000). Selenium volatilization from the soil — *Salicornia bigelovii* Torr. treatment system for the remediation of contaminated water and soil in the San Joaquin Valley. *J. Environ. Qual.* 29: 1048–1056.

Lombi, E., Zhao, F.J., Dunham, S.J., and McGrath, S.P. (2000). Cadmium accumulation in populations of *Thlaspi caerulescens* and Thlaspi goesingense. *New Phytol.* 145: 11–20.

Ma, L.Q., Komar, K.M., Tu, C. et al. (2001). A fern that hyperaccumulates arsenic. *Nature* 409: 579.

Macnair, M.R., Bert, V., Huitson, S.B. et al. (1999). Zinc tolerance and hyperaccumulation are genetically independent characters. *Proc. R. Soc. London, Ser. B* 266: 2175–2179.

Mariani, C., Debeuckeleer, M., Truettner, J. et al. (1990). Induction of male-strility in plants by a chimeric ribonuclease gene. *Nature* 347: 737–741.

Marmiroli, N. and McCutcheon, S.C. (2003). Making phytoremediation a successful technology. *Phytoremediation Transform. Control Contam.* 26: 85–119.

Marschner, H. (1995). *Mineral Nutrition in Higher Plants*. San Diego, CA: Academic Press.

Marschner, H., Roemheld, V., and Kissel, M. (1987). Localization of phytosiderophore release and of iron uptake along intact barley roots. *Physiol. Plant.* 71: 157–162.

McCormick, A.A., Kumagai, M.H., Hanley, K. et al. (1999). Rapid production of specific vaccines for lymphoma by expression of the tumor-derived single-chain Fv epitopes in tobacco plants. *Proc. Natl. Acad. Sci. U. S. A.* 96: 703–708.

McCutcheon, S.C. and Schnoor, J.L. (2003). Overview of phytotransformation and control of wastes. *Phytoremediation Transform. Control Contam.* 26: 1–58.

McGrath, S.P., Dunham, S.J., and Correll, R.L. (2000). Potential for phytoextraction of zinc and cadmium from soils using hyperaccumulator plants. In: *Phytoremediation of Contaminated Soil and Water* (ed. N. Terry and G. Bañuelos), 109–128. Boca Raton, FL: CRC Press.

Meagher, R.B. (2000). Phytoremediation of toxic elemental and organic pollutants. *Curr. Opin. Plant Biol.* 3: 153–162.

Meagher, R.B., Rugh, C.L., Kandasamy, M.K. et al. (2000). Engineered phytoremediation of mercury pollution in soil and water using bacterial genes. In: *Phytoremediation of Contaminated Soil and Water* (ed. N. Terry and G. Bañuelos), 201–221. Boca Raton, FL: Lewis.

Meharg, A.A. and Macnair, M.R. (1992). Suppression of the high affinity phosphate uptake system, a mechanism of arsenate tolerance in *Holcus lanatus* L. *J. Exp. Bot.* 43: 519–524.

Meyers, B., Zaitsman, A., Lacroix, B. et al. (2010). Nuclear and plastid genetic engineering of plants: comparison of opportunities and challenges. *Biotechnol Adv.* 6: 747–756.

Misra, S. and Gedamu, L. (1989). Heavy metal tolerant transgenic *Brassica napus* L. and *Nicotiana tabacum* L. plants. *Theor. Appl. Genet.* 78: 161–168.

Moon, H.S., Abercrombie, J.M., Kausch, A.P., and Stewart, C.N. Jr. (2010a). Sustainable use of biotechnology for bioenergy feedstocks. *Environ. Manage.* 46: 531–538.

Moon, H.S., Li, Y., and Stewart, C.N. Jr. (2010b). Keeping the genie in the bottle: transgene biocontainment by excision in pollen. *Trends Biotechnol.* 28: 3–8.

Moon, H.S., Abercrombie, L.L., Eda, S. et al. (2011). Transgene excision in pollen using a codon optimized serine resolvase CinH-RS2 site-specific recombination system. *Plant Mol. Biol.* 75: 621–631.

Mukhopadhyay, R., Li, J., Bhattacharjee, H., and Rosen, B.P. (1998). Metalloid resistance mechanisms. *Adv. Exp. Med. Biol.* 456: 159–181.

Murphy, A., Zhou, J., Goldsbrough, P.B., and Taiz, L. (1997). Purification and immunological identification of metallothioneins 1 and 2 from Arabidopsis thaliana. *Plant Physiol.* 113: 1293–1301.

Neuhierl, B. and Böck, A. (1996). On the mechanism of selenium tolerance in selenium-accumulating plants: purification and characterization of a specific selenocysteine methyltransferase from cultured cells of *Astragalus bisulcatus*. *Eur. J. Biochem.* 239: 235–238.

Nies, D.H. (1999). Microbial heavy-metal resistance. *Appl. Microbiol. Biotechnol.* 51: 730–750.

Nies, D.H., Nies, A., Chu, L., and Silver, S. (1989). Expression and nucleotide sequence of a plasmid-determined divalent cation efflux system from *Alcaligenes eutrophus*. *Proc. Natl. Acad. Sci. U. S. A.* 86: 7351–7355.

Noctor, G., Arisi, A.C.M., Jouanin, L. et al. (1998). Glutathione: biosynthesis, metabolism and relationship to stress tolerance explored in transformed plants. *J. Exp. Bot.* 49: 623–647.

Oliver, M.J. and Hake, K. (2012). Seed-based gene containment strategies. In: *Plant Gene Containment* (ed. M.J. Oliver and Y. Li), 113–124. Oxford, UK: Blackwell Publishing Ltd.

Oliver, M.J., Quisenberry, J.E., Trolinder, N. and Keim, D.L. (1998). Control of plant gene expression. US patent 5723765.

Oliver, M.J., Quisenberry, J.E., Trolinder, N. and Keim, D.L. (1999a). Control of plant gene expression. US patent 5925808.

Oliver, M.J., Quisenberry, J.E., Trolinder, N. and Keim, D.L. (1999b). Control of plant gene expression. US patent 5977441.

Pan, A., Tie, F., Zhenwen, D. et al. (1994a). Alpha-domain of human metallothionein I-A can bind to metals in transgenic tobacco plants. *Mol. Gen. Genet.* 242: 666–674.

Pan, A., Yang, M., Tie, F. et al. (1994b). Expression of mouse metallothionein-I gene confers cadmium resistance in transgenic tobacco plants. *Plant Mol. Biol.* 24: 341–352.

Pence, N.S., Larsen, P.B., Ebbs, S.D. et al. (2000). The molecular physiology of heavy metal transport in the Zn/Cd hyperaccumulator *Thlaspi caerulescens*. *Proc. Natl. Acad. Sci. U. S. A.* 97: 4956–4960.

Pilon-Smits, E.A.H. and Pilon, M. (2000). Breeding mercury-breathing plants for environmental clean-up. *Trends Plant Sci.* 5: 235–236.

Pilon-Smits, E. and Pilon, M. (2002). Phytoremediation of metals using transgenic plants. *Crit. Rev. Plant Sci.* 21 (5): 439–456. https://doi.org/10.1080/0735-260291044313.

Pulford, I.D. and Watson, C. (2003). Phytoremediation of heavy metalcontaminated land by trees - a review. *Environ. Int.* 29: 529–540.

Raskin, I. (1996). Plant genetic engineering may help with environmental clean-up. *Proc. Natl. Acad. Sci. U. S. A.* 93: 3164–3166.

Raskin, I., Smith Robert, D., and Salt David, E. (1997). Phytoremediation of metals: using plants to remove pollutants from the environment. *Curr. Opin. Biotechnol.* 8: 221–226.

REN21. (2001). Renewables 2011: global status report. 13–4. http://www.ren21.net/Portals/97/documents/GSR/GSR2011_Master18.pdf.

Rogers, E.E., Eide, D.J., and Guerinot, M.L. (2000). Altered selectivity in an *Arabidopsis* metal transporter. *Proc. Natl. Acad. Sci. U. S. A.* 97: 12356–12360.

Rosegrant, M.W. and Cavalieri, A.J. (2008). Bioenergy and agro-biotechnology commissioned paper submitted to the German Council on global change. International Food Policy Research Institute (IFPRI), Washington, DC.

Ruf, S., Hermann, M., Berger, I.J. et al. (2001). Stable genetic transformation of tomato plastids and expression of a foreign protein in fruit. *Nat. Biotechnol.* 19: 870–875.

Rugh, C.L., Wilde, H.D., Stack, N.M. et al. (1996). Mercuric ion reduction and resistance in transgenic *Arabidopsis thaliana* plants expressing a modified bacterial merA gene. *Proc. Natl. Acad. Sci. U. S. A.* 93: 3182–3187.

Rugh, C.L., Senecoff, J.F., Meagher, R.B., and Merkle, S.A. (1998). Development of transgenic yellow poplar for mercury phytoremediation. *Nat. Biotechnol.* 16: 925–928.

Rugh, C.L., Bizily, S.P., and Meagher, R.B. (2000). Phytoreduction of environmental mercury pollution. In: *Phytoremediation of Toxic Metals — Using Plants to Clean up the Environment* (ed. I. Raskin and B.D. Ensley), 151–171. Wiley, New York.

Saier, M.H. (2000). A functional-phylogenetic classification system for transmembrane solute transporters. *Microbiol. Mol. Biol. Rev.* 64: 354–411.

Salt, D.E. and Krämer, U. (2000). Mechanisms of metal hyperaccumulation in plants. In: *Phytoremediation of Toxic Metals: Using Plants to Clean-up the Environment* (ed. I. Raskin and B.D. Ensley), 231–246. New York: Wiley.

Salt, D.E., Blaylock, M., Kumar Nanda, P.B.A. et al. (1995). Phytoremediation: a novel strategy for the removal of toxic metals from the environment using plants. *Bio/Technology* 13: 468–474.

Sandhu, S., Blount, A.R., Quesenberry, K.H., and Altpeter, F. (2010). Apomixis and ploidy barrier suppress pollen-mediated gene flow in field grown transgenic turf and forage grass (Paspalum notatum Flugge). *Theor. Appl. Genet.* 121: 919–929.

Schillberg, S., Twyman, R.M., and Fischer, R. (2005). Opportunities for recombinant antigen and antibody expression in transgenic plants—technology assessment. *Vaccine* 23 (15): 0–1769. https://doi.org/10.1016/j.vaccine.2004.11.002.

Schmoger, M.E., Oven, M., and Grill, E. (2000). Detoxification of arsenic by phytochelatins in plants. *Plant Physiol.* 122: 793–801.

Sexton, S. and Zilberman, D. (2011). How agricultural biotechnology boosts food supply and accomodates biofuels. Working paper 16699. http://www.nber.org/papers/w16699.

Shaul, O., Hilgemann, D.W., de Almeida-Engler, J. et al. (1999). Cloning and characterization of a novel Mg^{2+}/H^{+} exchanger. *EMBO J.* 18: 3973–3980.

Shrift, A. (1969). Aspects of selenium metabolism in higher plants. *Annu. Rev. Plant Physiol.* 20: 475–494.

Staub, J.M., Garcia, B., Graves, J. et al. (2000). High-yield production of a human therapeutic protein in tobacco chloroplasts. *Nat. Biotechnol.* 18: 333–338.

Stewart, C.N. Jr., Halfhill, M.D., and Warwick, S.I. (2003). Transgene introgression from genetically modified crops to their wild relatives. *Nat. Rev. Genet.* 4: 806–817.

Stewart, C.N. Jr. (2008). Pharming in crop commodities. *Nat. Biotechnol.* 26: 1222–1223.

Sun, S.S.M. (2008). Application of agricultural biotechnology to improve food nutrition and healthcare products. *Asia Pac. J. Clin. Nutr.* 17: 87–90.

The Biology Refugia (2011). Blogging about ecology, evolution, taxonomy and biodiversity from Singapore.

Thomine, S., Wang, R., Ward, J.M. et al. (2000). Cadmium and iron transport by members of a plant metal transporter family in Arabidopsis with homology to Nramp genes. *Proc. Natl. Acad. Sci. U. S. A.* 97: 4991–4996.

Twyman, R.M., Stoger, E., Schillberg, S. et al. (2003). Molecular farming in plants: host systems and expression technology. *Trends Biotechnol.* 21: 570–578.

Van der Zaal, B.J., Neuteboom, L.W., Pina, J.E. et al. (1999). Overexpression of a novel *Arabidopsis* gene related to putative zinc-transporter genes from animals can lead to enhanced zinc resistance and accumulation. *Plant Physiol.* 199: 1047–1055.

Vaquero, C., Sack, M., Chandler, J. et al. (1999). Transient expression of a tumor-specific single-chain fragment and a chimeric antibody in tobacco leaves. *Proc. Natl. Acad. Sci. U. S. A.* 96: 11128–11133.

Vázquez, M.D., Barceló, J., Poschenrieder, C. et al. (1992). Localization of zinc and cadmium in *Thlaspi caerulescens* (Brassicaceae), a metallophyte that can hyperaccumulate both metals. *J. Plant Physiol.* 140: 350–355.

Verch, T., Yusibov, V., and Koprowski, H. (1998). Expression and assembly of a full-length monoclonal antibody in plants using a plant virus vector. *J. Immunol. Methods* 220: 69–75.

Warwick, S.I., Beckie, H.J., and Hall, L.M. (2009). Gene flow, invasiveness, and ecological impact of genetically modified crops. *Ann. N. Y. Acad. Sci.* 1168: 72–99.

Williams, L.E., Pittman, J.K., and Hall, J.L. (2000). Emerging mechanisms for heavy metal transport in plants. *Biochim. Biophys. Acta* 1465: 104–126.

von Wirén, N., Marschner, H., and Roemheld, V. (1996). Roots of ironefficient maize also absorb phytosiderophore-chelated zinc. *Plant Physiol.* 111: 1119–1125.

Wolfenbarger, L.L. and Phifer, P.R. (2000). The ecological risks and benefits of genetically engineered plants. *Science* 290: 2088–2093.

Wolt, J.D. (2009). Advancing environmental risk assessment for transgenic biofeedstock crops. *Biotechnol. Biofuels* 2: 27.

Xiang, C. and Oliver, D.J. (1998). Glutathione metabolic genes coordinately respond to heavy metals and jasmonic acid in *Arabidoposis*. *Plant Cell* 10: 1539–1550.

Yamaguchi, T. and Blumwald, E. (2005). Developing salt-tolerant crop plants: challenges and opportunities. *Trends Plant Sci.* 10: 615–620.

Yeargan, R., Maiti, I.B., Nielsen, M.T. et al. (1992). Tissue partitioning of cadmium in transgenic tobacco. Seedlings and field grown plants expressing mouse metallothionein I gene. *Transgenic Res.* 1: 261–267.

Yeo, A.R., Yeo, M.E., and Flowers, T.J. (1987). The contribution of an apoplastic pathway to sodium uptake by rice roots in saline conditions. *J. Exp. Bot.* 38: 1141–1153.

Zhu, Y.L., Pilon-Smits, E.A.H., Jouanin, L., and Terry, N. (1999a). Overexpression of glutathione synthetase in Indian mustard enhances cadmium accumulation and tolerance. *Plant Physiol.* 119: 73–79.

Zhu, Y.L., Pilon-Smits, E.A.H., Tarun, A.S. et al. (1999b). Cadmium tolerance and accumulation in Indian mustard is enhanced by overexpressing γ-glutamylcysteine synthetase. *Plant Physiol.* 121: 1169–1177.

Zimmermann, M.B. and Hurrell, R. (2002). Improving iron, zinc and vitamin A nutrition through plant biotechnology. *Curr. Opin. Biotechnol.* 13: 142–145.

20

Comprehensive Note on Various Wastewater Treatment Strategies

Amna Aqeel and Javaria Zafar

Institute of Industrial Biotechnology, Government College University Lahore, Lahore, Pakistan

20.1 Introduction

Water is an important ingredient of our environment and a key player for subsistence on the planet earth. Amid the sociocultural growth, the meteoric economic boom in Asia has put the scarcity of water and its limited accessibility to the fore. The modernization in agriculture, drift toward urbanization, and a swift industrial revamping have made water a main victim in the changing global dynamics (Visvanathan 2018). Water has been a portent of life for humans and all other living beings, an important ingredient in the health of people but the changing world's priorities have made insignificant changes in the physical and chemical properties of water. Water makes up about 70% of the world and is a major element in accomplishing poverty reduction, food security, suitable living standards and great harmony with the essential ecosystem of the Earth. While the population and consumption of water in various forms is increasing exponentially, the production of wastewater has experienced a multifold rise, but the appropriate sanitation and access to safe water are in extremely dire straits. According to the World Health Organization (WHO), almost 40% of the population of the world has no access to primary sanitation and the worst-case scenario is observed mostly in rural areas (WHO 2013). Moreover, it was reported by WHO in 2019 that about 785 million people are a global population of about 6.8 billion people using at least a basic water service. About two billion people worldwide consume water contaminated with feces that pose serious hazards to health and contribute to various waterborne diseases like cholera, diarrhea, polio, and typhoid. According to an estimation, about 485,000 deaths occur annually due to diarrhea, a product of the consumption of contaminated water. By 2025, it is assumed that half of the world's population will be experiencing water stress. The third-world countries have health care facilities in which 22% have no water service, 21% have no sanitation service along with 22% that have no service of waste management. In addition, in spite of the fact that there is an abundance of water resources on the earth, the natural causes and anthropogenic activities polluting the water have made it a competitive resource (WHO 2019).

The vital essence of life, water has been in use for domestic, industry, as well as irrigation purposes and wastewater obtained as a result is an amalgamation of residential waste, human and animal waste, i.e. sewage, effluents from industries, urban runoff, storm runoff as well as commercial firms (West 2003). It comprises the waste that demands a high amount of oxygen with inorganic chemicals, nutrients for plant growth, toxic compounds, minerals as well as sediments (Metcalf & Eddy Inc. 1991). Sewage water is mainly organic in nature while industrial waste is a consortium of relatively different and noxious organic compounds. The treatments for their remediation consist of various physicochemical and biological methods (Lettinga and Hulshoff 1986). For an extended period of time, the main focus was to reduce the content of oxygen-demanding compounds, suspended materials, dissolved inorganic materials as well as harmful pathogens. But, now there is a shift in prioritizing improvement in modes of waste disposal (Sonune and Rupali 2004). The last decade marks the use of anaerobic digestion for the treatment of sludge and the formation of biogas. This anaerobic mode of digestion is relatively slow and it was identified by Shimizu et al. (1996) that waste-activated sludge hydrolysis is a rate-limiting step and a pretreatment was therefore developed to improve the anaerobic digestion later on. The pretreatment can be mechanical (Chio et al. 1997; Tiehm et al. 2001; Chu et al. 2002), chemical as well as thermal (Camacho et al. 2002).

Aquatic Contamination: Tolerance and Bioremediation, First Edition. Edited by Rouf Ahmad Bhat, Gowhar Hamid Dar, Fernanda Maria Policarpo Tonelli, and Saima Hamid.

This present era makes the usage of primary, secondary, and tertiary methods. The primary method employs mechanical methods, the secondary treatment makes use of oxidation of organic waste by microbes and the tertiary is the advanced stage of treatment in which appropriate application of new physicochemical and biological methods are deprived of the basic drinking water with 90% of the users. The aerobic systems include extended aeration, oxidation ditch, MBR, fixed film processes, or hybrid processes. The anaerobic systems include continuously stirred tank reactors, upflow anaerobic sludge blanket (UASB), ultra-high-rate fluidized bed reactors, etc., with a low carbon footprint as compared to aerobic methods. The anaerobic method also carries a plus point of having relatively low capital investment with payback as compared to aerobic, which has a high investment with minimal payback (Arun et al. 2020). It is assumed that it is not the individuality of these processes that gives the maximal efficiency in wastewater treatment but rather the combination of these two in the advanced treatment technologies where the contamination of organic compounds is relatively high (Rasalingam et al. 2014).

20.2 Treatment Strategies

The treatment strategies of wastewater consist of three categories. These include physical, chemical, as well as biological methods. The level of contaminants varies in the consecutive stages of wastewater treatment. The basic scheme that is followed in a wastewater treatment plant is categorized into three phases. These include primary, secondary, and tertiary. There are various rigorous methods for the treatment of waste along with the control of essential nutrients. This chapter explains the various conventional as well as advanced treatment strategies for the treatment of surplus waste and describes the application for effective remediation (Asano and Levine 1996).

A typical wastewater treatment involves the following phases.

1) Preliminary treatment
2) Primary treatment
3) Secondary treatment
4) Tertiary treatment or advanced treatment

20.2.1 Preliminary Treatment

The first stage is named the preliminary treatment. It is used to remove the screening and grit that get entry from the sewer system into the wastewater treatment plant. The pathogens in the wastewater are least affected by the preliminary treatment. Primary treatment removes the coarse particles and some suspended solids (SS). The pathogens are still in the wastewater. So, the downstream is necessary for their proper removal to meet the requirements of the discharge or for reuse in various applications. It is expected that a suitable amount of pathogens could be reduced while they get attached to the particles that are removed.

The preliminary stage of wastewater treatment includes these basic steps:

1) Screening
2) Grit removal
3) Comminution of large objects

The aforementioned methods 1 and 2 are employed to guard the tools from any harm, to intercept unacceptable solids, and to change the solids to physically acceptable forms for the treatment. These two are mostly used for military domestic wastewaters (Metcalf & Eddy Inc. 1991).

20.2.1.1 Screening

The screening is done to detain the floating objects in the passing wastewater as well as intercept the suspended debris with the help of some attached equipment so that the screenings are removed and the organic matter simply moves to the sewage flow with the final screenings hacked and compacted into final form for dispatching out to the site for disposal (US EPA 1995). The elimination of screenings is not just the name of the physical removal of solids aided with screens but it also includes the in-flow or side-stream disintegration of the solids. Their removal cannot be done in separate ancillary but washing, dewatering, and incineration can be done for final disposal (Sidwick 2007) There are mainly two types of screens that are employed in screening. These are fine as well as coarse. The coarse one comprises manually

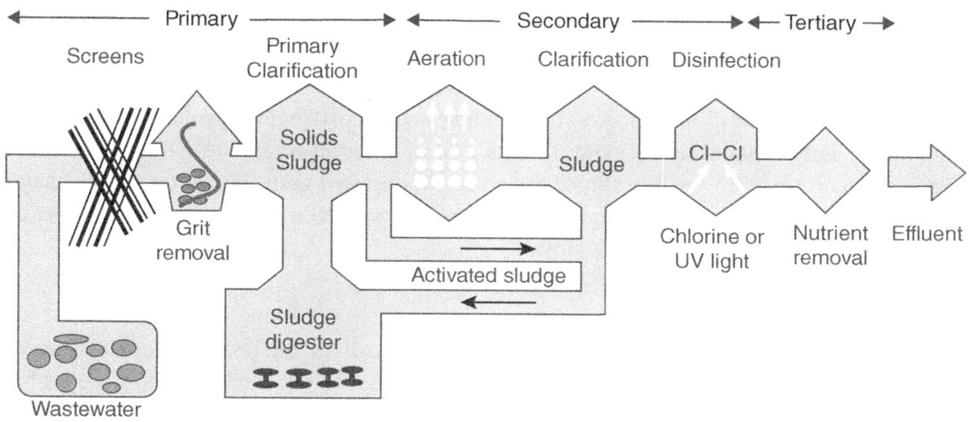

Figure 20.1 Stages of wastewater treatment.

as well as mechanically operated bar screens along with racks for trash. Bar screens comprise inclined or vertical steel bars that are distributed throughout the channel and separate the particles from the wastewater when it flows through the channel (Raouf et al. 2019). Screening precedes the mechanically operated systems like raw sewage pumps, grit chambers along with tanks used for primary sedimentation. Amid the exit from the regular bar screens or comminutors, wastewater remains move to the trash racks. These are made up of round or parallel rectangular steel bars with openings. The criteria for making coarse screens include size along with angle and spacing from the vertical. Whereas, the fine screens are made up of screen media of varying nature comprising slotted perforated plates, wire mesh, wedge-like wire and woven wire cloth. They have very fine openings that need to be scrubbed perpetually with scrapers, brushes, forced air or water spray. The fineness of the openings and the flow of sewage through them determine the efficiency of the fine screens (Bhargava 2016).

20.2.1.2 Grit Removal

The waste after passing through the screen gets entry into the grit chambers. There the larger particles are detained from moving in the remaining wastewater that exits the grit removal channels. The assembly consist of grit hoppers and an aerated system that causes the aeration of the wastewater along with cumulating the residual material of heavier inorganic nature in the hoppers attached at the bottom of the grit removal system. It intercepts the inorganic particles and comprises the grit washing and storage capability. The grit removal makes the process of downstreaming very easy and wears down any hindrance and potential damage in the pumps, dewatering plant, sludge as well as machinery (US EPA 1995). It is an economical process with various advantages:

1) Eliminates the chance of potential damage and abrasive wear to the equipment, i.e. sludge scrapers and pumps.
2) Ward of any deposition in the openings, channels, basins, and pipes.
3) Intercept any capacity reduction in sludge handling systems.

These facilities can be used individually for separate sanitary systems and in combined sewer systems that may have large amounts of inert material. These are present subsequent to the bar screens and communitors and precede raw sewage drives. Moreover, there are certain cases in which their detection in advance of the raw sewage pumps becomes difficult due to the presence of a deep influent line. Hence, it is necessary in those cases to employ a pump for the wastewater carrying grit with it. The pumps that have an ability to cater for the grit and handle must be used in these circumstances.

20.2.1.3 Communitor

The first assembly into which the wastewater enters is either a communitor or a screen. After the introduction of the wastewater into the communitor, it acts as a cutter of the bar, a width of the channel, a velocity of upcoming wastewater as well as a screen. The communitors and grinding devices can scrap solids in a range of 6–19 mm (0.25–0.75 in). These are mounted in a wastewater treatment channel and allow the passage of wastewater through them. Blades are responsible for cutting the rags, debris, and larger particles and reduce their sizes until they can pass freely through the slots or openings (US EPA 2003).

The speed of most shredders or grinders may range from 34 rpm (60 rpm) in the case of macerators to about 2500 rpm for high-speed hammer mills. Some units comprise channels that are specially designed for the proper hydraulic environments and hence are more expensive than usual. There is a bypass conduit in the communitors to maintain the equipment. They are usually installed at the points where the handling of screenings renders inappropriate most probably between the grit chambers and primary settling tanks. They ameliorate the odor, flies as well as unsightliness. These may carry either rotating or oscillating cutters. The former comprises a separate stationary screen juxtaposed with the cutlers or the former with the combined. There is a barminutor, a type of communitor, that consists of a bar screen along with rotating cutters (Bhargava 2016). The comminutors are usually provided in large plants.

20.2.1.4 Grease Removal

These are the equipments that are located between the grit chambers and primary clarifiers and are used for the removal of oil, grease, and fats from the wastewater. The free form of fat that has an ability to come on top is removed in the attached rectangular tanks and is discharged. This method has its foundation in the principle of lowering the flow rate of water. The fat collected is discharged manually in a specially designed area along with siphoning of water. These equipments allow the segregation of lubricant and emollient remains at the surface of inland wastewaters and are positioned between grit chambers and primary clarifiers. Free molecules of fat, that tend to appear on the water surface are alienated into rectangular vats, and the process is based on dropping the flow rate of water. The fat present in a colloidal state of emulsified droplets do not have the capability to rise on the top for separation and hence is removed by blustering air at minimum pressure in the divided tanks and then removed (Craciun et al. 2004). The recently made grinders can compact the screenings after chopping and washing. The communitors when used in the cold weather banish the need for prevention of screenings from freezing. They have a lower profile as compared to screens which are cost-effective when the units are enclosed (US EPA 2003).

20.2.1.5 Primary Clarifier

The next step is the primary clarifier or settling tank that has an ability to retain the bulk of suspensions present in the wastewater. This is based on the sedimentation principle that takes place with the gravitational force. They are large-sized containers composed of concrete or cement with a height of about 30 m and a width of approximately 9 m. They are present in a variety of shapes, either rectangular or round. The two are mentioned as follows:

1) **Rectangular primary clarifier**:
   The type of primary clarifier is rectangular in shape and in which water stays for approximately 1.5–2 hours. The deposits that are detained are collected in the hoppers several times a day. The collection is done with the use of scrapers that are mounted on the bridge and are helpful in preventing their fermentation (Rojanschi and Ognean 1997).
2) **Round primary clarifier**:
   This is a round-type primary clarifier in which the wastewater gets entry into the mid-section and after treatment, elopes through the trough present on the periphery in the outer region. There is a bridge mounted on the upper surface of the tank that has attached sludge scrapers with it and rotates slowly while the primary clarifier functions. These scrapers are responsible for moving the sludge such that it moves toward the bottom and then forwarding it into the sludge funnel present in the centre where it is eliminated from the clarifier. The bottom with a slide slope also has floating scum scrapers that withhold the floating solids and push them into the scum trough for separation (Zabava et al. 2015).

20.2.2 Primary Sedimentation

The next stage is the primary sedimentation that is a centralized or semicentralized treatment in the conventional mode of wastewater treatment in which the settleable solids and floating materials are removed for further treatment of wastewater. It works on the primary objective of eliminating organic solids as well as floating organic substances known as scum so that the SS load is reduced in order to downstream the wastewater in the next stages of treatment (Metcalf & Eddy 2014). The primary sedimentation tanks are rectangular or circular in shape and have a hydraulic detention period of 1.5–3 hours based on the average daily flow rate. The scum that is collected is disposed of separately or combined with sludge in the plants dedicated to wastewater treatment. It does not have any pathogen concentration in it but assumptions are made regarding some significant concentrations in it. So, it must be handled carefully. The putrifiable solids that are present in the primary sludge are detached unceasingly from the tank's bottom and are sent for stabilization under anaerobic and aerobic conditions (Wen et al. 1999). The primary sludge obtained as a result carries 2–5% total solids along with 60–80% organic solid load. The treatment at the primary stage can remove up to 70% of total suspended solids (TSS) and about 45% of biological oxygen demand BOD5 (Metcalf & Eddy 2014). The primary sludge is sent for further reduction in the organic load in the secondary treatment containers that are aerobic technologies or natural systems (Jasim 2020). The chemical addition can be done after the primary sedimentation but sometimes it is not necessary at this stage (Muller et al. 1995). There are other auxillary units that are employed in the primary treatment for sludge treatment, disposal, and disinfection steps. For several years, the primary treatment was only used as a sole treatment for wastewater but later on, the tables turned and now it is no longer considered acceptable. The secondary treatment stage is a must-employed treatment to meet the criteria of the regulations available for wastewater treatment. According to a report published by Metcalf and Eddy in 2014, primary sedimentation produces about 110–170 kg of dry solids per 1000 m^3 of the treated wastewater and the sludge is then stabilized and treated for pathogens (Andreoli et al. 2007).

20.2.3 Secondary Treatment

It involves the treatment of sewage obtained from primary treatment to eradicate the lingering organics as well as SS. The solids are distributed in three categories with 30% suspended that are removed at the primary stage along with 6% colloidal and 65% dissolved solids that need removal in the secondary stage. The primary stage has already removed a large amount of SS and the remaining effluent is composed of mainly colloidal along with organic as well as inorganic solids in dissolved form. Recent trends in effluent treatment require the secondary treatment of organics from the wastewater that can be done with the help of aerobic and anaerobic methods of removal (US EPA 1995). A typical secondary plant can eliminate about 85–95% of the BOD and suspended entities from the influent. This tributary treatment mainly focuses on biological methods and the use of microbes under controlled environment (Singh and Shrivastava 2011). Several methods are employed at the secondary stage with variations in the way the oxygen is provided for consumption by microbes to effectively degrade and metabolize the organic matter (Amoatey and Richard 2011).

20.2.4 Tertiary Treatment/Advanced Wastewater Treatment

This stage is mainly divided into three categories based on the process devised and employed. These are:

- Tertiary treatment
- Physicochemical treatment
- Combined biological-physical treatment/Hybrids

The first kind is tertiary treatment in which additional units are combined in the secondary treatment assembly and modification is done for enhanced treatment. It could be simply an insertion of a filter for targetting SS are can be a small consortium of various units that detain organic solids and suspended particles alongwith the removal of phosphorus and nitrogen (Kuba et al. 1996). The physicochemical treatment combines the physical and chemical units alternatively to procure the desired effluent that could be recycled or reused (Gupta et al 2012). The biological-physical–chemical treatments in combination with each other bring a new category named combined biological-physical–chemical systems or hybrid systems (Wang et al. 2006). These are distinct from the tertiary system because it is an intermingling of the three processes and tertiary is only the scale-up of the conventional biological or secondary-stage treatment. There is another classification of advanced treatment that is done on the basis of desired treatment goals. It is used for:

a) Surplus removal of organic solids and suspended particles
b) Elimination of nitrogenous oxygen demand (NOD)
c) Amputation of nutrient
d) Elimination of toxic substances

The conventional secondary treatment gives a suitable BOD along with the removal of SS. The advanced treatment is necessary to generate effluent that can be reprocessed unswervingly or circuitously and reuse to meet the current needs of native water supply. A BOD of 20 mg/l and a COD of 60–100 mg/l are obtained in an average secondary effluent. 65% of the COD of influent is removed already in the secondary plant. Hence, for procuring high-quality effluent, additional methods of removal must be utilized. There is also an additional demand for oxygen from the present nitrogen in wastewater in most secondary effluents.

20.3 Methods of Wastewater Treatments

There are three categories of wastewater treatment. These include

Physical/Mechanical methods
Chemical Methods
Biological Methods

20.3.1 Physical Methods and Mechanical Methods

There are certain methods in the treatment strategies that are of physical-mechanical nature. These include sedimentation, floatation, filtration, as well as centrifugal methods and disinfection. Most of these are used at the primary stage of wastewater treatment and some are used in the secondary stage. The mechanical method reduces the BOD of the wastewater by at least 20% and TSS by at least 50% before it makes an exit from the plant. These systems employ an amalgamation of physical, biological, as well as chemical processes to complete their objectives of effective treatment. The natural processes are performed under artificial environmental conditions and utilize a chain of screens, grinders, tanks, pumps, and blowers along with other mechanical units. Two of the most important things that fall into the category of mechanical methods are screenings and filters (Rosa and Rubio 2005; Rubio 1998). Screens are solely of mechanical nature but filters fall into either the mechanical or biological domains. Flow meters and the control systems for flow rate determination, activated sludge systems, sequencing batch reactors (Artan et al. 1995) or SBRs, oxidation ditches, and extended aeration systems are all mechanical and biological in nature. The physical methods are those that employ physical forces to remove contaminants from the wastewater (Wastewater Technology Fact Sheet 2000, Betty 1999). They form the basis of most of the process flow systems that are used for wastewater treatment (Cheremisinoff 2002; Bettley 1992).

20.3.2 Flow Equalization

It is a method to minimize the variations in the flow rates and composition of wastewater to effectively treat it in the facility. Each unit is designed accordingly for the specified characteristics of wastewater. The efficiency and control can be improved when all the units operate at a uniform flow criterion. If, at any instance, there is an occurrence of an

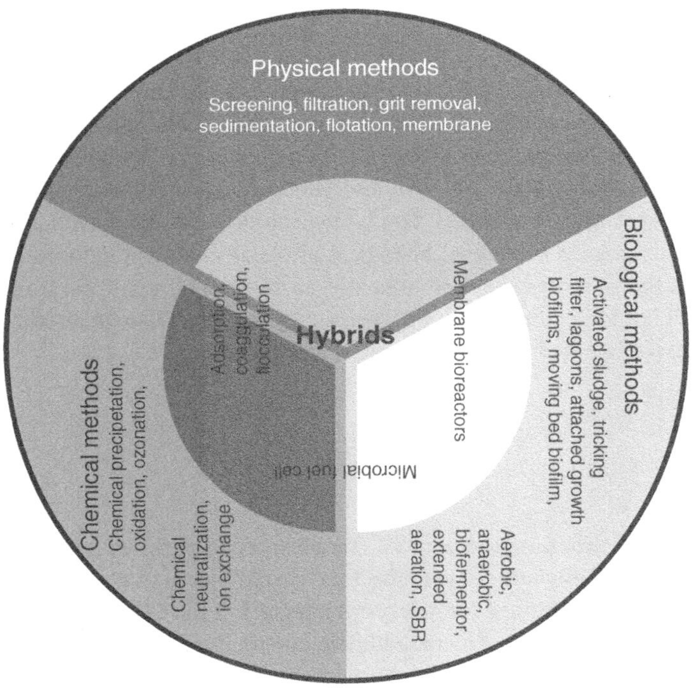

Figure 20.2 Various strategies of wastewater treatment. *Source:* Adapted from Tee et al. (2016).

extensive disparity in the flow configuration with respect to time, the performance of the overall process can be hampered (Wang et al. 2005). The variations in the flow composition are the product of several industrial processes or seasonal variations (Sincero and Sincero 2016). Equalization basins come to the rescue in these circumstances. The influent gets entry into the equalization basin at first and then forward toward the next stages of treatment. The basins serve more purpose as compared to storage tanks (Pons and Corriou 2002). These tanks serve various purposes. They accumulate smaller volumes of wastewater for the operation in the batch reactors and are also used in continuous systems for equalization of the waste flow to discharge the effluent at a uniform rate. They improve hydraulic detention time and increase the efficiency of a biological process because of the minimum shock loads and toxins during the further secondary treatment operations (Nielsen et al. 1996). The manual as well as automated control of operations depending on flow rate like disinfection, chemical feeding as well as sludge pumping, are also simplified (Wang et al. 2005). Other benefits also include the dilution of the inhibiting substances and lessening the temperature shocks in the tanks treating stormwater (Metcalf & Eddy Inc. 1991).

There are two types of the diurnal flow equalization method. These include:

1) In-line flow in which whole wastewater passes directly through the equalization basin.
2) Off-line flow also known as stream basin that deals with the wastewater flow above some predetermined flow rate passing through the equalization basin.

The pumping requirements are reduced in case of off-line flow but the pollutant concentration is not reduced much as compared to an in-line method (Charlton 1994). This technique thus helps in equalizing certain constraints like flow, temperature, and the level of pollutants over a period for an effective secondary treatment (Zhang et al. 2009).

20.3.3 Sedimentation

A mechanophysical treatment that basically involves sedimentation, a technique that employs gravitational force to settle down heavy particles in a mixture and removes grit and particulate substance in the primary resolving basin, biotic accumulation, or floc in case of activated sludge tank as well as the chemical floc in case of chemical coagulation treatment (Boari et al. 1997; Ahmed et al. 2009).

The process allows the removal of adjourned artifacts by permitting the water to stay undisturbed for different ranges of time intervals in altered tanks. The suspended objects settle down under the influence of gravity (Franklin 1991; Nemerow and Dasgupta 1991). The time required for settling is contingent upon various factors like the magnitude of the entities, their density as well as the velocity of the water when it is in motion. Moreover, various substances are also added to augment the competence of the process. The separation aided by gravity reduces the SS by up to 60%. It is a very important method in the treatment of the effluents from paper, refinery industries, and water employed for ion exchange as well as membrane processes (Marmagne and Coste 1996).

20.3.4 Flotation/Flocculation

The bubbles are trapped inside the particles resulting in increasing the buoyant force of the bubble combined with the particle or the other case may be the gas bubble simply holding the particles onto the surface (Rubio et al. 2007; Parekh and Miller 1999).

With the help of this method, the SS, greases, oils, or other biological particles form agglomerates that are then skimmed off effortlessly (Praksh et al. 2008). There are some chemicals that are useful in floatation. These are alum, activated silica, or sometimes compressed air. Compressed air is allowed to pass through the water, which helps in the flotation process. An advanced technique known as electroflotation (electroflocculation) has also been in use for effective wastewater treatment and removes up to 75% of SS along with 99% of oil as well as grease.

The treatment of wastewater follows a combination of a coagulant and a flocculent in a subsequent manner. An approximate amount of inexpensive coagulant is used to lessen the external charge of suspended particles The flocculent is mainly of cationic nature and has a high mass with low charge density (Ahmed et al. 2009; Wang et al. 2011; Zarkovic et al. 2011). The flocculants that are expensive when used with an excess of negative charges can be wasted. There are some systems in which sensitization is done with a suitable amount of cationic coagulant that makes the agglomerates and anionic flocculant favor their precipitation (Wang et al. 2020). The nonionic polyacrylamide is ineffective in the treatment of wastewater from the paper mill (Ganjidoust et al. 1997).

20.3.5 Chemical Treatment

The other method of wastewater treatment is the chemical method that employs the use of chemicals for clarification and elimination of the residual particles from the wastewater. This mode of treatment strategy was developed long ago however its implementation got a little delayed with respect to the biological methods that seemed to be more pocket-friendly than the substances at that period (Culp 1967; Culp et al. 1978). However, there are certain circumstances that does not favor the use of biological methods like intermittent waste flows, and the wastes that carry compounds toxic and fatal to the microbes as well as those containing impurities that are nonbiodegradable in nature (Harleman and Morrissey 1990). In this span of time, the shift has been seen toward chemically enhanced treatment strategies as the revolution is made in polymer chemistry, with increased concern in the control of nutrients and elevating costs in construction and energy industries. This has also occurred in response to various obstacles operation and regulation of wastewater dischargers along with plants dedicated to treating the wastewater (Eilbeck and Mattock 1987). Nowadays, there are many countries that depend on chemical methods to remediate their raw wastewaters, be it as the only advanced primary method or a secondary staged strategy or as a pre-treatment method preceding the next stage of biological treatment (Morrissey and Harleman 1992; Odegaard 1992). Likewise, the shifting toward chemical strategies has in turn improved the quality of effluent obtained after the treatment. This improvement has narrowed down the size, complexity as well as cost of the secondary treatment (Harleman et al. 1991). These methods are of physicochemical nature and withdraw large quantities of fine particles that are arduous to eliminate with classical mechanical methods. The chemical processes are either employed at the primary stage like coagulation or also in the last stage like disinfection making the quality effluent at the end (Zabava et al. 2015). Their increasing popularity is due to the amount of benefit they confer while remediating wastewater (Poon and Brueckner 1975; Stukenberg 1975; Eilbeck and Mattock 1987):

- They are further expectable and easy to control with simple techniques. They can be operated manually or automated (Odegaard 1987, 1992).
- They are very tolerant to environmental changes like temperature and toxic substances and are easy to operate. They work on an on–off basis without the need for the prerequisite period for acclimatization.
- As compared to the biological systems, they feel free to adjust variable input piles and flows, like seasonal currents, along with recurrent and multifaceted expulsions that occur in the case of the wastewaters obtained from industrial sites (Harleman et al. 1991).
- The chemical plants are easily molded and are flexible when needed for any additional installation with negligible change in their basic chemical structure.
- They usually require less space and have a minimal cost of installation. But it should be kept in mind that their cost can be increased with large-scale treatment plants (Semerjian and Ayoub 2003).

20.3.6 Various Strategies for Physicochemical Treatment

The Primary Physical–Chemical Processes Included are Described as Follows:

1) Activated carbon adsorption
2) Chemical oxidation
3) Solids removal (clarification and precipitation)

20.3.6.1 Activated Carbon Adsorption

Adsorption is a technique that is an attachment of a compound to a solid support. It can be of two types: physical and chemical adsorption. Chemical adsorption simply deals with the chemical interactions between the solid and the adsorbed solute whereas physical adsorption relies on strong van der Waals forces and therefore is a reversible reaction (Wang et al. 2005). Physical adsorption is favored when the attractive molecular forces present between the solute and adsorbent become stronger than those present between the solute and the solvent. Carbon is a vital compound in the adsorption of toxic compounds and other organic compounds in wastewater treatment. The columns with carbon in the granular form are utilized in the purification of wastewater and its powdered form is also applied for the carbon-activated sludge system for remediation of wastewater. Another modified type is the granular activated carbon (GAC) adsorption that is notable for the tertiary treatment stage for effective remediation of the organic pollutants present in the secondary effluent obtained from the treatment strategy. Lately, GAC has also been in use at the secondary stage of treatment in the physical–chemical treatment (PCT) plants (Rim et al. 1997). This is due to their ability to eliminate biodegradable particles from wastewater along with refractory substances that are organic in nature. The alternate physical–chemical strategy involves coagulation with succeeding precipitation to treat raw municipal and industrial wastewaters with the adsorption of activated carbon at the last to eliminate soluble and insoluble organic substances harmful to the environment. Another application of GAC is observed in anaerobic filter systems that treat the industrial wastewaters comprising the organic and toxic compounds that are difficult to degrade.

20.3.6.2 Chemical Oxidation

I) **Advanced oxidation process**:

It is categorized as the advanced oxidation process (AOP) because it deals with reactive oxygen species like the hydroxyl radical. Examples include photocatalysis, sonocatalysis and the Fenton reaction.

i) **Photocatalysis**:

It is a type of AOP that has been discovered for its application in gas-phase treatments. A semiconductor is used as a photocatalyst whose electron gets excited with the exposure to light and jump from valence to conduction band, with the resulting pair of hole (h^+ in valence band) and electron (e^- in conduction band) (Herrmann et al. 1999; Czaplicka et al. 2006). This pair is highly unstable and can subsequently oxidize and reduce different species. The oxidation at the surface of the semiconductor by oxygen there is a generation of various species of oxidative nature (Descorme 2017). These include hydroperoxyl radicals, hydroxyl radicals as well as superoxide radicals, etc. They are involved in the conduction of the reactions by photocatalysts. Metal nanoparticles are also employed to improve the photocatalytic activity of the semiconductors. They make trap centers to prevent recombination. An example is given as a series of silver nanoparticles attached to a ferrous oxide-forming complex of metal with ferrous oxide (MFe_2O_4). The metal or M can be copper, cobalt, zinc, or nickel (Zhu et al. 2015). Other examples are Ag-Ag_2O composite and a heterostructured Ag-$BiVO_4$/$InVO_4$ photocatalyst (Hu et al. 2015; Lin et al. 2015).

ii) **Fenton**:

The second type of AOP is Fenton which is known for its function in the exclusion of toxic compounds and contaminants from wastewater (Lu et al. 2002). This process uses hydrogen peroxide (H_2O_2) along with the salts of ferrous or cuprous (Fe^{2+}, Cu^{2+}). For a long time, hydroxyl radical is contemplated as the only one counted in Fenton especially when homogeneous catalysts are involved whereas the heterogenous catalyst favors the assumption of involvement of other reactive species like ferryl ion ($Fe[IV]O^{2+}$) (Bull and McManamon 1999). There are different factors that also contribute to the Fenton process. These include optimum pH, that is 3 for a homogeneous catalyst, while pH below 2.5 favors $Fe(II)(H_2O)^{2+}$ species formation and has a little reactivity and scavenging power for H^+ ions. The free iron radicals are decreased at pH_4 with the precipitate formation of the ferric oxyhydroxides. Meanwhile, the oxidation potential and concentration of hydrogen peroxide are also important parameters to be considered in the Fenton process (Descorme 2017, Kuo et al. 1996). Hydrogen peroxide in large amounts can be lethal to microbes in the environment and might be responsible for collecting hydroxyl radicals that are produced. This process reduces sludge production and improves effective oxidation (Xu et al. 2019).

iii) **Sonocatalysis**:

The ultrasonic waves are used for irradiation of water aided with OH· radicals in sonolysis (Choi et al. 2020). The formation of microbubbles arises with ultrasonic irradiation that oscillates, grows as well as collapses (regarded as acoustic cavitation), which allows the breakdown of water vapor to form hydroxyl radicals in the area of concentrated and localized energy and hydrogen atoms form as a result (Suslick et al. 1999).

Those niches of concentrated energy have temperatures of about 5000 K and pressures of about 1000 atm (Suslick et al. 1999). Hence, the process is not only utilized to decompose hard substances but also to dissociate larger organic compounds into smaller ones (Skorb et al. 2011; Cravotto et al. 2013). The examples of sonolytic materials are Al_2O_3, ZnO, and TiO_2. It is inferred that the nucleation sites are favored by solid catalysts (Tiehm et al. 1999). However, the mechanism undergoing sonocatalysis is not completely comprehended (Qiu et al. 2018).

II) **Direct oxidation with strong oxidants**:

The direct oxidation technique is performed by employing a strong oxidant directly. These strong oxidants include ozone and hydrogen peroxide. For the oxidation with hydrogen peroxide, the wet air oxidation is a modified version that is utilized.

i) **Oxidation with ozone**:
As a disinfectant, ozone is used as primary mechanism to prevent the spread of various waterborne diseases (Chiang et al 2006). It is utilized for the oxidation of several inorganic compounds, humic substance as well as contaminants that are toxic to the humans while water is used for drinking purposes (Rice et al. 1981). The reaction with ozone follows two pathways. It can be direct by only an ozone mole or an indirect one with hydroxyl radical as an oxidant (Hoigne and Bader 1976). The former is highly selective but proceeds at a slow pace with their targets like unsaturated bonds of amines, aromatics and olefins (Hoigne and Bader 1983) while the latter indirectly has a fast pace with slow selectivity (Prado et al. 1994; Staehelin and Hoigne 1985; Hoigne 1997). Low pH favors direct ozone reactions while indirect hydroxyl radical reactions prevail at high pH. Self-decomposition of ozone increases at basic pH and results in faster decomposition of contaminants in dye solution treated with ozonation at high levels of pH (Arslan-Alaton and Akmehmet Balcioglu 2002). There is less amount of hydroxyl radicals formed at an acidic pH that is reported by Chu and Ching in 2003 to treat 2,4-dichlorophoxyacetic acid present in acidic water. It should be kept in a close monitor that in special cases of ozone treatment in ozone-based systems (i.e. O_3/H_2O_2 or O_3/UV). O_2 is not exhausted during the process since the $HO^{\bullet}{}_2/O^{\bullet-}$ radicals formed in several reactions with peroxyl radical produce more $HO^{\bullet}$ radicals as a consequence of the speedy reaction of $O_2{}^{\bullet-}$ radical with O_3 (Gogate and Pandit 2004; Flaherty and Huang 1992).

ii) **Oxidation with H_2O_2**:
The reaction system with O_3/H_2O_2-system as the main key player is considered to be a vital and practical system because it deals with the incorporation of hydrogen peroxide into a solution with already present ozone. At this point, various other niches are being investigated that contribute to much higher yields. An example includes the multistage injection systems that bring forth better outcomes as compared to conventional modes of techniques. The additions are adjusted with each step by employing reactors in series. The concentration of the organic products as well as the remaining pollutants are the defining factors that affect the additions (Silva and Jardim 2006). Other advanced methods are UV irradiation and chlorination which ensure better consequences in terms of remediated water but are costly (Lazarova et al. 2000, 2003). H_2O_2 shows a decrease in the value of COD, BOD, and offensive odor along with foaminess in wastewater from different sources (Geissen 2000). Alongside, it has also been a powerful and improved treatment with respect to the existing physical or biological strategies. It can also employ a catalyst in certain cases like iron (Fe^{2+} or Fe^{3+} ions), UV light or ozone (O_3) for the catalytic autoxidation as well as the decrease of the BOD/COD of wastewater with the incorporation of alkali or can be used alone under many circumstances (Millero et al. 1989; Ksibi 2006).

20.3.6.3 Solids Removal

III) **Clarification**:
The matter present in a suspended form in the wastewater must be eliminated by various methods. The matter may be large solids that are settleable and the ones that are colloidal in nature and comes under the category of nonsettleable. Their removal is achieved by using coagulation, flocculation as well as sedimentation. Their combination is referred to as conventional clarification.

i) **Coagulation**:
This process combines the smaller particles and makes their aggregates or flocs and these also adsorb dissolved organic matter to their surface and, in turn, remove the impurities by separation. Aluminum sulfate and ferric chloride are notable compounds for the modern coagulation process (Jiang 2015). Coagulants can mostly be of inorganic nature or prehydrolyzed metal-ion coagulants that are based on aluminum (i.e. poly-aluminum chloride) or the ferric iron dependants (i.e. polyferric sulfate) or the coagulant with mixed polymeric nature (i.e. polyaluminum-iron-sulfate). These coagulants give better results as compared to conventional coagulants like aluminum sulfate or ferric sulfate (Jiang and Graham 1996). The process of coagulation and flocculation takes place subsequently allowing particle collision as well as the floc growth, and is then succeeded with the sedimentation process (Prakash et al. 2014).

ii) **Precipitation**:
Alkalization at high pH for chemical clarification
There is the involvement of calcium and magnesium compounds in the efficient wastewater treatment that was reported in the late 1920s and its contribution to high pH makes the treatment more feasible (Dziubek and Kowal 1984). The coagulation process is favored at high pH using various sources of magnesium ions like

magnesium chloride, magnesium hydroxide, magnesium carbonate, processed dolomite, etc. According to Leentvaar and Rebhun (1982), low pH (9) lime treatment also shows a significant decrease in TSS and organic carbon content. Whereas at high-pH (11.0–11.5), the treatment was more effective because of the presence of both calcium carbonate and magnesium hydroxide precipitation (Merril and Jorden 1975). Alkalizing agents like lime or caustic soda are also utilized for the traditional chemical clarification process while raising pH.

I) **Lime alkalization**:
Lime includes quicklime (CaO), hydrated lime Ca(OH) x, as well as dolo-2 mitic lime under general terminology. The two kinds that are predominant in wastewater treatment are quick lime and hydrated lime (Forstner and Leirde 1981). It has a plus point of no contribution to a rise in salinity as compared to alum or iron salts (Dziubek and Kowal 1984). Along with it, lime is also a precipitant for phosphates, bacteria, and trace metals, and a coagulant utilized for eliminating suspended matter and colloidal particles in wastewater (Semerjian and Ayoub 2003; Parker 1972; Stukenberg 1975). Algae are also employed for the removal of 80% suspended matter comprising bicarbonate alkalinity, hardness as well as phosphates (Semerjian and Ayoub 2003; Culp et al. 1978).

II) **Caustic soda alkalization**:
Sodium hydroxide is also utilized for increasing pH levels for calcium and magnesium-mediated precipitation. The elimination of turbidity, suspended matter and organic solids aided with NaOH was reported by Merril and Jordan in 1975. Moreover, its usage is less favored as compared to lime alone.

20.4 Electrochemical Methods of Wastewater Treatment

There are various applications of electrochemical treatments in wastewater treatment strategies. These include electrodeposition, electrocoagulation (EC), electroflotation (EF) as well as electrooxidation (Szpyrkowicz et al. 1994). The first one is electrodeposition that is important in wastewaters rich in heavy metals as they recover heavy metals from them. EC has been an important strategy aided with aluminum, iron or hybrid Al/Fe electrodes for wastewater treatment (Stuart 1946; Bonilla 1947). Another technique is electroflocculation that removes flocs from the sludge. It can eliminate colloidal particles, oil or grease along with other organic pollutants (Nenno et al. 1988). It is efficient as compared to dissolved air flotation, sedimentation, or impeller flotation (IF). Now, with the adoption of newly developed active and stable electrodes, this technology has been improved. Electrooxidation is responsible for efficiently degrading the pollutants that are refractive in nature and found on the surface of a few electrodes. The electrodes titanium-based boron-doped diamond film electrodes (Ti/BDD) are highly stable and active in the industries (Chen 2004).

20.4.1 Ion-Exchange Technique

This method is used in wastewater treatment and interchange one ion with the other for the demineralization purpose. The two types of exchange resins are used in this method. These are cation resins as well as anion exchange resins. Their type is not important when it comes to membrane fouling as it is easy to get this problem with untreated water. Hence, a separate part is also necessary to eliminate any suspended matter if present in wastewater. This should be ensured before the influent undergoes any treatment strategy. If there is any organic matter present, it should be eliminated so that a little load is there for an ion exchange unit. Sodium zeolite softening depicts the example of the typical ion exchange process that removes the hardness due to calcium and magnesium. The process is utilized in steam boilers as well as industrial water treatment facilities. The bed completely gets saturated with Ca and Mg ions and is regenerated afterward. Any how, there are very limitations to this process (Kansara et al. 2016).

20.4.2 UV Disinfection

It is the most important method that is applied for disinfection throughout the world. This system is highly efficient for its removal capability for microbes and results in minimal byproducts with less cost (Lazarova et al. 1999). The main hindrance associated with this is the resistance of certain microbes against UV radiation which needs higher doses to be effectively eliminated (Liberti et al. 2002). The efficiency of this process depends upon the quality of the influent and the performance of the system also varies with respect to characteristics of water like turbidity, concentration of SS as well as transmittance (Harleman and Murcott 1999). Moreover, this system of UV is linked to pretreatments for a better quality of influent (Gomez et al. 2007).

20.5 Biological Treatment

These include both physical and biochemical strategies. The biological strategy itself has two divisions – aerobic and anaerobic. The former is the degradation of organic matter with oxygen aided by aerobic bacteria feeding on these compounds while the latter deals with anaerobic bacteria under negligible oxygen (Zabava et al. 2015). The neutralization of toxic compounds in wastewater results in treated waters with varying properties either use for irrigation or sludge to be processed, stored, degraded or recovered (Robescu 2000). After these processes, the treated water reaches an acceptable quality as proposed by the general guidelines for wastewater (Zabava et al. 2015). The generated biomass is removed by secondary sedimentation. The biological treatment is carried out mostly by heterotrophic bacteria and less often by fungi (Gray 2005; Samer 2015).

20.6 Strategies for Biological Treatment

The classification of biological systems is done on the basis of microbes present in them. Aerobic and facultative are present in aerobic treatment while anaerobic and facultative favor anaerobic mode of treatment. There are two categories into which the processes are divided. These are:

- Suspended growth processes.
- Fixed growth processes.

20.6.1 Suspended Growth Processes

They are referred to as the processes in which the wastewater and related microbes are confined in a container. The oxygen is then incorporated to let the biological activity take place in the reactor. Their examples include ponds, lagoons as well as activated sludge systems (Ebrahimi and Najafpour 2016).

20.6.2 Fixed Growth Processes

This name is given to the systems where biological mass is fixed onto a growth medium and wastewater is made to come in contact with it. The mass then stabilizes and treats the wastewater as it is passed through it. Examples include trickling filters as well as rotating biological contractors (Ebrahimi and Najafpour 2016).

20.6.2.1 Lagoons

These are continuously operating bioreactors that contain wastewater and natural microbes in them to treat wastewater. In lagoons, the continuous flow and volume are maintained for effective aeration and mixing in order to treat wastewater. These may be up to thousands of m^3. Anyhow, good biodegradation leads to increasing biomass which is a disadvantage in this process. The lagoons may be aerobic, anaerobic, or facultative (Cicek et al. 1998).

20.6.2.2 Facultative Lagoons

These are the mostly used forms of lagoons. These come under the aquatic treatment methods. In these cases, the water gets divided into three layers. The one near the surface is aerobic, while the intermediate layer is facultative with the bottom anaerobic layer. The intermediate layer is aerobic on its top while anaerobic close to its bottom, and forms a facultative zone.

20.6.2.3 Aerated Lagoons

These are smaller in volume and deeper than the facultative lagoons. These are modifications of stabilization ponds in which the devices are installed for aeration and this removes the odors from the septic conditions. The mechanical or diffused air systems are used for aeration purposes. The primary demerit of lagoons is their high solid content of effluent, which could be 100 mg/l (Fonade et al. 2000).

1) **Activated sludge**:
It is a vital process for meeting the criteria of secondary treatment effluent limits. The design is made on the basis of overflow rate along with solids loading. These are used for sedimentation purposes. There are various modifications that are made with the basic principle of activated sludge systems. These are employed for achieving greater treatability and lowering the capital cost. The variations have their basis in completely mixed systems, step aeration, contact stabilization, extended aeration as well as pure oxygen systems. Most of the modifications are able to achieve an effective BOD removal. Extended aeration is a flexible method but is costly. The two methods – extended aeration as well as contact stabilization – are also favorable as package plants (Wisconsin Department of Natural Resources 2016).

20.6.3 Sequencing Batch Reactors

The varying characteristics of the wastewater influent and the rigorous regulations regarding effluents have been a bulwark against effective wastewater treatment. Contrarily, another reliable, lucrative as well as highly efficient technology is developed known as SBR. Being a modified version of the activated sludge system, it functions on a fill-and-draw technique. It brings together two phases aerobic and anaerobic in a single unit and results in reserving about 25% cost of aeration accompanying less production of sludge. Concomitantly, the removal of nitrogen and phosphorus can be obtained by modifying the actual operating cycle. SBR is a typical batch reactor which functions over a chain of reactions comprising the SBR cycle. Fill, react, settle, decant and idle periods are the stages of the SBR cycle. The system could achieve a higher degree of biological removal of nutrient with alternations of aerobic and anaerobic cycles that exceeds two in the count with varying desired outcomes. The removal of nitrogen involves two stages of nitrification that occurs aerobically with the conversion of ammonia to nitrates and the other denitrification stage that is the anaerobic transformation of nitrate into the nitrogen gas. Nitrifying bacteria perform nitrification and pH gets increase in this step while heterotrophic bacteria conduct denitrification and alkaline production gets increased in this step (Yang et al. 1999). SBR is definitely a quality alternative in place of continuous-flow systems of biological nutrient removal (BNR) because of the incorporation of anaerobic, anoxic along with aerobic reactions in a single treatment tank during the SBR cycle. The laboratory scale SBR has shown a high degree of BNR with low COD along with great removal results for N and P carrying wastewater. The introduction of the anoxic stage in anaerobic–aerobic SBR results in an increasing ratio of phosphate uptake in the anoxic stage as compared to aerobic uptake (Lee et al. 2001). Moreover, the removal of nitrogen and phosphorus can also be obtained by using control parameters like pH and ORP (Akin and Ugurlu 2005).

20.6.4 Oxidation Ditches

Another modification of the activated sludge system is the oxidation ditch (Metcalf & Eddy Inc. 1991). These are used for the removal of organic solids that are biodegradable under long solids retention times (SRTs) that ranges from 2.5 to 3.5 lb./Hp-hour (A. Adams, personal communication). These consist of an intermixed system and can be manipulated for the plug flow conditions that need diffused air for mixing purposes with less time as compared to the usual oxidation ditch. It is composed of a ring-, oval-, or horseshoe-shaped basin that works individually or makes a multichannel configuration (Allen et al. 1995). These are regarded as "racetrack-type" systems with aerators that are mounted on top either horizontally or vertically to entertain the ditch with oxygen transfer and circulation. It has the advantage of partial denitrification, a constant level of water for continuous discharge that reduces the rate of overflow and eradicates the periodic surge of effluent that is common to all other biological methods like SBRs (Ettlilch 1978). It also has a long time of hydraulic retentions with a minimal footprint of shock load or hydraulic surge. It also generates little amount of sludge in comparison to other biological methods because of extended biological activity during the process.

20.6.5 Fixed Film Processes

20.6.5.1 Trickling Filters

The attached growth systems are used effectively for the treatment of wastewater for may years. Their applications are increased with their implications in remote locations as well as various areas of third-world countries for BOD and toxic content removal (Naz et al. 2015). The trickling filter is based on an aerobic mode of treatment. The effluent is passed over the fixed bed of the support medium. This makes the biological film grow on the surface of the attached medium and the microbes thus start acting on the wastewater passing through them. The biofilm is partially exposed to the surface of wastewater and partially to the air for oxygen. The stabilization obtained through this method depends on various factors like volumetric or organic loading, ventilation, type of medium used for support and temperature. This system hardly attains values beyond 85% for BOD reduction.

20.6.5.2 Rotating Biological Contractors

The rotating biological contractors are made up of a set of large flat or corrugated disks that are placed together in a close manner and mounted on a common horizontal shaft with partially or completely sinking in the effluent requiring a treatment. The continuous movement of the shaft is aided by a mechanical motor and results in the formation of a biofilm on the entire surface of the supporting medium and degrading, the organic content present in wastewater (Flickinger 2009). The aerobic processes require the rotation of media to encourage the transfer of oxygen. This rotation also gives rise to turbulence along with shearing forces to remove excessive microbes from the attached medium (Patwardhan 2003; Rodgers and Zhan 2003). The excessive biomass is removed in a clarifier and is separated from treated wastewater. The RBC can have a single compartment or several ones joining together with disks and packed medium in every stage.

20.6.6 Other Approaches or Hybrid Systems

There are various modifications with chemical-biological-physical methods incorporation and others with one category removed from the major three. The wastewater carrying a large content of organic compounds experiences serious complications. So, the intermixing of two or more methods can overlook the weakness of one and improve the overall efficiency (Tee et al. 2016). There are some approaches to attain maximum efficiency with less cost. These are as follows:

20.6.6.1 Membrane Bioreactors

These are the combined aerobic and anaerobic systems with membrane organisms that can be used in suspended growth biological systems (Amoatey and Richard 2011). The recent developments that are made in the following respect include the installation of new architecture to improve the strains that are responsible for the biological treatment and include nitrifiers along with providing a supported field for an increase in performance and maintaining reasonable adsorbents to cater to membrane fouling (Dhaouadi and Marrot 2008, Ueda and Horan 2000). MBR systems are used for the effective and instantaneous removal of pollutants and contaminants (Aslam et al. 2017). There are two categories of MBR that differ on the basis of location of membranes. These may be submerged in the bioreactor or located externally (Yamamoto et al. 1989; Ueda et al. 1996, 1997). The submerged-type membranes carrying MBR are notable for their better performance in procuring quality effluent as compared to the activated sludge systems (Yamamoto et al. 1989; Karadag et al. 2015; Fraga et al. 2017). The primary demerit of MBR is that the membrane fouling control is very difficult in these systems in contrast to the external membrane system. Other methods like intermittent suction method as well as back flushing are incorporated to decrease the membrane fouling (Visvanathan et al. 1997; Randall and Barnard 1998).

20.6.6.2 Anaerobic RBC and Aerobic SBR Systems

The RBC process is known for its short time for retention, low requirement of energy, less cost, and outstanding process control and handling of an extensive range of flows, whereas, its disadvantage is its vulnerability to the characteristics of wastewater. This results in restricted operational flexibility against various conditions of operation and loading. The fill and draw AS systems are improved to the new aerobic SBR which is made up of one or more tanks. Every tank has the capacity to undergo the separation of solids as well as stabilization (Wang and Wu 2009). The merits of the SBR process were reported according to which it is concluded that it is flexible while treating variable flows, giving aerobic and anaerobic conditions in the same compartment or tank, has an efficient contact between oxygen and microbes, has a great ability to remove contaminants and occupies a small space. These pros have made it an appropriate and highly used process on wastewater from industrial sites (Yang et al. 1999). The intermixing of anaerobic RBC and aerobic SBR will result in the efficient generation of bioenergy and a feasible treatment system for wastewater because a large amount of methane is produced by anaerobic RBC and can be treated effectively with aerobic SBR. This integration is employed for treating screened dairy manure as well as for its mixture with cheese whey. It can achieve up to 98% decrease in COD and produce a significant amount of methane gas.

20.6.6.3 UASB and Aerobic Fluidized Bed System

The liquid circulation is used in the fluidized bed reactors to fluidize the particles covered with biofilm. As a result, the problem of substrate diffusion is ameliorated which is most often in stationary bed reactors. Shieh and Keenan (1986) gave the advantage of AFB reactors that have a high concentration of biomass, small HRT, high organic loading, less resistance for small external mass transfer, and a large surface area for transfer of mass with no clogging of bed. Whereas the demerits were their issue of oxygen distribution and bed expansion (Lazarova and Manem 1994; Saravanane and Murthy 2000). Along with it, more energy is also required as it requires a high ratio of circulation of liquid. According to Tavares et al. (1995), AFB can attain up to 82% removal of average COD at a low average HRT of about 30 minutes. Moreover, a removal efficiency of 75% was reported for synthetic textile-polluted wastewater. This outcome was obtained by integrating UASB with the AFB reactor at an HRT of 14 hours. The integrated system produced 45% lower sludge volume as compared to the aerobic system. Anaerobic biomass incorporated into the AFB reactor improves the overall COD removal and results in more content of SS. However, a minimum amount of cell mass must be ensured in the UASB reactor prior to the treatment in the AFB reactor. The treatment is highly applicable for the biological treatment of wastewater from industrial sites with moderate levels of pollution because the UASB-AFB system has less sludge formation, high tolerance for pH change and stable performance in COD removal. It is also ideal for its economic and technical implications and areas where less space is needed (Goli et al. 2019).

20.6.6.4 ANR and Ozonation

The technique of ozonation is notable for its applications in the treatment of bacteria and other organic pollutants present in drinking water. It is an important disinfectant that has its foundation in the utilization of (Li et al. 2015) active molecules of ozone gas (O_3). According to the new studies, the autotrophic nitrogen removal (ANR) process is found to be effective when used along with ozonation–adsorption (activated carbon) as a hybrid system and can eliminate up to 82.6% of the COD with 78.4% of total nitrogen content or TN in the landfill leachate. The ANR system alone is much less efficient as compared to the hybrid system (Tee et al. 2016)

20.6.6.5 Integrated Reverse Osmosis (RO) Process with the Sorption Membrane Filtration (MF)

This hybrid system is important for its usage in the removal of boron from seawater (Guler et al. 2011). The ion exchange resin used in the process has a size of about 0–20 mm for the boron elimination from the RO permeates. The sorption of boron was done on an ion exchange resin selective for very fine powdered boron and the resin loaded with boron is separated with the help of submerged membranes afterward. The process is beneficial as it uses very fine particles of resin for sorption membrane separation and in turn results in specific surfaces as well as faster kinetics. RO-ion exchange-MF hybrid can remove boron by almost 95–98% with microfiltration membrane along with the removal of 92.8–93.8% with ultrafiltration membrane (Li et al. 2013).

20.6.6.6 Microbial Fuel Cell (MFC)

This system employs both chemical and biological treatment methods. Microbial fuel cells (MFCs) are under the spotlight for transforming organic waste into electricity. Many studies have been reported on MFC as an alone system but there is no much disclosure on its integration with other systems. The wastewaters are highly populated with organic matter that could be transformed into energy. MFC is quite a distinctive technology as it treats wastewater and this process is a method of capturing energy and consequentially making electricity or hydrogen gas, as compared to drain on electrical energy (Weld and Singh 2011).

20.6.6.7 MCABR

This hybrid system is made up of an integration of membrane coagulation with an adsorption bioreactor (MCABR) that includes the incorporation of polyaluminum chloride (PACl) that acts as coagulant with powdered activated carbon (PAC) that plays its part as an adsorbent into the bioreactor (Tian et al. 2010). This removes a great amount of organic matter from slightly polluted surface water. MCABR incorporates four types of mechanisms. These include separation aided with membrane, biodegradation with the help of microbes, PACl-mediated coagulation and adsorption with PAC in conjunction with each other contributing to the elimination of dissolved organic matter (Tee et al. 2016).

References

Ahmed, B., Mohamed, H., Limem, E., and Nasr, B. (2009). Degradation and mineralization of organic pollutants contained in actual pulp and paper mill wastewaters by a UV/H_2O_2 process. *Ind. Eng. Chem. Res.* 48 (7): 3370–3379.

Akin, B.S. and Ugurlu, A. (2005). Monitoring and control of biological nutrient removal in a sequencing batch reactor. *Process Biochem.* 40 (8): 2873–2878.

Arslan-Alaton, I. and Akmehmet Balcioglu, I. (2002). Biodegradability assessment of ozonated raw and biotreated pharmaceutical wastewater. *Arch. Environ. Contam. Toxicol.* 43: 0425–0431. https://doi.org/10.1007/s00244-002-1235-y.

Allen, S.J., Khader, K.Y.H., and Bino, M. (1995). Electrooxidation of dyestuffs in waste waters. *J. Chem. Technol. Biotechnol.* 62: 111–117.

Amoatey, P. and Richard, B. (2011). Chapter 20: wastewater management. In: *Waste Water Evaluation and Management* (ed. F.S.G. Einschlag), 1–22. IntechOpen.

Andreoli, C.V., von Sperling, M., and Fernandes, F. (2007). *Sludge Treatment and Disposal*, vol. 6, 14–17. IWA Publishing https://doi.org/10.2166/9781780402130.

Artan, N., Wilderer, P., Orhon, D. et al. (2002). Model evaluation and optimisation of nutrient removal potential for sequencing batch reactors. *Water SA* 28 (4): 423–432.

Arun, S., Sinharoy, A., Pakshirajan, K., and Lens, P.N.L. (2020). Algae based microbial fuel cells for wastewater treatment and recovery of value-added products. *Renew. Sustain. Energy Rev.* 132, 110041, ISSN 1364-0321, https://doi.org/10.1016/j.rser.2020.110041.

Asano, T. and Levine, A.D. (1996). Wastewater reclamation, recycling and reuse: past, present, and future. *Water Sci. Technol.* 33 (10): 1–14.

Aslam, M., Charf, A., Lesage, G. et al. (2017). Membrane bioreactors for wastewater treatment: a review of mechanical cleaning by scouring agents to control membrane fouling. *Chem. Eng. J.* 307: 897–913.

Bettley, A. (1992). Physical methods of liquid effluent treatment separation processes. An outline of the role of the separation processes in effluent treatment including pros and cons of selected electrically intensive techniques in comparison to gravity settlement. *Build. Res. Inf.* 20 (2): 102–108.

Betty, A.C. (1999). *Baker Process Personal communication with Senior Process Engineer.* Memorandum to Parsons Engineering Science Inc.

Bhargava, A. (2016). Physico-chemical waste water treatment technologies: an overview. *Int. J. Sci.* 4: 08–19.

Boari, G., Mancini, I.M., and Trulli, E. (1997). Technologies for water and wastewater treatment. In: *Aspects économiques de la gestion de l'eau dans le bassin méditerranéen* (ed. B. Dupuy), 261–287. Bari: CIHEAM (Options Méditerranéennes: Série A. Séminaires Méditerranéens; n. 31).

Bonilla, C.F. (1947). Possibilities of the electronic coagulator for water treatment. *Water Sewage* 21–85.

Bull, A.R. and McManamon, J.T. (1999). Destruction of inorganic pollutants in wastewater with hydrogen peroxide. In: *Emerging Technologies in Hazardous Waste Management* (ed. W. Tedder and G. Pohland), 52–66. Washington, DC: American Chemical Society.

Camacho, P., Deleris, S., Geaugey, V. et al. (2002). A comparative study between mechanical, thermal and oxidative disintegration techniques of waste activated sludge. *Water Sci. Technol.* 46 (10): 79–87.

Casellas, M., Dagot, C., and Baudu, M. (2006). Set up and assessment of a control strategy in a SBR in order to enhance nitrogen and phosphorus removal. *Process Biochem.* 41 (9): 1994–2001.

Charlton, J. (1994). Biological nutrient removal applied to weak sewage. *Water Sci. Technol.* 29 (12): 41–48.

Chen, G. (2004). Electrochemical technologies in wastewater treatment. *Sep. Purif. Technol.* 38 (1): 11–41.

Cheremisinoff, N.P. (2002). *Handbook of Water and Wastewater Treatment Technologies*. Boston: Butterworth Heinemann.

Chiang, Y.P., Liang, Y.Y., Chang, C.N., and Chao, A.C. (2006). Differentiating ozone direct and indirect reactions on decomposition of humic substances. *Chemosphere* 65 (11): 0–2400.

Chio, H.B., Hwang, K.Y., and Shin, E.B. (1997). Effect on anaerobic digestion of sewage sludge pretreatment. *Water Sci. Technol.* 35 (10): 207–211.

Choi, Y., Daein, L., Hong, S. et al. (2020). Investigation of the synergistic effect of sonolysis and photocatalysis of titanium dioxide for organic dye degradation. *Catalysts* 10 (5): 500.

Chu, C.P., Lee, D.J., You, C.S., and Tay, J.H. (2002). Weak ultrasonic pre-treatment on anaerobic digestion of flocculated activated biosolids. *Water Res.* 36: 2681–2688.

Cicek, N., Winnen, H., Suidan, M.T. et al. (1998). Effectiveness of the membrane bioreactor in the biodegradation of high molecular weight compounds. *Water Res.* 32 (5): 1553–1563.

Craciun, V., Chiorescu, E., and Balan, O. (2004). *Recycling of Waste and Residue from Agriculture and Food Industry*. Iaşi: Cermi Publishing.

Cravotto, G., Gaudino, E.C., and Cintas, P. (2013). On the mechanochemical activation by ultrasound. *Chem. Soc. Rev.* 42: 7521–7534.

Culp, G.L. (1967). Chemical treatment of raw sewage. *Water Wastewater Eng.* 4: 61–63.

Culp, R.L., Wesner, G.M., and Culp, G.L. (1978). *Handbook of Advanced Wastewater Treatment*, 2nde. New York: Van Nostrand Reinhold Company.

Czaplicka, M., Hazard, J., and Mater, B. (2006). Review photo-degradation of chlorophenols in the aqueous solution. *J. Hazard. Mater.* 134 (1–3): 45–59.

Descorme, C. (2017). Catalytic wastewater treatment: oxidation and reduction processes recent studies on chlorophenols. *Catal. Today* 297: 324–334.

Dhaouadi, H. and Marrot, B. (2008). Olive mill wastewater treatment in a membrane bioreactor: process feasibility and performances. *J. Chem. Eng.* 145 (2): 225–223.

Dziubek, A.M. and Kowal, A.L. (1984). Effect of magnesium hydroxide on chemical treatment of secondary effluent under alkaline conditions. *Proceedings of Water Reuse Symposium III, American Waterworks Association Research Foundation*, San Diego. Elsevier.

Ebrahimi, A. and Najafpour, G.D. (2016). Biological treatment processes: suspended growth vs. attached growth. *IJEE* 7 (2): 114–123.

Eilbeck, W. and Mattock, G. (1987). *Chemical Processes in Waste Water Treatment*. Chichester, Sussex: Ellis Horwood Ltd.

Environmental Finance Center Network. Chapter 1: introduction to wastewater management. *Handbook of Wastewater Management for Local Representatives*. 2e, Syracuse University.

Ettlilch, W.F. (1978). *A Comparison of Oxidation Ditch Plants to Competing Processes for Secondary and Advanced Treatment of Municipal Wastes*. Springfield, VA: Environmental Protection Agency.

Fahad, A., Mohamed, S., Maya, R. et al. (2019). Wastewater and its treatment techniques: an ample review. *Indian J. Sci. Technol.* 12 (25): 1–13.

Flaherty, K.A. and Huang, C.P. (1992). *Chemical Oxidation*, vol. 2 (ed. W.W.B.A.R. Eckenfelder and J.A. Roth), 58. Lancaster: Technomic Inc.

Flickinger, M.C. (2009). *Encyclopedia of Industrial Biotechnology (Bioprocess, Bioseparation, and Cell Technology)*. Bioreactors: Rotating Biological Contactors.

Fonade, C., Rols, J.L., Goma, G. et al. (2000). Improvement of industrial wastewater treatment by aerated lagoon: case studies. *Water Sci. Technol.* 42 (5–6): 193–200.

Forstner, U. and Leirde, J.H. (1981). Trace metals in water purification processes. In: *Metal Pollution in the Aquatic Environment* (ed. U. Forstner and G.T.W. Wittmann). Berlin: Springer-Verlag Publishers.

Fraga, F.A., Garcia, H.A., Hooijmans, C.M. et al. (2017). Evaluation of a membrane bioreactor on dairy wastewater treatment and reuse in Uruguay. *Int. Biodeterior. Biodegrad.* 119: 552–564.

Franklin, L.B. (1991). *Wastewater Engineering: Treatment*. New York: Disposal and Reuse, Inc.

Ganjidoust, H., Tatsumi, K., Yamagishi, T., and Gholian, R. (1997). Effect of synthetic and natural coagulant on lignin removal from pulp and paper wastewater. *Water Sci. Technol.* 35 (2–3): 291–296. https://doi.org/10.1016/S0273-1223(96)00943-2.

Geissen, S.U. (2000). Water demand in arid countries: an opportunity to reduce environmental pollution. *Mediterranean Area Proceedings of the Int Conf WATRAMA*, Tunis, 1.

Gogate, P.R. and Pandit, A.B. (2004). A review of imperative technologies for wastewater treatment I: oxidation technologies at ambient conditions. *Adv. Environ. Res.* 8: 553.

Goli, A., Shamiri, A., Khosroyar, S. et al. (2019). A review on different aerobic and anaerobic treatment methods in dairy industry wastewater. *J. Environ. Treat. Tech.* 6 (1): 113–141.

Golil, A., Shamiri, A., Khosroyar, S. et al. (2019). A review on different aerobic and anaerobic treatment methods in dairy industry wastewater. *J. Environ. Trat. Tech* 6 (1): 113–141.

Gomez, M., Plaza, F., Garralon, G. et al. (2007). A comparative study of tertiary wastewater treatment by physico-chemical UV process and macrofiltration–ultrafiltration technologies. *Desalination* 202 (1–3): 369–376.

Gray, N.F. (2005). *Water Technology: An Introduction for Environmental Scientists and Engineers*, 2nde. Amsterdam, The Netherlands: Elsevier.

Guidelines for Water Reuse (1992–2004). *US EPA*. DC: Washington.

Guler, E., Kabay, N., Yuksel, M. et al. (2011). Integrated solution for boron removal from seawater using RO process and sorption-membrane filtration hybrid method. *J. Membr. Sci.* 375 (1): 249–257.

Gupta, K.V., Ali, I., Saleh, T.A. et al. (2012). Chemical treatment technologies for waste-water recycling - an overview. *RSC Adv.* 2 (16): 6380–6388.

Harleman, D.R. and Morrissey, S. (1990). Chemically enhanced treatment: an alternative to biological secondary treatment for ocean outfalls. *Proceedings of the National Conference on Hydraulic Engineering*, American Society of Civil Engineers, San Diego, CA.

Harleman, D.R.F. and Murcott, S. (1999). The role of physical-chemical wastewater treatment in the mega-cities of the developing world. *Water Sci. Technol.* 40: 75–80.

Harleman, D.R., Morrissey, S., and Murcott, S. (1991). *The Case for Chemically Enhanced Primary Treatment in a New Boston Harbor Clean-up Plan*. Cambridge, MA: Ralph M Parsons Laboratory, Massachusetts Institute of Technology.

Herrmann, J.M., Matos, J., Disdier, J. et al. (1999). Solar photocatalytic degradation of 4- chlorophenol using synergistic effect between titania and activated carbon in aqueous suspension. *Catal. Today* 54: 255–265.

Hoigne, J. (1997). Inter-calibration of OH radical sources and water quality parameters. *Water Sci. Technol.* 35: 1–8.

Hoigne, J. and Bader, H. (1976). The role of hydroxyl radical reactions in ozonation processes in aqueous solutions. *Water Res.* 10: 377–386.

Hoigne, J. and Bader, H. (1983). Rate constants of reactions of ozone with organic and inorganic compounds in water. II. Dissociation organic compounds. *Water Res.* 17: 185–194.

Hu, X., Hu, C., and Wang, R. (2015). Enhanced solar photodegradation of toxic pollutants by long-lived electrons in Ag–Ag_2O nanocomposites. *Appl. Catal. B Environ.* 637: 176–177.

Jasim, N.A. (2020). The design for wastewater treatment plant (WWTP) with GPS X modelling. *Cogent Eng.* 7 (1): 1723782. https://doi.org/10.1080/23311916.2020.1723782.

Jiang, J.Q. (2015). The role of coagulation in water treatment. *Curr. Opin. Chem. Eng.* 8: 36–44.

Jiang, J.Q. and Graham, N.J.D. (1996). Enhanced coagulation using Al/ Fe(III) coagulants: effect of coagulant chemistry on the removal of color-causing NOM. *Environ. Technol.* 17: 937–950.

Kansara, N., Bhati, L., Narang, M., and Vaishnavi, R. (2016). Critical review wastewater treatment by ion exchange method: a review of past and recent researches trade science Inc. *ESAIJ* 12 (4): 143–150.

Karadag, D., Koroglu, O.E., Ozkaya, B., and Cakmakci, M. (2015). A review on anaerobic biofilm reactors for the treatment of dairy industry wastewater. *Process Biochem.* 50 (2): 262–271.

Ksibi, M. (2006). Chemical oxidation with hydrogen peroxide for domestic wastewater treatment. *J. Chem. Eng.* 119 (2–3): 161–165.

Kuba, T., van Loosdrecht, M.C.M., and Heijnen, J.J. (1996). Phosphorus and nitrogen removal with minimal COD requirement by integration of denitrifying dephosphatation and nitrification in a two-sludge system. *Water Res.* 30: 1702–1710.

Kuo, W.G. (1992). Decolorizing dye wastewater with Fenton's reagent. *Water Res.* 26 (7): 881–886.

Lazarova, V. and Manem, J. (1994). Advances in biofilm aerobic reactors ensuring effective biofilm activity control. *Water Sci. Technol.* 29 (10–11): 319–327.

Lazarova, V., Savoye, P., Janex, M.X. et al. (1999). Advanced wastewater disinfection technologies: state of the art and perspectives. *Water Sci. Technol.* 40: 203–213.

Lazarova, V., Cirelli, G., Jeffrey, P. et al. (2000). Enhancement of integrated water management and water reuse in Europe and the Middle East. *Water Sci. Technol.* 42 (1/2): 193–202.

Lazarova, V., Sheilds, P., Levine, B. et al. (2003). Production of high quality water for reuse purposes: the West Basin experience. *Water Supply* 13 (3): 167–175.

Lee, D.S., Jeon, C.O., and Park, J.M. (2001). Biological nitrogen removal with enhanced phosphate uptake in a sequencing batch reactor using single sludge system. *Water Res.* 30 (2): 277–284.

Leentvaar, J. and Rebhun, M. (1982). Effect of magnesium and calcium precipitation on coagulation-flocculation with lime. *Water Res.* 16 (5): 655–662. https://doi.org/10.1016/0043-1354(82)90087-2.

Lettinga, G. and Hulshoff, L. (1986). New technologies for anaerobic wastewater treatment. *Water Sci. Technol.* 18: 41–53.

Li, W., Krantz, W.B., Cornelissen, E.R. et al. (2013). A novel hybrid process of reverse electrodialysis and reverse osmosis for low energy seawater desalination and brine management. *Appl. Energy* 104: 592–602.

Li, W., Lu, X., Xu, K. et al. (2015). Cerium incorporated MCM-48 (Ce-MCM-48) as a catalyst to inhibit bromate formation during ozonation of bromidecontaining water: efficacy and mechanism. *Water Res.* 87: 49–58.

Liberti, L., Notarnicola, M., and Petruzzelli, D. (2002). Advanced treatment for municipal wastewater reuse in agriculture. UV disinfection: parasite removal and by-product formation. *Desalination* 152: 315–324.

Lin, X., Guo, X., Shi, W. et al. (2015). Ternary heterostructured Ag–$BiVO_4$/$InVO_4$ composites: synthesis and enhanced visible-light-driven photocatalytic activity. *J. Alloys Compd.* 635: 256.

Lu, M.C., Chen, J.N., and Huang, H.H. (2002). Role of goethite dissolution in the oxidation of 2-chlorophenol with hydrogen peroxide. *Chemosphere* 46: 131.

Marmagne, O. and Coste, C. (1996). Colour removal from textile plant effluents. *Am. Dyest. Rep.* 85: 15–20.

Merril, D.T. and Jorden, R.M. (1975). Lime-induced reactions in municipal waters. *J. Water Pollut. Control Fed.* 47: 2783–2808.

Metcalf & Eddy Inc. (1991). *Wastewater Engineering: Treatment, Disposal, and Reuse*, 3ee, vol. 4 (ed. G. Tchobanoglous and F.L. Burton). New York: McGraw-Hill.

Metcalf & Eddy (2014). *Wastewater Engineering: Treatment and Resource Recovery*, 5ee. New York: McGraw-Hill.

Millero, F.J., LeFerriere, A., Fernandez, M. et al. (1989). Oxidation of H_2S with H_2O_2 in natural waters. *Environ. Sci. Technol.* 23: 209–213.

Morrissey, S.P. and Harleman, D.R.F. (1992). Retrofitting conventional primary treatment plants for chemically enhanced primary treatment in the USA. In: *Chemical Water and Wastewater Treatment II* (ed. R. Klute and H. Hahn). Berlin, Heidelberg: Springer https://doi.org/10.1007/978-3-642-77827-8_26.

Muller, E., Stouthamer, A., Van Verseveld, H., and Eikelboom, D. (1995). Aerobic domestic waste water treatment in a pilot plant with complete sludge retention by crossflow filtration. *Water Res.* 29 (4): 1179–1189.

Naz, I., Saroj, D.P., Sadia, M. et al. (2015). Assessment of biological trickling filter systems with various packing materials for improved wastewater treatment. *Environ. Technol.* 36 (4): 424–434.

Nemerow, N. and Dasgupta, A. (1991). *Industrial and Hazardous Waste Treatment*. New York: Van Nostrand Reinhold.

Nenno, V.E., Zelentsov, V.I., Melnichuk, E.V. et al. (1988). Experience in operating a device for concentration of mineral raw material combining electroflotation and separation in a froth layer. *EOM* 6: 77–79.

Nielsen, M.K., Carstensen, J., and Harremoes, P. (1996). Combined control of sewer and treatment plant during rainstorm. *Water Sci. Technol.* 34 (3–4): 181–187.

Odegaard, H. (1987). Particle separation in wastewater treatment. *Documentation on the 7th EWPCA Symposium*, Munich, Germany, 351–400.

Odegaard, H. (1992). Norwegian experiences with chemical treatment of raw wastewater. *Water Sci. Technol.* 25: 255–264.

Parekh, B.K. and Miller, J.D. (1999). *Advances in Flotation Technology*. Littleton, CO: Society for Mining, Metallurgy, and Exploration. Inc.

Parker, D.S. (1972). *Process Development for Nitrogen Removal of the CCCSD Water Reclamation Plant*. Advanced Waste Treatment Design Seminar: University of California, Riverside.

Patwardhan, A.W. (2003). Rotating biological contactors: a review. *Ind. Eng. Chem. Res.* 42: 2035–2051.

Pons, M.N. and Corriou, J.P. (2002). Implementation of storage tanks on the cost 624 benchmark. *Water Sci. Technol.* 45 (4–5): 159–166.

Poon, C.P. and Brueckner, T.G. (1975). Physico-chemical treatment of wastewater–seawater mixture by electrolysis. *J. Water Pollut. Control Fed.* 47: 66–78.

Prado, J., Arantegui, J., Chamarro, E., and Esplugas, S. (1994). Degradation of 2,4-D by ozone and light. *Ozone Sci. Eng.* 16: 235–245.

Prakash, N.B., Sockan, V., and Jayakaran, P. (2014). Waste water treatment by coagulation and flocculation. *Int. J. Eng. Sci. Innov. Technol.* 3 (2): 479.

Prakash, O., Talat, M., Hasan, S.H., and Pandey, R.K. (2008). Factorial design for the optimization of enzymatic detection of cadmium in aqueous solution using immobilized urease from vegetable waste. *Bioresource technology* 99 (16); 7565–7572. https://doi.org/10.1016/j.biortech.2008.02.008.

Qiu, P., Park, B., Choi, J. et al. (2018). A review on heterogeneous sonocatalyst for treatment of organic pollutants in aqueous phase based on catalytic mechanism. *Ultrason. Sonochem.* 45: 29–49.

Randall, C.W. and Barnard, J.L. (1998). *Design and Retrofit of Wastewater Treatment Plants for Biological Nutritient Removal*. CRC Press.

Raouf, M.E.A., Maysour, N.E., Farag, R.K. et al. (2019). Wastewater treatment methodologies, review article. *Int. J. Environ. Agric. Sci.* 3: 01–08.

Rasalingam, S., Peng, R., and Koodali, R.T. (2014). Removal of hazardous pollutants from wastewaters: applications of TiO_2-SiO_2 mixed oxide materials. *J. Nanomater.* 42.

Rice, R.G., Robson, C.M., Miller, G.W., and Hill, A.G.M. (1981). Use of ozone-bromide reactions. *J. Am. Water Works Assoc.* 85: 63–72.

Rim, Y.T., Yang, H.T., Yoon, C.H. et al. (1997). Chemical treatment technologies for waste-water recycling, an overview. *RSC Adv.* 2 (16): 6380.

Robescu, D. (2000). *Technlogies, Installations and Equipment for Wastewater Treatment*. Bucharest: Technical Publishing.

Rodgers, M. and Zhan, X.M. (2003). Moving medium biofilm bioreactors. *Rev. Environ. Sci. Biotechnol.* 2: 213–224.

Rojanschi, V. and Ognean, T. (1997). *Book of Operator in Wastewater Treatment Plants*. Bucharest: Technical Publishing.

Rosa, J.J. and Rubio, J. (2005). The FF (flocculation–flotation) process. *Miner. Eng.* 18 (7): 701–707.

Rubio, J. (1998). *Environmental Applications of the Flotation Process. Effluent Treatment in the Mining Industry*, vol. 9, 335–364. Chile: Departamento de Engenharia Metalúrgica, Universidade de Concepcion.

Rubio, J., Carissimi, E., and Rosa, J.J. (2007). Flotation in water and wastewater treatment and reuse: recent trends in Brazil. *Int. J. Environ.* 30 (2): 197–212.

Samer, M. (2015). *Biological and Chemical Wastewater Treatment Processes*. InTech.

Saravanane, R. and Murthy, D. (2000). Application of anaerobic fluidized bed reactors in wastewater treatment: a review. *Environ. Manage. Health* 11 (2): 97–117.

Semerjian, L., & Ayoub, G. (2003). High-pH–magnesium coagulation–flocculation in wastewater treatment. *Adv. Environ. Res.*, 7(2), 389–403. https://doi.org/10.1016/S1093-0191(02)00009-6

Shieh, W.K. and Keenan, J.D. (1986). Fluidized bed biofilm reactor for wastewater treatment. In: *Bioproducts* (ed. T. Scheper and R. Ulber), 131–169. Berlin, Heidelberg: Springer.

Shimizu, Y., Okuno, Y.-I., Uryu, K. et al. (1996). Filtration characteristics of hollow fiber microfiltration membranes used in membrane bioreactor for domestic wastewater treatment. *Water Research* 30 (10): 2385–2392, ISSN 0043-1354. https://doi.org/10.1016/0043-1354(96)00153-4.

Sidwick, J.M. (2007). The preliminary treatment of wastewater. *J. Chem. Technol. Biotechnol.* 52 (3): 291–300.

Silva, L.M.D. and Jardim, W.F. (2006). Trends and strategies of ozone application in environmental problems. *Química Nova* 29 (2): 310–317.

Sincero, A.P. and Sincero, A.G. (2016). Review paper on industrial wastewater treatment processes. Research gate.

Singh, M. and Srivastava, R.K. (2011). Sequencing batch reactor technology for biological wastewater treatment: a review. *Asia Pac. J. Chem. Eng.* 6 (1): 3–13.

Skorb, E.V., Fix, D., Shchukin, D.G. et al. (2011). Sonochemical formation of metal sponges. *Nanoscale* 3: 985–993.

Sonune, A. and Rupali, G. (2004). Water and land management institute. *Desalination* 167: 55–63.

Staehelin, J. and Hoigne, J. (1985). Decomposition of ozone in water in the presence of organic solutes acting as promoters and inhibitors of radical chain reactions. *Environ. Sci. Technol.* 19 (12): 1206–1213.

Stuart, E.F. (1946). Electronic water purification progress report on the electronic coagulator a new device which gives promise of unusually speedy and effective results. *Water Sewage* 84: 24–26.

Stukenberg, J.R. (1975). Physical-chemical wastewater treatment using a coagulation-adsorption process. *J. (Water Pollut. Control Federation)* 47 (2): 338–353. http://www.jstor.org/stable/25038633.

Suslick, K.S., Didenko, Y., Fang, M.M. et al. (1999). Acoustic cavitation and its chemical consequences. *Philos. Trans. R Soc. A* 357: 335–353.

Szpyrkowicz, L., Naumczyk, J., and Zilio-Grandi, F. (1994). Application of electrochemical processes for tannery wastewater treatment. *Toxicol. Environ. Chem.* 44: 189–202.

Tavares, C., Sant'anna, G., and Capdeville, B. (1995). The effect of air superficial velocity on biofilm accumulation in a three-phase fluidized-bed reactor. *Water Research* 29 (10): 2293–2298. https://doi.org/10.1016/0043-1354(95)00043-K.

Tee, P.F., Abdullah, O.M., Tan, I.A.W. et al. (2016). Review on hybrid energy systems for wastewater treatment and bio-energy production. *Renew. Sust. Energ. Rev.* 54: 235–246.

Tian, J.Y., Chen, Z.L., Nan, J. et al. (2010). Integrative membrane coagulation adsorption bioreactor (MCABR) for enhanced organic matter removal in drinking water treatment. *J. Membr. Sci.* 352 (1): 205–212.

Tiehm, A. (1999). Combination of ultrasound and biodegradation: enhanced bioavailability of polycyclic aromic hydro carbons. In: *Ultrasound in environmental engineering, TU Hamburg-Harburg Reports on Sanitary Engineering*, vol. 25 (ed. A. Tiehm and U. Neis), 167–180. GFEU-Verlag.

Tiehm, A., Nickel, K., Zellhorn, M., and Neis, U. (2001). Ultrasonic waste activated sludge disintegration for improving anaerobic stabilization. *Water Res.* 35 (8): 2003–2009. https://doi.org/10.1016/s0043-1354(00)00468-1. PMID: 11337847.

Tilche, A., Bacilieri, E., Bortone, G. et al. (1999). *Water Sci. Technol.* 40 (1): 199–206.

Ueda, T. and Horan, N.J. (2000). Fate of indigenous bacteriophage in a membrane bioreactor. *Water Res.* 34 (7): 2151–2159.

Ueda, T., Hata, K., and Kikuoka, Y. (1996). Treatment of domestic sewage from rural settlements by a membrane bioreactor. *Water Sci. Technol.* 34 (9): 189–196.

Ueda, T., Hata, K., Kikuoka, Y., and Seino, O. (1997). Effects of aeration on suction pressure in a submerged membrane bioreactor. *Water Res.* 31 (3): 489–494.

US EPA (1995). Treatment and Storage. AP-42, vol. I, CH 4.3. https://www3.epa.gov. (accessed 21 May 1995)

US EPA (2003). Environmental Protection Agency Wastewater Technology fact sheet - US EPA. https://www3.epa.gov/npdes/pubs/final_sgrit_removal.pdf.

Visvanathan, C. (2018). *Eighth Regional 3R Forum in Asia and the Pacific Water Quality and Security in Asia and the Pacific, What 3R and Circular Economy Can Offer? United Nations Centre for Regional Development (MoHUA)*. Government of India and Ministry of the Environment, Government of Japan.

Visvanathan, C., Yang, B.S., Muttamara, S., and Maythanukhraw, R. (1997). Application of air backflushing technique in membrane bioreactor. *Water Sci. Technol.* 36 (12): 259–266.

Wang, Z. and Wu, Z. (2009). Distribution and transformation of molecular weight of organic matters in membrane bioreactor and conventional activated sludge process. *J. Chem. Eng.* 150 (2): 396–402.

Wang, L.K., Hung, Y.T., and Shammas, N.K. (2005). *Physicochemical Treatment Processes Granular Activated Carbon Adsorption*, vol. 15, 573–633. Springer.

Wang, L.K., Hung, Y.T., and Shammas, N.K. (2006). *Handbook of Environmental Engineering, Physicochemical Treatment Processes*, 3rde. Totowa: The Humana Press Inc.

Wang, J.P., Chen, Y.Z., Wang, Y. et al. (2011). Optimization of the coagulation-flocculation process for pulp mill wastewater treatment using a combination of uniform design and response surface methodology. *Water Res.* 45 (17): 5633–5640. https://doi.org/10.1016/j.watres.2011.08.023.

Wang, L., Ali, J., Zhang, C. et al. (2020). Simultaneously enhanced photocatalytic and antibacterial activities of TiO2/Ag composite nanofibers for wastewater purification. *J. Environ. Chem. Eng.* 8 (1): 1021040. https://doi.org/10.1016/j.jece.2017.12.057.

Waste Water Treatment Manuals Preliminary Treatment (1995). *EPA*, Ireland.

Wastewater Technology Fact Sheet Oxidation Ditches (2000). *US EPA*, Washington, DC.

Weld, R.J. and Singh, R. (2011). Functional stability of a hybrid anaerobic digester/microbial fuel cell system treating municipal wastewater. *Bioresour. Technol.* 102 (2): 842–847.

Wen, C., Huang, X., and Qian, Y. (1999). Domestic wastewater treatment using an anaerobic bioreactor coupled with membrane filtration. *Process Biochem.* 35 (3): 335–340.

West, S. (2003). Innovative on-site and decentralised sewage treatment, Recycling and management systems in Northern Europe and the USA: Report of a study tour. February to November. Australia.

Wisconsin Department of Natural Resources (2016). *Introduction to Activated Sludge Study Guide*. Madison.

World Health Organization (WHO) (2013). Research for Universal Coverage: World Health Report.

World Health Organization (WHO) (2019). Research for Universal Coverage: World Health Report.

Xu, Z., Song, X., Li, Y. et al. (2019). Removal of antibiotics by sequencing-batch membrane bioreactor for swine wastewater treatment. *Sci. Total Environ.* 684: 23–30. https://doi.org/10.1016/j.scitotenv.2019.05.241.

Yamamoto, K., Hiasa, M., Mahmood, T., and Matsuo, T. (1989). Direct solid-liquid separation using hollow fiber membrane in an activated sludge aeration tank. *Water Sci. Technol.* 21 (4–5): 43–54.

Yang, H.J., Shin, E.B., Chung, Y.C., and Ryu, J.K. (1999). *J. Environ. Sci. Health* 34 (5): 1105–1116.

Zabava, B.S., Voicu, G., Ungureanu, N. et al. (2015). Basic equipment for the mechanical treatment of wastewater. *International Symposium ISB-INMA TEH, Agricultural and Mechanical Engineering*, Bucharest, Romania.

Žarković, D. B., Rajaković-Ognjanović, V. N., & Rajaković, L. V. (2011). Conservation of resources in the pulp and paper industry derived from cleaner production approach. *Resources, Conservation and Recycling*, 55(12), 1139–1145. https://doi.org/10.1016/j.resconrec.2011.07.003

Zhang, R., Khorshed, C., Vigneswaran, S., and Kandasamy, J. (2009). Submerged microfiltration coupled with physcio-chemical processes as pretreatment to sea water desalination. *Desalin. Water Treat.* 11 (1–3): 52–57.

Zhu, Z., Liu, F., Zhang, H. et al. (2015). Photocatalytic degradation of 4-chlorophenol over Ag/MFe_2O_4 (M = Co, Zn, Cu, and Ni) prepared by a modified chemical co-precipitation method: a comparative study. *RSC Adv.* 5: 55499–55512.

21

Case Studies of Aquatic Contamination and Bioremediation

Younis A. Hajam[1] *and Diksha*[2]

[1] *Department of Life Sciences and Allied Health Sciences, Sant Baba Bhag Singh University, Jalandhar, Punjab, India*
[2] *Department of Biosciences, Division Zoology, Career Point University, Hamirpur, Himachal Pradesh, India*

21.1 Introduction

The uninterrupted efforts to improve lifestyle have resulted in the continuous progression of the human race over the last decades. Whereas, advancement in the life of humans has induced a negative impact on the biosphere. Different anthropogenic activities such as transportation, industrialization, urbanization, and agriculture are inducing harsh effects on the environment. Even though industrialization is fundamental for the economic stability of developing nations, this industrial development has produced a negative effect on the economy by increasingly depleting natural resources and environmental pollution. The frequency of increase in natural disasters due to environmental and mostly climate change troubled the economy of developing nations. Likewise, the prerequisite of a safe environment in terms of water and air for the well-being of inhabitants also becomes difficult due to the addition of various pollutants. Likewise, the speculation of a safe environmental with respect to water and air for the well-fare of organisms has become difficult because of the continuous generation of various contaminants and pollutants to the environment. The use of polluted water or breathing polluted air produces pressure on an economy by increasing the health and well-being cost. The term "environmental pollution" is used to explain the situation of any unwanted substance thrown into the environment in the form of materials and energy. The addition of these materials to the soil, water, or air could produce an acute or chronic stress on the ecological systems of the biosphere resulting in the loss of ecosystem services. These damages can be categorized into primary or secondary pollution; for example, if the effect is direct and observable in the area, the damage is primary and the damage caused by this pollutant is known as a primary pollutant. However, in a case where the level of pollution is small leads to damage to other entities or affects the ecological balance, and becomes visible after some time period, it can be considered secondary damage and a pollutant is known as a secondary pollutant. Secondary pollutants are also those entities which do not produce an direct harm to the environment. However, they enter the environment along with other creatures or pollutants and produce toxic substances that typically damage the environment. For decades, environmental pollution has become more difficult. The addition of new substances continuously in the environment and their effects varies at a very high rate from region to region. One more challenge to tackling the environmental pollution is that the impact frequently looks different from the source. Hence, a global agreement is required to resolve the issue of environmental pollution. Environmental pollution is frequently demonstrated in three aspects of the biosphere, i.e. water, air, and soil. In the recent era, contamination of water has become a serious issue.

21.2 Water Contamination

The addition of toxic chemicals to water bodies has created a serious problem in the whole world. There is continuous addition of pollutants to the water released from industries, agricultural runoff, and municipal supplies. Although different treatment plants exist for the treatment of wastewater, the frequency of the problem is more than the potential of these

Aquatic Contamination: Tolerance and Bioremediation, First Edition. Edited by Rouf Ahmad Bhat, Gowhar Hamid Dar, Fernanda Maria Policarpo Tonelli, and Saima Hamid.

treatment plants. About two million tons of waste from sewage and industries are released daily into open water channels. Likewise, annual production of wastewater is about 1500 km^3, which is six times higher as compared to the overall water of rivers in the whole world. The distribution of freshwater resources is not equally distributed in the entire world that makes the accessibility of freshwater a challenge to the deprived societies. Thus, people sometimes use unhealthy and unsafe water.

Water is known as a universal solvent. Hence, in fact, clear water could be home to an entire number of toxic contaminants or pathogens. Availability of water and sanitation has become a challenge in the world, whereas the condition has become more serious in developing and underdeveloped countries where about 80% of people are devoid of the basic facilities of safe water for drinking. Moreover, groundwater is also being contaminated because of the continuous addition of pollutants from the leachate of surface pollutants including municipal solid waste, food processing waste, and wastewater containing toxic materials, etc. This increased pollution has created pressure on the water ecosystem. It has been reported that one-third of the amphibians and half of the freshwater fishes are on the verge of extinction due to pollution.

Other than the remarkable level of the modern revolution, the ongoing industrial development at high speed has increased the ecological contamination issues by delivering a variety of harmful toxins into the environment. For example, chemically dynamic buildups, antitoxins, poisonous weighty components, harmful phenols, and related dangerous mixtures all unfavorably influence the whole living environment with their direct/backhanded free development in water networks (Bilal et al. 2020; Morsi et al. 2020; Rasheed et al. 2020). Consequently, there is a critical need to successfully forestall, moderate, or decrease the generation of toxins to support a better and green climate (Liu et al. 2019). However, various remediation processes have been explored out and are used for the remediation of harmful and poisonous pollutants (Bilal et al. 2019; Aleya et al. 2019a,b; Ławniczak et al. 2020). However, most of these remediation and recycling techniques are very costly, inappropriate for the remediation of very poisonous synthetic compounds, does not reduces the age of harmful effects/buildups or optional ecological toxins (Dasgupta et al. 2015; Liu et al. 2019). Therefore, bioremediation technologies are now in a race to remediate the harmful ecological poisons for various benefits, like response productivity, significant expense viable proportion, boundless mineralization, and negligible/no side contamination age issues (García-García et al. 2016; Rasheed et al. 2019). In these bioremediation technologies, various bacterial species are used including "*Achromobacter*", "*Dehalococcoides*", "*Pseudomonas*", "*Burkholderia, Rhodococcus*", "*Comamonas*", "*Alcaligenes, Sphingomonas*", and "*Ralstonia*" can transform the highly harmful pollutants into less/no harmful byproducts, thus decreasing or eliminating contaminations from the sea-going air (Lloyd et al. 2003). Basically, microorganisms used in bioremediation secrete reactant catalysts, which convert the destructive products into nor-harmful products (Kumar and Dhote 2015; Dangi et al. 2019; Liu et al. 2019; Iqbal and Bilal 2020). Various methodologies have been developed taking into consideration the advantage of microbial strains. Momentarily, (i) transmission and cloning of chosen microbial strains are able to reduce the toxicity, (ii) articulation of proteins of interest which have novel applications and economically viable, (iii) articulation of corruption qualities for various contaminations, and (iv) protoplast combination for contamination debasement.

On planet earth, three major oceans, five widely large seas, and five smaller seas cover about 71% surface (Garrison 2012). The sea has captured human interest for ages; it is the main source of the natural pecking order and is a habitat for a variety of amphibian species (Abeyratne 2014; Soma et al. 2014; Weis 2015). Some studies have reported that the planet earth is at risk "due to the bioaccumulation and biomagnification of hazardous compounds in petroleum hydrocarbons (Rogers and Laffoley 2011; Weis 2015). Water contamination has become a serious issue and requires a suitable strategy and procedures to screen and execute systems and approaches for accomplishing arrangements. The yearly wastewater released has accounted for about 1500 km^3 (World Water Assessment Programme [United Nations] 2003). To decrease the contamination of freshwater and treatment of wastewater, there is increasing demand to develop some treatment approaches. In recent years various compounds have been identified in water due to anthropogenic or natural climates, which has increased the concerns about the state of the world's surface waters. Rodriguez et al. (2016) reported that the xenobiotics compounds occur naturally in nature at concentrations ranging from parts per trillion (ppt or ng/L) to parts per billion (ppb or g/L). These mixtures of compounds are known as emerging toxins (ECs)", micropollutants, foreign substances of arising concern (CEC), or follow-natural mixtures (TrOCs). The US Land Study characterizes these ECs are synthesized by engineering or microorganisms that are not generally checked in the climate, but rather that might unfavorably affect the climate or human wellbeing.

Release of harmful pollutants from human-made sources brings about exploitation of different regular resources like water and soil (Prescott 2002; Kumari et al. 2013). It is important to reduce the contamination or to change harmful products into less harmful ones. This can be accomplished by utilizing a procedure known as bioremediation; this obliterates the impurities without affecting the environment with the assistance of organic means (Maksymiv 2015). Bioremediation can be characterized as an innovative technique using different organisms. It is a more secure, cleaner, eco-accommodating, and savvy innovation (Pandey and Fulekar 2012). This cycle includes the oxidation of parent compounds and the transport of carbon dioxide and water that provided energy to the microorganism (Vidali 2001). This cycle brings about complete corruption of the objective contamination as this interaction does not include the exchange of pollutants starting with one spot and then onto the next (for instance, from water to soil or the other way around) (Walsh 1999). Successful biodegradation depends to select appropriate microorganisms perfectly positioned with the right natural circumstances like pH, temperature, and a lot more is important (Walsh 1999). According to the process of digestion, there are two types of substrates, viz. essential substrates and optional substrates (Vidali 2001). Essential substrates are those substrates which upon digestion provided energy to the cell, and the impurities are alluded to as the essential substrate. Optional substrates are those substrates when the cell uses them, they provide energy but does not hinder the progression of the process (Shishir et al. 2019). Bioremediation methods have been utilized for the purification of surface and subsurface soils, freshwater, and marine frameworks, soils, and groundwater and sullied land biological systems. In some cases, bioremediation initially generates oil hydrocarbon defilement to immobilize pollutants or change them to synthetic items at this point not perilous to human wellbeing and the climate.

A general rule is that debasement includes the utilization of microorganisms and associated compound to breakdown, modulate, detoxify, or change contamination potential. Detoxification of contaminants is possible by using particular microorganisms as well as specific conditions (Liu et al. 2019). Thus, it is similarly critical to plan and foster various remediation courses of action according to the accessible contaminations in the climate by developing microorganisms with designed reactant possibilities. Polychlorinated biphenyls (PCBs) are generated from chlorinated fragrant hydrocarbons widely distributed on the whole planet. PCBs have created serious "to the environment as well as human beings" because they remain in the environment for a prolonged time and have strong contamination potential and cause various diseases (Aldhafiri et al. 2018). Microbial communities developed through genetic modification are essentially capable to eliminate PCBs. Important catalyst responsible for the breakdown of PCBs is biphenyl 2,3-dioxygenase (BDO), which is encoded by high-quality BPHA (Shumkova et al. 2015). According to the Natural Confirmation Association, 45% of all pesticides used worldwide are often organophosphorus chemicals that can affect the climate (Shabbir et al. 2018). *Escherichia coli* are used for the degradation of a mixture of pesticides such as organochlorines (OCs) and organophosphates (Tasks) (Yang et al. 2012).

Petro-based hydrocarbon spills into natural grids such as shorelines, oceanfront seas, and land, are mostly the result of human activity and have a detrimental, destructive impact on biological organizations all over the world. Petro-based hydrocarbon spills can be persistently toxic and depend on the overall amount and transparency duration (Ndimele 2017). According to Roy et al. (2014), all nations should take seriously the convincing degradation of ocean oil contamination. Remembering that the oil from the spill is a mixture of different hydrocarbons (including n-alkanes, polycyclic sweet-smelling hydrocarbons, and fragrant hydrocarbons), the development of naturally altered microorganisms with additional opportunities to direct or ruin various oil hydrocarbons is a proactive step to control offshore oil pollution (Gallo et al. 2012). A close genomic analysis of 26 Sphingomonas and Sphingobium strains was performed by Zhao et al. (2017) to determine the limits of bioremediation, the capacity for biodegradation, and the level of quality trade. Catechol was considered to be an important factor in the detoxification of scented mixes. Genomic evaluations demonstrated that the test strains' key biodegradation routes were the -ketoadipate and homogentisate pathways (Zhao et al. 2017).

The most common source of environmental toxins is crude petroleum-based hydrocarbons. According to several studies (Varjani 2017; Al-Khalid and El-Naas 2018; Zhang et al. 2019), the hydrocarbon oil industry's cleaning procedures release hazardous odorous normal combinations into the environment, including polyaromatic hydrocarbons (PAHs), phenolic substances that are hardly ever degradable, chlorophenols, and cresols harms from hydrocarbons. Of course, oil contamination problems during transportation have increased crude oil spills. Strong clean-up and treatment procedures are necessary for marine crude oil spills. Various physicochemical and ecological treatment methods have generally been used to remove crude oil pollution from saltwater. Because of its feasibility and affordability, a typical treatment method that uses organisms, advancements, and green development for the biodegradation of crude oil has actually received a lot of attention. As part of the therapy, researchers have used bioaugmentation, or single strain and group small animals, to decontaminate the most outrageous portion of the spilt raw oil. Currently, a popular example of agricultural waste is increasing day by day. However, it can be used for the development of beneficial products. There are many low-cost substrates available, including molasses, solid waste dates, whey from paneer, corn steep liquor, and soybean waste oil for the remediation of agricultural waste. All of these agricultural modern wastes are not only used for the development of improved natural element, but also may serve as the main for the development of sustainable agriculture at smaller-scale (Silva et al. 2014; El Mahdi et al. 2016; Patowary et al. 2016; Souza et al. 2016). A couple of experts have used various strategies by applying N:P:K supplement extents (Amenaghawon et al. 2014; Suja et al. 2014) and food wastes (Joo et al. 2008). A couple of researchers have applied it in the field (Al-Sayegh et al. 2016; Souza et al. 2016; El Mahdi and Aziz 2017; Mohajeri et al. 2017; Wu et al. 2017).

The oil spill from sea water should be removed through proper practical bioremediation guidelines. These guidelines should address issues like when to use bioremediation, which bioremediation experts to use, how to use them, and how to monitor and survey the results (Sayed et al. 2021). In the current effort to promote green development, the use of smaller than typical natural materials for the bioremediation of oil hydrocarbon damages is vital (Wu et al. 2017; Cai et al. 2021). One of the biggest barriers now standing in the way of "oil bioremediation in marine water is the lack" of regulations governing how and when to exploit this advancement. The ability of incorporating bioremediation techniques in areas where water improvement is less of a concern is one remarkable and valuable aspect. This should be possible with floating oil guideline impacts, which are typically used to stop moving water advancement with oil spills that create an enclosed region, also known as the flourishing method.

21.2.1 Different Cases Studies

21.2.1.1 Ramifications for Marine Animals

The oil leak crisis involving petroleum hydrocarbons has an impact on the maritime ecosystem and climate (Fingas 2019; Kim et al. 2019). Oil hydrocarbons spread out in a thin, smooth layer on the surface of the water as TPH is supplied clearly into aquatic bodies by oil spills. Direct destruction and cover-up occur when a transparent substance fills the body of a natural substance with petroleum hydrocarbon oil (Farrington 2014). The normal climate of the shallow coral reefs has been affected by oil spills involving hydrocarbons. Coral deterioration and death as a result of petroleum hydrocarbon oil transparency have been observed frequently (Zhang et al. 2019). Warm-blooded animals and "birds are covered with petroleum oil", which reduces the species' resilience to extra-biological "stresses including temperature changes", severe illnesses, and other poisons (Farrington 2014). Seabirds are specific, defenseless because oil contact impairs their flight. The subsequent confirmation of contaminated food, internal breath, and recurrent interactions with the oil-water association site, according to Zhang et al. (2019), results in significant individual injury with high passing rates. Through layers of the body, such as the gill surfaces, ingested or broken down oil generates direct destructive noxiousness, sublethal consequences, and conceptive frustration in marine life structures (Farrington 2014). Turtles caught in oil spills are exposed to delayed real contact with both floating oils and food contaminated by oil or tar balls, as well as petroleum-saturated air that is typically breathed. Due to petroleum hydrocarbons blocking their throats, old and active turtles were thought to be starving to death (Zhang et al. 2019). The lack of financial resources brought on by coordinated mortality, the disappearance of regular environmental conditions, and the termination of fisheries has an influence on business and aquaculture endeavors (Zhang et al. 2019). Results on the advancement of economically significant species in the aquaculture industry are not good. In essence, oil contamination renders products unsuitable for sale. Another problem is that things become intolerable when there are excessive concentrations of oil-manufactured stressors (Farrington 2014).

The lack of knowledge regarding when and how to employ this development is now one of the most significant obstacles to its implementation in marine water (Sayed et al. 2021). Using bioremediation techniques in situations where the quality of water in the enclosed area is not as good is advantageous and profitable. In order to create this kind of scenario, oil control shoots, also known as shooting, which are frequently "used to contain oil spills in moving water and" restrict improvements of moving water with oil spills achieving an enclosed environment, can be set up on the external layer of the water. The best method of oil spill control, after floating impacts and impediments, is oil variety of materials, which have been attempted by and large. The usage of oil spill impacts as floating deterrents should adjust to natural, mechanical and utilitarian restrictions. Numerical impact direct exhibiting strategies may be used to design or affirm impacting frameworks that meet these cutoff points (Muttin et al. 2017). The waiting oil (tainting) obsession after physicochemical treatment in seawater is not permanently set up by on-the-spot TPH analyzers (Sayed et al. 2021). Miniature creatures are either separated from seawater with the goal that they can be utilized used for daily uses. Studied have conducted using bioaugmentation (BA), biostimulation (BS)," or a combination of BA-BS in watery "media considering the problem of bioavailability of toxic substances, protozoan predation, or contention from local microbiota, etc.

Bioremediation frequently continues from weeks to months after the oil spill and is typically employed as a cleaning stage after the usage of conventional mechanical clean-up selections (Sayed et al. 2021). Bioremediation helps discard explicit hazardous materials, which requires manpower and is also cost-effective (Prince 1997; Gogoi et al. 2003). In addition to these bioremediation technologies (BA and BS or a combination of both) are used to clean harmful petroleum hydrocarbon mixtures and restrict them to enter the environment. Complex cycles applicable in all contamination processes cannot produce a significant momentary result and ought not to be adjusted exclusively to each contaminated site as a defensive first measure in the process that high centralizations of oil are available (Gogoi et al. 2003). It has been studied that clearance of specific oil contaminant bioremediation is a practical strategy for treatment (Sayed et al. 2021; Rajaoalison et al. 2022).

Vidali (2001) has studied that the use of microorganisms in remediation processes such as "*Pseudomonas*", "*Alcaligenes, Sphingomonas*", "*Rhodococcus*", and "*Mycobacterium*" increases the degradation process. These organisms have frequently been accounted to decontaminate pesticides and hydrocarbons, alkanes and polyaromatic compounds. A large number of these microscopic organisms use the pollutant as a source of carbon and energy. Anaerobic microorganisms cause detoxification without oxygen. Demand to use anaerobic microorganisms in bioremediation in order to remediate PCBs in waterway silt, dechlorination of the dissolvable "trichloroethylene (TCE)", and chloroform. Ligninolytic fungi are growths, for example, the white decay parasite Phanaerochaetechrysosporium and can corrupt a very assorted scope of persevering or harmful ecological toxins. Methylotrophs are oxygen-consuming microscopic organisms that develop using methane for carbon and energy. The underlying catalysts used in the pathway for vigorous decontamination, methane monooxygenase, have an expansive substrate range and are having dynamic nature against many inorganic mixtures. Progresses in bioremediation outfit subatomic, microbial science and protein designing apparatuses and depend on the use of novel metal-sequestering peptides, judicious, and silly pathway designing, and compound plans.

Natural synthetic compounds are reproduced by using these standard or designed substances, associated with significant changes in their patterns of species including fishes (Braga et al. 2005; Brown et al. 2006). Scientists have developed for the remediation of medical wastes such as analgesics and sanitizers, hostile to microbial and various engineered substances (Wilson and Ashraf 2018). Assessment percentage for different contaminants varies viz. pesticides 57%, solutions 17%, and phthalates 5% (Gani and Kazmi 2017). Reported that emerging contaminants were found in 14 countries and investigated occasionally for their fate and advancement. Moreover, it has also been reported that more than 80 different types of PCPs, EDCs, and pharmaceuticals were found in treated as well as in untreated sewage, streams, lakes, seas, sediment, and, shockingly, normal water (Vickers 2017). The risk associated with the level of emerging pollutants (ECs) and other xenobiotics in amphibians has created a significant issue. Once these synthetic pollutants enter into water, they undergo physicochemical changes and then become more poisonous and harmful than their native forms. Photolysis and Drug Contextual analyses are principally two unique and free ways to evaluate the risk associated with TP arrangement. According to Escher and Fenner (2011), it is possible to develop methodologies for the detoxification of TP.

21.2.1.2 Remediation of Amiloride (AMI)

Amiloride (AMI) is a potassium-sparing diuretic. The phototransformation technique was used for the remediation of AMI in water (Calza et al. 2008). Direct photolysis and photocatalyzed degradation (with TiO_2 as a photocatalyst) directly degrade the drug. Complete degradation was observed after four hours (direct photolysis) and after 30 minutes (photocatalysis). Photoproducts were isolated by HPLC coupled with HRMS through an ESI interface. After 30 minutes of illumination under sun light, roughly half of AMI was detoxified and along with AMI, two photoproducts at m/z 212.0869 (AMI1) and 213.0709 (AMI2) were separated. Mass spectra analysis demonstrated that the primary cycles are dechlorination and hydroxylation of the heteroaromatic ring (Li et al. 1999). In the photocatalytic light, it is feasible to distinguish two different TPs beginning from the unit of the guanidine moiety and the cleavage of the heteroaromatic ring (AMI 3 and AMI 4).

21.2.1.3 Remediation of Chlorpromazine

Degradation of chlorpromazine (CRP) by applying natural sunlight, various studies analyzed the photosensitive characteristics of this drug, under aerobic situations, 12 TPs have been isolated by chromatographic technique and it has been observed that the production of TPs are associated with the production of brown and then violet color significant variation in the photochemistry were noted between the presence and absence of oxygen in the examined solution, with the production of polymers under anaerobic conditions. Illumination with short UVB (270 nm) leads to photoionization; however, exposure with long UVA (320–400 nm) generates free radical intermediates. The photochemical activities of CPR under environmental conditions were examined in the recent past. LC–MS investigation of photolysis samples displayed first-order kinetics with about whole irradiation of CPR prior 256 minutes of illumination. A significant reduction in the peak at the retention time of CPR can be noticed and new peaks (mostly at earlier times) come out during elimination with the Xenon lamp. Whereas, DOC level was found constant at a higher level revealing no mineralization. At the last stage of examination, a DOC irradiation of 0.1% was observed. But only residuals of CPR were identified as observed in LC–MS assessment, constituting for an initial irradiation of 99.7% and transformation of CPR but no mineralization of CPR but no mineralization during the period of the analysis. The pH values changed from neutral (pH 7.4 at 0 minute) to acidic (pH 3.9 at 256 minutes). This was observed to be due to photolytic dechlorination and 2-hydroxylation of CPR with resulting proton discharge into the test solution. At the final stage of the test mostly multiple hydroxylated and tentatively dimer species were observed. These analytical results are correlated with the change in color during the experiment from clear to ruby as dimmers are mostly responsible for the absorption of light".

The LC–MS data revealed that 57 TPs were tentatively detected at the end of the test; most of them were dimmers with highest retention times than CRP itself. And 28 TPs and MSm experiments of higher order were conducted, hence permitting structural illumination of these compounds. "Fragmentation pathways and proposals for the molecular structures were provided for the three highly abundant ions, which are produced during the process of the test. 317.3 m/z at tr, 1¼6.5 minutes was detected as 2-hydroxypromazine sulfoxide (CPR 1), which is the product of dechlorination, followed by hydroxylation and sulfoxidation. 335.4 m/z at tr, 2¼14 minutes was noted as CPR sulfoxide (CPR 2) 301. 1 m/z at 16.8 minutes was produced in the initial illumination time phases and was displayed to be 2-hydroxypromazine (CPR 3), hence being a possible photoproduct precursor of 317 m/z.

Trautwein and Kümmerer (2012) studied that breakdown of VPR was examined according to the international standard guidelines (4–7). These tests provide an indicator of biodegradability, abiotic and anaerobic breakdown of molecules in wastewater treatment plants instead of complete examination of environmental toxicity. The statement is that substances clearing these tests are easily degraded and then do not reach or deposit in the aquatic systems. The findings revealed that CPR was not biodegraded under aerobic OECD tests or in the anaerobic test. However, CPR does not showed any toxicity toward the aerobic bacteria used in the test whereas few of them were found toxic toward the aerobic sludge. CPR or its most forms of TPs (335 m/z and 301 m/z) were categorized as toxic to anaerobic sludge. It has been reported that toxicity of CPR incubated under the dark followed by its photodegradation with UVA and visible light was examined Nałęcz-jawecki et al. Two bioassays were used viz. the Spirotox test with the ciliated protozoan Spirotomum ambiguum, a short-term acute toxicity test that considers deformation (EC50) and lethal response (LC50) as endpoints, and second one is the Thamnotoxkit F with the anostracan crustacean *Thamnocephalus platyurus* that indicates its lethality (LC50) after 24 hours' incubation as the endpoint as per the standard operational protocols. A value of LC50 equal to 0.38 mg/L and an EC50 of 0.35 mg/L for the protozoan after 24 hours of exposure to CPR were observed. Test samples of CPR illuminated with UVA and VIS for seven days were still found toxic for the protozoan instead the reduction in the parent compound. An acute LC50 value equal to 0.62 mg/L was recorded after testing the pure compounds on the crustacean *T. platyurus*, whereas the same organism was not found sensitive to the TPs with an extreme alteration in toxicity demonstrated by the reduction in the concentration of parent compound. CPR was not found mutagenic or genotoxic. The UV illumination CPR does not bind with DNA and does not cause genotoxicity by the production of transitory radicals. However, photoactivated CPR shows a mutagenic effect in the Ames test at 10 mg/ml concentration.

21.2.1.4 Diclofenac (DCF)

Photolysis was observed as a very important transformation pathway of diclofenac (DCF) in surface waters; it causes the transformation of about 90% of DCF in a few hours. Reverse-phase HPLC fractionation of the solid phase extract of the DCF solution gives six fractions. Samples of the nonfractionated (NF) extract and every fraction was tested by taking three technical replicates to determine their inhibitory effect on the reproduction of algae of Scenedesmus vacuolatus. In NF, the algal reproduction was found suppressed considerably by about 80.5% in one fraction in comparison to other fractions not more than 30%. The active fraction was examined using GC/EI/MS in full-scan mode and 2-[(2 chlorophenyl) amino]benzaldehyde (DCF 1, was detected and confirmed by computer-aided structure elucidation from mass spectra.

The EC50 of DCF on S. vacuolatus was 48.1 mg/L and for DCF 1 4.8 mg/L. The method used for this evaluation was completely different, following the effect-directed analysis (EDA) of DCF to detect the TPs. In EDA assessment, the reason for the toxicity of environmental samples or mixtures was found by combining biotesting on one or more organisms from different taxonomic groups having separation and fractionation methods. The detection of chemicals and their confirmation of TPs are conducted following the analysis of the toxicity of the fractionated mixtures. This method is comparable "to the toxicity identification evaluation (TIE), it is one of the first standardized EDA procedures, developed by the

U.S. Environmental Protection Agency (EPA), as a practical tool to detect and identify the chemical reason of measurable impact by a sequential decrease in the complexity of environmental mixtures eventually to single toxicant.

21.2.1.5 Dipyrone

A prodrug is called dipyrone (DIP). It actually hydrolyzes swiftly into 4-methylaminoantipyrine (4-MAA), which allows its enzymes to catalyze its transformation into 4-acetylaminoantipyrine (4-AAA) and 4-formylaminoantipyrine (4-FAA). All three DIP metabolites' photochemical behavior was studied in different contexts. Lighting experiment employing milli-Q water, synthetic seawater, and freshwater to analyze solutions (10 and 50 mg) in the sun simulator. The "degradation kinetics" and the identification of the TPs were both documented using a liquid chromatography system with a time-of-flight mass spectrometer (LC-TOF-MS). According to the LC-MS findings, 13 TPs representing the key metabolite (4-MAA) that the lighting route produced were assumed. Two major TPs (DIP2 and DIP3) from the 4MAA were found to be very stable even after 50 hours of illumination. N-phenylacetamide (DIP 3) was also the main TP for the metabolite 4-AAA, but 4-FAA was also found to be capable of producing very persistent dimeric products.

It has been observed that oral administration of DIP gets quickly hydrolyzed into 4-MAA and the measured load of its derivatives in the STP effluents and surface waters, the acute toxicity assessment was conducted to test the metabolites 4-MAA, 4-AAA and 4-FAA at an initial concentration of 10 mg/L, prior and after their treatment via photolysis, on *Daphnia magna* hatched number of immobile daphnids was evidenced after 24 and 48 of exposure to the toxicants to evaluate the percentage of immobilization. The findings have revealed that nonilluminated samples do not have a considerable toxic effect on *D. magna* (maximum value at 24 hours 20% and at 48 hours 40%), while following photolysis, an elevation in acute toxicity was found in parallel with the prompt decrease in initial concentration of compound and the simultaneous production of toxic compounds. Whereas, after photolysis, toxicity level of acute toxicity was found to increase, the median immobilization concentrations evidenced as EC50.

21.2.1.6 Propranolol

The profile of TPs of propranolol (PRO) against fate exposure time was examined in deionized water and river water samples. When testing samples of PRO at two different concentrations (1 and 10 mg/L) in deionized water (DIW) and river water (RW) were illuminated with the solar simulator (l¼295–800 nm)." The experiment was conducted at low concentrations, with samples exposed in RW and DIW at different duration of time 0, 6, 24, and 48 hours. The investigation was conducted by positive ion LC–MS operated in full scan mode. Even though about 20TPs were found, only five products were found abundantly and all of them were found less than 10% pf PRO after 70 hours of illumination under the Xe lamp. TPs were masses m/z at 292. 276, 308, and 264 were summarized, and the percentage of these TPs was counted according to their peak areas relative to the peak area of the original PRO compound from standard solutions.

The comparative chronic toxicities of PRO and photo-degraded mixtures were investigated using the freshwater green alga "*Pseudokirchneriella subcapitata*" and the "*rotifer Brachionus calyciflorus*". Exposure concentrations of 1 and 10 mg/L PRO in deionized water (DIW) and river water (RW), earlier sterilized to stop any biodegradation, were illuminated. Samples were compared to dark control solutions in the same situations. The suppression of the algal growth test with some modifications from procedures was run for 72 hours and fluorescence was detected indicating the measurement of algal biomass.

When using animals as excellent indicators of the bioremediation of novel substances in soil and water, it has been found that parasites have high metabolic rates that can be used for radionuclides. Parasites are the "best microorganisms to grow and develop in acidic" conditions and radionuclide-polluted settings due to their great metabolic control (Azman et al. 2017; Tkavc et al. 2018). Chitin and chitosan, are known as good biosorbents of large metal particles, are found on the parasite cell surface (Das et al. 2008). According to studies (Farhan and Khadom 2015; Cárdenas González et al. 2019) development of "*Penicillium spp*"., "*Aspergillus spp.*" "*Rhizopus spp.*," "*Mucor spp.*," "*Saccharomyces sp.*" as well as "*Fusarium sp.*" are amazing biosorbents of metal particles. Damaging metals including "U, Th, Sr, Ni, Zn, Pb, Cr, and As" can be used as bio-absorbed by the biomass of *Penicillium* sp., *Rhizopus* sp., as well as *Saccharomyces* sp. (Tan and Cheng 2003; Iram et al. 2015; Bano et al. 2018). The use of water and food contaminated with high concentrations of uranium creates the risk of lung and bone disease, while uranium molecules themselves can cause a number of risks to human prosperity, the most important of them affecting kidneys (disturbance) and changes in pee tone. Current assessments have shown how techniques can use a variety of pathways to decrease the contamination of radionuclides in the environment (Vázquez-Campos et al. 2015; Cai et al. 2016; Al-Hagara et al. 2020). However, previous studies conducted on the uranium up-take of parasitic biomass. The cycle can be upset in the field by variations in the physical-manufactured, climatic, and wastewater production of the soil and water in each place. This has created a need for bioremediation processes, such as parasite-mediated remediation and to adapt under unambiguous conditions (Cerny 1974; Gupta et al. 2017).

Diesel, a mixture of hydrocarbons is primarily been used in industry and daily life as a source of energy like oil, fuel, light oil, and combustible gas (Sarıkoç 2020). When 40% of oil is used as a major power source, a vast amount of land and aquatic bodies are predicted to produce negative effects (Van Hamme et al. 2003). Diesel that has leaked into groundwater or the ocean will persist for a longer time and spread quickly to create silver sheens and a far-reaching rainbow film. The isolating cycles include the processes of spreading, dispersing, dispersion, deterioration, biodegradation, and emulsification. In the larger body of water, the higher sub-nuclear segments will sink (Wade et al. 2016; Zhao et al. 2016; Ji et al. 2018). Furthermore, water contaminated with diesel may be more dangerous than initial spilled oil. The occurrence of Changes in physicochemical occurs after mixing the stuff with water increases its propulsion.

According to Lominchar et al. (2018), diesel contains 64% aliphatic compounds, primarily cycloalkanes and n-alkanes, 35% fragrant blends, and 1% olefinic compounds. Diesel is frequently linked to hydrocarbon contamination via spills, limit tanks, transportation, and pipelines. From the incidents that took place, different effects of oil contamination address a threat to the ecosystem and the continuation of life, including typically altered vegetation. The insoluble and suspended oil prevents sunlight and oxygen from entering the water, causing animals to perish and leading to the extinction of species and a decrease in microbial populations (Beyer et al. 2016).

During the two decades some nations, including China, Brazil, the United States, India, Australia, and Malaysia were extremely prosperous and had a large number of manufacturing shares. In general, their evaluation location indicated significant solid areas with the Regular Noticing and Assessment journal and was energetically weighted in polycyclic fragrant hydrocarbons (PAHs). Furthermore, China is the main nation emphasizing "reusability" of waste in the manufacturing of useable products.

21.3 Noxious and Hazardous Combinations in Diesel-Tarnished Water

Ramadass et al. survey from 2017 found that major harmful combinations, "such as benzene, toluene, ethyl benzene, and xylene were present" in diesel water that had been subjected to forced division, exceeding the safeguarded level endpoints for the maritime environment (BTEX). These combinations often have little nuclear weight as opposed to long-moored hydrocarbons. The US Environment Affirmation Office (USEPA) lists them among the extremely dangerous and significant toxins that need to be removed from present effluents before they are released into the environment (EPA 2021). BTEX will transform into smolder when exposed to environmental stress, whereas others brought together will separate in water (Mohammadi et al. 2020). Higher water dissolvability may be badly designed and conclude the speed of bioremediation take-up by specific life structures (Varjani and Upasani 2017).

Along with BTEX, the USEPA has identified 16 specific combinations of polycyclic aromatic sweet-smelling hydrocarbons (PAHs) that are present in the environment and need to be disposed of or bioremediate them. A survey carried out at the Yinma Stream Bowl in China showed the 16 required PAHs. A total of 16 mixtures were successfully filtered out of the water tests, which include "naphthalene, acenaphthene, acenaphthylene, fluorene, phenanthrene, anthracene, fluoranthene, pyrene, chrysene, benzo (a) pyrene, benz (a) anthracene, benzo (b) fluoranthene, benzo (k) fluoranthene, dibenz (a, h)". Because various PAHs are cytotoxic and mutagenic. Since different PAHs are cytotoxic, mutagenic, and disease-causing, their presence in water could seriously harm humans, higher living things, and microorganisms (Moubasher et al. 2015). According to Abdel-Shafy and Mansour (Abdel-Shafy and Mansour 2016), they disrupt the immune system and excretory system, PAHs became the basis for threatening development and change advancement (kidney). Aliphatic hydrocarbons are stored in the organic framework via a variety of cycles, including the release of diesel and gasoline, marine oil, gas exploration and oil spills, the beginning of non-environmentally friendly power sources, and even nature itself. As four essential classes of aliphatic hydrocarbons, alkanes (n-alkanes), cycloalkanes, alkenes, and alkynes are proposed (Madhukar and Rajanikanth 2019). The typical origins and ecotoxicological dangers of 26 aliphatic hydrocarbons have been studied in Khark Island, Iran. In light of receptiveness to present wastewater discharges, petroleum inputs, and ecological sworn statements loaded with heavy metals and common harmful compounds, petroleum spoiling was viewed as the primary driver (Akhbarizadeh et al. 2016; Khalid et al. 2021). However, due to their resistance to corruption, destruction, and cancer-causing character, chlorinated aliphatic hydrocarbons tended by trichloroethylene (TCE) were identified as conclusive contamination in the groundwater of the United States (Li et al. 2017). Early biodegradation studies that have concentrated on the oil-polluted ocean surface have revealed cyclic alkanes that have diverged from n-alkanes at a lower level of corruption. Meanwhile, the higher noxiousness of cyclic alkanes has been presented to be the reason for low corruption (Jaekel et al. 2015).

In light of this, bioremediation is an excellent practice or plan above conventional processes in such circumstances and it may be advised as a trademark purifier to recover spoiled conditions employing commonplace tools (Ahmed and Fakhruddin 2018). Additionally, bioremediation is the most secure biotechnology method because the results it produces (such as carbon dioxide, water, unsaturated fats, and inorganic salts) are less dangerous than unique mixtures (Abatenh et al. 2017). Different components are included in the unique bioremediation process (Arora 2018). The processes involve using microorganisms or plants to help immobilize hazardous compounds and remediate the water and soil. They include bio sequestration, biodegradation, phytohydraulics, regular extraction, and volatilization.

At this point, the utilization of the two creatures and parasites has had a significant impact by making a mix of an ideal and strong environment as the result (Quintella et al. 2019).

Dangi et al. (2019) reported that living creatures produce compounds and expect huge parts when metabolic reactions occur. Creatures will attack and change the poisons absolutely, subsequently improving microbial bioremediation. No matter what the triumphs uncovered, bioremediation needs a way that is immovably associated with natural and environmental factors (Xenia and Refugio 2016; Zhu et al. 2017). The natural components impacting the speed of bioremediation are pH, temperature, pressure, as well as supplement source (Xue et al. 2015). Further limitations of bioremediation consolidate the bioavailability, bioactivity and natural science of the systems (Luka et al. 2018).

As they meet the growing need for the reduction in water contamination, local microbiological organizations anticipate playing a significant role in diesel pollution treatment. The removal of harmful contaminants from agro-modern wastes, the treatment of polluted shorelines, and microbial-based bioremediation in seas have all been suggested as effective large-scale treatments for petroleum hydrocarbon pollution. Microbial remediation is generally designed to normally destroy oil through the reducing framework. However, the necessary and anticipated effects have been indisputably delayed due to the climate's healthy ungainliness (Macaulay 2015).

Lately, a couple of assessments highlighted the bioremediation through microorganisms offering normally secured, useful, and especially viable systems in degrading hydrocarbons as well as alkanes and polyaromatic compounds and at least one or two defilements. Microorganisms that sort out some way to degrade hydrocarbons at a high rate have been represented in the past investigation, specifically "*Acinetobacter haemolyticus*" (Onur et al. 2015), "*Mycobacterium sp.*" (Gong et al. 2015), "*Pseudomonas aeruginosa*" (Kiraye et al. 2016), "*Bacillus subtilis*", "*Klebsiella sp.*," "*Acinetobacter junii*", "*Acinetobacter sp.*" (Rehman et al. 2018), "*Rhodococcus sp.*," "*Azotobacter vinelandii*" (Chaudhary et al. 2019), and "*Bacillus megaterium*", "*Acinetobacter iwoffii*" (Fahid et al. 2020)." The vulnerability of hydrocarbons to microbial degradation was situated in a slipping solicitation. The solicitation starts with direct alkanes followed by extended alkanes, little aromatics, cyclic alkanes and polyaromatic hydrocarbons (Das and Chandran 2011).

21.4 Halophilic Tiny Creatures Expected to Work as Bioremediation Trained Professionals

In diesel bioremediation, contaminants from polluted streams are eliminated by using organisms that break down fuel. Microorganisms that love salt, or halophiles, are incredibly rare. Additionally, those that are found in bodies of water have higher salt concentrations than ocean water. They roughly constitute 70 genera and more than 150 have been documented (de Lourdes Moreno et al. 2013). The halotolerant microorganisms are the strains that have between 0% and 5% pungency, and the social events have mild halophiles with 2–5% NaCl, moderate halophiles with 5–20% NaCl, silly halophiles with 20–30% NaCl, and halophiles with a pungency of 20–30% or more (Mohammadipanah et al. 2015). They are having applications in biotechnology. One of the applications of these halophiles colonies is the defiler or alteration of a degree of regular and inorganic pollution. In areas with high levels of pollution, high bitterness and supplement availability (nitrogen and phosphorus) are likely the factors restricting the biodegradation cycle by microorganisms.

Transportation frequently causes oil spills in the water or sea. On account of the shifting NaCl centre, scientists and specialists conducted extensive studies on the microbial depletion of hydrocarbon in the marine environment. One of the marine environments that are spectacularly impacted by oil hydrocarbons in light of extensive human activity is the Red Sea in Jeddah, Saudi Arabia.

According to Jamal and Pugazhendi's report from 2018, the Red Sea water trial halophilic microorganism consortium of *Ochrobactrumhalosaudis*, *Stenotrophomonasmaltophilia*, *Achromobacter xylosoxidans*, and *Mesorhizobiumhalosaudis* had actually contaminated different PAHs (phenanthrene, fluorene, and pyrene) with different sub-nuclear burdens The majority of the bacteria isolated from diesel oil belonged to the following genera: *Dietzia*, *Arthrobacter*, and *Halomonas*, and about 40% of the diesel oil components were successfully" eliminated by the halophilic consortium (RezaeiSomee et al. 2018). For their ability to degrade hydrocarbons in bizarre environments, halomonas are obvious and strikingly halophilic little living creatures (Fathepure 2014; Budiyanto et al. 2018; CortiMonzón et al. 2018; Yanan et al. 2018; Tena-Garitaonaindia et al. 2019). Additionally, the ability of Halomonas, Arthrobacter sp., to contaminate aliphatic and fragrant mixtures has been discovered in earlier studies (Abdulrasheed et al. 2020).

Chrysene and pyrene are both difficult to decontaminate. However, *Hortaea* sp. managed to find a technique to increase their rates of remediation in the maritime environment by up to 77% and 92%, respectively (Al Farraj et al. 2019). In high pungency conditions, Wang et al. (2018) found that Marinobacter contributed to phenanthrene degradation at salinities ranging from 3% to 20% while keeping an eye on the surrounding area. The Marinobacter consortium, which also included *Marinobacter hydrocarbonoclasticus* and *Marinobacter* sp., used phenanthrene and pyrene in 4% saline conditions, according to Jamal (2022). Despite the modest clearing rate as the pungency increased, the phenanthrene was totally destroyed for a very long period.

Another halophile, *Vibrio alginolyticus*, also increases the breakdown of diesel to a maximum of 98.20% and converts the fundamental mixtures into carbon dioxide, water, and a few intermediates (Isiodu et al. 2016; Imron and Titah 2018). By using hydrocarbons as the primary carbon source, dispersing biosurfactants, and defining biofilms surrounding oil dots and the oil–water interface, the Alcanivorax genera controlled maritime conditions that were affected by oil pollution (Yakimov et al. 2019). The first hexadecane-tainting strain among the Halorientalis sort has been identified as *Halorientalis hydrocarbonoclasticus* sp., with productive corruption quantified at 57% inside observing 3.6 M NaCl at oil-dirtied hypersaline settings (Zhao et al. 2017). The high (pyrene, benzo (e) pyrene, and benzo (k) fluoranthene) and low (naphthalene, phenanthrene, anthracene, and fluorene) nuclear weight PAHs have also been successfully degraded (above 90%) by promising bacterial strains, such as *Ochrobactrum* and *Pseudomonas* under a range of saline conditions (4–30%). Unusually, various reports made sure that hydrocarbon biodegradation is changed when pungency increases (Edbeib et al. 2016).

21.5 Parts Drew in with Diesel Bioremediation by Organisms

Microorganisms can degrade hydrocarbons due to four significant breakthroughs in intracellular protein catalysis (Li et al. 2019). Microorganisms absorb impurities to begin the cycle's crucially fundamental phase, which then radiates surfactants to aid in the emulsification (Kachienga et al. 2018). The toxins are then adsorbed into the cell film's outside layer. After that, they will enter the cell layer directly through endocytosis as a distinct or segregated transport and undergo an enzymatic reaction with the particular mixtures acting as driving forces to finish the general cycle (Hua and Wang 2014).

Methane monooxygenases (mmo attributes, pmoA (Murrell et al. 2000), central film non-heme iron alkane hydroxylases (or alkane monooxygenases, alkB (Van Beilen and Funhoff 2007), and cytochrome P450-type alkane hydroxylases (Van Beilen and Funhoff 2007) are examples of organized alkane hydroxylase structures. Aldehyde dehydrogenases and alcohol also oxidize substances, and the resultant unsaturated lipids join the carboxyl-horrific (TCA) cycle. In the earthly stage, anaerobic degradation pathways unquestionably stand sufficiently apart to be seen. Sulfate or nitrate is most frequently used as an electron acceptor in the anaerobic oxidation of alkanes, and it has been hypothesized that the expansion of a C group is what causes alkane degradation.

Cleaning up spilt oil in sea water is essentially one of high-impact bioremediation's most significant applications. The kind and concentration of the toxins present have an overall impact on how quickly and completely the majority of common toxins deteriorate in oxygen-consuming conditions. The limited accessibility of microorganisms, either due to poor biodiversity of nearby living things or the absence of nearby gathered microorganisms at filthy complains, could have an additional significant impact. According to reports (Al-Hawash et al. 2018), these natural materials possess the significant substrate qualities anticipated for the corruption of various hydrocarbons at the location. Additionally, the group microorganisms have demonstrated a better likelihood of success in using hydrocarbons than the isolated social orders that share a location with something essentially indistinguishable (Varjani and Upasani 2016).

For instance, the consortium was extracted from diesel-dirty conditions by "*Pseudomonas sp.*," "*Micrococcus sp.*," "*Staphylococcus sp.*," "*Bacillus sp.*," "*Flavobacterium sp.*," *Achromobacter* sp.," "*Klebsiella sp.*," "*Actinomycetes sp.*," "*Acetobacter sp.*," *and* "*Rhodococcus*" To contain the hydrocarbons in diesel and utilize them as their carbon sources, they went through a few cycles (Cruz et al. 2017).

Polycyclic fragrant hydrocarbons (PAHs) are significant and considered most dangerous components of oil spills to the environment. Strangely, when oxygen is scarce, bacteria promote the degradation of PAHs. Hydrocarbons initiate the intracellular attack that leads to the high-impact bacterial PAHs degradation by being oxidized, started, and joined by oxygen (Khalid et al. 2021). A cis-dihydrodiol was created through the hydroxylation of a fragrant ring from the substrate. Dioxygenase is the catalyst for this event. A protein development called dehydrogenase eventually dehydrogenates cis-dihydrodiol to a diol place. Ring-disconnecting dioxygenases may then cleave the metabolites through the ortho- or meta-cleavage route. While the ongoing shift triggers the production of TCA cycle intermediates, the previous response triggers the advancement of other intermediates, such as salicylic disastrous and catechol (Mahmoud and Bagy 2018). Intermediates are undoubtedly not the only thing present at the end of the high-impact digestion of PAHs, but they are also present together with biomass and final metabolic products like oxygen and carbon dioxide.

n-alkanes and cycloalkane are degraded that exhibit a small partitioning mechanism. The protein known as hydroxylase effortlessly oxidizes the cycloalkane to create a cycloalkanol. When oxygen is liberated, a molecule responds by chasing after the alkyl pack, which is how cycloalkanol is produced. Cycloalkanol is then converted to naphthenic ketone by dehydrogenation. The naphthenic ketone then undergoes a transition to become caprolactone with the help of an oxygenase substance. The ring of caprolactone is broken by the enhancement of water particles (H_2O), which transforms adipic terrible into a universally appealing substance. Following beta-oxidation, the adipic horrendous produces Acetyl-CoA, which microbes can use to remember their life history (Salamanca and Engesser 2014; Karande et al. 2018). However, only a small number of microorganism species may use cycloalkanes as the only carbon point of convergence for their metabolism.

21.6 Conclusion

Water contamination is certainly a global problem rather than a local one. It is obvious that clean water will aid in maintaining awareness of persistence while lessening risks to human prosperity and the maritime environment. Diesel hydrocarbons, which mostly combine aromatic and aliphatic combinations stored in the water bodies, are therefore very important to sanitize. One of the likely solutions to restore impure water is bioremediation. It is obvious that under hypersaline or saline environments, highly pungent microbes could be effectively employed to corrupt many naturally hazardous compounds. Since they obtain the essential molecules (such as oxygenase and dehydrogenase) that can act as catalysts to corrupt and utilize the hydrocarbons as a carbon source, the involvement of organisms in the removal of spilt oil is conceivable. In view of past exploration, the benefits from the ramifications of organisms, prominently halophiles, are critical. A more profound and better comprehension of biodegradation instruments holds huge natural significance that is exceptionally reliant upon the activities of microorganisms to change over or mineralize the natural contaminations. All things considered, the elements that might impact the biodegradation cycle ought to be considered to forestall controlling variables from hoarding the interaction. All in all, in light of the current survey, microbial corruption could be a central participant in the cleaning measures for diesel hydrocarbon remediation.

References

Abatenh, E., Gizaw, B., Tsegaye, Z., and Wassie, M. (2017). The role of microorganisms in bioremediation-a review. *Open J. Environ. Biol.* 2 (1): 038–046.

Abdel-Shafy, H.I. and Mansour, M.S. (2016). A review on polycyclic aromatic hydrocarbons: source, environmental impact, effect on human health and remediation. *Egypt. J. Pet.* 25 (1): 107–123.

Abdulrasheed, M., Zulkharnain, A., Zakaria, N.N. et al. (2020). Response surface methodology optimization and kinetics of diesel degradation by a cold-adapted Antarctic bacterium, Arthrobacter sp. strain AQ5-05. *Sustainability* 12 (17): 6966.

Abeyratne, R. (2014). *Aviation and Climate Change: In Search of a Global Market Based Measure*. Montreal, QC: Springer International Publishing.

Ahmed, F. and Fakhruddin, A.N.M. (2018). A review on environmental contamination of petroleum hydrocarbons and its biodegradation. *J. Environ. Sci. Nat. Resour.* 11 (3): 1–7.

Akhbarizadeh, R., Moore, F., Keshavarzi, B., and Moeinpour, A. (2016). Aliphatic and polycyclic aromatic hydrocarbons risk assessment in coastal water and sediments of Khark Island, SW Iran. *Mar. Pollut. Bull.* 108 (1-2): 33–45.

Al Farraj, D.A., Hadibarata, T., Yuniarto, A. et al. (2019). Characterization of pyrene and chrysene degradation by halophilic Hortaea sp. B15. *Bioprocess Biosyst. Eng.* 42 (6): 963–969.

Aldhafiri, S., Mahmoud, H., Al-Sarawi, M., and Ismail, W.A. (2018). Natural attenuation potential of polychlorinated biphenyl-polluted marine sediments. *Pol. J. Microbiol.* 67 (1): 37–48.

Aleya, L., Grisey, E., Bourioug, M., and Bourgeade, P. (2019a). Performance assessment of Etueffont (France) lagooning treatment system: report from a 16-year survey. *Sci. Total Environ.* 648: 518–529.

Aleya, L., Grisey, E., Bourioug, M. et al. (2019b). Proposed changes for post-closure monitoring of Etueffont landfill (France) from a 9-year survey. *Sci. Total Environ.* 656: 634–644.

Al-Hagara, O.E., Bayoumib, R.A., Azizd, O.A.A., and Mousad, A.M. (2020). Biosorption and adsorption of some heavy metals by Fusarium sp. F6c isolate as affected by gamma irradiation and agricultural wastes. *Sci. Asia* 46 (1): 1–37.

Al-Hawash, A.B., Dragh, M.A., Li, S. et al. (2018). Principles of microbial degradation of petroleum hydrocarbons in the environment. *Egypt. J. Aquat. Res.* 44 (2): 71–76.

Al-Khalid, T. and El-Naas, M.H. (2018). Organic contaminants in refinery wastewater: characterization and novel approaches for biotreatment. In: *Recent Insights in Petroleum Science and Engineering*, 371. IntechOpen.

Al-Sayegh, A., Al-Wahaibi, Y., Joshi, S. et al. (2016). Bioremediation of heavy crude oil contamination. *Open Biotechnol.* 10 (1).

Amenaghawon, A.N., Osunbor, O., and Obahiagbon, K.O. (2014). Impact of nutrients, aeration and agitation on the bioremediation of crude oil polluted water using mixed microbial culture. *Int. J. Sci. Res. Environ. Sci.* 2 (2): 43.

Arora, N.K. (2018). Bioremediation: a green approach for restoration of polluted ecosystems. *Environ. Sustain.* 1 (4): 305–307.

Azman, A.S., Othman, I., Fang, C.M. et al. (2017). Antibacterial, anticancer and neuroprotective activities of rare Actinobacteria from mangrove forest soils. *Indian J. Microbiol.* 57 (2): 177–187.

Bano, A., Hussain, J., Akbar, A. et al. (2018). Biosorption of heavy metals by obligate halophilic fungi. *Chemosphere* 199: 218–222.

Beyer, J., Trannum, H.C., Bakke, T. et al. (2016). Environmental effects of the deepwater horizon oil spill: a review. *Mar. Pollut. Bull.* 110 (1): 28–51.

Bilal, M., Rasheed, T., Nabeel, F. et al. (2019). Hazardous contaminants in the environment and their laccase-assisted degradation–a review. *J. Environ. Manage.* 234: 253–264.

Bilal, M., Mehmood, S., Rasheed, T., and Iqbal, H.M. (2020). Antibiotics traces in the aquatic environment: persistence and adverse environmental impact. *Curr. Opin. Environ. Sci. Health* 13: 68–74.

Braga, O., Smythe, G.A., Schäfer, A.I., and Feitz, A.J. (2005). Fate of steroid estrogens in Australian inland and coastal wastewater treatment plants. *Environ. Sci. Technol.* 39 (9): 3351–3358.

Brown, K.D., Kulis, J., Thomson, B. et al. (2006). Occurrence of antibiotics in hospital, residential, and dairy effluent, municipal wastewater, and the Rio Grande in New Mexico. *Sci. Total Environ.* 366 (2-3): 772–783.

Budiyanto, F., Thukair, A., Al-Momani, M. et al. (2018). Characterization of halophilic bacteria capable of efficiently biodegrading the high-molecular-weight polycyclic aromatic hydrocarbon pyrene. *Environ. Eng. Sci.* 35 (6): 616–626.

Cai, C.X., Xu, J., Deng, N.F. et al. (2016). A novel approach of utilization of the fungal conidia biomass to remove heavy metals from the aqueous solution through immobilization. *Sci. Rep.* 6 (1): 1–12.

Cai, Y., Wang, R., Rao, P. et al. (2021). Bioremediation of petroleum hydrocarbons using Acinetobacter sp. SCYY-5 isolated from contaminated oil sludge: strategy and effectiveness study. *Int. J. Environ. Res. Public Health* 18 (2): 819.

Calza, P., Massolino, C., Monaco, G. et al. (2008). Study of the photolytic and photocatalytic transformation of amiloride in water. *J. Pharm. Biomed. Anal.* 48 (2): 315–320.

Cárdenas González, J.F., Rodríguez Pérez, A.S., Vargas Morales, J.M. et al. (2019). Bioremoval of cobalt (II) from aqueous solution by three different and resistant fungal biomasses. *Bioinorg. Chem. Appl.* 2019.

Cerny, P. (1974). The present status of the analcime-pollucite series. *Can. Mineral.* 12 (5): 334–341.

Chaudhary, D.K., Bajagain, R., Jeong, S.W., and Kim, J. (2019). Development of a bacterial consortium comprising oil-degraders and diazotrophic bacteria for elimination of exogenous nitrogen requirement in bioremediation of diesel-contaminated soil. *World J. Microbiol. Biotechnol.* 35 (7): 1–11.

CortiMonzón, G., Nisenbaum, M., Herrera Seitz, M.K., and Murialdo, S.E. (2018). New findings on aromatic compounds' degradation and their metabolic pathways, the biosurfactant production and motility of the halophilic bacterium Halomonas sp. KHS3. *Curr. Microbiol.* 75 (8): 1108–1118.

Cruz, A., Cavaleiro, A.J., Paulo, A. et al. (2017). Microbial remediation of organometals and oil hydrocarbons in the marine environment. In: *Marine Pollution and Microbial Remediation* (ed. M.M. Naik and S.K. Dubey), 41–66. Singapore: Springer.

Dangi, A.K., Sharma, B., Hill, R.T., and Shukla, P. (2019). Bioremediation through microbes: systems biology and metabolic engineering approach. *Crit. Rev. Biotechnol.* 39 (1): 79–98.

Das, N. and Chandran, P. (2011). Microbial degradation of petroleum hydrocarbon contaminants: an overview. *Biotechnol. Res. Int.* 2011: 941810.

Das, N., Vimala, R., and Karthika, P. (2008). Biosorption of heavy metals–an overview. *Indian J. Biotechnol.* 7: 159–169.

Dasgupta, J., Sikder, J., Chakraborty, S. et al. (2015). Remediation of textile effluents by membrane based treatment techniques: a state of the art review. *J. Environ. Manage.* 147: 55–72.

Edbeib, M.F., Wahab, R.A., and Huyop, F. (2016). Halophiles: biology, adaptation, and their role in decontamination of hypersaline environments. *World J. Microbiol. Biotechnol.* 32 (8): 1–23.

El Mahdi, A.M. and Aziz, H.A. (2017). Hydrocarbon biodegradation using agro-industrial wastes as co-substrates. In: *Handbook of Research on Inventive Bioremediation Techniques* (ed. J.N. Bhakta), 155–185. IRMA.

El Mahdi, A.M., Aziz, H.A., Abu Amr, S.S. et al. (2016). Isolation and characterization of *Pseudomonas* sp. NAF1 and its application in biodegradation of crude oil. *Environ. Earth Sci.* 75 (5): 1–11.

EPA (2021). Toxic and priority pollutants under the clean water act. US EPA. https://www.epa.gov/eg/toxic-and-priority-pollutants-under-clean-water-act. Accessed on 24 January 2021.

Escher, B.I. and Fenner, K. (2011). Recent advances in environmental risk assessment of transformation products. *Environ. Sci. Technol.* 45 (9): 3835–3847.

Fahid, M., Arslan, M., Shabir, G. et al. (2020). Phragmitesaustralis in combination with hydrocarbons degrading bacteria is a suitable option for remediation of diesel-contaminated water in floating wetlands. *Chemosphere* 240: 124890.

Farhan, S.N. and Khadom, A.A. (2015). Biosorption of heavy metals from aqueous solutions by *Saccharomyces Cerevisiae*. *Int. J. Ind. Chem.* 6 (2): 119–130.

Farrington, J.W. (2014). Oil pollution in the marine environment II: fates and effects of oil spills. *Environ. Sci. Policy Sustain. Dev.* 56 (4): 16–31.

Fathepure, B.Z. (2014). Recent studies in microbial degradation of petroleum hydrocarbons in hypersaline environments. *Front. Microbiol.* 5: 173.

Fingas, M. (ed.) (2019). *Marine Oil Spills 2018*. MDPI.

Gallo, G., Lo Piccolo, L., Renzone, G. et al. (2012). Differential proteomic analysis of an engineered Streptomyces coelicolor strain reveals metabolic pathways supporting growth on n-hexadecane. *Appl. Microbiol. Biotechnol.* 94 (5): 1289–1301.

Gani, K.M. and Kazmi, A.A. (2017). Contamination of emerging contaminants in Indian aquatic sources: first overview of the situation. *J. Hazard. Toxic Radioact. Waste* 21 (3): 04016026.

García-García, J.D., Sánchez-Thomas, R., and Moreno-Sánchez, R. (2016). Bio-recovery of non-essential heavy metals by intra-and extracellular mechanisms in free-living microorganisms. *Biotechnol. Adv.* 34 (5): 859–873.

Garrison, T.S. (2012). *Oceanography: An Invitation to Marine Science*. Cengage Learning.

Gogoi, B.K., Dutta, N.N., Goswami, P., and Mohan, T.K. (2003). A case study of bioremediation of petroleum-hydrocarbon contaminated soil at a crude oil spill site. *Adv. Environ. Res.* 7 (4): 767–782.

Gong, X., Xu, X., Gong, Z. et al. (2015). Remediation of PAH-contaminated soil at a gas manufacturing plant by a combined two-phase partition system washing and microbial degradation process. *Environ. Sci. Pollut. Res.* 22 (16): 12001–12010.

Gupta, S., Wali, A., Gupta, M., and Annepu, S.K. (2017). *Plant-Microbe Interactions in Agro-Ecological Perspectives*. Springer.

Hua, F. and Wang, H.Q. (2014). Uptake and trans-membrane transport of petroleum hydrocarbons by microorganisms. *Biotechnol. Biotechnol. Equip.* 28 (2): 165–175.

Imron, M.F. and Titah, H.S. (2018). Optimization of diesel biodegradation by Vibrio alginolyticus using Box-Behnken design. *Environ. Eng. Res.* 23 (4): 374–382.

Iqbal, H.M. and Bilal, M. (2020). Time to automate the microbial detection and identification: the status quo. *J. Pure Appl. Microbiol.* 14 (1): 1–3.

Iram, S., Shabbir, R., Zafar, H., and Javaid, M. (2015). Biosorption and bioaccumulation of copper and lead by heavy metal-resistant fungal isolates. *Arab. J. Sci. Eng.* 40 (7): 1867–1873.

Isiodu, G.G., Stanley, H.O., Ezebuiro, V., and Okerentugba, P.O. (2016). Role of plasmid-borne genes in the biodegradation of polycyclic aromatic hydrocarbons (PAHs) by consortium of aerobic heterotrophic bacteria. *J. Pet. Environ. Biotechnol.* 7 (264): 2.

Jaekel, U., Zedelius, J., Wilkes, H., and Musat, F. (2015). Anaerobic degradation of cyclohexane by sulfate-reducing bacteria from hydrocarbon-contaminated marine sediments. *Front. Microbiol.* 6: 116.

Jamal, M.T. (2022). Enrichment of potential halophilic Marinobacter consortium for mineralization of petroleum hydrocarbons and also as oil reservoir indicator in Red Sea, Saudi Arabia. *Polycycl. Aromat. Compd.* 42 (2): 400–411.

Ji, H., Gong, Y., Duan, J. et al. (2018). Degradation of petroleum hydrocarbons in seawater by simulated surface-level atmospheric ozone: reaction kinetics and effect of oil dispersant. *Mar. Pollut. Bull.* 135: 427–440.

Joo, H.S., Ndegwa, P.M., Shoda, M., and Phae, C.G. (2008). Bioremediation of oil-contaminated soil using Candida catenulata and food waste. *Environ. Pollut.* 156 (3): 891–896.

Kachienga, L., Jitendra, K., and Momba, M. (2018). Metagenomic profiling for assessing microbial diversity and microbial adaptation to degradation of hydrocarbons in two South African petroleum-contaminated water aquifers. *Sci. Rep.* 8 (1): 1–6.

Karande, R., Salamanca, D., Schmid, A., and Buehler, K. (2018). Biocatalytic conversion of cycloalkanes to lactones using an in-vivo cascade in Pseudomonas taiwanensis VLB120. *Biotechnol. Bioeng.* 115 (2): 312–320.

Khalid, F.E., Lim, Z.S., Sabri, S. et al. (2021). Bioremediation of diesel contaminated marine water by bacteria: a review and bibliometric analysis. *J. Mar. Sci. Eng.* 9 (2): 155.

Kim, H., Choe, Y., and Huh, C. (2019). Estimation of a mechanical recovery system's oil recovery capacity by considering boom loss. *J. Mar. Sci. Eng.* 7 (12): 458.

Kiraye, M., John, W., and Gabriel, K. (2016). Bioremediation rate of total petroleum hydrocarbons from contaminated water by Pseudomonas aeruginosa case study: Lake Albert, Uganda. *J. Bioremed. Biodegrad.* 7: 335.

Kumar, A. and Dhote, M.C. (2015). Microbe associated phytoremediation technology for management of oil sludge: phytoremediation for oil sludge management. In: *Handbook of Research on Uncovering New Methods for Ecosystem Management through Bioremediation*, 1–28. IGI Global.

Kumari, R., Kaur, I., and Bhatnagar, A.K. (2013). Enhancing soil health and productivity of *Lycopersicon esculentum* Mill. Using *Sargassum johnstonii* Setchell & Gardner as a soil conditioner and fertilizer. *J. Appl. Phycol.* 25 (4): 1225–1235.

Ławniczak, Ł., Woźniak-Karczewska, M., Loibner, A.P. et al. (2020). Microbial degradation of hydrocarbons—basic principles for bioremediation: a review. *Molecules* 25 (4): 856.

Li, Y.N.B., Moore, D.E., and Tattam, B.N. (1999). Photodegradation of amiloride in aqueous solution. *Int. J. Pharm.* 183 (2): 109–116.

Li, J., de Toledo, R.A., and Shim, H. (2017). Multivariate optimization for the simultaneous bioremoval of BTEX and chlorinated aliphatic hydrocarbons by *Pseudomonas plecoglossicida*. *J. Hazard. Mater.* 321: 238–246.

Li, X., Li, H., and Qu, C. (2019). A review of the mechanism of microbial degradation of petroleum pollution. *IOP Conf. Ser.: Mater. Sci. Eng.* 484 (1): 012060.

Liu, L., Bilal, M., Duan, X., and Iqbal, H.M. (2019). Mitigation of environmental pollution by genetically engineered bacteria—current challenges and future perspectives. *Sci. Total Environ.* 667: 444–454.

Lloyd, J.R., Lovley, D.R., and Macaskie, L.E. (2003). Biotechnological application of metal-reducing microorganisms. *Adv. Appl. Microbiol.* 53: 85–128.

Lominchar, M.A., Santos, A., De Miguel, E., and Romero, A. (2018). Remediation of aged diesel contaminated soil by alkaline activated persulfate. *Sci. Total Environ.* 622: 41–48.

de Lourdes Moreno, M., Pérez, D., García, M.T., and Mellado, E. (2013). Halophilic bacteria as a source of novel hydrolytic enzymes. *Life* 3 (1): 38–51.

Luka, Y., Highina, B.K., and Zubairu, A. (2018). Bioremediation: a solution to environmental pollution-a review. *Am. J. Eng. Res.* 7 (2): 101–109.

Macaulay, B.M. (2015). Understanding the behaviour of oil-degrading micro-organisms to enhance the microbial remediation of spilled petroleum. *Appl. Ecol. Environ. Res.* 13 (1): 247–262.

Madhukar, A. and Rajanikanth, B.S. (2019). Cascaded plasma–ozone injection system: a novel approach for controlling total hydrocarbon emission in diesel exhaust. *Plasma Chem. Plasma Process.* 39 (4): 845–862.

Mahmoud, G.A.E. and Bagy, M.M.K. (2018). Microbial degradation of petroleum hydrocarbons. In: *Microbial Action on Hydrocarbons*, 299–320. Singapore: Springer.

Maksymiv, I. (2015). Pesticides: benefits and hazards. *J. Vasyl Stefanyk Precarpath. Nat. Univ.* 2 (1): 70–76.

Mohajeri, L., Zahed, M.A., Abdul Aziz, H., and Hasnain Isa, M. (2017). Assessment of bioaugmentation and biostimulation efficiencies for petroleum contaminated sediments. *Environ. Energy Econ. Res.* 1 (1): 89–98.

Mohammadi, L., Rahdar, A., Bazrafshan, E. et al. (2020). Petroleum hydrocarbon removal from wastewaters: a review. *Processes* 8 (4): 447.

Mohammadipanah, F., Hamedi, J., and Dehhaghi, M. (2015). Halophilic bacteria: potentials and applications in biotechnology. In: *Halophiles*, 277–321. Cham: Springer.

Morsi, R., Bilal, M., Iqbal, H.M., and Ashraf, S.S. (2020). Laccases and peroxidases: the smart, greener and futuristic biocatalytic tools to mitigate recalcitrant emerging pollutants. *Sci. Total Environ.* 714: 136572.

Moubasher, H.A., Hegazy, A.K., Mohamed, N.H. et al. (2015). Phytoremediation of soils polluted with crude petroleum oil using Bassiascoparia and its associated rhizosphere microorganisms. *Int. Biodeterior. Biodegrad.* 98: 113–120.

Murrell, J.C., Gilbert, B., and McDonald, I.R. (2000). Molecular biology and regulation of methane monooxygenase. *Arch. Microbiol.* 173 (5): 325–332.

Muttin, F., Campbell, R., Ouansafi, A., and Benelmostafa, Y. (2017). Numerical modelling and full-scale exercise of oil-spill containment boom on Galician coast. *Int. J. Comput. Methods Exp. Meas.* 5 (5): 760–770.

Ndimele, P.E. (ed.) (2017). *The Political Ecology of Oil and Gas Activities in the Nigerian Aquatic Ecosystem*. Academic Press.

Onur, G., Yilmaz, F., and Icgen, B. (2015). Diesel oil degradation potential of a bacterium inhabiting petroleum hydrocarbon contaminated surface waters and characterization of its emulsification ability. *J. Surfactant Deterg.* 18 (4): 707–717.

Pandey, B. and Fulekar, M.H. (2012). Bioremediation technology: a new horizon for environmental clean-up. *Biol. Med.* 4 (1): 51.

Patowary, R., Patowary, K., Kalita, M.C., and Deka, S. (2016). Utilization of paneer whey waste for cost-effective production of rhamnolipid biosurfactant. *Appl. Biochem. Biotechnol.* 180 (3): 383–399.

Prince, R.C. (1997). Bioremediation of marine oil spills. *Trends Biotechnol.* 15 (5): 158–160.

Quintella, C.M., Mata, A.M., and Lima, L.C. (2019). Overview of bioremediation with technology assessment and emphasis on fungal bioremediation of oil contaminated soils. *J. Environ. Manage.* 241: 156–166.

Rajaoalison, H., Knez, D., and Zamani, M.A.M. (2022). A multidisciplinary approach to evaluate the environmental impacts of hydrocarbon production in Khuzestan Province, Iran. *Energies* 15 (22): 8656.

Rasheed, T., Bilal, M., Nabeel, F. et al. (2019). Environmentally-related contaminants of high concern: potential sources and analytical modalities for detection, quantification, and treatment. *Environ. Int.* 122: 52–66.

Rasheed, T., Bilal, M., Hassan, A.A. et al. (2020). Environmental threatening concern and efficient removal of pharmaceutically active compounds using metal-organic frameworks as adsorbents. *Environ. Res.* 185: 109436.

Rehman, K., Imran, A., Amin, I., and Afzal, M. (2018). Inoculation with bacteria in floating treatment wetlands positively modulates the phytoremediation of oil field wastewater. *J. Hazard. Mater.* 349: 242–251.

RezaeiSomee, M., Shavandi, M., Dastgheib, S.M.M., and Amoozegar, M.A. (2018). Bioremediation of oil-based drill cuttings by a halophilic consortium isolated from oil-contaminated saline soil. *3 Biotech* 8 (5): 1–13.

Rodrigueza, O., Peralta-Hernandeza, J.M., Goonetillekeb, A., and Bandalac, E.R. (2016). Treatment technologies for emerging contaminants in water: a review. *Chem. Eng. J.* 323: 361–380.

Rogers, A. and Laffoley, D.D.A. (2011). *International Earth System Expert Workshop on Ocean Stresses and Impacts: Summary Workshop Report*. DIANE Publishing.

Roy, A.S., Baruah, R., Borah, M. et al. (2014). Bioremediation potential of native hydrocarbon degrading bacterial strains in crude oil contaminated soil under microcosm study. *Int. Biodeterior. Biodegrad.* 94: 79–89.

Salamanca, D. and Engesser, K.H. (2014). Isolation and characterization of two novel strains capable of using cyclohexane as carbon source. *Environ. Sci. Pollut. Res.* 21 (22): 12757–12766.

Sarıkoç, S. (2020). Fuels of the diesel-gasoline engines and their properties. In: *Diesel and Gasoline Engines* (ed. R. Viskup), 31. IntechOpen.

Sayed, K., Baloo, L., and Sharma, N.K. (2021). Bioremediation of total petroleum hydrocarbons (TPH) by bioaugmentation and biostimulation in water with floating oil spill containment booms as bioreactor basin. *Int. J. Environ. Res. Public Health* 18 (5): 2226.

Sayed, K., Baloo, L., Yekeen, S.T. et al. (2021). Determination of Total petroleum hydrocarbons concentration in coastal seawater of Teluk Batik Beach, Perak, Malaysia. In: *Key Engineering Materials*, vol. 888, 119–128. Trans Tech Publications Ltd.

Shabbir, M., Singh, M., Maiti, S. et al. (2018). Removal enactment of organo-phosphorous pesticide using bacteria isolated from domestic sewage. *Bioresour. Technol.* 263: 280–288.

Shishir, T.A., Mahbub, N., and Kamal, N.E. (2019). Review on bioremediation: a tool to resurrect the polluted rivers. *Pollution* 5 (3): 555–568.

Shumkova, E.S., Egorova, D.O., Boronnikova, S.V., and Plotnikova, E.G. (2015). Polymorphism of the bphA genes in bacteria destructing biphenyl/chlorinated biphenils. *Mol. Biol.* 49 (4): 569–580.

Silva, E.J., e Silva, N.M.P.R., Rufino, R.D. et al. (2014). Characterization of a biosurfactant produced by *Pseudomonas cepacia* CCT6659 in the presence of industrial wastes and its application in the biodegradation of hydrophobic compounds in soil. *Colloids Surf. B: Biointerfaces* 117: 36–41.

Soma, K., Ramos, J., Bergh, Ø. et al. (2014). The "mapping out" approach: effectiveness of marine spatial management options in European coastal waters. *ICES J. Mar. Sci.* 71 (9): 2630–2642.

Souza, A.F., Rodriguez, D.M., Ribeaux, D.R. et al. (2016). Waste soybean oil and corn steep liquor as economic substrates for bioemulsifier and biodiesel production by *Candida lipolytica* UCP 0998. *Int. J. Mol. Sci.* 17 (10): 1608.

Suja, F., Rahim, F., Taha, M.R. et al. (2014). Effects of local microbial bioaugmentation and biostimulation on the bioremediation of total petroleum hydrocarbons (TPH) in crude oil contaminated soil based on laboratory and field observations. *Int. Biodeterior. Biodegrad.* 90: 115–122.

Tan, T. and Cheng, P. (2003). Biosorption of metal ions with Penicilliumchrysogenum. *Appl. Biochem. Biotechnol.* 104 (2): 119–128.

Tena-Garitaonaindia, M., Llamas, I., Toral, L., and Sampedro, I. (2019). Chemotaxis of halophilic bacterium Halomonasanticariensis FP35 towards the environmental pollutants phenol and naphthalene. *Sci. Total Environ.* 669: 631–636.

Tkavc, R., Matrosova, V.Y., Grichenko, O.E. et al. (2018). Prospects for fungal bioremediation of acidic radioactive waste sites: characterization and genome sequence of Rhodotorulataiwanensis MD1149. *Front. Microbiol.* 8: 2528.

Trautwein, C. and Kümmerer, K. (2012). Degradation of the tricyclic antipsychotic drug chlorpromazine under environmental conditions, identification of its main aquatic biotic and abiotic transformation products by LC–MSn and their effects on environmental bacteria. *J. Chromatogr. B* 889: 24–38.

Van Beilen, J.B. and Funhoff, E.G. (2007). Alkane hydroxylases involved in microbial alkane degradation. *Appl. Microbiol. Biotechnol.* 74 (1): 13–21.

Van Hamme, J.D., Singh, A., and Ward, O.P. (2003). Recent advances in petroleum microbiology. *Microbiol. Mol. Biol. Rev.* 67 (4): 503–549.

Varjani, S.J. (2017). Microbial degradation of petroleum hydrocarbons. *Bioresour. Technol.* 223: 277–286.

Varjani, S.J. and Upasani, V.N. (2016). Biodegradation of petroleum hydrocarbons by oleophilic strain of Pseudomonas aeruginosa NCIM 5514. *Bioresour. Technol.* 222: 195–201.

Varjani, S.J. and Upasani, V.N. (2017). A new look on factors affecting microbial degradation of petroleum hydrocarbon pollutants. *Int. Biodeterior. Biodegrad.* 120: 71–83.

Vázquez-Campos, X., Kinsela, A.S., Collins, R.N. et al. (2015). Uranium binding mechanisms of the acid-tolerant fungus Coniochaetafodinicola. *Environ. Sci. Technol.* 49 (14): 8487–8496.

Vickers, N.J. (2017). Animal communication: when i'm calling you, will you answer too? *Curr. Biol.* 27 (14): R713–R715.

Vidali, M. (2001). Bioremediation: an overview. *Pure Appl. Chem.* 73 (7): 1163–1172.

Wade, T.L., Sericano, J.L., Sweet, S.T. et al. (2016). Spatial and temporal distribution of water column total polycyclic aromatic hydrocarbons (PAH) and total petroleum hydrocarbons (TPH) from the deepwater horizon (Macondo) incident. *Mar. Pollut. Bull.* 103 (1–2): 286–293.

Walsh, J.B. (1999). A feasibility study of bioremediation in a highly organic soil. Master of Science Thesis, Worcester Polytechnic Institute.

Wang, C., Huang, Y., Zhang, Z., and Wang, H. (2018). Salinity effect on the metabolic pathway and microbial function in phenanthrene degradation by a halophilic consortium. *AMB Express* 8 (1): 1–13.

Weis, J.S. (2015). *Marine Pollution: What Everyone Needs to Know*. Oxford University Press.

Wilson, M. and Ashraf, M.A. (2018). Study of fate and transport of emergent contaminants at waste water treatment plant. *Environ. Contam. Rev.* 1: 01–12.

Willey, Joanne M., Linda Sherwood, and Christopher J. Woolverton (2002). Prescott's microbiology. Vol. 7. New York: McGraw-Hill.

World Water Assessment Programme (United Nations) (2003). *Water for People, Water for Life: The United Nations World Water Development Report: Executive Summary*. Unesco Pub.

Wu, M., Li, W., Dick, W.A. et al. (2017). Bioremediation of hydrocarbon degradation in a petroleum-contaminated soil and microbial population and activity determination. *Chemosphere* 169: 124–130.

Xenia, M.E. and Refugio, R.V. (2016). Microorganisms metabolism during bioremediation of oil contaminated soils. *J. Bioremed. Biodegrad.* 7 (2): 1000340.

Xue, J., Yu, Y., Bai, Y. et al. (2015). Marine oil-degrading microorganisms and biodegradation process of petroleum hydrocarbon in marine environments: a review. *Curr. Microbiol.* 71 (2): 220–228.

Yakimov, M.M., Golyshin, P.N., Crisafi, F. et al. (2019). Marine, aerobic hydrocarbon-degrading *Gammaproteobacteria*: the family *Alcanivoracaceae*. In: *Taxonomy, Genomics and Ecophysiology of Hydrocarbon-Degrading Microbes* (ed. T.J. McGenity), 167–179. Springer.

Yanan, W., Yingying, L., Jianliang, X. et al. (2018). Exploring the degradation potential of Halomonas bacteria from oil-contaminated marine environment. *China Pet. Process. Petrochemical Technol.* 20 (4): 91.

Yang, J., Liu, R., Song, W. et al. (2012). Construction of a genetically engineered microorganism that simultaneously degrades organochlorine and organophosphate pesticides. *Appl. Biochem. Biotechnol.* 166 (3): 590–598.

Zhang, B., Matchinski, E.J., Chen, B. et al. (2019). Marine oil spills—oil pollution, sources and effects. In: *World Seas: An Environmental Evaluation*, 391–406. Academic Press.

Zhao, X., Liu, W., Fu, J. et al. (2016). Dispersion, sorption and photodegradation of petroleum hydrocarbons in dispersant-seawater-sediment systems. *Mar. Pollut. Bull.* 109 (1): 526–538.

Zhao, Q., Yue, S., Bilal, M. et al. (2017). Comparative genomic analysis of 26 *Sphingomonas* and *Sphingobium* strains: dissemination of bioremediation capabilities, biodegradation potential and horizontal gene transfer. *Sci. Total Environ.* 609: 1238–1247.

Zhao, D., Kumar, S., Zhou, J. et al. (2017). Isolation and complete genome sequence of Halorientalishydrocarbonoclasticus sp. nov., a hydrocarbon-degrading haloarchaeon. *Extremophiles* 21 (6): 1081–1090.

Zhu, X., Chen, B., Zhu, L., and Xing, B. (2017). Effects and mechanisms of biochar-microbe interactions in soil improvement and pollution remediation: a review. *Environ. Pollut.* 227: 98–115.

Glossary

a

Activated sludge
Aerobic granular sludge (AGS)
Aerobic rotating biological contactor (RBC)
Ammonia oxidizing microorganisms
Anaerobic fixed film reactor (AFFR)
Anaerobic treatment
Aquaporins
Arabidopsis thaliana
Arbuscular mycorrhizal fungi (AMF)
Arsenate reductase

b

Bacillus subtilis polysaccharide
Benzoic acid
Bioabsorbents
Biocalcation
Bioconcentration Factor (BCF)
Biofilm
Biological oxygen demand (BOD)
Biosensors

c

Capsular polysaccharide (CPS)
Carbamates
Carbonfiltration
Chemical oxygen demand (COD)
Colanic acid
Copper zinc tin sulfide (CZTS) technique
Constructed wetlands
Cyanoremediation
Cypermethrin (CYP)

d

Dark septate fungi (DSF)

Aquatic Contamination: Tolerance and Bioremediation, First Edition. Edited by Rouf Ahmad Bhat, Gowhar Hamid Dar, Fernanda Maria Policarpo Tonelli, and Saima Hamid.

e
Ectomycorrhizal fungi (EMF)
Endophytic fungi (EF)
Environs
Eucalyptus sp.

f
Free-floating macrophytes

g
Gene annotation
Granulated activate carbon (GAC)
Grassed filter strips
Green synthesized nanomaterials
Glycosyl transferase (*rhlB*)

h
Hydrocarbonoclastic bacteria
Hydrogenation

i
Iron oxide NPs

j
Jatropha curcas L.

l
Lemon fruits

m
Membrane bioreactor (MBR)
Metallothioneins
Microbial amylase
Microbial cellulase
Microbial fuel cells (MFCs)
Microbial lipase
Microbial pollulanase
Microbial protease
Microplastic degradation
Molecularly imprinted polymers
Moving bed biofilm reactor (MBBR)
Myconanotechnology
Mycoremediation

n
Nanochitosan
Nanofiltration

o
Oligonucleotide probes
Organochlorines

Organophosphates
Operons
Oxidation ditches (OD)

p
Permeable reactor barrier (PRB)
Phenyl amides
Phragmites sp.
Phytochelatins
Phytodegradation
Phytoextraction
Phytomanagement
Phytoremediation
Phytosiderophores
Phytostabilization
Phytovolatilization
Phtalimides
Point sources
Polysaccharides
Polysaccharide intercellular adhesion
Pseudomonas putida
Pyrethroids

r
Radioactive elements
Reactive oxygen species (ROS)
Rhizodegradation
Rhizofiltration
Root zone

s
Sesbania grandiflora
Sorghum bran
Sublethal effects
Superoxide dismutase (SOD)
Synthetic dyes

t
TiO_2 NPs
Tangerine
Terminalis chebula
Thin film composite (TFC)
Thlaspi caerulescens
Transcriptomics
Transgenic plants
Translocation factor (TF)
Trazines
Trichloroethylene
Trihalomethanes
Trinitrotoluene (TNT)

w
Water hyacinth
Whole-genome sequencing

v
Vegetated filter strips
Volatile organic compounds (VOCs)

z
Zinc oxide nanoparticles

Index

Aquatic Contamination: Tolerance and Bioremediation, First Edition. Edited by Rouf Ahmad Bhat, Gowhar Hamid Dar, Fernanda Maria Policarpo Tonelli, and Saima Hamid.